GLENCOE

Health

About the Author

Mary H. Bronson, Ph.D., has taught health education in grades K–12, as well as health education methods classes at the undergraduate and graduate levels. As health education specialist for the Dallas School District, Dr. Bronson developed and implemented a district-wide health education program, *Skills for Living*, which was used as a model by the state education agency. She has assisted school districts throughout the country in developing local health education programs. She is also the author of Glencoe's *Teen Health* textbook series.

connectED.mcgraw-hill.com

Copyright © 2015 McGraw-Hill Education

Send inquiries to:
McGraw-Hill Education
8787 Orion Place
Columbus, OH 43240

ISBN: 978-0-02-140707-1
MHID: 0-02-140707-X

Printed in the United States of America.

7 8 9 10 11 QVS 21 20 19 18 17

Health and Educational Consultants

Unit 1: A Healthy Foundation

Lisa M. Carlson, MPH, C.H.E.S.
Academic Program Director
Emory Transplant Center
Atlanta, Georgia

Betty M. Hubbard, Ed.D., C.H.E.S.
Professor of Health Education
Department of Health Sciences
University of Central Arkansas
Conway, Arkansas

Deborah L. Tackmann, M.E.P.D.
Health Education Instructor
North High School
Eau Claire, Wisconsin

Unit 2: Mental and Emotional Health

Rani Desai, Ph.D., M.P.H.
Associate Professor
Yale University
New Haven, Connecticut

Unit 3: Healthy and Safe Relationships

Jill English
Health Education and Evaluation Consultant
Orange, California

Unit 4: Nutrition and Physical Activity

Roberta Duyff, R.D., C.F.C.S.
Food and Nutrition Education Consultant
St. Louis, Missouri

Don L. Rainey
Lecturer and Director
Physical Fitness and Wellness Program
Texas State University
San Marcos, Texas

Unit 5: Personal Care and Body Systems

Dyan Campbell, R.N., M.P.H.
Campbell Consulting L.L.C.
Parksville, New York

Ismael Nuño, M.D.
Chief, Cardiac Surgery
LAC+USC Medical Center
Los Angeles, California

Unit 6: Growth and Development

Susan Giarratano Russell, Ed.D., M.S.P.H., C.H.E.S.
Health Education and Evaluation Consultant
Valencia, California

Unit 7: Drugs

Donna Breitenstein
Health Educator
Boone, North Carolina

Jeanne Title
Coordinator, Prevention Education
Napa County Office of Education and
Napa Valley Unified School District
Napa, California

Unit 8: Diseases and Disorders

Donna Breitenstein
Health Educator
Boone, North Carolina

Ismael Nuño, M.D.
Chief, Cardiac Surgery
LAC+USC Medical Center
Los Angeles, California

Unit 9: Safety and Environmental Health

Kelly Cartwright
College of Lake County
Department of Biological and Health Sciences
Grayslake, Illinois

Ismael Nuño, M.D.
Chief, Cardiac Surgery
LAC+USC Medical Center
Los Angeles, California

Greg Stockton
American Red Cross
Washington, DC

Teacher Reviewers

Mark Anderson
Supervisor of Health and
Physical Education
Cobb County Schools
Marietta, Georgia

Nita Auer
Health Educator
North Side High School
Ft. Wayne, Indiana

Mike Beasley
Health Educator
Blackville-Hilda High School
Blackville, South Carolina

Melissa Broussard
English and Health Educator
Iota High School
Iota, Louisiana

Julie Brumfield
Health Educator
Cabell Midland High School
Ona, West Virginia

Kristen Bye
Health and Physical Education
Teacher
Bedford North Lawrence High
School
Bedford, Indiana

Theresa Despino
Health and Physical Education
Teacher
Alexandria Senior High School
Alexandria, Louisiana

Rick Duffield
Health Educator
Valley High School
Pine Grove, West Virginia

Colette Dux
Health Educator
El Camino Real High School
Woodland Hills, California

Kelly Gamble
Health and Physical Education
Teacher
Avon High School
Avon, Indiana

Charles Harrison
Health Educator
Berkeley High School
Moncks Corner, South Carolina

Anna Harvley
Health Educator
Lugoff-Elgin High School
Elgin, South Carolina

Cindy Henderson
Health Teacher
Putnam City North High School
Oklahoma City, Oklahoma

Mimi Herald
K–12 Health Education
Coordinator
Bethel Public Schools
Bethel, Connecticut

David Horton
Health and Physical Education
Department Chair
Bedford North Lawrence High
School
Bedford, Indiana

Marita Hunt
Health Educator
Captain Shreve High School
Shreveport, Louisiana

Jia Oliver Jordan, M.Ed.
Health Educator
Booker T. Washington Magnet
High School
Montgomery, Alabama

Deborah Larson
Health Educator
Pendleton High School
Anderson, South Carolina

Kathy Marlowe
Healthful Living Educator
A. C. Reynolds High School
Asheville, North Carolina

Randall Nitchie
Department Chair and Health
Educator
Osseo Area Schools
Maple Grove, Minnesota

Sabra Papich
Health Teacher
Albuquerque High School
Albuquerque, New Mexico

Gretchen Shafer
Health/Nutrition Department
Chair
Fishers High School
Fishers, Indiana

Jason Simmons
Health and Physical Education
Department Chair
Ben Davis High School
Indianapolis, Indiana

Phyllis Simpson
Consultant
DeSoto, Texas

Tammy Smith
Assistant Athletic Director
Tulsa Public Schools
Tulsa, Oklahoma

Cindy Williams
Health Educator
Russellville Jr. High
Russellville, Arkansas

Tom Williams
Health Educator
Fayetteville High School
Fayetteville, Arkansas

Contents

UNIT 1 A Healthy Foundation

UNIT 2 Mental and Emotional Health

UNIT 3 Healthy and Safe Relationships

UNIT **4** Nutrition and Physical Activity

UNIT 5 Personal Care and Body Systems

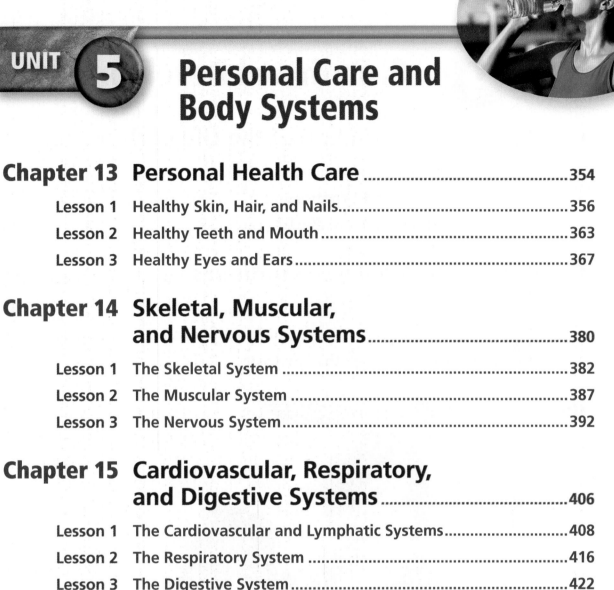

©liquidlibrary/PictureQuest

UNIT 6 Growth and Development

UNIT 7 Drugs

UNIT 8 Diseases and Disorders

UNIT 9 Safety and Environmental Health

Health Skills Activity

Real World CONNECTION

TEENS Making a Difference

Hands-On HEALTH

Fitness Handbook

Physical fitness is for everyone, regardless of a person's skill level. Non-athletes who choose not to join organized sports can develop a personal fitness plan to stay in shape. Even athletes can use some of the tips to cross train for their favorite sport.

Planning a Routine: The Fitness Handbook helps you plan a fitness routine that helps your body adjust slowly to activity. Over time, you will increase both the length of time you spend and the number of times that you are physically active each week. Teens should aim to get at least one hour of physical activity each day. These periods of physical activity can be divided into shorter segments, such as three 20 minute segments each day.

Before You Start Exercising: Before we begin, let's explain what we mean by the word *exercise*. Exercise can include any physical activity, such as completing a fitness program, playing individual or group sports, or even helping clean at home. The key is to keep your body moving.

Every activity session should begin with a warm-up to prepare your body for exercise. Warm-ups raise your body temperature and get your muscles ready for physical activity. Easy warm-up activities include walking, marching, and jogging, as well as basic calisthenics or stretches. A warm-up can also be an easier version of the exercise you have chosen. For example, if you have decided to run five miles, as a warm-up you may choose to jog for the first 10 minutes.

As you increase the time you spend doing a fitness activity, you should increase the time you spend warming up. Check the Sample Physical Fitness Plan to see how much time you should dedicate to warm-ups and fitness activities. For additional fitness activities, visit the Fitness Zone site in ConnectEd.

The Five Elements of Fitness

When developing a fitness plan, it's helpful to have a goal. Maybe your goal is to comfortably run three miles each day to stay in shape, or maybe you want to run in a marathon in the future. Regardless of the reasons why you develop a fitness plan, focusing on the five elements of fitness, or health-related fitness, will help you achieve overall physical fitness. The five elements are:

1. Cardiorespiratory Endurance

The ability of your heart, lungs, and blood vessels to send fuel and oxygen to your tissues during long periods of moderate to vigorous activity. Examples are jogging, walking, bike riding, and swimming.

2. Muscular Endurance

The ability of your muscles to perform physical tasks over a period of time without tiring. Many activities that build cardiovascular endurance also build muscle endurance, such as jogging, walking, and bike riding.

3. Muscular Strength

The amount of force your muscles can exert. Activities that can help build muscular endurance include push-ups, pull-ups, lifting weights, and running stairs.

4. Flexibility

The ability to move your body parts through their full range of motion. You can improve your flexibility by stretching before and after exercise.

5. Body Composition

The ratio of fat to lean tissue in your body. A healthy body is made up of more lean tissue and less fat. Body composition is a result of diet, exercise, and heredity.

Skill-Related Fitness

Skill-related fitness can enhance your ability to complete daily tasks unrelated to exercise. For health-related and skill-related fitness activities, visit the Fitness Zone site in ConnectEd. Skill-related fitness requires that you consider six elements. These include:

1. Agility

The ability to change and control the direction and position of the body while maintaining a constant, rapid motion. Sports that require a high level of agility include football, soccer, basketball, baseball, and softball.

2. Balance

The ability to control or stabilize the body while standing or moving. Examples of sports that require balance include gymnastics, golf, and ice skating.

3. Coordination

The ability to use the senses to determine and direct the movement of your limbs and head. Gymnastics, cheerleading, and juggling demand a high level of coordination.

4. Speed

The ability to move your body, or parts of it, swiftly. Foot speed is measured over a short and straight distance, usually less than 200 meters. Other speed evaluations might include hand and arm speed. The baseball pitcher, boxer, sprinter, and volleyball spiker all require specific kinds of speed.

5. Power

The ability to move the body parts swiftly while simultaneously applying the maximum force of your muscles. The long jump, power lifting, and swimming all require high levels of power.

6. Reaction Time

The ability to react or respond quickly to what you hear, see, or feel. Good reaction time is important to sprinters and swimmers, who must react to starts. The tennis player, boxer, and hockey goalie all require quick reaction times as well.

Fitness Handbook

Creating a Fitness Plan

When planning a personal activity program, choose activities that you enjoy and that you can realistically do. For example, think about what type of activity can realistically fit into your schedule. If your schedule is already full of after-school activities will you be you be tempted to skip workouts?

Another factor to consider when choosing a type of exercise is whether or not the exercise will help the social and mental/emotional sides of your health triangle. If meeting new people is one of your goals, will playing the sport help you meet people with whom you share interests? Also, your cultural background may impact your choices. In the U.S., football, basketball, and baseball are all popular sports. In most of the world, soccer is the most popular sport. You may choose to play soccer because it is a popular sport in the country of your ancestors, and you want to learn more about their lifestyle. Learning about a sport that commonly played in another country may help you learn more about that culture through the sport.

Most importantly, pick an activity that you enjoy. If you do not enjoy the activity, chances are you will find excuses not to exercise. The list below offers other factors may affect your activity choices:

- **Cost.** Some activities require expensive equipment. It may make sense to borrow or rent equipment, rather than buying it, when you try a new sport.
- **Where you live.** Is your local area flat or hilly? What is the climate like? Factors like these will affect the activities that you can do close to home.
- **Your schedule.** If you like to sleep late, planning to jog every morning will probably fail. Choose activities that fit your schedule and habits.
- **Your health and fitness level.** Do you have a health condition that may affect your exercise plan, such as asthma? If so, talk to your doctor before starting a new activity.
- **Personal safety.** When choosing activities, make sure that you have a safe environment to perform them in. For instance, you should not go running on busy streets with no sidewalks.

	Monday		Tuesday		Wednesday		Thursday		Friday	
Week	*Warm Up*	*Activity*	*Warm Up*	*Activity*	*Warm Up*	*Activity*	*Warm Up*	*Activity*	*Warm Up*	*Activity*
1	5 min	5 min	---	---	5 min	5 min	---	---	5 min	5 min
2	5 min	7 min	---	---	5 min	7 min	---	---	5 min	7 min
3	5 min	10 min	---	---	5 min	10 min	---	---	5 min	10 min
4	5 min	12 min	---	---	5 min	12 min	---	---	5 min	12 min
5	7 min	15 min	---	---	7 min	15 min	---	---	7 min	15 min
6	7 min	17 min	---	---	7 min	17 min	---	---	7 min	17 min
7	10 min	20 min	---	---	10 min	20 min	---	---	10 min	20 min
8	10 min	20 min	10 min	20 min	10 min	20 min	---	---	10 min	20 min
9	10 min	20 min	10 min	20 min	10 min	20 min	10 min	20 min	10 min	20 min

Sample Physical Fitness Plan

Fitness Circuit

Many public parks have Fitness Circuits (sometimes called Par Courses) with exercise stations located throughout a park. You walk or run between stations as part of your workout. You may also consider creating your own par course at home. Fitness Circuits can be adapted to a person's individual skill level and ability.

What Will I Need?

- Access to a public park or a home-made Fitness Circuit course.
- Comfortable workout clothes that wick away perspiration.
- Athletic shoes.
- Stopwatch (optional).
- Jump rope, dumbbells, elastic exercise bands, or check out the Fitness Zone Clipboard Energizer Activity Cards, Circuit Training for ideas.

How Do I Start?

- Warm-up with a 5 minute walk and stretching
- Read the instructions at each exercise station and perform the exercises as shown. Use the correct form. Try to do as many repetitions as you can for 30 seconds.
- After you finish the exercise, walk or run to the next station and complete that exercise.

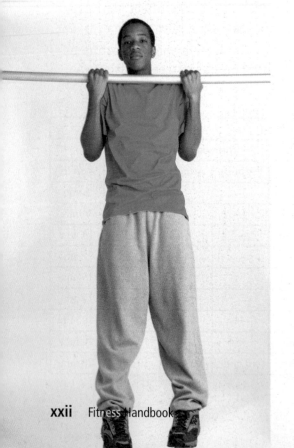

- Check your heart rate to see how intensely you exercised at the end of the Fitness Circuit.
- Cool-down by walking, standing in place and moving your feet up and down, or jogging slowly. End your cool-down with 3 to five minutes of stretching.
- Every month or so, consider adding a new exercise.

How Can I Stay Safe?

- Be alert to your surroundings in a public park. It is best to have a friend with you and it makes exercising even more fun.
- Leave enough room between stations at home to allow you to move and exercise freely. Avoid clutter in your exercise area.
- Perform the exercises correctly and at your own pace.

For more circuit training ideas, visit Cardiovascular Fitness – Circuit Training Activity 7 on the Fitness Zone site in ConnectEd.

Walking

By walking for as little as 30 minutes each day you can reduce your risk of heart disease, manage your weight, and even reduce stress. Walking requires very little equipment and you can do it almost anywhere. Walking is also something you can do by yourself or with friends and family.

What Will I Need?

- Running or walking shoes. Many athletic shoe stores sell both.
- Loose comfortable clothes that wick away perspiration. Layering is also a good idea. Consider adding a hat, sunglasses, and sunscreen if needed.
- Stopwatch and water bottle unless there are water fountains on your route.
- A pedometer or GPS to track your distance.

How Do I Start?

- Warm-up for 10 minutes by walking slowly and stretching.
- Walk upright with good posture. Do not exaggerate your stride or swing your arms across your body.
- Build your time and distance slowly. One mile or 20 minutes every other day may be enough for the first couple weeks. Eventually you will want to walk at least 30–60 minutes five days a week.
- End your walk with a five minute cool-down to stretch your muscles.

How Can I Stay Safe?

- Let your parents know where you will be walking and how long you will be gone.
- Avoid wearing headphones if by yourself or if walking on a road or street.

For more information about using a pedometer, visit the What Do You Need to Be Fit? – Activity 5 on the Fitness Zone site in ConnectEd.

Running

Running uses the large muscles of the legs thereby burning lots of calories and also gives your heart and lungs a good workout in a shorter amount of time. Running also helps get you into condition to play team sports like basketball, football, or soccer. Running can be done on your schedule although it's also fun to run with a friend or two.

What Will I Need?

- A good pair of running shoes. Ask your physical education teacher or an employee at a specialist running shop to help you choose the right pair.
- Socks made of cotton or another type of material that wicks away perspiration.
- Bright colored or reflective clothing.
- A stopwatch or watch with a second hand to time your runs or track your distance.
- Optional equipment might include a jacket or other layer depending on the weather, sunscreen, and sunglasses.

How Do I Start?

Your ultimate goal is to run at least 20 to 30 minutes at least 3 days a week. Use the training schedule shown below. Start by walking and gradually increasing the amount of time you run during each exercise session. Starting slowly will help your muscles and tendons adjust to the increased work load. Try spacing the three runs over an entire week so that you have one day in-between runs to recover. Here is a plan to get you started as a runner:

- Start each run with a brisk 3–5 minute walk to warm-up.
- Take some time to slowly stretch the muscles and areas of the body involved in running. Avoid "bouncing" when stretching or try to force a muscle or tendon to stretch when you start to feel tightness.
- Begin slowly and gradually increase your distance and speed. The running plan included in this section can give you some tips on how to train for a 5K run.
- Use the "talk test." Can you talk in complete sentences during your training runs? If not, you are running too fast. For more information about the talk test, visit Cardiovascular Fitness – Activity 2 on the Fitness Zone site in ConnectEd.

Training Schedule		
M W F	Split Schedule	Duration
Week 1	Brisk 5 min. walk Walk: 60–90 seconds Run: 60 seconds	Repeat for 20 minutes
Week 2 and 3	Brisk 5 min. walk Walk: 60 seconds Run: 60–90 seconds	Repeat for 20 minutes
Week 4	Brisk 5 min. walk Walk: 60 seconds Run: 3–5 minutes	Repeat for 20 minutes
Week 5+	Brisk 5 min. walk Run: 20–30 minutes	

Activities and Sports

Preparing for Sports and Other Activities

Are you are thinking about playing a sport? If so, think about developing a fitness plan for that sport. Some of the questions to ask yourself are: Does the sport require anaerobic activity, like running and jumping hurdles? Does the sport require aerobic fitness, like cross-country running? Other sports, such as football and track require muscular strength. Sports like basketball require special skills like dribbling, passing, and shot making. A workout plan for that sport will help you get into shape before organized practice and competition begins.

What Will I Need?

Talk to a coach or physical education teacher about how to get ready for your sport. You can also conduct online research to learn what type of equipment you will need, such as:

- Proper footwear and workout clothes for a specific sport.
- What facilities are available for training and practice, such as a running track, tennis court, football or soccer field, or other safe open area.
- Where you can access weights and others form of resistance training as part of your training program.

How Do I Start?

Now that your research is done, you can create your fitness plan. Include the type of exercises you will do each training day. Use the Sample Physical Fitness plan on page xxi to create your plan.

- Include a warm-up in your plan.
- List the duration of time that you will work out.
- Plan to exercise 3–5 days a week doing at least one kind of exercise each day. Remember to include stretching before every workout.

How Can I Stay Safe?

- Get instruction on how to use free weights and machines.
- Make sure you start every exercise activity with a warm-up.
- Ease into your Fitness Plan gradually so you do not pull a muscle or do too much too soon.
- Practice good nutrition and drink plenty of water to stay hydrated.

For more individual and team sport ideas, visit Individual and Team Sports – Activities 1–4 on the Fitness Zone site in ConnectEd.

John Flournoy/McGraw-Hill Education

Interval Training

Interval training consists of a mix of activities. First you do a few minutes of intense exercise. Next, you do easier, less-intense activity that enables your body to recover. Interval training can improve your cardiovascular endurance. It also helps develop speed and quickness. Intervals are typically done as part of a running program. However, intervals can also be done riding a bicycle or while swimming. On a bicycle, alternate fast pedaling with easier riding. In a pool, swim two fast laps followed by slower, easier laps.

What Will I Need?

- A running track or other flat area with marked distances like a football or soccer field.
- If at a park, 5–8 cones or flags to mark off distances of 30 to 100 yards.
- A training partner to help you push yourself (optional).

How Do I Start?

- After warming up, alternate brisk walking (or easy jogging). On a football field or track, walk 30 yards, jog 30 yards, and then run at a fast pace for 30 yards. Rest for one minute and repeat this circuit several times. If at a park, use cones or flags to mark off similar distances.

- Accelerate gradually into the faster strides so you stay loose and feel in control of the pace.

- If possible, alternate running up stadium steps instead of fast running on a track. This will help your coordination as well as your speed. Running uphill in a park would have similar benefits.

How Can I Stay Safe?

- Interval training should only be done once or twice a week with a day off between workouts.

- Check with your doctor first if you have any medical condition like high blood pressure or asthma.

For more interval training ideas, visit Cardiovascular Fitness – Activity 6 on the Fitness Zone site in ConnectEd.

Safety and Sportsmanship

Preventing Injuries

While you may be ready to jump right in to a fitness program, it is important to take proper caution to avoid injury. To prevent or safely treat injuries, follow these guidelines:

- Pay attention to your body. If you feel unusually sore or fatigued, postpone activity or exercise until you feel better.
- Include a proper warm-up and cool-down in your personal fitness program.
- Monitor the frequency, intensity, time, and type of your exercise closely. Progress slowly but steadily.
- If you run or walk along busy streets, always face oncoming traffic.
- Wear reflective clothing during night physical activities or exercise, such as walking or jogging.
- Use proper safety equipment for activities with a higher injury risk, such as skateboarding, snowboarding, and cycling.
- Always seek medical attention when you have an injury.

Being a Good Sport

What does it mean to be a good sport? Sportsmanship means you play fairly and follow the rules, respect your opponents, and show polite behavior to coaches, officials, teammates, and opponents. Practicing good sportsmanship also means using appropriate language and not trash-talking your opponents. Supporting your team is a great idea, and can add to the fun. However, using insulting and cruel references to another team or individual does not show good sportsmanship. Participating in sports can be exciting and emotions can be strong, however remember to take a deep breath and think before you say something out of anger. Whether you win or lose, remember that everyone deserves to be treated fairly.

UNIT 1 A Healthy Foundation

Chapter 1
Understanding Health and Wellness

Chapter 2
Taking Charge of Your Health

Building Healthy Communities

Using Visuals America on the Move Foundation is a national nonprofit organization dedicated to promoting healthful eating and active living among individuals, families, and communities. Every year during the month of September, America on the Move sponsors *Step*tember, a campaign promoting physical activity and good nutrition.

Get Involved. Conduct research to identify nonprofit organizations that work to promote healthy living in your community. Contact one organization and find out how teens can volunteer to help.

"Health is a state of complete physical, mental and social well-being, and not merely the absence of disease or infirmity."
— from the constitution of the World Health Organization

CHAPTER **1**

Understanding Health and Wellness

Activating Prior Knowledge

Using Visuals Look at the picture on this page. Write three sentences beginning with the words *These people appear healthy because . . .* Discuss your ideas about what makes someone healthy.

Chapter Launchers

Health in Action

Discuss the **BIG** Ideas

Think about how you would answer these questions:

▶ What is *health*?

▶ Why would you want to be healthy?

▶ Who is most responsible for your health?

Assess Your Health

Read each statement. On a separate sheet of paper, write "yes," "sometimes," or "no" based on your typical behavior.

1. I get between eight and ten hours of sleep each night.

2. I eat at least three nutritionally balanced meals each day, beginning with breakfast.

3. I do at least 20 minutes of aerobic physical activity three or more days a week.

4. I avoid using tobacco, alcohol, and other drugs.

5. I think about my health influences and try to make healthy choices most of the time.

A "yes" response shows that you practice healthy behaviors. "Sometimes" indicates that you should analyze and possibly modify your behavior. A "no" response means that you should modify the behavior.

Digital Vision/Getty Images

GUIDE TO READING

BIG Idea *Being in the best of health throughout your life means making healthy choices and practicing healthful behaviors.*

Before You Read

Create a Cluster Chart. Draw a circle and label it "Health." Use surrounding circles to define and describe this term. As you read, continue filling in the chart with more details.

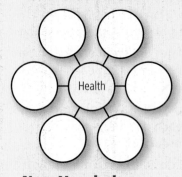

Health

New Vocabulary

▶ health
▶ spiritual health
▶ wellness
▶ chronic disease

Your Total Health

Real Life Issues ·······················

Being Healthy. On the first day of spring semester, Keisha comes home thinking about the challenge her health teacher had presented during class: What does *health* mean to you and your future? Keisha opens her journal and pauses. There are so many things to write about. She starts thinking about what she wants to accomplish in life, and the role that good health will play in helping her achieve her goals.

Writing *Write Keisha's journal entry for this day.*

Take Charge of Your Health

Main Idea You are responsible for your own health.

You probably have several ideas about what makes somebody healthy. Do you picture a healthy person as someone who is physically active and involved in sports? Do you think that a healthy individual gets along well with others and generally feels good about himself or herself? These images are all part of the "big picture" of **health**, *the combination of physical, mental/emotional, and social well-being.* Every day, you make decisions that shape your health. That's what this book is all about: giving you the knowledge and skills you need to take charge of your health for a lifetime.

Your Health Triangle

Main Idea It's important to balance your physical, mental/emotional, and social health.

When you are in good health, you have the energy to enjoy life and pursue your dreams. So, what can you do to stay healthy? Start by understanding the three areas of health.

These include your physical health, mental/emotional health, and social health. It's important to pay attention to all three areas of your health triangle. If you concentrate too much or too little on one area, the triangle can become unbalanced.

Physical Health

Physical health is all about how well your body functions. Having a high level of physical health means having enough energy to perform your daily activities, deal with everyday stresses, and avoid injury.

What does it take to get and keep a healthy body? Here are five important actions you can take:

- Get eight to ten hours of sleep each night.
- Eat nutritious meals and drink eight cups of water each day.
- Engage in 30 to 60 minutes of physical activity every day.
- Avoid the use of tobacco, alcohol, and other drugs.
- Bathe daily, and floss and brush your teeth every day.

Mental/Emotional Health

Mental/emotional health is about your feelings and thoughts. It's a reflection of how you feel about yourself, how you meet the demands of your daily life, and how you cope with the problems that occur in your life.

(b) Lawrence M. Sawyer/Getty Images, (l)©Blend Images/Alamy, (r)Digital Vision/Getty Images

FITNESS ZONE

I keep hearing that we should try to walk 10,000 steps every day. So my best friend and I wear pedometers and walk as much as possible. Instead of taking a bus, we walk. We also use the stairs rather than an escalator or elevator. We've been doing this for three months now, and we feel great! For more physical activity ideas, visit the Fitness Handbook and Fitness Zone sites in ConnectEd.

■ **Figure 1.1** Your health triangle is made up of three equally important areas. *What do you do to stay in good health?*

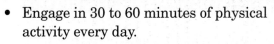

Social

Physical

Mental/Emotional

■ Figure 1.2 Each person strives to keep his or her health triangle in balance throughout life. *How might the health triangles of these two individuals be alike or different?*

People who are mentally and emotionally healthy

- enjoy challenges that help them grow.
- accept responsibility for their actions.
- have a sense of control over their lives.
- can express their emotions in **appropriate** ways.
- usually can deal with life's stresses and frustrations.
- generally have a positive outlook.
- make thoughtful and responsible decisions.

Spiritual Health Mental/emotional health also includes **spiritual health**, *a deep-seated sense of meaning and purpose in life*. Being spiritually healthy does not necessarily mean that you belong to a religious group, although it could include being a member of a spiritual community. Spiritual health involves having a feeling of purpose and a sense of values.

Social Health

Getting along with others, also known as *social health*, is as important to your overall health and wellness as having a fit body and mind. Your social network includes your family, friends, teachers, and other members of your community. You don't need to have lots of friends to have good social health. Sometimes just having a few special people with whom you can share your thoughts and feelings is enough. Maintaining healthy relationships is one way of caring for your social health. This involves

- seeking and lending support when needed.
- communicating clearly and listening to others.
- showing respect and care for yourself and others.

READING CHECK

Identify List the three components of health.

Keeping a Balance

When your health triangle is balanced, you have a high degree of **wellness**, *an overall state of well-being or total health*. Wellness comes from making decisions and practicing behaviors that are based on sound health knowledge and healthful attitudes. Maintaining wellness means keeping a balance among the three components of health.

Think about someone whose friends are the most important part of her life. She enjoys spending a lot of time with them, but she doesn't always get the rest she needs. Her social health is fine, but her physical and mental/emotional health are suffering.

What about a teen who spends all his time working out? He may be physically fit, but he doesn't have many friends, so he sometimes feels lonely and depressed. This teen pays too much attention to his physical health at the expense of his mental/emotional and social health.

Ignoring any area of your health triangle affects your total health. To keep a balance, you need to pay equal attention to all three areas of your health. Throughout this book, you will learn how to make responsible decisions and practice healthful behaviors that will help you keep a balance and maintain your wellness.

■ **Figure 1.3** When you feel your best, you perform at your best. *Which areas of this teen's health triangle are receiving attention?*

The Health Continuum

Main Idea Healthful behaviors will promote your wellness.

Your health and wellness are always changing. For instance, you may feel great one day and catch a cold the next. Your health at any moment can be seen as a point along a *continuum,* or sliding scale, such as the one in **Figure 1.4** on page 10. The continuum spans the complete range of health, from a loss of health and wellness at one end to high-level wellness at the other.

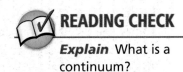

READING CHECK

Explain What is a continuum?

Figure 1.4 The Health Continuum

Your health can be measured on a sliding scale. *Where would you place your health along the continuum right now?*

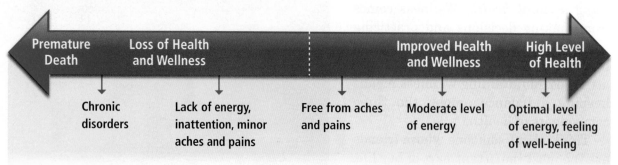

Premature Death	Loss of Health and Wellness		Improved Health and Wellness	High Level of Health
Chronic disorders	Lack of energy, inattention, minor aches and pains	Free from aches and pains	Moderate level of energy	Optimal level of energy, feeling of well-being

As you mature, your position on the continuum continues to change. Many Americans, unfortunately, start moving toward the lower end of the continuum. One-half of all American adults live with a **chronic disease,** *an ongoing condition or illness* such as heart disease, obesity, and cancer. The leading risk factors for many chronic diseases are smoking, lack of physical activity, poor nutrition, being overweight, and lack of health screenings. However, by making a lifelong commitment now to practice healthful behaviors, you will be more likely to maintain a high level of wellness and stay at the higher end of the continuum.

LESSON 1 ASSESSMENT

After You Read

Reviewing Facts and Vocabulary

1. Define the word *health*.

2. List important steps you can take to promote your physical health.

3. What is the health continuum? Describe the continuum's endpoints.

Thinking Critically

4. **Relate.** How can poor mental/emotional health affect physical health?

5. **Predict.** How might your behaviors today affect your health now? How might they affect your health in the future?

Applying Health Skills

6. **Communication Skills.** Create a poster that explains the three areas of health to fourth or fifth graders.

Writing Critically

7. **Persuasive.** Write an editorial for the school newspaper about why each person is responsible for his own health, and why this is important.

Real Life Issues ....

After completing the lesson, review and analyze your response to the Real Life Issues question on page 6.

What Affects Your Health?

Real Life Issues

Too Much Sun. Jason enjoys being outdoors, and he spends a lot of time in the sun as a member of the cross-country team. He always uses sunscreen. During the summer months, he and his friends enjoy swimming and boating at the lake. He has invited his cousin Sean to come to the lake for a week. The first day, when Jason offers Sean some sunscreen, Sean says no. He tells Jason he doesn't like the sticky feeling it leaves on his skin.

Writing *Write a brief dialogue between Jason and Sean. Have Jason try to convince Sean to protect his health by using sunscreen.*

Influences on Your Health

Main Idea Heredity, environment, attitude, behavior, media, and technology can all influence your health.

It is your responsibility to make healthy decisions and take actions to ensure your well-being. Factors such as heredity, environment, attitude, behavior, media, and technology can influence how you live. Understanding these influences will help you make informed decisions about your health.

Heredity

Your **heredity** refers to *all the traits that were biologically passed on to you from your parents.* LaToya inherited her brown eyes, black hair, and tall body type from her parents. LaToya also inherited genes that put her at risk for diabetes, a serious disorder that prevents the body from converting food into energy. Both of LaToya's parents have diabetes.

BIG Idea *Understanding how heredity, environment, and other factors affect your health can help you make healthy decisions.*

Before You Read

Create a K-W-L Chart. Make a three-column chart. In the first column, list what you **k**now about influences on your health. In the second column, list what you **w**ant to know about this topic. As you read, use the third column to summarize what you **l**earned.

K	W	L

New Vocabulary

▶ heredity
▶ environment
▶ peers
▶ culture
▶ media
▶ technology

They have taken steps to control their diet and started an after-dinner walking program to keep their condition from getting worse. By watching her parents, LaToya has learned to eat healthfully, maintain a normal weight, get adequate rest, and stay active. She's learned from her parents' example that these healthful behaviors may help her avoid getting diabetes herself.

It's important to understand the influences heredity has on your health. Ask your parent or grandparent questions about what health conditions and diseases run in your family. Knowing this information can help you take actions to stay well and healthy.

Environment

Your **environment** is *the sum of your surroundings,* including the physical places in which you live and the people who make up your world. The culture you live in is part of your environment as well.

Physical Environment You may not have much control over your physical environment at this time in your life. However, it's still important to recognize how your physical environment can impact all aspects of your health. Some environmental factors that can affect your health include

- neighborhood and school safety.
- air and water quality.
- availability of parks, recreational facilities, and libraries.
- access to medical care.

There are some things in your environment over which you *do* have control. For instance, you can keep your room clean and help reduce litter at your school. How else can you improve your physical environment?

■ **Figure 1.5** Your physical environment influences your health in several ways. *Identify some positive and negative influences in your physical environment. How do you use the positive influences to protect your health? How can you overcome the negative influences?*

■ **Figure 1.6** These teens are being influenced by their physical and social environments. *What are some environmental influences that affect you physically? What are some that affect you socially?*

Social Environment Your social environment is made up of all the people around you, including your family and peers. Your **peers**, *people of the same age who share similar interests,* also include your friends. All these people can be positive role models who support your healthful decisions, or they can increase your health risks. For example, Brandon promised his dad that he wouldn't drink. Hanging out with peers who drink, though, made it hard to keep that promise. Ultimately, Brandon decided to honor his commitment to his dad and found a new group of friends. Peers can have a positive influence on you, too. If your friends are involved in community service, chances are good that you'll join them in such activities.

Culture **Culture** refers to *the collective beliefs, customs, and behaviors of a group.* This group may be an ethnic group, a community, a nation, or a specific part of the world. Culture may include the language you speak, the foods you eat, your spiritual beliefs, and the traditions you practice. These **factors** can be a big influence on your health. For instance, some cultures enjoy a diet based on vegetables, fruits, grains, and very little meat. People from these backgrounds may be less likely to develop high cholesterol and may be better able to maintain their weight than those who eat a higher-fat diet.

Academic Vocabulary

factor *(noun):* an element that contributes to a particular result

Attitude

Your attitude, or the way you view situations, can have a big effect on your health. If you believe that adopting healthful habits will influence your health in positive ways, then you're more likely to make the decision to practice them.

READING CHECK

Analyze Why is it important to understand the influences on your health?

In addition, optimists—people who "see the glass as half full"—are usually in better health than pessimists, who "see the glass as half empty." Even if you have a natural tendency toward pessimism, you can remind yourself to look at challenging situations positively.

Behavior

Although you can't choose your heredity and may have only limited control over your environment, you have total control over your own behaviors. You can choose to avoid high-risk behaviors in favor of healthful behaviors, like choosing low-fat, nutritious foods and participating in daily physical activity.

Media and Technology

Every day you encounter one of the most powerful influences on your health—the media. **Media** are the *various methods for communicating information*. This content is delivered via **technology,** such as *radio, television, and the Internet,* and through print media, like newspapers and magazines. The constant presence of media messages has a significant influence on your decisions.

Media personalities and celebrities might be seen as role models because they get a lot of attention. Sometimes this attention is related to positive achievements, such as excelling in athletics or contributing time and money to help others in need. Sometimes the attention is related to negative behaviors. For example, some actors or models may practice unsafe eating habits in an attempt to maintain an extremely thin appearance. You also know about athletes who take drugs to help them perform better. Characters in movies or TV programs may drive dangerously, use alcohol or other drugs, smoke, or engage in sexual activity. They never seem to face any consequences associated with these behaviors. When you see these behaviors in the media, you may get the false impression that everyone is doing it.

More powerful and far reaching than radio, television, newspapers, and magazines is the Internet, which surpasses all other forms of media as an information source. Thousands of pages of health information from all over the world are available at the click of a mouse. Unfortunately, not all health messages and sources are valid. Some Web sites are sponsored by advertisers who only want you to buy their products. For valid health information, stick to Web sites that have *.gov* and *.edu* in their addresses, or sites maintained by professional health organizations, such as the American Medical Association and the Centers for Disease Control and Prevention (CDC).

■ **Figure 1.7** The HONcode seal tells you that a Web site's information is of high quality. *Why is it important to know whether media messages are trustworthy?*

One way to tell whether a Web site has reliable information is to look for the HONcode. HONcode is run by the Health On the Net Foundation, which is dedicated to improving the quality of online health information. Web sites accredited by the HONcode must follow a strict code of conduct.

Understanding Your Influences

Main Idea You can take control of your health by understanding the factors that influence it.

Think about all the factors that influence your health, including your heredity, your physical and social environments, your culture, your attitudes, your behaviors, and the media. Understanding these influences and committing to a healthy lifestyle are the first steps toward taking charge of your health.

In the pages that follow, you will learn more about risks and behaviors that are harmful to your health. You will gain the knowledge and skills you need to take responsibility for promoting your own health by avoiding these risks. You will also learn ways to promote the health of others. Learning these health skills and knowledge will help you achieve and maintain wellness.

 READING CHECK

Explain How would understanding the influence of media and technology make a difference in your health?

LESSON 2 ASSESSMENT

 After You Read

Reviewing Facts and Vocabulary

1. What does *heredity* mean?
2. Define *environment*. Identify three types of environment.
3. Evaluate two ways that media and technology may influence your health.

Thinking Critically

4. **Evaluate.** How does the environment in which you live affect your health?
5. **Synthesize.** Oliver's family has a history of heart disease. What steps might he take to protect his health?

Applying Health Skills

6. **Communication Skills.** With a classmate, role-play a scenario in which a teen tries to persuade a friend to adopt a positive health behavior.

Writing Critically

7. **Narrative.** Write a short story about Jesse, who just moved from a small town to a large city. Choose one of the possible influences on his health and describe how his well-being might be affected by this influence.

Real Life Issues ....

After completing the lesson, review and analyze your response to the Real Life Issues question on page 11.

LESSON 3

GUIDE TO READING

BIG Idea *Risk behaviors can harm your health, but there are steps you can take to avoid or reduce these risks.*

Before You Read

Create a Cluster Chart. Draw a circle and label it "Health Risks." Use surrounding circles to define and describe this term. As you read, continue filling in the chart with ways to reduce these risks.

Health Risks

New Vocabulary

▶ risk behaviors
▶ cumulative risks
▶ prevention
▶ abstinence
▶ lifestyle factors

Health Risks and Your Behavior

Real Life Issues ·······························

Worrying About a Friend. Jenna and her best friend, Madison, are discussing their plans for the weekend. Jenna is excited because Jackson, a classmate, has invited her to a party on Saturday night. The party is at the home of a classmate whose parents will be away. Madison suspects there will be alcohol at the party and no adult supervision. Jackson has a reputation for being wild, and Madison is worried for Jenna.

Writing *Write a dialogue in which Madison discusses with Jenna the potential dangers of going to the party.*

Identifying Health Risks

Main Idea Engaging in risk behaviors can harm your health.

Every day you are faced with some degree of risk. Simple events, such as crossing a street or using electrical appliances, carry a degree of risk. Being aware of certain risks to your health is part of becoming an adult.

Risk behaviors are *actions that can potentially threaten your health or the health of others*. It's important to recognize that you can control most risk behaviors. By understanding the risks associated with certain behaviors, you can make safe and responsible decisions about which risks to avoid. In this way, you actively protect and promote your health.

Recognizing Risk Behaviors

The Centers for Disease Control and Prevention (CDC) has identified six risk behaviors that account for most of the deaths and disability among young people under age 24.

These risk behaviors can lead to heart disease, cancer, and other serious illnesses later in life:

- Tobacco use
- Unhealthy dietary behaviors
- Inadequate physical activity
- Alcohol and other drug use
- Sexual behaviors that may result in HIV infection, other sexually transmitted diseases, and unintended pregnancies
- Behaviors that contribute to unintentional injuries and violence

Real World CONNECTION

Teen Risk Taking

To track patterns of risk taking among teens, the CDC developed the Youth Risk Behavior Survey (YRBS). It is administered every two years to a sample of high school students across the country. The information gathered in this survey is used in a variety of ways to influence change and improve the health and well-being of teens. Some of the major risk behaviors, with key findings in each category, appear in the graph below. When you analyze the data, you may be surprised. Despite the headlines, most teens are not drinking or using drugs. Most wear automobile safety belts, and two-thirds are physically active.

Activity Mathematics

Use the graph to answer these questions.

1. Approximately what percent of teens did *not* participate in vigorous physical activity three or more days per week?

2. Which risk behaviors did 90 percent of teens avoid?

3. **Writing** Write a paragraph explaining how you think these statistics can be used to promote teens' health and well-being.

Concept **Measurement and Data: Data Analysis** A bar graph represents data using shaded bars to show each value. The legend explains what each shade represents.

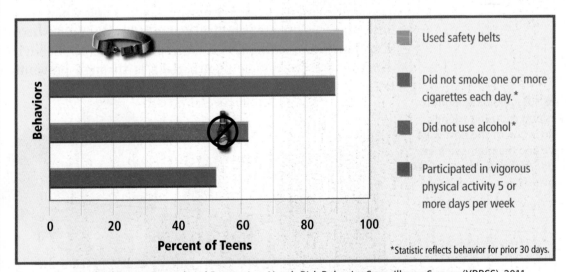

Legend:
- Used safety belts
- Did not smoke one or more cigarettes each day.*
- Did not use alcohol*
- Participated in vigorous physical activity 5 or more days per week

*Statistic reflects behavior for prior 30 days.

Source: Centers for Disease Control and Prevention, Youth Risk Behavior Surveillance Survey (YRBSS), 2011.

Risks and Consequences

Risk behaviors can have a serious impact on your health. In other words, these behaviors carry significant consequences. Both the short-term and long-term consequences can harm your health and well-being. Some risk behaviors can even be fatal. Before you engage in risk behaviors, it's important to evaluate the consequences. For example, smoking can have immediate health consequences, such as bad breath, yellow teeth, and headaches. If a person continues to smoke, the long-term consequences can include lung cancer, emphysema, and heart disease.

Risks can also add up over time. **Cumulative risks** are *related risks that increase in effect with each added risk.* Eating an occasional high-fat meal at a fast-food restaurant probably won't permanently affect your overall health. If you regularly eat high-fat meals, though, the negative effects accumulate over time and may lead to serious health problems.

Cumulative risks also increase when several risk factors are combined. For example, using a cell phone while driving carries risks. So does speeding. If an individual engages in both of these risk behaviors, the chance of getting into a car accident becomes even greater. The more risk behaviors you participate in, the more likely you are to experience negative consequences.

READING CHECK

Evaluate Why is it important to understand risk behaviors?

How to Avoid or Reduce Risks

Main Idea You can take action to reduce your exposure to health risks.

You can protect your health and minimize the possibility of risk by practicing positive health behaviors. Many of your automatic safety checks—wearing a safety belt when you get into a car, checking the depth of water before diving, or wearing a helmet when riding a bike—are positive health behaviors. Another way to reduce health risks is through **prevention**.

■ **Figure 1.8** Avoiding high-risk behaviors and choosing friends who do so is one of the best ways to achieve and maintain wellness. *How might friends help you avoid risk behaviors?*

■ **Figure 1.9** Participating in positive health behaviors benefits all three sides of your health triangle. *What are some positive health behaviors you and your friends enjoy?*

This means *taking steps to keep something from happening or getting worse.* Prevention includes getting regular medical and dental checkups. Checkups can detect health problems early, thus preventing them from getting worse.

Abstaining from High-Risk Behaviors

One of the most effective strategies for protecting your health is practicing abstinence. **Abstinence** is *a deliberate decision to avoid high-risk behaviors, including sexual activity and the use of tobacco, alcohol, and other drugs.*

All areas of your health triangle benefit when you choose to abstain from high-risk behaviors. For example, when you avoid tobacco, alcohol, and other drugs, you protect yourself from the chronic diseases associated with using these substances. You also feel good about yourself, which strengthens your mental/emotional health and your social relationships.

When you abstain from high-risk behaviors, you show that you value your well-being. You demonstrate maturity by taking responsibility for your health and playing an active role in maintaining your wellness.

Promoting Your Health

Main Idea Regularly participating in health-promoting behaviors will help you reach a high level of wellness.

Every day you make decisions, large and small, that **affect** your health. For example, if you choose to play a sport after school, you are likely to have fun and feel energized. If you choose to play video games instead, you may end up feeling sluggish because you didn't get enough physical activity. Understanding how your decisions impact your health will inspire you to adopt healthful behaviors that can promote wellness and prevent the development of disease.

Academic Vocabulary

affect *(verb):* to produce an effect upon

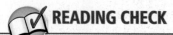

Lifestyle Factors

Lifestyle factors are *the personal habits or behaviors related to the way a person lives.* Scientists have found that these habits make a difference in people's overall health, happiness, and longevity. In other words, people who practice positive health habits regularly tend to be healthier and live longer. Lifestyle factors that can improve a person's level of health include

- getting eight hours of sleep each night.
- starting each day with a healthy breakfast.
- eating a variety of nutritious foods each day.
- being physically active for 30 to 60 minutes most days of the week.
- maintaining a healthy weight.
- abstaining from smoking or using other tobacco products.
- abstaining from the use of alcohol and other drugs.

Think about your daily habits. Do you regularly practice the lifestyle factors listed above? Can you think of ways to incorporate more of these behaviors into your daily routine? Remember, you have control over your lifestyle. By making the best possible decisions for yourself, you can achieve a high level of wellness now and into adulthood.

LESSON 3 ASSESSMENT

 After You Read

Reviewing Facts and Vocabulary

1. Define the term *risk behavior*.
2. Why is cumulative risk a serious concern?
3. How might changes in lifestyle factors influence your health in positive ways?

Thinking Critically

4. **Explain.** How might monitoring risk behaviors affect the well-being of teens?
5. **Synthesize.** Consider a risk behavior teens are exposed to, and predict how lifestyle factors can positively influence teens to avoid that risk.

Applying Health Skills

6. **Accessing Information.** Research organizations that offer after-school programs to help teens avoid risk behaviors. Write a short description of one such organization in your community.

Writing Critically

7. **Expository.** Using the data shown on page 17, write an article about the results of recent research on how many teens avoid risk behaviors.

Real Life Issues ....

After completing the lesson, review and analyze your response to the Real Life Issues question on page 16.

Promoting Health and Wellness

Real Life Issues •••••••••••••••••••••••••

Learning from Experience. Taylor is taking an elective class called Intergenerations. Students in this class are paired with older adults. Taylor's "classmate" is an active 89-year-old man named Harry. Harry remembers riding in a horse-drawn cart from his family farm to church on Sunday mornings. He also remembers there was no television while he was growing up, and that his family ate what they grew on the farm. Taylor's assignment is to interview Harry about his secrets to a long, healthy, and happy life.

Writing *Write a short questionnaire listing what Taylor might ask Harry. Cover all the factors you think might contribute to a long, healthy life.*

The Importance of Health Education

Main Idea Individual, family, community, and national health require planning and responsible behavior on everyone's part.

Achieving a high level of wellness means a higher quality of life for each individual. It means more time in which to feel physically and mentally healthy, to enjoy family and friends, and to achieve your personal goals.

Keeping people healthy is also a good investment. In 2012, the cost of U.S. healthcare reached $2.8 trillion, or $8,915 per person. Much of that expense could be avoided if people made healthier decisions about the way they live, adopted health-promoting habits, and took responsibility for maintaining their wellness.

GUIDE TO READING

BIG Idea *Staying healthy takes knowledge, a plan, and practicing healthful behaviors.*

Before You Read

Create Vocabulary Cards. Write each new vocabulary term on a separate note card. For each term, write a definition based on your current knowledge. As you read, fill in additional information related to each term.

Health Education

New Vocabulary

▶ health education
▶ *Healthy People*
▶ health disparities
▶ health literacy

Educating the public is the key to creating a healthier nation. **Health education** includes *providing accurate health information and teaching health skills to help people make healthy decisions*. Understanding health information and learning health skills empower people to live healthfully and improve their quality of life.

The Nation's Health Goals

Helping you to maintain good health is an important goal of the federal government. The U.S. government releases national health goals and objectives through the *Healthy People* program, *a nationwide health promotion and disease prevention plan designed to serve as a guide for improving the health of all people in the United States*. The *Healthy People* plan changes every 10 years. It outlines goals to improve the health status of the U.S. population within a 10 year period of time. The U.S. government is currently focusing on the goals established in the *Healthy People 2020* plan. As you can see in **Figure 1.10**, one of the goals of Healthy People 2010 is to reduce the overweight and obesity rates in the U.S.

By developing programs to promote health and prevent disease, *Healthy People* provides a common plan for everyone to follow. National, state, and local health agencies across the country carry out programs based on the plan's goals. The government tracks health behaviors and outcomes to measure their progress in achieving the national objectives.

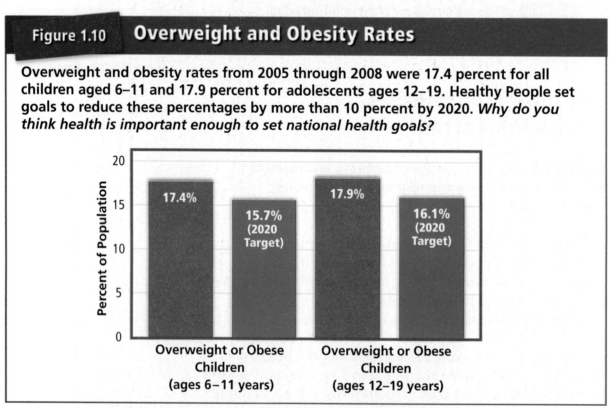

Figure 1.10 Overweight and Obesity Rates

Overweight and obesity rates from 2005 through 2008 were 17.4 percent for all children aged 6–11 and 17.9 percent for adolescents ages 12–19. Healthy People set goals to reduce these percentages by more than 10 percent by 2020. *Why do you think health is important enough to set national health goals?*

Source: CDC | NCHS, National Health and Nutrition Examination Survey (NHANES) 2007–2008.

TEENS   Making a Difference

"I'm helping to make things happen."

Mentoring Others

Matt M. from Georgia has found a way to save teens from boredom. "I heard about the Youth Council of Fayette County (YCFC) and their mission . . . to change the lives of teens in our county."

YCFC schedules all kinds of events, including the Battle of the Bands competition, holiday caroling at a nursing home, monthly trivia nights at a local pizzeria, and volunteering at the local thrift store.

Since joining YCFC three years ago, Matt has had little time to be bored. "Our long-term goal is to build a teen center where middle school students can go after school to get help with homework and play games. High school students would be mentors," says Matt. "Instead of complaining about my town, I'm helping to make things happen."

Activity Write your answers to the following questions in your personal health journal.

1 What is the purpose of the Youth Council of Fayette County?

2 How will the teen center help teens become leaders?

3 What steps would you take to create a Youth Council in your community?

Goals of *Healthy People*. *Healthy People 2010* established two general goals for the future: increase the quality and length of a healthy life for all Americans, and remove differences in health outcomes that result from factors such as gender, race, education, disability, and location. These *differences in health outcomes among groups* are called **health disparities**. Working toward these two goals will ensure that more people can enjoy the benefits of a healthy life.

Healthy People 2020 offers a blueprint that will shape the nation's health priorities until 2020. The health goals that guided early development of *Healthy People 2020* include the following:

1. Promote the best possible health in order to end preventable death, illness, injury, and disability.

2. Eliminate health disparities.

3. Make wellness a way of life and enhance quality of life for individuals and communities.

4. Promote healthy places and environments.

READING CHECK

List What are some of the goals for *Healthy People 2020*?

Becoming Health Literate

Tim Fuller Photography

Main Idea A health-literate person knows how to find and use reliable health information.

Every day people all across the country have to make important decisions that affect their health. To become an informed individual who can make sound health decisions, one must

- know where to find health information.
- decide if the information is correct.
- assess the risks and benefits of treatment.
- figure out how much medicine to take.
- understand test results.

What You Can Do

In order to increase your knowledge and take steps to improve your wellness, you need to develop health literacy.

■ **Figure 1.11** People who are informed know how to interpret the information they need to make good health decisions. *Where do you find information to make your daily health decisions?*

Health literacy refers to *a person's capacity to learn about and understand basic health information and services, and to use these resources to promote one's health and wellness.* Experts believe that poor health literacy influences a person's health more than age, income, and education.

Qualities of a health-literate individual include being

- **a critical thinker and problem solver**—a person who can develop evaluation criteria for health information before making decisions. This person knows how to apply these criteria to make responsible, healthy choices.

- **a responsible, productive citizen**—someone who acts in a way that promotes the health of the community. This person chooses safe, healthful, and legal behaviors that are consistent with family guidelines and that show respect for the individual and others.

- **a self-directed learner**—someone who searches for health information to make health-related decisions. This person knows how to evaluate health information to determine if it is reliable, accurate, and current. Such information is available on television and radio, on the Internet, and from health care professionals.

- **an effective communicator**—a person who is able to express health knowledge in a variety of ways.

READING CHECK

Explain What are the attributes of a health-literate person?

Academic Vocabulary

consistent *(adjective):* free from variation or contradiction

---

LESSON 4 ASSESSMENT

 After You Read

Reviewing Facts and Vocabulary

1. Why is health education important?
2. What are *health disparities*?
3. List three criteria that are needed for an individual to make sound health decisions.

Thinking Critically

4. **Analyze.** How does *Healthy People* hope to help the United States become a healthier country?
5. **Synthesize.** What are some steps you can take to become a health-literate individual?

Applying Health Skills

6. **Accessing Information.** Work with classmates to compile a list of resources in your community that support healthy lifestyle behaviors. Examples might include parks, libraries, and health organizations.

Writing Critically

7. **Expository.** Write an essay explaining what individuals, families, and communities can do to promote wellness.

Real Life Issues ....

After completing the lesson, review and analyze your response to the Real Life Issues question on page 21.

Hands-On HEALTH

 Activity **Reducing the Risks**

This activity encourages you to advocate for healthy change. Your group will research one of the top six risk behaviors that result in death of people under age 24. Risks include: alcohol and drug use, injury and violence, tobacco use, nutrition, physical inactivity, and sexual activity.

What You'll Need

- 2 sheets of 8½" × 11" paper
- poster board
- markers or paints

What You'll Do

Step 1

Research and identify six ways that one of the risk behaviors can affect physical, mental/emotional, and social health.

Step 2

Using your research, list ten lifestyle choices that could reduce the risk. Write your list on the second sheet of paper. Create a poster presenting your findings.

Step 3

Present your poster to the class. Urge the class to make healthy choices in support of *Healthy People 2010*.

Apply and Conclude

Advocate for reducing risk behaviors. Write a letter to the editor encouraging others to make healthful choices.

Checklist: Advocacy

☑ Did we take a clear, health-enhancing stand?

☑ Can we support our position with reliable sources?

☑ Did we demonstrate an awareness of our target audience?

☑ Did we deliver the message with enough passion and conviction?

LESSON 1

Your Total Health

Key Concepts

▶ Health is the combination of physical, mental/emotional, and social well-being.

▶ It is important to balance the three components of health.

▶ Making a lifetime commitment to practice healthful behaviors can improve your long-term well-being.

Vocabulary

▶ health (p. 6)
▶ spiritual health (p. 8)
▶ wellness (p. 9)
▶ chronic disease (p. 10)

LESSON 2

What Affects Your Health?

Key Concepts

▶ Your heredity plays a role in your health and wellness.

▶ You cannot always control your physical environment, but you can look for ways to overcome its negative influences.

▶ A positive attitude and healthful behaviors promote wellness.

Vocabulary

▶ heredity (p. 11)
▶ environment (p. 12)
▶ peers (p. 13)
▶ culture (p. 13)
▶ media (p. 14)
▶ technology (p. 14)

LESSON 3

Health Risks and Your Behavior

Key Concepts

▶ Risk behaviors can harm your health and the health of others.

▶ Risk behaviors that contribute to illness and disability include tobacco use, unhealthy dietary behaviors, inadequate physical activity, and alcohol and other drug use.

▶ Abstaining from high-risk behaviors will protect your health.

Vocabulary

▶ risk behaviors (p. 16)
▶ cumulative risks (p. 18)
▶ prevention (p. 18)
▶ abstinence (p. 19)
▶ lifestyle factors (p. 20)

LESSON 4

Promoting Health and Wellness

Key Concepts

▶ Health education is the key to creating a healthier nation.

▶ The national health goals of *Healthy People* provide guidelines for promoting health and preventing disease.

▶ A health-literate person has the necessary skills to function in today's health promotion and disease prevention environment.

Vocabulary

▶ health education (p. 22)
▶ *Healthy People* (p. 22)
▶ health disparities (p. 23)
▶ health literacy (p. 25)

LESSON 1

Vocabulary Review

Use the vocabulary terms listed on page 27 to complete the following statements.

1. _____ is the combination of physical, mental/emotional, and social well-being.

2. _____ provides people with a deep-seated sense of meaning and purpose in life.

3. A person with a balanced health triangle is said to have a high degree of _____.

Understanding Key Concepts

After reading the question or statement, select the correct answer.

4. Which of the following is *not* an aspect of physical health?
   a. Eating well and drinking water
   b. Making and keeping friends
   c. Being physically active
   d. Getting enough sleep

5. Which statement is true about a person who is in good social health?
   a. She spends a lot of time alone.
   b. She has few friends at school.
   c. She may not be in good physical health.
   d. She gets along with others.

6. You are likely to move in a negative direction on the health continuum if
   a. you engage in physical activity daily.
   b. you accept responsibility for your health.
   c. you fail to practice healthful behaviors.
   d. you regularly eat a healthful diet.

Thinking Critically

After reading the question or statement, write a short answer using complete sentences.

7. **Describe.** What are some characteristics of a person with good mental/emotional health?

8. **Explain.** How can your health triangle become unbalanced, and how can this imbalance affect your health?

9. **Identify.** List specific actions that teens can take to improve their wellness.

Digital Vision/Getty Images

10. **Explain.** How do your health behaviors affect your position on the health continuum?

LESSON 2

Vocabulary Review

Correct the sentences below by replacing the italicized term with the correct vocabulary term.

11. Your *environment* consists of traits that are biologically passed on to you by your parents.

12. A person's ethnicity, religion, and language are part of her *peers*.

13. *Technology* personalities may become our role models for how to behave.

Understanding Key Concepts

After reading the question or statement, select the correct answer.

14. What technique can you use to locate valid health information on the Internet?
   a. Find sites that are the most popular.
   b. Use only sites belonging to manufacturers of health care products.
   c. Locate sites that use *.gov* or *.edu* in their addresses.
   d. All of the above

15. Cultural influences on your health include
    a. biologically inherited traits.
    b. beliefs, customs, and behaviors.
    c. the health continuum and triangle.
    d. all of the above.

16. The media is a powerful influence because it
    a. encourages teens to live healthy lives.
    b. is constantly present.
    c. provides healthy role models.
    d. warns the audience of risk behaviors.

Thinking Critically

After reading the question or statement, write a short answer using complete sentences.

17. **Describe.** What are some ways that peers can influence your health both positively and negatively?

18. **Analyze.** Think about your own culture, including your ethnic background, spirituality, language, and community. What are some practices within your culture that influence your health?

19. **Describe.** What are some ways that your attitudes influence your health?

LESSON 3

Vocabulary Review

Choose the correct term in the sentences below.

20. *Risk behaviors / Prevention* means taking steps to keep something from happening or getting worse.

21. *Resiliency / Abstinence* is a deliberate decision to avoid high-risk behaviors.

22. *Cumulative / Serious* risks are risks that add up over time.

Understanding Key Concepts

After reading the question or statement, select the correct answer.

23. Which of the following statements is true?
    a. Risk behaviors are illegal for everyone.
    b. Risk behaviors can harm your health.
    c. Most teens engage in risk behaviors.
    d. There is no way to avoid risk behaviors.

24. A person who practices multiple risk behaviors at the same time is likely to
    a. be unaware of what he is doing.
    b. be a role model for his friends.
    c. face more negative consequences.
    d. show sound judgment.

25. Personal habits and behaviors that relate to the way a person lives are called
    a. negative consequences.
    b. risk participation.
    c. health promotion.
    d. lifestyle factors.

Thinking Critically

After reading the question or statement, write a short answer using complete sentences.

26. **Identify.** What are two risk behaviors that pose a threat to the health of teens today?

27. **Discuss.** How are teens' perceptions of risk behaviors influenced by what they believe others are doing?

28. **Synthesize.** What is abstinence, and what are the effects of practicing abstinence?

LESSON 4

Vocabulary Review

Use the vocabulary terms listed on page 27 to complete the following statements.

29. _____ empowers people to live healthfully and improve their quality of life.

30. Health goals for the United States may be found in _____.

31. A person who lacks _____ finds it difficult to obtain, understand, and use valid health information.

Understanding Key Concepts

After reading the question or statement, select the correct answer.

32. Health education provides
    a. medical health coverage.
    b. accurate health information.
    c. a wellness guarantee.
    d. none of the above.

33. Experts think that poor health literacy influences a person's health more than
    a. critical thinking and problem solving.
    b. attitude, environment, and income.
    c. education, income, and attitude.
    d. age, income, and education.

Thinking Critically

After reading the question or statement, write a short answer using complete sentences.

34. **Explain.** Why is it a good investment for the United States to keep its citizens healthy?

35. **Evaluate.** What role does the individual play in helping the nation achieve the goals of *Healthy People*?

36. **Describe.** How can health education help the nation achieve the *Healthy People* goals?

37. **Analyze.** How does being a self-directed learner affect a person's health literacy?

Technology   PROJECT-BASED ASSESSMENT

A Health Initiative

Background

Healthy People is an initiative set forth by the Department of Health and Human Services. The initiative establishes guidelines and goals that various people, states, communities, and professional organizations can use to improve the health of all Americans.

Task

Conduct an online search for the goals and guidelines of the initiative, and prepare a multimedia presentation that applies the guidelines to your school.

Audience

Students at your school

Purpose

Create a set of specific recommendations that can help the students in your school be healthier.

Procedure

1. Review the information in Chapter 1 regarding general health and wellness.

2. With your group, develop a policy for improving school health. Create an outline of the multimedia presentation.

3. Divide the main task into smaller tasks, including the Internet search, creating the slides, and delivering the presentation. Assign tasks to each group member.

4. After the research is complete, meet as a group to discuss the possible applications of *Healthy People* in your school.

5. Show your multimedia presentation to the class.

Math Practice

Analyze Geometric Properties. Read the passage, and use the equilaterial triangle ABC to answer the questions.

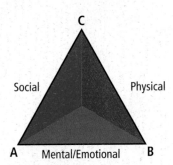

Natalie is doing a report on the health triangle. She sketched a model, but now she needs to make a larger version on poster board so that everyone in the classroom can see it. In her sketch, side AC is 9.5 cm long. Natalie decides to make the triangle on the poster board 3.25 times larger than her scale model.

1. When Natalie finishes drawing the large triangle on the poster board, what will be the approximate measure of side BC? Round to the nearest centimeter.

2. Side AB of the health triangle measures $5x$. Side BC measures $x + 20$. Which of the following statements explains why the equation $5x = x + 20$ can be used to solve for x?
   A. The angle measures of an equilateral triangle are never equal.
   B. Equilateral triangles have unequal sides.
   C. All sides of an equilateral triangle are always equal.
   D. Only two sides of an equilateral triangle are equal.

3. How might Natalie visually represent the health triangle of someone who neglects one or more aspects of health?

Reading/Writing Practice

Understand and Apply. Read the passage below, and then answer the questions.

(1) Veronica was late for her soccer game. (2) On her way to the field, she realized she had forgotten to pack her water bottle. (3) She did not think to get a drink from a nearby water fountain.
(4) Near the end of the first half, Veronica's leg cramped up. (5) After resting and drinking a bottle of water, she began to feel better. (6) But her coach told her that she had become dehydrated and refused to let her back into the game. (7) Veronica hadn't even felt thirsty before she got the cramp.
(8) Veronica didn't recognize you cannot count on thirst for knowing when you need water. (9) People can become dehydrated without feeling thirsty.
(10) Before playing any sport, make sure you drink plenty of noncarbonated fluids.

1. Which sentence includes details that support the author's point of view?
   A. Sentence 2   C. Sentence 9
   B. Sentence 7   D. Sentence 10

2. How does the writer show that the purpose of this essay is to persuade?
   A. The writer emphasizes the use of proper-fitting protective gear.
   B. The writer contrasts the different types of soccer gear people use.
   C. The writer explains that soccer's increasing popularity has led to more injuries.
   D. The writer describes what happened to someone who did not drink enough water.

3. Write a paragraph explaining the importance of drinking plenty of water before, during, and after sports activities.

National Education Standards

Math: Geometry
Language Arts: LACC.910.L.3.6

CHAPTER 2

Taking Charge of Your Health

Ed-Imaging

Activating Prior Knowledge

Using Visuals Look at what is happening in this photo. Why is it important to comparison shop when buying health-related products? Explain your thoughts in a short paragraph.

Chapter Launchers

Health in Action

Discuss the **BIG Ideas**

Think about how you would answer these questions:

▶ What is a health skill?

▶ How can health skills help you achieve wellness?

▶ Why should you be a health-literate consumer?

Assess Your Health

Read each statement. On a separate sheet of paper, write "yes," "sometimes," or "no" based on your typical behavior.

1. I use "I" messages to express my feelings.

2. I manage my stress in healthful ways.

3. I use refusal skills to say no to unhealthy behaviors.

4. I think about creating short-term and long term goals and how to reach my goals.

5. I know how to use consumer skills to make good decisions when buying health products.

A "yes" response shows that you practice healthy behaviors. "Sometimes" indicates that you should analyze and possibly modify your behavior. A "no" response means that you should modify the behavior.

Ryan McVay/Lifesize/Getty Images

GUIDE TO READING

BIG Idea *You can develop skills that will help you manage your health throughout your life.*

Before You Read

Create Vocabulary Cards. Write each new vocabulary term on a separate note card. For each term, write a definition based on your current knowledge. As you read, fill in additional information related to each term.

Health Skills

New Vocabulary

▸ health skills
▸ interpersonal communication
▸ refusal skills
▸ conflict resolution
▸ stress
▸ stress management skills
▸ advocacy

Building Health Skills

Real Life Issues

Fitting in Fitness. Alejandro is carrying a full schedule of advanced courses this semester. He also plays an instrument in the school jazz band and has a part-time job at the grocery store. Alejandro wants to add some physical activity to his schedule, but can't figure out how to fit it in. He asks his good friend Phil for suggestions.

Writing *Write a conversation in which Alejandro explains his situation to Phil. Phil should be supportive and offer possible strategies that can help Alejandro add physical activity into his schedule.*

Learning Health Skills

Main Idea Health skills help you manage your health.

Health skills are *specific tools and strategies to maintain, protect, and improve all aspects of your health.* Health skills are also called *life skills*, because once you've developed these skills, you can use them throughout your life to stay healthy. **Figure 2.1** lists the health skills you will learn in this chapter. You will have opportunities to practice them throughout the rest of the book.

Communication Skills

Main Idea Good communication is a vital health skill.

Three health skills—interpersonal communication, refusal skills, and conflict resolution—deal with how you give and receive information. Communication is more than just talking.

Figure 2.1   The Health Skills

Health skills are tools that help you take responsibility for your health.

Health Skill	Benefit to Your Health
Communication	You share your ideas and feelings, and listen carefully when others express theirs.
Refusal	You say no to unhealthy behaviors.
Conflict Resolution	You resolve problems with others in healthy ways.
Accessing Information	You locate valid sources of health information, products, and services.
Analyzing Influences	You understand the many influences on your health, including peers, family, culture, media, and technology.
Practicing Healthful Behaviors	You act to reduce risks and protect yourself against illness and injury.
Stress Management	You use healthy ways to reduce and manage stress in your life.
Advocacy	You work to improve your own health and the health of your family and your community.
Decision Making	You use a step-by-step process to evaluate your options and make healthy choices.
Goal Setting	You set goals and develop a plan to achieve those goals.

It involves carefully choosing your words and expressions to clearly say what you really mean. It also involves listening closely to others. **Interpersonal communication**—*the exchange of thoughts, feelings, and beliefs between two or more people*—helps you build strong relationships with others.

You can strengthen your interpersonal communication skills by doing the following:

- **Use "I" messages to express your feelings.** Saying "I feel upset when I'm left out of our plans" focuses on your emotions rather than placing blame.

- **Communicate with respect and caring.** Keep your voice calm and use a respectful tone when talking to another person.

- **Be an active listener.** Pay attention to what the other person is saying. Let him say what he has to say without interrupting. Try to understand the other person's point of view.

You will learn more about interpersonal communication in Chapter 6.

Figure 2.2 Refusal Strategies

These refusal strategies can help you say no to potentially harmful activities.

- **Say NO in a Firm Voice.** Do this calmly and clearly. Use expressions such as "No, I'd rather not."
- **Explain Why.** State your feelings. Tell the other person that the suggested activity or behavior goes against your values or beliefs.
- **Offer Alternatives.** Suggest a safe, healthful activity to do instead of the one offered.
- **Stand Your Ground.** Make it clear that you don't intend to back down from your position.
- **Leave If Necessary.** If the other person continues to pressure you, or won't take no for an answer, simply walk away.

Refusal Skills

Refusal skills are *communication strategies that can help you say no when you are urged to take part in behaviors that are unsafe or unhealthful, or that go against your values.* Someone may ask you to ride in a car with a driver who has been drinking. Or, someone may offer you a cigarette even though you don't smoke. Developing strong refusal skills helps you say no firmly, respectfully, and effectively.

Figure 2.2 lists important refusal strategies that will help you the next time a person tries to influence you to engage in an activity that you don't want to do. You may use one or several of these strategies in your refusal. Chapter 8 provides additional information on refusal strategies.

Conflict-Resolution Skills

Think of a recent argument you had. How was it resolved? Were all the people involved satisfied with the outcome? If so, you probably used the skill of **conflict resolution**, *the process of ending a conflict through cooperation and problem solving.* This health skill can help people resolve problems in ways that are agreeable to everyone involved.

Conflict-resolution skills include stepping away from an argument, allowing the conflict to subside, using good interpersonal communication skills, and maintaining an attitude of respect for yourself as well as for the other person. Sometimes, individuals must make a compromise in order to resolve the conflict. In a compromise, both parties give up something but still gain a desired result. You will learn more about conflict resolution in Chapter 9.

READING CHECK

Identify Name three communication skills that help protect your health.

Accessing Information

Main Idea Use reliable sources of health information.

Knowing how to find and evaluate health information will help you make decisions that benefit your well-being. To decide whether health information is valid, you need to determine the reliability of the group or individual sharing the information. Some valid sources include

- health care providers and professionals.
- valid Internet sites, such as those of government agencies and professional health organizations.
- parents, guardians, and other trusted adults.
- recently published material written by respected, well-known science and health professionals.

READING CHECK

Explain Why is the ability to find valid health information an important tool for protecting your health?

Analyzing Influences

Main Idea Understanding what influences you helps you to make more healthful choices.

Do you ever stop and think about *why* you do the things you do? Many factors can influence our decisions and actions. **Figure 2.3** identifies some examples of influences on our behaviors. The more aware you are of the various influences in your life and how they affect *you,* the better able you are to make informed choices about your health.

Figure 2.3 Influences on Your Health

Many factors influence your health. *Which sources have the most influence on you?*

Personal Values • Things I think are important • Likes and dislikes • Skills and talents	**Your Family and Culture** • Beliefs, behaviors, and habits • Family traditions • Food served at home
Personal Beliefs • Plans for the future • Goals • Hopes and dreams	**Media and Technology** • TV and movies • Magazines • Internet
Perceptions • Behaviors that I think are common or accepted	**Friends and Peers** • Behaviors and opinions of my friends and classmates
Curiosity/Fears • Things I wonder about • Things that scare or frighten me • Things I want to try • Things I never want to try	**School and Community** • Place where I live • School I attend • Air quality • Sources of recreation

Figure 2.4

Health Behaviors Checklist

Developing good health habits is important to maintaining your health.
Which of these health habits do you practice every day?

- ❑ I eat well-balanced meals, including breakfast, and I choose healthful snacks.
- ❑ I get regular daily physical activity and at least eight hours of sleep every night.
- ❑ I avoid using tobacco, alcohol, and other drugs.
- ❑ I floss and brush my teeth regularly.
- ❑ I wear a safety belt every time I ride in a car.
- ❑ I stay within 5 pounds of my healthy weight.
- ❑ I practice good personal hygiene habits.
- ❑ I get regular physical checkups.
- ❑ I keep a positive attitude.

- ❑ I express my emotions in healthy ways.
- ❑ I take responsibility for my actions.
- ❑ I think of my mistakes as chances to learn.
- ❑ I relate well to family, friends, and peers.
- ❑ I have one or more close friends.
- ❑ I treat others with respect.
- ❑ I use refusal skills to avoid risk behaviors.
- ❑ I get along with many kinds of people.
- ❑ I can put myself in other people's place and understand their problems.
- ❑ I volunteer to help others whenever I can.

Self-Management Skills

Main Idea Practicing healthy habits will protect your health.

Academic Vocabulary

promote *(verb):* to contribute to the growth of

Self-management means taking charge of your own health. When you manage your behaviors, you act in ways that protect your health and **promote** your own well-being. There are two self-management skills:

- **Practicing healthful behaviors.** You practice healthful behaviors when you make good health habits part of your everyday life. Take a look at the checklist shown in **Figure 2.4.** These positive behaviors can contribute to all aspects of your health.

- **Managing Stress.** Do you get nervous just before a test? Do you get stage fright? These are signs of **stress**, *the reaction of the body and mind to everyday challenges and demands.* Stress is a normal part of life, but too much unrelieved stress can lead to illnesses. That's why it's important to learn **stress management skills**, *skills that help you reduce and manage stress in your life.* Exercising, relaxation, and managing time efficiently are some effective ways to manage stress. You will learn more about stress management in Chapter 4.

Figure 2.5 Self-management skills help you stay healthy. *What skill are these teens practicing?*

Advocacy

(Main Idea) **Advocacy lets you share your health knowledge.**

At the beginning of this unit, on pages 2–3, you saw a picture of teens participating in a community event. This is an example of **advocacy**, *taking action to influence others to address a health-related concern or to support a health-related belief.* Participating in such activities allows you to encourage others to practice healthful behaviors. You can also advocate for better health by obeying laws that protect community health, sharing health information with family and friends, and developing and sending out health messages.

LESSON 1 ASSESSMENT

 ### After You Read

Reviewing Facts and Vocabulary

1. Define the term *health skills*.
2. What are two interpersonal communication skills that can reduce your health risk?
3. What is *advocacy*?

Thinking Critically

4. **Synthesize.** Why is it important to recognize and analyze the various influences on your behavior?
5. **Analyze.** How can advocacy help you with health issues that are important to you?

Applying Health Skills

6. **Stress Management.** List all the healthful strategies you used in the past week to relieve stress. Which ones were most helpful?

Writing Critically

7. **Narrative.** Marcos and Sarah disagree about which movie to see. Write a dialogue in which they resolve their disagreement using effective interpersonal communication strategies.

Real Life Issues ....

After completing the lesson, review and analyze your response to the Real Life Issues question on page 34.

Mike Watson Images/SuperStock RF

GUIDE TO READING

BIG Idea *You can actively promote your well-being by making healthful choices and setting positive goals.*

Before You Read

Create a K-W-L Chart. Make a three-column chart. In the first column, list what you **k**now about decision making and goal setting. In the second column, list what you **w**ant to know about this topic. As you read, use the third column to summarize what you **l**earned.

K	W	L

New Vocabulary

▶ values
▶ decision-making skills
▶ goals
▶ short-term goal
▶ long-term goal
▶ action plan

Making Responsible Decisions and Setting Goals

Real Life Issues ....

Making Decisions. Tara has been playing soccer since elementary school. Tryouts for the varsity soccer team are coming up, and she's having trouble deciding whether to try out. Tara loves soccer, but she's not sure she's good enough to make the team. If she *does* make the team, she might not have enough time to study and do well in school.

(Writing) *Write a conversation in which Tara explains her situation to her school counselor. The counselor should help Tara figure out what the potential outcomes of her choices might be.*

Decisions, Goals, and Your Health

(Main Idea) **Achieving good health begins with making responsible decisions.**

Now that you're in high school, do you have more freedom than you did when you were younger? Maybe you're allowed to stay out later on weekends and have more control over your schedule and activities. You may have a wider circle of friends than you did in middle school. Having more freedom is an exciting benefit of growing up.

As you're probably finding out, the freedom you gain as you grow older comes with more responsibility. For example, you may have to make tough decisions. You'll also have to set goals for yourself and plan how to reach them. Making decisions and setting goals means you're taking responsibility in determining your life's purpose and direction.

Decision Making

> **Main Idea** Decision-making skills help you make successful, responsible choices.

Life is filled with decisions. You make plenty of them every day. Some decisions are small, like what to wear to school or what to eat for breakfast. Other choices may be life changing, like deciding which college to attend or which career to pursue. Developing good decision-making skills will help you make responsible choices that contribute to your health and quality of life.

Your Values

The decisions you make reflect your personal values and the values of your family. **Values** are *the ideas, beliefs, and attitudes about what is important that help guide the way you live.* For example, you may value a strong, healthy body. The decisions you make about how to take care of your body will reflect this value. If you value your relationships with family and friends, you will make choices that show your caring and respect.

Because you first learned your values from your family, it's often a good idea to talk with family members about a decision that is troubling you. You share important values with them, so they can provide you with helpful feedback.

The Decision-Making Process

Have you ever thought about what actually goes into making a decision? **Decision-making skills** are *steps that enable you to make a healthful decision.* **Figure 2.7** on page 42 illustrates the six steps in making good decisions. Notice that one of the steps involves the HELP strategy. This strategy includes asking yourself the following questions:

- **H** *(Healthful)* Does this choice present any health risks?

- **E** *(Ethical)* Does this choice reflect what you value?

- **L** *(Legal)* Does this option violate any local, state, or federal laws?

- **P** *(Parent Approval)* Would your parents or guardians approve of this choice?

READING CHECK

Analyze Why is it important to develop good decision-making skills?

■ **Figure 2.6** Family members often know you better than anyone else knows you. *Why is it a good idea to talk over important decisions with family members?*

Figure 2.7   Steps of the Decision-Making Process

STEP 1 State the Situation.
Clearly identify the situation. Ask yourself: What decision do I need to make? Who is involved? Am I feeling pressure to make a decision? How much time do I have to decide?

STEP 2 List the Options.
What are all the possible choices you could make? Remember that sometimes it is appropriate *not* to take action. Share your options with parents or guardians, siblings, teachers, or friends. Ask for their advice.

STEP 3 Weigh the Possible Outcomes.
Weigh the consequences of each option. Use the HELP strategy to guide your choice.

STEP 4 Consider Values.
A responsible decision will reflect your values.

STEP 5 Make a Decision and Act on It.
Use everything you know at this point to make a responsible decision. You can feel good that you have carefully thought about the situation and your options.

STEP 6 Evaluate the Decision.
After you have made the decision and taken action, reflect on what happened. What was the outcome? How did your decision affect your health and the health of those around you? What did you learn? Would you take the same action again? If not, how would your choice differ?

Goal Setting

Main Idea Working toward goals helps you achieve your hopes and dreams.

How do you see yourself in the future? What would you like to accomplish? What are your hopes and dreams? The answers to these questions form your **goals**, *those things you aim for that take planning and work*. Whether you reach your goals—and how successfully you reach them—depends on the plans you make now. Suppose your goal is to go to college. To reach that goal, you'll plan what courses to take in high school so that you meet the entrance requirements of the college you choose. You'll also work hard to earn the grades that will get you in.

Just as you set life goals because you have dreams for the future, you also set goals for your health in order to stay well. For instance, you may set a goal to drink more water and fewer soft drinks. To reach this goal, you need to plan how to make water available instead of soda when you're thirsty. You might plan to carry a refillable water bottle in your backpack, and to order water instead of soda when you're eating out with friends.

Types of Goals

Time is a consideration when you're setting goals. How long do you think it will take to reach your goal? A **short-term goal**, like finishing a term paper by Friday, is *a goal that you can reach in a short period of time*. A **long-term goal** is *a goal that you plan to reach over an extended period of time*.

Sometimes short-term goals become stepping stones to long-term goals. For example, making a high school sports team can be a stepping stone to your goal of becoming a professional athlete.

Short-Term Goals You can accomplish a short-term goal fairly quickly. Let's say your goal is to find and read three articles on an assigned topic over the weekend. On Saturday you search the Internet, locate, and print out your articles. On Sunday you read the articles so you're ready to discuss them in class on Monday.

Long-Term Goals Long-term goals call for more time as well as more planning. If you want to run a 10K (6.2-mile) race, you know you need to train for several months to build up your endurance and speed. A series of short-term goals can help you achieve this. You can practice running shorter distances until you are able to run a mile in a reasonable time.

FITNESSZONE

My teacher said that when you set a goal, you need to be specific and choose things that can be measured. Goals like "lose weight" or "gain muscle" are too general. Be specific, like, "I want to finish a 5K race," or "I want to eat at least five servings of fruits and vegetables a day." That way you can track your success. For more physical activity ideas, visit the Fitness Handbook and Fitness Zone sites in ConnectEd.

READING CHECK

Describe Identify and describe two types of goals.

■ **Figure 2.8** Many teens set health-related goals based on personal assessments of their health. *What steps can you take to improve your health? What strategies could you use?*

■ **Figure 2.9** This teen trained hard to reach the State Finals. *What other types of long-term goals might you set that can be reached by setting short-term goals?*

Then, you work up to running 5K (3.1 miles). Working short-term goals into the planning of your long-term goal helps you feel good each week as you run faster and farther.

Reaching Your Goals

To reach your goal, you need an action plan. An **action plan** is *a multistep strategy to identify and achieve your goals.* You can turn your dreams into reality by following the steps in **Figure 2.10**.

| Figure 2.10 | **Developing an Action Plan** |

Set a specific, realistic goal and write it down. You need to have a destination so you'll know where you're going.

List the steps you will take to reach your goal. Think of short-term goals as the steps that lead you to the long-term goal.

Identify sources of help and support. Who are your team members? They might include friends, family members, teachers, or community leaders.

Set a reasonable time frame for achieving your goal. Write down your time frame next to your goal statement.

Evaluate your progress by establishing checkpoints. Your checkpoints could be particular dates within your time frame, or they could be short-term goals that you plan to accomplish.

Reward yourself for achieving your goal. Celebrate with family and friends when you reach your goal. Plan smaller rewards along the way as you accomplish each short-term goal. This will help keep you motivated.

Health Skills Activity

Decision-Making Skills

Making New Friends

Justine made a new friend, Michelle, in biology class. Justine really likes Michelle, and the two girls have a lot in common. However, Justine knows that Michelle hangs out with a group of friends who smoke.

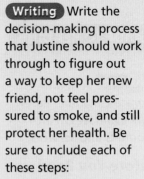

During class this morning, Michelle asked if Justine wanted to go to the mall with her and some friends on Saturday. Justine wants to go, but she doesn't want to be pressured about her choice not to smoke. She also doesn't want to lose Michelle as a friend. Justine wonders how she should handle the situation.

Writing Write the decision-making process that Justine should work through to figure out a way to keep her new friend, not feel pressured to smoke, and still protect her health. Be sure to include each of these steps:

1. State the situation.
2. List the options.
3. Weigh possible outcomes.
4. Consider values.
5. Make a decision and act.
6. Evaluate the decision.

LESSON 2 ASSESSMENT

 After You Read

Reviewing Facts and Vocabulary

1. How can decision-making skills improve your health?
2. Why would you set a health goal?
3. Give an example of one short-term and one long-term goal related to improving physical fitness.

Thinking Critically

4. **Evaluate.** What might happen if a teen made a decision that went against her personal values?
5. **Analyze.** How can responsible decision making help you achieve your health goals?

Applying Health Skills

6. **Goal Setting.** Choose a short-term goal that you personally would like to achieve. Write an action plan to accomplish your goal.

Writing Critically

7. **Descriptive.** Recall a time when you had to make a health-related decision. Describe how you made the decision, what happened as a result, and how you might change your decision-making process based on what you learned in this lesson.

Real Life Issues ....

After completing the lesson, review and analyze your response to the Real Life Issues question on page 40.

Being a Health-Literate Consumer

Real Life Issues ....

Analyzing Product Labels. Brad's skin is starting to break out, and he decides to buy an acne cream. In the skin care section at the drugstore, he finds 20 different products. Brad starts to read the package labels. After the fifth label, he feels overwhelmed and doesn't know how to sort out all the information.

Writing *If you were Brad, how would you respond to this situation? Write a paragraph explaining which criteria you use to select health products.*

Making Informed Choices

Main Idea You can learn to make good consumer choices.

Are you a smart buyer? Do you try to get a good-quality product at a reasonable price when you shop? Being a smart shopper is especially important when it comes to making choices about health products and services. It's up to you, as a **health consumer**—*someone who purchases or uses health products or services*—to make informed buying decisions.

Probably the most important influence you need to be aware of as a consumer is advertising. **Advertising** is *a written or spoken media message designed to interest consumers in purchasing a product or service.* Although advertising can provide useful information, its primary purpose is to get you to buy the product.

Advertisers use various techniques and hidden messages to promote their products and services. **Figure 2.11** shows some of these common techniques. Being a health-literate consumer means being aware of these messages and knowing how to evaluate them.

Figure 2.11   Hidden Messages in Advertising

Recognizing advertising techniques will help you make informed purchasing decisions.

Technique	Example	Hidden Message
Bandwagon	Group of people using a product or service	Everyone is using it, and you should too.
Rich and famous	Product displayed in expensive home	It will make you feel rich and famous.
Free gifts	Redeemable coupons for merchandise	It's too good a deal to pass up.
Great outdoors	Scenes of nature	If it's associated with nature, it must be healthy.
Good times	People smiling and laughing	The product will add fun to your life.
Testimonial	People for whom a product has worked	It worked for them, so it will work for you, too.

Evaluating Products

There are two effective ways to sharpen your consumer skills when buying health products: read product labels, and do some comparison shopping before you buy.

Product Labels Labels give you important information about what a product contains. Product labels carry the product's name, its intended use, directions, warnings, manufacturer's information, and the amount in the container. You will usually find the product's ingredients listed by weight in descending order. The active ingredients are the most important: they're the ones that make the product effective. By comparing the amount of active ingredients in different acne medications, for example, you can figure out which one contains more of the active ingredient, or whether a less expensive product contains the same active ingredient as a brand-name product.

Comparison Shopping A second great tool for smart health consumers is **comparison shopping**, or *judging the benefits of different products by comparing several factors, such as quality, features, and cost.* Here are some criteria you can use to judge health products and services:

- **Cost and quality.** Generic products may work the same as brand-name products. Compare the quality of lower-cost items, and look for products that meet your needs but cost less.

READING CHECK

Describe How does advertising influence your decision to purchase a health-related product?

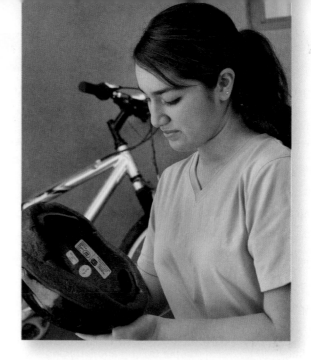

■ **Figure 2.12** Product labels contain useful health-related information. *How can you tell if a product has been safety tested?*

- **Features.** Figure out which product features are most important to you so that you don't waste money on features you don't want.

- **Warranty.** Many products come with a **warranty**, *a company's or a store's written agreement to repair a product or refund your money if the product doesn't function properly.* Ask about warranties before buying expensive products, and read them carefully to make sure you understand what they cover.

- **Safety.** When you are evaluating sports, recreation, and home-safety products, look for logos from well-known, reputable organizations that show the product has been tested for safety. For example, the **Underwriters Laboratories (UL)** tests and certifies products such as electrical appliances and fire extinguishers. **Snell,** a nonprofit foundation, and the **American National Standards Institute (ANSI)** monitor safety standards for helmets and other protective equipment.

- **Recommendations.** Listen to the opinions of people you trust who have used the product or service that you are considering. Also, check the consumer product ratings from organizations such as Consumers Union.

Evaluating Information and Services To evaluate health information or services, ask yourself these questions:

- Does this information come from a **valid** source? Does the service come from a respected provider?

- If the source is a Web site, who pays for the site? Is it a reputable organization? What is the purpose of the site?

Keep in mind that your doctor, nurse, and pharmacist are also great sources for reliable health care information.

Academic Vocabulary

valid *(adjective):* well-grounded or justifiable

Real World CONNECTION

Comparing Products

Teens spend $155 billion each year on clothing, music, personal care items, and other products. As a result, retailers pay close attention to teens' consumer behaviors and promote products directly to them.

It's important to look past the glossy advertising and fancy packaging to evaluate a product carefully. One feature to consider is cost and how to get the most for your money. Compare the following products.

Product	Size	Price
A. Celebrity-brand purifying gel	5.5 oz.	$25.99
B. Foaming acne cleanser	6 oz.	$6.99
C. Generic-brand acne cleanser	6 oz.	$4.99

Activity Mathematics

Use the chart to answer these questions.

1. What is the unit price of each cleanser?
2. If you use one bottle of product B every four months, how much will you pay for this product in a year?
3. **Writing** Write a paragraph describing two or more features you would compare when shopping for a facial cleanser.

Concept Measurement and Data: Unit Price To calculate unit price, divide the cost of the product by the volume, or total ounces. This will yield the cost per ounce.

LESSON 3 ASSESSMENT

 After You Read

Reviewing Facts and Vocabulary

1. Who is a *health consumer*?
2. How does comparison shopping make you a smart consumer?
3. Define *warranty*.

Thinking Critically

4. **Synthesize.** Susie wants to buy an expensive pedometer after seeing her favorite actress use it in an ad. The ad says the product is the most accurate on the market. What hidden messages in the ad is Susie responding to?

5. **Analyze.** How does a warranty help you become a smarter consumer?

Applying Health Skills

6. **Accessing Information.** Your friend takes a multivitamin, and you're wondering if you should too. You check a few Web sites to learn more about multivitamins before you ask your parents for permission to use them. How would you evaluate the validity of the information you find?

Writing Critically

7. **Expository.** Find three ads in a teen magazine. Write a short essay explaining the advertising techniques used in each ad. Refer to Figure 2.11.

Real Life Issues

After completing the lesson, review and analyze your response to the Real Life Issues question on page 46.

GUIDE TO READING

BIG Idea *Knowing how to handle consumer problems is an important skill to learn.*

Before You Read

Create Vocabulary Cards. Write each new vocabulary term on a separate note card. For each term, write a definition based on your current knowledge. As you read, fill in additional information related to each term.

Health Fraud

New Vocabulary

▶ consumer advocates
▶ malpractice
▶ health fraud

Review Vocabulary

▶ warranty (Ch.2, L.3)

Managing Consumer Problems

Real Life Issues

Health Fraud Terms. A number of claims and terms are commonly used to sell fraudulent health products.

Natural
Non-toxic
Money-back guarantee
Scientific breakthrough
Testimonials from people who claim amazing results
No-risk
Ancient remedy
Miraculous cure

Writing *Think of an advertisement that you have seen that uses one or more of these gimmicks. Write a paragraph describing the product and its potentially false or misleading claims.*

Resolving Consumer Problems

Main Idea Take action to correct consumer problems.

Have you ever bought a product and been dissatisfied with it after you used it? Maybe the product didn't work the way it should, didn't work at all, or a part was missing or broken. What can you do when this happens?

When searching for a product, find out what the store's return policy is *before* making a purchase. If the product has a warranty, check that it's in the package or ask for a copy. Scan the warranty and read the store's return instructions. After you get home and open the product, save the packaging, along with your receipt and warranty.

If the product comes with instructions, read them carefully. Pay particular attention to the directions for skin and hair care products. Follow all the steps for the product's use or assembly. Make sure you use the product exactly the way it was designed and made to be used.

Ingram Publishing

■ Figure 2.13 Most products have instructions that tell you how to use them correctly. *What are some products you use that have specific directions to follow?*

If you are using the product correctly and it isn't working the way you expected, read the warranty to learn how the manufacturer requires you to return it. You may be able to return it to the store where you bought it. However, some manufacturers want the product returned directly to them. Put the product back in its original packaging, and follow the manufacturer's return instructions. You may be asked to write a letter describing the problem and requesting a replacement or your money back. Date your letter and keep a copy for your files, along with a shipping receipt to prove you returned the product.

If you are not satisfied with the response to your efforts, ask for help from one of the following organizations:

- The **Better Business Bureau** handles complaints about local merchants. Its basic services are dispute resolution and truth-in-advertising complaints.

- **Consumer advocates** are *people or groups whose sole purpose is to take on regional, national, and even international consumer issues.* Some, like Consumers Union, test products and inform the public about potential problems. Others keep an eye out for consumer concerns about products and services.

- **Local, state, and federal government agencies** work to protect consumers' rights. The federal agencies most concerned with consumer health issues are the FDA, which is responsible for ensuring that medicines are safe, effective, and properly labeled, and the Consumer Product Safety Commission, which recalls dangerous products.

READING CHECK

Describe What would you do if you purchased a product that didn't work?

Tim Fuller Photography

Sometimes people run into problems with their health care providers, such as difficulty in scheduling appointments, or their health insurance won't cover nontraditional **approaches** such as acupuncture or herbal treatments. Often, people can avoid more serious problems by changing health care professionals or insurance companies. To make sure they are getting the best care possible, many people get a second opinion from another doctor for any major health concern, especially if it involves surgery or other serious treatment.

Occasionally, health care professionals may fail to provide adequate treatment and may be guilty of **malpractice**, *failure by a health professional to meet accepted standards*. If you experience a serious problem with a health care professional, you can contact organizations such as the American Medical Association or a state licensing board for help.

Health Fraud

Main Idea Protect yourself from health fraud.

Have you ever seen an ad on TV or in a magazine that promises an instant cure for a health problem? Did you think that sounded too good to be true? You were probably right. Such ads are a kind of **health fraud**, *the sale of worthless products or services that claim to prevent disease or cure other health problems*. Health fraud is often called *quackery*.

Weight-loss and beauty products are two areas particularly susceptible to health fraud. Read ads for these products very carefully before deciding to buy. Look out for claims like the following:

- "Secret formula"
- "Miracle cure"
- "Overnight results"
- "All natural"
- "Hurry, this offer expires soon"

■ **Figure 2.14** You can consult a registered pharmacist if you have questions about a product's health claims. *What other reliable sources can you go to for medical advice?*

Health clinics that provide "miracle" cures for ailments or questionable treatments, such as "microwaving" cancer cells, are also guilty of health fraud. Some fraudulent clinics have been shut down after it was discovered that the people running them were not the doctors they claimed to be; some even had criminal records. These clinics often take advantage of people who are very ill and desperate for a cure.

To protect yourself from health fraud, you can do the following:

- Check out the product's or service's claim with a doctor or other health professional.

- Talk to family and friends to get their opinion.

- Check with the Better Business Bureau to see if there have been complaints about the product or service.

- Check with a professional health organization about the claim. The American Heart Association, for example, will be familiar with health frauds related to heart disease treatment.

Remember, you have the power and the responsibility to protect your health and well-being!

 **READING CHECK**

Describe How can you protect yourself from health fraud when buying a health-related product?

LESSON 4 ASSESSMENT

After You Read

Reviewing Facts and Vocabulary

1. How do consumer advocates help you to be a better health consumer?

2. What is *malpractice*?

3. Define *health fraud*.

Thinking Critically

4. **Analyze.** Why is it important for health consumers to take an active role if they are dissatisfied with a product or service?

5. **Evaluate.** Review the list of claims that are often associated with health fraud. What do these claims have in common?

Applying Health Skills

6. **Advocacy.** Two of your friends are thinking about purchasing products that promise immediate weight loss. You know these claims are unrealistic and that the products may be unsafe. Create a text message that warns your friends about unrealistic claims in weight-loss products.

Writing Critically

7. **Narrative.** Write a story about a teen who encounters a consumer problem. Your story should describe the problem and demonstrate how the teen handles it.

Real Life Issues

After completing the lesson, review and analyze your response to the Real Life Issues question on page 50.

Hands-On
HEALTH

Activity — All About You

This activity is all about you! You will write a letter describing yourself and assessing your health habits. In the letter, you will set a health goal and develop a plan to reach that goal.

What You'll Need

- paper
- pen or pencil

What You'll Do

Step 1

Review Chapter 2. Then write a letter to yourself describing your personality, your likes and dislikes, and your values. In your letter, identify a health habit or skill you want to improve.

Step 2

Develop a health-related goal statement and an action plan for reaching that goal. Identify people who can provide help and support.

Step 3

Share your letter with a peer and ask for feedback. Revise your goal and action plan if necessary.

Apply and Conclude

At the end of one week, reread your letter. Write a reflection, and identify influences that affected your progress. These might include family, peers, culture, media, and personal values. Continue challenging yourself to reach your goal.

Checklist: Goal Setting

- ✓ Identification of realistic goal
- ✓ Clear goal statement
- ✓ Plan for reaching the goal
- ✓ List of people who can provide help and support
- ✓ Evaluation or reflection on the plan

LESSON **1**

Building Health Skills

Key Concepts

▶ Health skills are tools that help you manage your health.

▶ Good interpersonal communication helps you build strong relationships with others.

▶ When you are aware of how influences such as family, peers, culture, media, and personal values affect you, you are better able to make informed choices about your health.

Vocabulary

▶ health skills (p. 34)
▶ interpersonal communication (p. 35)
▶ refusal skills (p. 36)
▶ conflict resolution (p. 36)
▶ stress (p. 38)
▶ stress management skills (p. 38)
▶ advocacy (p. 39)

LESSON **2**

Making Responsible Decisions and Setting Goals

Key Concepts

▶ Use the steps in the decision-making process to make safe and responsible decisions.

▶ Short-term goals can help you reach long-term goals.

▶ To accomplish your goals, create an action plan.

Vocabulary

▶ values (p. 41)
▶ decision-making skills (p. 41)
▶ goals (p. 42)
▶ short-term goal (p. 43)
▶ long-term goal (p. 43)
▶ action plan (p. 44)

LESSON **3**

Being a Health-Literate Consumer

Key Concepts

▶ To be a smart consumer, read labels, comparison shop, and evaluate advertisements for hidden messages.

▶ Evaluate health information and services carefully to make sure they come from valid sources or respected providers.

Vocabulary

▶ health consumer (p. 46)
▶ advertising (p. 46)
▶ comparison shopping (p. 47)
▶ warranty (p. 48)

LESSON **4**

Managing Consumer Problems

Key Concepts

▶ Always read and follow the instructions for products you buy.

▶ Consumer and health organizations help fight health fraud.

Vocabulary

▶ consumer advocates (p. 51)
▶ malpractice (p. 52)
▶ health fraud (p. 52)

LESSON 1

Vocabulary Review

Use the vocabulary terms listed on page 55 to complete the following statements.

1. _____ are tools that you can use to maintain all aspects of your health.

2. If you influence another person to adopt a healthful behavior, that's called _____.

3. A person who goes for a brisk walk when feeling overwhelmed by a busy schedule is practicing a health skill called _____.

Understanding Key Concepts

After reading the question or statement, select the correct answer.

4. Sofia is angry that Alisa interrupts her. If Sofia says, "I'm upset because my ideas are not being heard," she is
   a. using an "I" statement to express her feelings.
   b. blaming Alisa for interrupting her.
   c. practicing poor interpersonal communication skills.
   d. all of the above.

5. Joe and Tony are having a heated argument. Tony decides to cool off before continuing the discussion. Tony is practicing a health skill called
   a. advocacy.
   b. conflict resolution.
   c. stress management.
   d. accessing information.

Thinking Critically

After reading the question or statement, write a short answer using complete sentences.

6. **Compare and Contrast.** How are interpersonal communication and conflict-resolution skills similar? How are they different?

7. **Evaluate.** What types of information could you use to evaluate the validity of health information?

8. **Explain.** How does technology influence your health choices?

9. **Analyze.** What does it take to advocate for health?

Ryan McVay/Lifesize/Getty Images

LESSON 2

Vocabulary Review

Mark the following sentences as True (T) or False (F).

10. The decisions you make about your health should reflect your values.

11. Decision making is a random process, depending on your mood.

12. Breaking a long-term goal into several short-term goals can make the long-term goal easier to achieve.

Understanding Key Concepts

After reading the question or statement, select the correct answer.

13. It's a good idea to talk over major decisions with your family because
    a. they always know what is best for you.
    b. they have a right to know everything you do.
    c. they share your values, the basis for making decisions.
    d. they will tell you what you should do.

14. Setting health-related goals
    a. ensures that you will obtain your goals.
    b. helps you plan and safeguard your well-being.
    c. takes a lot of time and creates stress.
    d. is a one-time event when you set healthy goals.

15. An action plan should include
    a. a written statement of your goal.
    b. the steps you will take to accomplish it.
    c. neither a nor b.
    d. both a and b.

Thinking Critically

After reading the question or statement, write a short answer using complete sentences.

16. **Analyze.** Why is decision making a key health skill? How can this skill contribute to your safety and well-being?

17. **Evaluate.** What criteria are important in weighing the possible consequences of your choices?

18. **Analyze.** Why is it important to set health-related goals?

LESSON 3

Vocabulary Review

Use the vocabulary terms listed on page 55 to complete the following statements.

19. Each of us is a(n) _____ because we all buy health products and services.

20. Evaluating the features of two similar products is called _____.

21. _____ techniques include bandwagon, rich and famous, free gifts, great outdoors, good times, and testimonial.

Understanding Key Concepts

After reading the question or statement, select the correct answer.

22. To be a critical thinker about advertising,
    a. note how often you see the same ad.
    b. remember product names.
    c. look for hidden messages in ads.
    d. compare ads for similar products.

23. Smart health consumers
    a. read product labels.
    b. listen to infomercials.
    c. write to product manufacturers.
    d. try out different products.

Thinking Critically

After reading the question or statement, write a short answer using complete sentences.

24. **Explain.** What strategies do smart consumers use to protect their health when purchasing health products?

25. **Discuss.** How does an understanding of advertising help you become a smarter consumer?

26. **Evaluate.** What are some effective strategies for evaluating health information and services?

LESSON 4

Vocabulary Review

Choose the correct term in the sentences below.

27. Someone who tests products and informs the public about potential problems is a *product expert / consumer advocate*.

28. Selling a worthless weight loss product is an example of *health fraud / trickery*.

29. When a doctor fails to live up to professional standards in medicine, he may be accused of *arrogance / malpractice*.

Assessment

Understanding Key Concepts

After reading the question or statement, select the correct answer.

30. If you buy a product and are not satisfied with it, you should
    a. read the warranty to find out how to return it.
    b. throw it away.
    c. immediately buy a replacement.
    d. all of the above.

31. Which of the following is *not* a good way to protect yourself from health fraud?
    a. Checking out claims with a health care professional
    b. Trying the product or service for yourself
    c. Talking to others who have used the product or service
    d. Consulting with the Better Business Bureau

Thinking Critically

After reading the question or statement, write a short answer using complete sentences.

32. **Synthesize.** How are consumers empowered to protect their health and well-being?

33. **Explain.** What are the best steps to take if you are dissatisfied with the health-related product that you have purchased?

34. **Analyze.** Why are some people attracted to products that make fraudulent claims?

35. **Discuss.** What criteria would you use to evaluate health services?

Technology PROJECT-BASED ASSESSMENT

Volun-teen

Background

Teens can advocate for healthy living on the local, national, and international levels. You can help make a difference by getting involved in this effort too!

Task

Conduct an online search for volunteer opportunities that promote healthy living and that are available to teens in your community. Write a blog entry or create a Web page that describes the volunteer opportunities and encourages teens to get involved. Ask other teens to add information about their favorite volunteer activities

Audience

Students in your class and teens in your community

Purpose

Inform teens of health-related volunteer opportunities, and encourage their involvement.

Procedure

1. Conduct an online search for volunteer opportunities with several health organizations. Examples include hospitals, the Red Cross, and the American Cancer Society.

2. Gather details about the volunteer opportunities. Find out the type of work involved, the minimum age requirement, the length of the assignment, and any necessary contact information.

3. Review several of the organization's Web sites to familiarize yourself with their writing style.

4. Write a blog entry or create a Web page that describes the volunteer opportunities you researched.

5. The blog or Web page should encourage teens to pursue volunteer opportunities.

Math Practice

Interpret Graphs. The graph below shows the percentages of young adults who volunteer each year. Use the graph to answer Questions 1–3.

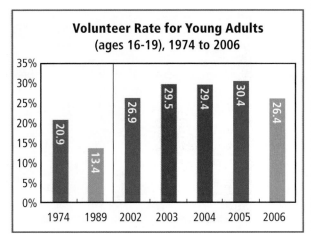

Volunteer Rate for Young Adults (ages 16-19), 1974 to 2006

1974: 20.9
1989: 13.4
2002: 26.9
2003: 29.5
2004: 29.4
2005: 30.4
2006: 26.4

Adapted from "Volunteering in America: 2007 State Trends and Rankings in Civic Life," Corporation for National and Community Service, April 2007.

1. Which years show an increase from the previous year in the volunteering rate among young adults?
   A. 1974 and 2002
   B. 1989 and 2006
   C. 2003 and 2005
   D. 2004 and 2006

2. Between which two years is there a change of only one-tenth of a percent?
   A. 2002 and 2003
   B. 2003 and 2004
   C. 2004 and 2005
   D. 2005 and 2006

3. According to the graph, how do volunteering rates in the 2000s compare to rates in the late 1900s?
   A. Fewer young adults volunteer in this century.
   B. More young adults volunteer in this century.
   C. There is no change in the number of young adults who volunteer.
   D. More adults volunteer than young adults during both centuries.

Reading/Writing Practice

Understand and Apply. Read the passage below, and then answer the questions.

> Good communication is a skill you can use every day. One way you can demonstrate effective communication skills is to be a good listener. The speaker may think that you are not listening if your eyes wander around the room. Closed body language, such as crossing your arms, conveys that you may not be open to hearing what others have to say.
>
> To show that you are listening attentively, make eye contact and let the other person finish what he or she is saying before you speak. Use body language such as nodding your head to show that you are interested in what the person is saying. A good listener makes statements that encourage the speaker to explain his or her views, such as "What do you mean by that?" It can also be helpful to restate what the person tells you to make sure that you understand what is being said.

1. Which of the following is a behavior that characterizes good listening skills?
   A. Crossed arms
   B. Making eye contact
   C. Wandering eyes
   D. Interrupting

2. Which statement best summarizes the main point of the article?
   A. Closed body language is negative.
   B. You can demonstrate effective communication by being a good listener.
   C. Always ask speakers to explain their views.
   D. Never restate what you hear.

3. Describe the effects of attentive listening and poor listening on communication.

National Education Standards

Math: Measurement and Data
Language Arts: LACC.910.L.3.6

CAREER CORNER — Health Careers

Health Teacher

Health teachers help students understand how to maintain good health. These teachers provide information on nutrition, fitness, and social issues. They can also give basic information on how the human body functions. Health teachers may need to spend time organizing lesson plans, as well as group and individual activities.

A health teacher must have a four-year teaching degree. Public school health teachers must be licensed, but this is not a requirement for private school health teachers. High school classes that can help prepare a student to become a health teacher include biology, science, fitness, and communications.

Medical Writer

If you like to write and are interested in health care, becoming a medical or health writer might be a good career for you. These professionals organize and write complex medical information for the public in lay-person's terms. Medical writers may work for publishing companies, hospitals, radio and TV stations, universities, government agencies, and pharmaceutical companies.

Medical writers need a bachelor's degree, with courses in science and English or journalism. To prepare for this career, high school students should take classes in biology, chemistry, math, and English.

Health Information Technician

If you think of yourself as well organized and detail oriented, you might enjoy a career as a health information technician. These technicians assemble patients' health information and assign codes to medical diagnoses and procedures. Some use computer programs to tabulate and analyze data for research.

Health information technicians are employed at hospitals, clinics, doctors' offices, and insurance companies. You need a two-year degree in health informatics to become a health information technician. To get into a qualified program, take biology, chemistry, health, and computer science courses in high school.

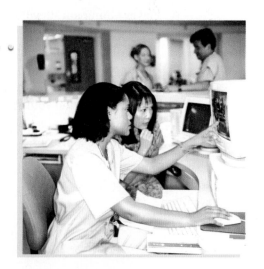

CAREER SPOTLIGHT

Health Promoter

Elizabeth Pedroza became a health promoter with the Hispanic Health Projects before she knew anything about diabetes or women's health. Today, she spends her days teaching community members about diabetes and other health issues, developing diabetes education materials, and translating medical information for Spanish-speakers.

Q. How can you teach others when you are not a health expert?

A. *I work with experts who teach me what I need to know. The experts know the content and I know the culture. You don't need a college degree for my job. It's more about asking questions, listening, and sharing important information.*

Q. What specific training have you received?

A. *I'm a skilled interviewer with good communications skills, and I am comfortable talking to people with different backgrounds.*

I've also been trained on how to get better information out of people, how to present myself in a respectful way, and how to foster trust. These are important talents to have when you are trying to improve someone's health.

Q. What attracted you to this job?

A. *It provides the opportunity to go into the community and help people. It isn't about what I would gain from it, but what I could give.*

Activity Beyond the Classroom

Writing Health Education Careers. Identify organizations in your community where health educators work. Don't forget to include your school! Interview three or four health educators to find out more about what they do and what they find challenging about their jobs.

Based on what you learn, write a newspaper recruitment ad for a health educator's job.

UNIT 2 Mental and Emotional Health

UNIT PROJECT

Mental and Emotional Support

Using Visuals Big Brothers Big Sisters of America is the oldest and largest mentoring program in the United States. The program matches children with teen and young adult volunteers from the community. The one-to-one relationships that form have a long-lasting, positive effect on both the mentors and the children.

Get Involved. Which organizations in your community provide mentoring for children and teens? Research community resources to find out how you can contribute or participate. Based on your findings, create a brochure advertising these mentoring programs.

"*Learn from yesterday, live for today, hope for tomorrow.*
The important thing is not to stop questioning."
— *Albert Einstein, 20th-century physicist*

CHAPTER 3

Achieving Mental and Emotional Health

Lesson 1

Developing Your Self-Esteem

BIG Idea *Good mental and emotional health helps you develop healthy self-esteem.*

Lesson 2

Developing Personal Identity and Character

BIG Idea *Healthy identity is based on being a person of good character.*

Lesson 3

Expressing Emotions in Healthful Ways

BIG Idea *Managing your emotions allows you to express them in healthful ways.*

Activating Prior Knowledge

Using Visuals Look at the picture on this page. How is this teen expressing his identity? List three characteristics that help define who we are.

Chapter Launchers

Health in Action

Discuss the BIG Ideas

Think about how you would answer these questions:

▸ How would you describe your level of self-esteem?

▸ How is your self-esteem related to your identity?

▸ How does the way you express emotions reflect your mental health?

Assess Your Health

Read each statement. On a separate sheet of paper, write "yes," "sometimes," or "no" based on your typical behavior.

1. I am happy and enjoy my life.
2. I have the confidence to try new things, and believe that setbacks are temporary.
3. I choose friends who value and respect me.
4. I learn from my mistakes.
5. I am a trustworthy and honest person.

A "yes" response shows that you practice healthy behaviors. "Sometimes" indicates that you should analyze and possibly modify your behavior. A "no" response means that you should modify the behavior.

Developing Your Self-Esteem

BIG Idea *Good mental and emotional health helps you develop healthy self-esteem.*

Before You Read

Create an Outline. Preview this lesson by scanning the pages. Then organize the headings and subheadings into an outline. As you read, fill in your outline with important details.

I.
A.
1.
2.
B.
II.

New Vocabulary

▸ mental/emotional health
▸ resilient
▸ self-esteem
▸ competence
▸ hierarchy of needs
▸ self-actualization

Real Life Issues

Staying Positive. Kevin has been swimming since he was 5 years old. When he recently didn't make the swim team, he thought there had been a mistake. He did well in swimming competitions throughout elementary school. He even took time to work out every day after school prior to the tryouts. He feels embarrassed about being cut from the team and doesn't know how he can face his friends.

Writing *Imagine you are Kevin's close friend. Write a dialogue between yourself and Kevin discussing how he should deal with his disappointment.*

What Is Mental and Emotional Health?

Main Idea Mental and emotional health helps you function effectively each day.

Do you see yourself in a positive way? Are you able to handle challenges and setbacks well? The ability to answer "yes" to these questions is one sign of mental and emotional health. **Mental/emotional health** is *the ability to accept yourself and others, express and manage emotions, and deal with the demands and challenges you meet in your life.* Having good mental/emotional health is an important part of your total health.

Most people have ups and downs throughout their lives. For example, you may have felt proud because you performed well during a school play, but were disappointed when you didn't make the varsity team at school. Such ups and downs are normal, especially during the teen years when you are adapting to many changes in your life.

The Importance of Mental and Emotional Health

Mentally healthy people are, in general, happy and enjoy their lives. They feel confident and comfortable spending time alone or with others. They're also flexible and can cope with a wide variety of feelings and situations.

Good **mental** and emotional health influences your physical and social health too. For example, if you're worried, you might eat an unhealthful diet, not get enough sleep, or stop exercising regularly. If you're worried and become irritable, your relationships with friends and others may suffer.

Academic Vocabulary

mental *(adjective):* of or relating to the mind

Characteristics of Good Mental and Emotional Health

How do you know if you have good mental and emotional health? Here are some general characteristics.

- **Sense of belonging.** Feeling close to family members, friends, teachers, and others provides you with support.
- **Sense of purpose.** Recognizing that you have value and importance as a person lets you set and reach goals.
- **Positive outlook.** Seeing the bright side of life reduces stress and increases your chances of success.
- **Self-sufficiency.** Having the confidence to make responsible decisions promotes your sense of independence and self-assurance.
- **Healthy self-esteem.** Having healthy self-esteem helps you accept and recover from difficulties and failures.

Everyone has to manage difficult and stressful situations. Mentally and emotionally healthy people handle stresses in positive ways. These people are **resilient**—they have *the ability to adapt effectively and recover from disappointment, difficulty, or crisis.*

 READING CHECK

Name What are the characteristics of good mental/emotional health?

■ **Figure 3.1** Close friends encourage one another. *How might this kind of encouragement affect a person's mental health?*

©Robert Daly/age fotostock

FITNESS ZONE

My team lost the last game of the season, and I thought about quitting the sport. Then my dad told me that Michael Jordan was cut from the varsity basketball team when he was in 10th grade. Imagine if he had quit! He wouldn't have become one of the greatest NBA basketball players in history. I decided to be like Mike and not to give up on exercise or sports. For more physical activity ideas, visit the Fitness Handbook and Fitness Zone site in ConnectEd.

READING CHECK

Describe Identify several benefits of healthy self-esteem.

Self-Esteem

Main Idea Healthy self-esteem is necessary for good mental/emotional health.

Developing **self-esteem**, or *how much you value, respect, and feel confident about yourself,* influences the other characteristics of good mental health. If you feel valued, loved, and accepted by others, and you value, love, and accept yourself your overall attitude and outlook will be good. Having good self-esteem will also affect your overall attitude and the health choices you make. Taking healthful risks can raise your self-esteem. Trying new challenges can also raise your sense of **competence**, or *having enough skills to do something.*

How You Develop Self-Esteem

You probably remember a time when your family praised you for doing something well, or reassured you and gave you advice on tasks you hadn't yet mastered. When you are praised for mastering a task or reassured when you do not, your self-esteem increases. Self-esteem also increases when you believe that you can succeed, or when you master new challenges.

No one succeeds at new tasks and activities all the time. If you don't succeed at a new task or activity, think about the reasons why you may not have succeeded. How did you prepare? Was it realistic to expect to succeed on your first try? Also, everyone has unique abilities. You may have to work harder at a new task or activity if you do not have the unique abilities to master that task or activity easily.

How you react emotionally to situations also affects your self-esteem. *Self-talk,* the encouragement or criticism that you give yourself, can affect your self-esteem. Using positive self-talk will strengthen your self-esteem. Try to replace negative thoughts by using positive self-talk.

Benefits of Healthy Self-Esteem

Healthy self-esteem helps you feel proud of yourself and your abilities, skills, and accomplishments. You believe that setbacks are temporary. You have the confidence to confront challenges and overcome them.

Healthy self-esteem also gives you the confidence to try new things. You're not afraid to try a new sport, or join a club at school, or even get a job and learn new tasks. People with healthy self-esteem know that they may not be as good at some tasks as they are with others. They don't see themselves as a failure if they don't succeed at something.

Figure 3.2 Important people in your life play a role in shaping your self-esteem. *How has someone important to you affected your self-esteem?*

Improving Your Self-Esteem

Main Idea You can improve your self-esteem and your overall mental and emotional health.

You can control many things that affect your self-esteem. Avoid criticizing yourself, or spending time with people who criticize you. Set realistic expectations, and don't expect everything to be "perfect." Expecting perfection can prevent you from enjoying your successes. You may judge a success as a failure if it doesn't meet your critera for perfection. These additional suggestions can help you improve your self-esteem.

- Choose friends who value and respect you.
- Focus on positive aspects about yourself.
- Replace negative self-talk with supportive self-talk.
- Work toward accomplishments rather than perfection.
- Consider your mistakes learning opportunities.
- Try new activities to discover your talents.
- Write down your goals and the steps you will take to achieve them.
- Exercise regularly to feel more energized.
- Volunteer your time to help someone.
- Accept the things you can't change, and focus your energy on changing the things you can.

Developing Self-Awareness

Main Idea Understanding your needs and meeting them in healthy ways will help you reach your highest potential.

As infants, we rely on others to meet our basic needs. Those needs include food, clothing, and physical safety and comfort. As we grow, our needs become more complex. The psychologist Abraham Maslow created a theory that explains human development and motivation (see **Figure 3.3**). The **hierarchy of needs** is *a ranked list of those needs essential to human*

Figure 3.3

Maslow's Hierarchy of Needs

Maslow's model helps us understand our needs. Meeting these needs in healthy ways strengthens our mental/emotional health.

LEVEL 5 REACHING POTENTIAL
Need for self-actualization

LEVEL 4 FEELING RECOGNIZED
Need to achieve, need to be recognized

LEVEL 3 BELONGING
Need to love and be loved, need to belong

LEVEL 2 SAFETY
Need to be secure from danger

LEVEL 1 PHYSICAL
Need to satisfy basic needs of hunger, thirst, sleep, and shelter

growth and development, presented in ascending order, starting with basic needs and building toward the need to reach your highest potential.

Maslow's hierarchy shows that our earliest motivations are to satisfy our physical needs. Once these basic needs are met, we become interested in meeting the need to belong and be loved, the need to be valued and recognized, and the need to reach our potential or to achieve **self-actualization** or *to strive to be the best you can.*

As well as understanding needs, you need to learn how to meet them in healthy ways. Meeting a need in a high-risk way will not lead to healthful development. For example, some teens may join gangs in order to belong to a group. Joining a gang is a high-risk behavior that affects all sides of the health triangle.

Try using Maslow's model to evaluate your personal development. What needs are you focused on right now? How are you meeting those needs? As your self-awareness grows, you can begin to take more control of your personal growth. Reaching out to others can help you develop deeper relationships and a stronger support group.

■ **Figure 3.4** These teens have found a way to contribute to their community. *Why do you think taking time to attend to the needs of others is beneficial to self-esteem?*

 READING CHECK

Explain Why is joining a gang an unhealthy way to meet the need to be valued?

LESSON **1** ASSESSMENT

📖 After You Read

Reviewing Facts and Vocabulary

1. List the characteristics of good mental/emotional health.
2. Define the term *self-esteem.*
3. Identify the five levels of Maslow's hierarchy of needs.

Thinking Critically

4. **Analyze.** Explain how being mentally and emotionally healthy contributes to the quality of your life.
5. **Identify.** What are three ways that you can demonstrate healthy self-esteem and good mental/emotional health?

Applying Health Skills

6. **Practicing Healthful Behaviors.** Keiko just found out that she didn't make the track team. Write a script showing how she can use positive self-talk to deal with this disappointment.

Writing Critically

7. **Descriptive.** Imagine a teen whose suggestions are ignored during a group project. With healthy self-esteem, how might the teen respond?

Real Life Issues ....

After completing the lesson, review and analyze your response to the Real Life Issues question on page 66.

©Ariel Skelley/Blend Images/Corbis

GUIDE TO READING

BIG Idea *Healthy identity is based on being a person of good character.*

Before You Read

Create a Cluster Chart. Draw a center circle and label it "Character." Draw circles around it and use these to define and describe this term. As you read, continue filling in the chart with more details.

Character

New Vocabulary

▶ personal identity
▶ role model
▶ personality
▶ character
▶ integrity
▶ constructive criticism

Review Vocabulary

▶ values (Ch.2, L.2)

Developing Personal Identity and Character

Real Life Issues ••••••••••••••••••••••••••

Choosing a Path. Casey is in the process of deciding what to do after he graduates next year. He visits the school guidance counselor to discuss his options. The counselor begins by asking, "What are you interested in? What are your talents?"

Writing *What are your interests and talents? Write an essay describing how these interests and talents might play a role in your future.*

Your Personal Identity

Main Idea Your personal identity describes who you are.

When you first meet someone, you tell that person your name. As you get to know the person better, you may share more information. These attributes are your **personal identity**, *your sense of yourself as a unique individual.*

Your personal identity depends a lot on your age and circumstances. Other parts of your personal identity are unique to you. Identity development is one of the most important tasks you will accomplish during your teen years.

How Identity Forms

Identity is partly formed by recognizing your likes and dislikes. Your relationships and experiences with family and friends also influence your personal identity. As you mature, you'll meet a greater number and variety of people, and will develop your own opinions. As your experiences broaden, you develop likes and dislikes based on how your experiences fit with your values and beliefs.

You may identify a **role model**, *someone whose success or behavior serves as an example for you*. Your identity will change throughout your life as your interests change. You will struggle at times with alternatives and choices, but eventually you will develop a clear sense of your own values, interests, beliefs, occupational goals, and relationship expectations.

Aspects of Identity

One aspect of your identity is your **personality**, *a complex set of characteristics that makes you unique*. Your personality sets you apart from other people and determines how you will react in certain situations. Although it plays a big role in defining your identity, it isn't the only thing. Other relationships, such as those with your family, your ethnic group, and even your close friends, also define who you are. These shared characteristics and your unique qualities form your identity.

The Importance of Good Character

Main Idea Character plays a significant role in your decisions, actions, and behavior.

An important aspect of your identity is your **character**, *the distinctive qualities that describe how a person thinks, feels, and behaves*. Good character is an outward expression of inner values and is a vital part of healthy identity. A person of good character demonstrates *core ethical values,* such as responsibility, honesty, and respect. Such values are held in high regard across all cultures and age groups.

READING CHECK

List What are some aspects of personal identity?

■ **Figure 3.5** Friends share similar values. *What values do you share with your friends?*

©Brand X Pictures/PunchStock

I've noticed that there's definitely a connection between having a positive outlook on life and feeling healthy. One day last week, no one could go skateboarding with me. I was bored and started to feel sad. Then I remembered that a family with a girl about my age moved in down the street. I stopped by and introduced myself. She and I went for a walk. We got some fresh air and exercise and I had a chance to make a new friend—it felt great! For more physical activity ideas, visit the Fitness Handbook and the Fitness Zone sites in ConnectEd.

READING CHECK

Identify What are the six traits of good character?

■ **Figure 3.6** Courtesy is a sign of respect. *How do you show respect for others?*

Traits of Good Character

Six traits are commonly used to describe good character. By demonstrating these traits consistently in your actions and behaviors, you show others that you have **integrity**, or *a firm observance of core ethical values*.

- **Trustworthiness.** You are honest, loyal, and reliable—you do what you say you'll do. For example, if you tell a friend that you'll meet at a certain time, you try your best to be on time. You have the courage to do the right thing, and you don't lie, cheat, or steal.

- **Respect.** You are considerate of others and accept their differences. You make decisions that show you respect your health and the health of others. Even if you disagree with another person's point of view, you use good manners in your dealings with people. You treat them and their property with care and respect.

- **Responsibility.** You use self-control—you think before you act and consider the consequences. You are accountable for your choices and decisions, and don't blame others for your actions. You try your best and complete projects you start, even when things don't go as planned.

- **Fairness.** You play by the rules, take turns, and share. You are open minded, and you listen to others. You don't take advantage of others, and you don't blame others.

- **Caring.** A caring person is kind and compassionate. You express gratitude, are forgiving toward others, and want to help people in need.

- **Citizenship.** Demonstrating good citizenship means you advocate for a safe and healthy environment at school and in your community. You take an interest in the world around you. You obey rules and laws, and show respect for authority.

Tim Fuller Photography

Working Toward a Positive Identity

Main Idea You can develop a healthy identity.

You may think that your family and your circumstances form your identity. This is partly true, but *you* control who you *become*. As you mature, you will make more personal choices and decisions. For example, you will choose a career. The list in **Figure 3.7** can help you develop a positive identity.

Recognize Your Strengths and Weaknesses

To begin to understand your identity, analyze your strengths and weaknesses. Be honest and realistic. If you are a trustworthy friend or a talented singer, be proud of yourself. At the same time, evaluate your weaknesses without being too critical, and set realistic goals to improve. For instance, if you tend to put things off, such as homework, set a goal to develop new habits. With planning and commitment, you can improve habits.

Demonstrate Positive Values

Practicing good character is not always easy, but it helps you build a positive identity. For instance, if you are an honors student who feels pressured to cheat on an exam, that action could harm your self-esteem, your reputation, or both.

Develop a Purpose in Your Life

A sense of purpose helps you set goals and work to achieve them. It also provides you with a framework to build a healthy identity. Some of your goals will be short term, like studying for and passing an exam. Others will be long term, such as planning for higher education and acquiring job skills.

READING CHECK

Explain How is developing a purpose for your life helpful?

Figure 3.7   **Tips for Promoting a Healthy Identity**

▸ **List your skills and strengths.** Include physical, mental/emotional, and social strengths. Read the list when you're feeling down.

▸ **Surround yourself with positive, supportive people.** Choose friends who support and respect your rights and needs.

▸ **Find something that you love to do, and do it frequently.** If you're always too busy to do the things you enjoy, you're not taking care of yourself.

▸ **Stop making life a contest.** Recognize that there will always be people more and less able than you in areas of life. Be content with doing the best you can in all areas that matter to you.

▸ **Help someone else.** One way to feel good about yourself is to see the positive effects of your own words or actions on someone else's life.

Hill Street Studios/Getty Images

Form Meaningful Relationships

Meaningful relationships, such as those with family, friends, and others, are **crucial** to the development of your identity. Relationships provide a support system that can help you build confidence and develop a sense of security and belonging. Within a meaningful relationship, family, friends, or others may give you **constructive criticism**, or *nonhostile comments that point out problems and encourage improvement.* For example, when a friend doesn't do well at a task, you might make helpful suggestions without judging the way your friend performed.

Avoid Unhealthful High-Risk Behaviors

Risk taking is part of life. Playing sports, taking part in artistic or creative activities, public speaking, and making friends all involve some risk. These risks are healthful. They challenge you to develop skills and to mature in new ways. However, high-risk behaviors, such as using tobacco, alcohol, or other drugs, reckless driving, or joining a gang, are dangerous and harmful.

Contribute to the Community

Your community is your extended support system. It provides services and resources to meet many of your needs. For a community to remain strong, however, all of its members must participate in making it work. Giving back to the community in the form of volunteering is part of being a good citizen. Volunteering within your community improves the quality of people's lives, gives you a sense of accomplishment and belonging, and increases your self-esteem.

Academic Vocabulary

crucial *(adjective):* important or essential

■ **Figure 3.8** This player relies on his coach for honest feedback. *Whom else might a teen rely on for honest feedback?*

Real World CONNECTION

Your Sources of Support

People around you regularly provide support. Some fulfill material needs, others offer comfort and honest opinions, and others provide information. One thing that all of these people have in common is a genuine concern for you. Together they form your support system.

Identify your support system. Make a chart with the different types of support: **Material** (providers of money, transportation, physical help), **Emotional** (providers of comfort, sympathy, encouragement), **Information** (providers of knowledge and referrals), and **Appraisal** (providers of feedback, praise, suggestions). Under each category, identify who you count on and why. List those people whose names appear often. These individuals make up your core support group.

Activity Technology

Pick one person from your core support group.

1. Create a streaming video to express your appreciation for his or her support and encouragement.
2. Identify ways that the person has supported you and helped build your identity.
3. Tell that person how his or her support has shaped your goals for the future.
4. Describe to the person how his or her support has influenced you to support others throughout your life.

LESSON 2 ASSESSMENT

After You Read

Reviewing Facts and Vocabulary

1. Define the term *personal identity*.
2. Identify the six traits of good character.
3. Explain the benefit of constructive criticism.

Thinking Critically

4. **Analyze.** Describe how role models help in forming identity.
5. **Describe.** Explain how healthful risk taking can help you mature in new ways.

Applying Health Skills

6. **Communication Skills.** With a classmate, role-play situations where constructive criticism is given.

Writing Critically

7. **Expository.** If you were to choose a role model, who would it be? Write a short essay explaining your choice.

Real Life Issues ....

After completing the lesson, review and analyze your response to the Real Life Issues question on page 72.

Expressing Emotions in Healthful Ways

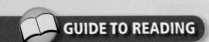

GUIDE TO READING

BIG Idea *Managing your emotions allows you to express them in healthful ways.*

Before You Read

Create a K-W-L Chart. Make a three-column chart. In the first column, list what you **k**now about emotions and ways to express them. In the second column, list what you **w**ant to know about this topic. As you read, use the third column to summarize what you **l**earned about the topic.

K	W	L

New Vocabulary

▶ emotions
▶ hormones
▶ hostility
▶ empathy
▶ defense mechanisms

Real Life Issues

Expressing Feelings. Learning how to manage anger and other strong emotions can reduce the risk of violence.

Source: Centers for Disease Control and Prevention; National Center for Injury Prevention and Control; Youth Risk Behavior Surveillance Survey

> **In 2010, over 563,863 people ages 10 to 24 were treated in emergency rooms because of a violent act.**

> **29% of teens reported feeling sad or hopeless that they stopped doing some usual activities.**

Writing *Write a paragraph describing ways to manage strong emotions and reduce the risk of violence.*

Understanding Your Emotions

Main Idea Recognizing and acknowledging your emotions is a sign of good mental and emotional health.

Have you ever seen a movie that made you feel happy, sad, or even scared? These feelings are examples of **emotions**, *signals that tell your mind and body how to react.* Many times, the most intense emotions you feel will be related to an event in your life. How you respond to your emotions can affect your mental/emotional, physical, and social health.

Changes during puberty are caused by **hormones**—*chemicals produced by your glands that regulate the activities of different body cells.* These hormones can make you feel as if your emotions are swinging from one extreme to another. It's normal for teens to feel overcome by emotions during puberty.

Learning to recognize your emotions and to understand their effects on you will help you learn to manage them in healthful ways. Below are some common emotions.

- **Happiness.** Being satisfied and feeling positive are good descriptions of happiness. When you are happy, you usually feel energetic, creative, and sociable.

- **Sadness.** Feeling sad is a normal, healthy reaction to difficult life events. These feelings may be mild, like being disappointed because you didn't do well on a test, or they may be deep and long lasting, such as the grief you feel when a pet or family member dies.

- **Love.** Strong affection, deep concern, and respect are expressions of love. Loving someone means that you support the needs and growth of that person and respect the person's feelings and values.

- **Fear.** When you are startled by someone or something, you may feel fear. Feelings of fear can increase your alertness and help you escape from possibly harmful situations. Some people let fear of imagined threats prevent them from taking healthful risks.

- **Guilt.** Guilt is the feeling of shame and regret that occurs when you act against your values. Sometimes people feel guilt about situations that they have no control over. For instance, some children and teens may blame themselves if their parents divorce.

- **Anger.** Anger is a normal reaction to being emotionally hurt or physically harmed. Anger that is not handled in a constructive way can lead to violence. Another form of anger is **hostility**, *the intentional use of unfriendly or offensive behavior.* Hostility can hurt others, as well as the hostile person. Often, anger is complicated because it can hide another emotion, such as hurt or guilt.

READING CHECK

Explain How do hormones affect emotions?

■ **Figure 3.9** Actors portray strong emotions by using body language and changing their tone of voice. *How might an actor use body language to convey each of the emotions described above?*

Managing Your Emotions

Main Idea Knowing how to recognize your emotions can help you manage them in healthful ways.

Emotions are neither good nor bad. The way you express your emotions, however, can produce good or bad consequences. Learning to express emotions in a healthful way will not only help you cope with emotional upsets, but also helps those around you to better handle their emotions.

Dealing with Emotions in Positive Ways

As a young child, you learned from parents, teachers, and friends how to express your emotions. Some emotions, such as happiness and love, are expressed through facial expressions like smiles and glances, and through behaviors like laughing and hugging. **Empathy**, or *the ability to imagine and understand how someone else feels*, is expressed by supporting a friend who is going through a difficult time. You may also have learned that emotions are private if you know people who are uncomfortable expressing their feelings.

No doubt you also learned that people sometimes deal with their feelings in harmful ways. They may exaggerate their emotions, pretend they have no feelings, or even hurt another person deliberately while expressing emotions.

To help you recognize your emotions and express them in positive ways, ask yourself these questions:

- Why do I feel the way I do about this event?
- Will this event matter later on in my life?
- Why should I wait before responding?
- What can I do to feel better?
- Who can I ask to help me deal with my negative feelings?

■ **Figure 3.10** Healthful expression of feelings lets you enjoy life more. *What are some positive ways to express emotions?*

Figure 3.11 Common Defense Mechanisms

▶ **Repression.** Involuntarily pushing unpleasant feelings out of one's mind.

▶ **Regression.** Returning to behaviors characteristic of a younger age, rather than dealing with problems in a mature manner.

▶ **Denial.** Unconscious lack of recognition of something that is obvious to others.

▶ **Projection.** Attributing your own feelings or faults to another person or group.

▶ **Suppression.** Consciously and intentionally pushing unpleasant feelings out of one's mind.

▶ **Rationalization.** Making excuses to explain a situation or behavior, rather than taking responsibility for it.

▶ **Compensation.** Making up for weaknesses and mistakes through gift giving, hard work, or extreme efforts.

Responding to Difficult Emotions

Feeling bad, or emotional, when things happen in your life is normal. These feelings, however, can be managed. Some techniques to reduce the intensity of your emotions include taking several deep breaths, relaxing your muscles, getting away from the situation until you calm down, analyzing your emotions by writing about them in a private journal, or talking to someone you trust about the way you feel.

Some people choose to manage difficult emotions by avoiding situations that make them uncomfortable. **Defense mechanisms** are *mental processes that protect individuals from strong or stressful emotions and situations*. **Figure 3.11** lists some common defense mechanisms used to respond to difficult emotions.

Sometimes you may use defense mechanisms unconsciously as a way to protect yourself from intense emotional pain. You may not even be aware you are using them. Although defense mechanisms can help you deal with emotions for a short time, eventually you will need to work through the problem. Relying on defense mechanisms too long can keep you from facing—and solving—what's upsetting you.

Some emotions, such as fear, guilt, and anger, can be very damaging. People may respond to these emotions without thinking about the consequences. By analyzing the cause of these feelings, you can learn to manage them.

Handling Fear Most people are afraid of something. You can overcome some fears by recognizing that you're afraid and figuring out what is causing this fear. For example, you may be afraid to speak in front of a group, but need to give a presentation as part of a group assignment. For this type of fear, try talking to a friend or an adult who can suggest ways to organize your material and prepare for the presentation.

FITNESS ZONE

Some days I just don't feel like exercising. On those days, I use positive self-talk to remind myself how good I feel after a workout. Now, I congratulate myself every time I take a step toward one of my goals. After I finish a workout, I think, "Awesome! I did it!" For more physical activity ideas, visit the Fitness Handbook and Fitness Zone sites in ConnectEd.

Academic Vocabulary

resource (noun): a source
of supply or support

Other fears, such as the fear of going to college or learning to drive a car, may require the help of **resources** within your community. If you're unable to control your fears, consider seeking the help of a mental health professional.

Dealing with Guilt Guilt is another very destructive emotion. If it is not managed, it can harm your self-esteem. If you feel guilty about something, think about the cause. Have you hurt someone? Admitting a mistake, apologizing, and promising to be more thoughtful in the future can help manage feelings of guilt. Keep in mind that you may not be able to control some situations. Look at the circumstances realistically and honestly. Some situations are out of your control. For instance, if your parents are divorcing, it may upset you, but it's not your fault.

Managing Anger Anger is one of the most difficult emotions to handle. As with guilt, it is best to figure out what is causing your anger, and then deal with it in a healthy way. When you first feel anger building up inside you, take time to calm down. You might try deep breathing or slowly repeating a calming word or phrase. If this doesn't work, physically remove yourself from the situation. Then try one of these strategies:

READING CHECK

Explain How do people use defense mechanisms?

- **Do something to relax.** Listen to soothing music, read a book, or imagine sitting on a beach or walking through the woods.

- **Channel your energy in a different direction.** Use the energy generated by your anger to do something positive. Take a walk, go for a bike ride, play the piano or guitar, or write your feelings down in a private journal.

- **Talk with someone you trust.** Sharing your thoughts and feelings with a trusted friend or family member may help you see the situation from the viewpoint of another person. Not only will you feel better, but the listener also may be able to give you some tips on how to deal with the situation.

■ **Figure 3.12** Physical activity is a healthy way to use the energy that can build up with anger. *Which strategy for dealing with anger would you most likely use?*

Health Skills Activity

Practicing Healthful Behaviors

Managing Your Anger

When Tina took out her favorite sweater to wear, she saw a big stain on the sleeve. Furious, she marched into her sister Judy's room. "I never said you could borrow my sweater! Look what you did to it! It's ruined!"

"It was clean when I put it back in your closet!" Judy shot back.

"I don't believe you," Tina said. "Now I don't have anything to wear tonight. Don't ever touch my things again!"

"But when I put it back, it was clean," said Judy.

Tina stormed back to her room and slammed the door.

Writing How might Tina have better dealt with her immediate feelings of anger? Rewrite this scene having Tina deal with her anger using the following steps:

1. Use strategies to reduce anger such as taking deep breaths or relaxing your muscles.
2. Analyze your feelings to recognize your emotions.
3. Talk to a parent about your feelings.
4. Write a letter to Judy expressing your feelings to her.

LESSON 3 ASSESSMENT

After You Read

Reviewing Facts and Vocabulary

1. What are *emotions*? How can emotions affect your behavior?
2. What are five common defense mechanisms?
3. List three strategies for handling anger in a healthful way.

Thinking Critically

4. **Analyze.** What role do hormones play in affecting a teen's emotions?
5. **Explain.** Describe what can happen when you take time to think before you respond to a strong emotion. How can this help you stay healthy?

Applying Health Skills

6. **Communication Skills.** Write a one-page script describing how a teen helps a friend manage an emotion, such as fear or excitement.

Writing Critically

7. **Descriptive.** Write a poem describing a situation that was emotional for you. Tell how you managed your emotions in a healthful way.

Real Life Issues ....

After completing the lesson, review and analyze your response to the Real Life Issues question on page 78.

Hands-On HEALTH

Activity "I feel . . . "

Managing your emotions helps you express them in a healthful way, set personal boundaries, and create healthy relationships with others. This activity will allow you to identify how you felt in certain situations and how to manage your feeling(s).

What You'll Need

- three 3" x 5" index cards per person
- markers
- notebook paper and pen or pencil

What You'll Do

Step 1

Using a marker, write in large print one emotion in the middle of each card and place face down.

Step 2

Using the statement **"I feel . . . when . . . and I need . . ."** one person chooses an index card, reads the emotion out loud, and then completes the rest of the statement.

Step 3

Continue until all group members have shared and used all the cards.

Apply and Conclude

After completing the activity, reflect in writing how it felt to share and listen to your peers. Explain how it feels and describe how this might impact a person's mental/emotional health.

Checklist: Communication Skills

- ☑ Clear, organized message
- ☑ Listening skills
- ☑ Use of "I" messages
- ☑ Respectful tone
- ☑ Appropriate body language

LESSON **1**

Developing Your Self-Esteem

Key Concepts

▸ Mentally healthy people sometimes have mental and emotional problems, but can cope with their emotions as well as know when to seek help.

▸ Healthy self-esteem involves having a sense of personal worth and a sense of competence.

▸ You can improve your self-esteem.

Vocabulary

▸ mental/emotional health (p. 66)
▸ resilient (p. 67)
▸ self-esteem (p. 68)
▸ competence (p. 68)
▸ hierarchy of needs (p. 70)
▸ self-actualization (p. 71)

LESSON **2**

Developing Personal Identity and Character

Key Concepts

▸ You develop your personal identity by developing a clear sense of your values, beliefs, skills, and interests.

▸ A person of good character demonstrates core ethical values.

▸ You can build a healthy identity from both the good and bad influences in your life.

Vocabulary

▸ personal identity (p. 72)
▸ role model (p. 73)
▸ personality (p. 73)
▸ character (p. 73)
▸ integrity (p. 74)
▸ constructive criticism (p. 76)

LESSON **3**

Expressing Emotions in Healthful Ways

Key Concepts

▸ Recognizing and understanding your emotions will provide you with ways to maintain your emotional health.

▸ Learning to manage your feelings is an important part of being mentally and emotionally healthy.

▸ When you deal with difficult emotions, such as fear, guilt, and anger, you may need to carefully consider the situation and use specific strategies to handle your feelings.

Vocabulary

▸ emotions (p. 78)
▸ hormones (p. 78)
▸ hostility (p. 79)
▸ empathy (p. 80)
▸ defense mechanisms (p. 81)

LESSON 1

Vocabulary Review

Use the vocabulary terms listed on page 85 to complete the following statements.

1. Having enough skills to do something is called _____.

2. Valuing, respecting, and feeling confident about yourself describes _____.

3. Having the ability to adapt successfully and recover from disappointment, difficulty, or crisis is called being _____.

Understanding Key Concepts

After reading the question or statement, select the correct answer.

4. Which of the following is *not* a characteristic of good mental and emotional health?
   a. Sense of purpose
   b. Pessimistic outlook
   c. Autonomy
   d. Healthy self-esteem

5. Which statement about self-esteem is *not* true?
   a. Self-esteem is always the same.
   b. Self-esteem develops over time.
   c. Self-talk affects self-esteem.
   d. Feedback from others affects self-esteem.

6. Which need within Maslow's hierarchy is the highest-level need?
   a. A safety need
   b. An esteem need
   c. A physical need
   d. The need to reach your potential

Thinking Critically

After reading the question or statement, write a short answer using complete sentences.

7. **Explain.** Why do you think one person can be considered mentally healthier than another when neither has a serious mental problem?

8. **Describe.** How can poor mental health affect your physical health?

9. **Synthesize.** Select one of the suggestions for improving self-esteem. Explain a practical way to make the action or behavior part of your life.

10. **Compare and Contrast.** How might a person with healthy self-esteem respond to a difficult challenge differently than a person with poor self-esteem?

11. **Explain.** Describe how self-esteem can affect your ability to reach your potential.

LESSON 2

Vocabulary Review

Correct the sentences below by replacing the italicized term with the correct vocabulary term.

12. A *friend* is someone whose success or behavior serves as an example for you.

13. *Trustworthiness* is a firm observance of core ethical values.

14. *Judgment* involves positive comments that point out problems and encourage improvement.

15. Your *personal identity* is the complex set of characteristics that make you unique.

Understanding Key Concepts

After reading the question or statement, select the correct answer.

16. A person's unique characteristics and group affiliations are known as
    a. features of character.
    b. strengths and weaknesses.
    c. features of identity.
    d. examples of core ethical values.

17. Which quality of good character reflects the importance of community concerns, such as obeying laws and voting?
    a. Responsibility
    b. Trustworthiness
    c. Caring
    d. Citizenship

18. Choosing not to cheat is an example of
    a. recognizing your strengths and weaknesses.
    b. demonstrating positive values.
    c. developing a purpose in your life.
    d. forming meaningful relationships.

19. The groups you belong to help you define
    a. characteristics that you share with other people.
    b. ways to get along with other people.
    c. how people are different.
    d. none of the above.

Thinking Critically

After reading the question or statement, write a short answer using complete sentences.

20. **Discuss.** Name some values that parents likely pass on to their children.

21. **Synthesize.** Why do you think responsibility, honesty, and respect are values that exist across cultures?

22. **Explain.** How might unhealthful risk behaviors affect your health and identity?

23. **Identify.** What are some examples of healthful risk behaviors?

24. **Describe.** What are some ways that good character is related to healthy identity?

LESSON 3

Vocabulary Review

Use the vocabulary terms listed on page 85 to complete the following statements.

25. A chemical produced by your glands that regulates the activities of different body cells is a(n) _____.

26. The intentional use of unfriendly or offensive behavior is called _____.

27. The ability to imagine and understand how someone else feels is called _____.

28. Mental processes that you use to protect yourself from strong or stressful emotions or situations are called _____.

Understanding Key Concepts

After reading the question or statement, select the correct answer.

29. Which of the following is *not* true about anger?
    a. It can result in violence.
    b. Often another emotion is involved.
    c. You become angry as you think about a situation.
    d. It causes little emotional harm.

30. A cause of guilt is
    a. acting against your values.
    b. doing a good deed.
    c. repressing an unpleasant feeling.
    d. recognizing you are not the cause of a negative situation.

31. Which is *not* a positive way to express an emotion?
    a. Hugging
    b. Smiling
    c. Yelling
    d. Laughing

32. Which of the following defense mechanisms uses excuse-making to explain a situation?
    a. Repression
    b. Rationalization
    c. Denial
    d. Compensation

Thinking Critically

After reading the question or statement, write a short answer using complete sentences.

33. **Analyze.** How do peers, family, and friends influence the way you express and manage emotions?

34. **Evaluate.** What might the effects of changing hormone levels during the teen years have on emotions?

35. **Explain.** Why are emotions neither good nor bad?

36. **Identify.** Name one positive characteristic that can be developed when you learn to recognize and express emotions in healthful ways. Discuss how acquiring this characteristic might affect your relationships.

37. **Evaluate.** What are possible consequences to everyone involved when a person responds violently to anger?

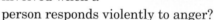

Technology — PROJECT-BASED ASSESSMENT

Watching for Signs of Mental Illness

Background

Good mental health is important to the well-being of everyone. The signs of mental illness, however, are sometimes easy to miss. Recognizing the early signs can help address and treat these problems.

Task

Conduct an online search and create a podcast describing the early warning signs of one or two mental illnesses.

Audience

Students at your school

Purpose

Help students recognize the warning signs of mental illness. Encourage them to seek help for themselves and others.

Procedure

1 Work in groups to review the information in Chapter 3 regarding mental health. Assign tasks and responsibilities to each group member.

2 Visit various Web sites that discuss the early warning signs of the mental illnesses the group has chosen.

3 Create a podcast which includes clear examples of the warning signs.

4 Be sure to explain how mental illness affects teens, and give resources for help.

5 Present your podcast to the students in your class.

6 Ask your teacher for help to create a unified podcast that represents the entire class, to be presented to the principal and possibly added to the school's Web site.

Math Practice

Interpret Graphs. A survey of 500 U.S. teens ages 14 to 17 shows that participating in afterschool activities can improve grades. Use the graph to answer Questions 1–3.

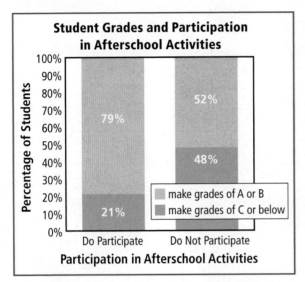

Student Grades and Participation in Afterschool Activities

79%
52%
48%
21%

make grades of A or B
make grades of C or below

Do Participate   Do Not Participate

Participation in Afterschool Activities

Percentage of Students

Adapted from "The YMCA's Teen Action Agenda" by Nels Ericson, *Office of Juvenile Justice and Delinquency Prevention Fact Sheet,* May 2001.

1. Among 1,000 teens who do not participate in afterschool activities, how many receive grades of A or B?
   - **A.** 210
   - **B.** 480
   - **C.** 520
   - **D.** 790

2. Choose the fraction of teens who participate in afterschool activities and receive grades of A or B.
   - **A.** 1/8
   - **B.** 1/7
   - **C.** 1/4
   - **D.** 4/5

3. Using the information in the chart, what could you conclude about teens between the ages of 14 and 17?
   - **A.** Most make grades of A or B.
   - **B.** Most make grades of C or below.
   - **C.** Most are in afterschool activities.
   - **D.** Most are not in afterschool activities.

Reading/Writing Practice

Understand and Apply. Read the passage below, and then answer the questions.

Rob had a social studies project due on Monday. "I've got plenty of time," he decided on Thursday. On Friday, he waited for a brilliant idea before giving up. Saturday he went to a baseball game. Rob finally sat down Sunday and worked late into the night. His grade reflected the lack of time and planning he spent on the project.

Rob experienced two common reasons why people procrastinate, or put things off: waiting for inspiration to strike and lack of planning. There are several ways to overcome procrastination. Break large tasks into smaller, more manageable parts. Make a list of everything that needs to be done. Work on each item separately. Tell friends and family your deadlines to help reduce distractions.

1. Which statement best sums up the main point? Procrastination
   - **A.** can be overcome by approaching a task in a variety of ways.
   - **B.** prevents you from starting projects.
   - **C.** creates negative consequences.
   - **D.** causes delays starting for fear of not doing a good enough job.

2. Which of the following does not summarize a suggestion for overcoming procrastination?
   - **A.** Break down the task into smaller, more manageable parts.
   - **B.** Make a list of everything you need to do.
   - **C.** Just sit down and do the work.
   - **D.** Set a deadline for completing.

3. Write a brief essay describing three ways you can avoid Rob's dilemma.

National Education Standards

Math: Measurement and Data, Data Analysis
Language Arts: LACC.910.L.3.6

CHAPTER 4

Managing Stress and Coping with Loss

Activating Prior Knowledge

Using Visuals Look at the picture on this page. How is this teen managing her stress? What other types of activities can help you manage stress? How do you deal with the stressors in your life? Explain your thoughts in a paragraph.

Chapter Launchers

Health in Action

Discuss the BIG Ideas

Think about how you would answer these questions:

▶ What is stress?

▶ Can you avoid stress?

▶ In what ways can other people help you deal with stress?

Assess Your Health

Read each statement. On a separate sheet of paper, write "yes," "sometimes," or "no" based on your typical behavior.

1. I react to stress in a positive way.

2. I recognize the stressors in my life and take action to manage stress.

3. I plan my time wisely to reduce stress at school.

4. I practice relaxation techniques to reduce the effects of stress.

5. I understand that using alcohol, tobacco, and other drugs, can harm the body and cause more stress.

A "yes" response shows that you practice healthy behaviors. "Sometimes" indicates that you should analyze and possibly modify your behavior. A "no" response means that you should modify the behavior.

LESSON 1

BIG Idea *Stress can affect you in both positive and negative ways.*

Before You Read

Create a K-W-L Chart. Make a three-column chart. In the first column, list what you **k**now about stress. In the second column, list what you **w**ant to know about this topic. As you read, use the third column to summarize what you **l**earned.

K	W	L

New Vocabulary

▶ perception
▶ stressor
▶ psychosomatic response

Review Vocabulary

▶ stress (Ch.2, L.1)

Understanding Stress

Real Life Issues ••••••••••••••••••••••••••••••

Stage Fright. Cari woke up this morning with a vague sense of dread. Now, sitting at her desk at school, she has butterflies in her stomach and her palms are sweaty. Today is oral report day, and Cari is next. She is nervous about speaking in front of her classmates.

Writing *Why do you think Cari is experiencing these symptoms? How might she try to calm herself for the presentation? Explain your thoughts in a paragraph.*

What Is Stress?

Main Idea How you think about a challenge determines whether you will experience positive or negative stress.

Feeling stress is a natural part of life. Stress is the reaction of the body and mind to everyday challenges and demands. It might appear quickly, like when you are late and running to catch the bus. Stress can also slowly build for days, like when you feel the pressure to perform well in your next basketball game or on a final exam.

Often, situations associated with stress are unavoidable. How much the stress of an event affects you, however, depends in part on your perception of it. **Perception** is *the act of becoming aware through the senses.* For example, based on your perception, you might believe that a disagreement with a friend has ruined your relationship. Your friend, on the other hand, might believe that you'll eventually work out the issue. Because of your perception of the event, you are more likely to experience a higher level of stress about the situation than your friend is.

Your reaction to stressful events depends on your previous experiences. If you enjoy playing in a band, performing a solo may not make you nervous. However, if you've made a mistake during a band performance, you might worry about how well you'll play during a solo.

Reacting to Stress

Some people believe that stress is always unhealthy. Stress can have both a positive and a negative effect. Positive stress can motivate you. For example, this type of stress can inspire you to work harder if you have a deadline approaching.

Stress has a negative effect, however, when it interferes with your ability to perform. It might cause you to feel distracted, overwhelmed, impatient, frustrated, or even angry. Negative stress can harm your health. Understanding the causes of stress and how you respond to it will help you develop effective stress-management skills.

Causes of Stress

> **Main Idea** Stressors vary among individuals and groups.

A **stressor** is *anything that causes stress*. Stressors can be real or imagined, **anticipated** or unexpected. People, objects, places, events, and situations are all potential stressors. Certain stressors, like sirens, affect most people the same way—causing heightened alertness.

As you've learned, the specific effects of most stressors will depend on your experiences and perceptions. What causes stress for you may not cause stress for someone else. **Figure 4.2** on page 94 identifies some common teen stressors.

READING CHECK

Explain How can your perception of an event affect the amount of stress you feel?

Academic Vocabulary

anticipate *(verb):* to expect

■ **Figure 4.1**
Meeting the demands of an active schedule can be stressful. *How do you deal with the stresses of a regular school day?*

Your Body's Response to Stressors

Main Idea Stressors activate the nervous system and specific hormones.

When you perceive something to be dangerous, difficult, or painful, your body automatically begins a stress response. For example, if you walk by your neighbors' house and their dog barks, you would likely feel startled and your heart might start racing. The sudden, loud barking is a stressor that affects you automatically, without any thought.

Both your nervous system and endocrine system are active during your body's response to stress. This physical response is largely involuntary, or automatic. The stress response, which occurs regardless of the type of stressor, involves three stages:

- **Alarm.** Your mind and body go on high alert. This reaction, illustrated in **Figure 4.3**, is sometimes referred to as the "fight-or-flight" response because it prepares your body either to defend itself or to flee from a threat.

- **Resistance.** If exposure to a stressor continues, your body adapts and reacts to the stressor. You may perform at a higher level and with more endurance for a brief period.

- **Fatigue.** If exposure to stress is prolonged, your body loses its ability to adapt. You begin to tire and lose the ability to manage other stressors effectively.

READING CHECK

List What are the three stages of the body's stress response?

Figure 4.2	Stressors for Teens			
Life Situations	**Environmental**	**Biological**	**Cognitive (Thinking)**	**Personal Behavior**
• School demands	• Unsafe neighborhood	• Changes in body	• Poor self-esteem	• Taking on a busy schedule
• Problems with friends, bullying	• Media (TV, magazines, newspapers, Internet)	• Illness	• Personal appearance	• Relationship issues
• Peer pressure		• Injury	• Not fitting in	• Smoking
• Family problems, abuse	• Natural disasters	• Disability		• Using alcohol or other drugs
• Moving or changing schools	• Threat of terrorist attacks			
• Breaking up with a girlfriend or boyfriend	• War			
	• Global warming			

FUTURE LOVE PRESSURE RACISM FAMILY DANGER DRUGS POLLUTION VIOLENCE HATE GRADES WAR

Figure 4.3   **The Alarm Response**

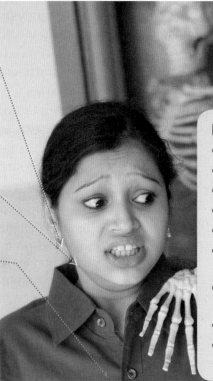

 Alarm begins when the *hypothalamus,* a small area at the base of the brain, receives danger signals from other parts of the brain. The hypothalamus releases a hormone that acts on the pituitary gland.

 The pituitary gland secretes a hormone that stimulates the adrenal glands.

 The adrenal glands secrete adrenaline. *Adrenaline* is the "emergency hormone" that prepares the body to respond to a stressor.

Physical Symptoms
- Dilated pupils
- Increase in perspiration
- Faster heart rate and pulse
- Rise in blood pressure
- Faster respiration rate
- Narrowing of arteries to internal organs and skin
- Increased blood flow to muscles and brain
- Increase in muscle tension
- Release of blood sugar, fats, and cholesterol

Stress and Your Health

Main Idea Ongoing stress affects all aspects of your health.

The physical changes that take place in your body during the stress response can take a toll on your body. Prolonged stress can lead to a **psychosomatic response**, *a physical reaction that results from stress rather than from an injury or illness*. Some of the physical effects of stress include

- headache,
- a weakened immune system,
- high blood pressure,
- bruxism, clenching the jaw or grinding the teeth, and
- digestive disorders.

Mental/emotional and social effects of stress include difficulty concentrating, irritability, and mood swings. Using alcohol or drugs to relieve stress may create more problems, if the person begins abusing these substances.

Real World CONNECTION

How Stressed Out Are You?

School is a cause of stress for many teens. In a study that examined what worried teens most about going back to school, nearly a third named schoolwork. Almost as many teens reported that they were worried about social concerns and physical appearance issues. The results of the study found that

▸ 32 percent reported schoolwork issues.

▸ 30 percent reported social issues.

▸ 25 percent reported physical appearance issues.

▸ 3 percent reported extracurricular issues.

▸ 10 percent reported no worries about returning to school.

Identifying the causes of stress in your life is the first step to handling it. If you know the cause, you can figure out how to prevent it or at least reduce its effects on you.

Activity Mathematics

The study received completed surveys from 600 teens.

1. How many teens felt that issues other than schoolwork caused them stress?

2. How many teens experienced no worries about returning to school?

3. **Writing** What worries you about returning to school? How do you cope with the stress of new classes?

Concept Ratios and Proportional Relationships: Percents A percent is a ratio comparing a number to 100. It can also be represented as a fraction with 100 as the denominator. To find the percent of a number, change the percent to a fraction or decimal, then multiply by the number.

LESSON 1 ASSESSMENT

 After You Read

Reviewing Facts and Vocabulary

1. Define the word *perception*.

2. What are three cognitive stressors for teens?

3. Identify the two body systems involved in the stress response.

Thinking Critically

4. **Synthesize.** Identify one way that stress has had a positive effect on your performance.

5. **Analyze.** Explain how a person in an extremely high-stress situation is able to accomplish an incredible feat of strength, such as lifting a car to free a person trapped underneath.

Applying Health Skills

6. **Analyzing Influences.** Describe ways that peer influence might increase the amount of stress that teens experience.

Writing Critically

7. **Expository.** Write a paragraph describing the positive and negative effects that stress has on your emotions.

Real Life Issues ....

After completing the lesson, review and analyze your response to the Real Life Issues question on page 92.

Managing Stress

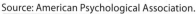

Real Life Issues

Ways to Handle Stress. Common ways to manage stress are:

Source: American Psychological Association.

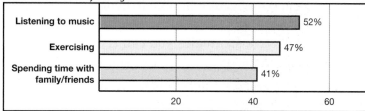

Listening to music	52%
Exercising	47%
Spending time with family/friends	41%

Writing *Write a paragraph describing how you manage stress.*

When Stress Becomes a Problem

Main Idea **Identifying what is stressful is the first step in learning how to manage stress.**

You are keenly aware of stress when its cause is obvious, such as when you're late for an appointment, your computer crashes while you're doing homework, or when you realize that you've left the materials you need to complete a project at home. When you know the source of stress, you can find ways to resolve the problem. Unfortunately, people often don't recognize the stressors in their lives. Many times, people recognize that they're feeling stressed only after the stress has begun to affect their health.

The effects of stress are *additive,* meaning they build up over time. Unless you find ways of managing stress, it will take a physical and mental toll on you. An increasing number of teens are experiencing **chronic stress**, *stress associated with long-term problems that are beyond a person's control.* For these individuals, stress has become a constant burden that can last for months.

Fortunately, there are positive actions you can take to deal with stress. Although you can't eliminate all stress from your life, you can manage it. The trick is to learn strategies to keep stress from building up and to deal with individual stressors effectively.

GUIDE TO READING

BIG Idea *You can manage stress by learning skills to reduce the amount and impact of stress in your life.*

Before You Read

Create a Cluster Chart. Draw a circle and label it "Stress-Management Skills." Use surrounding circles to define and describe this term. As you read, continue filling in the chart with more details.

Stress Mgmt.

New Vocabulary

▸ chronic stress
▸ relaxation response

Review Vocabulary

▸ stress management skills (Ch.2, L.1)
▸ resilient (Ch.3, L.1)

Tim Fuller Photography

Stress-Management Techniques

(Main Idea) You can develop strategies to both avoid and reduce your stress.

Stress-management skills help you manage stressors in a healthful, effective way. Some skills involve strategies to prevent stress. Others focus on coping with the impact of stress.

Avoiding and Limiting Stress

Avoiding situations that cause stress is the easiest way to reduce its effects. If you're unable to avoid a stressor, you can try to restrict or limit the amount of stress you're exposed to. These are effective strategies you can try:

- **Use refusal skills.** Determine whether you have time for a new activity before agreeing to take it on. If the new activity will add to your stress, use refusal skills to say no. You will learn about refusal skills in Chapter 8.

- **Plan ahead.** Manage your time wisely by planning ahead. Think about how stressed you feel before a test. **Figure 4.4** lists ways to reduce stress when studying for and taking tests.

- **Think positively.** We can't control everything in our lives, but we *can* control how we respond to events. A positive outlook limits stress by shifting your perception and the way you react to a stressor. For example, try viewing a typical stressor, like a job interview, as a learning opportunity instead of a threat.

FITNESS ZONE

With school, work, and everything else, my friends and I can get really stressed out. I found that working out is the best stress reliever. When I'm feeling really stressed, I go for a run, swim, or just shoot some baskets on my own. Afterward, I always feel less stressed out. For more physical activity ideas, visit the Fitness Handbook and Fitness Zone sites in ConnectEd.

| Figure 4.4 | **Overcoming Test Anxiety** |

Relaxation techniques, such as deep breathing and stretching, can reduce stress.

- Plan for tests well in advance, studying a little each night.

- Learn to outline material, highlighting and numbering important points to learn them quickly.

- During a test, do some deep breathing. Get comfortable in your chair. Use positive self-talk such as "I can do this!" or "Way to go!"

- Answer all the questions you are sure of, then go back and answer the ones that are more difficult.

- After getting your corrected test back, examine your mistakes. If you don't understand the correction, ask your teacher.

■ **Figure 4.5** Planning ahead can help you avoid or limit stress. *What other actions can you take to manage stress?*

- **Avoid tobacco, alcohol, and other drugs.** Using tobacco, alcohol, and other drugs in an attempt to relieve stress will actually harm the body and cause more stress.

Handling Stress and Reducing Its Effects

Some stressors may be unavoidable. Some days you may be running late for school because the weather is bad, or the bus had a flat tire. If you have a part-time job, your boss might be stressed himself on some days, which makes your workday stressful. For stressors that are unavoidable, try to find ways to reduce their negative effects. To lower the impact of stress on your health, try these tips:

- **Practice relaxation techniques.** Deep breathing, thinking pleasant thoughts, stretching, taking a warm bath, getting a massage, and even laughing can relieve your stress. Practicing these techniques regularly can help you achieve a **relaxation response**, *a state of calm.* **Figure 4.6** describes a relaxation technique.

- **Redirect your energy.** When intense energy builds up from stress, the best thing to do is use that energy in a constructive way. You can put your nervous energy to good use by working on a creative project, going for a walk or a swim, jogging, riding your bike, or playing a game of pickup basketball.

READING CHECK

Explain How can refusal skills help you avoid stress?

Academic Vocabulary

technique *(noun):* a method of accomplishing a desired aim

Figure 4.6	**Progressive Muscle Relaxation**

By practicing these relaxation techniques daily, you can prepare yourself to manage stress when it occurs.

Breathe deeply and slowly throughout the process.
1. Loosen your clothing and get comfortable. Lie down or relax in a comfortable chair.
2. Tighten the muscles in your toes. Hold for a count of 10. Relax.
3. Flex the muscles in your feet. Hold for a count of 10. Relax.
4. Move slowly up your body, tensing and then relaxing the muscles in your legs, abdomen, back, shoulders, arms, neck, and face.

Health Skills (Activity)

Stress Management

When Demands Are Too High

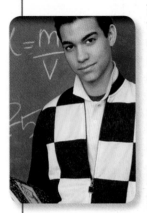

Juan has too many commitments and is feeling the effects of stress. After his parents leave for work, he helps his younger brother get ready for school and waits until his brother's school bus arrives. At school, he volunteers on the yearbook committee, plays on the basketball and soccer teams, and is taking extra classes to boost his grades and qualify for a college scholarship. At home, he helps his family by doing chores. On weekends, he works at a local bookstore to earn his own money. Juggling all of his responsibilities has become more difficult. Now his girlfriend wants him to spend more time with her.

Writing Write a letter to Juan suggesting ways that he could manage his stress. Use the following questions to guide your thinking.

1. What are Juan's stressors?
2. Which stressors, if any, can he avoid or prevent? How?
3. Which stressors can he limit? How?
4. Which stress-management techniques can help Juan deal with his stress?

- **Seek support.** Sometimes just talking about your problem can make you feel better. When you feel stressed, try confiding in someone you trust, such as a parent, guardian, sibling, teacher, or close friend. They can provide you with an objective view and valuable advice.

Staying Healthy and Building Resiliency

Main Idea Taking care of your health is essential to stress management.

READING CHECK

Explain What three self-maintenance habits can reduce your level of stress?

In addition to learning stress-management skills, developing habits that maintain your general health will also help reduce the effects of stress. These self-maintenance habits help you deal with stress in positive ways. They can also play a role in preventing stress, reducing stress, and helping your mind and body recover from stress.

Get Adequate Rest Too little sleep can affect your ability to concentrate. This can affect schoolwork, athletics, and even relationships. By contrast, adequate sleep can help you face the challenges and demands of the next day. Using time-management skills will allow you to get the eight to nine hours of sleep that you need each night.

Get Regular Physical Activity Participating in regular physical activity benefits your overall health whether or not you are feeling the effects of stress. Physical activity can release pent-up energy and clear your mind. Done regularly, exercise increases your energy level and your endurance. It helps you sleep better, too.

Eat Nutritious Foods Eating a variety of healthful foods and drinking plenty of water not only helps your body function properly, but it also reduces the effects of stress. In contrast, poor eating habits can contribute to stress, causing weakness, fatigue, and a reduced ability to concentrate. Overeating and undereating can also put your body under stress. Beverages high in caffeine and sugar, such as coffee drinks or quick-energy drinks, can increase the effects of stress. You'll learn more about good nutrition in Chapters 10 and 11.

By including self-maintenance and stress-management strategies in your daily routine, you can become more resilient. This means you're able to adapt effectively and recover from disappointment, difficulty, or crisis. For example, you would probably feel disappointed if you didn't win the part you wanted in the school play. A resilient teen would bounce back from this disappointment and work harder for the next audition. Resiliency helps you handle difficulties and challenges in healthful ways and achieve long-term success in spite of negative circumstances.

LESSON 2 ASSESSMENT

After You Read

Reviewing Facts and Vocabulary

1. What is *chronic stress*?
2. Identify four strategies to avoid or limit stress.
3. Identify three relaxation techniques.

Thinking Critically

4. **Synthesize.** It's Wednesday, and Ariana's biology test is on Friday. As she sits down to study, her friend Conner calls and asks her to go out. How might Ariana balance her activities and manage her stress?
5. **Describe.** Explain the role of positive thinking as a stress-management strategy.

Applying Health Skills

6. **Practicing Healthful Behaviors.** Some of the habits that you practice to maintain overall health can also help manage stress. Design a poster illustrating the habits that can help you manage stress.

Writing Critically

7. **Personal.** Evaluate your own wellness in regard to stress. Write a paragraph to explain your assessment.

Real Life Issues ....

After completing the lesson, review and analyze your response to the Real Life Issues question on page 97.

GUIDE TO READING

BIG Idea *Understanding the grieving process helps you cope with loss and manage your feelings in healthy ways.*

Before You Read

Create Vocabulary Cards. Write each new vocabulary term on a separate note card. For each term, write a definition based on your current knowledge. As you read, fill in additional information related to each term.

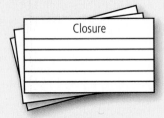

Closure

New Vocabulary

▶ stages of grief
▶ closure
▶ coping
▶ mourning
▶ traumatic event

Coping with Loss and Grief

Real Life Issues •

Losing a Close Relative. Kelly has always been close to her grandfather. Every weekend they would spend time together, taking walks, watching movies, playing chess, or just talking. He has just passed away at the age of 92. Kelly misses him terribly and feels there is a big hole in her life.

Writing *If you were Kelly's friend, how might you comfort her as she tries to cope with the loss of her grandfather? Write a dialogue between you and Kelly in which you offer support and sympathy.*

Acknowledging Loss

Main Idea Acknowledging a loss is one way to help begin the healing process.

You have probably experienced a loss that left you feeling sad. Perhaps you moved to a new city and left behind good friends. You may have even experienced the death of someone you love. Everyone experiences loss during their lives and the grief that it brings. For example, you may have felt the pain of rejection, the breakup of a relationship, or the death of a pet, friend, or family member. Maybe you had to move or change schools and miss the friends you left behind.

Grieving is a common and natural reaction to any loss that brings on strong emotions. Loss feels hurtful, but it does not have to be harmful. Immediately after the loss, you may feel that your life will never be the same, and that you may never recover. Again, these feelings are natural. Acknowledging and understanding your grief will help you begin the healing process. This in turn will help you to cope with the loss and manage your feelings.

Expressing Grief

Main Idea The grieving process can help people accept the loss and start to heal.

Feelings of loss are very personal. Some people feel sadness, guilt, or even anger. Some may talk about their loss; others may want to be alone. Sometimes people experience several or all of these emotions.

The Grieving Process

While everyone grieves in their own way, Swiss-American psychiatrist Elisabeth Kübler-Ross noted that the grieving process includes **stages of grief**, *a variety of reactions that may surface as an individual makes sense of how a loss affects him or her.* Not everyone goes through each stage, and the order may be different for each person. Here are the stages:

- **Denial or Numbness.** It may be difficult to believe the loss has occurred.
- **Emotional Release.** The loss is recognized. This stage often involve periods of crying.
- **Anger.** The person uses anger because he or she feels powerless and unfairly deprived.
- **Bargaining.** As the reality of the loss sets in, the person may promise to change if what was lost can be returned.
- **Depression.** Beyond the feelings of sadness, feelings of isolation, alienation, and hopelessness may occur.
- **Remorse.** The person may become preoccupied with thoughts about how the loss could have been prevented.
- **Acceptance.** The person faces the reality of the loss, and experiences **closure**, or the *acceptance of a loss.*
- **Hope.** Remembering becomes less painful, and the person begins to look ahead to the future.

Experiencing and accepting your feelings during grieving is necessary for healing. These feelings are part of **coping**, or *dealing successfully with difficult changes in your life.*

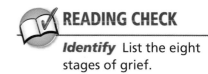

READING CHECK

Identify List the eight stages of grief.

■ **Figure 4.7** Grieving is a process that you need not experience alone. *How might receiving comfort and support help you through a loss?*

Figure 4.8 Memorial services and sites help people grieve and show respect. *What are other ways to remember a loved one?*

Coping with Death

Main Idea Coping with death involves receiving and showing support.

Death is one of the most painful losses we can experience. Even if a person dies after a long illness, it's likely that the survivors will grieve. If the death was sudden or traumatic, the survivors may also experience shock.

Most people respond to loss by **mourning**, *the act of showing sorrow or grief.* Mourning includes talking about the person, experiencing the pain of the loss, and searching for meaning. It may be difficult for some people to move out of the mourning process. Dwelling on things that can't be changed will only add to your hurt. Instead, try to think about how the relationship was positive in your life.

Showing Empathy

Grieving alone makes the process more difficult. The friendship and support of others who are also grieving may make the process easier. If you can't talk to family and other loved ones, try talking to a supportive friend.

If you know someone who is grieving, there are ways you can show support.

- Help the person to recall happy, positive memories.

- Be a sympathetic listener, and use silence when appropriate. Sometimes, just nodding your head shows that you understand what the person is saying.

- Don't rush the grieving process or attempt to **resolve** the person's grief in one day. Remember, no one can lead another person through this process or hurry through it.

Academic Vocabulary

resolve *(verb):* to deal with successfully

Community Support

A person's cultural background also influences grieving. Common mourning rituals, such as memorial services, wakes, and funerals are events that celebrate the life of the person who has died. Telling stories or describing why the person was special can help you move through the grieving process. The clergy and mental health professionals who specialize in grief can also provide support.

READING CHECK

Explain What are ways to support someone who is grieving?

Coping with Traumatic Events

Main Idea Support from family, friends, and community resources can help individuals recover from a traumatic event.

A **traumatic event** is *any event that has a stressful impact sufficient to overwhelm your normal coping strategies.* Traumatic events are sudden and shocking, such as accidents, violent assaults, suicides, and natural disasters. After a traumatic event, you may question your sense of security and confidence. Seek support from family members, friends, and community groups and agencies to help you manage your shock and grief. Also, trying to resume your normal activities can help you through the grieving process.

LESSON 3 ASSESSMENT

After You Read

Reviewing Facts and Vocabulary

1. Identify the stages of grief.
2. Define the term *coping*.
3. List three examples of a traumatic event.

Thinking Critically

4. **Analyze.** How might coping with a death resulting from a long-term illness differ from coping with a sudden death caused by an accident?

5. **Apply.** Recall a story of personal loss that you read about in a book or saw in a movie. Write a paragraph that describes the process of grieving that the main character went through.

Applying Health Skills

6. **Communication Skills.** Write a letter expressing caring and empathy to a friend who is grieving for a loved one.

Writing Critically

7. **Expository.** Write a paragraph describing ways that people in the community and community support groups can help someone who is coping with a loss.

Real Life Issues ....

After completing the lesson, review and analyze your response to the Real Life Issues question on page 102.

Hands-On
HEALTH

Activity Juggling Stress

Do you sometimes feel overwhelmed and exhausted trying to juggle everything in your life—school, homework, activities, family and household responsibilities—all at once? This activity will help you identify your stressors and find ways to manage stress.

What You'll Need

- 3 large index cards
- pen or pencil

What You'll Do

Step 1

In the middle of each index card, draw a 1-inch circle. Write one stressor that is currently affecting your life.

Step 2

Your teacher will demonstrate the tennis ball activity. Each of your stressors is like a tennis ball you juggle as you try to balance your daily activities.

Step 3

Select one of your index cards. List four signs or symptoms describing how this particular stressor is currently affecting your health.

Apply and Conclude

Think about the stressors that you identified. List stress-management techniques that you can practice to reduce or manage the stressful situations you identified. Implement your plan and evaluate its effectiveness.

Checklist: Stress-Management Skills

- ✓ Identification of situations that cause stress
- ✓ Techniques you can use to avoid stressful situations
- ✓ Ways to manage stress
- ✓ Evaluate the effectiveness of the techniques you use to manage stress

Medioimages/Photodisc/Getty Images

Understanding Stress

Key Concepts

▶ Stress is a natural part of life; everyone experiences stress.
▶ The specific effects of stressors on your life depend on your experiences and perception.
▶ Too much stress can be unhealthful.

Vocabulary

▶ stress (p. 92)
▶ perception (p. 92)
▶ stressor (p. 93)
▶ psychosomatic response (p. 95)

Managing Stress

Key Concepts

▶ You can manage stress by using refusal skills, planning ahead, thinking positively, and avoiding tobacco, alcohol, and other drugs.
▶ Stress-management techniques include relaxation, redirecting your energy, and seeking support.
▶ Taking care of your health can help you prevent and reduce stress, as well as recover from its effects.

Vocabulary

▶ chronic stress (p. 97)
▶ stress-management skills (p. 98)
▶ relaxation response (p. 99)
▶ resilient (p. 101)

Coping with Loss and Grief

Key Concepts

▶ Grief is caused by many kinds of loss.
▶ Each person's response to a loss is unique to the situation and to the individual.
▶ Each person goes through stages of grief, although not in any particular order.
▶ Coping with death requires closure.

Vocabulary

▶ stages of grief (p. 103)
▶ closure (p. 103)
▶ coping (p. 103)
▶ mourning (p. 104)
▶ traumatic event (p. 105)

Assessment

LESSON 1

Vocabulary Review

Use the vocabulary terms listed on page 107 to complete the following statements.

1. _____ is the reaction of the body and mind to everyday challenges and demands.

2. Anything that causes stress is called a(n) _____.

3. A physical reaction that results from stress rather than from an injury or illness is called a(n) _____.

Understanding Key Concepts

After reading the question or statement, select the correct answer.

4. The amount of stress that you experience mostly relates to
   a. the type of friends that you have.
   b. where you go to school.
   c. your perception of stressors.
   d. how your parents respond to stress.

5. During which stage of the body's stress response are hormones released?
   a. Resistance
   b. Alarm
   c. Fatigue
   d. Recovery

6. Which of the following is *not* a physical symptom of the alarm response?
   a. Faster heart rate and pulse
   b. Decreased blood flow to muscles and brain
   c. Increase in perspiration
   d. Increase in blood pressure

Thinking Critically

After reading the question or statement, write a short answer using complete sentences.

7. **Identify.** What are the five categories of stressors?

8. **Infer.** How would trying an activity for the first time affect your stress level?

9. **Synthesize.** Suppose you've been assigned to work on a project with three students you don't know. How might this affect your perception of doing the project?

10. **Analyze.** Describe how mental fatigue that results from stress can affect your ability to study.

11. **Describe.** List three physical symptoms of stress.

12. **Infer.** You've spent three months feeling stressed while studying for an important exam. What impact could this stress have on your health?

LESSON 2

Vocabulary Review

Correct the sentences below by replacing the italicized term with the correct vocabulary term.

13. Using refusal skills, planning ahead, and practicing relaxation techniques are examples of *chronic stress.*

14. You are *relaxed* if you are able to adapt effectively and recover from disappointment, difficulty, or crisis.

15. Practicing stress-management techniques can help you achieve a state of calm, or a *chronic stress,* when stressed.

16. Stress associated with long-term problems beyond one's control is known as *resilient.*

Understanding Key Concepts

After reading the question or statement, select the correct answer.

17. Which is not a relaxation technique?
    a. Taking a warm bath
    b. Laughing
    c. Eating a comfort food
    d. Deep breathing

18. One way to prevent taking on an activity that will add to your level of stress is to
    a. procrastinate.
    b. use refusal skills.
    c. think positively.
    d. redirect your energy.

19. Which of the following is *not* a way to redirect energy that may build up as a result of stress?
    a. Going for a walk
    b. Watching TV
    c. Riding your bike
    d. Working on a creative project

20. Which is *not* an effect of physical activity on stress?
    a. Clears your head
    b. Increases energy level
    c. Helps you sleep better
    d. Helps you avoid stress

Thinking Critically

After reading the question or statement, write a short answer using complete sentences.

21. **Explain.** How does identifying personal stressors help in the development of a stress-management plan?

22. **Analyze.** Why are the effects of stress additive? How can additive stressors affect your health?

23. **Synthesize.** What might be some important considerations when planning ahead for a research project that is due in three weeks?

24. **Analyze.** Why shouldn't people smoke cigarettes as a way to relieve stress?

25. **Analyze.** Explain how having resiliency can help you manage your stress.

LESSON 3

Vocabulary Review

Choose the correct term in the sentences below.

26. *Closure/Coping* is acceptance of a loss.

27. *Coping/Mourning* is the act of showing sorrow or grief.

28. A stressful event that overwhelms your coping strategies is called a *traumatic event/stage of grief.*

Understanding Key Concepts

After reading the question or statement, select the correct answer.

29. Which is *not* a stage of grief?
    a. Remorse
    b. Empathy
    c. Acceptance
    d. Denial

30. The needed outcome of grieving is
    a. anger.
    b. sympathy.
    c. remorse.
    d. closure.

31. You can show support to someone who is grieving by
    a. helping the person recall happy memories.
    b. being a sympathetic listener.
    c. not rushing the grieving process.
    d. all of the above.

32. During which stage of grief do people make a promise to change if what was lost can be returned?
    a. Denial
    b. Depression
    c. Bargaining
    d. Hope

33. Which of the following strategies can help someone cope with a traumatic event?
    a. Spending time alone
    b. Delaying getting back to a daily routine
    c. Putting off grieving
    d. Seeking support from the community

Thinking Critically

After reading the question or statement, write a short answer using complete sentences.

34. **Describe.** What are four examples of loss that could cause someone to experience the grieving process?

35. **Evaluate.** How do you think the ability or inability to remain open to relationships could affect the way a person responds to loss?

36. **Analyze.** What is necessary in order for healing to occur after a loss?

37. **Explain.** At which stage of the grieving process might a person become unable to move on? What should people do if they have difficulty moving through the stages of grief?

38. **Explain.** How do mourning rituals following a death help individuals during the grieving process?

39. **Identify.** Who is available in a community to respond to the needs of survivors of a traumatic event?

Technology PROJECT-BASED ASSESSMENT

The Stages of Grief

Background

We all have to cope with a significant loss some time in our lives. A best friend may move to another city, or a close family member may pass away. The grieving process occurs in stages, and understanding those stages will help you cope with loss.

Task

Conduct an online search about the grieving process. Create a podcast describing the grieving process.

Audience

Students in your class

Purpose

Help other students learn about the grieving process and how they can help someone who is grieving.

Procedure

1. Form small groups. Divide the tasks. Some members may want to conduct the Internet search or write the script for the podcast. Others may want to perform in the podcast.

2. Using the information in Chapter 4, as well as additional Internet resources, research the stages of grief.

3. Write the script for a podcast about a student who has suffered a significant loss. The podcast should describe the student moving through all the stages of grief.

4. Record the podcast to be played in class.

Math Practice

Interpret Tables. Angelika conducted a survey of 296 students at her school to determine what caused them the most stress. The results of her survey are shown in the table below. Use the table to answer Questions 1–3.

Student Stressors at Washington High	
Greatest Stressor	**Number of Students**
Grades	93
Peer conflict	81
Family issues	64
Work responsibilities	24
After-school activities	8
Personal health	8
Other	18

1. What percentage of students did not feel that work responsibilities were a stressor?
   A. 88%
   B. 76%
   C. 19%
   D. 92%

2. What number of students reported that family issues caused them the most stress?
   A. 64
   B. 78
   C. 192
   D. 53

3. What could you conclude from the information given in the table?
   A. Most students are stressed about work responsibilities.
   B. After-school activities are stressors for many teens.
   C. Peer conflict ranked second as the greatest stressor for students surveyed.
   D. Personal health issues are stressful for students.

Reading/Writing Practice

Understand and Apply. Read the letter below, and then answer the questions.

Dear Maya,
  Things have changed since you moved away. The factory closed and more than 3,000 people are unemployed. I'm sure more people will be leaving town like your family did. Manuel and his family moved, too. Now both of my best friends have left.
  My parents don't want to tell me, but I can tell that things are not good. I overhear them talking, but they clam up whenever I ask anything. Dad has a new job, but he is making less money.
  Please write back and let me know how things are going for you. Do your parents have jobs? Have you made new friends? I miss you!
    Love,
    Isabel

1. Which word *best* describes the tone of this letter?
   A. Angry
   B. Worried
   C. Bitter
   D. Resigned

2. What does this letter reveal most about Isabel?
   A. It's hard for her to make new friends.
   B. She understands life outside her town.
   C. The factory closing led to lost jobs and wages.
   D. She feels stress due to all the changes.

3. Pretend you are Maya and write a reply to Isabel's letter. As Maya, explain what life is like now for you and your family, and how you are dealing with the stress of moving.

National Education Standards

Math: Statistics and Probability
Language Arts: LACC.910.L.3.6; LACC.910.RL.2.4

Edi-Imaging

CHAPTER 5

Mental and Emotional Problems

Activating Prior Knowledge

Using Visuals Some mental health professionals recommend art therapy as one strategy for coping with problems. Why do you think creating art might help an individual deal with difficult emotions?

Chapter Launchers

Health in Action

Discuss the **BIG** Ideas

Think about how you would answer these questions:

▶ What are some reasons that teens might feel anxiety?

▶ What mental health disorders can you name?

▶ What are some sources of help for people with mental health disorders?

Assess Your Health

Read each statement. On a separate sheet of paper, write "yes," "sometimes," or "no" based on your typical behavior.

1. I find ways to manage anxiety by planning ahead to reduce stress.

2. I recognize that everyone feels depressed at times.

3. I understand that mental health disorders are health disorders.

4. I recognize that suicidal thoughts require the help of mental health professionals.

5. I know how to seek help for depression and other mental health disorders.

A "yes" response shows that you practice healthy behaviors. "Sometimes" indicates that you should analyze and possibly modify your behavior. A "no" response means that you should modify the behavior.

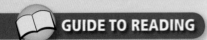
Dealing with Anxiety and Depression

GUIDE TO READING

BIG Idea *Anxiety and depression are treatable mental health problems.*

Before You Read

Create an Outline.
Look through the lesson to find the headings and subheadings. Write down the headings to make an outline. As you read, fill in details beneath each heading or subheading.

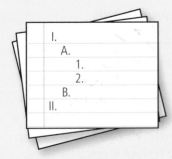

I.
  A.
    1.
    2.
  B.
II.

New Vocabulary

▶ anxiety
▶ depression
▶ apathy

Review Vocabulary

▶ emotions (Ch.3, L.3)

Real Life Issues

Difficult Times. Tony's parents are separating. He's not surprised because they have been arguing a lot lately. However, he still feels hurt by their decision. Tony is also worried about how his family will get by financially without both of his parents living at home. The constant feelings of sadness and uncertainty are starting to affect other aspects of his life. Tony is tired all the time, has withdrawn from his friends, and has lost his appetite. All he wants to do is stay home alone and sleep.

Writing *Write a dialogue between you and Tony, as if you were Tony's friend. In your conversation, show him empathy and support during this difficult time. What advice might you give him?*

Understanding Anxiety

Main Idea Occasional anxiety is a normal, manageable reaction to many short-term, stressful situations.

Experiencing difficult emotions is a normal part of life. They occur for a variety of reasons, including hormonal changes, relationship issues, grief, or stress. A common feeling is **anxiety**, *the condition of feeling uneasy or worried about what may happen.* You may, for example, feel anxious about an important class presentation.

Occasional anxiety is a natural response to life events. Brief feelings of worry, insecurity, fear, self-consciousness, or even panic are common responses to stress. Usually, once the stressful situation is over, so is the anxiety it created.

■ **Figure 5.1** Depression can cause a person to withdraw and suffer alone. *Why might this symptom be dangerous?*

Coping with Anxiety

Knowing that anxiety is common doesn't make it easier to manage. Think about the situations that have caused you to be anxious in the past. What can you do in the future to plan ahead so that stress will not build and cause anxiety? The stress-management techniques described in Chapter 4 can help reduce anxiety. Some people use substances such as alcohol or drugs to escape from anxiety. These substances produce a temporary, false sense of relaxation, but can cause other physical, mental/emotional, social, or legal problems.

Understanding Depression

(**Main Idea**) **Depression can linger or be severe enough to disrupt daily activities.**

Depression is *a prolonged feeling of helplessness, hopelessness, and sadness.* Feelings of sadness affect everyone, but depression usually lasts longer and may produce symptoms that do not go away over time. Depression is a serious condition that may **require** medical help. About 11 percent of all teens will display some signs of depression. It's one of the most common mental health concerns among teens. Types of depression include *major depression,* which is intense and can last for weeks or months. *Mild depression* has less severe symptoms, but can last for years. *Adjustment disorder* is a reaction to a specific life event. For example, a person may have trouble reaching closure when grieving.

 READING CHECK

Identify What is the benefit of using stress management techniques to manage anxiety?

Academic Vocabulary

require *(verb):* to demand as necessary

Figure 5.2

Warning Signs of Depression

Five or more of these symptoms must persist for two or more weeks before a diagnosis of major depression is indicated.

▶ Persistent sad or irritable mood

▶ Loss of interest in activities once enjoyed

▶ Change in appetite or body weight

▶ Difficulty sleeping or oversleeping

▶ Restlessness or irritability

▶ Loss of energy

▶ Feelings of worthlessness or inappropriate guilt

▶ Difficulty concentrating or making decisions

▶ Recurrent thoughts of death or suicide

▶ Feeling hopeless

Causes and Effects of Depression

Depression can be caused by physical, psychological, or social reasons. A medical condition or illness may cause depression. It may also be caused by psychological reasons, such as surviving a traumatic life event. Finally, social or environmental factors, such as living in poverty or in a physically or emotionally harmful environment may cause depression. **Figure 5.2** lists warning signs of depression. Other symptoms include the following:

- **Changes in thinking.** People who are depressed may have trouble concentrating and making decisions. They may have self-destructive thoughts.

- **Changes in feelings.** People who are depressed may experience **apathy**, or *a lack of strong feeling, interest, or concern*. They may not feel pleasure in things they once enjoyed. They may be sad, or irritable and angry.

- **Changes in behavior.** People with depression may become emotional, and they may begin eating too little or too much. The person may have trouble sleeping and may seem tired. The person might also neglect basic hygiene and withdraw from social situations.

Getting Help for Depression

Main Idea Depression is a treatable illness.

READING CHECK

Identify Who can a depressed teen ask for help?

If you recognize signs of depression in yourself or a friend, discuss your concerns with someone you trust. Depression is serious, but it is treatable. If a friend asks you not to tell anyone that he or she is depressed, it's okay to break that promise. Health professionals can develop a plan to treat depression that may include taking medication, making changes in the home or school environment, or counseling. Treating depression takes time, persistence, and patience.

Health Skills Activity

Accessing Information

Recognizing Reliable Resources

"Hey, Con, how's it going?" Devin asked his close friend Connor as they met up in the hallway. Connor smiled slightly at Devin. "I'm feeling better, thanks," he replied. "I met with my family doctor last week. He was concerned that I was depressed."

"Yeah?" Devin looked interested. "What did he say?"

"He prescribed some medication, and I'm meeting with a mental health specialist tomorrow.

Devin didn't know what to say at first. He knew that Connor had been unhappy, but didn't know much about depression.

"I'm sorry Connor. I didn't realize you were feeling that bad. I hope things get better soon."

The friends said goodbye and went to their classes. In study hall, Devin thought more about depression and realized he didn't know much about it. He decided to do research on the Web to find out if there is anything he can do—or shouldn't do—to help his friend.

Writing List sources of information for depression and its treatment that Devin might use to help Connor. Evaluate the information using the criteria listed below.

1. What are the qualifications of the authors?
2. Is the material backed by a nationally recognized and respected mental health organization?
3. Can the information be confirmed by other sources?

LESSON 1 ASSESSMENT

 After You Read

Reviewing Facts and Vocabulary

1. Define the term *anxiety*.
2. What are the causes of depression?
3. Describe changes in thinking that might be effects of depression.

Thinking Critically

4. **Analyze.** Explain the difference between "feeling down or depressed" and "having depression." Provide examples to show the difference.
5. **Synthesize.** If you believe a friend might be depressed, what can you do to help?

Applying Health Skills

6. **Analyzing Influences.** Divide a sheet of paper into three columns. Label the columns "Family," "Friends," and "School." Use this chart to describe how depression can affect each aspect of your life.

Writing Critically

7. **Expository.** Write a paragraph discussing why it is important for someone with depression to get professional help.

Real Life Issues

After completing the lesson, review and analyze your response to the Real Life Issues question on page 114.

Lesson 1 Dealing with Anxiety and Depression **117**

GUIDE TO READING

BIG Idea *Gaining an understanding of mental health disorders builds insight and empathy.*

Before You Read

Create Vocabulary Cards. Write each new vocabulary term on a separate note card. For each term, write a definition based on your current knowledge. As you read, fill in additional information related to each item.

Mental Disorder

New Vocabulary

▸ mental disorder
▸ stigma
▸ anxiety disorder
▸ mood disorder
▸ conduct disorder

Mental Disorders

Real Life Issues

Cutting. Bree is having a difficult time at school and is constantly arguing with her parents at home. She began cutting herself to release emotional pain. She wears long-sleeved shirts to hide her scars because she doesn't want anyone to lecture her about it. Bree's friend, Kris, sees the cuts and asks, "Bree, are you okay? What's happening?" Bree mumbles a response and walks away.

Writing *Pretend that you are Bree's friend. You are scared by the cuts on her arms. Write a letter to Bree telling her you care for her and want her to get help.*

Understanding Mental Disorders

Main Idea **Mental disorders are medical conditions that require diagnosis and treatment.**

Each year, approximately 57.7 million people in the United States are affected by some form of **mental disorder**—*an illness of the mind that can affect the thoughts, feelings, and behaviors of a person, preventing him or her from leading a happy, healthful, and productive life.* That's about one in every four Americans. Many do not seek treatment because they feel embarrassed or ashamed. Others worry about the stigma associated with mental disorders. A **stigma** is *a mark of shame or disapproval that results in an individual being shunned or rejected by others.*

Many people don't understand that mental disorders are medical conditions, and require diagnosis and treatment just like any physical illness or injury. Learning about mental and emotional problems will help erase the stigma associated with these disorders, and will help encourage people to seek medical help early. Many times, mental and emotional problems cannot be solved without professional help.

Types of Mental Disorders

Main Idea Mental disorders can be identified by their symptoms.

Mental disorders are medical conditions that can begin as early as childhood. Many times, these problems require help from health professionals.

Anxiety Disorders

An **anxiety disorder** is *a condition in which real or imagined fears are difficult to control.* It is one of the most common mental health problems among children and teens. Reports have shown that as many as 25 **percent** of children between ages 13 to 18 will experience an anxiety disorder in his or her lifetime. People with anxiety disorders try to avoid situations that make them feel anxious or fearful. **Figure 5.3** describes five types of anxiety disorders.

Impulse Control Disorders

People with impulse control disorders cannot resist the urge to hurt themselves or others. Impulse control disorders may begin in childhood or the teen years, and can continue into adulthood. People with this disorder may cause physical harm to themselves and others. They may also cause financial harm by overspending and gambling. People with impulse control disorder may also behave in ways that cause them to lose friends. **Figure 5.4** on page 120 provides examples of these disorders.

Academic Vocabulary

percent *(noun):* one part in a hundred

Figure 5.3 Anxiety Disorders

Phobia	A strong, irrational fear of something specific, such as heights or social situations.
Obsessive-Compulsive Disorder	Persistent thoughts, fears, or urges (obsessions) leading to uncontrollable repetitive behaviors (compulsions). For example, the fear of germs leads to constant hand washing.
Panic Disorder	Attacks of sudden, unexplained feelings of terror. "Panic attacks" are accompanied by trembling, increased heart rate, shortness of breath, or dizziness.
Post-Traumatic Stress Disorder (PTSD)	A condition that may develop after exposure to a terrifying event. Symptoms include flashbacks, nightmares, emotional numbness, guilt, sleeplessness, and problems concentrating.
Generalized Anxiety Disorder (GAD)	Exaggerated worry and tension for no reason. People with GAD startle easily and have difficulty concentrating, relaxing, and sleeping.

Figure 5.4 Impulse Control Disorders

Kleptomania	Unplanned theft of objects
Cutting	Repetitive cutting on parts of the body that can be hidden
Pyromania	Setting fires to feel pleasure or release tension
Excessive Gambling	Continuing to gamble despite heavy losses
Compulsive Shopping	Spending money on items that you can't afford and don't need

Eating Disorders

Eating disorders commonly occur during the teen years. As teens reach puberty, body changes and media images may cause some teens to put pressure on themselves to look a certain way. These teens may develop symptoms of anorexia nervosa, bulimia nervosa, or binge eating disorder. Eating disorders are more common among girls, but can affect boys too. Eating disorders can lead to unhealthful weight loss and death. You will learn more about eating disorders in Chapter 11.

Mood Disorders

A **mood disorder** is *an illness that involves mood extremes that interfere with everyday living*. These extremes are more severe than the normal highs and lows everyone experiences. Mood disorders include depression and bipolar disorder. *Bipolar disorder*, or manic-depressive disorder, is marked by extreme mood changes, energy levels, and behavior.

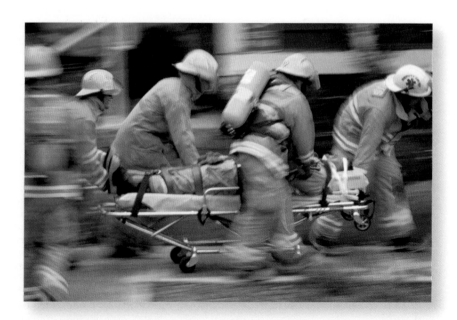

■ **Figure 5.5** Post-traumatic stress disorder may occur in the aftermath of a crisis. *What can community members do to support one another during a crisis?*

Conduct Disorder

Children and teens with **conduct disorder** engage in *patterns of behavior in which the rights of others or basic social rules are violated.* Examples include stealing, violence, and vandalism. Treatment includes learning to adapt to the demands of everyday life.

Schizophrenia

Schizophrenia (skit-suh-FREE-nee-uh) is a mental disorder in which a person loses contact with reality. Symptoms include delusions, hallucinations, and thought disorders. Schizophrenia affects about one percent of adults in the U.S.

People with this disorder behave unpredictably. Professional help and medication are needed to treat the illness successfully.

Personality Disorders

Teens with personality disorders are unable to regulate their emotions. They may feel distressed in social situations or may behave in ways that are distressing to others. The cause of personality disorders is unknown. It is felt that genetics and environmental factors lead to personality disorder.

READING CHECK

List What are examples of conduct disorder?

 LESSON 2 ASSESSMENT

After You Read

Reviewing Facts and Vocabulary

1. Define the term *stigma*. How can a stigma affect your health?

2. Identify the five types of anxiety disorders.

3. Which mental disorder can cause a person to have hallucinations?

Thinking Critically

4. **Evaluate.** Explain why mental disorders should be viewed like any other physical illness. Why is it important not to stigmatize someone with a mental disorder?

5. **Analyze.** Why are eating disorders both a mental health problem and a physical health problem?

Applying Health Skills

6. **Advocacy.** Teens suffering from mental disorders often feel confused, isolated, scared, or ashamed. Create a poster promoting awareness of and empathy toward mental illnesses. Focus on specific ways to be supportive, patient, and understanding.

Writing Critically

7. **Expository.** Choose one of the mental disorders you read about in this lesson. Explain why you chose this particular disorder, and how you would learn more about it.

Real Life Issues

After completing the lesson, review and analyze your response to the Real Life Issues question on page 118.

LESSON 3

📖 GUIDE TO READING

BIG Idea *Professional intervention and support from friends and family can often help prevent suicide.*

Before You Read

Create a K-W-L Chart.
Make a three-column chart. In the first column, list what you **k**now about the prevention of suicide. In the second column, list what you **w**ant to know about the topic. As you read, use the third column to summarize what you **l**earned.

K	W	L

New Vocabulary

▶ alienation
▶ suicide
▶ cluster suicides

Suicide Prevention

Real Life Issues ·····························

Helping a Friend. Nick's friend Ryan has been feeling down lately. Even though Ryan is a good student, he failed several important tests. As a result, he was suspended from the baseball team. To make matters worse, his girlfriend broke up with him. To Nick, Ryan seems depressed all the time and doesn't care about anything anymore. When Nick tries to talk with him, Ryan says he just wants to get away from it all.

Writing *Write a brief paragraph describing how you might respond to Ryan if you were in Nick's position.*

Knowing the Facts About Suicide

Main Idea Certain risk factors increase thoughts of suicide and suicide attempts.

Most people learn to manage stress in a healthful way. For some people, however, stress can cause **alienation**, *feeling isolated and separated from everyone else.* These people may be unable to cope with difficult life experiences. They may lack the support from family and friends, and be unable to access community resources for help. They may seek to escape from the pain and consider ending their lives.

Suicide is *the act of intentionally taking one's own life.* It is the third leading cause of death for teens ages 12 to 19. Each year, approximately 4,000 teens commit suicide.

Suicide Risk Factors

Among those who commit suicide, two risk factors are common. More than 90 percent are suffering from depression or another mental disorder, or have a history of abusing alcohol or other drugs. Sometimes, both risk factors are present.

Tim Fuller Photography

Some people use alcohol or other drugs to relieve their depression. Alcohol and drugs, however, have a depressant effect and lower one's inhibitions, making self-destructive behavior more likely. These people usually have more than one risk factor, such as a stressful situation or loss; previous suicide attempts; family history of mental disorders, substance abuse, or suicide; and access to guns.

Exposure to other teens who have died by suicide is a risk factor that can lead to **cluster suicides**, *a series of suicides occurring within a short period of time and involving several people in the same school or community.* Cluster suicides account for about 5 percent of all teen suicides. Some cluster suicides result from pacts made among peers. In other cluster suicides, the teens may not know one another, but may share an environmental stressor, such as a tragic event in their school or community. Some teens may learn of suicides through the news media.

Strategies to Prevent Suicide

Main Idea Recognizing the signs of suicide may help prevent it.

Most suicidal thoughts, behaviors, and actions are expressions of extreme distress. The warning signs of suicide are described in **Figure 5.7** on page 124. A person **displaying** only a few signs may not necessarily be considering suicide. When someone talks about committing suicide—whether it's done in a serious, casual, or even humorous way—*take it seriously.* Any discussion or suggestion about suicide requires immediate attention. Never agree to keep a secret if a friend says he or she is considering suicide. Tell an adult without delay.

 READING CHECK

Describe What are some behaviors that might indicate a person is thinking about suicide?

Academic Vocabulary

display *(verb):* to make evident

Figure 5.7   Recognizing the Warning Signs of Suicide

The warning signs of suicide should be taken seriously. The more signs exhibited, the more likely it is that the person is thinking about suicide.

- Direct statements such as "I wish I were dead."
- Indirect statements such as "I can't take it anymore."
- Writing poems, song lyrics, or diary entries that deal with death
- Direct or indirect suicide threats
- An unusual obsession with death
- Withdrawal from friends
- Dramatic changes in personality, hygiene, or appearance
- Impulsive, irrational, or unusual behavior

- A sense of guilt, shame, or rejection; negative self-evaluation
- Deterioration in schoolwork or recreational performance
- Giving away personal belongings
- Substance abuse
- Complaints about physical symptoms, such as stomachaches, headaches, and fatigue
- Persistent boredom and indifference
- Violent actions, rebellious behavior, or running away
- Intolerance for praise or rewards

Sources: American Academy of Child and Adolescent Psychiatry; National Mental Health Association

How You Can Help

READING CHECK

Describe What steps can you take to help someone who may be considering suicide?

■ **Figure 5.8** Suicide survivor support groups are available in most communities. *How might such support groups prevent suicides?*

People who are considering suicide often believe that their death will not matter to anyone. Showing empathy when talking with that person will let him or her know you are concerned. If someone you know may be considering suicide, that person needs help. Try the following:

- **Initiate a meaningful conversation.** Show interest, compassion, patience and understanding. Don't respond by saying "You really don't want to do that," or "Everyone feels sad sometimes."

- **Show support and ask questions.** Remind the person that all problems have solutions, and that suicide is *not* the answer. Tell your friend that most suicide survivors later express gratitude that they did not die.

- **Try to persuade the person to seek help.** Encourage the person to talk with a parent, counselor, or other trusted adult. Offer to go with him to get help.

If you believe a friend may be suicidal, tell an adult, and find out what steps the adult will take. If the adult doesn't seem to believe the threat is serious, talk to other adults until someone takes action. You can also contact community resources, such as a crisis center or suicide hotline. You will learn more about community resources in Lesson 4.

Real World CONNECTION

Depression and Suicide

Untreated depression is the leading cause of suicide. People who consider suicide feel that they don't matter to others. People who appear to have a mental health problem and may be considering suicide need to be encouraged to seek help.

What can you, as a friend, do to help prevent suicide? Should you tell someone that a friend has mentioned suicide, even if that friend asked you to keep the information private? What can people in the community do to help prevent suicides? In small groups, conduct an online search for reliable information about seeking help for depression.

Activity Writing

Once you have gathered your information as a group, complete the following activity:

1. Write a script for a skit that urges teens to seek help if they are depressed and considering suicide.

2. Record a video of the skit and include a statement encouraging teens to get help.

3. Remind the audience that all problems can be solved and that depression is treatable.

4. Encourage teens who may be depressed to talk to a parent, teacher, or other trusted adult.

5. Provide contact information for local crisis centers and suicide hotlines.

LESSON 3 ASSESSMENT

 After You Read

Reviewing Facts and Vocabulary

1. Define the term *alienation*.
2. What two risk factors have the strongest association with suicide?
3. Name five warning signs of suicide.

Thinking Critically

4. **Synthesize.** Make a list of three direct statements and three indirect statements that could indicate a teen is considering suicide.
5. **Explain.** Define the term *cluster suicides* and explain why they happen.

Applying Health Skills

6. **Decision Making.** Imagine that you have a friend who shares negative comments about herself. Use the six steps of decision making to determine what actions to take.

Writing Critically

7. **Descriptive.** Write a note to a teen who has exhibited some suicidal warning behaviors. Use the suggestions listed in the lesson to help the teen rethink his or her situation.

Real Life Issues ....

After completing the lesson, review and analyze your response to the Real Life Issues question on page 122.

LESSON 4

📖 GUIDE TO READING

BIG Idea *Mental health professionals and related agencies provide treatment and support for people with mental health problems.*

Before You Read

Create a Cluster Chart. Draw a circle and label it "Getting Help." Use surrounding circles to identify professionals in the community who can help individuals with mental health problems. As you read, continue filling in the chart with more details.

New Vocabulary

▶ psychotherapy
▶ behavior therapy
▶ cognitive therapy
▶ family therapy
▶ group therapy
▶ drug therapy

Getting Help

Real Life Issues ••••••••••••••••••••••••

No One to Turn To. Angie is desperate. She manages to get through each day, but inside she feels as though her life is spiraling out of control. She's confused and tries not to let her fears and frustrations show. Angie knows she needs help, but doesn't know who to ask. She's afraid of what the person will think of her.

Writing *Write a letter to Angie encouraging her to talk to a trusted adult and ask for help. Make sure the tone of your letter is understanding and considerate.*

When Help Is Needed

Main Idea The first step to getting help for a mental health problem is being aware that help is needed.

Many teens with mental health problems do not recognize the seriousness of their condition, or understand that help is available. In fact, most adult mental health disorders have their roots in untreated childhood and adolescent problems. More than half of suicidal youths had symptoms of a mental disorder for more than a year prior to their deaths.

Mental health influences every aspect of a person's life. No one should ever feel embarrassed to talk with someone about mental or emotional problems. Teens should seek help if they experience any of the following:

• Feeling trapped or worrying all the time

• Feelings that affect sleep, eating habits, schoolwork, job performance, or relationships

• Becoming involved with alcohol or other drugs

• Becoming increasingly aggressive, violent, or reckless

Often, friends and family are the first to recognize that a problem is affecting the teen's life and relationships. Their concern may encourage the individual to seek help.

Overcoming the Stumbling Blocks

Main Idea The benefits of treatment encourage people to overcome a reluctance to get help.

Seeking help for mental health problems can be difficult. However, these problems are *not* easily managed without help. Initially, talking about problems may make a person feel more vulnerable. When asking for help, remember these facts:

- Asking for help is a sign of inner strength. It shows responsibility for one's own wellness.
- Serious disorders, compulsions, and addictions are complex and require professional intervention.
- Sharing your thoughts with an objective, helpful individual can be a great relief.
- Financial help to pay for care may be available.

READING CHECK

Explain What does asking for help from a mental health professional show?

Where to Go for Help

Main Idea People in your community are available to help.

It takes courage to confront a problem and try to solve it. Talking with a trusted adult, such as a parent, guardian, teacher, or school nurse, can get you started.

Many teens receive help for a mental health problem at school. A counselor or the school nurse can identify and contact support services. Other options for community help are talking with the clergy, and crisis hotlines. Crisis hotlines allow people to talk anonymously. The workers are trained to deal with difficult mental and emotional situations.

Treatment for mental health problems is unique to each individual. Sometimes, a treatment plan may not work. If that happens, talk to someone else. It may be necessary to try several different treatments. People with mental health problems should continue to seek help until they feel better.

■ **Figure 5.9** Adults working at a crisis hotline are usually volunteers motivated by the desire to help people who are suffering. *How might a person who is caring and yet objective be helpful during an emotional crisis?*

Mental Health Professionals

READING CHECK

List Name some people who can help teens with mental health problems.

Help is available from a variety of professionals who work in your community's schools, clinics, hospitals, and family agencies. These specialists are trained to help people with mental and emotional problems, and include the following:

- **Counselor**—a professional who handles personal and educational matters
- **School psychologist**—a professional who specializes in the assessment of learning, emotional, and behavioral problems of schoolchildren
- **Psychiatrist**—a physician who diagnoses and treats mental disorders and can prescribe medications
- **Neurologist**—a physician who specializes in physical disorders of the brain and nervous system
- **Clinical psychologist**—a professional who diagnoses and treats emotional and behavioral disorders with counseling. Some can prescribe medications
- **Psychiatric social worker**—a professional who provides guidance and treatment for emotional problems in a hospital, mental health clinic, or family service agency

Treatment Methods

Main Idea Several methods can be helpful in treating a mental health problem.

Mental health professionals may use several treatments depending on their expertise and the needs of the patient. The following are the most commonly used therapy methods.

- **Psychotherapy** is *an ongoing dialogue between a patient and a mental health professional.* The dialogue is designed to find the cause of a problem and devise a solution.
- **Behavior therapy** is *a treatment process that focuses on changing unwanted behaviors through rewards and reinforcements.*

■ **Figure 5.10** A mental health specialist respects a patient's concern for confidentiality. *What are other benefits of seeking help from a mental health specialist?*

©Lisa F. Young/Alamy

- **Cognitive therapy** is *a treatment method designed to identify and correct distorted thinking patterns that can lead to feelings and behaviors that may be troublesome, self-defeating, or self-destructive.*

- **Family therapy** focuses on *helping the family function in more positive and* **constructive** *ways by exploring patterns in communication and providing support and education.* Family therapy is most successful when every member of the family attends the sessions.

- **Group therapy** involves *treating a group of people who have similar problems and who meet regularly with a trained counselor.* Group members agree that whatever is said in the group is private. They agree not to discuss information heard during the group with others.

- **Drug therapy** is *the use of certain medications to treat or reduce the symptoms of a mental disorder.* It is sometimes used alone, but is often combined with other treatment methods such as those listed above.

Sometimes a mental health problem is serious enough to require hospitalization. In a hospital, a patient can receive intensive care and treatment from doctors, nurses, and a variety of mental health specialists. When someone is receiving care after being hospitalized, these specialists are available 24 hours a day.

Academic Vocabulary

constructive *(adjective):* promoting improvement or development

READING CHECK

Explain What types of treatment methods can help those with a mental disorder?

LESSON 4 ASSESSMENT

After You Read

Reviewing Facts and Vocabulary

1. What is *behavior therapy*?

2. Which mental health professional treats physical disorders of the brain?

3. Who might a teen reach out to at school about a mental health problem?

Thinking Critically

4. **Analyze.** What protective factors do you have or can you develop to help you deal with stress in your life?

5. **Synthesizing.** How does developing a positive outlook strengthen your resiliency?

Applying Health Skills

6. **Accessing Information.** Compile a list of local resources for mental health problems. Include mental health professionals, school counselors, hospital emergency rooms, and hotlines.

Writing Critically

7. **Persuasive.** Write an editorial about the importance of seeking help for mental health problems. Include strategies for getting help.

Real Life Issues ....

After completing the lesson, review and analyze your response to the Real Life Issues question on page 126.

Hands-On HEALTH

Activity Life-Saving Resources

Many people want to help someone who has a mental/emotional problem. Develop an accurate and reliable health resource list that anyone can access in your school or in your community. You must reference your sources, access your sources to confirm they are appropriate, and identify the type of help available from the sources.

What You'll Need

- access to the Internet
- library or media center
- phone book and phone
- other resources
- paper and writing utensil

What You'll Do

Step 1

Conduct an Internet search and create a list of valid resources for a mental/emotional problem.

Step 2

When possible, order free pamphlets and other information.

Step 3

Create an action plan suggesting ways a friend can help a person with a mental/emotional problem.

Apply and Conclude

Discuss with the other members of your class and create a wallet card of a full list of resources.

Checklist: Accessing Information

✓ List all of your sources of information.

✓ Evaluate the sources to determine their reliability.

✓ Judge the appropriateness of your sources.

Dealing with Anxiety and Depression

Key Concepts

▶ Seek help if thoughts, emotions, or behaviors affect daily life.

▶ Causes of depression include stressful life events, unhappy family environments, social conditions, and illness.

▶ Depressed people need treatment from a medical or mental health professional.

Vocabulary

▶ anxiety (p. 114)
▶ emotions (p. 114)
▶ depression (p. 115)
▶ apathy (p. 116)

Mental Disorders

Key Concepts

▶ Education can overcome the stigma of mental illness.

▶ If left untreated, many mental disorders that begin in childhood or adolescence can continue into adulthood.

▶ Anxiety disorders are common disorders that teens experience.

Vocabulary

▶ mental disorder (p. 118)
▶ stigma (p. 118)
▶ anxiety disorder (p. 119)
▶ mood disorder (p. 120)
▶ conduct disorder (p. 121)

Suicide Prevention

Key Concepts

▶ The two risk factors most associated with suicide are depression and abusing alcohol or other drugs.

▶ Most suicidal thoughts, behaviors, and actions are expressions of extreme distress.

▶ A suicidal teen needs immediate adult intervention.

Vocabulary

▶ alienation (p. 122)
▶ suicide (p. 122)
▶ cluster suicides (p. 123)

Getting Help

Key Concepts

▶ Help for mental health problems may be available at school or through community resources.

▶ The reluctance to get help can be overcome by recognizing the benefits of treatment.

▶ Mental health professionals can diagnose a mental health problem and devise an appropriate treatment plan.

Vocabulary

▶ psychotherapy (p. 128)
▶ behavior therapy (p. 128)
▶ cognitive therapy (p. 129)
▶ family therapy (p. 129)
▶ group therapy (p. 129)
▶ drug therapy (p. 129)

Assessment

Design Pic/Don Hammond

Vocabulary Review

Use the vocabulary terms listed on page 131 to complete the following statements.

1. Prolonged feelings of helplessness, hopelessness and sadness can be a sign that you are suffering from _____.

2. Feelings of unease or worrying about what may happen are signs of _____.

3. A lack of strong feeling, interest, or concern is called _____.

Understanding Key Concepts

After reading the question or statement, select the correct answer.

4. Which of the following is *not* a change in behavior one might expect with depression?
   a. Trouble sleeping
   b. Eating too much
   c. Eating too little
   d. Running alone

5. Which of the following is *not* a warning sign of depression?
   a. Boredom
   b. Increased interest in life
   c. Poor concentration
   d. Increased irritability

Thinking Critically

After reading the question or statement, write a short answer using complete sentences.

6. **Identify.** What are three feelings you may experience when you are depressed?

7. **Explain.** How might depression affect your sleep?

8. **Analyze.** How might community violence cause someone to become depressed?

9. **Discuss.** While being treated for depression, what else can you do to help the healing process?

Vocabulary Review

Choose the correct term in the sentences below.

10. Illnesses that involve mood extremes that interfere with everyday living are called *mood disorders / stigma.*

11. Patterns of behavior in which the rights of others or basic social rules are violated are typical of a *mental disorder / conduct disorder.*

12. Conditions in which real or imagined fears are difficult to control are called *stigma / anxiety disorder.*

Understanding Key Concepts

After reading the question or statement, select the correct answer.

13. Which is *not* an anxiety disorder?
    a. Phobia
    b. Pyromania
    c. Panic disorder
    d. Post-traumatic stress disorder

14. Kleptomania is
    a. an anxiety disorder.
    b. a mood disorder.
    c. an impulse control disorder.
    d. a conduct disorder.

SW Productions/Getty Images

15. Bipolar disorder is
- **a.** a conduct disorder.
- **b.** a personality disorder.
- **c.** an anxiety disorder.
- **d.** a mood disorder.

Thinking Critically

After reading the question or statement, write a short answer using complete sentences.

16. **Explain.** Describe how misconceptions of mental illness can be overcome.

17. **Analyze.** Explain why some people with a mental disorder may not seek help for their problem.

18. **Describe.** Identify several examples of anxiety triggers for teens.

19. **Infer.** Consider the types of problems that people with impulse control disorders have. Explain what problems people with this disorder may face before getting treatment.

LESSON 3

Vocabulary Review

Correct the sentences below by replacing the italicized term with the correct vocabulary term.

20. The act of intentionally taking one's own life is called *alienation*.

21. A series of suicides occurring within a short period of time and involving several people in the same school or community is referred to as *suicide*.

Understanding Key Concepts

After reading the question or statement, select the correct answer.

22. Suicide is the _____ leading cause of teen deaths.
- **a.** first
- **b.** second
- **c.** third
- **d.** fourth

23. Of the following risk factors for teen suicide, which should probably be of most concern?
- **a.** A stressful situation or loss
- **b.** Substance abuse
- **c.** Family history of mental disorders
- **d.** Exposure to other teens who have died by suicide

24. Which is *not* a warning sign of suicide?
- **a.** Withdrawal from friends
- **b.** An overwhelming sense of guilt
- **c.** Persistent indifference
- **d.** Preoccupation with buying new things

Thinking Critically

After reading the question or statement, write a short answer using complete sentences.

25. **Analyze.** Explain why drinking alcohol is not an effective way to try to relieve depression.

26. **Describe.** What are five warning signs of suicide?

27. **Explain.** Why might cluster suicides occur in a community where the individuals may not even know one another?

28. **Evaluate.** Explain why it is important never to keep secret a person's threat to commit suicide.

LESSON 4

Vocabulary Review

Use the vocabulary terms listed on page 131 to complete the following statements.

29. A treatment method designed to identify and correct distorted thinking patterns is known as _____.

30. The use of certain medications to treat symptoms of a mental disorder is called _____.

Assessment

Understanding Key Concepts

After reading the question or statement, select the correct answer.

31. A mental health professional who handles personal and educational matters is a
    a. counselor.
    b. school psychologist.
    c. psychiatrist.
    d. neurologist.

32. A treatment method that uses ongoing dialogue between a patient and a mental health professional is
    a. family therapy.
    b. cognitive therapy.
    c. psychotherapy.
    d. group therapy.

33. Which is *not* true regarding crisis-hotline workers?
    a. They are trained to deal with difficult mental/emotional situations.
    b. They are usually volunteers.
    c. They know about your personal situation.
    d. They allow you to remain anonymous.

Thinking Critically

After reading the question or statement, write a short answer using complete sentences.

34. **Describe.** Identify the behaviors that help you recognize that a friend needs help.

35. **Analyze.** What are the possible consequences of not getting help for an adolescent mental disorder?

36. **Synthesize.** What criteria would be important to you when choosing someone to talk with about a mental health problem?

Technology PROJECT-BASED ASSESSMENT

Phobias

Background

A phobia is a strong fear of something specific. For example, arachnophobia is a fear of spiders. Other phobias include agoraphobia, the fear of being in an open space, or claustrophobia, the fear of being in a closed space.

Task

Conduct an Internet search to learn about different types of phobias. Create a Web page describing types of phobias and how they are treated.

Audience

Students at your class

Purpose

Develop awareness of one kind of mental illness that affects many children, teens, and adults.

Procedure

1. Conduct research on the Internet to learn more about phobias.

2. Select three to four phobias that you will describe on your Web page.

3. Identify the kinds of professional help and solutions available for treating specific phobias.

4. Research online about what might occur if a person's activities should put him or her near the object or situation that is the source of the phobia.

5. Create sections on your Web page for each type of phobia, giving as much information as possible.

6. Present your Web page to your class.

Math Practice

Understand and Apply. Read the paragraph below, and then answer the questions.

> *Nearly everyone is mildly depressed at some time, but 16 percent of the U.S. population will suffer from major depression in their lifetime. A study was conducted on more than 9,000 people ages 18 and older. Fifty-seven percent of those who had major depression sought help. This rate is almost 40 percent higher than the rate reported 20 years before the study. Even though the number of patients treated is increasing, it is estimated that only 21 percent are receiving adequate care.*

1. If the size of the general population is 200 million people, how many people will experience major depression at some time during their lives?
   - **A.** 14 million
   - **B.** 42 million
   - **C.** 75 million
   - **D.** 92.8 million

2. What function can be used to find the number of people who are seeking help for depression if you know the size of the population with depression? (Hint: The variable N is the number of people seeking help, and P is the size of the population.)
   - **A.** $N = P$
   - **B.** $N = (0.57)(0.07)P$
   - **C.** $N = 0.57P$
   - **D.** $P = 0.07N$

3. Examine the percentages reflecting how many people have major depression, how many of these people seek help, and how many who seek help receive adequate care. Of 20,000 people, how many people would you expect to be receiving adequate care for major depression? Justify your answer.

Reading/Writing Practice

Analyze and Infer. Read the passage below, and then answer the questions.

> *John F. Nash Jr. is known for his work as a creative mathematician. He is also an example of how one person can succeed in his chosen field even if he is battling a difficult mental health challenge: paranoid schizophrenia.*
>
> *While working at Princeton University in the 1950s, Nash made great strides in a field of mathematics called game theory. This research later earned him a share in the 1994 Nobel Prize in Economics. However, soon after completing this work, he began to suffer what was later diagnosed as paranoid schizophrenia. After taking a break for nearly 30 years, Nash returned to mathematics and now continues to do research and write at Princeton.*

1. What information supports the claim that Nash is successful?
   - **A.** Nash was born in West Virginia.
   - **B.** Nash has paranoid schizophrenia.
   - **C.** Nash stopped his research for 30 years.
   - **D.** Nash won a Nobel Prize in Economics.

2. Why did the author write this passage?
   - **A.** To cite examples of famous people with various mental disorders
   - **B.** To describe how a mathematician came up with his prize-winning research
   - **C.** To explain how schizophrenia affects mental and physical health
   - **D.** To show how a person can be successful in spite of a mental disorder

3. Write a paragraph describing the effects of schizophrenia on a person's mental and emotional health.

National Education Standards

Math: Measurement and Data, Problem Solving
Language Arts: LACC.910.L.3.6

TEENS *Speak Out*

©Corbis

Are Teens Overscheduled?

*T*he high school years can bring many different kinds of stress. Teens today are busier than ever as they try to balance school, athletics and other extracurricular activities, part-time jobs, friendships, dating relationships, and family responsibilities. All these demands can cause a great deal of stress, possibly leading to health problems. Do teens have too many responsibilities? Are they overscheduled? Should parents help teens include free time in their schedule to pursue interests such as reading, art, or just relaxing? Read on to find out two teens' viewpoints about this issue.

Benefits of a Full Schedule

Having a full schedule of different activities can help teens develop new interests and skills because they're always trying new things. Many of these skills, such as multitasking, may help them succeed in college and in the work world. Meeting the challenges of a full schedule can also give teens a sense of accomplishment and build self-esteem.

" *I have a busy schedule, and I like it that way. It keeps me challenged, and I don't feel bored. Juggling school, baseball, a job, and time with family and friends also helps prepare me for the real world. It can get stressful sometimes, but that's part of life."*

–Jeff Z., age 17

Benefits of a Relaxed Schedule

Having a more relaxed schedule allows teens to devote attention to a few important activities that they really enjoy instead of stretching themselves too thin. A relaxed schedule can help them manage their stress level and avoid stress-related health problems. Also, by not overloading their schedules, teens can better explore their creative interests.

" *Some of my friends are stressed all the time because they're trying to do too much. After trying to keep up with schoolwork, studying for SATs, being on sport teams, holding down jobs, and doing chores at home, they don't have any time for themselves. It's important to keep a balance so you don't burn out."*

–Alison R., age 16

Activity Beyond the Classroom

1. **Summarize** your thoughts on this issue. Do you think some teens are overscheduled? Why do you think they are trying to do so much? How might this affect their mental and emotional health?

2. **Synthesize** your ideas. Imagine that you are a columnist at a teen magazine. Write an article about balancing responsibilities and activities. Discuss how teens can tell if they are overscheduled, and provide strategies for maintaining an appropriate activity load.

Healthy and Safe Relationships

UNIT PROJECT

A Positive Place

Using Visuals Boys & Girls Clubs of America calls itself "The Positive Place for Kids." Four thousand clubs across America offer programs and activities to children and teens in all age groups. Volunteers provide key support to staff.

Get Involved. Use community or online resources to locate youth centers in your area. Call or visit one center to find out what kinds of programs and activities it offers. Summarize your findings in a brief report.

©Monashee Frantz/age fotostock

"**The only way to have a friend is to be one.**"
— *Ralph Waldo Emerson, 19th-century writer and poet*

Skills for Healthy Relationships

Lesson 1
Foundations of a Healthy Relationship

BIG *Idea* *Building strong relationships is important to your overall health.*

Lesson 2
Respecting Yourself and Others

BIG *Idea* *You can promote healthy relationships by showing respect for yourself and others in your life.*

Lesson 3
Communicating Effectively

BIG *Idea* *Effective communication is a key to building healthy relationships.*

Activating Prior Knowledge

Using Visuals Look at the photo on this page. What kind of relationship do these teens seem to have? Discuss the characteristics that you think are most important to a healthy and strong relationship.

Chapter Launchers

Health in Action

Discuss the **BIG** Ideas

Think about how you would answer these questions:

▶ What relationships in your life are most important?

▶ What makes these relationships special?

▶ What do you do to keep these relationships strong?

Assess Your Health

Read each statement. On a separate sheet of paper, write "yes," "sometimes," or "no" based on your typical behavior.

1. I understand that strong relationships have a positive influence on my health.

2. I treat others with kindness and consideration.

3. I am honest and open with others.

4. I treat others with respect.

5. I accept that everyone is different and I value diversity in others.

A "yes" response shows that you practice healthy behaviors. "Sometimes" indicates that you should analyze and possibly modify your behavior. A "no" response means that you should modify the behavior.

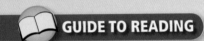

LESSON 1

GUIDE TO READING

BIG Idea *Building strong relationships is important to your overall health.*

Before You Read

Create a K-W-L Chart. Make a three-column chart. In the first column, list what you **k**now about relationships. In the second column, list what you **w**ant to know about this topic. As you read, use the third column to summarize what you **l**earned.

K	W	L

New Vocabulary

- relationship
- friendship
- citizenship
- role
- cooperation
- compromise

Review Vocabulary

- interpersonal communication (Ch.2, L.1)

Foundations of a Healthy Relationship

Real Life Issues

Important Relationships. Andrew's classmate has an extra ticket to a sold-out concert on New Year's Eve. Andrew really wants to go but he already promised his best friend that he would attend his New Year's Eve party. Andrew doesn't want to let his best friend down but he may not have another chance to see this band in concert.

Writing *If you were Andrew, how would you respond? How can you reassure the people in your life of their importance to you? Write a paragraph explaining your choice and how it might impact your friendship.*

Relationships in Your Life

Main Idea You have many types of relationships in your life, and you play different roles in all of them.

As you learned in Chapter 3, one of the most basic human needs is the need to belong and to feel loved. Building and maintaining healthy relationships can help you meet this need. A **relationship** is *a bond or connection you have with other people.*

Although some people use the word *relationship* to refer to a romantic involvement, there are actually all kinds of relationships that can be important in your life. For instance, you have relationships with family members, friends, teachers, classmates, and people in your community. All of these relationships can affect your health in ways that may be positive or negative.

Stockbyte/PunchStock

Relationships with Family

Some of the most important relationships in your life are with the family members who share your home, such as parents or guardians, brothers, and sisters. You also have family relationships with other relatives, such as grandparents, aunts, uncles, and cousins. One thing that makes family relationships special is that they last your entire life. The friends you have in high school may not be your friends ten years from now, but your family is your family for life.

Healthy family relationships strengthen every side of your health triangle. Parents or guardians take care of your physical needs for food, clothing, and shelter. The love, care, and encouragement they provide are important to your mental and emotional health. They also help build your social health by teaching you the values and social skills that will guide you in all your other relationships.

Relationships with Friends

A **friendship** is *a significant relationship between two people that is based on trust, caring, and consideration.* Although you probably have many friends your own age, friendships can form between people of any age. You may choose your friends because you have similar interests, because they share your values, or maybe just because they live nearby. Whatever the reason, good friends can benefit your health in many ways. They have a positive influence on your self-esteem and can help you resist harmful behaviors.

■ **Figure 6.1** Strong friendships have a positive influence on your health. *What friendships are most important in your life?*

 READING CHECK

Explain How can friends benefit your health?

Relationships in Your Community

Being part of a strong community has a positive impact on every aspect of your health. It can promote healthful behaviors and also provide resources to help you when you're in trouble. You reinforce your ties to the community through good **citizenship**—*the way you conduct yourself as a member of the community*. Good citizens work to strengthen their communities by obeying laws, being friendly to neighbors, and helping to improve the places where they live.

Roles in Relationships

A **role** is a *part you play in your relationships*. In the course of a single day, you may play many roles with different people. You might be a son or daughter at home, a student at school, a friend when you're hanging out with your buddies, a teammate during gym class, and an employee at an after-school job. For an illustration of how a single person can play many roles, see **Figure 6.2**.

Figure 6.2 **Intersecting Relationships**

This diagram illustrates the many overlapping roles that one teen, Felicia, plays in her relationships at home, at school, and in her community. *What different roles do you play in your life?*

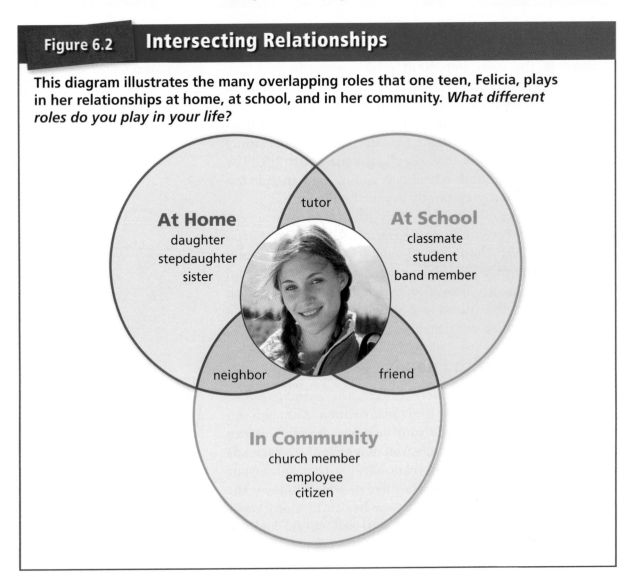

At Home
daughter
stepdaughter
sister

tutor

At School
classmate
student
band member

neighbor

friend

In Community
church member
employee
citizen

Sometimes you can even play more than one role with the same person. For instance, when you babysit a younger brother or sister, you temporarily take on the role of caregiver in that relationship. Your role in a relationship can also change over time. For example, someone you work with might become a friend.

Balancing all the different roles in your life can be tricky. You may feel at times that you can't handle all the demands being made on you. In such situations, you may decide that you need to focus more on one relationship right now.

Traits of Healthy Relationships

Main Idea In a healthy relationship, people respect and support each other.

Healthy relationships nurture you. They bring out the best in you and encourage you to make healthful choices in your life. Qualities of healthy relationships include

- **mutual respect.** You treat other people with respect, and they respect you in return. You accept each other's opinions, tastes, and traditions, even if they are different. At times you may agree to disagree instead of trying to force your opinions on each other.

- **caring.** You treat other people with kindness and consideration. During difficult times, you show empathy and support. You're also willing to help out others.

- **honesty.** You are honest and open with others, rather than concealing your thoughts, feelings, or actions.

- **commitment.** You contribute to the relationship and work to keep it strong, even if it means making some sacrifices. You deal with problems in a positive way and are able to overcome them.

Skills for Building Healthy Relationships

It takes work to maintain a healthy relationship. The people involved need to make an effort to understand each other and get along in different situations. Three skills that can help are communication, cooperation, and compromise—sometimes known as the three Cs of healthy relationships.

Communication As you learned in Chapter 2, *interpersonal communication* is the exchange of thoughts, feelings, and beliefs between two or more people. In relationships, people need to understand each other. It's important to learn effective communication skills so you can express your thoughts, feelings, and expectations to others and understand theirs in return. You will learn more about how to communicate well with others in Lesson 3.

READING CHECK

Identify What are the characteristics of a healthy relationship?

■ **Figure 6.3** Communication is more than just talking. It's getting your message across to others and hearing their response. *Describe three different ways you might communicate with someone.*

Cooperation Have you ever had to move a heavy desk or other large piece of furniture? Tasks like this are nearly impossible without **cooperation**, *working together for the good of all.* Cooperating with others to reach a common goal can strengthen your relationships. For example, when Jonah and his mom worked together to build a set of shelves for his room, they learned to interact better together and shared a sense of accomplishment in the project.

Compromise Sometimes, when people in relationships want different things, they may decide to compromise. **Compromise** is *a problem-solving method in which each participant gives up something to reach a solution that satisfies everyone.*

The give-and-take of effective compromise strengthens relationships. It allows you to resolve disagreements in a way that everyone can accept. Remember, though, that compromise works only when all the people involved are happy with the solution. You should *not* compromise on things that really matter to you, like your values and beliefs. The art of getting along with others involves knowing when it's appropriate to compromise and when you need to stand your ground.

Character and Relationships

In Chapter 3, you learned about the six traits of good character. Each of these traits contributes in its own way to healthy relationships. Here are examples of how each trait can strengthen a relationship.

- **Trustworthiness.** Maribel has promised to go to the movies with her friend Noriko on Friday night. On Friday afternoon, a guy Maribel likes asks her out for that evening. She'd like to say yes, but she doesn't want to break her promise to Noriko. By keeping her promise, Maribel shows that Noriko can count on her.

- **Respect.** Kyle's parents have taught him always to listen when someone else is talking and to avoid interrupting.

■ **Figure 6.4** Demonstrating the traits of good character can strengthen your relationships with others. *Which traits of good character is this teen showing?*

By extending this courtesy to each other whenever they talk, Kyle and his parents improve their communication as a family.

- **Responsibility.** While at a party, Tara accidentally knocks over a glass and breaks it. She immediately apologizes to the host, helps clean up the broken glass, and offers to pay to replace it. Her responsible action improves her host's opinion of her.

- **Fairness.** Enrique and his brother take turns using their computer to do schoolwork, send e-mail, and play games. Sharing the **computer** fairly keeps Enrique and his brother from fighting over it.

- **Caring.** When her friend Carl is having trouble with math, Alison offers to tutor him. Carl considers her a good friend for caring about him and helping him out.

- **Citizenship.** Ruby's family bought a snow blower to help keep the sidewalk and driveway clear. Now, in addition to clearing her own driveway, Ruby offers to do her neighbor's as well. Her actions improve her relationship with her neighbor and her reputation in the community.

Academic Vocabulary

computer *(noun):* a device that can store, retrieve, and process data

---

LESSON 1 ASSESSMENT

 After You Read

Reviewing Facts and Vocabulary

1. Identify three kinds of relationships you have in your life.

2. Define *citizenship* and give an example of good citizenship.

3. What are the three Cs of healthy relationships?

Thinking Critically

4. **Analyze.** Explain how relationships with family members are important to all three sides of your health triangle.

5. **Synthesize.** Think about your interactions with other people over the course of a day. Analyze these interactions to identify what roles you played in your relationships that day. Summarize your findings.

Applying Health Skills

6. **Advocacy.** Brainstorm a list of ways that students can demonstrate good citizenship at school. Based on these ideas, create a flyer encouraging students to be good citizens of the school community.

Writing Critically

7. **Narrative.** Write a conversation between two characters who share a relationship. In your narrative, show how the communication between these characters affects their relationship in a positive or negative way.

Real Life Issues ....

After completing the lesson, review and analyze your response to the Real Life Issues question on page 142.

BIG Idea *You can promote healthy relationships by showing respect for yourself and others in your life.*

Before You Read

Create a T-Chart. Draw a two-column chart. Label the left column "Self-Respect" and the right column "Respect for Others." As you read, fill in the columns with information about how each type of respect can improve your relationships.

Self-Respect	Respect for Others

New Vocabulary

▶ prejudice
▶ stereotype
▶ tolerance
▶ bullying
▶ hazing

Review Vocabulary

▶ personal identity (Ch.3, L.2)
▶ values (Ch.2, L.2)

Respecting Yourself and Others

Real Life Issues •

Bullying. Many teens have experienced bullying.

Source: National Center for Education Statistics, Indicators of School Crime and Safety, 2012.

> **In 2011, 28% of teens were bullied at school during the school year.**

> **In 2011, 9% of teens reported being cyber-bullied anywhere during the school year.**

> **4% of students reported they were afraid of attack or harm at school.**

Writing *Write a journal entry describing how being bullied can make someone feel.*

Respect for Yourself

Main Idea Self-respect will strengthen your relationships.

Having self-respect is an important foundation for developing and maintaining healthy relationships. When you respect yourself, you're more likely to seek out relationships with people who treat you with respect. Self-respect makes you less likely to let other people talk you into taking risks that could harm your health.

The Need for Strong Values

During your teen years, you may be searching for your *personal identity*—your sense of who you are and where you belong in the world. You may struggle to develop your personal values system. *Values*, as you have learned, are the beliefs, ideas, and attitudes about what is important that help guide the way you live.

Being unsure of your values can complicate your relationships. If you aren't clear about your values, it's much more difficult to communicate them to others. As a result, the people around you may not be able to tell what is important to you. This could increase the chance that you'll face pressure to participate in unhealthful behaviors.

In contrast, when you are clear about your values, you strengthen your relationships. Other people will know what you believe in and understand what's important to you. Upholding your values shows that you respect yourself, and communicating your values to others can help them respect you too.

Respect for Others

Main Idea It's important to treat people with respect.

You can strengthen your relationships with all the people in your life by treating them with the same respect you'd like them to show you. With strangers and casual acquaintances, you can show respect through common courtesy. You might hold a door open for someone or say "Thank you" to the checker at the grocery store. With close friends and family members, you can show respect in more significant ways:

- **Listen to other people.** Be willing to hear and consider their points of view, even if you disagree with them.

- **Be considerate of others' feelings.** Before you act or speak, consider how it might make the other person feel.

- **Develop mutual trust.** Let others know they can trust you by being honest and dependable. Show that you trust them by believing what they say and confiding in them.

- **Be realistic in your expectations.** For example, you can't expect friends and family members to always make you their top priority.

READING CHECK

Explain Why might being unsure about values complicate relationships?

■ **Figure 6.5** Lending your MP3 player to your brother is one way to show that you trust him. *What are some other ways to demonstrate trust?*

READING CHECK

Evaluate Why is prejudice a barrier to healthy relationships?

Tolerance

Sometimes people treat others with disrespect because of prejudice. **Prejudice** is *an unfair opinion or judgment of a particular group of people.* For example, a teen might decide he dislikes all cheerleaders because a cheerleader once turned him down for a date. Some forms of prejudice involve stereotypes. A **stereotype** is *an exaggerated or oversimplified belief about people who belong to a certain group.* Assuming that all boys like sports is an example of a gender stereotype.

Prejudice is a barrier to healthy relationships. It can keep people from getting to know others as individuals. In contrast, demonstrating tolerance can help you build healthy relationships. **Tolerance** is *the ability to accept others' differences.* People who are tolerant value diversity and can appreciate differences in other people's cultures, interests, and beliefs.

Disrespectful Behaviors

Has a fellow student ever picked on you for no reason? Perhaps this person called you names, or even threatened you with physical violence. This disrespectful behavior is an example of **bullying**—*deliberately harming or threatening other people who cannot easily defend themselves.* Bullies may tease their victims, spread rumors about them, or try to keep them out of a group. They may even attack others physically by pushing, shoving, or hitting them.

Some bullies push other people around because it makes them feel superior. They may also do it as a way to feel they are part of a group or to keep from being bullied themselves. As many as one out of four students in the United States gets bullied on a regular basis. Kids and teens who are bullied at school may stay home out of fear. They may even try to harm themselves because the bullying has seriously damaged their self-esteem. Bullying behavior is also harmful to the bullies themselves. They are more likely to drop out of school and to have problems with alcohol or violence.

Hazing, a related problem, means *making others perform certain tasks in order to join the group.* Hazing activities may be physically or emotionally harmful. Examples include yelling or swearing, forcing new group members to go without sleep, physically beating them, or forcing them to drink alcohol. Severe hazing incidents have been known to result in death. Hazing is often meant to humiliate new members or prove that they are inferior to existing members.

■ **Figure 6.6** Bullies may intimidate others through verbal attacks, malicious rumors, or even physical force. *How would you feel if you were facing the bully in this picture?*

Real World CONNECTION

Dealing with a Bully

Stop Bullying Now! is one example of a Web site that recommends various ways to deal with a bully. In recent years, technology has given teens a new way of bullying each other. Cyberbullying can involve sending mean or threatening messages through e-mails or Web pages. Conduct an Internet search for information on other ways to deal with a bully. Look for information on the definitions of bullying, the effects it has on those who are bullied, whom to talk to about bullying, and how to prevent it. Use only information from credible sources, and list the Web sites where you obtain information.

Activity Technology

Work with a small group to conduct your research. Then write, perform, and record a video of a skit using the information you gathered. The skit should show what the bullied teens and the teens who oppose bullying can do to prevent it.

LESSON 2 ASSESSMENT

 After You Read

Reviewing Facts and Vocabulary

1. Identify four ways to show respect in your relationships.

2. What are *stereotypes*?

3. List three reasons some teens bully others.

Thinking Critically

4. **Synthesize.** Give an example of how demonstrating strong values can strengthen your relationships with others.

5. **Analyze.** How is bullying different from hazing?

Applying Health Skills

6. **Decision Making.** Ahmed has just made the swim team, but he's concerned about reports that the varsity swimmers haze the new members. Use the decision-making process to analyze how Ahmed might deal with this problem.

Writing Critically

7. **Persuasive.** Write an editorial about the problem of bullying in schools. Your article should encourage students to help create a positive climate in which bullying is not tolerated.

Real Life Issues ....

After completing the lesson, review and analyze your response to the Real Life Issues question on page 148.

BIG Idea *Effective communication is a key to building healthy relationships.*

Before You Read

Create a Word Web.
Write the phrase "Effective Communication" in the center of a piece of paper. Jot down characteristics of effective communication. As you read, add more notes to your word web.

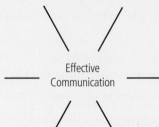

Effective
Communication

New Vocabulary

▶ aggressive
▶ passive
▶ assertive
▶ "I" message
▶ active listening
▶ body language

Review Vocabulary

▶ constructive criticism (Ch.3, L.2)

Communicating Effectively

Real Life Issues ....

A Pushy Friend. Erin's friend Louise is the kind of person who always wants to have her own way. When they go out together, Louise decides where they'll go and what they'll do. If Erin offers a suggestion, Louise dismisses it. Erin wants to stay friends with Louise, but she's tired of being pushed around. She wishes she knew how to stand up for herself without being rude.

Writing *Write a letter from Erin to Louise. In it, Erin should discuss how Louise's behavior makes her feel and explain that she'd like to have a say in what they do.*

Communication Styles

Main Idea There are three types of communication styles.

Do you know someone who's like Louise, always insisting on doing things her own way? How about someone who's like Erin, always going along with what other people suggest? These two characters reflect two of the three major styles of communication:

- **Aggressive.** Being **aggressive** means being *overly forceful, pushy, or hostile.* It may involve bullying or intimidation. People with an aggressive communication style may not pay attention to others' thoughts, feelings, or needs.

- **Passive.** Being **passive** means being *unwilling or unable to express thoughts and feelings in a direct or firm manner.* This involves putting others' needs ahead of your own. People may adopt a passive communication style because they dislike conflict and will go out of their way to avoid an argument.

- **Assertive.** Being **assertive** means *expressing your views clearly and respectfully.* Assertiveness involves standing up for your rights and beliefs while also respecting those of others. Dealing with a disagreement in an assertive way can involve negotiating with others to find the best solution to the problem.

Using an assertive communication style will improve your relationships with others. It will help ensure that your own needs are met, along with those of other people.

READING CHECK

Identify What are the three styles of communication?

Ways to Communicate

Main Idea To communicate effectively, you need to learn speaking skills, listening skills, and nonverbal communication.

Communication is a two-way street. It's not enough just to get your messages across to the other person. You also need to understand the messages being sent to you. This includes both verbal and nonverbal messages.

Speaking Skills

The key to good communication is to say what you mean. It's not reasonable to expect other people to read your mind or be able to pick up on subtle hints. If something's on your mind, you need to say what it is. For example, if a friend has hurt your feelings, you need to let that person know as clearly and directly as you can.

■ **Figure 6.7** Communicating assertively makes your relationships run more smoothly. *Which teen in this photo would you say is communicating assertively? How can you tell?*

Figure 6.8 — "You" Messages and "I" Messages

Compare the messages in these two columns. *How might a listener react to each message?*

"You" Messages 	"I" Messages 
"Why can't you ever show up on time?"	"I really don't like to be left waiting—it makes me feel like you don't think I'm important."
"You never listen to anything I say."	"I feel like my suggestions aren't being taken seriously."
"I said I'd take out the trash, and I will! You don't have to nag me about it every five minutes!"	"I'm feeling stressed because I have a big project due tomorrow. I'll take out the trash as soon as I finish working on this."
"You're always taking my CDs without asking."	"It bothers me when I get home and find all my CDs in your room."
"You always ignore me when your other friends are around."	"I feel hurt when I'm left out of a conversation."

Of course, being clear and direct doesn't mean being disrespectful. When people feel that they are being attacked, they may be less willing to listen. One way to make sure you don't sound disrespectful when talking about a touchy subject is to use "I" messages. An **"I" message** is *a statement that focuses on your feelings rather than on someone else's behavior.* Using "I" messages helps you communicate your feelings in a positive way without placing blame on someone else. **Figure 6.8** gives some examples of "I" messages.

Listening Skills

To communicate effectively, listening is just as important as speaking. You can make sure other people's messages get through to you by practicing **active listening**, *paying close attention to what someone is saying and communicating.* Here are some ways to practice active listening:

- **Don't interrupt.** Give your full attention to what the speaker is saying.
- **Show interest.** Face the speaker and make eye contact to show that you are paying attention. You can also encourage the speaker by nodding or making comments, such as "I see," "Go on," or "I understand," at appropriate times.

- **Restate what you hear.** Rephrase or summarize the speaker's words to make sure you understand what you're hearing.

- **Ask questions.** Asking questions can help you understand what the speaker is saying. It can also help the speaker clarify her own thoughts and feelings.

- **Show empathy.** Let the other person know that you can relate to his feelings. Try not to pass judgment on the speaker's attitudes and actions.

Nonverbal Communication

Sometimes what you say isn't as important as how you say it. A comment such as "Nice outfit" can actually sound like an insult if it's delivered in a sarcastic tone. Your tone of voice is an example of nonverbal communication.

Your body language can also affect the meaning of the messages you send. **Body language** is *nonverbal communication through gestures, facial expressions, behaviors, and posture.* It includes everything from nodding, which shows that you agree, to turning away, which shows that you aren't listening. **Figure 6.9** shows different examples of body language.

Sometimes you send messages through body language without even realizing it. If you're feeling embarrassed, you may look at the ground instead of at the person you're talking to. In some cases, your body language may even **contradict** what you're saying. For instance, saying "I'm fine" in an angry tone will probably make people think you're anything but fine. Being aware of your body language can help you avoid sending mixed messages that may confuse your listeners.

READING CHECK

Describe What does active listening involve?

Academic Vocabulary

contradict *(verb):* to imply the opposite of

| Figure 6.9 | **Using Body Language** |

Body language is an important part of nonverbal communication.

"I'm really interested in what you're saying."

"I don't want to talk to you."

"I'm worried."

Offering Useful Feedback

Main Idea Offering constructive feedback can improve your relationships with others.

Even in a strong relationship, every now and then people say or do things that bother other people. You might have a friend who's lots of fun to be around, except for the fact that he always interrupts you when you're talking. If you want your friend to change his behavior, you have to let him know how you feel—but in a way that doesn't come across as a personal attack. In other words, you need to offer constructive criticism, nonhostile comments that point out problems and encourage improvement.

The goal of constructive criticism is to bring about positive changes. Thus, it's counterproductive to give it in an aggressive way. Attacking someone isn't going to encourage him to change. Instead, use "I" messages that focus on the problem, not on the person. To offer constructive criticism, point out a specific problem, explain why it bothers you, and suggest a solution. You might say, "I feel that sometimes my ideas don't get heard. I would like to finish what I'm saying. Then I'll be glad to listen to any responses."

Letting people know how their actions make you feel isn't something you should do only when there's a problem. It's also important to let people know you appreciate what they do for you. If a friend goes out of her way to help you and doesn't get so much as a "thank you," she might feel that her actions went unnoticed or unappreciated. Let the people in your life know you value them. Tell your dad how much you enjoyed a meal that he prepared, or compliment a friend on her artistic skills. A little appreciation can go a long way in strengthening a relationship.

READING CHECK

Explain What is the goal of constructive criticism?

■ **Figure 6.10** Encouraging your friends to pursue their interests is one way to show your appreciation for them. *What are other ways to let people in your life know you value them?*

©Paul Bradbury/age fotostock

Health Skills Activity

Communication Skills

Coping with Criticism

Dennis and Vince are in the same history class, but are not close friends and have never worked together before. Their history teacher has asked them to complete an important project together. Both teens want the extra credit points they'll receive if they do a good job.

However, Vince criticizes all of Dennis's work. Most recently, Vince complained that he doesn't like Dennis's writing style. Dennis is beginning to feel frustrated. Nothing he does satisfies Vince. Dennis is beginning to feel his self-esteem weaken.

Writing Write a dialogue between Dennis and Vince. Dennis might suggest that they talk about what outcome they both want before they begin the work. He should also ask Vince to offer only constructive advice. Remember to use the following rules for good communication:

1. Speak calmly and clearly.
2. Use "I" messages.
3. Show respect and empathy.
4. Listen carefully and ask appropriate questions.

LESSON 3 ASSESSMENT

 After You Read

Reviewing Facts and Vocabulary

1. What are the three main styles of communication?

2. List three ways to show interest in what another person is saying.

3. Define the term *body language* and give an example.

Thinking Critically

4. **Evaluate.** Leah is an aggressive communicator. When somebody says something she disagrees with, she always says, "You're wrong!" How could Leah's communication style affect her relationships with others?

5. **Synthesize.** In a paragraph, discuss how having strong communication in your relationships can contribute to your personal health and safety.

Applying Health Skills

6. **Analyzing Influences.** Think about different factors that can influence how you communicate with others. Factors may include family, peers, culture, or personality. In a paragraph, discuss how one of these factors affects your personal communication style.

Writing Critically

7. **Narrative.** Write a dialogue in which one character offers constructive criticism to another. Follow the guidelines for giving constructive criticism.

Real Life Issues ....

After completing the lesson, review and analyze your response to the Real Life Issues question on page 152.

Hands-On HEALTH

Activity **The New Employee**

Rachel has an after-school job at a local store. A new employee started working there recently, and his behavior is often aggressive and inappropriate. Rachel wants to talk to the manager, but she doesn't know where to begin.

What You'll Need

- paper and pen or pencil
- 5 or more index cards

What You'll Do

Step 1

Review the communication skills outlined in the chapter. On the front of each index card, write one skill that Rachel will need to communicate effectively with her manager. On the back of each card, describe how to use that communication skill in this situation.

Step 2

Arrange the cards in the order that Rachel should use them.

Step 3

Use the index cards to help you write a script in which Rachel discusses her concerns with her manager.

Apply and Conclude

Rehearse and role-play the script with classmates. Make sure that Rachel's needs and feelings are clearly communicated to her manager.

Checklist: Communication Skills

- ✓ Clear, organized message
- ✓ "I" messages
- ✓ Active listening
- ✓ Respectful tone
- ✓ Appropriate body language
- ✓ Reason(s) for the request

CHAPTER 6 Review

LESSON 1

Foundations of a Healthy Relationship

Key Concepts

- Relationships in your life affect all sides of your health triangle.
- Important relationships in your life may include family relationships, friendships, and relationships in your community.
- Healthy relationships involve mutual respect, caring, honesty, and commitment.
- Communication, cooperation, and compromise are important skills for building healthy relationships.
- Demonstrating the six traits of good character strengthens your relationships.

Vocabulary

- relationship (p. 142)
- friendship (p. 143)
- citizenship (p. 144)
- role (p. 144)
- interpersonal communication (p. 145)
- cooperation (p. 146)
- compromise (p. 146)

LESSON 2

Respecting Yourself and Others

Key Concepts

- Having self-esteem and demonstrating strong values can improve your relationships with other people.
- You can strengthen your relationships with all the people in your life by treating them with respect.
- Demonstrating tolerance, or the ability to accept others' differences, can help you build healthy relationships.
- Bullying and hazing are disrespectful and harmful behaviors.

Vocabulary

- personal identity (p. 148)
- values (p. 148)
- prejudice (p. 150)
- stereotype (p. 150)
- tolerance (p. 150)
- bullying (p. 150)
- hazing (p. 150)

LESSON 3

Communicating Effectively

Key Concepts

- You can improve your relationships by communicating assertively, rather than aggressively or passively.
- Communication involves speaking, listening, and nonverbal communication such as body language.
- Using "I" messages helps you communicate your feelings in a positive way without placing blame on someone else.
- Active listening involves paying close attention to what someone is saying and communicating.
- Constructive criticism can bring about positive changes by pointing out problems in a nonhostile way.

Vocabulary

- aggressive (p. 152)
- passive (p. 152)
- assertive (p. 153)
- "I" message (p. 154)
- active listening (p. 154)
- body language (p. 155)
- constructive criticism (p. 156)

LESSON 1

Vocabulary Review

Choose the correct word in the sentences below.

1. A bond or connection you have with other people is called a *relationship / friendship*.

2. *Communication / Citizenship* lets you express your thoughts, feelings, and expectations to others.

3. In *cooperation / compromise*, each participant gives up something to reach a solution that satisfies everyone.

Understanding Key Concepts

After reading the question or statement, select the correct answer.

4. Which of the following is an example of good citizenship?
   a. Looking after a younger brother or sister
   b. Helping a friend study for a test
   c. Taking part in an effort to clean up a local river
   d. Getting good grades in school

5. Jeanne and her father have very different political views. However, they accept their differences and do not try to change each other's opinions. Which quality of strong relationships does this action show?
   a. Mutual respect
   b. Caring
   c. Honesty
   d. Commitment

Thinking Critically

After reading the question or statement, write a short answer using complete sentences.

6. **Describe.** Identify and describe three roles you play in your relationships with others.

7. **Explain.** How does cooperation strengthen your relationships?

8. **Evaluate.** Give an example of a situation in which you should *not* be willing to compromise.

9. **Synthesize.** Identify one trait of good character, and give an example of how it can strengthen a relationship.

LESSON 2

Vocabulary Review

Use the vocabulary terms listed on page 159 to complete the following statements.

10. _____ is an unfair opinion or judgment of a particular group of people.

11. People display _____ when they recognize and appreciate the differences among people.

12. Deliberately harming or threatening other people who cannot easily defend themselves is known as _____.

Understanding Key Concepts

After reading the question or statement, select the correct answer.

13. Teens who respect themselves probably will choose friends who
    a. are the smartest students in the class.
    b. participate in a wide variety of school activities.
    c. share all of their beliefs, tastes, and values.
    d. treat them with respect.

14. Which of the following statements is *not* an example of prejudice?
    a. I get nervous when I'm in large groups of strangers.
    b. I think kids from private schools are stuck-up.
    c. I don't want girls on our baseball team because they can't throw.
    d. I want to be friends only with Asian American kids because they're such good students.

15. Which of the following is *not* an example of bullying?
    a. Repeatedly making fun of the way a classmate talks
    b. Threatening to beat up a student if he doesn't hand over his lunch money
    c. Accepting someone into your group regardless of the clothing brand she wears
    d. Shoving a younger kid on the school bus

16. If you are being bullied at school, you should
    a. just ignore the bully until she goes away.
    b. stay home from school to avoid the bully.
    c. get a group of friends to gang up on the bully and attack him.
    d. tell a trusted adult about the problem and ask for help.

Thinking Critically

After reading the question or statement, write a short answer using complete sentences.

17. **Discuss.** How might you demonstrate respect in your relationship with a teacher?

18. **Explain.** In what ways can prejudice harm relationships?

19. **Analyze.** Explain how bullying can be harmful both to victims and to bullies themselves.

20. **Evaluate.** At Corinne's school, students who are new to the drama club usually get assigned to sing in the chorus or paint the sets. In contrast, more experienced students get the lead roles in the plays. Is this an example of hazing? Explain why or why not.

LESSON 3

Vocabulary Review

Correct the sentences below by replacing the italicized term with the correct vocabulary term.

21. Trying to get your own way through bullying or intimidation is an example of *passive* communication.

22. Someone using *assertive* communication is unwilling or unable to express his thoughts and feelings.

23. When you use *body language,* you focus on your own feelings rather than on someone else's behavior.

Understanding Key Concepts

After reading the question or statement, select the correct answer.

24. Some of the friends in your group want to go out for Thai food, but you don't like Thai food. Which of the following would be an assertive way to respond?
    a. Tell them Thai food is not your favorite and politely suggest an alternative.
    b. Go with them for Thai food, but don't eat anything.
    c. Just go along with whatever the group wants.
    d. Insist on going for Mexican food instead.

25. Which of the following skills is *not* a part of active listening?
    a. Listening without interrupting
    b. Making eye contact with the speaker
    c. Asking questions for clarification
    d. Using "I" messages

Assessment

26. If you stand with your hands on your hips and your lips in a frown while someone is talking, what message might you be sending with your body language?

   a. "I'm really interested in everything you're saying."

   b. "I'm feeling angry."

   c. "I'm embarrassed about something."

   d. "I am respectfully considering your opinion."

27. Which of the following is an example of constructive criticism?

   a. "Do you always have to leave all your junk out in the hall?"

   b. "I can't stand the way you interrupt me all the time."

   c. "Haven't you ever heard of knocking? You're always barging into my room!"

   d. "Next time, would you mind calling first to let me know you're coming over?"

Thinking Critically

After reading the question or statement, write a short answer using complete sentences.

28. **Evaluate.** Which communication style will most enhance your relationships? Explain how.

29. **Analyze.** Describe the roles that speaking, listening, and body language play in communication.

30. **Synthesize.** A friend calls to tell you that she just broke up with her boyfriend. Explain how you might use active listening in this situation.

31. **Explain.** Why are "I" messages an important part of constructive criticism?

Technology PROJECT-BASED ASSESSMENT

Effective Communication Skills

Background

As you've learned, effective communication involves strong speaking and listening skills. The use of technology can play a major role in communication.

Task

Show two scenarios of people communicating to one another through e-mail. The first scenario should show how someone with weak communication skills may end up sending confusing messages. The second scenario should demonstrate how using effective communication skills can prevent the problems shown in the first example.

Audience

Other students in your class

Purpose

Demonstrate and practice communication skills

Procedure

1 Create an example of communication between two people through e-mail. The topic can involve going to the movies, going on a date, or something similar.

2 The first example of communication should show how a teen can experience problems in communication, such as sending an unclear message.

3 Create the second communication example in which the teen avoids communication problems by using effective strategies.

4 Present your communication examples to the class.

Math Practice

Calculate Percentages. Bullying occurs more frequently among middle school students, but it also happens in high school. This diagram shows the estimated number of students affected by bullying.

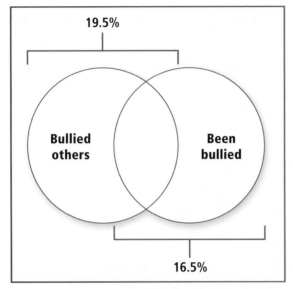

19.5%

Bullied others

Been bullied

16.5%

1. About 6% of all students have both bullied other students and have been the victim of a bully. From this information and the diagram, what percentage of students have either been bullied or have bullied another student, but not both?
   A. 3.5% **C.** 30%
   B. 6% **D.** 36%

2. Suppose there are 842 students in your school. Using the percentages shown in the diagram, estimate the number of students in your school who have bullied another student.
   A. 25 **C.** 139
   B. 51 **D.** 164

3. Using the percentages given above, explain how you would determine the number of students in your school who have neither bullied another student nor been bullied.

Reading/Writing Practice

Understand and Apply. Read the passage below, and then answer the questions.

1. **Will:** *Hey, Jim, are you busy on Sunday? We could catch the afternoon showing of that new action movie.*
2. **Jim:** *Actually, I saw that last week with my brother.*
3. **Will:** *Well, that's the only movie I want to see.*
4. **Jim:** *Oh, it's okay. I don't mind seeing it again.*
5. **Will:** *Good. Meet me there at one.*
6. **Jim:** *Um, I have church that day, so that will be cutting it close . . . but don't worry about it.*
7. **Will:** *Who's worried?*

1. What does line 3 reveal about Will's personality?
   A. He does not like movies.
   B. He treats his friends with respect.
   C. He is used to getting his own way.
   D. He is unwilling to express his thoughts and feelings.

2. Jim could best be described as
   A. enthusiastic.
   B. passive.
   C. inconsiderate.
   D. uncommunicative.

3. Rewrite this dialogue with Jim using a more assertive communication style. Your new dialogue should show how being assertive results in an outcome that is more acceptable for both teens.

National Education Standards

Math: Ratios and Proportional Relationships
Language Arts: LACC.910.L.3.6; LACC.910.RL.2.4

CHAPTER 7

Family Relationships

Ed-Imaging

Lesson 1

Healthy Family Relationships

BIG Idea *Your relationships with family members influence your total health.*

Lesson 2

Strengthening Family Relationships

BIG Idea *Family members support and care for one another, especially during difficult times.*

Lesson 3

Help for Families

BIG Idea *Families may require outside assistance to deal with serious problems.*

Activating Prior Knowledge

Using Visuals Look at the photo on this page. Based on what you have learned about relationships, write a paragraph explaining how these family members are strengthening their relationship. Discuss the ways their interactions might contribute to their physical, mental/emotional, and social health.

Chapter Launchers

Health in Action

Discuss the **BIG** Ideas

Think about how you would answer these questions:

▸ How do you and your family depend on each other?

▸ What helps you and your family through tough times?

▸ Where would you go for help with a family problem?

Assess Your Health

Read each statement. On a separate sheet of paper, write "yes," "sometimes," or "no" based on your typical behavior.

1. I recognize that there are many types of families.

2. I know that our families provide guidance, values, and cultural traditions.

3. I understand that family members can help each other cope with change.

4. I recognize that abuse within a family may require help from professionals and law enforcement.

5. I know where to seek help within my community for help with family problems.

A "yes" response shows that you practice healthy behaviors. "Sometimes" indicates that you should analyze and possibly modify your behavior. A "no" response means that you should modify the behavior.

LESSON 1

Tim Fuller Photography

Healthy Family Relationships

GUIDE TO READING

BIG Idea *Your relationships with family members influence your total health.*

Before You Read

Organize Information. Draw a triangle. Label the sides "Physical," "Mental/Emotional," and "Social." As you read, record information about how families affect each area of health.

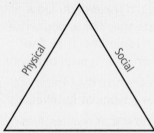

Physical

Social

Mental/Emotional

New Vocabulary

▸ siblings
▸ nuclear family
▸ blended family
▸ extended family
▸ foster care
▸ affirmation

Review Vocabulary

▸ role (Ch.6, L.1)

Real Life Issues ••••••••••••••••••••••••••

A Family Discovery. Recently, Jack found out that he's adopted. The discovery came as a shock, and now he's starting to question everything he's ever known about himself and his family. Even though he loves his parents and brother and knows they love him too, he still feels confused. He doesn't know whether to think of them as his "real" family anymore, and he also wonders about his biological parents.

Writing *Write a journal entry from Jack's point of view. In it, Jack should express his feelings about being adopted and reflect on how this discovery could affect his life.*

The Family Unit

Main Idea There are many kinds of families, but all family members have certain responsibilities toward each other.

What is a family? The question isn't as simple as it sounds. There are many different types of families. Family members may be related to each other by birth, marriage, or adoption. People in the same family may live together or separately.

No matter who is in your family, your relationships with them are some of the most important in your life. Family relationships have a strong influence on your total health. Healthy families provide support to their members and help children and teens develop the values and skills to become successful members of society. Being part of a strong family can also be an important *protective factor* for teens by helping them avoid behaviors that may put their health at risk. Ultimately, healthy families are the foundation of a healthy society.

Types of Families

When you think about families, you may picture your own parents or **siblings**, your *brothers and sisters*. To some people, the word *family* suggests a **nuclear family**—*two parents and one or more children living in the same place*. Although this is a common family structure in America, increasing numbers of children live in other types of families:

- **Single-parent families.** These are families with one parent caring for one or more children. A child may live with one parent after a divorce or the other parent's death.

- **Blended families.** These families form when a single parent remarries. A **blended family** consists of *a married couple and their children from previous marriages*. The new couple may also decide to add to their blended family by having more children.

- **Extended families.** An **extended family** is *a family that includes additional relatives beyond parents and children*. Relatives include grandparents, aunts, uncles, and cousins. Some people live with extended family members.

- **Adoptive families.** These families consist of a parent or parents and one or more adopted children. Some families have both biological children and adopted children.

- **Foster families.** **Foster care** is *the temporary placement of children in the homes of adults who are not related to them*. Children may be placed in foster care because of problems like abuse (discussed in Lesson 3). In some cases, foster parents may decide to adopt a child who has been living with them.

READING CHECK

Compare and Contrast How do adoptive families differ from foster families?

■ **Figure 7.1** Families may include members who are related by birth, by adoption, or both. *What do you think makes someone part of a family?*

Real World CONNECTION

America's Families

The structure of American families has shifted over the past several decades. The table below shows how the living arrangements of American children under 18 have changed since 1960. Review the statistics, and use the table to answer the questions.

Activity Mathematics

1. In the year 1960, what percentage of all children lived in each type of household?

2. In the year 2013, what percentage of all children lived in each type of household?

3. What factors do you think contributed to the shift in family structure during this 53-year period?

Concept Number and Operations: Percents A percent is a ratio comparing a number to 100. To calculate percentage, divide the given amount by the total amount. Then multiply the answer by 100 and add a percent sign (%).

Living Arrangements of American Children Under 18

Year	Total children under 18	Two parents	Mother only	Father only	Other relatives	Non-relatives
1960	63,727	55,877	5,105	724	1,601	420
1970	69,162	58,939	7,452	748	1,546	477
1980	63,427	48,624	11,406	1,060	1,949	388
1990	64,137	46,503	13,874	1,993	1,421	346
2000	72,012	49,795	16,162	3,058	2,160	837
2007	73,746	52,153	16,658	2,389	2,545	535
2013	73,910	50,646	17,532	2,999	2,121	612

Source: U.S. Bureau of the Census

Family Interactions

In a family, each member plays certain roles and has certain responsibilities. In general, parents or guardians are in charge of meeting the family's basic needs, such as food and shelter. Parents also serve as teachers in the family, establishing rules and setting limits to protect their children's health and safety. They teach their children about the reasons for these rules, and teach the values and skills that will guide them in the future.

Children and teens, meanwhile, also have roles and responsibilities. When they are young, their main job is to respect the **authority** of parents or guardians. As they get older, they may take on more responsibilities, such as doing chores or caring for younger siblings. By taking on such tasks, teens can help the family run more smoothly and boost their own self-esteem.

Other relatives play a role in the family as well. For example, grandparents may help care for children and teach them about the family's history. Aunts and uncles may serve as mentors and role models. Cousins who are close in age may be playmates and friends.

Academic Vocabulary

authority *(noun):* the right to make decisions and give commands

Your Family and Your Health

Main Idea Your family members contribute to your health.

Being part of a family helps you meet your most basic needs. Beyond that, being part of a healthy family can strengthen all three aspects of your health.

Promoting Physical Health

The most obvious way your family promotes your physical health is by providing for your basic physical needs. Your parents or guardians make sure that you receive food, clothing, and shelter. They also promote your physical health by

- **providing medical care.** When you were young, your parents or guardians took you to the doctor when you were sick. They also made sure you got medical and dental checkups and necessary immunizations.

- **setting limits on behavior.** Do your parents set rules, such as how late you can stay out at night? The purpose of these rules is to promote your safety and health. For instance, setting a curfew can protect you from risky situations and also help make sure you get enough sleep.

- **teaching health skills.** In addition to setting limits on your behavior, your parents helped teach you the skills you needed to control your own behavior as you got older. They may have taught you basic safety skills, such as wearing a helmet when you ride a bike. They may also have encouraged you to develop healthy habits, like eating nutritious foods and engaging in physical activity.

FITNESS ZONE

It feels good to do something nice for someone, and it can be good for your health too. That's why I like to take my little brother and sister to the park to play catch or basketball. I want to be a good role model and teach them just how important exercise is. Besides, the smiles on their faces make it all worthwhile. For more physical activity ideas, visit the Fitness Handbook and Fitness Zone sites in ConnectEd.

■ **Figure 7.2** By encouraging healthful behaviors such as physical activity, parents and other family members can promote physical health. *What are other ways your family influences your physical health?*

Promoting Mental and Emotional Health

As you get older, you may rely less on your family to meet your physical needs. However, it is likely that your family still plays an important role in meeting your mental and emotional needs. For example, your family can provide a safe environment for you to express and deal with your emotions. Family members can also give you love and support, helping to meet your need to feel that you belong. This sense of belonging, in turn, can help boost your self-esteem.

Your family can also help meet your need to feel valued and recognized by providing **affirmation**. This is *positive feedback that helps others feel appreciated and supported*. For instance, they can celebrate your achievements with you or show appreciation for the ways you help out at home.

Promoting Social Health

Your family also contributes to your social development. In the first few years of your life, family members helped you learn how to communicate and get along with others. As you grew, your family may have helped you learn other important social skills, such as how to cooperate with others and how to resolve conflicts. The social skills you learned from your family will help you make your own way in the world as an independent adult.

Values One of the most important ways families promote social health is by instilling values. Parents play a significant role in helping children develop core ethical values, including responsibility, honesty, and respect. Learning these values is a key to developing strong character.

■ **Figure 7.3** Family members can provide affirmation by celebrating each other's achievements. *What are other ways family members can support each other mentally and emotionally?*

Families can teach values in different ways. One way is by explanation. For instance, if two siblings are fighting over a toy, a parent might sit down with them and explain why it's important to share. Teaching by example can be an even more powerful way to promote good values. Let's say a parent is shopping with a child and receives too much change back for a purchase. By immediately returning the extra money, the parent teaches the child about honesty and fairness. Likewise, parents who demonstrate kindness and respect in their daily behaviors reinforce these same values in their children. By being positive role models, parents and other family members help children develop strong values.

Cultural Heritage Families also promote social health by sharing their culture and traditions. For example, families may light candles together at Kwanzaa or enjoy a barbecue and fireworks on the Fourth of July. Sharing their culture enriches the lives of family members and helps them develop a sense of cultural identity. This awareness of being part of a larger culture can create important social bonds that extend beyond the family.

READING CHECK

Identify Name two ways families can teach values.

LESSON 1 ASSESSMENT

After You Read

Reviewing Facts and Vocabulary

1. What is a *sibling*?
2. Name three kinds of families.
3. Identify four ways in which families promote the physical health of children and teens.

Thinking Critically

4. **Synthesize.** Explain how the role you play within your family has changed over time.
5. **Analyze.** How does providing affirmation within the family promote mental and emotional health?

Applying Health Skills

6. **Communication Skills.** Work with a classmate to write and perform a scene that shows family members supporting each other mentally and emotionally. The scene should include "I" messages, active listening, and appropriate body language.

Writing Critically

7. **Personal.** Write a personal essay about your family. Describe how you interact, and discuss how family members contribute to each other's total health.

Real Life Issues ....

After completing the lesson, review and analyze your response to the Real Life Issues question on page 166.

LESSON 2

BIG Idea *Family members support and care for one another, especially during difficult times.*

Before You Read

Create a T-Chart.
Make a two-column table. Label the columns "Change in Family Structure" and "Change in Circumstances." As you read, fill in each column with examples of changes that can affect families, and strategies strong families can use to deal with these changes.

Change in Structure	Change in Circumstances

New Vocabulary

▸ separation
▸ divorce
▸ custody

Review Vocabulary

▸ stress (Ch.2, L.1)

Strengthening Family Relationships

Real Life Issues

Dealing with Divorce. Beth has just learned that her parents are getting a divorce. Her father will be moving across town, and she knows she's going to be asked who she wants to live with. Beth is close to both her parents, and she doesn't want to have to choose between them. She'd like to tell her parents how she feels, but she doesn't want to add to their problems.

Writing *Write a dialogue in which Beth discusses her feelings with one or both of her parents. Each character should demonstrate good communication skills.*

Characteristics of Strong Families

Main Idea Strong families support their members in a variety of ways.

Different families interact together in different ways. For example, Joyce's family tends to be reserved around each other. They express their feelings calmly and rationally. When Joyce goes to her friend Ted's house, she's amazed at how openly his family expresses emotions. Ted and his family laugh and cry easily together. They tease each other and get into arguments, but they always make up.

This doesn't mean Ted's family is healthier than Joyce's, or vice versa. They just interact in different ways. The important thing is that both Ted and Joyce feel secure and loved. Both of their families demonstrate traits of strong families:

- **Good communication.** Healthy families share their thoughts and feelings honestly with each other. They listen to each other and demonstrate empathy.

■ **Figure 7.4** Spending time together strengthens family relationships. *What are other ways for family members to show their commitment to each other?*

©Hero/Corbis/Glow Images

- **Caring and support.** Family members show they love each other through their words and actions. They express appreciation for each other and help each other through difficult times.

- **Respect.** Family members respect each other's opinions, tastes, and abilities. They show consideration by sharing, being courteous, respecting each individual's privacy, and helping out with household tasks.

- **Commitment.** Healthy families make time for each other. They work together to solve problems, and they're willing to make sacrifices for the good of the family.

- **Trust.** In a healthy family, parents earn their children's trust by being honest and keeping their promises. Children show that they are worthy of trust by being honest, loyal, and reliable.

 **READING CHECK**

Identify How can family members show respect for each other?

Coping with Change

Main Idea Family members can help each other cope with changes in the family's structure or circumstances.

Families can face a variety of problems, both major and minor. Many of these problems have to do with changes in the family's structure or circumstances. A parent losing a job, for example, or a grandparent's serious illness can lead to long-term stress for the whole family. Even positive events, such as a move or the marriage of a relative, can create stress. Because change is a normal part of life, healthy families must be prepared to deal with changes and help each other cope.

Changes in Family Structure

The structure of a family changes when someone new joins the family or when a family member moves out of the home. Examples of such changes include birth, adoption, separation, divorce, remarriage, and the death of a family member.

Birth and Adoption Welcoming a new baby or an adopted child into the family is a joyful event. However, adjusting to the new situation isn't always easy. Making room for the new child means that everyone else has to make do with less space at home. Also, as parents devote time and energy to the new child, they may have less time for the other children—and for each other. All these changes can create stress for everyone. Family members can help each other through this time by sharing the responsibility for taking care of the new child. They can also make an effort to find time for each other.

Separation and Divorce Separation and divorce are difficult, especially since they result in a family member leaving the home environment. **Separation** is *a decision by two married people to live apart from each other.* Couples who separate may hope to eventually work out their differences and live together again. **Divorce**, by contrast, is *a legal end to a marriage contract.*

READING CHECK

Compare and Contrast What is the difference between separation and divorce?

When parents divorce, they need to come to an agreement about where the children will live. **Custody** is *the legal right to make decisions affecting children and the responsibility for their care.* Custody may be granted to only one parent (sole custody) or divided so that both parents share in the child rearing (joint custody). Adapting to either arrangement can be difficult for the children. They may find it hard to go for long periods without seeing one of their parents. In the case of joint custody, they may find it stressful to move back and forth between two homes.

Parents can help their children get through this difficult period by reminding them that both parents still love them. They can also reassure the children that the divorce was not their fault. Children may find it easier to cope if they discuss their feelings with parents and other trusted adults. In some cases, they may want to consider joining a support group for children of divorce. Being part of such a group may help them realize that they are not alone.

Remarriage After a divorce, one or both parents may decide to marry again. A parent may also remarry after the death of a spouse. When a parent remarries, the children must adjust to having, or living with, a stepparent. If the stepparent has children from a previous marriage, all members of the blended family will need time to adjust. Good communication and mutual respect will make this process easier.

Figure 7.5 The remarriage of a parent can bring mixed feelings. *How can teens show support for a parent's decision to remarry?*

Death of a Family Member Perhaps the most difficult change a family can go through is the death of a family member. In Chapter 4, you learned about the feelings of grief that can accompany a death or other loss. Family members can help each other through this difficult time by sharing their feelings and memories about the person they've lost. It's also important for family members to respect each other's feelings and remember that the process of grieving is different for everyone. Joining a support group or seeking help from a counselor may also help those who have lost a loved one recover from their pain.

Changes in Family Circumstances

Changes in a family's **circumstances** can also be a source of stress. Family members can help each other deal with these changes by communicating honestly and showing as much support as possible. Here are some examples of changes in family circumstances:

- **Moving to a new home.** When a family moves, especially over a long distance, family members may miss their old friends and familiar surroundings. Teens may be anxious about making new friends and adjusting to a new school. When a move results from the breakup of a marriage, it can add to the stress already caused by the divorce.

- **Changes in the family's financial situation.** Financial problems can result from the loss of a job, a medical emergency, poor planning, or uncontrolled spending. Not having enough money to pay the bills can be stressful.

Academic Vocabulary

circumstance *(noun):* an event that influences another event

It can also lead to arguments about how the family's limited funds should be used. Interestingly, a sudden financial gain can also be a source of stress. Unaccustomed wealth can trigger anxiety and confusion as people wonder what to do with the money and whether it's going to change the way people see them.

- **Illness and disability.** A serious illness or disability can disrupt a family's normal routine. One or more family members may need to change their schedules to care for the sick or disabled person. Coping with this situation may be easier if each family member plays a role in caring for the sick or disabled person.

- **Alcohol or other drug abuse.** Substance abuse is one of the most serious problems a family can face. Family members must seek outside help to deal with the situation. Teens may wish to consult teachers, other trusted adults, or organizations such as Alateen. You will learn more about confronting the problem of substance abuse in Chapters 21 and 22.

Coping with Changes

One of the most important strategies for coping with changes in the family is to talk honestly and openly with each other. Just talking about your feelings can help reduce stress. Letting family members know about your needs and wants can also make it easier for them to help you.

You, in turn, can make an effort to support your family members during a difficult period. For example, you can offer to take on more chores and responsibilities at home. You can also make a point of being there for family members if they want to talk.

If this strategy is not enough, family members may find it helpful to talk with someone outside the family, such as a counselor, teacher, or member of the clergy. They can also try to learn more about the situation they're dealing with, either by reading books or by talking with people who have been through similar experiences. Finally, families should be willing to seek professional help if they need it. Lesson 3 discusses resources that can help.

■ **Figure 7.6** Talking with a parent or other trusted adult can help you deal with the stress of family changes. *To whom do you turn when you need to talk?*

Health Skills Activity

Communication Skills

Family Finances

A month ago, Kenny's dad lost his job. Ever since then, his parents have been tense and anxious. They have whispered conversations so the children can't hear them, but Kenny knows they're talking about money. Kenny is worried about how the family is going to manage financially, but he's even more upset that his parents don't trust him enough to talk to him about the problem.

Writing Write a dialogue between Kenny and his parents. In it, Kenny should express his concerns about the family's financial situation and his feelings about being left out of his parents' discussions. Follow these guidelines for good communication:

1. Speak calmly and clearly.
2. Use "I" messages.
3. Show respect and empathy.
4. Listen carefully, and ask appropriate questions.

LESSON 2 ASSESSMENT

 After You Read

Reviewing Facts and Vocabulary

1. How can family members demonstrate good communication?

2. What are the two main types of changes that cause stress in families?

3. Identify three situations that can lead to a change in family structure.

Thinking Critically

4. **Analyze.** Compare and contrast the difficulties sole custody and joint custody can pose for teens whose parents are divorced.

5. **Synthesize.** Give an example of a positive or negative event that could cause stress within a family. Explain what strategies the family might use to deal with this stress.

Applying Health Skills

6. **Stress Management.** Think of a stressful family situation. Then list five stress-management techniques that can help you handle this stress.

Writing Critically

7. **Narrative.** Children sometimes go through stages of grief (denial, anger, bargaining, depression, and acceptance) in response to their parents' divorce. Write a story about a teen who goes through several of these stages. Describe how the teen expresses his or her feelings at each stage.

Real Life Issues ....

After completing the lesson, review and analyze your response to the Real Life Issues question on page 172.

📖 GUIDE TO READING

BIG Idea *Families may require outside assistance to deal with serious problems.*

Before You Read

Create a Word Web.
In the center of a sheet of paper, write the phrase "Sources of Support." As you read, add information about sources of help for families in trouble.

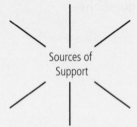

Sources of
Support

New Vocabulary

▸ abuse
▸ domestic violence
▸ spousal abuse
▸ child abuse
▸ neglect
▸ elder abuse
▸ cycle of violence
▸ crisis center

Help for Families

Real Life Issues ••••••••••••••••••••••••••••••

Worried About a Friend. Mark is concerned about his friend Sofia. Sofia says her parents argue a lot, and she thinks her dad hits her mom. One evening, Mark called Sofia to ask her a question about a homework assignment. He heard her parents arguing in the background. Another time, Mark thought he saw Sofia crying at school.

Writing *Write a dialogue in which Mark encourages Sofia to confide in him and seek help. Make sure Mark uses good communication techniques.*

Violence in Families

Main Idea **Violence in families can cause lasting harm.**

All families have problems from time to time, and that's normal. In most cases, families can work through their problems with the help of good communication and mutual support. However, some problems are too serious for family members to handle on their own. One of the most dangerous problems a family can face is **abuse**, *the physical, mental, emotional, or sexual mistreatment of one person by another.* When abuse results in *acts of violence involving family members,* it is called **domestic violence**. You will learn more about the different forms of abuse in Chapter 9.

Spousal Abuse

Domestic violence or any other form of abuse directed at a spouse is called **spousal abuse**. This form of violence can occur in all kinds of families, regardless of income, ethnicity, or education level. Spousal abuse can involve physical or sexual violence as well as emotional abuse. Abusers may threaten or intimidate their victims and try to cut them off from family or friends.

Spousal abuse is a criminal act that can be prosecuted by law. However, this crime often goes unreported. Victims may blame themselves for their partners' abusive behavior, thinking that they somehow deserve the mistreatment. They may also be unwilling to tear the family apart by leaving an abusive spouse. Many fear they will be unable to support themselves or their children if they leave. In some cases, the abuser may threaten to hurt or kill the victim or their children if the spouse attempts to leave.

Child Abuse

Child abuse is *domestic abuse directed at a child.* It includes any action that harms or threatens a child's health and development. Like spousal abuse, child abuse can be physical, emotional, or sexual. Child abuse may also involve **neglect**, *the failure to provide for a child's basic needs.* Neglected children may lack adequate food, clothing, shelter, or medical support. Leaving children alone and unsupervised for long periods of time is also a form of neglect.

Parents who abuse their children don't always want to hurt them. Sometimes they simply don't know how to take care of children. Many abusive parents were abused themselves as children and don't know any other way for a family to function. Alcohol and drug abuse also increase the risk of violence in the home. Whatever the reasons behind it, abusing a child is always unacceptable and dangerous.

Figure 7.7 Abuse can harm children emotionally as well as physically. *What forms can child abuse take?*

Elder Abuse

Elder abuse, *the abuse or neglect of older family members,* is a growing problem that often goes unnoticed. Elder abuse can occur both within the family and in institutional settings such as nursing homes. Like children, older family members may suffer physical, emotional, and sexual abuse, as well as neglect. Elder abuse can also be financial. For instance, caregivers may take advantage of elders by manipulating or pressuring them into handing over control of their money and other assets.

Academic Vocabulary

domestic *(adjective):*
of or relating to the
household or the family

Effects of Abuse

Victims of **domestic** abuse may suffer physical injuries, such as bruises, burns, or broken bones. In the worst cases, physical abuse can lead to permanent injury or death. For many victims, however, the emotional scars left by abuse last even longer than the physical injuries. Victims often experience feelings of shame and worthlessness. Abused children may be anxious or depressed and have difficulty in school. Without treatment, abused children often grow up to become abusers themselves. The *pattern of repeating violent or abusive behaviors from one generation to the next* is known as the **cycle of violence**.

Children who live in abusive homes may try to escape by running away. Others are thrown out of their homes by an abusive parent or guardian. Many runaways and "thrown-aways" end up living on the street or in the company of predatory adults. They are at risk for drug problems, crime, and continuing physical or sexual abuse.

To avoid these risks, children suffering abuse at home need to seek help from an adult they can trust, such as a relative, teacher, medical professional, or religious adviser. The police can also connect these teens with social services that can help them. Short-term shelters, for instance, can provide a safe place to stay. "Drop-in" services can provide food, clothing, medical attention, and crisis counseling.

Stopping Domestic Abuse

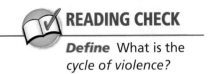

READING CHECK

Define What is the *cycle of violence?*

Stopping domestic violence depends on the three Rs: *recognize, resist,* and *report.* The first step is to *recognize* the problem. Victims and others need to be aware that child abuse and domestic violence are crimes. Any claim of abuse should be taken seriously, even if it sounds unbelievable.

Victims of domestic abuse can also *resist* their abusers. If someone tries to harm you, you can try to escape or to prevent the attack. Once you escape, seek help from a trusted adult.

However, resistance may not always be possible. That's why *reporting* the abuse is the third step in putting a stop to it. If you or someone you know is being abused, report the problem to someone who can help you. Try talking to a trusted adult, such as a family member or a school nurse. You can also contact an abuse hotline or a crisis center. Finally, you can go directly to the police. The victim may also require counseling and medical care.

Victims of domestic violence need help. Their abusers need help, too. Through counseling and other strategies, they can learn to manage their feelings and break the cycle of violence. You will learn more about sources of help for victims and abusers in Chapter 9.

TEENS Making a Difference

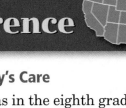

"It's your own choice to do good."

Taking Charge of a Family's Care

Ashleigh R., of Ohio, was in the eighth grade when her mother was diagnosed with multiple sclerosis (MS). Ashleigh responded by taking charge of not only her mother's care, but the care of her younger siblings as well. She also did all the housework and kept up with her studies at school.

Today, Ashleigh's mom is in remission. As a result of her mother's courageous battle, Ashleigh wants to become a doctor and find a cure for MS. After finishing high school, Ashleigh plans to attend Ohio State University, and then go on to medical school.

Through it all, Ashleigh has kept a positive attitude. She says, "It's your own choice to do good—it's all in your attitude toward life. Don't just go through life—go for the positive!"

Activity Write your answers to the following questions in your personal health journal.

1. What motivated Ashleigh to "take charge"?
2. List three ways you can help a family member who has a health problem.
3. How do you think maintaining a positive attitude has helped Ashleigh's health?

Sources of Support

Main Idea Communities offer many forms of support to families in crisis.

There are many community resources to help families deal with a variety of problems, including abuse. What type of help they need depends on the seriousness of the problem. Some problems, such as domestic violence, may require notifying the police. Others, such as substance abuse, may require medical help. Some sources of help for families facing difficulties include

- family counseling services.
- support groups.
- community services, such as shelters or hotlines.
- law enforcement officials.
- hospitals or clinics.
- faith communities.

Figure 7.8 **Support Groups**

These are just a few of the many support groups in the country.

Problem	Organization	Provides Support For
Substance abuse	Alcoholics Anonymous	Alcoholics
	Al-Anon	Family members and friends of alcoholics (subgroup, Alateen, is specifically for younger members)
	Narcotics Anonymous	Drug abusers
Eating disorders	Overeaters Anonymous	Compulsive overeaters
	Eating Disorders Anonymous	People with anorexia, bulimia, or binge eating disorder
Domestic violence	SAFE (Stop Abuse For Everyone)	Victims of domestic violence and abuse
Dealing with grief	Bereaved Parents of the USA	Parents who have lost a child

Counseling

Family counseling is therapy to restore healthy relationships in a family. Families come in as a group to meet with a counselor, discuss their problems, and seek solutions. Counseling can help some families deal with changes such as separation or divorce. It can also help in cases when one member has a problem that affects the entire family. Such problems may include anger, depression, or substance abuse. Sometimes individual counseling is also beneficial.

Support Groups

A support group is several people who are all coping with the same problem. The group meets regularly to discuss their problems and get advice from each other. Support groups can help many people just by reassuring them that they are not alone. **Figure 7.8** lists support groups that deal with various personal or family health issues.

Community Services

Families seeking help may also turn to resources in their community. Troubled family members may seek help from a **crisis center**, *a facility that offers advice and support to people dealing with personal emergencies.* People might turn to a crisis center to help them get through problems such as substance abuse or domestic violence. Some communities also have crisis hotlines. These are special telephone numbers people can call to receive help 24 hours a day.

READING CHECK

Identify Name three places that can provide help for families in crisis.

Communities also provide a variety of other services to families in need. For instance, public or private agencies may offer classes on parenting and conflict resolution. Social services can help provide food, clothing, shelter, and medical care. Public agencies can also help adults find a job or receive job training.

Finally, community services offer help for victims of domestic abuse. Social agencies can remove children from abusive homes and place them in foster care. Victims can also seek help by contacting an organization that deals with domestic violence. Many communities provide shelters where spouses and children can go to escape an abusive home. They may also help victims obtain counseling and legal services.

■ **Figure 7.9** In family counseling, the family meets with the counselor as a group to learn ways to resolve their problems. *Give an example of a problem that a family might seek to solve through family counseling.*

LESSON 3 ASSESSMENT

 After You Read

Reviewing Facts and Vocabulary

1. Identify four different forms of child abuse.

2. Describe the physical and emotional effects of abuse.

3. What is family counseling?

Thinking Critically

4. **Analyze.** Explain how neglect might affect each part of a child's health triangle.

5. **Evaluate.** Hector's dad recently moved out of the house. Hector feels lonely and guilty about his parents' separation. He believes no one understands how he feels. What source of support do you think would be most helpful for Hector, and why?

Applying Health Skills

6. **Accessing Information.** Consult phone directories, bulletin boards, and Web sites to find resources in your community that help families in crisis. Based on your findings, create a brochure that describes sources of support and how to contact them.

Writing Critically

7. **Expository.** Write an article discussing the problem of domestic abuse. Describe the effects of abuse and identify ways victims can seek help.

Real Life Issues ....

After completing the lesson, review and analyze your response to the Real Life Issues question on page 178.

Hands-On HEALTH

Activity Healthy Families in the Future

Congratulations! You are going to create your future healthy family! During this activity, you will identify characteristics that you believe are necessary for a healthy family. Consider the characteristics of a strong family, including traits that promote physical, social, and mental/emotional health.

What You'll Need

- 1 sheet of paper
- pen or pencil

What You'll Do

Step 1

Review Chapter 7 and identify ten characteristics that you believe are necessary for a physically, socially, and mentally/emotionally healthy family. Write each characteristic on your sheet of paper.

Step 2

Rank the characteristics from 1 to 10 with number 1 being the most important.

Step 3

Compare and contrast your list to another classmate's list, giving evidence to support your top five characteristics.

Apply and Conclude

Identify lifestyle choices and behaviors you can do now to help you achieve your healthy family in the future.

Checklist: Self-Management Skills

☑ Demonstrate healthful behaviors, habits, and techniques

☑ Identify protective behaviors (such as first-aid techniques, safety steps, or strategies) to help you avoid and manage unhealthy or dangerous situations

☑ List steps in correct order

LESSON **1**

Healthy Family Relationships

Key Concepts

▶ Your relationships with family members have a strong influence on your total health.

▶ Family members may be related by birth, marriage, or adoption.

▶ All members of a family share responsibility for the family's health.

▶ Families promote physical health by meeting basic physical needs, providing medical care, setting limits on behavior, and teaching health skills.

▶ Family members promote mental and emotional health by giving each other love, support, and affirmation.

▶ Families promote social health by teaching social skills, instilling values, and sharing cultural traditions.

Vocabulary

▶ siblings (p. 167)
▶ nuclear family (p. 167)
▶ blended family (p. 167)
▶ extended family (p. 167)
▶ foster care (p. 167)
▶ role (p. 168)
▶ affirmation (p. 170)

LESSON **2**

Strengthening Family Relationships

Key Concepts

▶ Strong families demonstrate good communication, love and support, respect, commitment, and trust.

▶ Changes in family structure or circumstances can be a major source of stress within families.

▶ Family members can help each other cope with change by talking about their feelings and offering help and support.

Vocabulary

▶ stress (p. 173)
▶ separation (p. 174)
▶ divorce (p. 174)
▶ custody (p. 174)

LESSON **3**

Help for Families

Key Concepts

▶ Abuse in families can be physical, emotional, or sexual.

▶ Victims of abuse include spouses, children, and older relatives.

▶ Families in crisis can seek support from counselors, support groups, crisis centers, and other community services.

Vocabulary

▶ abuse (p. 178)
▶ domestic violence (p. 178)
▶ spousal abuse (p. 178)
▶ child abuse (p. 179)
▶ neglect (p. 179)
▶ elder abuse (p. 179)
▶ cycle of violence (p. 180)
▶ crisis center (p. 182)

LESSON 1

Vocabulary Review

Correct the sentences below by replacing the italicized term with the correct vocabulary term.

1. A(n) *single-parent family* consists of a married couple and their children from previous marriages.

2. Two parents and one or more children living in the same place form a(n) *extended family.*

3. *Adoption* is the temporary placement of children in the homes of adults who are not related to them.

Understanding Key Concepts

After reading the question or statement, select the correct answer.

4. Relatives such as aunts, uncles, and grandparents are part of a person's
   a. nuclear family.
   b. blended family.
   c. extended family.
   d. foster family.

5. In a family, children are often responsible for
   a. meeting the family's basic needs, such as food and shelter.
   b. setting limits on family members' behaviors.
   c. teaching values and skills.
   d. performing household chores.

6. Parents promote their children's mental and emotional health by
   a. providing for basic needs, such as food, clothing, and shelter.
   b. providing medical care.
   c. providing affirmation.
   d. sharing cultural traditions.

Thinking Critically

After reading the question or statement, write a short answer using complete sentences.

7. **Describe.** What is one purpose of foster care?

8. **Explain.** How can setting limits on children's behavior promote physical health?

9. **Give Examples.** Name two healthful behaviors children may learn from their parents.

10. **Evaluate.** Why might teaching values by example be more powerful in some cases than teaching by explanation?

LESSON 2

Vocabulary Review

Use the vocabulary terms listed on page 185 to complete the following statements.

11. During a(n) _____, a couple may attempt to work out their problems so that they can live together again.

12. A(n) _____ is a legal end to a marriage contract.

13. After a divorce, sole or joint _____ of the children may be granted to one or both parents.

Understanding Key Concepts

After reading the question or statement, select the correct answer.

14. Helping a younger sibling with a difficult school assignment is an example of
   a. good communication.
   b. support.
   c. respect.
   d. trust.

15. Which of the following is an example of a change in family structure?
   a. The birth of a new baby
   b. The loss of a parent's job
   c. A family member's serious illness
   d. Moving to a new home

16. Joint custody is an arrangement in which
   a. the children live with their mother.
   b. the children live with their father.
   c. both parents share responsibility for the children.
   d. the children are placed in foster care.

17. Which of the following is *not* a helpful way to cope with changes in the family?
   a. Talking openly with other family members
   b. Making more of an effort to help out with chores and other responsibilities
   c. Keeping feelings to yourself to avoid worrying family members
   d. Showing empathy for family members' feelings

Thinking Critically

After reading the question or statement, write a short answer using complete sentences.

18. **Describe.** What are five traits of a healthy family?

19. **Compare and Contrast.** Explain how families in movies and TV shows may differ from real families.

20. **Infer.** Why might a divorced parent's remarriage cause mixed feelings for a teen?

21. **Evaluate.** Why can financial gains, as well as losses, be a source of stress for families?

LESSON 3

Vocabulary Review

Choose the correct term in the sentences below.

22. *Cycle of violence / Abuse* is the physical, mental, emotional, or sexual mistreatment of one person by another.

23. Any act of violence involving family members is known as *domestic violence / spousal abuse.*

24. Child *violence / neglect* is the failure to provide for a child's basic needs.

Understanding Key Concepts

After reading the question or statement, select the correct answer.

25. Yelling at or threatening a child is an example of
   a. physical abuse.
   b. emotional abuse.
   c. sexual abuse.
   d. neglect.

26. Older family members are much more likely than young children to suffer
   a. physical abuse.
   b. sexual abuse.
   c. emotional abuse.
   d. financial abuse.

27. If a friend confides that he is being abused, you should
   a. assume the person is just exaggerating.
   b. confront the abuser face-to-face.
   c. keep quiet for fear of putting the victim at further risk.
   d. seek help from a trusted adult.

28. Which type of organization can provide families in need with food, shelter, and medical care?
    a. Counseling services
    b. Support groups
    c. Crisis hotlines
    d. Social services

Thinking Critically

After reading the question or statement, write a short answer using complete sentences.

29. **Analyze.** Explain why some victims of spousal abuse are unwilling to leave their abusers.

30. **Explain.** Why are people who were abused as children more likely to become abusive parents?

31. **Evaluate.** James lives with an abusive, alcoholic parent. He has considered running away from home. What are the possible consequences he might face if he does so?

32. **Compare and Contrast.** What is the main difference between counseling and support groups as a way to deal with family problems?

33. **Explain.** How can community services offer help and support for victims of domestic abuse? Give specific examples.

Technology PROJECT-BASED ASSESSMENT

Coping During Times of Stress

Background

A family is a team. For a family to work as a single unit, everyone needs to communicate clearly and carry out their responsibilities. Successful families care for, support, and help each other.

Task

Create a blog about a fictional family. This family just survived a natural disaster, such as a hurricane or tornado. Some family members live in other areas that were not affected by the disaster. The family members are working together and supporting each other through this difficult time.

Audience

Students in your class

Purpose

Help students learn how family support is especially important in times of stress.

Procedure

1. Organize into small groups. Review the information in Chapter 7 about family relationships.

2. Conduct an online search on families that have survived natural disasters. How does each family member function independently and as part of a group? Obtain examples of how they support each other.

3. Identify four or five key points to make in the blog.

4. Work together to create a blog about a fictional family supporting each other after a natural disaster. Have each member of your group play the role of a family member. Make sure key points are clearly explained and supported.

Math Practice

Interpret Statistics. The chart below provides marriage and divorce statistics for the U.S. population in 2004 and 2005. Use the statistics to answer Questions 1–3.

Number of marriages in 2004:	2,279,000
Number of marriages in 2005:	2,230,000
Marriage rate in 2004:	7.8 per 1,000 people
Marriage rate in 2005:	7.5 per 1,000 people
Divorce rate in 2004:	3.6 per 1,000 people
Divorce rate in 2005:	3.7 per 1,000 people

Divorce rates include only 46 states and D.C. Adapted from "Births, Marriages, Divorces, and Deaths: Provisional Data for 2005, table A," *National Center for Health Statistics,* July 2006.

1. By how much did the number of marriages decrease from 2004 to 2005?
   A. 3,000
   B. 30,000
   C. 49,000
   D. 79,000

2. What proportion could be used to estimate the total U.S. population in 2005?
   A. $1,000 - 7.5 = x - 2,230,000$
   B. $1,000 - 7.5 = 2,230,000 - x$
   C. $7.5/1,000 = x/2,230,000$
   D. $7.5/1,000 = 2,230,000/x$

3. According to the marriage statistics, which figure best estimates the total U.S. population in 2005?
   A. 16,725
   B. 223,000,000
   C. 297,300,000
   D. 16,725,000,000

Reading/Writing Practice

Understand and Apply. Read the passage below, and then answer the questions.

There are four people in my family: me, my mom, and my two older sisters. My mom adopted all three of us when I was very young. I don't remember my birth parents, so this is the only family I've ever known.

Like any other family, we get into arguments sometimes. But we also take care of each other. My sisters help me with homework, and my mom is always there for us.

My mom thinks it's important for my sisters and me to be in touch with our Korean heritage. Even though she's not Korean, she learned to make Korean foods for us. Every year we go to the local heritage festival to celebrate our traditions as a family. It makes me feel valued to know that Mom respects our heritage and doesn't want us to change.

1. In the first sentence, the part after the colon should be changed to read:
   A. me and my mom and my two sisters.
   B. my two older sisters and I and my mom.
   C. my mom, my two older sisters, and me.
   D. I, my mom, and my two older sisters.

2. The author mentions arguments to show that
   A. his family is not healthy.
   B. he gets along with his mother but not with his sisters.
   C. his family is supportive and caring.
   D. his family is much like any other.

3. Think about a cultural tradition that you and your family share. Write a short essay describing this tradition and how it helps bring you together as a family.

National Education Standards

Math: Measurement and Data, Statistics
Language Arts: LACC.910.L.3.6

CHAPTER 8

Peer Relationships

Lesson 1

Safe and Healthy Friendships

BIG Idea *Mutual respect and honesty are important characteristics of healthy friendships.*

Lesson 2

Peer Pressure and Refusal Skills

BIG Idea *Learning effective refusal skills will help you deal with negative peer pressure.*

Lesson 3

Practicing Abstinence

BIG Idea *Setting dating limits and practicing abstinence will benefit all three sides of your health triangle.*

Activating Prior Knowledge

Using Visuals The teens in this picture are friends who share similar interests. Write five sentences, each beginning with the words *A true friend is someone who . . .* Discuss your ideas of what characteristics make a true friend.

BananaStock/Jupiterimages

Chapter Launchers

Health in Action

Discuss the **BIG** Ideas

Think about how you would answer these questions:

▶ Who are your peers?

▶ Why are peer relationships important?

▶ How can peer relationships affect your health?

Assess Your Health

Read each statement. On a separate sheet of paper, write "yes," "sometimes," or "no" based on your typical behavior.

1. I support and encourage my friends and peers.

2. I manage feelings of jealousy and envy in healthful ways.

3. I practice refusal skills to help me deal with negative peer pressure.

4. I set limits regarding relationships.

5. I practice abstinence and avoid risky behaviors.

A "yes" response shows that you practice healthy behaviors. "Sometimes" indicates that you should analyze and possibly modify your behavior. A "no" response means that you should modify the behavior.

Safe and Healthy Friendships

GUIDE TO READING

BIG Idea *Mutual respect and honesty are important characteristics of healthy friendships.*

Before You Read

Create a Cluster Chart. Draw a circle and label it "Friendship." Use surrounding circles to define and describe this term. As you read, continue filling in the chart with more details.

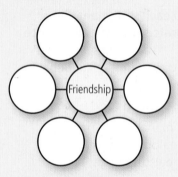

New Vocabulary

▸ platonic friendship
▸ clique

Review Vocabulary

▸ peers (Ch.1, L.2)
▸ friendship (Ch.6, L.1)
▸ prejudice (Ch.6, L.2)
▸ stereotypes (Ch.6, L.2)

Real Life Issues

Maintaining Friendships. Tom and Jarod have been friends since the sixth grade. They promised to join the same clubs and sports teams in high school to stay close friends. Now that they're sophomores, Tom is meeting new friends, and his interests have changed. He wants to try new things but wants to remain friends with Jarod, too.

Writing *If you were Tom, how might you express your concerns? Write a brief letter to Jarod explaining your thoughts and feelings.*

Peer Relationships

Main Idea We will all have many types of friends.

During adolescence, you continue to develop and strengthen your personal identity. The development of your identity will be influenced by many factors, including your peers. Peers are people of similar age who share similar interests. Peer relationships can play an important role in your health and well-being. Your friends and peers may influence you to try new activities, such as joining the debate club or learning to play tennis. These activities, in turn, can promote all aspects of your health.

As you get older, your social groups expand. You may also get a part-time job where you'll meet new people. These opportunities to meet people from different age groups, cultures, races, and religions contribute to your social development. Some of the people you meet during your high school years may become lifelong friends.

Friendships

You will form many kinds of friendships throughout your life. As you learned in Chapter 6, a friendship is a significant relationship between two people. Friends not only enjoy spending time together, they also care for, respect, trust, and show consideration for each other. They also share interests, hobbies, and other friends. Friendships have several common **attributes**:

- Similar values, interests, beliefs, and attitudes
- Open and honest communication
- Sharing of joys, disappointments, dreams, and concerns
- Mutual respect, caring, and support
- Concern about each other's safety and well-being

You probably have several types of friendships, including casual, close, and platonic friends. With the widespread use of the Internet today, many teens are also forming online friendships.

Casual and Close Friendships A casual friend is someone with whom you share interests but not deep emotional bonds. As you get to know a casual friend better, your relationship may develop into a close friendship. Close friends have strong emotional ties to each other. You feel comfortable sharing your thoughts, feelings, and experiences with a close friend.

When something is bothering you, a close friend offers support and encouragement. She or he listens to your concerns without passing judgment. Close friends also feel comfortable talking about problems that may arise in the friendship.

©Image Source/PunchStock

Academic Vocabulary

attribute *(noun):* a quality or characteristic

■ **Figure 8.1** These teens have a casual friendship based on a common interest. *What interests do you share with the peers you think of as casual friends?*

Platonic Friendships Your friends can include both males and females. A **platonic friendship** is *a friendship with a member of the opposite gender in which there is affection, but the two people are not considered a couple*. Platonic friendships can help you understand and become comfortable with members of the opposite gender.

Online Friendships The Internet has created opportunities for new kinds of friendships. Online friendships can be rewarding because you can get to know people in other parts of the world and learn about other cultures and traditions.

Social networking sites have created a great way to interact with others. Online friendships, however, can be dangerous. For example, online friends may not be truthful. A person who claims to be a teen may really be an adult. These sites can expose you to people who may harm you, including sexual predators. When communicating with online friends, keep these guidelines in mind:

- Don't share personal information or pictures of yourself.
- Don't offer your phone number or street address.
- Never arrange a face-to-face meeting.
- Always tell a trusted adult if an online friend suggests you do something that makes you feel uncomfortable.

Building Strong Friendships

Main Idea Good friends offer loyalty, support, and motivation.

As friends grow closer and share more serious thoughts and feelings, friendships may become complex. It's natural for friendships to grow and change, but always remember that healthy relationships are based on:

- mutual respect
- caring
- honesty
- commitment

Friends can reinforce your values and motivate you. Additional traits of a positive friendship include the following:

- **Empathy.** Does your friend consider your needs and feelings? Does she or he demonstrate understanding?
- **Fairness.** Does your friend treat you fairly?
- **Shared interests.** Do you enjoy the same things?
- **Acceptance.** Do you and your friend accept and appreciate each other's differences?
- **Support.** Does your friend support you during difficult times?
- **Loyalty.** Does your friend keep your confidences? Does he or she stay true to your friendship?

FITNESS ZONE

My friends and I aren't "sports nuts," but we want to be more active. We asked our PE teacher to suggest some activities. He said we should try noncompetitive activities like walking in the park, playing a round of miniature or disc golf, or just hitting tennis balls (but not keeping score). For more physical activity ideas, visit the Fitness Handbook and Fitness Zone sites in ConnectEd.

SW Productions/Getty Images

■ **Figure 8.2** Friendships can contribute positively to your well-being and enrich your life. *Identify some qualities of strong and healthy friendships.*

Recognizing Problems in Friendships

Main Idea It's important that you know how to recognize problems in a friendship and how to resolve those problems.

Friendships can have a positive or negative effect on you. They are positive when they offer support and encouragement. They have a negative effect if they influence you to engage in harmful activities. Cyberbullying is an example of how some friendships can be harmful. You will learn more about this in Chapter 26. To avoid unhealthy friendships, you need to recognize and resolve problems that arise.

Cliques

A **clique** is *a small circle of friends, usually with similar backgrounds or tastes, who exclude people viewed as outsiders.* Often members of a clique share interests, dress similarly, and behave in the same way. Being part of a clique may provide members with a sense of belonging. However, cliques may discourage individual members from thinking for themselves or acting as individuals.

Sometimes, clique members may **exclude** others by showing prejudice. They may make assumptions or judgments about an individual without really knowing him or her. These judgments may include stereotypes, exaggerated or oversimplified beliefs about an entire group of people, such as an ethnic or religious group or a gender.

Academic Vocabulary

exclude *(verb):* to prevent or restrict the entrance of

Managing Feelings of Envy or Jealousy

Another problem that can occur in friendships is envy or jealousy. Such feelings may arise if one friend compares himself or herself to another friend. Envy and jealousy can harm a friendship. If you feel jealous of a friend, ask yourself the following questions:

- What is making me feel jealous?
- Is my friend deliberately trying to make me feel this way?
- What can I do to manage or reduce these feelings? How can I feel better about myself?
- Are these feelings of jealousy more important than our friendship?
- What positive qualities make this person a good friend?

To overcome feelings of envy or jealousy, remind yourself of your unique talents and the positive aspects of your life. Friendships can survive the occasional feelings of jealousy if individuals focus on the reasons why they became friends.

When Friendships Change

As you grow older, you and a close friend might spend less time together and develop new interests. When close friends grow apart, the friendship may become casual. Other times, you may decide to end a friendship because it is becoming harmful. Here are some reasons for ending a friendship:

- A friend pressures you to do something that is unsafe or goes against your values.
- A friend says hurtful and insulting things to you.
- A friend constantly tries to get you to change your beliefs or actions.

If you decide to end a friendship, communicate your feelings to that friend in a clear and respectful way. Use "I" messages to explain your feelings and reasons for ending the friendship. Let your friend respond with his or her point of view. Although it may be difficult, sometimes ending a friendship is the best decision for both individuals. Remember, you can always talk to another friend or ask a trusted adult for advice about dealing with these situations.

READING CHECK

Explain What is one way of dealing with feelings of jealousy in a friendship?

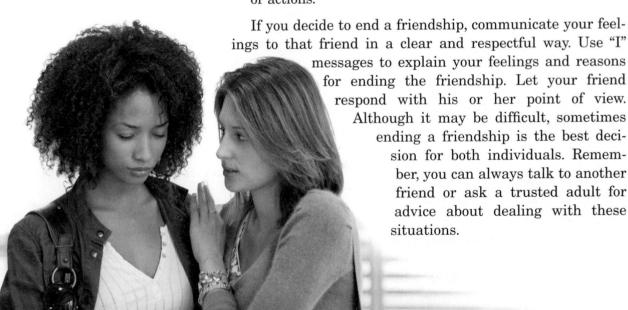

■ **Figure 8.3** Talking with a trusted friend can help you deal with difficulties in other peer relationships. *What are some other strategies for handling problems in friendships?*

Health Skills Activity

Communication Skills

When Friendships Change

Allen saw Dave on the way to school. "Dave, what are you doing after school? I'm going out with Eric's group tonight, and they invited you along."

"Sorry, I've got a project to finish," said Dave. "Why do you want to hang out with those guys, anyway? They cause a lot of trouble."

"They're fun to be around," Allen said. Then he added angrily, "Why are you so uptight?"

Dave looked at Allen. They had been friends since grade school, but lately Allen seemed to lose interest in doing the things that they used to enjoy together. Allen quit the tae kwon do class and the math study group they were in. Now Allen was spending time with a rough crowd and getting into trouble at school. Dave wondered how he should respond.

Writing Write an ending to the scenario in which Dave communicates to Allen that he thinks their friendship is changing. Make sure Allen knows that Dave has decided to pursue healthier friendships. Use the following tips as a guideline.

1. Use "I" messages.
2. Speak calmly and clearly.
3. Use a respectful tone.
4. Listen carefully and ask appropriate questions.
5. Use appropriate body language.

LESSON 1 ASSESSMENT

 After You Read

Reviewing Facts and Vocabulary

1. Define the word *peers*.
2. Define *friendship*. Identify four traits of healthy friendships.
3. List two problems that may affect friendships.

Thinking Critically

4. **Evaluate.** What actions can you take to promote safe and healthy friendships?
5. **Describe.** Name two possible outcomes of lying to a friend. How might this affect the friendship?

Applying Health Skills

6. **Communication Skills.** With a classmate, role-play a scenario in which close friends communicate needs, wants, and emotions in healthful ways.

Writing Critically

7. **Expository.** Write a dialogue in which peers express disagreement about an issue while still showing respect for self and others.

Real Life Issues ....

After completing the lesson, review and analyze your response to the Real Life Issues question on page 192.

©BananaStock/PunchStock

Peer Pressure and Refusal Skills

GUIDE TO READING

BIG Idea *Learning effective refusal skills will help you deal with negative peer pressure.*

Before You Read

Create Vocabulary Cards. Write each new vocabulary term on a separate note card. For each term, write a definition based on your current knowledge. As you read, fill in additional information related to each term.

Peer Pressure

New Vocabulary

▶ peer pressure
▶ harassment
▶ manipulation

Review Vocabulary

▶ assertive (Ch.6, L.3)
▶ refusal skills (Ch.2, L.1)
▶ passive (Ch.6, L.3)
▶ aggressive (Ch.6, L.3)

Real Life Issues

Peer Pressure. Kelly failed her driving test today. All of her friends have their driver's licenses. She doesn't want to go back to school where she will have to explain to her friends what happened. Kelly worries that some of her friends will make fun of her. She is tired of being dropped off at school by her mom while all of her friends have their own cars.

Writing *Kelly is experiencing unspoken peer pressure. Write a paragraph about a time when you experienced either spoken or unspoken peer pressure. How did you handle the situation?*

Peer Pressure

Main Idea Peers can influence how you think, feel, and act.

You are hanging out with friends on a Friday night when someone suggests going to a movie. Everyone agrees, but you aren't sure because the movie ends late. If you stay for the entire movie, you'll miss your curfew. "Come on! It's the weekend!" your friends say. You hesitate, trying to decide what to do.

During the teen years, it's common to experience pressure situations like the one presented above. How you respond to these situations can impact your health and safety. That's why it's important to learn ways to handle pressure from peers. *The influence that people your age may have on you* is called **peer pressure**. Peer pressure can have a positive or negative influence on your actions and behaviors. Evaluating forms of peer pressure and developing strategies for responding to it will help you maintain healthy relationships.

Paul Burns/Getty Images

■ Figure 8.4 Positive peer pressure can motivate you to try new activities that can benefit all sides of your health triangle. *What are some examples of positive peer pressure that you have experienced?*

Positive Peer Pressure

Peers can influence you in many positive ways. Your peers might inspire you to try a new activity, like an art class, or to try ethnic foods that you've never tasted before. They may also encourage you to participate in community projects, such as a cleanup campaign. Agreeing to work with your peers on a volunteer project benefits your social health because you have the opportunity to interact with others in a positive way. It also benefits the community by providing a cleaner environment. Volunteering to serve food at a homeless shelter or working at a Special Olympics event because a friend does are other examples of positive peer pressure.

Sometimes, positive peer pressure involves *not* participating in risky behaviors or activities. For instance, having friends who do not use tobacco, alcohol, or other drugs may positively influence you to avoid these harmful substances. You can also use positive peer pressure to influence others in healthful ways. You might encourage a peer to try out for the softball team or to study hard for an important test.

Negative Peer Pressure

Peers sometimes pressure others to take part in behaviors or accept beliefs with negative consequences. The members of a clique, for example, may be disrespectful toward people they do not consider acceptable to their group. Such behavior may involve **harassment**, or *persistently annoying others*. Harassment may include hurtful behaviors such as name-calling, teasing, or bullying.

READING CHECK

List What are the differences between positive and negative peer pressure?

Negative peer pressure may also lead some teens to engage in behaviors that go against their values. For instance, a peer might pressure a classmate to help him or her cheat on a test.

Another way that some people exert negative peer pressure is through **manipulation**. This is *an indirect, dishonest way to control or influence other people*. Take a look at **Figure 8.5**, which lists some examples of how people manipulate one another. It is important to discourage this kind of hurtful behavior and to encourage the victim to report the problem to a trusted adult.

Resisting Negative Peer Pressure

Main Idea Practicing refusal skills will help you deal with negative peer pressure.

Peer pressure does not stop at the end of your teen years. Throughout your life, you will experience instances in which peers, including friends and co-workers, try to influence you to behave in a particular way. They might even make direct requests or demands of you.

In some cases, your responses to these siutations will directly affect your health. For example, getting into a car with friends who have been drinking can lead to serious injury or even death. To protect your health and safety, you need to learn effective strategies for resisting negative peer pressure.

One way to resist negative peer pressure is to develop friendships with people who share your values and interests. Friends who have respect for your health and well-being will be less likely to pressure you into doing something that goes against your values. You will also find that it is much easier to resist negative peer pressure when you have supportive friends who stand by you and respect your decision.

Figure 8.5	**Common Methods of Manipulation**

- **Making threats**—promising violence or some other negative consequence if the person does not do what is asked
- **Blackmail**—threatening to reveal some embarrassing or damaging information if the person does not do what is asked
- **Mocking or teasing**—making fun of another person in mean or hurtful ways
- **"Guilt trips"**—making a person feel guilty to get desired results
- **Bargaining**—offering to make a deal to get what one wants
- **Flattery**—using excessive praises to influence another person
- **Bribing**—promising money or favors if the person does what is asked

TEENS Making a Difference

Saving Her Future

"We should be better people and live better lives."

Shawntae B., of Louisiana, is president of Students Against Violence Everywhere (SAVE). "If a violent act happened to me, I would want my friends to do something about it. So I joined SAVE to help reduce violence in my school and community. We've helped clean the school and pick up trash in and around the building. These tasks help to make the community look nice which helps to reduce violence."

Volunteers with SAVE also visit a women's and men's penitentiary. They see how the prisoners live and the freedoms that are lost. "We know that we should be better people and live better lives because we don't want to end up in that situation." Shawntae believes SAVE volunteers are helping the community. "As a member of SAVE, you are putting your community first and making it a better place to live in."

Activity Write your answers to the following questions in your personal health journal.

1. What is purpose of SAVE?
2. How does cleaning up the environment help to reduce violence?
3. What three things could you specifically do in your school or community to help reduce violence?

Sometimes, however, the pressure to participate in unsafe or potentially harmful activities can be difficult to resist. When pressured by a friend, many teens worry jeopardizing the relationship. They may agree to an activity that goes against their values in an attempt to maintain a friendship or to make new friends.

Another concern that teens have is that refusing to go along with a group may make them appear "uncool." They fear that peers will tease or make fun of them for their decision. Even though these situations may be difficult, it is important to remain firm and stay true to yourself.

Remember, you have the responsibility to make decisions that have the best possible effect on your well-being. In making decisions that involve potentially risky consequences, your health and safety come first. To help you say "no" to the pressure to participate in an activity that is unsafe or goes against your values, practice refusal skills. Rehearsing assertive refusals will make it easier to say no when pressure situations arise.

■ **Figure 8.6** Practicing refusal skills will help you deal with negative peer pressure. *How would you say no if someone pressured you to participate in an unsafe activity?*

Assertive Refusal

When you practice assertive communication, you state your position and stand your ground while acknowledging the rights of others. This is the most effective approach when facing negative peer influences.

Refusal Skills An important aspect of being assertive is the ability to demonstrate appropriate refusal skills. Refusal skills are communication strategies that can help you say no when you are urged to take part in behaviors that are unsafe or unhealthy, or that go against your values. Effective refusal skills involve a three-step process:

- **Step 1: State Your Position.** The first step in resisting negative peer pressure is to say no. You need to state your position simply and firmly. When you say no, make sure you really mean it. Combining your words with nonverbal messages, such as those shown in **Figure 8.7**, will make your statement more effective. Having said no, give an honest reason for your response. Your reason may be as simple as, "It goes against my values." Offering a legitimate reason will help strengthen your refusal.

- **Step 2: Suggest Alternatives.** When a peer asks you to take part in an activity with which you are uncomfortable, try suggesting another activity. For example, if a friend wants to go to a party where there is no adult supervision, you might say, "No, let's go to the movies instead." By offering an alternative, you create an opportunity to spend time with your friend in a way that makes you comfortable. Keep in mind that your suggestion is most effective if it takes you away from the dangerous or unpleasant situation.

- **Step 3: Stand Your Ground.** Even after you refuse, some peers may continue trying to persuade you to join in. Make it clear that you mean what you said. Use strong body language and maintain eye contact, but do not touch the other person or become physical in any way. If this doesn't work, remove yourself from the situation. Simply say, "I'm going home," and walk away.

When you're faced with negative peer pressure, refusal skills can help you avoid unsafe situations. Learning and practicing these steps will help you deal with high-pressure situations in a way that keeps you safe and healthy. Knowing that you made the decision to protect your safety and uphold your values will make you feel good about resisting negative peer pressure.

Figure 8.7 Body Language and Assertive Refusal

Reinforce the meaning of your words with appropriate body language.

Shaking your head is one way to communicate no.

Raising your hands in a "Stop" or "No way" signal tells others that you are not interested.

If the other person continues to pressure you, you can walk away from the situation.

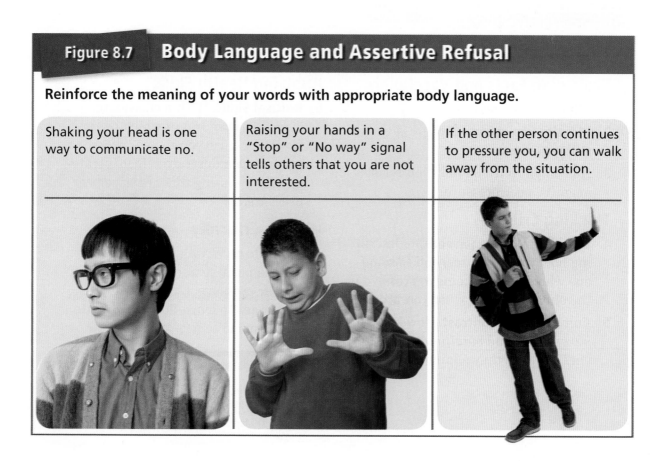

Passive and Aggressive Responses

READING CHECK

Explain Why is it important to use assertive refusal skills, rather than using passive or aggressive responses?

Being assertive may take some practice. To some people, a passive response to negative peer pressure seems easier. Passive communicators are unwilling or unable to express their thoughts and feelings in a direct or firm manner. Teens who respond passively to peer pressure may believe they are making friends by going along. However, being passive may cause others to view them as pushovers who aren't worthy of respect.

Some people may feel more comfortable with an aggressive response. Such responses are overly forceful, pushy, or hostile. An aggressive way of resisting peer pressure may involve yelling, shouting, shoving, or other kinds of verbal or physical force. Aggressive people may get their way, but most people react to aggressive behavior by avoiding the individual or by fighting back. Either reaction can result in emotional or physical harm to both parties.

Practicing assertive communication is the most effective way to deal with peer pressure. Being assertive shows that you will stand up for your rights, beliefs, and needs. It shows that you respect yourself and those around you.

LESSON 2 ASSESSMENT

After You Read

Reviewing Facts and Vocabulary

1. What is *peer pressure*?
2. Identify two examples of manipulation.
3. How might a friend help you resist negative peer pressure?

Thinking Critically

4. **Describe.** Write a paragraph describing how you would respond to someone who says that being aggressive is the only way to get what you want.
5. **Compare and Contrast.** How are *harassment* and *manipulation* different? How are they similar?

Applying Health Skills

6. **Refusal Skills.** With a classmate, develop a scenario in which peers try to pressure you to use tobacco or alcohol. Demonstrate refusal strategies for resisting this negative peer pressure.

Writing Critically

7. **Expository.** Write an essay analyzing the positive and negative effects of peer pressure. Explain why it is important to learn how to evaluate and respond to peer pressure.

Real Life Issues ....

After completing the lesson, review and analyze your response to the Real Life Issues question on page 198.

Practicing Abstinence

Real Life Issues ....

Thinking About Dating. Kayla has a close group of friends. Dan, one of her good friends, recently told her that he wants to date her exclusively. Kayla knows that Dan's been sexually active in the past. She likes Dan but doesn't think she's ready for a serious relationship.

Writing *Write a dialogue in which Kayla expresses her feelings to Dan. Both individuals should be honest and respectful.*

Dating Decisions

Main Idea Personal values and priorities will influence your dating decisions.

During the teen years, you may start thinking about dating. Dating can be a great way to get to know another person. It also provides opportunities to develop social skills, discover new interests, and reaffirm personal values.

Some teens, however, may decide not to date for personal reasons. They might not feel ready or they may have other priorities. **Priorities** are *the goals, tasks, values, and activities that you judge to be more important than others.* Priorities can include focusing on school or spending time with family.

Talking to a parent or other trusted adult can help you decide if you're ready to date. If you decide to date, try to establish healthful dating expectations. Keep the following in mind:

- You and your date deserve to be treated with consideration and respect.

- Be yourself and communicate your thoughts and feelings honestly.

- Never feel pressured to do anything that goes against your values or your family's values.

BIG Idea *Setting dating limits and practicing abstinence will benefit all three sides of your health triangle.*

Before You Read

Create a K-W-L Chart. Make a three-column chart. In the first column, list what you **k**now about dating and abstinence. In the second column, list what you **w**ant to know about this topic. As you read, use the third column to summarize what you **l**earned.

K	W	L

New Vocabulary

▶ priorities
▶ intimacy
▶ infatuation
▶ self-control
▶ sexually transmitted diseases (STDs)

Review Vocabulary

▶ abstinence (Ch.1, L.3)

Setting Limits

Your parents or guardians may set limits regarding your dating relationships. Such limits are intended to protect your health and safety. For example, many parents insist on a curfew, a set time at which teens must be home at night. As you mature, you'll need to set your own limits. Your parents or guardians can guide you through this **process**. It's a good idea to set a limit on the age of the people you date. You'll also need to set limits with your date regarding where you will go, how you will get there, and what you will do when you get there. Setting limits and making them clear before a date will help ensure safe and positive dating experiences.

The most important limit you can set is to practice abstinence. As you learned in Chapter 1, abstinence is a deliberate decision to avoid high-risk behaviors, including sexual activity and the use of tobacco, alcohol, and other drugs. Choosing abstinence will safeguard your health and future.

Abstinence

Main Idea There are many strategies that can help you commit to abstinence.

By choosing abstinence from sexual activity, you are taking responsibility for your well-being. Abstinence does not mean doing without intimacy or physical contact in a close, special friendship. **Intimacy** is *a closeness between two people that develops over time*. You can still express affection and develop intimacy while practicing abstinence. For example, you can hold hands, hug, kiss, and share your thoughts, feelings, and dreams. Keep in mind that it's important not to confuse genuine affection and intimacy with **infatuation**, or *exaggerated feelings of passion*.

Academic Vocabulary

process *(noun):* a series of actions geared toward an end result

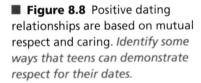

READING CHECK

Explain Why is it important to set dating limits?

■ **Figure 8.8** Positive dating relationships are based on mutual respect and caring. *Identify some ways that teens can demonstrate respect for their dates.*

Sexual Content on TV

Media messages can play an important role in a teen's decisions regarding sexual activity. Studies have shown that adolescents with higher exposure to sexual content on TV are more likely to engage in sexual activity. Consider these statistics:

▶ 64 percent of all television programs contain sexual content.

▶ Of programs with sexual content, 15 percent show abstinence or risk of sexual activity.

Analyzing media messages and comparing them to real-life situations is an important skill to develop. It will help you resist external pressures and stay committed to abstinence.

Activity Mathematics

Assume that 1,500 TV programs were surveyed.

1. How many programs had sexual content?
2. How many programs depicted abstinence or risk of sexual activity?
3. **Writing** Write a short essay describing how the higher rate of sexual content on TV influences teen behavior.

Concept Ratios and Proportional Relationships: Percents A percent is a ratio comparing a number to 100. It can also be represented as a fraction with 100 as the denominator. To find a decimal equivalent, divide the percent by 100. To convert a decimal to a percent, multiply it by 100.

Practicing abstinence requires planning and self-control. **Self-control** is *a person's ability to use responsibility to override emotions*. It's normal and healthy to have sexual feelings. You cannot prevent those feelings from occurring, but you can control how you react to those feelings. The following tips can help you maintain self-control and stay firm in your decision to practice abstinence:

• **Set limits for expressing affection.** Think about your priorities and set limits for your behavior before you are in a situation where sexual feelings may build.

• **Communicate with your partner.** Discuss your limits for expressing affection with your dating partner. Clear and honest communication will help your dating partner understand and respect your limits.

• **Talk with a trusted adult.** Ask a trusted adult, such as a parent or guardian, for suggestions on ways to manage your feelings.

• **Seek low-pressure dating situations.** Choose safe dating locations and activities. For example, attend parties only where an adult is present. Try group dating, which can eliminate the pressure to engage in sexual activity.

• **Date someone who respects and shares your values.** A dating partner who respects you and has similar values will understand your commitment to abstinence.

READING CHECK

Identify List three behaviors that can help you maintain self-control in a dating situation.

Avoiding Risk Situations

Some dating situations may increase your chances of being pressured to participate in sexual activity or other high-risk behaviors. Before you go on a date, know where you're going and what you will be doing. Find out who else will be there, and discuss with your parents or guardians what time they expect you home. Here are additional precautions:

- **Avoid places where alcohol and other drugs are present.** The use of alcohol or other drugs impairs judgment. People under the influence of these substances are more likely to engage in high-risk behaviors. Prevent such situations by not using alcohol or other drugs and by avoiding people who use these substances.

- **Avoid being alone with a date at home or in an isolated place.** You may find it more difficult to maintain self-control when you are home alone or in an isolated place with a date. These situations also increase the risk of being forced into a sexual act against your will.

Considering the Consequences

Main Idea Abstinence from sexual activity has a positive effect on all sides of your health triangle.

Sexual activity carries serious consequences. It is illegal for an adult to have sexual contact with someone under the age of consent. Consent laws, which vary from state to state, make it illegal for an unmarried minor to engage in sexual activity. For example, if a state's age of consent is 18, two 17-year-olds who engage in sexual activity would be breaking the law. Sexual activity can also harm a teen's physical, mental/emotional, and social health.

Effects on Physical Health

Many teens make the decision to practice abstinence because it is the only 100 percent effective method to eliminate health risks associated with sexual activity. These risks include unplanned pregnancy and sexually transmitted diseases. Also known as sexually transmitted infections (STIs), **sexually transmitted diseases (STDs)** are *infectious diseases spread from person to person through sexual contact.*

■ **Figure 8.9** Going out in a group can reduce some of the pressures of dating. *What are other benefits of group dates or double dates?*

Unplanned Pregnancy Every year in the United States, about one million teenage girls become pregnant. Female teens who have begun to ovulate are physically able to have babies. A pregnancy can result even if teens are engaging in sexual activity for the first time. A teen who becomes pregnant may not obtain the prenatal care that protects her life and that of the growing baby. Her partner may lack the maturity needed to support her during the pregnancy.

Sexually Transmitted Diseases Each year, about half of the diagnosed cases of STDs occur among teens and young adults between the ages of 15 and 24. Although many STDs can be treated and cured if diagnosed early, some STDs have no cure. If left untreated, some STDs can cause sterility in males and infertility in females. This means that a person may never be able to have a child. Other STDs, such as the herpes virus and HIV/AIDS, have no cure. In the case of AIDS, the disease can be fatal.

Effects on Mental/Emotional Health

In general, teens are not prepared for the emotional demands of a sexual relationship. Teens who engage in sexual activity before reaching emotional maturity may experience

- hurt because partners are not committed as in a marital relationship.
- guilt because teens are usually not truthful to their parents about being sexually active.
- loss of self-respect because sexual activity goes against personal and family values.
- regret and anxiety, if sexual activity results in an unplanned pregnancy or an STD.

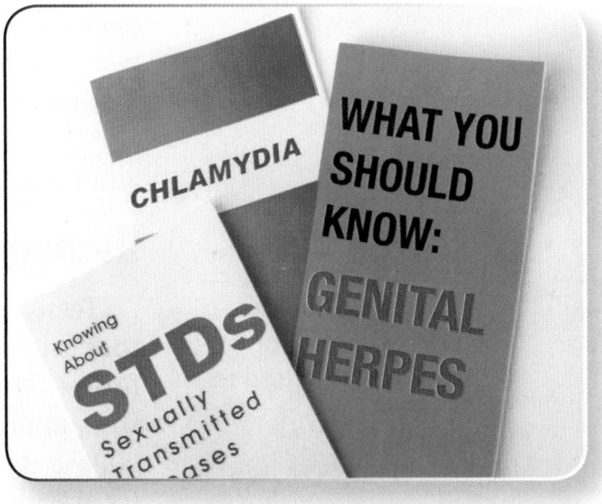

Effects on Social Health

Engaging in sexual activity can negatively affect a teen's relationships with other people. Sexually active teens may deprive themselves of the opportunity to pursue new interests or friendships. The decision to engage in sexual activity can also harm a teen's relationships with family members. Parents who discover that their teen is sexually active may express disappointment and worry. These feelings can cause tension in the family. In addition, teens who are sexually active risk an unplanned pregnancy. Teen parents face many challenges, such as providing financial and emotional support for their child. Teens who become parents may have to put their own education and career plans on hold.

READING CHECK

Recall What are two health risks associated with sexual activity?

■ **Figure 8.10** Learning the facts about STDs and other negative consequences of sexual activity will help you make informed dating decisions. *How might contracting an STD affect a teen's mental/emotional and social health?*

Ken Karp/McGraw-Hill Education

Committing to Abstinence

Main Idea Honest communication with your dating partner will help you stay committed to abstinence.

To stay firmly committed to abstinence, continue to remind yourself of the reasons that you chose abstinence. Make sure that you communicate your values and decisions to your dating partner. It can be difficult to talk about abstinence, but the following tips can help make the conversation go more smoothly:

- Choose a relaxed and comfortable time and place.
- Begin on a positive note, perhaps by talking about your affection for the other person.
- Be clear in your reasons for choosing abstinence.
- Be firm in setting limits in your physical relationship.

Using Refusal Skills

Committing to abstinence means not letting a partner, peers, or the media pressure you to do something you don't want to do. Practice the refusal skills that you learned in Lesson 2 to help you stand firm in your decision. Resist pressure to engage in sexual activity by using refusal statements similar to those shown in **Figure 8.11**.

Recommitting to Abstinence

Teens who have been sexually active in the past may feel that they cannot choose to abstain from sexual activity in the future. They may feel pressured to remain sexually active. It is important to understand that choosing abstinence is *always* an option regardless of past experiences. Returning to abstinence is a positive alternative to previous sexual behavior. Teens who recommit to abstinence will feel good about their decision to protect their health and well-being.

READING CHECK

Identify What are two strategies for staying committed to abstinence?

| Figure 8.11 | **Using Refusal Skills to Say No to Peer Pressure** |

Practicing effective refusal statements will help you resist the pressure to engage in sexual activity.

Pressure Line	Your Response
▶ Everybody does it.	▶ No. Not everybody is doing it.
▶ I thought you were cool.	▶ I *am* cool, and the answer's still no.
▶ No one will know.	▶ I'll know, and I'm the one who matters.
▶ If you loved me, you'd do it.	▶ If you loved me, you'd respect my decision.

■ **Figure 8.12** Careful consideration of the negative consequences associated with sexual activity will reinforce the decision to practice abstinence. *What are some of your reasons for practicing abstinence?*

LESSON 3 ASSESSMENT

 After You Read

Reviewing Facts and Vocabulary

1. How is *intimacy* different from *infatuation*?

2. What are three negative consequences of teen sexual activity?

3. Identify ways of resisting persuasive tactics regarding sexual involvement.

Thinking Critically

4. **Synthesize.** What are the benefits of practicing abstinence?

5. **Analyze.** How can teen parenthood harm an individual's social development?

Applying Health Skills

6. **Refusal Skills.** Write a scenario in which a teen is being pressured to engage in sexual activity. The teen should demonstrate effective refusal skills to resist the pressure.

Writing Critically

7. **Personal.** Write an essay describing what your life will be like in ten years. Include an explanation of how practicing abstinence will help you achieve your goals.

Real Life Issues ....

After completing the lesson, review and analyze your response to the Real Life Issues question on page 205.

Hands-On HEALTH

Activity — Assert Yourself

Learning to be assertive can help you maintain your commitment to a healthful lifestyle. By practicing assertiveness, you will find it easier to resist negative peer pressure and live according to your personal values. In this activity, you will role-play assertive communication skills.

What You'll Need

- large index cards
- pen or pencil
- paper

What You'll Do

Step 1

With a partner, think of a realistic scenario in which you are being pressured by one or more peers to do something against your values.

Step 2

Write your scenario on an index card, and then trade cards with another pair of students.

Step 3

Role-play the scenario you've received. Use the checklist on this page to make sure you include the elements of assertive communication.

Apply and Conclude

Write a short reflective paper describing how being assertive can help protect your physical, mental/emotional, and social health.

Checklist: Communication Skills

- ✓ "I" messages
- ✓ Respectful but convincing tone of voice
- ✓ Alternative to the action
- ✓ Clear, simple statement
- ✓ Appropriate body language

LESSON **1**

Safe and Healthy Friendships

Key Concepts

▸ The friendships you form will depend on several common attributes that you and your friends share.

▸ Friendships can change as you develop new interests and expand your social group.

▸ Using "I" messages when explaining your feelings to a friend is the best way to talk about problems in a friendship.

Vocabulary

▸ peers (p. 192)
▸ friendship (p. 193)
▸ platonic friendship (p. 194)
▸ clique (p. 195)
▸ prejudice (p. 195)
▸ stereotypes (p. 195)

LESSON **2**

Peer Pressure and Refusal Skills

Key Concepts

▸ Peer pressure can have a positive or negative influence on your actions and behaviors.

▸ You can resist negative peer pressure by learning to use refusal skills.

▸ Assertive refusal is a positive way to resist negative peer pressure.

▸ Passive and aggressive responses are not effective ways to handle negative peer pressure.

Vocabulary

▸ peer pressure (p. 198)
▸ harassment (p. 199)
▸ manipulation (p. 200)
▸ assertive (p. 202)
▸ refusal skills (p. 202)
▸ passive (p. 204)
▸ aggressive (p. 204)

LESSON **3**

Practicing Abstinence

Key Concepts

▸ Some teens may decide that they want to postpone dating.

▸ Group dating gives you the opportunity to socialize with the opposite gender without the pressure of one-on-one dating.

▸ Becoming sexually active can affect your physical, mental/ emotional, and social health.

▸ Abstinence from sexual activity is the best choice for teens.

Vocabulary

▸ priorities (p. 205)
▸ abstinence (p. 206)
▸ intimacy (p. 206)
▸ infatuation (p. 206)
▸ self-control (p. 207)
▸ sexually transmitted diseases (STDs) (p. 208)

Vocabulary Review

Correct the sentences below by replacing the italicized term with the correct vocabulary term.

1. When two teens of the opposite gender have affection for each other but are not considered a couple, they are said to have a *romantic friendship.*

2. A *peer* excludes people viewed as outsiders.

3. An exaggerated or oversimplified belief about a group of people is called a *judgment.*

Understanding Key Concepts

After reading the question or statement, select the correct answer.

4. Which of the following attributes is *not* necessary for a friendship to work?
   a. Mutual respect
   b. Concern about each other's safety
   c. Identical beliefs and values
   d. Open, honest communication

5. Which of the following statements is true of online friendships?
   a. They're not *real* friends unless you meet them face-to-face.
   b. It's okay to assume that people are who they say they are.
   c. You should exchange photos and personal information.
   d. They can be a rewarding way to meet people from around the world.

6. If you need to end a friendship, how should you handle the situation?
   a. Give the friend a detailed list of what exactly he or she did wrong.
   b. Give the friend the "silent treatment" until he or she gets the message.
   c. Talk about your own feelings and reasons for ending the friendship.
   d. Have someone else tell the friend that you no longer want to be friends.

Thinking Critically

After reading the question or statement, write a short answer using complete sentences.

7. **Identify.** Name one positive effect and one negative effect of belonging to a clique.

8. **Explain.** How would you tell a friend that your friendship has changed?

9. **Discuss.** Name ways that peers might influence your identity as a teen.

10. **Compare and Contrast.** What qualities do casual friends, close friends, and platonic friends share? How do these social groups differ?

Vocabulary Review

Use the vocabulary terms listed on page 213 to complete the following statements.

11. Name-calling and bullying are examples of _____.

12. When you stand up for your rights in a firm and positive way, you are being _____.

13. If you are urged to take part in unhealthy behaviors, _____ will help you say no.

Understanding Key Concepts

After reading the question or statement, select the correct answer.

14. Which of the following is an example of positive peer pressure?
    a. Offering friendship in exchange for a favor
    b. Encouraging friends to become volunteers at a homeless shelter
    c. Smoking cigarettes to win approval
    d. Persuading a friend to bully another teen

15. Which of the following behaviors does *not* use manipulation?
    a. Flattering a person to influence her actions
    b. Teasing a person in a hurtful way
    c. Using a "guilt trip" to get desired results
    d. Asking a person to tell you honestly what she thinks

16. Which of the following is *not* part of the three-step process of refusal skills?
    a. Try it once before saying no.
    b. Suggest alternatives.
    c. Stand your ground.
    d. State your position.

17. Shaking your head and raising your hand in a "Stop" signal are examples of:
    a. Peer pressure
    b. Manipulation
    c. Nonverbal assertive refusal
    d. Aggressive behavior

Thinking Critically

After reading the question or statement, write a short answer using complete sentences.

18. **Explain.** What are the risks of responding passively to peer pressure?

19. **Describe.** What behaviors do people use when responding aggressively to peer pressure?

20. **Analyze.** Suppose a group of friends constantly teases a student in your school. How can you show disapproval of this inconsiderate and disrespectful behavior?

21. **Evaluate.** Analyze the similarities and differences between passive, aggressive, and assertive forms of communication.

LESSON 3

Vocabulary Review

Choose the correct term in the sentences below.

22. *Responsibility / Abstinence* is a deliberate decision to avoid high-risk behaviors.

23. The ability to practice responsible behaviors even when you are faced with temptation is called *self-control / priority*.

24. *Infatuation / Intimacy* is the closeness that grows over time between two people who care about each other.

Understanding Key Concepts

After reading the question or statement, select the correct answer.

25. Which of the following statements is true?
    a. STDs can be cured with over-the-counter medications.
    b. Teens under the age of 18 are immune to STDs.
    c. Some STDs have no cure, and some can cause infertility or even death.
    d. The symptoms of all STDs go away after a few months.

26. Which of the following behaviors will help you maintain self-control while dating?
    a. Date someone who respects and shares your values.
    b. Return home from dates before midnight.
    c. Avoid dating someone who goes to your own school.
    d. Limit the number of parties you attend.

27. A person trying to pressure you into sexual activity would probably *not* say:
   a. "Don't worry, no one will ever know."
   b. "If you feel uncomfortable with this, then we shouldn't do it."
   c. "If you love me, then show it."
   d. "Everyone in school is doing it."

28. Approximately how many teenage girls become pregnant every year in the United States?
   a. 1,000
   b. 10,000
   c. 100,000
   d. 1,000,000

Thinking Critically

After reading the question or statement, write a short answer using complete sentences.

29. **Identify.** When might sexual activity result in a loss of self-respect?

30. **Identify.** What are two *risk situations* that could increase your chances of being pressured into sexual activity?

31. **Describe.** How might engaging in sexual activity have a negative effect on your social health?

32. **Describe.** What are some ways that teens and their parents can set limits for dating relationships?

33. **Discuss.** How are teens' attitudes toward sexual activity influenced by the media and popular culture?

34. **Analyze.** Why is abstinence the best choice for teens? Include information on the social, mental/emotional, and physical benefits of abstinence.

Technology PROJECT-BASED ASSESSMENT

Friendship Survey

Background

Your peers are important to your development as an individual. During the teen years, you develop a variety of relationships. Some relationships are casual friendships, while others are close friendships. Dating relationships also develop during this time.

Task

Use an online survey tool to conduct a survey of students in your school to find out what characteristics they consider essential for a close friendship.

Audience

Students in your class

Purpose

Find out what qualities your peers believe differentiate a close friendship from other kinds of relationships.

Procedure

1 Use the information in Chapter 8 to develop a list of qualities that are essential for a close friendship.

2 Create a ten-question survey using an online survey program. The questions might gauge the importance of each quality as "very important," "important," or "not important."

3 Select 15 or more students for the survey. Use the online survey tool to tally the responses.

4 Examine the results, and make a table showing the responses to each question. Determine the qualities that are considered to be most essential for a close friendship.

5 Discuss the results with your class.

Math Practice

Interpret Graphs. The table below shows 12-year trends related to sexual activity among teens in the ninth through twelfth grades. Use the graph to answer the questions.

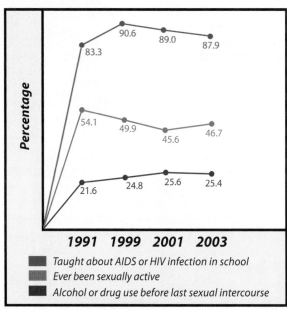

1991 1999 2001 2003

■ Taught about AIDS or HIV infection in school
■ Ever been sexually active
■ Alcohol or drug use before last sexual intercourse

Adapted from the Youth Risk Behavior Survey: 1991–2003, Centers for Disease Control and Prevention.

1. By what percentage has sexual activity among teens decreased between 1991 and 2003?
 A. 10.2   C. 4.3
 B. 7.4   D. 12.6

2. Which year shows the highest rate of teens reporting alcohol use before engaging in sexual activity?
 A. 2001   C. 1991
 B. 1999   D. 2003

3. Identify the years with the lowest and highest percentage of AIDS/HIV education. What is the difference in percentage between the two years?
 A. 3.2   C. 7.3
 B. 6.0   D. 4.5

Reading/Writing Practice

Understand and Apply. Read the passage below, and then answer the questions.

> *I have been best friends with Tamara since the first grade. Tamara is nice to me and to other people.*
> *Tamara's kindness shows in many ways. Once she gave her circus tickets to some kids who had never been to the circus. Tamara visits a nearby nursing home at least once a month. She worries about some of the people she has met there because they have no family.*
> *Tamara was a good friend to me when my parents divorced. She listened to me for hours as I talked about how upset I was. She also let me cry and never told me that I was overreacting. I knew that she couldn't do anything to change the situation, but she always made me feel better.*

1. How does the author show that Tamara is a good friend?
 A. By comparing Tamara's actions to those of her other friends
 B. By pointing out that Tamara once helped with a canned-food drive
 C. By citing examples of Tamara's kindness
 D. By saying that Tamara does not gossip

2. When Tamara listened to her friend talk about divorce, what characteristics of friendship did she show?
 A. Mutual respect   C. Support
 B. Caring   D. All of the above

3. Describe the qualities that you think make someone a good friend. Give examples and details to support your opinions.

National Education Standards

Math: Measurement and Data
Language Arts: LACC.910.L.3.6; LACC.910.RL.2.4

CHAPTER 9

Resolving Conflicts and Preventing Violence

Lesson 1
Causes of Conflict

BIG Idea *Knowing why conflicts occur can help you prevent them.*

Lesson 2
Resolving Conflicts

BIG Idea *Conflicts can be resolved through negotiation or mediation.*

Lesson 3
Understanding Violence

BIG Idea *Teens need to know about forms of violence and ways to protect themselves.*

Lesson 4
Preventing and Overcoming Abuse

BIG Idea *Abuse can cause physical, mental, and emotional damage.*

Activating Prior Knowledge

Using Visuals Describe what is happening in the scene shown on this page. What are some causes of violence in society? What can citizens do to reduce violence and crime?

Chapter Launchers

Health in Action

Discuss the **BIG Ideas**

Think about how you would answer these questions:

▶ What are some conflicts that you've had with people?

▶ How did you resolve these conflicts?

▶ Why do you think some conflicts result in violence?

Assess Your Health

Read each statement. On a separate sheet of paper, write "yes," "sometimes," or "no" based on your typical behavior.

1. I know many of the causes of conflicts and try to avoid them.

2. I walk away from a conflict that I think may become dangerous.

3. I recognize that negotiation is a good way to resolve a conflict.

4. I avoid gangs and other peer pressures that might result in violence.

5. I practice tolerance of others.

A "yes" response shows that you practice healthy behaviors. "Sometimes" indicates that you should analyze and possibly modify your behavior. A "no" response means that you should modify the behavior.

McGraw-Hill Education

Causes of Conflict

BIG Idea *Knowing why conflicts occur can help you prevent them.*

Before You Read

Organize Information. In the center of your paper, write "Conflict" and circle it. Label the space above this circle "Causes" and the space below it "Effects." As you read, list causes of conflict on the top half of the page, and effects of conflict on the bottom half of the page.

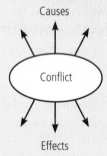

Causes

Conflict

Effects

New Vocabulary

▶ conflict
▶ interpersonal conflicts
▶ escalate

Real Life Issues ••••••••••••••••••••••••••••

Conflict in Communication. Jen has been looking for her new coat all evening, and now she is running late for a movie. As she rushes out the door, her sister, Lynn, walks in wearing the coat. Jen is furious. She yells, "If you can take my clothes without permission, I'll take yours whenever I feel like it!" Jen walks into Lynn's room, grabs an armful of clothes, and walks out the front door with them.

Writing *Write a paragraph explaining what you think caused this conflict and how it might have been avoided.*

Understanding Conflicts

Main Idea Conflicts can arise for a variety of reasons.

The term **conflict** refers to *any disagreement, struggle, or fight.* Some conflicts are fairly trivial, such as a squabble between two siblings over control of the TV remote. Others can be serious or even deadly, such as turf wars between rival gangs. *Conflicts between people or groups of people* are known as **interpersonal conflicts**. They tend to arise when one party's needs, wishes, or beliefs clash with those of another party. Interpersonal conflicts can involve groups of any size, from individual people to entire nations. *Internal conflicts,* by contrast, take place within an individual. For example, if your best friend's birthday party and your sister's championship soccer game fell on the same day, you might go through an internal conflict over which event to attend.

Common Causes of Conflict

Interpersonal conflicts can arise for a variety of reasons. Some arise out of misunderstandings. For instance, Lauren and Jesse got angry at each other over a miscommunication.

Sometimes, misunderstandings occur when an individual **misinterprets** another person's language, gestures, or sense of humor. This type of conflict might occur between people of different cultures or age groups. In other cases, someone deliberately starts a conflict—for example, by insulting or shoving someone else. Causes of conflict include

Academic Vocabulary

misinterpret *(verb):* to understand wrongly

- **power struggles.** A teen and her parent might have a conflict over how late she is allowed to stay out at night.

- **personal loyalties.** A teen might be angry with his best friend for taking another person's side in an argument.

- **jealousy and envy.** A teen might be upset when her friend starts going out with a boy she likes.

- **property disputes.** A teen might be angry with his brother for borrowing his MP3 player without permission.

- **conflicting attitudes and values.** Two friends might have an argument because one wants to hang out only with the "cool" crowd, while the other wants to be friendly to everyone.

- **lack of respect.** A teen might be rude to a classmate because of a prejudice against that student's ethnic group.

READING CHECK

Compare and Contrast How are interpersonal conflicts different from internal conflicts?

Understanding these causes of conflict may help you avoid some conflicts before they start. If it looks like a conflict is developing, you may be able to keep it from escalating. **Escalate** means *to become more serious.* Conflicts can escalate into fights when emotions get out of control. Feelings such as hurt pride, embarrassment, or the desire for revenge can turn a simple conflict into a situation that could be unsafe for everyone involved. In some cases, it's best to walk away before the conflict escalates.

■ **Figure 9.1** Conflicts between people can occur for many different reasons. *Which type of conflict does this picture show?*

"Prevent violence before it starts."

A Voice Against Violence

Rafael G., of North Carolina, knows how it feels to be bullied. That's why he decided to join Students Against Violence Everywhere (SAVE). In his work with SAVE, Rafael raised money to install security cameras and floodlights in hard-to-see areas on campus. His school's SAVE chapter also produced a television segment describing positive ways to respond to violence and crime.

Rafael is proud of the role that SAVE has played in raising awareness about violence. He also recognizes the importance of prevention. "It's better to have a strong force in place to prevent crime or violence before it starts," he says.

Activity Write your answers to the following questions in your personal health journal:

1. What motivated Rafael to join SAVE?
2. What can you do to help reduce or resolve conflicts in your school?
3. What programs does your school have in place to prevent violence?

Results of Conflict

Conflict is a normal part of life. Because each individual is different, it's inevitable that people will disagree sometimes. Learning to manage conflicts and deal with them before they get out of hand will strengthen all aspects of your health.

Sometimes conflict can actually bring about positive results. Working to resolve a conflict can help people improve their communication and problem-solving skills. It can also improve their social health by teaching them how to get along with people who disagree with them. In addition, dealing with conflicts can strengthen relationships. When two people make the effort to work through a conflict together, it shows their commitment to each other.

Unfortunately, conflicts can also have negative effects. They can be a major source of stress, resulting in problems such as headaches and lost sleep. Conflicts can harm your emotional and social health if they lead to anger, frustration, fear, and emotional pain. In addition, conflicts in the workplace can cause people to lose their jobs. In the worst cases, conflicts can escalate to violence, resulting in serious injury or even death.

Preventing Conflicts

It's often easier to prevent a conflict than it is to resolve it. For instance, if you know someone who is always trying to provoke you into an argument, you might decide to avoid that person. If you get involved in a minor disagreement with someone, you can remind yourself that the argument isn't that important in the long run. It's not worth damaging your relationship over something trivial.

Sometimes you can prevent conflicts by adjusting your own behavior. Suppose you have a friend who always forgets to bring money when you go out. Instead of feeling annoyed every time, you might just make a point of reminding this person to stop at the ATM beforehand. Adjusting your attitude can also help. If you tend to interpret any kind of personal remark as an attack on you, you might try to relax and not be bothered so much by what other people say.

■ **Figure 9.2** Reminding a friend who often borrows money to get some cash before you go out is one way to prevent a conflict. *What other ways can you think of?*

LESSON 1 ASSESSMENT

 After You Read

Reviewing Facts and Vocabulary

1. Identify two common causes of interpersonal conflicts.

2. How can conflicts be positive?

3. Give an example of how conflicts can negatively affect one's health.

Thinking Critically

4. **Analyze.** How might adapting your behavior help prevent conflicts?

5. **Evaluate.** Discuss the benefits and drawbacks of walking away from a developing conflict.

Applying Health Skills

6. **Analyzing Influences.** How might influences such as environment, culture, media, and personal values affect a conflict between two people?

Writing Critically

7. **Expository.** Write an essay about common teen conflicts. Explain how to prevent some of these conflicts.

Real Life Issues ....

After completing the lesson, review and analyze your response to the Real Life Issues question on page 220.

LESSON 2

GUIDE TO READING

BIG Idea *Conflicts can be resolved through negotiation or mediation.*

Before You Read

Create a Venn Diagram. Draw two overlapping circles and label them "Negotiation" and "Mediation." As you read, fill in the circles with information about these two methods of resolving conflicts. Traits the two methods have in common should go in the overlapping area.

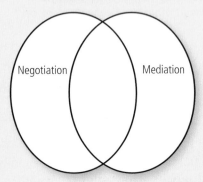

New Vocabulary

▶ negotiation
▶ mediation
▶ confidentiality
▶ peer mediation

Review Vocabulary

▶ conflict resolution (Ch.2, L.1)
▶ compromise (Ch.6, L.1)

Resolving Conflicts

Real Life Issues

Two Teens, One Bathroom. Joe bangs furiously on the bathroom door. "Maggie! You've been in there for half an hour! I need to get ready for school too, you know!" His sister, Maggie, flings open the door. "You know it takes me longer to get ready," she says. "I have to wash and blow-dry my hair every morning. Give me a break!" Joe sighs. "Look, we can't go through this every morning. Can't we work out some kind of deal?"

Writing *Continue the dialogue between Joe and Maggie. Have them brainstorm a solution to their conflict.*

Responding to Conflict

Main Idea **There are various ways to deal with a conflict.**

When you have a conflict with someone, you have two choices: walk away or respond to it. If you think the conflict could escalate and become dangerous, getting out is the best approach. This is also true if you are having trouble managing your own anger. Walking away will give you a chance to calm down so that you can approach the conflict rationally.

However, in many cases, walking away from a conflict will not make it go away. You may stop it from escalating, but the same issue is likely to come up again. Sooner or later, you will need to practice *conflict resolution,* the process of ending a conflict through cooperation and problem solving.

Compromise

You can often resolve minor conflicts with a compromise. If you and your brother disagree about what to watch on TV, you might agree to watch one show and record the other.

Ken Karp/McGraw-Hill Education

However, it can be difficult to reach a compromise when both parties have strong opinions about an issue. In addition, it's unwise to compromise when doing so could have harmful consequences or would go against your values.

Effective Negotiation

Main Idea Negotiation involves finding a solution that both sides can accept.

When conflicts are not resolved, they can get worse, sometimes even resulting in violence. It's important to understand that violence does not solve conflicts. One group may be able to force another to do what it wants, but that does not address the cause of the conflict. As a result, the same conflict is likely to occur again. This may cause the violence to escalate, harming more people. A better strategy is **negotiation**, *the use of communication and, in many cases, compromise to settle a disagreement.*

The Negotiation Process

The negotiation process involves talking, listening, and considering the other party's point of view. Mutual respect is an important factor in a successful negotiation. These are the steps of the negotiation process:

1. Take time to calm down and think over the situation.

2. Let each party take turns explaining its side of the conflict without interruption. Apply good communication skills, such as active listening and using "I" messages.

3. If necessary, ask for clarification to make sure that each party understands the other's position.

4. Brainstorm solutions to the conflict.

5. Discuss the advantages and disadvantages of each solution.

6. Agree on a solution that is acceptable to both sides. The ideal outcome will be a win-win solution. If this is not possible, the two parties may need to compromise.

7. Follow up to see whether the solution has worked for each party.

FITNESS ZONE

Playing sports is a great way to learn how to deal with conflict. In sports, there are disagreements as to whether there was a foul or if someone was out of bounds. When I play ball with my friends, we try to stay cool and discuss the issue. We usually decide to replay the point. After all, we're all friends and we'd like to keep it that way. For more fitness tips, visit the Fitness Handbook and Fitness Zone sites in ConnectEd.

■ **Figure 9.3** Compromise can help you resolve simple conflicts, such as whose CD to listen to. *When is it a bad idea to compromise?*

READING CHECK

Describe What are the steps involved in negotiation?

Preparing for Negotiation Successful negotiations require careful planning. Taking these steps ahead of time will increase the chances that negotiation will work:

- **Choose the time and place carefully.** The negotiation should take place at a time when both parties are calm, not impatient or rushed. Arrange to meet in a quiet place on neutral ground—not at the home of either party or in any other place that "belongs" to one side.

- **Check your facts.** Make sure your understanding of the situation is based on accurate information.

- **Plan what you will say.** Think about how to word your statement respectfully. You may wish to rehearse or write down your statement.

Tips for Successful Negotiation Staying calm is an important key to successful negotiation. Getting angry or upset could throw off the negotiations. Attack the problem, not each other. Avoid blaming and name-calling.

As you discuss the problem, try to keep an open mind. Listen attentively to what the other side has to say, and try to understand the other party's point of view. Be willing to take responsibility for your role in the conflict, and apologize if you have done something to hurt the other person. Remember, your goal is not to "win," but to find a solution that everyone can accept. Make sure to provide a way out of the conflict that will allow the other person to save face.

The Mediation Process

Main Idea Bringing in a neutral third party to mediate can help resolve some conflicts.

When two parties cannot reach a solution through negotiation, they may consider mediation. **Mediation** is *bringing in a neutral third party to help others resolve their conflicts peacefully.* The word *mediation* literally means "being in the middle." Having a third party "in the middle" helps put some distance between the two opposing parties.

■ **Figure 9.4** Mediation can help people settle interpersonal conflicts. *What qualities would an effective mediator need?*

Health Skills Activity

Conflict Resolution

Negotiating with Parents

Chloe's parents have a strict rule: she isn't allowed to be out past 9 P.M. on a school night and 11 P.M. on weekends. When she was younger, Chloe thought this was reasonable. Now that she's 16, it just doesn't seem fair. All her friends get to stay out later, and it's embarrassing for her to have to leave in the middle of a party. She wants to try negotiating a new curfew with her parents.

Writing Write a dialogue between Chloe and her parents in which they work out an acceptable solution. Follow these negotiation steps:

1. Let each party explain its side.
2. Brainstorm solutions.
3. Discuss the pros and cons of each solution.
4. Agree on an acceptable solution.

The presence of someone who is not on either side reduces the level of confrontation. Mediation can be especially useful for dealing with conflicts that go on for a long time and threaten to disrupt everyday life.

Mediation can be formal or informal. Formal mediation involves the help of a mediator who has special training in resolving conflicts. Informal mediation can be as simple as asking a teacher to help settle a dispute with a classmate. Effective mediation depends on these basic principles:

- **Neutrality.** The mediator must always be an outsider who has no stake in the dispute. The mediation session should also take place in a neutral location.

- **Confidentiality. Confidentiality** means *respecting the privacy of both parties and keeping details secret.* The mediator promises not to reveal anything said by either party during the process.

- **Well-defined ground rules.** Both parties must agree to the rules set by the mediator. In some cases, the mediator may ask the two parties to sign an agreement to work out the problem within a given time frame.

In a typical mediation, each party gets a chance to present its side of the argument. The mediator then summarizes the points made by each side and leads a discussion between the two parties. The mediator does not make judgments or impose solutions. Instead, the solutions must come from the two parties. However, the mediator can help them see the advantages and disadvantages of certain ideas.

READING CHECK

Compare and Contrast Explain how negotiation and mediation are similar and how they are different as strategies for resolving conflicts.

Peer Mediation

Many schools have started peer mediation programs to help resolve conflicts. **Peer mediation** is *a process in which specially trained students help other students resolve conflicts peacefully.* Typically, peer mediation involves

- **making introductions.** The mediator explains that she will remain neutral and that the session is confidential.

- **establishing ground rules.** Both parties must agree to such rules as listening without interrupting, telling the truth, and addressing each other with respect.

- **hearing each side.** Each student tells his story in turn. The mediator may ask questions and take notes.

- **exploring solutions.** The two sides discuss the situation and propose possible solutions. Each solution is discussed, and both parties try to find an acceptable solution.

- **wrapping it up.** The mediator sums up the agreement. In some cases, both sides sign a written **contract**.

Keep in mind that mediation is not an appropriate solution for every kind of problem in schools. Violence and other crimes, for instance, require action from school administration or legal authorities.

Academic Vocabulary

contract *(noun):* an agreement between two or more parties

LESSON 2 ASSESSMENT

 After You Read

Reviewing Facts and Vocabulary

1. List three steps you could take to prepare for a negotiation.

2. When might it be necessary to bring in a mediator to settle a conflict?

3. Give two examples of ground rules in a peer mediation process.

Thinking Critically

4. **Evaluate.** Suppose a friend wants to copy answers off your paper during a test. When you refuse, she gets angry. Explain whether this conflict could be resolved through compromise.

5. **Make Inferences.** Why might peer mediation for students work better than bringing in an adult mediator?

Applying Health Skills

6. **Conflict Resolution.** Luke wants to go to a basketball game, but his parents want him to help out with spring cleaning. Write a dialogue in which they use conflict-resolution techniques to settle this problem.

Writing Critically

7. **Narrative.** Write a short story that centers on a conflict between two teens. Show how the characters resolve their conflict through compromise, negotiation, or mediation.

Real Life Issues ....

After completing the lesson, review and analyze your response to the Real Life Issues question on page 224.

Understanding Violence

GUIDE TO READING

Real Life Issues ·······························

The Costs of Violence. There are various forms of violence that may occur on or off school property.

Source: Centers for Disease Control and Prevention, Youth Risk Behavior Survey, 2011.

> **32.8% of teens had been in a physical fight.**

> **3.9% of teens had been injured in a physical fight requiring medical care.**

Writing *Write a paragraph explaining how violence in schools can be avoided.*

Causes of Violence

Main Idea **Weapons, drugs, and gangs are some factors that can contribute to violence.**

Violence is *the threatened or actual use of physical force or power to harm another person or to damage property.* Some acts of violence result from interpersonal conflicts that escalate out of control. However, violence can also be random. People may commit violent acts for many reasons. Causes can include

- uncontrolled anger or frustration.
- a need to control others.
- hatred or prejudice against a particular group.
- retaliation or revenge for some past harm, whether real or perceived.

BIG Idea *Teens need to know about forms of violence and ways to protect themselves.*

Before You Read
Create a Word Web. Write "Violence" in the center of a sheet of paper. As you read the lesson, add information about causes of violence and forms that violence can take.

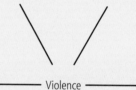

Violence

New Vocabulary
▶ violence
▶ assault
▶ random violence
▶ homicide
▶ sexual violence
▶ sexual assault
▶ rape

Review Vocabulary
▶ prejudice (Ch.6, L.2)
▶ refusal skills (Ch.2, L.1)

Certain risk factors make children and teens more likely to be involved in violence. Children are at a greater risk if their families are poor, have low levels of education, or are involved in illegal activities. For teens, friends and peers play a greater role. Having friends who are involved in violence and crime greatly increases teens' risk of committing violent acts themselves. Fortunately, there are factors that can help protect teens from participating in violence. Teens who are committed to school and have a negative attitude toward crime are less likely to commit acts of violence, even if they have several risk factors.

Alcohol and Drug Use

Studies have found that alcohol, in particular, plays a role in many violent crimes. There are several possible reasons for this connection:

- Drinking and drug use can lower people's self-control. As a result, they may be less likely to restrain their violent impulses.
- Drinking and drug use can damage people's judgment. They may overreact to something they see as a threat or fail to consider the consequences of their actions.
- Teens may engage in violent crimes as a way to get money to buy drugs.
- People who use drugs and drink alcohol are more likely to engage in other high-risk behaviors, such as fighting, carrying weapons, and engaging in unsafe sexual activity.

Some teens actually become involved with violence before they start to use alcohol or drugs. In other words, for some teens, it isn't using drugs and alcohol that makes them violent. Instead, their violent lifestyle puts them at risk for other problems, including substance abuse.

Mental and Emotional Problems

Academic Vocabulary

insecure *(adjective):* not confident or sure

Low self-esteem is another risk factor for violence among teens. Insecure teens may try to use violence to prove themselves. Teens who have had little success in life may use violence as a way of getting back at a system that they think has caused them to fail. In addition, teens with low self-esteem may be more likely to join gangs as a way to belong. Gang membership puts teens at much higher risk for violence.

Stress, depression, and strong emotions such as anger can lead some teens to become violent. Learning to control anger effectively can greatly reduce the risk of violence. Anger-management workshops and counseling can help people learn to deal with anger and avoid lashing out at others.

Availability of Weapons

A 2011 government survey revealed that nearly one in five high school students reported having carried a weapon within the past 30 days. Five percent of all students said they have carried a gun. Access to weapons can increase violence. To protect yourself from the dangers associated with weapons, follow these strategies:

- Do not carry a weapon. People who carry guns are twice as likely to become victims of gun violence.

- If you know that another teen is carrying a weapon, tell a trusted adult, such as a parent or teacher. If necessary, you can contact the authorities anonymously.

- If your parents keep a gun at home, encourage them to equip it with a trigger lock and to store it unloaded in a locked cabinet. (See Chapter 26.)

Violence in the Media

Every day, children and teens are exposed to violent words and images in television, movies, song lyrics, and video games. A 2007 study of violence in the media indicated that more than 60 percent of all television shows and nearly 90 percent of top-rated video games contain some violence. In addition, scenes that feature violence often fail to show its harmful consequences. In many cases, the characters who commit violent acts suffer no punishment as a result.

Exposure to violence in the media can influence the way people think about violence. Some young people who view scenes of violence may begin to perceive it as normal or even positive. Studies have found that children and teens act more aggressively right after watching violent scenes. Also, children and teens who are aggressive tend to watch more violent television than their less aggressive peers.

READING CHECK

Describe Name three ways for teens to protect themselves from situations involving weapons.

■ **Figure 9.5** Children and teens are exposed to violence in the media every day. *How might this exposure influence their behavior?*

Real World CONNECTION

Violence Among Teens

The National Youth Risk Behavior Survey keeps track of behaviors that put teens' health and safety at risk. The chart below shows how some behaviors related to violence have increased or decreased over time.

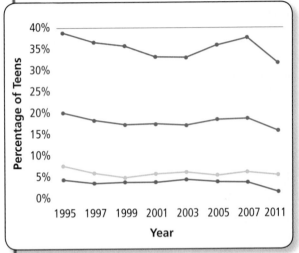

Source: Centers for Disease Control and Prevention, National Youth Risk Behavior Survey, 1995–2011

Activity — Mathematics

Use the graph to answer the following questions:

1. What percent of teens were involved in a physical fight in 2003?

2. What overall trend can you detect for all four risk behaviors over the 16-year period shown?

3. Which of these behaviors do you think poses the greatest danger to teens? Why?

Concept Measurement and Data: Interpreting Graphs A line graph compares the relationship between two variables. It is an effective tool for showing trends.

- Being in a physical fight (at least once in the past year)
- Carrying a weapon (at least once in the past month)
- Carrying a gun (at least once in the past month)
- Requiring treatment for injuries from a physical fight (at least once in the past year)

Gang Violence

Youth gangs are groups of teens or young adults who are involved collectively in violent or illegal activity. Gangs are often involved in drug dealing, robbery, and violent attacks on members of rival gangs. Teens who join gangs may be seeking protection from violence or looking for a way to fit in.

Teens who belong to gangs are much more likely than their peers to commit serious or violent crimes. They are also much more likely to become victims of violence. Being part of a gang reduces a teen's chances of graduating from school and finding a steady job. As a result, teen gang members may end up as career criminals.

To avoid gang influence, be aware of gang activity in your area, including the colors and symbols used by various gangs. Doing so will enable you to recognize and steer clear of gang members. It will also help you avoid dressing in a way that could cause you to be mistaken for a gang member. Seek out positive alternatives to gang membership, such as sports or after-school programs. Be prepared to use refusal skills if anyone ever tries to recruit you into a gang.

Types of Violence

Main Idea Violence may be physical or sexual.

In nearly half of all violent crimes, the victims know their attackers. This rate is higher for certain types of crimes. Victims of sexual attacks, for instance, are very likely to know their attackers, while robberies are typically random.

Assault and Homicide

An **assault** is *an unlawful physical attack or threat of attack.* Assaults range from minor threats to attacks that cause life-threatening injuries. Each year, more than 4 million assaults take place in the United States, and more than one million of those incidents result in injury. Roughly half of all assaults occur between people who know each other. However, assaults may also take the form of **random violence**—*violence committed for no particular reason.*

If the victim of an assault dies, the crime becomes a **homicide**, *the willful killing of one human being by another.* Teens can protect themselves from assault and homicide by avoiding the risk factors associated with violence in general. That means avoiding drugs, alcohol, weapons, and gangs. You can also work on developing your protective factors. For instance, strengthening your ties to your family and your school can lower your overall risk of violence.

■ **Figure 9.6** Teens who are involved in school activities are at less risk of violence, including assault and homicide. *Why do you think involvement with school can lower teens' risk of violence?*

Sexual Violence

You've probably heard of sexual harassment, or unwelcome sexual conduct. This may include jokes, gestures, or physical contact. Sometimes, sexual harassment may escalate to **sexual violence**, *any form of unwelcome sexual contact directed at an individual.* Sexual violence can include

- **sexual assault**, which means *any intentional sexual attack against another person.* In 2012, more than 73,000 sexual assaults were reported in the United States.

- **rape**, which is *any form of sexual intercourse that takes place against a person's will.* In 2010, more than 84,000 forcible rapes were reported to law enforcement in the U.S. Rape is one of the crimes least likely to be reported to the police. Survivors of rape may be unwilling to report the crime because of shame or fear.

Sexual violence can affect anyone. However, most victims are female, and most rapists are male. Of all violent crimes, rape and sexual assault are the ones in which victims are most likely to know their attackers.

Avoiding Sexual Violence A sexual attack can happen anywhere. To help protect yourself, be aware of your surroundings wherever you go. Refuse to go anywhere alone with someone you don't know or trust. Attend parties with friends so you can all watch out for each other. Avoid alcohol and drugs, which can make you an easier target. Finally, trust your instincts. If a situation feels unsafe, don't hesitate to get out of it. For more safety tips, see Chapter 26.

Responding to a Sexual Attack If you are ever sexually attacked, your goal is to survive. In some cases, that may mean resisting the attacker, while in other cases, it may be safer to submit. You may try to stall for time, distract the attacker, or scream to attract attention. Do whatever you need to do to survive the situation.

Reporting a sexual attack right away gives you the best chance of bringing the attacker to justice. To preserve evidence of the attack, do not bathe or brush your teeth until you have been examined. Seek medical help for any injuries and, if appropriate, get tested for pregnancy and sexually transmitted diseases (STDs).

Survivors of rape and sexual assault need time to heal physically and emotionally. They may suffer feelings of fear, guilt, and shame.

■ **Figure 9.7** Counseling can help survivors of sexual violence recover from the experience. *What steps are important to take right after a sexual attack?*

Many mistakenly blame themselves for the attack. Counseling can help survivors of a sexual attack recover from the experience.

READING CHECK

Identify What are some ways of responding to a sexual attack?

Hate Crimes

A hate crime is any crime motivated chiefly by hatred of or prejudice against a particular group. People may be targeted because of their race, religion, culture, sexual orientation, or other difference. Hate crimes can take many forms:

- **Harassment.** This can include racial slurs, stalking, or attempts to exclude a targeted group from community life.
- **Vandalism.** Perpetrators may use offensive messages or symbols to deface buildings.
- **Arson.** Criminals may blow up or set fire to buildings.
- **Assault and Homicide.** Criminals may physically attack or even kill members of the targeted group.

Hate crimes affect everyone, spreading fear, distrust, and anger throughout the community. The best way to stop hate crimes is to change the attitudes behind them. Practicing and teaching tolerance of other groups can go a long way toward ending these crimes. When a hate crime occurs, community members can condemn the crime and express support for the targeted group. This may prevent the hate violence from escalating.

LESSON 3 ASSESSMENT

 After You Read

Reviewing Facts and Vocabulary

1. Identify two factors that can contribute to violence.
2. What is *random violence*?
3. Identify two steps you can take to protect yourself from sexual violence.

Thinking Critically

4. **Evaluate.** Why might survivors of rape be reluctant to tell others about the crime?
5. **Analyze.** How can practicing and promoting tolerance help prevent violence?

Applying Health Skills

6. **Refusal Skills.** Write a dialogue between two teens at a party. One teen is trying to persuade the other to go somewhere alone together. The other teen uses refusal skills to avoid the threat of sexual violence.

Writing Critically

7. **Persuasive.** Write an editorial promoting tolerance and condemning hate crimes.

Real Life Issues

After completing the lesson, review and analyze your response to the Real Life Issues question on page 229.

Radius Images/Getty Images

GUIDE TO READING

BIG Idea *Abuse can cause physical, mental, and emotional damage.*

Before You Read

Organize Information. Make a chart with three columns. Label the columns "Physical Abuse," "Emotional Abuse," and "Sexual Abuse." As you read, fill in the columns with examples of each type of abuse, possible effects, and ways to prevent or respond to it.

Physical Abuse	Emotional Abuse	Sexual Abuse

New Vocabulary

▶ physical abuse
▶ emotional abuse
▶ verbal abuse
▶ sexual abuse
▶ stalking
▶ date rape

Review Vocabulary

▶ abuse (Ch.7, L.3)
▶ cycle of violence (Ch.7, L.3)

Preventing and Overcoming Abuse

Real Life Issues

A Dangerous Date. Elena smiles at Matt as she hops into his car for their date. On the drive, Matt turns and heads in a different direction. "Where are you going?" Elena asks. "The concert's that way."
"Change of plans," Matt says with a sly grin. "There's a party on the other side of town. Some guys figured out how to get into that old abandoned house. I thought it would give us time to be alone." Elena begins to worry. She's afraid that if she and Matt are alone, he might try to take advantage of her sexually.

Writing *Write a conclusion to this story that shows how Elena escapes from this potentially dangerous situation.*

Abuse in Relationships

Main Idea All forms of abuse are extremely harmful.

Abuse is the physical, mental, emotional, or sexual mistreatment of one person by another. In Chapter 7, you learned about abuse in families. However, abuse can also occur in other types of relationships, including dating relationships. A dating relationship may be abusive if one partner

- tries to pressure the other into sexual activity.
- tries to make the relationship serious or exclusive right away.
- acts jealous or possessive.
- tries to control the other's behavior.
- yells, swears, or otherwise emotionally attacks the other.
- threatens the other with physical violence.

Forms of Abuse

Abuse in relationships can take several forms. The most common forms include the following:

- **Physical abuse** is *a pattern of intentionally causing bodily harm or injury to another person.* Examples include hitting, kicking, shoving, biting, pulling hair, and throwing objects at another person. Physical abuse can result in serious injuries. It can also leave the victim emotionally scarred. Victims of physical abuse may respond with violence of their own.

- **Emotional abuse** is *a pattern of attacking another person's emotional development and sense of worth.* One form of emotional abuse is **verbal abuse**, *the use of words to mistreat or injure another person.* Examples include yelling, swearing, and making insults. Abusers may also humiliate their victims, attempt to control their behavior, threaten physical harm, or cut the person off from friends and family members. Emotional abuse can damage self-esteem and lead the victim to feel worthless or helpless. Victims may even come to feel that they deserve the abuse.

- **Sexual abuse** is *a pattern of sexual contact that is forced upon a person against the person's will.* Sexual assault, rape, and trying to pressure someone into sexual activity are examples of sexual abuse. Sexual abuse can harm victims physically and emotionally. It may also put them at risk for pregnancy or disease.

- **Stalking** is *repeatedly following, harassing, or threatening an individual.* Stalkers may follow their victims physically from place to place. They may also harass them by calling or e-mailing repeatedly and sending letters or gifts. More than 3.3 million people are stalked each year in the United States, and most of them know the stalker.

READING CHECK

Synthesize Give an example of verbal abuse.

■ **Figure 9.8** Communicating your sexual limits clearly to the people you date can help protect you from being in an abusive relationship. *What are some harmful effects of abuse?*

©Image100/Corbis

Sometimes teens who are in abusive relationships don't realize there is a problem. A boyfriend may think that being jealous and possessive of his girlfriend shows how much he loves her. The girlfriend may accept his efforts to control her as normal. It's important to understand that trying to control a **partner** is not a normal or healthy part of a relationship.

Protecting Yourself from Abuse

There are several steps you can take to avoid abusive relationships. For starters, you can hang out with others who share your values and treat you with respect. You can also know your own limits with regard to sexual activity and communicate those limits clearly to anyone you date. Avoiding drugs and alcohol is another important step. These substances can impair your ability to make sound decisions.

Know the warning signs of abuse in relationships. If you feel a relationship might be turning dangerous, trust your instincts and get out. If necessary, seek help from someone you trust, such as a parent or teacher. If you feel you are in immediate danger, you can contact the police. Remember that no matter what happens, you are not to blame for anyone else's behavior. You can control only your own actions.

Date Rape and Acquaintance Rape

Main Idea Rape that occurs in dating relationships is a form of abuse.

Sometimes abuse in dating relationships can take the form of sexual violence. **Date rape** occurs when *one person in a dating relationship forces the other person to take part in sexual intercourse.* More than 40 percent of female rape victims and more than 10 percent of male victims are romantically involved in some way with their attackers. In *acquaintance rape,* the attacker is someone the victim knows casually or considers a friend. This is the form of rape that affects male victims most often.

All forms of rape can harm survivors physically, mentally, and emotionally. Minor injuries, such as scratches and bruises, are common. A smaller percentage of survivors suffer major injuries such as broken bones. Long-term effects include chronic pain, headaches, or stomach problems. Survivors are also at risk for pregnancy and STDs. Rape can trigger feelings of shock, anxiety, guilt, and distrust of others. In the long term, survivors may develop mental and emotional problems, such as depression, eating disorders, or post-traumatic stress disorder (see Chapter 5).

READING CHECK

Identify What form of rape most often affects male victims?

■ **Figure 9.9** Getting your own drink and always keeping an eye on it can help you avoid date rape drugs. *What other strategies can help prevent date rape?*

Alcohol, Drugs, and Date Rape

Alcohol often plays a role in date rape. Drinking lowers people's inhibitions and impairs their judgment. Both females and males are more likely to be sexually attacked when they've been drinking. In addition, males are more likely to commit sexual attacks when under the influence of alcohol.

Some rapists use drugs to subdue their victims. Substances like Rohypnol ("roofies"), GHB, and ketamine are sometimes called "date rape drugs" because they can make someone an easier target. Mixed with food or drink, these drugs are difficult to detect. They work quickly, with effects ranging from drowsiness and dizziness to loss of consciousness. People who've been drugged often cannot remember what happened to them, making it difficult for them to identify attackers.

Avoiding Date Rape

The tips you learned in Lesson 3 on avoiding sexual violence also apply to date rape and acquaintance rape. Specific strategies for avoiding date rape include the following:

- Avoid being alone with a dating partner you don't trust or know well, or with anyone who makes you feel uneasy.

- Avoid alcohol and drugs. Stay sober and aware of what's going on around you.

- Be clear about your sexual limits with dating partners.

- Always get your own beverage at parties, and never leave it uncovered or unattended. Don't drink anything that smells or tastes strange.

- Make sure you have a way to get home. Don't depend only on your date for a ride.

- If you start to feel dizzy, disoriented, or otherwise unwell, tell someone you trust and ask for help getting home.

Overcoming Abuse

Eclipse Studios/McGraw-Hill Education

Main Idea Counseling can help survivors of abuse recover from its effects.

Victims of abuse or rape may be reluctant to tell others about what has happened to them. Recognizing that they are not to blame can be the first step in recovering from the experience. People need to understand that all forms of abuse, including rape, are illegal. Reporting the incident to authorities can help prevent future abuse.

Help for Survivors

People who have survived rape or abuse may feel angry, confused, or ashamed. They may withdraw from friends and family or develop symptoms of depression or anxiety. The traumatic experience can lead to a fear of intimacy and an inability to trust others. In the long term, these individuals may be at risk for problems such as alcohol or drug abuse, eating disorders, self-injury, and suicide.

Seeking professional help is the best way to work through these feelings and avoid long-term health consequences. It can be difficult to talk about something as traumatic as rape or abuse. However, talking about the experience in a safe, supportive environment is the best way to move toward healing. Survivors can seek support from sources such as

- parents, guardians, or other trusted adults.
- teachers, coaches, school nurses, or guidance counselors.
- members of the clergy.
- police.
- private physicians or hospital emergency rooms.
- shelters for victims of domestic violence.
- rape crisis centers.
- therapists, counselors, or support groups.

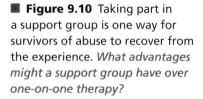

■ **Figure 9.10** Taking part in a support group is one way for survivors of abuse to recover from the experience. *What advantages might a support group have over one-on-one therapy?*

Counseling can take several forms. Some survivors of abuse prefer one-on-one sessions with a trained therapist. Others feel more comfortable with support groups where they can share their experiences with other survivors. Being in such a group may help them understand that they are not alone.

Help for Abusers

In cases of abuse, the victim isn't the only one who needs help. Abuse is a learned behavior, and many abusers were once victims of abuse themselves. They need help to break the cycle of violence. Some abusers may see their behavior as normal or justified. They need to recognize that abusing others is wrong and to learn healthier social behaviors. Other abusers understand that their behavior is wrong but feel powerless to stop it. They need to learn that they are responsible for their own behavior, and that with help, they can control their violent impulses.

Counseling can help abusers learn to cope with their emotions in healthier ways. Abusers should recognize that asking for help is an act of courage. Without it, their violent behavior may increase until it destroys their relationships or causes serious harm to someone they love. By getting the help they need, they may be able to save their relationships and stop the cycle from continuing to the next generation.

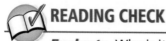

READING CHECK

Evaluate Why is it important to report all forms of abuse to authorities?

LESSON 4 ASSESSMENT

 After You Read

Reviewing Facts and Vocabulary

1. What is *verbal abuse*?
2. Identify two warning signs that a dating relationship may be abusive.
3. Identify two strategies for avoiding date rape.

Thinking Critically

4. **Analyze.** Why should you make sure you have a way to get home from a party or social event other than depending on your date for a ride?
5. **Evaluate.** Why is it beneficial for abusers, as well as survivors of abuse, to seek counseling?

Applying Health Skills

6. **Communication Skills.** Suppose a friend has just confided to you about being in an abusive relationship. Write a dialogue between you and your friend in which you show support and encourage your friend to seek help.

Writing Critically

7. **Creative.** Write a poem or song about the problem of abuse in relationships.

Real Life Issues ....

After completing the lesson, review and analyze your response to the Real Life Issues question on page 236.

Hands-On HEALTH

Activity ## Stop the Hurt

Every few seconds a person is sexually, physically, or emotionally abused. This activity will help you understand the consequences of abuse as well as the steps you can take to stop the hurt.

What You'll Need

- notebook paper
- markers
- poster board

What You'll Do

Step 1

Brainstorm all the behaviors, actions, or situations that you believe to be sexually, physically, or emotionally abusive (for example, racial slurs, rumors, or unwanted touching). Write each behavior on notebook paper.

Step 2

Discuss with your group each situation, placing them in sequential order from most hurtful to least hurtful.

Step 3

Develop specific steps that the victim can use in each abusive situation. Write them on your poster board and present them to your class. (For example: Using "I" messages and reporting the behavior.)

Apply and Conclude

Identify steps you can take to prevent physical, mental, and sexual abuse in your school and community.

Checklist: Self-Management Skills

☑ Demonstrate healthful behaviors, habits, and techniques

☑ Identify protective behaviors (such as first-aid techniques, safety steps, or strategies) to avoid and manage unhealthy or dangerous situations

☑ List steps in correct order

Review

LESSON 1

Causes of Conflict

Key Concepts

- Causes of conflict include misunderstandings, power struggles, property disputes, and lack of respect.
- Working together to resolve conflicts strengthens relationships.
- Conflicts can lead to stress, negative emotions, damaged relationships, financial losses, and violence.

Vocabulary

- conflict (p. 220)
- interpersonal conflicts (p. 220)
- escalate (p. 221)

LESSON 2

Resolving Conflicts

Key Concepts

- Compromise is one way to resolve a conflict.
- Negotiation is a step-by-step process used to resolve conflicts.
- Mediation involves bringing in a neutral third party to help resolve a conflict.

Vocabulary

- conflict resolution (p. 224)
- compromise (p. 224)
- negotiation (p. 225)
- mediation (p. 226)
- confidentiality (p. 227)
- peer mediation (p. 228)

LESSON 3

Understanding Violence

Key Concepts

- Substance use, availability of weapons, violence in the media, and gang presence may contribute to violence.
- Violent crimes include assault, homicide, and sexual assault.
- Following a sexual attack, the most important steps to take are reporting the crime and seeking medical help.

Vocabulary

- violence (p. 229)
- assault (p. 233)
- random violence (p. 233)
- homicide (p. 233)
- sexual violence (p. 234)
- sexual assault (p. 234)
- rape (p. 234)

LESSON 4

Preventing and Overcoming Abuse

Key Concepts

- Violence in dating relationships is dangerous and unhealthy.
- Abuse in relationships can be physical, emotional, or sexual.
- Alcohol and drug use can increase the risk of date rape.
- Both survivors of abuse and abusers need counseling to overcome the experience and break the cycle of violence.

Vocabulary

- physical abuse (p. 237)
- emotional abuse (p. 237)
- verbal abuse (p. 237)
- sexual abuse (p. 237)
- stalking (p. 237)
- date rape (p. 238)

McGraw-Hill Education

LESSON **1**

Vocabulary Review

Use the vocabulary terms listed on page 243 to complete the following statements.

1. The term _____ refers to any disagreement, struggle, or fight.

2. Disagreements can _____ into fights when emotions get out of control.

3. _____ can arise when one party's needs, wishes, or beliefs clash with those of another party.

Understanding Key Concepts

After reading the question or statement, select the correct answer.

4. Which of the following is *not* an example of an interpersonal conflict?
   a. Two friends disagree over which movie to see.
   b. Two rival gangs fight over control of a neighborhood.
   c. Two political parties clash over tax policy.
   d. A teen feels torn between loyalties to two friends who are not on good terms.

5. Which of the following is a positive result of conflict?
   a. Improved problem-solving skills
   b. Stress
   c. Damaged relationships
   d. Violence

Thinking Critically

After reading the question or statement, write a short answer using complete sentences.

6. **Discuss.** How might cultural differences contribute to interpersonal conflict? What steps could people take to avoid this problem?

7. **Analyze.** Explain how conflicts can both help and harm relationships.

8. **Synthesize.** How could you prevent a friend's frequent forgetfulness from becoming an ongoing source of conflict?

LESSON **2**

Vocabulary Review

Correct the sentences below by replacing the italicized term with the correct vocabulary term.

9. Solving a disagreement in a way that satisfies everyone involved is called *escalation*.

10. *Discussion* involves bringing in a neutral third party to help resolve conflicts.

11. *Neutrality* means respecting the privacy of both parties and keeping details secret.

Understanding Key Concepts

After reading the question or statement, select the correct answer.

12. Which of the following conflicts most likely could be resolved through compromise?
   a. The school drama club is split over which of two plays to put on.
   b. One group of parents wants the school to adopt a dress code, but the other doesn't.
   c. A teen wants to buy a used car, but his father objects because the car is unsafe.
   d. One friend makes fun of a boy whom the other likes.

13. Which of the following is *not* a good strategy for negotiation?
    a. Choose the time and place carefully.
    b. Refuse to compromise.
    c. Listen attentively to what the other side has to say.
    d. Be willing to take responsibility for your role in the conflict.

14. For a conflict between two teens who are friends, the best peer mediator would be
    a. the parent of one of the teens.
    b. a teen who is friends with one of them.
    c. a teacher.
    d. a student who does not know either teen well.

Thinking Critically

After reading the question or statement, write a short answer using complete sentences.

15. **Analyze.** Compromise is not always a good way of resolving a conflict. In what type of situation is it unwise to compromise?

16. **Evaluate.** Is a compromise between two parties the ideal outcome of a negotiation? Why or why not?

17. **Explain.** Why is it important for mediators in a conflict to be neutral?

LESSON 3

Vocabulary Review

Choose the correct term in the sentences below.

18. *Violence/Assault* is the use of physical force to harm people or damage property.

19. The willful killing of one human being by another is called *homicide/rape*.

20. The term *sexual violence/sexual assault* refers to any form of unwelcome sexual contact directed at an individual.

Understanding Key Concepts

After reading the question or statement, select the correct answer.

21. Teens are *less* likely to be involved in violence if they
    a. have friends or family who are involved in crime.
    b. use alcohol or drugs.
    c. are committed to school.
    d. have an underprivileged background.

22. Which of the following accurately describes the term *youth gang*?
    a. Teens who hang out together
    b. A major drug ring run by young adults
    c. Two teens who have shoplifted
    d. Teens who commit acts of vandalism and assault as a group

23. In most cases of rape, the victim
    a. does not know the rapist.
    b. is female.
    c. is over 18 years old.
    d. reports the crime to the police.

Thinking Critically

After reading the question or statement, write a short answer using complete sentences.

24. **Analyze.** What are the possible consequences of retaliating for a violent act?

25. **Explain.** How can media violence influence behavior?

26. **Evaluate.** What is the advantage of reporting a sexual attack to the police right away, without bathing or showering first?

LESSON 4

Vocabulary Review

Choose the correct term in the sentences below.

27. *Physical abuse/Emotional abuse* is a pattern of intentionally causing bodily harm or injury to another person.

28. Repeatedly following, harassing, or threatening an individual is known as *verbal abuse / stalking.*

29. *Sexual abuse / Date rape* is a pattern of sexual contact that is forced upon a person against the person's will.

Understanding Key Concepts

After reading the question or statement, select the correct answer.

30. Trey insists that his girlfriend ask his permission before she goes out with her friends. He also calls to check up on her whenever she's out. His behavior is an example of
    a. physical abuse.
    b. emotional abuse.
    c. sexual abuse.
    d. stalking.

31. Which of the following behaviors can *reduce* your risk of date rape?
    a. Using alcohol or other drugs
    b. Being alone with a date you don't trust or know well
    c. Relying on your date for a ride home
    d. Going to a party with a group of friends

Thinking Critically

After reading the question or statement, write a short answer using complete sentences.

32. **Explain.** Why is verbal abuse considered a form of emotional abuse?

33. **Analyze.** How does knowing your sexual limits and communicating them clearly to the people you date protect you from abuse?

34. **Evaluate.** Why are some survivors of rape or abuse reluctant to seek counseling? What might convince them that counseling can help?

Technology PROJECT-BASED ASSESSMENT

Recognizing the Warning Signs of Violence

Background
Recognizing the warning signs of violence can prevent a situation from escalating and becoming violent. Knowing how to prevent a conflict from becoming violent can reduce the risk of injury.

Task
Create a podcast or a short streaming video public service announcement (PSA) describing the warning signs of violence.

Audience
Students in grades 6–8

Purpose
Help students learn how to stay safe by recognizing the warning signs of violent behavior.

Procedure
1. Conduct an online search for public service announcements on other subjects. Create a list of the characteristics that make these PSAs effective.
2. Review the information in Chapter 9 on conflicts and violence.
3. Write the script for a PSA podcast or video. Have it reviewed by your teacher or a school counselor.
4. Record or film your PSA.
5. Present your PSA to a middle school class.

Math Practice

Interpret Tables. A study surveyed teachers at schools that implemented a peer-mediation program. The table below lists some of the problems that teachers reported and the percentage decrease in incidents of those problems since the program began. Use the table to answer Questions 1–3.

Problem	Percentage decrease since implementing program
Expulsions	73%
Assaults	90.2%
Discipline referrals	57.7%

Adapted from *Safe and Drug-Free Schools Program Inventory*, 2002.

1. Which problem decreased the most since the peer-mediation program began?
   A. assaults
   B. conflict
   C. discipline referrals
   D. expulsions

2. If the number of assaults reported before the program began was 150 per year, about how many assaults per year occurred after the program began?
   A. 15
   B. 60
   C. 90
   D. 135

3. A student looking at this table concluded that since the peer-mediation program began, problems decreased by 220.9% (the sum of all three percentages). Why would you question his reasoning?
   A. The percentages do not add up to 100%.
   B. 220.9% is not the sum of all three percentages.
   C. The percentage decreases are not likely related to the peer-mediation program.
   D. There might be overlap in the percentages. For example, assault may lead to expulsion.

Reading/Writing Practice

Understand and Apply. Read the passage below and then answer the questions.

If you have a friend in an abusive relationship, you may be tempted to "rescue" her or him. You may try to persuade your friend to leave the relationship by criticizing the abuser. However, criticizing the abuser may simply make the victim less willing to confide in you.

Ask your friend how she or he feels. Express your concerns in a way that focuses on the abusive behavior rather than on the abuser. You might say, "I'm worried that you're getting hurt," rather than "That jerk doesn't deserve you."

Let your friend know that she or he can always count on you for sympathy and support. That way, when your friend does feel ready to leave the relationship, she or he will be more likely to turn to you for help.

1. The purpose of this piece is
   A. to describe the consequences of abuse.
   B. to discuss ways to prevent abuse.
   C. to let teens know how to help a friend in an abusive relationship.
   D. to list sources of help for abused teens.

2. Which sentence should be added at the beginning of the second paragraph?
   A. Instead, the best approach is to listen without criticizing.
   B. The victim may deny the abuse.
   C. Abuse can destroy self-worth.
   D. Abuse is never acceptable.

3. Write a letter to a fictitious friend who is involved in an abusive relationship. Use the guidelines provided in this passage to offer sympathy and support.

National Education Standards

Math: Measurement and Data, Problem Solving
Language Arts: LACC.910.L.3.6

CAREER CORNER — Social Services Careers

Child Welfare Worker

Child welfare workers help children whose health and well-being are in jeopardy because the child's parents or guardians are not able to take care of them. If a parent or guardian abuses alcohol or other drugs, or abuses a child, a child welfare worker may remove a child from the family and work with the parents to solve their problems.

To help prepare for this career, you can take psychology, sociology, and communications courses in high school. To become a child welfare worker, a bachelor's or master's degree in social work and a license or certificate from your state may be required.

Family Counselor

Family counselors help families find ways to work out their problems and communicate more openly and honestly. They work with family members to overcome issues such as depression, marital problems, and parent-child conflicts. Counselors can help individuals build stronger relationships with their families.

To learn more about becoming a family counselor, take psychology and communications courses in high school. Family counselors are required to have a master's degree in counseling and a license in marriage and family therapy.

Social and Human Services Assistant

A social and human services assistant works under the direction of a social worker, nurse, psychiatrist, psychologist, or a physical therapist. They provide support by assessing a patient's needs and helping that person learn to solve problems.

Social and human services assistants need good communication skills and should demonstrate empathy and understanding. To learn more about this field, take communications, psychology, and sociology classes in high school. A college degree usually is not required, but having a background in the social services field is helpful. In some states, social and human services assistants earn a certificate or an associate's degree.

CAREER SPOTLIGHT

Health Educator

Elizabeth Jenkins decided to become a health educator when she was in high school. During a teen parenting course, she thought "I could teach this, and teach it better." After completing an internship, she was hired by a high school pregnancy prevention program.

Q. What training do you have?

A. I completed a Georgia Campaign for Adolescent Pregnancy Prevention program. Then I became a certified nurse assistant so that I could work with school nurses. I earned an associate's degree and am finishing up my bachelor's degree in criminal justice and business administration.

Q. What are your health goals for the teens you teach?

A. My main goal is that the teen moms graduate from high school. I encourage these young women to go to college. I expect all the teens in my school to further their education.

Q. Are you making an impact?

A. I have teen mothers who have gone to college on scholarships. I know a teen who, after listening to me, left an abusive relationship. I also helped a teen rebuild a relationship with her mother.

Activity   Beyond the Classroom

Writing Social Services Careers. Visit your school counselor or someone else in the social services field. Ask that person to tell you about his or her career and other related occupations. Invite the person to describe how individuals with these careers help people. Learn what education requirements there are for these jobs. Find out what high school classes you could take now to help prepare for these careers.

Based on what you learn, create a brochure describing two or three social services careers. Include information about how teens can prepare for these careers. Share the brochure with your classmates.

UNIT 4 Nutrition and Physical Activity

BananaStock/JupiterImages

UNIT PROJECT

Raising Awareness

Using Visuals The Susan G. Komen for the Cure is the world's largest group of breast cancer survivors and activists fighting to save lives and promote quality care for all. The organization raises funds for research into the causes and cures of breast cancer, and also funds education programs that help people learn how they can reduce their risk for this disease.

Get Involved. Do research to learn about organizations in your community that raise funds to fight cancer. Contact one of these organizations to find out how teens can volunteer. Share your findings with your classmates.

"He who has health has hope; and he who has hope has everything."
—Arabian proverb

CHAPTER 10

Nutrition for Health

Lesson 1
The Importance of Nutrition

BIG Idea *Learning to make healthful food choices will keep you healthy throughout your life.*

Lesson 2
Nutrients

BIG Idea *Each nutrient in your diet plays a unique and essential role in keeping you healthy.*

Lesson 3
Healthy Food Guidelines

BIG Idea *MyPlate is a tool that can help you choose healthful foods for all your meals and snacks.*

Lesson 4
Nutrition Labels and Food Safety

BIG Idea *By reading food labels and handling foods safely, you can avoid many food-related health problems.*

Activating Prior Knowledge

Using Visuals Take a look at the photo on this page. How are these teens practicing healthful behaviors? Explain your thoughts in a short paragraph.

Chapter Launchers

Health in Action

Discuss the BIG Ideas

Think about how you would answer these questions:

▸ What influences your food choices?

▸ Are your eating habits healthful? Why or why not?

Assess Your Health

Read each statement. On a separate sheet of paper, write "yes," "sometimes," or "no" based on your typical behavior.

1. I eat a variety of fruits and vegetables each day.

2. I understand the difference between hunger and appetite.

3. I choose to eat foods that are low in saturated fats.

4. I use MyPlate to help me make healthful food choices.

5. I read food labels to understand the nutritional information in the foods I eat.

A "yes" response shows that you practice healthy behaviors. "Sometimes" indicates that you should analyze and possibly modify your behavior. A "no" response means that you should modify the behavior.

GUIDE TO READING

BIG Idea *Learning to make healthful food choices will keep you healthy throughout your life.*

Before You Read

Create a K-W-L Chart. Make a three-column chart. In the first column, list what you <u>k</u>now about nutrition. In the second column, list what you <u>w</u>ant to know about this topic. As you read, use the third column to summarize what you <u>l</u>earned.

K	W	L

New Vocabulary

▶ nutrition
▶ nutrients
▶ calorie
▶ hunger
▶ appetite

The Importance of Nutrition

Real Life Issues

Schools can play a major role in the nutrition of teens.

Percentage of Districts That Almost Always or Always Used Healthy Food Preparation Practices, 2000, 2006, and 2012			
Practice	2000	2006	2012
Used part-skim or low-fat cheese instead of regular cheese	34.1	50.3	69.4
Reducing the amount of fats and oils in recipes or using low-fat recipes	25.3	26.4	41.4
Roasting, baking, or broiling meat rather than frying	NA	86.7	76.2
Steaming or baking other vegetables	59.5	77.7	83.7

Source: Centers for Disease Control and Prevention, School Health Policies and Programs Study, 2012.

Writing *Write a paragraph describing why it's important for schools to offer students healthful food choices.*

Why Nutrition Matters

Main Idea The food you eat affects your health and quality of life.

Most people know what foods they like. They may not understand how the body uses food. The food you eat plays a significant role in your total health. To make healthful food choices, you must first learn about **nutrition**, *the process by which your body takes in and uses food.*

Your body relies on food to provide it with **nutrients**, *substances in food that your body needs to grow, to repair itself, and to supply you with energy.* The energy your body receives from food is measured in calories. A **calorie** is *a unit of heat used to measure the energy your body uses and the energy it receives from food.* The calories in the food you eat provide the energy your body needs for activities such as walking, doing chores, and playing sports.

During your teen years, choosing the right foods in the right amounts will give your body the nutrients it needs for healthy growth and development. Healthful foods provide fuel for physical activities, help you stay mentally alert, and keep you looking and feeling your best.

Nutrition also affects your lifelong health. Eating a variety of healthful foods can help you avoid unhealthful weight gain and diseases such as type 2 diabetes. It can also lower your risk of developing other conditions that can threaten your life as you age. These include the following:

- Cardiovascular disease
- Stroke
- Certain cancers
- Osteoporosis

READING CHECK

Explain In what ways do your eating habits affect your lifelong health?

What Influences Your Food Choices?

Main Idea A variety of factors influence food choices.

When you make food choices, you need to understand what influences you. Did you eat oatmeal with raisins for breakfast because you like the taste? Maybe you grabbed a snack for a quick energy boost.

Hunger and Appetite

People eat for two reasons: hunger and appetite. **Hunger** is *the natural physical drive to eat, prompted by the body's need for food.* When you're hungry, you may feel tired or lightheaded. Once you satisfy your hunger, you feel better. **Appetite** is *the* **psychological** *desire for food.* Think of how the smell of fresh-baked bread tempts you, even if you're full.

Academic Vocabulary

psychological *(adjective):* directed toward the mind

■ **Figure 10.1** Several factors can influence your food choices. *What might be influencing the teen in this photo?*

VStock/Alamy

📖 READING CHECK

Make Inferences
Why do advertisers want to influence your food choices?

Food and Emotions

Sometimes people eat in response to an emotional need, like when they feel stressed, frustrated, lonely, or sad. In other cases, people may snack out of boredom or use food as a reward. Some people engage in "mindless eating," which is snacking continuously while absorbed in another activity. They eat even when their body doesn't need food.

Using food to relieve tension or boredom can lead to weight gain, since you're eating when your body doesn't need food. On the other hand, if you lose your appetite because you're upset, your body may not get all the nutrients it needs. Recognizing how emotions affect your eating can help you break such patterns and reconnect your eating with real hunger.

Food and Your Environment

The people and things around you also affect what you choose to eat. Environmental influences include:

- **Family and culture.** If your family eats most meals at home, this will influence what you eat. You may prefer certain foods because of your family's cultural influence.

- **Friends.** If your friends always go for pizza after school, you'll probably eat pizza too. You might try new foods with friends, including foods from other cultures.

- **Time and money.** People with busy schedules may choose foods that are quick and easy to prepare, such as convenience foods and microwavable meals. Cost can also be a factor. For instance, you may not eat expensive steaks very often.

- **Advertising.** Advertisers try to influence your decisions about food. They hope that an ad for a juicy hamburger will send you out to the nearest fast-food window.

■ **Figure 10.2** Seeing what your friends and peers eat can influence your own food choices. *What are some of the ways your friends have influenced your eating habits?*

SW Productions/Getty Images

Health Skills — Activity

Analyzing Influences

Food Choices

Alex couldn't wait to get home from school the day his brother, Jeff, came home from college. When Alex got home, his brother's car was already parked in front of the house. They talked about school for a couple of minutes before Alex said, "Jeff, Mom said you get to pick what we have for dinner. Should we order a pizza?"

Jeff patted his stomach. "No pizza for me. I've already gained the 'freshman 15,'" referring to the weight gain that many college freshmen experience.

Jeff said that his schedule was different from when he was in high school, and he was finding it hard to start a new routine. His friends often encouraged him to eat even when he wasn't hungry. Alex wonders how Jeff can manage his diet.

Writing Pretend you are Jeff and use the following steps to create a plan analyzing the influences on your diet.

1. Keep a food diary for a week, noting what you eat, when you eat, and what influences your choices.
2. Analyze whether you're eating because of hunger or another reason.
3. Create a healthy eating plan that you can follow.

LESSON 1 ASSESSMENT

After You Read

Reviewing Facts and Vocabulary

1. Name three health problems that good nutrition can help you avoid.
2. What is the difference between *hunger* and *appetite*?
3. Identify two emotions that influence eating when someone isn't hungry.

Thinking Critically

4. **Analyze.** Explain how advertising can influence your food choices.
5. **Evaluate.** Emily can't resist home-made chocolate-chip cookies. What is influencing her behavior? Is this influence healthy?

Applying Health Skills

6. **Stress Management.** Eating when you're not hungry can be a response to stress. List three healthier ways to respond to stress.

Writing Critically

7. **Descriptive.** Write an essay describing two ways in which your family or friends influence your food choices.

Real Life Issues ....

After completing the lesson, review and analyze your response to the Real Life Issues question on page 254.

Nutrients

BIG Idea *Each nutrient in your diet plays a unique and essential role in keeping you healthy.*

Before You Read

Create a Cluster Chart. Draw a circle and label it "Nutrients." Draw circles around it and use these to define and describe this term. As you read, continue filling in the chart with more details.

New Vocabulary

▶ carbohydrates
▶ fiber
▶ proteins
▶ cholesterol
▶ vitamins
▶ minerals
▶ osteoporosis

Real Life Issues ·

Too Much (Nutrition) Information. Lately, Judy has been getting a lot of advice on what to eat—and all of it is different. First her friend Heather told her you need lots of carbohydrates and little fat. Judy's friend, Rob, said eating certain combinations of foods is a good idea. Then Judy read a magazine article stating that eating too many carbohydrates will cause weight gain. Judy is confused by all the information. She's beginning to feel she should just eat whatever she likes.

Writing *Write a journal entry from Judy's point of view. Have her describe how she will figure out whether the health information she's getting is valid or not.*

Giving Your Body What It Needs

Main Idea Each of the six nutrients has a specific job or vital function to keep you healthy.

Everything you eat contains nutrients. Nutrients perform specific roles in maintaining your body functions. Your body uses nutrients in many ways:

• As an energy source
• To heal, and build and repair tissue
• To sustain growth
• To help transport oxygen to cells
• To regulate body functions

There are six types of nutrients. Three of these types—carbohydrates, proteins, and fats—provide energy. The other three—vitamins, minerals, and water—perform a variety of other functions. Getting a proper balance of nutrients during the teen years can improve your health through adulthood.

Nutrients That Provide Energy

Main Idea Carbohydrates, proteins, and fats provide your body with energy and help maintain your body.

The energy in food comes from three sources: carbohydrates, proteins, and fats. Each gram of carbohydrate or protein provides four calories of energy, while each gram of fat provides nine calories. The body uses these nutrients to build, repair, and fuel itself.

Carbohydrates

Carbohydrates are *starches and sugars found in foods, which provide your body's main source of energy.* Most nutrition experts recommend getting 45 to 65 percent of your daily calories from carbohydrates.

Types of Carbohydrates There are three types of carbohydrates: simple, complex, and fiber. Simple carbohydrates are sugars, such as fructose (found in fruits) and lactose (found in milk). Sugars occur naturally in fruits, dairy products, honey, and maple syrup. They are also added to many processed foods, such as cold cereals, bread, and bakery products.

Complex carbohydrates, or starches, are long chains of sugars linked together. Common sources include grains, grain products such as bread and pasta, beans, and root vegetables such as potatoes.

The last type of carbohydrate is **fiber**, *a tough complex carbohydrate that the body cannot digest.* Fiber moves waste through your digestive system. Eating foods high in fiber can help you feel full, and may reduce the risk of cancer, heart disease and type 2 diabetes. Experts recommend that teen girls ages 14 to 18 eat 26 grams of fiber daily. Teen boys should eat 38 grams of fiber. Good sources of fiber include fruits and vegetables, whole grains, nuts, seeds, and legumes.

■ **Figure 10.3** These foods are good sources of carbohydrates. *How does your body use carbohydrates?*

The Role of Carbohydrates Your body uses carbohydrates by breaking them down into their simplest forms. Most of the carbohydrates you consume are turned into a simple sugar called glucose, which is the main source of fuel for the body's tissues. Glucose can be stored in your body's tissue and used later during periods of intense activity.

Benefits of Fiber Although the body cannot digest fiber, it still plays an important role by aiding digestion and reducing the risk of disease. Experts recommend eating 26 grams of total fiber daily for teen girls ages 14 to 18 years, and 38 grams daily for boys the same age.

Proteins

Proteins are *nutrients the body uses to build and maintain its cells and tissues.* They are made up of chemicals called amino acids.

Types of Proteins Your body uses about 20 amino acids that are found in foods. You produce, or synthesize, all but nine of the amino acids. These nine are called essential amino acids because the body must get them from food. The rest are known as nonessential amino acids.

Other proteins are from animal sources—such as meat, eggs, and dairy products—and from soy. They are sometimes called "complete" proteins because they contain all nine essential amino acids. Proteins from plant sources are usually missing one or more of the essential amino acids. However, you can get all the essential amino acids by eating a variety of plant-based foods that are rich in protein. Examples of these foods are grains, nuts, seeds, and legumes.

The Role of Proteins Protein is the basic building material of all your body cells. Muscles, bones, skin, and internal organs are all constructed of protein. Protein helps your body grow during childhood and adolescence. Throughout your life, protein will maintain muscles, ligaments, tendons, and all body cells.

Proteins also do a variety of other jobs in the body. For example, the protein hemoglobin in your red blood cells carries oxygen to all your body cells. Proteins may also function as hormones, chemicals that regulate the activities of your various body systems. Although protein does not supply energy to your body as quickly or easily as carbohydrates do, it can be used as an energy source.

Teen boys ages 14 to 18 should consume about 52 grams of protein per day, and teen girls ages 14 to 18 need 46 grams per day. Between 10 and 15 percent of your total daily calories should come from protein.

■ **Figure 10.4** All these foods are good sources of protein. *Which of these foods provide complete proteins?*

Fats

Most of what you hear about fats is how to avoid them. Does this mean you shouldn't eat any fat at all? No. Your body needs a certain amount of fat to function properly. You can, however, choose healthier fats.

Types of Fats Dietary fats are composed of fatty acids, which are classified as either unsaturated or saturated. Fatty acids that the body needs but cannot produce on its own are called essential fatty acids. The fat in all foods is a combination of unsaturated and saturated fats:

- **Unsaturated fats.** Vegetable oils, nuts, and seeds tend to contain larger amounts of unsaturated fats. Eating unsaturated fats in moderate amounts may lower your risk of heart disease.

- **Saturated fats.** Saturated fat is found mostly in animal-based foods such as meat and many dairy products. A few plant oils (palm, coconut, and palm kernel) also contain a lot of saturated fats. Consuming too many saturated fats may increase your risk of heart disease.

- **Trans fats.** These fats are formed by a process called hydrogenation, which causes vegetable oil to harden. As it hardens, the fats become more saturated. Trans fats can be found in stick margarine, many snack foods, and packaged baked goods, such as cookies and crackers. Trans fats can raise your total blood cholesterol level, which increases your risk for heart disease. As a result of the risk of trans fats, the USDA now requires that the amount of trans fats be listed on the nutrition label. Some cities have passed laws limiting or eliminating the use of trans fats in foods prepared in restaurants.

Health Issues of Fats Your body needs a certain amount of fat to carry out its basic functions, however, consuming too much fat can be harmful. Because fatty foods are generally high in calories, consuming a lot of them can lead to unhealthful weight gain and obesity.

The Role of Fats Fats provide a concentrated form of energy. The essential fatty acids are also important to brain development, blood clotting, and controlling inflammation. They also help maintain healthy skin and hair. Fats also absorb and transport fat-soluble vitamins (A, D, E, and K) through the bloodstream.

■ **Figure 10.5** Olive oil is a good source of healthful, unsaturated fat. *Why are unsaturated fats better for your health than saturated fats?*

The calories from fats that your body does not use are stored as body fat. Stored fat, known as adipose tissue, provides insulation for the body. However, carrying too much body fat increases the risk of health problems, such as type 2 diabetes and cardiovascular disease.

In addition, consuming saturated fats can increase the levels of **cholesterol**—*a waxy, fatlike substance*—in your blood. Cholesterol is needed to create cell walls, certain hormones, and vitamin D. However, excess cholesterol in your blood can build up on the insides of the arteries. This raises your risk of heart disease. Trans fats behave like saturated fats and promote cholesterol buildup in your arteries.

Nutrition experts recommend that teens consume less than 25 to 35 percent of their calories from fats because of the health risks associated with fats. Choose healthful unsaturated fats and limit your intake of saturated fats, including trans fats, to less than 10 percent of your total calories.

Other Types of Nutrients

Main Idea Vitamins, minerals, and water do not provide energy, but perform a wide variety of body functions.

Some nutrients do not supply calories but are still necessary for carrying out various body functions. These include vitamins, minerals, and water. Each vitamin and mineral performs a different function in the body.

Vitamins

Vitamins are *compounds found in food that help regulate many body processes*. There are several different vitamins that perform different functions in the body. (See **Figure 10.6** on page 263.)

Vitamin C, folic acid, and the B vitamins are water soluble, meaning they dissolve in water and pass easily into the bloodstream during digestion. The body doesn't store these vitamins; any unused amounts are removed by the kidneys. The fat-soluble vitamins (A, D, E, and K), by contrast, are stored in body fat for later use. If consumed in large amounts, these vitamins can build up in the body to the point where they become harmful.

Minerals

Minerals are *elements found in food that are used by the body*. Because your body cannot produce minerals, it must get them from food. **Figure 10.7** on page 264 lists some of the minerals your body needs and how it uses them.

READING CHECK

Cause and Effect
What are the benefits of choosing snacks labeled "no trans fat"?

FITNESS ZONE

My doctor says that a lot of teens don't get the nutrients they need because of poor eating habits. Teenagers tend to eat foods high in fat, like fast foods. We need to eat foods like fruits, vegetables, and low-fat dairy products. Now I try to eat the right amount of fruits and veggies every day. For more physical activity ideas, visit the Fitness Handbook and Fitness Zone sites in ConnectEd.

Figure 10.6 **Vitamins**

Vitamins in yellow boxes are fat soluble. Those in blue boxes are water soluble.

Vitamin/Amount Needed Per Day by Teens Ages 14 to 18	Role in Body	Food Sources
Fat-Soluble Vitamins		
A Teen female: 700 mcg Teen male: 900 mcg	needed for night vision; stimulates production of white blood cells; regulates cell growth and division; helps repair bones and tissues; aids immunity; maintains healthy skin and mucous membranes	carrots, sweet potatoes, tomatoes, fortified cereals, leafy green vegetables, fish, liver, fortified dairy products, egg yolks
D (calciferol) Teen female: 15 mcg Teen male: 15 mcg	helps body use calcium and phosphorus (needed for building bones); aids immune function; helps regulate cell growth	fortified cereals and dairy products, fatty fish such as salmon and tuna **Note:** Your skin naturally produces vitamin D when exposed to sunlight.
E Teen female: 15 mg Teen male: 15 mg	protects cells from damage; aids blood flow; helps repair body tissues	fish, milk, egg yolks, vegetable oils, fruits, nuts, peas, beans, broccoli, spinach, fortified cereals
K Teen female: 75 mcg Teen male: 75 mcg	essential for blood clotting, aids bone formation	green leafy vegetables, vegetable oils, cheese, broccoli, tomatoes
Water-Soluble Vitamins		
B$_1$ (thiamine) Teen female: 1.0 mg Teen male: 1.2 mg	helps the body use carbohydrates for energy; promotes health of nervous system	enriched and whole-grain cereal products, lean pork, liver
B$_2$ (riboflavin) Teen female: 1.0 mg Teen male: 1.3 mg	helps the body process carbohydrates, proteins, and fats; helps maintain healthy skin	lean beef, pork, organ meats, legumes, eggs, cheese, milk, nuts, enriched grain products
B$_3$ (niacin) Teen female: 14 mg Teen male: 16 mg	helps body process proteins and fats; maintains health of skin, nervous system, and digestive system	liver, poultry, fish, beef, peanuts, beans, enriched grain products
B$_6$ Teen female: 1.2 mg Teen male: 1.3 mg	helps body use proteins and fats; supports immune and nervous systems; helps blood carry oxygen to body tissues; helps break down copper and iron; prevents one type of anemia; helps maintain normal blood sugar levels	organ meats, pork, beef, poultry, fish, eggs, peanuts, bananas, carrots, fortified cereals, whole grains
B$_{12}$ (cobalamin) Teen female: 2.4 mcg Teen male: 2.4 mcg	maintains healthy nerve cells and red blood cells; needed for formation of genetic material in cells; prevents one type of anemia	liver, fish, poultry, clams, sardines, flounder, herring, eggs, milk, other dairy foods, fortified cereals
C (ascorbic acid) Teen female: 65 mg Teen male: 75 mg	protects against infection; promotes healthy bones, teeth, gums, and blood vessels; helps form connective tissue; helps heal wounds	citrus fruits and juices, berries, peppers, tomatoes, broccoli, spinach, potatoes
Folic acid (folate) Teen female: 400 mcg Teen male: 400 mcg	helps body form and maintain new cells; reduces risk of birth defects	dark green leafy vegetables, dry beans and peas, oranges, fortified cereals and other grain products

READING CHECK

Explain What is the difference between water- and fat-soluble vitamins?

One mineral that is especially important to your health is calcium. Calcium promotes bone health. Eating calcium-rich foods helps reduce your risk of developing **osteoporosis**, *a condition in which the bones become fragile and break easily.* Osteoporosis is most common in women over the age of 50. You can take action now to prevent the likelihood that you will develop osteoporosis when you're older. Bone mass builds up most rapidly between the ages of ten and 20, reaching its peak around age 30. Eating plenty of calcium-rich foods as a teen can protect your health years down the road.

Water

Water is essential for most body functions. All of the body cells contain water. Water's functions include

- moving food through the digestive system.
- digesting carbohydrates and protein, and aiding other chemical **reactions** in the body.
- transporting nutrients and removing wastes.
- storing and releasing heat.
- cooling the body through perspiration.
- cushioning the eyes, brain, and spinal cord.
- lubricating the joints.

Academic Vocabulary

reaction *(noun):* a response to a stimulus or influence

Figure 10.7 Minerals

Mineral/Amount Needed Per Day by Teens Ages 14 to 18	Role in Body	Food Sources
Calcium Teen female: 1,300 mg Teen male: 1,300 mg	forms bones and teeth; aids blood clotting; assists muscle and nerve function; reduces risk of osteoporosis	dairy products, calcium-fortified juice, calcium-fortified soy milk and tofu, corn tortillas, Chinese cabbage, broccoli, kale
Phosphorus Teen female: 1,250 mg Teen male: 1,250 mg	produces energy; maintains healthy bones	dairy products, peas, meat, eggs, some cereals and breads
Magnesium Teen female: 360 mg Teen male: 410 mg	maintains normal muscle and nerve function; sustains regular heartbeat; aids in bone growth and energy production	meat, milk, green leafy vegetables, whole grains, nuts
Iron Teen female: 15 mg Teen male: 11 mg	part of a compound in the red blood cells needed for carrying oxygen; aids in energy use; supports immune system	meat, poultry, beans, fortified grain products

Teen girls need about 9 cups of fluids a day, and teen boys need about 13 cups each day. About 20 percent of your total daily water intake comes from the foods you eat, since all foods contain some water. In most cases, drinking fluids with your meals and any other time you feel thirsty will supply your body with all the water it needs.

If you are very active, however, you will need to drink even more water to replace what your body loses when you sweat. Make sure to drink extra water before, during, and after exercise, even if you are not feeling thirsty. One important point to remember: if you feel thirsty, you waited too long to take in fluids. You should also drink extra fluids in hot weather to prevent dehydration. Limit your consumption of coffee, tea, and soft drinks that contain caffeine. Caffeine is a substance that eliminates water from your body, so caffeinated drinks can actually make you dehydrated.

■ **Figure 10.8** Water is essential for just about every function in your body. *When should you make sure to drink extra water?*

LESSON 2 ASSESSMENT

 After You Read

Reviewing Facts and Vocabulary

1. Which nutrients can your body use as sources of energy?

2. What are essential amino acids? From what source do you obtain essential amino acids?

3. How does eating calcium-rich foods as a teen protect your lifelong health?

Thinking Critically

4. **Analyze.** Explain how saturated fats and trans fats may cause illnesses later in life, like heart disease.

5. **Synthesize.** What are the health benefits of eating a variety of fruits and vegetables?

Applying Health Skills

6. **Goal Setting.** Examine your school's weekly lunch menu. List the most healthful food choices available each day. Then use the steps for goal setting to create a healthy eating plan.

Writing Critically

7. **Narrative.** Write a story from the point of view of a nutrient. Have the nutrient describe itself, what it does in the body, and why it is important for health.

Real Life Issues ....

After completing the lesson, review and analyze your response to the Real Life Issues question on page 258.

©Paul Bradbury/age fotostock

Healthy Food Guidelines

GUIDE TO READING

BIG Idea *MyPlate is a tool that can help you choose healthful foods for all your meals and snacks.*

Before You Read

Create an Outline. Preview this lesson by scanning the pages. Organize the headings and subheadings into an outline. As you read, fill in your outline with important details.

I.
A.
1.
2.
B.
II.

New Vocabulary

▶ Dietary Guidelines for Americans
▶ MyPlate
▶ nutrient-dense

Real Life Issues ·····························

No Time for Breakfast. Ever since she started high school, Tina never seems to have enough time for breakfast. Homework keeps her up late, so when she wakes up the next morning, she barely has time to get dressed and catch the bus. Most mornings in class, she feels weak and sluggish, and by lunchtime she's ravenous. Tina wants to find the time to eat breakfast so she has more energy throughout the day.

Writing *Pretend you are Tina. In a paragraph, write out a plan to fit breakfast into your busy schedule.*

Guidelines for Eating Right and Active Living

Main Idea **MyPlate helps you apply what you know about nutrients to choose healthful foods.**

The **Dietary Guidelines for Americans** are *a set of recommendations about smart eating and physical activity for all Americans.* These guidelines, published by the U.S. Department of Agriculture (USDA) and the Department of Health and Human Services (HHS), provide science-based advice for healthful eating. The guidelines also provide information on the importance of active living. This advice can be summed up in three key guidelines:

• Make smart choices from every food group.

• Find your balance between food and activity.

• Get the most nutrition out of your calories.

Making Smart Choices

Choosing a variety of foods from each food group will provide all the nutrients your body needs. There are five major food groups: Grains, vegetables, fruits, milk, and proteins.

MyPlate Use **MyPlate**—*an interactive guide to healthful eating and active living*—shown in **Figure 10.9**, to choose foods from all five of the food groups. MyPlate helps you put the Dietary Guidelines into action.

The MyPlate graphic shows a dinner plate covered with four differently colored triangles. The blue circle represents a glass of milk or other dairy product. On the plate, each of the triangles is a different size and is labeled. The size of the triangle represents a proportion. For example, the vegetables portion of the plate is larger than protein. This means that you should eat more vegetables than protein. The MyPlate Web site offers advice on how to choose healthful food sources for the fats you eat.

For more vocabulary practice, go to the Interactive Health Tutor at **glencoe.com**.

Figure 10.9	**MyPlate**

MyPlate uses a different color and size to represent each food group. Explain nutrition guidelines that the MyPlate graphic represents.

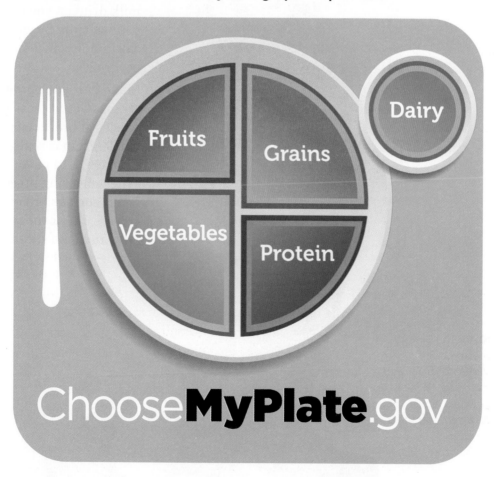

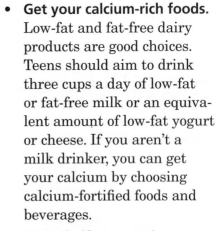

READING CHECK

Explain Why are the triangles on the plate different sizes?

Your level of physical activity should balance out the calories in the foods you eat. The MyPlate site provides individually tailored advice about your daily calorie needs based on your age, gender, and activity level.

Your Best Choices Within each food group, some choices are better than others. The Dietary Guidelines offer recommendations for choosing the most healthful foods from each food group:

- **Focus on fruits.** Eat a variety of fruits. Fresh whole fruits that provide fiber are a better choice than fruit juice.

- **Vary your veggies.** Vegetables fall into several categories. These categories include dark green vegetables, such as broccoli, kale, and spinach, and orange vegetables, such as carrots, pumpkin, and winter squash. Try to eat a good mix of different types of vegetables each day.

- **Get your calcium-rich foods.** Low-fat and fat-free dairy products are good choices. Teens should aim to drink three cups a day of low-fat or fat-free milk or an equivalent amount of low-fat yogurt or cheese. If you aren't a milk drinker, you can get your calcium by choosing calcium-fortified foods and beverages.

- **Make half your grains whole.** Get at least three ounces of brown rice or whole-grain cereals, breads, crackers, and pasta each day. When choosing processed foods, such as breads and cereals, check the ingredient label to make sure the grains are described as "whole."

■ **Figure 10.10** The Dietary Guidelines recommend choosing a variety of fruits and vegetables every day. *What fruits or vegetables would you choose for an afternoon snack?*

- **Go lean with protein.** Choose lean meats and poultry. Prepare them by grilling, baking, or broiling. Proteins, or any foods, that are prepared by frying in oil will add extra fat to your diet. This can increase the risk of overweight and obesity. Also, try getting more of your protein from fish, beans, peas, nuts, and seeds.
- **Limit certain foods.** Avoid foods that are high in fat—especially saturated fats and trans fats. Also, limit foods with salt and added sugars. Remember, it's okay to occasionally enjoy a few foods that are high in sugar, salt, or fat. If you enjoy eating a sweet snack each day, you can use physical activity to burn the extra calories.

Balancing Food and Physical Activity

Even if you eat the right amount and mix of healthful foods, you can still be overweight if you aren't getting enough physical activity. The Dietary Guidelines recommend that everyone balance the energy in the foods with regular physical activity.

The Dietary Guidelines recommend that teens should be physically active for 60 minutes almost every day to avoid unhealthy weight gain.

Getting the Most Nutrition Out of Your Calories

Every day your body needs a certain number of calories, depending on your age, your gender, and activity level. If you choose to spend your entire day's calorie needs with a single high-calorie fast-food meal, you may get the right amount of calories, but probably won't get the variety of nutrients your body needs. To make sure you get enough nutrients out of the foods you eat, choose **nutrient-dense** foods. These foods have *a high ratio of nutrients to calories.*

The more nutrient dense a food is, the more nutrients it packs into a given number of calories. For example, a single large carrot and a half ounce of potato chips have about the same number of calories, but the carrot is higher in nutrients. By eating more carrots and fewer potato chips, you will get more nutrients out of the same number of calories.

This doesn't mean that you have to give up all your favorite high-calorie foods. Any food that supplies calories and nutrients can be part of a healthful eating plan. You can plan to include them into your daily eating plan along with healthful, nutrient-dense foods. For example, try eating a small serving of potato chips with a lean, nutritious turkey sandwich with lettuce and tomato and some carrot or celery sticks. If your overall diet is nutrient dense, your eating plan can include an occasional treat.

FITNESS ZONE

Part of my fitness plan is to make healthier food choices. For me, that means just a few little changes. When I eat out, I choose healthy substitutes, like grilled chicken instead of fried chicken, or using ketchup and mustard instead of mayonnaise. It's not that hard to substitute, but it's made a big difference to my fitness level. For more physical activity ideas, visit the Fitness Handbook and Fitness Zone sites in ConnectEd.

Evaluate Your Eating Habits

"You are what you eat" is a common way of saying that your eating habits are important for your well-being. A diet that includes too much or too little of certain foods can affect your health now and in the future. The MyPlate Web site provides credible information about the dietary guidelines and how to get the nutrients you need. Keeping track of what you eat for a week can help you analyze your nutrient needs and make adjustments to your diet. Also, on the MyPlate Web site, you can create an eating plan based on your individual needs.

Activity Technology

To evaluate your eating habits, keep an online food diary for a week. Your food diary should record each food you eat and drink and the amount you consume.

Healthful Eating Patterns

Main Idea Use MyPlate and the Dietary Guidelines to plan all your meals and snacks.

Do you like to sit down to three meals a day, or do you prefer to eat six or more smaller meals throughout the day? MyPlate is flexible enough to adapt to just about any eating style. Some teens find it hard to make healthful choices in certain situations, such as breakfast time, eating on the go, or dining out. With a little planning, however, you can find ways to fit nutritious foods into any lifestyle.

Some people have trouble figuring out how to apply the Dietary Guidelines and MyPlate to their daily eating plan. One tool that can help is the plate diagram. With this tool, you can **visualize** how a healthful meal might look on your plate. **Figure 10.11** shows a plate diagram for one lunch or dinner. For breakfast, you might leave out the vegetables and high-protein foods and put the starchy food center stage. The colors on the plate match the colors on MyPlate. For example, orange represents grains, green shows a serving of vegetables, red is fruit, purple is for protein. Milk or other dairy products are shown in the glass.

Academic Vocabulary

visualize *(verb):* to form a mental image of

Figure 10.11 **What's on Your Plate?**

The plate diagram can help you visualize how much space you might devote to each type of food. *How might you adapt this diagram to a meal that has different types of foods mixed together, such as pasta with vegetables?*

Starting the Day Off Right

It's Monday morning, you've overslept, and you have just 20 minutes to get yourself out of bed and out the door. When you're hurried, it can be tempting to skip breakfast. However, you may pay the price later, when your stomach starts growling in the middle of a class. After eight hours of sleep, your body needs to refuel. If you force it to keep going, you will likely run short on energy.

Eating breakfast has many benefits for kids and teens. For example, children who eat breakfast typically do better in school and are less likely to be overweight. You may find it easier to fit breakfast into your schedule if you do some of the prep work the night before. For instance, you can set the table for breakfast before you go to bed. That way, all you have to do in the morning is fill your cereal bowl or put the bread in the toaster. Other ideas for quick and easy breakfasts are instant oatmeal or grits, hard-cooked eggs (which can be cooked the night before), and whole-grain muffins.

READING CHECK

Analyze a Graph
List ways that the plate diagram and the MyPlate guidelines are consistent.

■ Figure 10.12 Many different foods can be part of a healthful breakfast. *Name three nontraditional breakfast foods that you might like to try.*

If you simply don't care for traditional breakfast foods, there are plenty of other choices for starting your day off right. For instance, try a whole-grain bagel or toast with peanut butter or melted cheese. A breakfast burrito (eggs, cheese, and salsa on a tortilla) can also be a quick and healthy alternative. Another healthy choice may be to reheat last night's leftover spaghetti for breakfast.

Sensible Snacks

Healthful snacks can give you energy to keep you going between meals. Enjoying a sensible snack after school, for instance, can keep you from coming to the dinner table so hungry that you eat twice as much as you should. There are plenty of healthful foods that you can easily enjoy when you need a quick bite:

- Fresh fruit
- Cut-up vegetables, such as celery or carrot sticks
- String cheese
- Unsalted nuts
- Air-popped popcorn
- Fat-free yogurt
- Bread sticks

Eating Right When Eating Out

Making healthful food choices is just as important when you eat away from home. With a little effort, you can find the most healthful, nutrient-dense items on the menu. Here are a few tips to keep in mind:

- **Watch portion sizes.** Restaurant meals have grown larger over the years. If you think the serving size is more than you need, try splitting the meal with a friend or wrapping up the leftovers to take home.

- **Pay attention to how foods are prepared.** Anything fried is likely to be high in fat. Grilled, baked, and broiled foods are healthier choices.

- **Add fresh vegetables and fruits.** The salad bar can be a health-conscious eater's best friend. If the restaurant doesn't have one, order a salad off the menu or ask the server to provide extra lettuce and tomato for your sandwich.

- **Go easy on toppings.** High-fat sauces, mayonnaise, butter, and sour cream add fat and calories to a dish. You can make your meal lighter by asking the restaurant to leave these out or serve them on the side.

- **Don't drink your calories.** Choose water instead of soft drinks to satisfy your thirst without adding extra calories to your meal.

■ **Figure 10.13** Splitting a meal is one way to avoid overeating in a restaurant. *What are other strategies for choosing healthful foods when you go out to eat?*

LESSON 3 ASSESSMENT

After You Read

Reviewing Facts and Vocabulary

1. What are the five basic food groups?

2. What kinds of foods are best to avoid or limit?

3. Provide two examples of nutrient-dense foods.

Thinking Critically

4. **Analyze.** The Dietary Guidelines recommend regular physical activity. Why is this recommendation made?

5. **Synthesize.** Josh ate a cheeseburger, fries, and a soda for lunch. List the foods he could choose for dinner to balance out his lunch.

Applying Health Skills

6. **Accessing Information.** Search for information from credible sources that provide meal planning based on MyPlate.

Writing Critically

7. **Expository.** Write a description of a meal you had recently. Discuss what foods or cooking methods made this meal healthful or unhealthful.

Real Life Issues ....

After completing the lesson, review and analyze your response to the Real Life Issues question on page 266.

GUIDE TO READING

BIG Idea *By reading food labels and handling foods safely, you can avoid many food-related health problems.*

Before You Read

Organize Information. Fold a sheet of paper into quarters. Unfold it and label the four sections "Clean," "Separate," "Cook," and "Chill." As you read, fill in the sections with tips about the four steps in food safety.

Clean	Separate
Cook	Chill

New Vocabulary

▶ food additives
▶ foodborne illness
▶ pasteurization
▶ cross-contamination
▶ food allergy
▶ food intolerance

Nutrition Labels and Food Safety

Real Life Issues

Food Allergies. Alex is allergic to nuts. If he eats anything that contains nuts, his face swells up and he has to be taken to the hospital. He's learned to read food labels carefully to make sure nothing he eats has nuts in it. His friend Lauren has invited him to her house for dinner with her family. He'd like to say yes, but he knows that if anything they serve has nuts in it, he could be in serious trouble.

Writing *Write a paragraph explaining how you would handle this situation. How can Alex protect his safety and his friend's feelings at the same time?*

Nutrition Label Basics

Main Idea Food labels provide information about the ingredients and nutritional value of foods.

Whenever you buy a package of food, it has a label that tells you about the nutritional value of what's inside. The food label also lists all of the ingredients that were used to prepare the food. Among other things, the food label lists

- the name of the food product.
- the amount of food in the package.
- the name and address of the company that makes, packages, or distributes the product.
- the ingredients in the food.
- the Nutrition Facts panel, which provides information about the nutrients found in the food.

Ingredient List

The ingredients in a food appear on the label in descending order by weight. So, the ingredient that makes up the largest share of the weight comes first, followed by the one that makes up the next largest share of the weight, and so on. However, food labels that list several similar ingredients can be misleading. For example, a product that contains three kinds of sweeteners would list each one separately: *high-fructose corn syrup, corn syrup, sugar.* Therefore, the three sweeteners appear farther down on the list than they would if they were all listed as a single ingredient, *sugars.* This may give the impression that the product contains less added sugars than it really does.

Food Additives Some foods contain **food additives**, *substances added to a food to produce a desired effect.* Food additives may be used to keep a food safe for a longer period of time, to boost its nutrient content, or to improve its taste, texture, or appearance. Two food additives that concern some experts are aspartame, a sugar substitute, and olestra, a fat substitute. Many diet soft drinks are sweetened with aspartame. Some potato chips are made with olestra, which passes through the body undigested. Because olestra is not absorbed, some people experience gastrointestinal problems when eating it.

Nutrition Facts

The Nutrition Facts panel provides information about the nutrients found in the food. See **Figure 10.14** on page 276 for an example of a Nutrition Facts panel and the information it contains.

Nutritional Claims

Along with information about specific nutrients, food labels make other types of claims about nutritional value. Federal law gives uniform definitions for the following terms:

- **Free.** The food contains none, or an insignificant amount, of a given component: fat, sugar, saturated fat, trans fat, cholesterol, sodium, or calories. For instance, foods labeled as being "calorie-free" must have fewer than five calories per serving.

- **Low.** You can eat this food regularly without exceeding your daily limits for fat, saturated fat, cholesterol, sodium, or calories. Low-fat foods, for instance, must have three grams or less of fat per serving.

- **Light.** A food labeled as "light" must contain one-third fewer calories, one-half the fat, or one-half the sodium of the original version. On some packages, *light* may refer only to the color of the food, such as light brown sugar.

READING CHECK

Explain Why are additives used in foods?

FITNESS ZONE

After we learned to read food labels in health class, I started checking the label on everything I eat. I'm doing it as part of my overall fitness plan. I was surprised to see that fruit juice is high in calories! It's better to drink water or eat a piece of fruit. For more physical activity ideas, visit the Fitness Handbook and Fitness Zone sites in ConnectEd.

- **Reduced.** The food contains 25 percent fewer calories, or 25 percent less of a given nutrient, than the original version. This term may also be worded as *less* or *fewer*. Foods labeled as *reduced* may offer a much healthier option than the original version. The reduced version of a high calorie food may still contain a high number of calories.

- **High.** The food provides at least 20 percent of the daily value for a vitamin, mineral, protein, or fiber. Synonyms for this term include *rich in* and *excellent source of*.

- **Good source of.** The food provides 10 to 19 percent of the daily value for a vitamin, mineral, protein, or fiber. Synonyms for this term include *contains* and *provides*.

- **Healthy.** Foods described as healthy must be low in fat and saturated fat and contain limited amounts of cholesterol and sodium. They must also provide at least 10 percent of the daily value for vitamin A, vitamin C, iron, calcium, protein, or fiber.

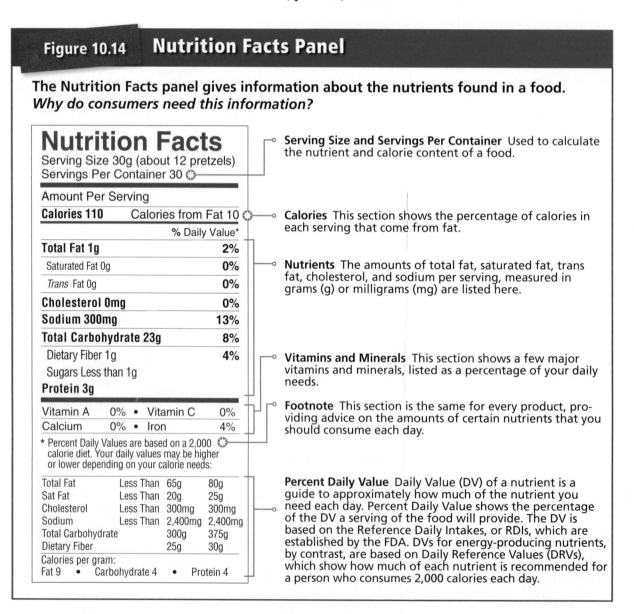

Figure 10.14   Nutrition Facts Panel

The Nutrition Facts panel gives information about the nutrients found in a food. *Why do consumers need this information?*

Nutrition Facts
Serving Size 30g (about 12 pretzels)
Servings Per Container 30 ⊙

Amount Per Serving	
Calories 110 Calories from Fat 10 ⊙	
	% Daily Value*
Total Fat 1g	**2%**
Saturated Fat 0g	**0%**
Trans Fat 0g	**0%**
Cholesterol 0mg	**0%**
Sodium 300mg	**13%**
Total Carbohydrate 23g	**8%**
Dietary Fiber 1g	**4%**
Sugars Less than 1g	
Protein 3g	

Vitamin A	0% •	Vitamin C	0%
Calcium	0% •	Iron	4%

* Percent Daily Values are based on a 2,000 ⊙ calorie diet. Your daily values may be higher or lower depending on your calorie needs:

Total Fat	Less Than	65g	80g
Sat Fat	Less Than	20g	25g
Cholesterol	Less Than	300mg	300mg
Sodium	Less Than	2,400mg	2,400mg
Total Carbohydrate		300g	375g
Dietary Fiber		25g	30g

Calories per gram:
Fat 9   •   Carbohydrate 4   •   Protein 4

Serving Size and Servings Per Container Used to calculate the nutrient and calorie content of a food.

Calories This section shows the percentage of calories in each serving that come from fat.

Nutrients The amounts of total fat, saturated fat, trans fat, cholesterol, and sodium per serving, measured in grams (g) or milligrams (mg) are listed here.

Vitamins and Minerals This section shows a few major vitamins and minerals, listed as a percentage of your daily needs.

Footnote This section is the same for every product, providing advice on the amounts of certain nutrients that you should consume each day.

Percent Daily Value Daily Value (DV) of a nutrient is a guide to approximately how much of the nutrient you need each day. Percent Daily Value shows the percentage of the DV a serving of the food will provide. The DV is based on the Reference Daily Intakes, or RDIs, which are established by the FDA. DVs for energy-producing nutrients, by contrast, are based on Daily Reference Values (DRVs), which show how much of each nutrient is recommended for a person who consumes 2,000 calories each day.

Organic Food Labels

In addition to nutritional claims, you may see one other notation on a food label: "USDA Organic." Foods labeled as *organic* are produced without the use of certain agricultural chemicals, such as synthetic fertilizers or pesticides. As well as not containing synthetic fertilizers or pesticides, these foods cannot contain genetically modified ingredients or be subjected to certain types of radiation. The USDA Organic label makes no claims, however, that organic foods are safer or more nutritious than conventionally grown foods.

Open Dating

Many food products have *open dates* on their labels. These dates help you determine how long the food will remain fresh. There are several types of open dates:

- **Sell by dates** show the last day on which a store should sell a product. After this date, the freshness of a food is not guaranteed.

- **Use by** or **expiration dates** show the last day on which a product's quality can be guaranteed. For a short time, most foods are still safe to eat after this date.

- **Freshness dates** appear on items with a short shelf life, such as baked goods. They show the last date on which a product is considered fresh.

- **Pack dates** show the day on which a food was processed or packaged. The pack date does not give the consumer an indication of the product's freshness.

(l)©USDA, (r)Andrew Resek/McGraw-Hill Education

Academic Vocabulary

item *(noun):* an object of concern or interest

■ **Figure 10.15** Foods bearing the USDA Organic label are produced without the use of certain agricultural chemicals. *Why might some consumers prefer these foods?*

Food Safety

Main Idea Handling food carefully can help you avoid foodborne illnesses and other hazards.

Have you ever seen a sign in a restaurant restroom reminding employees to wash their hands before returning to work? This restaurant policy helps prevent the spread of pathogens that can cause illness. It is one strategy for preventing **foodborne illness**, or *food poisoning.* About 76 million Americans become ill as a result of foodborne illnesses each year.

Foods can contain pathogens, or disease-causing organisms. Sometimes the pathogens produce disease. In other cases, it's the poisons that pathogens produce that cause illness. Some foods, such as certain mushrooms, that don't contain pathogens can still contain or produce poisonous chemicals. To protect yourself against foodborne illnesses learn what causes them and how to keep food safe.

How Foodborne Illness Occurs

Bacteria and viruses cause most cases of foodborne illness. The most common sources are the bacteria *Campylobacter, Salmonella, E. coli,* and a group of viruses known as the Norwalk and Norwalk-like viruses.

Some pathogens are naturally present in healthy animals. *Salmonella* bacteria can infect hens and enter their eggs. Shellfish may pick up bacteria that are naturally present in seawater. Fresh fruits and vegetables may become contaminated if they are washed with water that contains traces of human or animal wastes. Finally, infected humans who handle food can spread pathogens from their own skin to the food or from one food to another.

Some common symptoms of foodborne illness include cramps, diarrhea, nausea, vomiting, and fever. In most cases, people recover from foodborne illness within a few days. Occasionally, symptoms may be severe. Dehydration is one danger of foodborne illness. Fluids lost through vomiting and diarrhea can result in dehydration. If the following symptoms are present, consult a doctor:

- A fever higher than 101.5 degrees F
- Prolonged vomiting or diarrhea
- Blood in the stool
- Signs of dehydration, including a decrease in urination, dry mouth and throat, and feeling dizzy when standing

Keeping Food Safe to Eat

Food distributors and the U.S. government take steps to keep pathogens out of the food supply. One important process is pasteurization of milk and juices, which helps prevent *E. coli* infection. **Pasteurization** is *treating a substance with heat to kill or slow the growth of pathogens*. The Dietary Guidelines outline four basic steps for keeping food safe: clean, separate, cook, and chill.

Clean Wash and dry your hands frequently to keep pathogens on your skin from entering food. Be sure to wash your hands for at least 20 seconds with warm water and soap before and after handling food, as well as after using the bathroom, changing a diaper, or handling pets.

Clean utensils and surfaces carefully to prevent **cross-contamination**, *the spreading of pathogens from one food to another*. Wash cutting boards, dishes, utensils, and countertops with hot, soapy water after you finish preparing each food item. Mop up spilled food promptly using a paper towel or a clean cloth that has been washed in hot water.

Finally, wash the food itself. Rinse fresh fruits and vegetables under running water, and rub the surfaces of firm-skinned fruits and vegetables.

Separate The foods most likely to carry pathogens are raw meat, poultry, seafood, and eggs. To avoid cross-contamination, separate these from other foods. Store them separately when shopping and at home. Use separate cutting boards when preparing raw meats, poultry, and fish. After cooking meat, poultry, or fish, transfer the cooked food to a clean platter, rather than putting it back on the plate that held the raw food.

Cook Heating food to a high enough temperature will kill the pathogens that cause foodborne illness. To determine whether meat, poultry, and egg dishes are cooked thoroughly, use a food thermometer to measure the internal temperature (the temperature in the center of the food). **Figure 10.17** on page 280 shows the internal temperatures suggested for different foods.

READING CHECK

Explain Why is it best to always use warm water and soap when washing your hands?

■ **Figure 10.16** Washing hands, produce, utensils, and surfaces carefully is the first step in preventing foodborne illness. *How does this step prevent the spread of pathogens?*

For some foods, you can tell whether they are fully cooked by their appearance. Eggs should be firm, not runny; fish should be opaque and flake easily with a fork. When reheating soups, sauces, or gravy, bring the liquid to a boil. Heat all leftovers to 165 degrees F. When cooking food in a microwave oven, stir and rotate the food periodically to make sure there are no cold spots in which bacteria can survive.

Chill Refrigeration slows the growth of harmful bacteria. Refrigerate or freeze meat, poultry, and other perishable foods as soon as you bring them home from the store. Avoid overpacking the refrigerator; circulating air will help keep the food cool. Divide large amounts of food into small, shallow containers to help it cool more quickly.

Frozen foods should be thawed safely before cooking. Thaw frozen foods in the refrigerator, in a microwave, or under cold running water. Discard any food that has been sitting out at room temperature for two hours or longer—one hour when the temperature is above 90 degrees F.

Figure 10.17 **Safe Food Temperatures**

The top of this thermometer shows safe temperatures for cooking food, while the bottom shows safe temperatures for storing food. *Why is the area in the middle called the danger zone?*

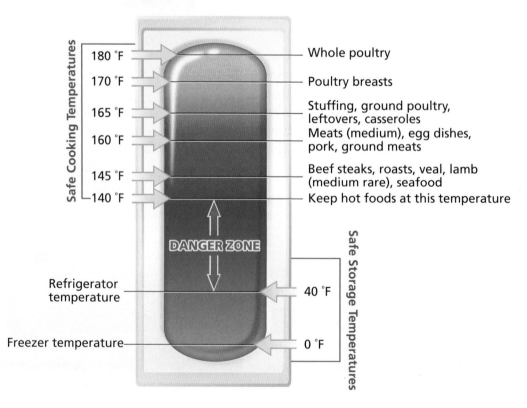

Food Sensitivities

Keeping pathogens out of food is important for everyone. Some people need to worry about specific foods. Food sensitivities—allergies and intolerances—can make some foods dangerous to eat. A **food allergy** is *a condition in which the body's immune system reacts to substances in some foods.* The most common allergens are found in milk, eggs, peanuts, tree nuts, soybeans, wheat, fish, and shellfish. Food labels are required telling whether a food product contains any of these ingredients or any protein derived from them.

The symptoms of food allergies vary from mild to life threatening. Some people experience skin irritations, such as rashes, hives, or itching while others develop gastrointestinal symptoms such as nausea, vomiting, or diarrhea. The most dangerous allergic reaction is anaphylaxis, a condition in which the throat swells up and the heart has difficulty pumping. Anaphylaxis can be life threatening and requires immediate medical attention.

A **food intolerance**—*a negative reaction to food that doesn't involve the immune system*—is more common than a food allergy. One of the most common is lactose intolerance, which occurs when a person's body does not produce enough of the enzyme needed to digest lactose, a sugar found in milk. People who are lactose intolerant may experience gas, bloating, and abdominal pain.

READING CHECK

Compare and Contrast What is the difference between a food allergy and a food intolerance?

LESSON 4 ASSESSMENT

After You Read

Reviewing Facts and Vocabulary

1. What does the term *light* mean when used on a food label?

2. What is the difference between a sell by date and a use by date?

3. What is another term that refers to *foodborne illness*?

Thinking Critically

4. **Evaluate.** An instant soup is very low in fat and calories but high in sodium. Can this food be labeled "healthy"? Explain why or why not.

5. **Synthesize.** What are the possible consequences of undercooked eggs?

Applying Health Skills

6. **Practicing Healthful Behaviors.** Summarize the steps for preventing foodborne illnesses. Post the steps in your kitchen as a reminder of food safety.

Writing Critically

7. **Persuasive.** Write an essay that convinces others of the importance of food safety.

Real Life Issues . . . .

After completing the lesson, review and analyze your response to the Real Life Issues question on page 274.

Hands-On HEALTH

Activity ## What's in the Bag?

Your group is opening a health food store and is looking for healthful foods to stock the store. Working in groups, pass around grocery bags containing nutrition labels. After reviewing the Nutrition Facts information in this chapter, analyze each label, and choose one food item from each bag to add to your store's inventory.

What You'll Need

- paper and pen or pencil
- one brown paper grocery bag per group
- seven to ten nutrition labels per bag

What You'll Do

Step 1

Choose a grocery bag for your group. Analyze the Nutrition Facts panel on each label and choose one.

Step 2

Write down the name of the food item and three reasons to support your group's choice.

Step 3

Exchange the bag with another group. Repeat steps 2 and 3 until you've selected one item from each bag.

Apply and Conclude

Describe your choices to the class. Include the reasons why your group selected each food item for your store.

Checklist: Accessing Information

✓ Did I access specific information from food labels?

✓ Did I use information on the labels to analyze the nutritional values of foods?

✓ Can I show that my choices are healthful?

LESSON **1**

The Importance of Nutrition

Key Concepts

▶ Nutrients supply your body with energy and help it to grow, repair itself, and function well.

▶ Hunger is a physical need for food. Appetite is a desire to eat.

▶ Family, culture, friends, time, money, and advertising can influence your food choices.

Vocabulary

▶ nutrition (p. 254)
▶ nutrients (p. 254)
▶ calorie (p. 254)
▶ hunger (p. 255)
▶ appetite (p. 255)

LESSON **2**

Nutrients

Key Concepts

▶ The six nutrients are carbohydrates, proteins, fats, vitamins, minerals, and water.

▶ Carbohydrates, proteins, and fats provide you with energy.

▶ Vitamins, minerals, and water do not provide energy but are necessary for many body functions and processes.

Vocabulary

▶ carbohydrates (p. 259)
▶ fiber (p. 259)
▶ proteins (p. 260)
▶ cholesterol (p. 262)
▶ vitamins (p. 262)
▶ minerals (p. 262)
▶ osteoporosis (p. 264)

LESSON **3**

Healthy Food Guidelines

Key Concepts

▶ The Dietary Guidelines for Americans provide recommenda-tions for healthy eating and regular physical activity.

▶ The five major food groups are grains, vegetables, fruits, milk, and meat and beans.

▶ It is important to eat nutrient-dense foods that have a high ratio of nutrients to calories.

Vocabulary

▶ Dietary Guidelines for Americans (p. 266)
▶ MyPlate (p. 267)
▶ nutrient-dense (p. 269)

LESSON **4**

Nutrition Labels and Food Safety

Key Concepts

▶ Food labels provide information about ingredients, nutritional value, serving sizes, and calories.

▶ Four steps to prevent foodborne illnesses are clean, separate, cook, and chill.

▶ People with food allergies or food intolerances must take special care about the foods they eat.

Vocabulary

▶ food additives (p. 275)
▶ foodborne illness (p. 278)
▶ pasteurization (p. 279)
▶ cross-contamination (p. 279)
▶ food allergy (p. 281)
▶ food intolerance (p. 281)

LESSON 1

Vocabulary Review

Use the vocabulary terms listed on page 283 to complete the following statements.

1. The process by which your body takes in and uses food is called _____.

2. Your body relies on food to provide it with the _____ it needs to grow, to repair itself, and to supply you with energy.

3. A _____ is a unit of heat used to measure the energy your body uses and the energy it receives from food.

Understanding Key Concepts

After reading the question or statement, select the correct answer.

4. Which of the following is not a way that choosing healthful foods affects your total health and wellness?
   a. It gives your body the nutrients it needs for growth and development.
   b. It helps you avoid unhealthful weight gain.
   c. It provides fuel for sports and other activities.
   d. It ensures that you will never get sick.

5. Preferring certain foods because you've grown up eating them is an example of the influence of
   a. family.
   b. friends.
   c. money.
   d. advertising.

Thinking Critically

After reading the question or statement, write a short answer using complete sentences.

6. **Analyze.** Why is emotional eating harmful?

7. **Synthesize.** How might the food choices of a high-powered business executive with a busy schedule differ from those of a part-time worker?

8. **Discuss.** Give an example of a way in which a person's cultural background could influence that person's food choices.

LESSON 2

Vocabulary Review

Choose the correct word in the sentences below.

9. Your body's main source of energy is *carbohydrates / proteins*.

10. Consuming saturated fats and trans fats can increase the levels of *fiber / cholesterol* in your blood.

11. *Vitamins / minerals* are elements found in food that are used by the body.

Understanding Key Concepts

After reading the question or statement, select the correct answer.

12. Which of the following is not one of the six basic nutrients?
    a. Carbohydrates
    b. Fiber
    c. Protein
    d. Vitamins

13. Your body uses carbohydrates by breaking them down into
    a. sugars.   c. fatty acids.
    b. amino acids.   d. water.

14. About what percentage of your daily calories should come from fat?
    a. 10 to 15 percent
    b. Less than 25 to 35 percent
    c. At least 30 percent
    d. 50 to 65 percent

Thinking Critically

After reading the question or statement, write a short answer using complete sentences.

15. **Describe.** How does fiber benefit your body?

16. **Explain.** Why is it dangerous to consume too much of a fat-soluble vitamin?

17. **Explain.** Why does your body need more water when you are very active?

LESSON 3

Vocabulary Review

Use the vocabulary terms listed on page 283 to complete the following statements.

18. The _____ contain recommendations about smart eating and physical activity for all healthy Americans.

19. An interactive guide to healthy eating and active living is the _____.

20. Foods that are _____ have a high ratio of nutrients to calories.

Understanding Key Concepts

After reading the question or statement, select the correct answer.

21. Which food group triangle in MyPlate is largest?
    a. Grains
    b. Fruits
    c. Vegetables
    d. Proteins

22. The Dietary Guidelines recommend that teens be physically active for
    a. 20 minutes, three or more times a week.
    b. 30 minutes a day.
    c. 50 minutes, five or more times a week.
    d. 60 minutes a day.

23. Which method of preparation tends to make food high in fat?
    a. Baking
    b. Broiling
    c. Frying
    d. Grilling

Thinking Critically

After reading the question or statement, write a short answer using complete sentences.

24. **Explain.** How can people who don't eat dairy products get enough calcium every day?

25. **Analyze.** Why is it important to include nutrient-dense foods in your daily eating?

26. **Identify.** Give two examples of healthful snacks.

LESSON 4

Vocabulary Review

Correct the sentences below by replacing the italicized term with the correct vocabulary term.

27. *Ingredients* may be used to keep a food fresh longer, to boost its nutrient content, or to improve its taste, texture, or appearance.

28. *Boiling* means treating a substance with heat to kill or slow the growth of pathogens.

29. It is important to clean utensils and surfaces carefully to prevent *foodborne illness,* the spread of pathogens from one food to another.

Assessment

Understanding Key Concepts

After reading the question or statement, select the correct answer.

30. Which of the following is *not* listed in the Nutrition Facts panel?
    a. The number of servings per container
    b. The number of calories per serving
    c. The vitamin and mineral content of the food
    d. The ingredients found in the food

31. Regular ice cream contains 7.5 grams of fat per serving. Ice cream that contains only 5 grams of fat per serving could be described as
    a. light.
    b. low-fat.
    c. reduced-fat.
    d. fat-free.

32. Which of the following is *not* one of the four basic steps for preventing foodborne illness?
    a. Clean
    b. Chop
    c. Cook
    d. Chill

Thinking Critically

After reading the question or statement, write a short answer using complete sentences.

33. **Compare and Contrast.** What is the difference between a food that is labeled "low-fat" and one that is labeled "reduced-fat"?

34. **Identify.** What are the usual symptoms of foodborne illness?

35. **Identify.** Name two foods that are common sources of allergens.

Technology PROJECT-BASED ASSESSMENT

The Importance of Nutrients

Background

Nutrients are the substances in food that your body needs. To have a healthful diet, your body needs six basic nutrients.

Task

You will work in a small group to create a wiki that a group of friends can use to help each other make healthy food choices.

Audience

Students in your class

Purpose

The purpose of the wiki is to help your peers learn to make healthy food choices based on the six basic nutrients and how the body uses them. The wiki will also show which foods provide the body with each of the nutrients.

Procedure

1. Review the information about the six groups of nutrients discussed in Chapter 10.

2. Create a wiki with your group that explains and gives examples and tips on how to make healthy food choices based on the six basic nutrients.

3. Be sure the healthy food choices are based on the six nutrients. Make any necessary revisions.

4. Obtain permission from your principal to post the wiki on the school's Web site.

Math Practice

Interpret Tables. To determine which food intake pattern to use, the following table gives an estimate of individual calorie needs. The calorie range for each age/sex group is based on physical activity level, from sedentary to active. *Sedentary* lifestyles include light physical activity. *Active* lifestyles include the equivalent to walking more than 3 miles per day at 3 to 4 miles per hour and the light physical activity typical of day-to-day life.

Calorie Range		
	Sedentary	Active
Females		
14–18	1,800	2,400
19–30	2,000	2,400
Males		
14–18	2,200	3,200
19–30	2,400	3,000

1. What are the approximate calorie needs of a sedentary 16-year-old male?
   - **A.** 1,800 calories
   - **B.** 2,000 calories
   - **C.** 2,200 calories
   - **D.** 2,400 calories

2. In 2000, the total number of active females age 14–18 in the United States was approximately 4,788,000. This was a 31 percent increase from the total number in 1975. What was the approximate number of active females age 14–18 in 1975?
   - **A.** 274,000
   - **B.** 398,000
   - **C.** 2,245,000
   - **D.** 3,655,000

3. About 35 percent of a 16-year-old male's calories should come from carbohydrates. Which most closely matches this number?
   - **A.** 1/4
   - **B.** 1/3
   - **C.** 3/5
   - **D.** 5/7

Reading/Writing Practice

Understand and Apply. Read the passage below and then answer the questions.

> Last weekend, after a game of basket-ball at the local community center, we all went to get a snack from the vending machine. Everything in the machine was high in fat, salt, or sugar. I put my money back in my pocket. My friends said, "Why don't you want anything?"
>
> Here's why. Last year, my dad found out he has high blood pressure. He's a bit overweight, so his doctor told him to cut out foods high in salt and fat. My parents didn't tell me to stop eating snacks like chips and cookies, but Dad's condition helped me understand that what I eat can affect my health.

1. What was the author's purpose in writing this piece?
   - **A.** To teach friends how to communicate better
   - **B.** To explain that eating better can affect your health
   - **C.** To persuade others to eat cookies rather than chips
   - **D.** To argue that healthy snacks taste better than unhealthy snacks

2. According to this text passage, high blood pressure may be related to
   - **A.** exercising occasionally.
   - **B.** choosing salty foods that are high in fat.
   - **C.** selecting low-fat foods that are salty.
   - **D.** eating foods that are high in fat and salt.

3. Write a paragraph giving your suggestions about how to improve eating habits. Provide details to support your main points.

National Education Standards

Math: Measurement and Data
Language Arts: LACC.910.L.3.6, LACC.910.RL.2.4, LACC.910.W.3.8

CHAPTER **11**

Managing Weight and Eating Behaviors

Ed-Imaging

Lesson 1

Maintaining a Healthy Weight

BIG Idea *Maintaining a healthy weight helps you protect your health and prevent disease.*

Lesson 2

Body Image and Eating Disorders

BIG Idea *Poor body image may lead to unhealthful and harmful eating behaviors.*

Lesson 3

Lifelong Nutrition

BIG Idea *Nutritional needs will change throughout your life.*

Activating Prior Knowledge

Using Visuals Look at the picture on this page. Based on what you see, how does the behavior of these teens contribute to their overall health? Write a paragraph explaining how these teens are taking care of their bodies.

Chapter Launchers

Health in Action

Discuss the **BIG** Ideas

Think about how you would answer these questions:

▶ What does it mean to have a healthy weight?

▶ Does your weight affect your self-image?

▶ How do your food needs differ from those of your friends?

Assess Your Health

Read each statement. On a separate sheet of paper, write "yes," "sometimes," or "no" based on your typical behavior.

1. I recognize that maintaining a healthy weight will help me stay healthy.

2. I know what body mass index (BMI) means and how it impacts my overall health.

3. I recognize that being underweight is as risky as being overweight.

4. I avoid fad diets.

5. I avoid using performance enhancing drugs.

A "yes" response shows that you practice healthy behaviors. "Sometimes" indicates that you should analyze and possibly modify your behavior. A "no" response means that you should modify the behavior.

BIG Idea *Maintaining a healthy weight helps you protect your health and prevent disease.*

Before You Read

Create a Venn Diagram. Draw two overlapping circles. Label them "Losing Weight" and "Gaining Weight." As you read, fill in the outer area of each circle with useful tips on the corresponding topic. Fill in the overlapping area with advice that is useful to everyone trying to maintain a healthy weight.

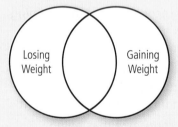

Losing Weight   Gaining Weight

New Vocabulary

▶ metabolism
▶ body mass index (BMI)
▶ overweight
▶ obese
▶ underweight

Maintaining a Healthy Weight

Real Life Issues ....

On the Unhealthy Track. Below are some examples of unhealthy habits. Source: Centers for Disease Control and Prevention, Youth Risk Behavior Surveillance, 2011.

> **13.8% of teens do not participate in at least 60 minutes of physical activity at least once a week.**

> **32.4% of teens watch three or more hours of TV on an average school day.**

Writing *Write a paragraph describing how a person can adopt a healthy habit.*

The Calorie Connection

Main Idea You maintain your weight by taking in as many calories as you use.

Calories are units used to measure the energy found in food. If you consume more calories than your body needs, you will gain weight. If you use more calories than you take in, you will lose weight. The balance between the calories you take in and those you burn is called energy balance.

Your Energy Balance

Your **metabolism**—*the process by which the body breaks down substances and gets energy from food*—converts the food you eat into fuel. It takes about 3,500 calories to equal 1 pound of body fat. Thus, if you consume 500 fewer calories than you use every day, you will lose 1 pound per week.

Figure 11.1

Calories in Common Snack Foods

High-Calorie Snack			Lower-Calorie Alternative		
Food Item	Serving Size	Calories	Food Item	Serving Size	Calories
Potato Chips	1 oz.	155	Pretzels	1 oz.	108
Cola	12 oz.	151	Water	16 oz.	0
Chocolate/caramel candy bar	1.6 oz.	208	Apple	1 medium	70
Chocolate sandwich cookies	6 cookies	282	Granola bar, raisin nut	1 oz.	127
Cream-filled snack cakes	2 (3 oz.)	314	Vanilla yogurt (low-fat)	8 oz.	193

How Many Calories?

As a rule, foods that are high in fat will also be high in calories. A gram of fat contains nine calories while a gram of protein or carbohydrate contains four. Some low-fat foods, however, may also be high in calories. Sugary foods contain more calories than fresh vegetables and fruits, which are higher in water and fiber.

Food preparation also plays a role in how many calories a food delivers. Fried foods, or those served with a cream sauce or otherwise prepared in a way that adds extra fats and sugars, are likely to be high in calories. To control your weight, eat less of high calorie foods or eat them less often. **Figure 11.1** compares the calories of common snack foods.

Maintaining a Healthy Weight

Main Idea Body mass index and body composition help you judge whether your weight is healthy.

To maintain a healthy weight, burn the same amount of calories that are consumed. The right weight for each person is based on several factors, including age, gender, height, body frame, and stage and rate of growth.

Body Mass Index

To learn if your body weight falls into a healthy **range**, calculate your **body mass index (BMI)**—*a measure of body weight relative to height.* Compare it to the charts on page 293 to determine if you're **overweight**—*heavier than the standard weight range for your height*—or at risk for being overweight.

READING CHECK

Predict What would probably happen if you increased your activity level without eating more food?

Academic Vocabulary

range *(noun):* the distance between possible extremes

It's important to remember that every teen grows at his or her own rate. It's normal that some of your friends will be taller or shorter than you, and that some will weigh more or less than you.

Body Composition

Although BMI is a quick, handy way to evaluate your weight, it doesn't tell the whole story. A person who is very muscular, for instance, may have a higher BMI but still be healthy. It's also important to consider your body composition—the ratio of fat to lean tissue in your body.

One commonly used method to measure your BMI is called skin-fold testing. It involves measuring the thickness of skin folds at different points on the body to figure out how much fat is stored beneath the skin. This test should be performed by a qualified professional.

Your Weight and Your Health

Main Idea Being either overweight or underweight carries health risks.

People whose weight does not fall into a healthy range are at a higher risk for various diseases. Weighing too much can increase your risk for health problems such as heart disease, cancer, asthma, osteoarthritis, gallbladder disease, or type 2 diabetes. Teens who weigh too little may feel weak, tire easily, or have trouble concentrating.

Weighing Too Much

You've probably heard that more than 15 percent of teens in the United States are overweight. This percentage has tripled since the 1980s. Teens who are overweight may be at risk of becoming **obese**—*having an excess of body fat*. Being obese carries serious health risks.

Photodisc/Getty Images

READING CHECK

Explain What does body mass index (BMI) measure?

■ **Figure 11.2** Staying active helps you maintain a healthy weight. *What might happen if these teens spent their afternoons playing video games instead of engaging in physical activity?*

Real World CONNECTION

Determining BMI

Here's an example for a 16-year-old male who is 6 feet tall and weighs 182 pounds.

182 ÷ 72 = 2.528

2.528 ÷ 72 = .035

0.035 × 703 = 24.6

BMI = 24.6

Use this formula to determine your BMI. First, convert your height into inches. Divide your weight in pounds by your height in inches. Divide that result by your height again, and multiply the result by 703.

Look at the charts to the right to determine if you're at risk for being overweight or underweight.

Activity **Quantities**

Some problems require more than one step to solve. Think through your approach before choosing an operation to use.

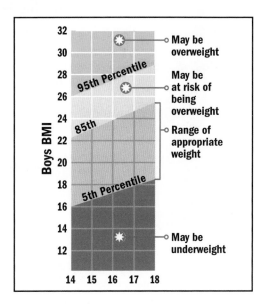

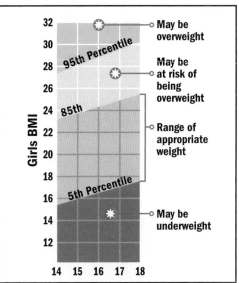

Some people are overweight or obese because of heredity or genetics. Some people may have a slow metabolism, which may lead to weight gain. However, many people who are overweight consume too many calories and get too little physical activity.

Weighing Too Little

Being **underweight**, or *below the standard weight range for your height,* also carries health risks. Some thin people may have trouble fighting off disease. Others are naturally thin because of genetics or because they have a fast metabolism.

■ **Figure 11.3** There is a wide range of weights that can be considered healthy. *Why should you avoid comparing your weight to that of your friends?*

Tim Fuller Photography

Teens may be thin because their bodies are growing very quickly. As their growth slows, their bodies may "fill out." For other teens, however, being too thin can mean that they aren't getting the calories and nutrients their growing bodies need, or that they are exercising excessively to burn calories.

If your BMI suggests that you might be underweight or overweight, get advice from a health care professional about healthful ways to gain or lose weight.

Managing Your Weight

Main Idea Stay physically active and eat healthful foods.

If your weight seems to be in a healthy range, then you probably don't need to worry too much about the number of calories you consume. If you want to lose or gain weight, however, you'll need to adjust either the number of calories you take in, the number you burn through physical activity, or both. The Dietary Guidelines for Americans does not recommend that teens diet. Instead, teens should try to eat a healthful, well-balanced diet every day to reach a healthy weight. Some healthful ways to manage your weight include the following strategies:

- **Target a healthy weight.** Learn your ideal weight range from a health care professional.
- **Set realistic goals.** Eat a consistently healthful diet and exercise regularly.
- **Personalize your plan.** Incorporate foods you enjoy into your daily eating plan.
- **Put your goals and plan in writing.** Write down your goals and your plan.
- **Evaluate your progress.** Track your weight on a weekly basis.

Healthful Ways to Lose Weight

MyPyramid provides information on food groups, recommended amounts, and the importance of physical activity. Here are some points to keep in mind:

- **Choose nutrient-dense foods.** Fruits, vegetables, and whole grains supply nutrients with fewer calories.
- **Watch portion sizes.** Stick to recommended portion sizes for each major food group.
- **Eat fewer foods that are high in fats and added sugars.** These add calories without many nutrients.

READING CHECK

Summarize What health problems may underweight teens have?

- **Enjoy your favorite foods in moderation.** Try enjoying a small scoop of ice cream less often.
- **Be active.** The information in **Figure 11.5** on page 296 compares the number of calories burned in different types of activity.
- **Tone your muscles.** Since muscle tissue takes more calories to maintain than fat, increasing your muscle mass means that your body will use more calories.
- **Stay hydrated.** Teens should drink between 9 and 13 cups of fluids a day.

Healthful Ways to Gain Weight

If you are trying to gain weight, the following strategies can help. Teens who want to gain weight should try to increase the amount of healthy muscle on their bodies, not fat. To gain healthy weight, continue a regular exercise program while using the strategies listed below.

- **Select foods from the five major food groups that are higher in calories.** Choose whole milk instead of low-fat or fat-free milk.
- **Choose higher-calorie, nutrient-rich foods.** Examples include nuts, dried fruits, cheese, and avocados.
- **Eat nutritious snacks.** Enjoy healthful snacks more often to increase your daily calorie intake.
- **Get regular physical activity.** If you're increasing your calorie intake to gain weight, don't forget exercise. Physical activity will ensure that most of the weight you gain is muscle rather than fat.

My dad always says, "If you always do what you always did, you'll always get what you always got." I realized he is saying that if you want to change the outcome, you have to change your behavior. I wanted to get into shape, but skipped workouts and ate junk food. When I changed my behavior, I got what I wanted. Making healthier food choices and exercising regularly improved my fitness level, and I felt a lot healthier. For more fitness tips, visit the Fitness Handbook and Fitness Zone sites in ConnectEd.

■ **Figure 11.4** Your food choices can help you either gain or lose weight. *List three nutritional qualities that make this lunch a good choice for someone trying to lose weight.*

Figure 11.5 | Calories Burned During Physical Activity

Physical activity may mean a brief workout. This graph shows how many calories a person weighing 125 to 175 pounds can burn doing each activity for 10 minutes.

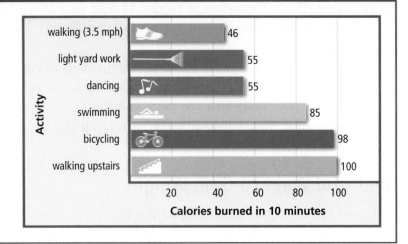

Activity

walking (3.5 mph) — 46
light yard work — 55
dancing — 55
swimming — 85
bicycling — 98
walking upstairs — 100

Calories burned in 10 minutes

Physical Activity and Weight Management

Physical activity can help you lose or maintain a healthy weight. Some added benefits of regular physical activity:

- It helps relieve stress.
- It promotes a normal appetite response.
- It increases self-esteem, which helps you keep your plan on track.
- It helps you feel more energetic.

LESSON 1 ASSESSMENT

 After You Read

Reviewing Facts and Vocabulary

1. What is *metabolism*?
2. Explain how to calculate your body mass index.
3. List three health problems associated with being overweight and obese.

Thinking Critically

4. **Analyze.** Explain how exercise that builds muscle can help promote loss of body fat.
5. **Synthesize.** Mike is 15 years old. He is 5 feet 9 inches tall and weighs 180 pounds. Explain whether his weight is in a healthy range.

Applying Health Skills

6. **Practicing Healthful Behaviors.** Write a plan describing the strategies you will use to maintain a healthy weight throughout your life.

Writing Critically

7. **Expository.** Write a short article aimed at middle school students describing the causes and effects of overweight and obesity problems among teens.

Real Life Issues ....

After completing the lesson, review and analyze your response to the Real Life Issues question on page 290.

Body Image and Eating Disorders

Real Life Issues

Warning Signs of an Eating Disorder. People with eating disorders such as anorexia nervosa have an irrational fear of gaining weight. Would you recognize the signs of anorexia in someone?

Could It Be Anorexia?
Obsessed with weight loss
Brittle hair and nails
Dry and yellowish skin
Constantly feeling cold

Writing *Write a dialogue of how you would talk to a friend you suspect may have an eating disorder. What would you say to show caring and concern?*

Your Body Image

Main Idea The media and other influences can affect your body image.

When you look in the mirror, do you like what you see? If the answer is yes, that means that your **body image**—*the way you see your body*—is positive. Though many teens like the way they look, many others feel insecure about their changing bodies. During your teen years, you will experience many physical changes at a rapid pace. You may feel unhappy with your body type and wish you were taller, shorter, thinner, shapelier, or more muscular.

Where does body image come from? Some teens may compare their bodies to those of models, athletes, or actors. You should keep in mind that the images shown in the media aren't always realistic. Peers can also influence body image. Overweight or underweight teens may feel pressured by their friends and others to look a certain way.

GUIDE TO READING

BIG Idea *Poor body image may lead to unhealthful and harmful eating behaviors.*

Before You Read

Create a Comparison Chart. Make a chart and label the rows "Anorexia," "Bulimia," and "Binge Eating." Label the columns "Symptoms" and "Health Risks." As you read, fill in the chart with information about these eating disorders.

	Symptoms	Health Risks
Anorexia		
Bulimia		
Binge Eating		

New Vocabulary

▶ body image
▶ fad diets
▶ weight cycling
▶ eating disorders
▶ anorexia nervosa
▶ bulimia nervosa
▶ binge eating disorder

Figure 11.6 Media images may influence teens' views about ideal body types. *Do you think trying to look like magazine models is a realistic or healthy goal?*

READING CHECK

Define What is body image?

Accepting Yourself

The rapid pace of physical changes you experience during your teen years may affect your body image. Growth spurts may cause some teens to look thin. Hormonal changes may cause other teens to gain weight. Try to accept yourself the way you are, or talk to a parent or other trusted adult about your feelings. You can't change your basic body type, and you could hurt your health if you try.

Fad Diets

> **Main Idea** Fad diets are neither safe nor reliable ways to lose weight.

Teens who want to lose weight may be tempted to try **fad diets,** *weight-loss plans that tend to be popular for only a short time.* Fad diets typically promise quick, easy weight loss. People on these diets may lose weight temporarily, but they usually regain it after going off the diet. As a result, they may fall into **weight cycling,** *a repeated pattern of losing and regaining body weight.*

Fad diets may restrict the types and amounts of foods that you eat, making it difficult to stay on them for a long time. Other fad diets use pills or supplements that seem to offer an easy solution to weight loss. Research has shown that they are not effective. Fad diets can **pose** serious health risks.

In fact, most teens should not diet at all. Teens who feel they may need to lose weight should talk to their doctor before starting any diet plan. In rare circumstances, teens with a serious weight problem may be advised to follow a low-calorie plan, but only under the supervision of a health care professional. In general, teens who want to maintain a healthy weight should follow the nutrition guidelines of MyPyramid and get regular physical activity.

Academic Vocabulary

pose *(verb):* to present or set forth

Types of Fad Diets

All fad diets lose their popularity once people realize that they're unhealthy and that they just don't work. Still, certain types of fad diets keep reappearing every few years in a different form. Here are some common ones:

- **Miracle foods.** These plans promise you can "burn fat" by eating lots of a single food or type of food. In reality, there is no single food that can destroy fat. Moreover, eating only certain types of food will not give your body all the nutrients it needs.

- **Magic combinations.** These plans promise that certain foods will trigger weight loss when they're eaten together. The food combinations may be safe to eat as part of an overall healthy diet, but there's no evidence that combining certain foods will lead to weight loss.

- **Liquid diets.** These plans replace solid food with ultra-low-calorie liquid formulas. These diets can lead to dangerous side effects if they are followed incorrectly. However, doctors may recommend them (with medical supervision) for people who are seriously obese.

- **Diet pills.** Some diet pills and supplements claim to suppress your appetite so that you eat less. Others claim to "block" or "flush" fat from the body. Diet pills can be addictive. In addition, they may cause drowsiness, anxiety, a racing heart, or other serious side effects.

- **Fasting.** Fasting deprives the body of needed nutrients and can result in dehydration. Some religious and cultural customs require people to fast for short periods, such as specific days or times of the day during certain months. This kind of short-term fasting is safe for most people.

Recognizing Fad Diets

How can you tell the difference between a fad diet and a legitimate weight-loss plan? Any plan that does not follow the MyPyramid guidelines may deprive your body of nutrients. Plans that promise ultra-fast weight loss (more than 2 pounds a week) are likely to be unsafe or ineffective. Plans that promise you can lose weight without boosting your physical activity also are likely to be unsafe or ineffective. Watch out for such words as *effortless, guaranteed, miraculous, breakthrough, ancient,* or *secret.* Diets that require you to buy certain products rather than choose healthful foods should also raise your suspicions. Finally, be skeptical about claims that "doctors don't want you to know about this weight-loss plan." Ask yourself, "Why would my doctor want to keep me from reaching a healthy weight?"

■ **Figure 11.7** Combining a healthful, lower-calorie eating plan with physical activity is a more reliable way to lose weight than any fad diet. *What are some drawbacks of fad diets?*

READING CHECK

Explain What are some typical characteristics of fad diets?

Eating Disorders

Main Idea Eating disorders are extreme and dangerous eating behaviors that require medical attention.

As you have learned, some types of weight-loss diets are unhealthy. Yet some people eat in ways that are more harmful to their health. They suffer from **eating disorders**—*extreme, harmful eating behaviors that can cause serious illness or even death*. Eating disorders are classified as mental illnesses, and they are often linked to depression, low self-esteem, or troubled personal relationships. Social and cultural forces that emphasize physical appearance can also play a role. Eating disorders often run in families. Research also suggests that genetics may be a factor in the development of eating disorders.

Anorexia Nervosa

Anorexia nervosa is *an eating disorder in which an irrational fear of weight gain leads people to starve themselves*. It mainly affects girls and young women. People with anorexia see themselves unrealistically as overweight even when they are dangerously thin. The disorder affects a person's self-concept and coping abilities. Outside pressures, high expectations, a need to be accepted, and a need to achieve are characteristics associated with anorexia nervosa. Medical specialists say that genetics and other biological factors may also play a role in the development of this disorder. Often, people with anorexia develop obsessive behaviors related to food, such as

- avoiding food and meals.
- eating only a few kinds of food in small amounts.
- weighing or counting the calories in everything they eat.
- exercising excessively.
- weighing themselves repeatedly.

Health consequences of anorexia nervosa are related to malnutrition and starvation. The bones of people with eating disorders may become brittle. Body temperature, heart rate, and blood pressure may drop, and there may be a reduction in organ size. Anorexia nervosa can lead to heart problems and sudden cardiac death.

■ **Figure 11.8** People with anorexia can be very thin and still see themselves as overweight. *What are other symptoms associated with anorexia?*

Bulimia Nervosa

Bulimia nervosa is *an eating disorder that involves cycles of overeating and purging, or attempts to rid the body of food.* People with bulimia regularly go on *binges*, eating a huge amount of food in a single sitting. During the binge they may feel out of control, often gulping down food too fast to taste it. After the binge they purge, forcing themselves to vomit or taking laxatives to flush the food out of their systems. Instead of purging, some people with bulimia may fast or exercise frantically after a binge.

Unlike people with anorexia, bulimia sufferers are typically in the normal weight range for their age and height. However, they share the same fear of weight gain and dissatisfaction with their bodies.

Health consequences of bulimia nervosa include dehydration, sore and inflamed throat, and swollen glands. The teeth of people with bulimia nervosa may become damaged by regular exposure to stomach acid from vomiting. They may also damage their stomach, intestines, or kidneys. In severe cases, the chemical imbalances that result from purging can lead to irregular heart rhythms, heart failure, and death.

Binge Eating Disorder

Binge eating disorder is *an eating disorder in which people overeat compulsively.* They binge in much the same way people with bulimia do, eating large amounts of food in a short period of time. These eating binges generally do not occur as frequently as binges associated with bulimia. During a binge, the person may feel guilty and disgusted about his or her behavior, but feel powerless to stop it. Binge eating disorder is more common in males than any other eating disorder, accounting for more than a third of all cases.

Consequences of binge eating disorder include becoming overweight or obese. People with binge eating disorder do not purge. They can also develop the health problems associated with obesity, including high blood pressure, type 2 diabetes, and cardiovascular disease.

Seeking Help

Eating disorders are serious and dangerous illnesses. People with these disorders need help to overcome them. Medical help may involve counseling, nutritional guidance, a doctor's care, and in extreme cases, a hospital stay.

For anorexia nervosa, the goal of treatment is to restore the patient's body weight to a healthy level. The patient also receives psychological and family therapy. Family members and friends can help by creating a supportive environment and helping the patient learn to eat normally again.

FITNESS ZONE

Some of my friends skip breakfast to lose weight. Some say they don't have the time to eat it. My dad says that he read a magazine article saying that people who eat breakfast every day are more likely to lose weight and keep it off. Eating breakfast gives you an energy boost. It also increases your metabolism, which means that you burn more calories. I think it's so cool that you can actually eat to maintain a healthy weight or lose weight. For more fitness tips, visit the Fitness Handbook and Fitness Zone sites in ConnectEd.

 READING CHECK

Compare and Contrast How is binge eating disorder different from bulimia nervosa?

Figure 11.9 Self-help groups in some communities provide ongoing support for people recovering from eating disorders. *How might being part of such a group be helpful?*

©Vibe Images/Alamy

The key to treating bulimia nervosa is to break the cycle of binging and purging. Behavioral therapy can sometimes help with this goal. After that, psychotherapy can address the emotional problems that led to the eating disorder. Similar treatments are used for binge eating disorder.

People with eating disorders often cannot admit that they have a problem. Family members and friends can help them recognize the problem and seek treatment. Some patients end up requiring long-term care to recover. Support groups can help with this process. If you think that you or someone you know may have an eating disorder, your first step might be to talk to a trusted adult, such as a parent, counselor, or school nurse.

LESSON 2 ASSESSMENT

 After You Read

Reviewing Facts and Vocabulary

1. List two factors that influence body image.
2. Define *fad diets*.
3. List three types of eating disorders.

Thinking Critically

4. **Synthesize.** How might a poor body image result in an eating disorder?
5. **Evaluate.** If you read an ad in a magazine promising you can lose up to 15 pounds in one month while still eating all your favorite foods, would you think this was a fad diet or a legitimate plan? Explain why.

Applying Health Skills

6. **Analyzing Influences.** Write an essay describing how teen magazines portray teens and their bodies. How might the magazine's pictures affect the body image of teens?

Writing Critically

7. **Narrative.** Write a story about a teen who seeks help for an eating disorder. Describe the symptoms and how the disorder affects the teen's life.

Real Life Issues ....

After completing the lesson, review and analyze your response to the Real Life Issues question on page 297.

Lifelong Nutrition

Real Life Issues ....

A Personal Choice. Miranda is facing a moral dilemma. After spending her summer at a camp where she helped care for farm animals, she's become uncomfortable with the idea of eating meat. She thinks of animals as friends, and she doesn't like the thought of eating them. However, food is an important part of her family's traditions. Miranda is afraid her parents won't understand if she tells them she doesn't want to eat meat anymore.

Writing *Write a dialogue in which Miranda tries to explain to her parents her desire to become a vegetarian.*

Lifelong Nutritional Needs

Main Idea Your age, gender, lifestyle, and health needs can affect your body's food needs.

An active 16-year-old girl has different nutritional needs than an 80-year-old man. Also, we all have individual nutritional preferences and considerations. Some people choose to eat only plant-based foods. A person with an allergy to nuts will avoid those foods. There are several factors that affect your nutritional needs, including the following:

- **Age.** During your teen years, your body's calorie needs increase to support your growth. As you get older, your needs will change based on your activity level.

- **Gender.** On average, females tend to need fewer calories than males. Throughout their lives, females have a greater need for some nutrients, like iron and calcium. Pregnant women also have special nutritional needs. To ensure the health of their babies, pregnant women need extra calcium, iron, and folic acid, and more calories.

GUIDE TO READING

BIG Idea *Nutritional needs will change throughout your life.*

Before You Read

Create a Cluster Chart. Draw a circle and label it "Special Nutritional Needs." Draw surrounding circles and use these to describe nutritional needs for vegetarians, athletes, and those with health conditions. As you read, continue filling in the chart with more details.

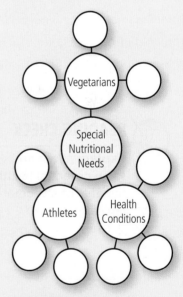

New Vocabulary

▶ vegetarian
▶ dietary supplements
▶ performance enhancers
▶ herbal supplements
▶ megadoses

■ **Figure 11.10** Vegetarians can choose from a wide variety of healthful and tasty foods to meet their protein needs. *Name the foods in each dish that are a source of protein.*

READING CHECK

Explain What are some potential health benefits of a vegetarian eating style?

• **Activity level.** The more active you are, the more calories your body needs. Very active people need to consume more calories, preferably from nutrient-dense foods, to maintain their weight.

Vegetarian Diets

A **vegetarian** is *a person who eats mostly or only plant-based foods.* There are several different types of vegetarianism. The strictest vegetarians, known as vegans, eat only plant-based foods. Other types of vegetarians include

• Lacto-ovo vegetarians who also eat dairy foods and eggs,

• Lacto vegetarians who add dairy foods to their diet, and

• Ovo vegetarians who include eggs in their diet.

People may choose a meatless diet for many reasons. Many believe a vegetarian diet is more healthful. Still others are vegetarians for religious, cultural, or economic reasons—or because they simply prefer vegetarian foods.

Advantages of the vegetarian eating style are that plant-based foods tend to be lower in saturated fat and cholesterol, and higher in fiber, than most animal-based foods. As a result, a well-planned meatless diet may help reduce the risk of cardiovascular disease and some types of cancer.

Drawbacks of the vegetarian eating style are that plant-based foods tend to be lower in certain nutrients, such as protein, iron, calcium, zinc, and some B vitamins. One nutrient, vitamin B^{12}, is found only in animal-based foods. Some vegetarians may need to take **dietary supplements**—*products that supply one or more nutrients as a supplement to, not a substitute for, healthful foods*—to obtain all the nutrients.

A healthful vegetarian diet contains a variety of foods, including plenty of vegetables, fruits, and whole-grain foods. Choices such as nuts and legumes, as well as eggs and dairy products, can help vegetarians consume enough protein.

Health Conditions

The foods people eat can **trigger** certain diseases or health conditions. People with these conditions may need to avoid or limit certain foods in order to avoid health problems. Below are some of the health conditions that can be affected by foods:

Academic Vocabulary

trigger *(verb):* to initiate or set off

- **Diabetes.** People with diabetes have to monitor their eating carefully to make sure their blood sugar stays in a healthy range. Diabetics who use insulin must tailor the amount of insulin they inject to the amount of carbohydrates in the foods they eat. Others may be able to control their diabetes without medication by carefully controlling the carbohydrates in foods and beverages they consume. Those who are overweight may find that losing weight helps them control their blood sugar.

- **Food allergies.** Food allergies can range from merely annoying to life threatening. People with severe food allergies must avoid the foods and food ingredients they are allergic to. This means checking ingredient lists on packaged foods and quizzing servers at restaurants. This can be difficult, but not nearly as difficult as being rushed to the emergency room.

- **Lactose intolerance.** People with this food intolerance can't easily digest the lactose in milk and some dairy products. Some people with lactose intolerance can control the problem by consuming smaller portions of milk or getting their calcium from cheese and yogurt, which contain less lactose. Some may take lactase (the enzyme needed to digest lactose) in liquid or tablet form when they eat dairy products.

- **Celiac disease.** Also known as gluten intolerance, this condition makes people unable to tolerate a protein called gluten, which is found in wheat, rye, and barley. Oats may also be harmful to those with celiac disease. The only treatment is to avoid these grains and anything made from them, including bread, pasta, and beer.

- **High blood pressure.** Consuming salt can raise a person's blood pressure. This effect is stronger in some individuals than in others. People with high blood pressure are often encouraged to keep their salt intake low.

- **High cholesterol.** People with high cholesterol may need to reduce their intake of saturated fats and trans fats. These fats increase cholesterol production in the body.

READING CHECK

Name What are three health conditions that are caused by foods?

Nutrition for Athletes

Eating right affects an athlete's performance. Like everyone else, athletes need a balanced diet that supplies enough nutrients to support health. The most important difference is that when you're very active, you need more calories to provide additional fuel. Teen athletes may need from 2,000 to 5,000 calories per day, depending on their sport and on the intensity, length, and frequency of their training.

Athletes need more protein and carbohydrates than inactive people. They may also need more calories from nutrient-dense foods and foods higher in carbohydrates. These types of foods will help student athletes maintain their energy and keep their weight up for athletic competition.

Making Weight In sports such as wrestling or boxing, your weight is important because you have to compete with others in the same weight class. If your sport requires you to "make weight," you must compete at a weight that's right for you. Your ideal weight will put your BMI in a healthy range and allow you to eat enough to meet your daily nutrient needs.

Some athletes try to compete in a weight class that's too low for them because they think they will have a better chance against smaller opponents. In reality, they are hurting themselves by trying to force their bodies down to a weight that isn't healthy. Extreme measures such as fasting or trying to sweat off extra weight can cause dehydration, harming performance as well as health. If you really do need to lose some weight, stick to a sensible plan that will take off ½ to 1 pound per week.

■ **Figure 11.11** Drinking plenty of water during a workout helps you avoid dehydration and keep performing at your peak. *What problems can dehydration cause?*

Hydration

Teen girls should try to drink about 9 cups of noncaffeinated fluids each day, and teen boys should try to drink 13 cups. Student athletes may need more fluids. When you sweat during exercise, your body loses fluids. These fluids must be replaced to avoid dehydration and heatstroke. Dehydration can lead to fatigue, dizziness or light-headedness, and cramping. Becoming dehydrated can lead to an imbalance of *electrolytes*—minerals that help maintain the body's fluid balance. The minerals sodium, chloride, and potassium are all electrolytes. To prevent dehydration, drink water before and after you exercise, and about every 15 minutes or so during a workout.

Avoiding Performance Enhancers

Some athletes try to gain an extra edge by using **performance enhancers**—*substances that boost athletic ability.* Many of these substances pose health risks, especially for teens. Using performance enhancers is illegal and has been banned under the rules of many sports organizations. Some of the best-known performance enhancers include the following:

- **Anabolic steroids.** These dangerous drugs, which are illegal without a doctor's prescription, have the same effect as male hormones (known as androgens) in the body. Athletes who take steroids disregard the many health risks in an effort to boost muscle growth. You will learn more about steroids and their risks in Chapter 22.

- **Androstenedione** (better known as "andro"). Andro is a weaker form of the androgens that the body produces naturally. Although some athletes take it to build muscle, its actual benefits are doubtful. Andro has many of the same side effects as steroids, and its use is now banned in professional sports.

- **Creatine.** This compound helps release energy. Some athletes take creatine supplements to give them a quick burst of power and reduce muscle fatigue. However, it can actually hurt athletic performance on account of its side effects, which include cramps and nausea. Using it at high doses may damage the heart, liver, and kidneys.

- **Energy Drinks.** These contain high amounts of caffeine. Energy drinks provide quick energy in an unhealthful way by increasing your heart rate. Using energy drinks to enhance your performance may actually hurt you. Drinking caffeinated beverages may cause your body to lose more fluids, leading to dehydration.

Using any type of performance enhancer, whether it's legal or illegal, is not worth the risk. You can perform at your peak without using performance enhancers by training, eating right, and getting enough rest.

Eating Before a Competition

Eating before a competition provides your body with the energy it needs to get through the competition. Try to eat about three to four hours before a competition so that your stomach is empty by the time you compete.

Before competing, try to choose meals that are high in carbohydrates and low in fat and protein. Fat and protein stay in the digestive system for a longer period of time. Good choices of foods to eat before a competition include pasta, rice, vegetables, breads, and fruits. Also, remember to drink plenty of water before, during, and after the competition.

FITNESSZONE

I carry healthy snacks in my backpack every day. When I get hungry, I'm not tempted to buy something that's not healthy. An apple or a small bag of carrots are easy to carry. They're healthier than a candy bar or chips, too. For more fitness tips, visit the Fitness Handbook and Fitness Zone sites in ConnectEd.

Using Supplements

Figure 11.12 Supplements can help you meet your needs for specific nutrients, but they cannot take the place of healthful foods. *What are some reasons people might use supplements?*

| Main Idea | Dietary supplements can help people meet their nutrient needs if they cannot do it with food alone.

Walk down the aisles of any drugstore, and you will see a huge array of dietary supplements. These supplements provide various combinations of vitamins, minerals, protein, and fiber. You may also see **herbal supplements**, which are *dietary supplements containing plant extracts.*

The most important thing to know about supplements is that they are no substitute for eating a variety of healthful foods. Some people, however, may not be able to get all the nutrients they need through food alone. For example, strict vegetarians may use supplements to provide the nutrients they are not getting from animal-based foods. Pregnant or nursing women may use them to make sure they get all the extra nutrients their bodies need. Supplements can also help people who are recovering from illness or taking medications that reduce the body's ability to absorb certain nutrients.

Concerns About Dietary Supplements

Most people who follow a nutritious, well-balanced eating plan, such as the one recommended in MyPyramid, will not need a multivitamin. However, multivitamin and mineral supplements are generally safe to use, as long as you use them correctly. For starters, do not take supplements that provide more than 100 percent of the Daily Values for any nutrient. Taking **megadoses**, or *very large amounts,* of any supplement can be dangerous. Some vitamins, such as A, D, E, and K, can build up in the body and become toxic.

Many herbal supplements raise additional concerns. Some people take these "natural" products because they believe they are a safe alternative to drugs for treating certain conditions. These supplements can still be dangerous. Using the herb ephedra, or ma huang, can lead to a heart attack or stroke. Products containing this herb were banned in 2004. Other herbs, such as kava and comfrey, have been linked to serious liver damage.

The National Institutes of Health (NIH) cautions that herbal supplements are just like drugs. Herbal supplements aren't regulated in the same ways as foods or drugs. However, the U.S. Food and Drug Administration (FDA) can take action to stop the sale of supplements that are unsafe or mislabeled. To be safe, treat supplements with the same caution you'd use with any drug. Check with your health care provider before using any herbal supplements, especially if you are already taking other medications.

READING CHECK

Identify List two kinds of people who might use dietary supplements.

©Plush Studios/Getty Images

Health Skills Activity

Accessing Information

Evaluating Supplements

Sanjay knows that teens may need more calcium than adults because their bodies are growing at a fast rate. He's thinking about taking a calcium supplement. Looking in the supplements aisle at the drugstore, he sees that the labels make many types of claims. In his health class, Sanjay learned that certain claims must be reviewed by the FDA. For instance, a product's manufacturers must get FDA approval to claim that their product can prevent or treat any disease. However, FDA approval is not needed for general claims about the product's benefits, such as "promotes circulatory health." Sanjay thinks about which of the Calci-Treat claims can be proven.

CALCI-TREAT

☀ Provides a full day's supply of calcium
☀ Builds strong bones
☀ Helps prevent osteoporosis
☀ Delicious chocolate flavor
☀ Easy to use

Writing Write a paragraph describing how Sanjay could evaluate the claims made on this supplement's label. Consider these questions:

1. Which of the claims can be proved? Which are opinions?
2. Which of these claims require FDA approval? Which might not have received FDA approval?
3. What sources can Sanjay check to confirm the claims made on this label?

LESSON 3 ASSESSMENT

 After You Read

Reviewing Facts and Vocabulary

1. What are *dietary supplements*?
2. List three factors that can affect your body's nutrient needs.
3. Why should teen athletes avoid performance enhancers?

Thinking Critically

4. **Analyze.** How would cutting back on food and water affect the performance of a student athlete?
5. **Evaluate.** Is it safe for a vegan to take a daily supplement that provides the recommended dose of iron, calcium, and B vitamins? Explain why.

Applying Health Skills

6. **Advocacy.** Create a flyer telling teens about the dangers of using dietary supplements. Include warnings about specific types of supplements.

Writing Critically

7. **Expository.** Write an article for the school newspaper. In the article, tell teens the right and wrong ways to improve athletic performance.

Real Life Issues ....

After completing the lesson, review and analyze your response to the Real Life Issues question on page 303.

Hands-On HEALTH

Activity The Match Game

This high-energy activity will test your ability to understand vocabulary terms and their definitions. It will also ask you to make a decision related to managing your weight in a healthy and positive way.

What You'll Need
- 34 index cards and a marker
- textbook

What You'll Do

Step 1

On each of 17 cards, list one key vocabulary term found in Chapter 11. On the remaining cards, write one definition per card. Mix the cards up and place face down in a stack.

Step 2

When the teacher says "go," turn the cards over and, as quickly as you can, put the vocabulary terms in a column to the left, matching each word to its definition on the right. First group to correctly match wins.

Step 3

Mix the cards up again and pass to the group on your left. Repeat matching.

Apply and Conclude

List the factors associated with one weight management issue or eating disorder and make a health-enhancing decision about it.

Checklist: Decision Making

- ✓ Gives a clear description of the situation
- ✓ Gives several options with possible outcomes of each
- ✓ Shows influence of values on possible decisions
- ✓ Shows a health-enhancing decision and evaluation of it

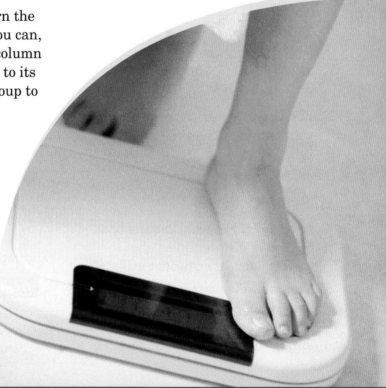

Maintaining a Healthy Weight

Key Concepts

▸ Body mass index and body composition can help determine whether your weight is healthy.

▸ Being overweight carries health risks such as hypertension, cardiovascular disease, and type 2 diabetes.

▸ Health risks of being underweight include nutrient deficiencies, difficulty fighting off disease, weakness, and tiring easily.

▸ Following nutrition guidelines on MyPyramid and being physically active will help you maintain a healthy weight.

Vocabulary

▸ metabolism (p. 290)
▸ body mass index (BMI) (p. 291)
▸ overweight (p. 291)
▸ obese (p. 292)
▸ underweight (p. 293)

Body Image and Eating Disorders

Key Concepts

▸ Influences on teens' body image include family, peers, and the media.

▸ Fad diets promise quick and easy weight loss, but they are not a safe and reliable way to lose weight.

▸ Eating disorders, such as anorexia, bulimia, and binge eating, are extreme and dangerous eating behaviors that require professional help.

Vocabulary

▸ body image (p. 297)
▸ fad diets (p. 298)
▸ weight cycling (p. 298)
▸ eating disorders (p. 300)
▸ anorexia nervosa (p. 300)
▸ bulimia nervosa (p. 301)
▸ binge eating disorder (p. 301)

Lifelong Nutrition

Key Concepts

▸ Your body's nutritional needs depend on such factors as your age, health, and lifestyle.

▸ A vegetarian eating style can offer health benefits if foods are chosen carefully to provide all the nutrients the body needs.

▸ Health conditions, such as diabetes, food allergies, and high blood pressure, can impact food choices.

▸ Athletes need to consume extra calories, drink extra water, and avoid harmful performance enhancers.

▸ In addition to a healthful eating plan, some people take dietary supplements to help meet their nutrient needs.

Vocabulary

▸ vegetarian (p. 304)
▸ dietary supplements (p. 304)
▸ performance enhancers (p. 307)
▸ herbal supplements (p. 308)
▸ megadoses (p. 308)

LESSON 1

Vocabulary Review

Use the vocabulary terms listed on page 311 to complete the following statements.

1. _____ is a measure of body weight relative to height.

2. Adults who have an excess of body fat may be considered _____.

3. People who are _____ have a BMI that is lower than the healthy range.

Understanding Key Concepts

After reading the question or statement, select the correct answer.

4. Measuring the thickness of skin folds at different points on the body is a way to determine your
   a. body mass index.
   b. body composition.
   c. metabolism.
   d. energy balance.

5. Which of the following is *not* a health risk associated with being overweight?
   a. Hypertension (high blood pressure)
   b. Type 2 diabetes
   c. Osteoarthritis (a joint disease)
   d. Anemia (a condition in which the blood cannot carry needed oxygen to the body)

6. A safe, reasonable rate of weight loss is
   a. 1 pound per day
   b. 5 pounds per week
   c. 1/2 to 1 pound per week
   d. 1 to 2 pounds per year

Thinking Critically

After reading the question or statement, write a short answer using complete sentences.

7. **Explain.** How is your weight related to your energy balance?

8. **Identify.** List three healthful steps you could take if you wanted to gain weight.

9. **Compare and Contrast.** What is the difference between being overweight and obese?

10. **Evaluate.** Why is physical activity important for all teens, regardless of weight?

LESSON 2

Vocabulary Review

Correct the sentences below by replacing the italicized term with the correct vocabulary term.

11. A repeated pattern of losing and regaining body weight is called *binge eating.*

12. *Fad diets* are extreme, harmful eating behaviors that can cause serious illness or even death.

13. *Anorexia nervosa* is an eating disorder in which people overeat compulsively.

Understanding Key Concepts

After reading the question or statement, select the correct answer.

14. Which of the following might cause teens to develop a negative body image?
    a. Focusing on their good qualities
    b. Being picked on at school because of the way they look
    c. Being physically active
    d. Having friends with positive attitudes toward their own bodies

15. Teens who think they need to lose weight should
   a. take diet pills.
   b. follow a liquid diet.
   c. begin a fast.
   d. consult a doctor.

16. Which of the following is *not* a behavior associated with anorexia nervosa?
   a. Avoiding food and meals
   b. Exercising excessively
   c. Eating a large amount of food in a single sitting
   d. Weighing oneself repeatedly

17. The first step in treating bulimia nervosa is to
   a. break the cycle of binging and purging.
   b. get the patient's weight back to a normal level.
   c. address the emotional problems that led to the eating disorder.
   d. provide nutritional guidance.

Thinking Critically

After reading the question or statement, write a short answer using complete sentences.

18. Explain. Why are fad diets generally not safe or reliable ways to lose weight?

19. Explain. What makes very-low-calorie diets dangerous for teens?

20. Analyze. Identify three signs that distinguish a fad diet from a legitimate weight-loss plan.

21. Explain. What are three risks associated with using diet pills?

22. Compare and Contrast. How are the eating disorders anorexia and bulimia similar? How are they different?

23. Compare and Contrast. How are bulimia and binge eating disorder similar? How are they different?

LESSON 3

Vocabulary Review

Use the vocabulary terms listed on page 311 to complete the following statements.

24. People who eat mostly or only plant-based foods are called _____.

25. _____ are dietary supplements containing plant extracts.

26. Taking a _____, or a very large amount, of any supplement can be dangerous.

Understanding Key Concepts

After reading the question or statement, select the correct answer.

27. Which of the following foods would all vegetarians refuse to eat?
   a. Eggs
   b. Milk
   c. Chicken
   d. Bread

28. People with celiac disease must avoid foods that contain
   a. sugar.
   b. lactose.
   c. gluten.
   d. fiber.

29. Which medical condition may require people to limit their salt intake?
   a. Allergies
   b. Diabetes
   c. High cholesterol
   d. High blood pressure

Assessment

Thinking Critically

After reading the question or statement, write a short answer using complete sentences.

30. Identify. List three factors that may influence a person's calorie and nutrient needs.

31. Analyze. How can good nutrition enhance your health throughout your life?

32. Evaluate. Describe the health advantages and disadvantages of a vegetarian eating style.

33. Explain. How are dehydration and electrolyte imbalance related?

34. Explain. Why are dietary supplements not a substitute for eating a variety of healthful foods?

35. Identify Problems and Solutions. Toby has a milk allergy and is unable to consume dairy products. How might he benefit from dietary supplements?

36. Describe. What health problems can result when athletes take performance-enhancement drugs or supplements?

Technology PROJECT-BASED ASSESSMENT

Helping a Friend

Background

Eating disorders are a serious medical problem that can result in lifelong health problems and even death. Often, people with eating disorders try to hide the disorder from their friends and family.

Task

With your group of three students, create a streaming video of a short skit that demonstrates how you would talk to a friend who you suspect may have an eating disorder. Encourage your friend to get help. Your video should also demonstrate ways to obtain help for the friend from a trusted adult.

Audience

Students in your class

Purpose

Practice communicating effectively with your peers and with adults

Procedure

1 Conduct an online search on symptoms of eating disorders and review those in your textbook. Choose one eating disorder to focus on in your video.

2 Collaborate as a group to write the script, film, and upload the video with these characters: a student with a suspected eating disorder, the student's friend, and an adult.

3 Be sure your chosen method of communication demonstrates the following: expressing concern and support, telling an adult about the problem, and making sure your friend receives help.

4 When creating the video, remember that a person with an eating disorder may not accept advice.

5 Show your video to the class. Ask permission to upload your group's video to the school's Web site.

Math Practice

Problem Solving. Some math problems require you to read a text passage. Read the text carefully and answer the questions that follow.

Mohammed's school started a fitness and nutrition program. Mohammed joined the program and developed food and physical activity plans. Exactly two weeks after starting the plan, Mohammed had lost 3 pounds and noticed that he had more energy.

Later that week, Mohammed left school feeling restless. He decided to go for a power walk. He burned 37 calories warming up before he went on the walk and 2.3 calories for every minute of walking once he got started. When Mohammed finished his power walk, he spent 10 minutes cooling down.

1. What was Mohammed's average weight loss per day in the first two weeks of his plan?
   A. 0.21 pounds/day   **C.** 1.50 pounds/day
   B. 0.67 pounds/day   **D.** 4/67 pounds/day

2. If x represents the number of minutes Mohammed power walked and y represents how many calories he usually burns cooling down, which expression below could be used to figure out how many total calories he burned on the power walk?
   A. $x(2.3 + 37 + y)$
   B. $(2.3 + 37)(x + y)$
   C. $(2.3 \times y) + 37 + x$
   D. $(2.3 \times x) + 37 + y$

3. Mohammed burned 37 calories during his warm-up, 20 calories during his cool down, and power-walked for 43 minutes. How many total calories did he burn?

Reading/Writing Practice

Understand and Apply. Read the passage below and then answer the questions.

About 50 million Americans begin weight-loss diets each year. Very few—perhaps 5 percent—will manage to keep the weight off. Why? Most people approach weight loss the wrong way. They look for "quick fixes" that will let them lose weight with as little effort as possible. They may put their faith in "magic" weight-loss formulas or combinations of food that will "melt away" the pounds.

The only sensible approach is to cut your calorie intake by following the guidelines in MyPyramid, and getting more physical activity. This plan may not be quick—and it may not be easy—but it will produce lasting results, a promise no other diet can live up to.

1. What is the main idea of this article?
   A. Millions of Americans try to lose weight each year.
   B. Weight-loss diets can be harmful.
   C. Nobody really loses weight by dieting.
   D. Only a sensible plan will result in permanent weight loss.

2. Which sentence could be added to introduce the second paragraph?
   A. There are no shortcuts to weight loss.
   B. Fad diets can help some people.
   C. A healthy weight has many benefits.
   D. For many people, losing weight is impossible.

3. Contrast the realities of fad diets with losing weight by following a plan of healthy eating and physical activity.

National Education Standards

Math: Number and Quantities; Quantities
Language Arts: LACC.910.L.3.6

CHAPTER 12

Physical Activity and Fitness

Lesson 1
Benefits of Physical Activity

BIG Idea *Being physically active benefits your total health in a variety of ways.*

Lesson 2
Improving Your Fitness

BIG Idea *Different types of exercise can help you evaluate and improve the various elements of fitness.*

Lesson 3
Planning a Personal Activity Program

BIG Idea *Planning your physical activity can help you achieve specific fitness goals.*

Lesson 4
Fitness Safety and Avoiding Injuries

BIG Idea *It is important to learn how to prevent injuries and respond to them when they occur.*

Activating Prior Knowledge

Using Visuals Look at the picture on this page. Based on what you have learned in previous chapters, write a paragraph describing how these teens are affecting their physical, mental/emotional, and social health.

Chapter Launchers

Health in Action

Discuss the **BIG** Ideas

Think about how you would answer these questions:

▶ What physical activities do you enjoy?

▶ How does physical activity fit into your daily life?

▶ Do you consider yourself fit?

Assess Your Health

Read each statement. On a separate sheet of paper, write "yes," "sometimes," or "no" based on your typical behavior.

1. I engage in physical activity at least 60 minutes each day.

2. I use the stairs instead of taking an elevator whenever possible.

3. I set fitness goals to improve my physical fitness.

4. I do a warm-up and cooldown after physical activity.

5. I use proper safety gear when participating in physical activities or sports.

A "yes" response shows that you practice healthy behaviors. "Sometimes" indicates that you should analyze and possibly modify your behavior. A "no" response means that you should modify the behavior.

Benefits of Physical Activity

BIG Idea *Being physically active benefits your total health in a variety of ways.*

Before You Read

Organize Information. Divide a sheet of paper into three columns. Label them "Physical," "Mental/Emotional," and "Social." As you read the lesson, fill in the chart by listing how physical activity benefits these three aspects of your health.

Physical	Mental/ Emotional	Social

New Vocabulary

▶ physical activity
▶ physical fitness
▶ exercise
▶ sedentary

Real Life Issues

Trying Something New. Nina and Marianne have been best friends since grade school. When Marianne took up kickboxing last year, Nina was disappointed that her friend had a new activity she didn't share. Marianne kept talking about how much she enjoyed her new sport and encouraging Nina to try it. In the beginning, Nina was hesitant, but she finally decided to take up the sport as well. Now the two friends have another activity they can enjoy together.

Writing *How do your friends affect your choice of sports and other physical activities? Describe an activity that someone you know has influenced you to try.*

Physical Activity and Your Health

Main Idea Physical activity benefits all aspects of your health.

Suppose you decided you were going to do just one thing to improve your health. What one action would you choose? Probably the single most important step you could take would be to lead a physically active life. **Physical activity** is *any form of movement that causes your body to use energy.* It benefits just about every system in your body.

Physical activity doesn't just mean "working out." It includes all kinds of activities that you do on a daily basis, such as walking to school, cleaning your room, or playing sports with your friends. There are lots of different ways to make physical activity a part of your life and enjoy its many benefits to your health. Turn to the Fitness Handbook to learn more about physical activity.

Physical Benefits

Being active on a regular basis improves your **physical fitness**, *the ability to carry out daily tasks easily and have enough reserve energy to respond to unexpected demands.* Teens should try for at least 60 minutes of physical activity every day. Depending on what kind of activity you do, it can strengthen your muscles and bones, boost your energy level, or improve your posture. You might feel that committing 60 minutes every day to physical activity will be difficult. Try dividing the time into smaller segments to get your 60 minutes throughout the day.

You can achieve specific fitness goals through **exercise**, *purposeful physical activity that is planned, structured, and repetitive, and that improves or maintains physical fitness.* All kinds of physical activity—not just exercise—will improve your health. Being physically active can help you maintain a healthy weight and may reduce your risk of many serious diseases. **Figure 12.1** shows the ways in which physical activity can benefit several different body systems and maintain your overall health.

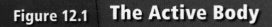

Figure 12.1 The Active Body

This illustration shows just a few of the ways physical activity makes your body stronger. *Which systems in your body benefit from regular physical activity?*

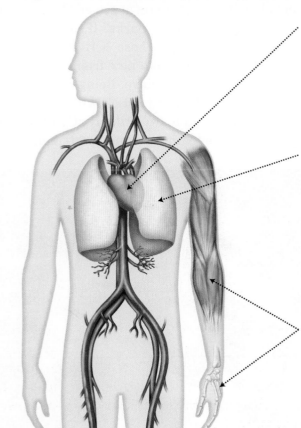

Cardiovascular System
Regular physical activity strengthens the heart muscle so that it pumps blood more efficiently. It reduces blood pressure and lowers the levels of artery-clogging cholesterol.

Respiratory System
As your activity level increases, your lungs begin to work more efficiently, pulling in larger amounts of air and increasing the amount of oxygen delivered to your body. As a result, you can do many activities more easily — for example, running a greater distance without becoming short of breath.

Musculoskeletal System
Physical activity strengthens muscles and bones, reducing your risk of developing fragile bones as you age. Strengthening your bones and muscles can also improve your balance and coordination.

Mental and Emotional Benefits

Being physically active maintains your physical health and has a positive effect on your mental and emotional health. It can provide the following:

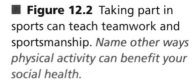

- **Stress relief.** Being active stimulates your body to produce chemicals called endorphins. This results in a feeling of well-being, aids relaxation, and relieves physical pain. Some types of physical activity, such as stretching, can ease muscle tension as well.

- **Mood enhancement.** Have you ever gone out for a walk when you were in a bad mood and returned feeling much better? Physical activity is a natural mood lifter. In addition to endorphins, it promotes the production of other brain chemicals that combat anxiety and depression. For this reason, people with anxiety and depression are advised to get regular physical activity.

- **Better sleep.** Moderate activity at least three hours before bedtime helps you relax and get to sleep more easily.

- **Improved self-esteem.** The physical fitness you develop through increased activity can translate into more self-confidence. It can give you a sense of accomplishment and also help you look and feel your best.

Social Benefits

Are you a member of a sports team at school? Do you enjoy hiking or exploring trails in nearby parks? If so, you've probably formed friendships through these activities. Physical activity can be a great way to make new friends and spend time with the friends you already have. Being active as part of a group can help motivate you to stick with your fitness program. It can also help you learn skills that will improve your relationships, such as teamwork and sportsmanship.

■ **Figure 12.2** Taking part in sports can teach teamwork and sportsmanship. *Name other ways physical activity can benefit your social health.*

TEENS Making a Difference

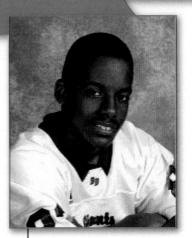

"I wanted to help kids get fit."

For the Love of the Game

James M., of Indiana, enjoys volunteering. As a football player on his high school team, he and 20 other players spend three hours after practice once a week mentoring elementary and middle school students in the inner city. "With the increase in obesity, I wanted to help kids get fit and teach them the game of football. Some who are overweight, as well as others, learn that they can do the same things as me."

James feels it is a lot of responsibility to volunteer. "You have to make time in your busy schedule. But it's so rewarding when you see them grasp a concept and do it correctly, like run a route, or block."

"It's fun having kids learn the game and get fit at the same time. It also helps the community—it keeps them off the streets and gives them something to do."

Activity Write your answers to the following questions in your personal health journal:

1. Why does James like volunteering?
2. What are some of the benefits of teaching others to be physically active?
3. What can you do to help people in your community become physically active?

The increased self-esteem that comes with physical fitness can help your social life as well. It can give you confidence when meeting new people or dealing with social situations. Finally, physical activity can help you manage stress, rather than letting it build up until it has a negative impact on your relationships.

Risks of Being Inactive

Main Idea An inactive lifestyle puts you at risk for a variety of health problems.

Despite the many benefits of physical activity, many teens still lead lives that are **sedentary**—*involving little physical activity*. Sedentary teens may spend their free time watching TV, playing video games, or surfing the Internet. All of us **devote** some time to sedentary activities, but being sedentary all the time puts you at risk for a variety of health problems.

Academic Vocabulary

devote *(verb):* to give time or effort to an activity

READING CHECK

Identify Problems and Solutions How can teens reduce their risk of obesity, cardiovascular disease, and type 2 diabetes?

Health problems that may result from being sedentary include

- unhealthful weight gain and obesity;
- cardiovascular disease, such as heart attack and stroke;
- type 2 diabetes;
- certain types of cancer;
- asthma and other breathing problems;
- osteoporosis, a condition in which the bones become porous and fragile, making them much more likely to break;
- osteoarthritis, a condition caused by the breakdown of cartilage and bone in the body's joints;
- psychological problems such as stress, anxiety, and depression; and
- premature death.

Increasing your level of physical activity lowers your risk of these health problems. Teens should aim for 60 minutes of physical activity every day, or at least most days.

Making Time for Physical Activity

Main Idea There are several ways to fit physical activity into your daily life.

Setting aside an hour a day for exercise may be difficult for some busy teens. You can get the same benefits from several shorter periods of activity spread out over the course of a day. For example, engaging in 10 minutes of physical activity six times a day provides the same benefits as an hour-long workout. **Figure 12.4** shows some ways you can fit physical activity into your life by choosing active alternatives to the things you do every day.

READING CHECK

Identify List two ways that you can make time for physical activity.

■ **Figure 12.3** Just by turning off the TV and getting out of the house for a little exercise, you can reduce your risk of health problems. *What are other advantages of participating in physical activities?*

Figure 12.4 Active Alternatives

Instead of this . . .	Try this . . .
• Taking the elevator	• Taking the stairs
• Using a snowblower	• Shoveling snow
• Getting a ride to school or to a friend's house	• Walking, skating, or riding your bike
• Using a shopping cart	• Carrying your groceries to the car
• Taking the car through a car wash	• Washing the car by hand
• Playing video or computer games	• Playing basketball, soccer, or tennis

LESSON 1 ASSESSMENT

 After You Read

Reviewing Facts and Vocabulary

1. What is the difference between *physical activity* and *exercise*?

2. Name three body systems that benefit from regular physical activity.

3. Identify two types of disease associated with a sedentary lifestyle.

Thinking Critically

4. **Analyze.** Explain how being physically active on a regular basis makes your body better able to respond to physical demands.

5. **Synthesize.** Camilla thinks there's no point in trying to improve her physical fitness because she doesn't have a free hour in her daily schedule for exercise. What advice would you give her?

Applying Health Skills

6. **Stress Management.** Raul is taking several honors classes that require a lot of homework, and often feels stressed. How can he incorporate physical activity into his schedule to help reduce stress?

Writing Critically

7. **Expository.** Write an essay describing what might influence some teens to choose a sedentary lifestyle. Suggest ways to encourage these teens to become more physically active.

Real Life Issues ....

After completing the lesson, review and analyze your response to the Real Life Issues question on page 318.

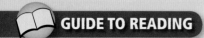
Improving Your Fitness

BIG Idea *Different types of exercise can help you evaluate and improve the various elements of fitness.*

Before You Read

Create a Comparison Chart. Draw a chart. Label the columns "Define," "Measure," and "Improve." Label the rows "C/E" (cardio endurance), "M/S" (muscular strength), "M/E" (muscular endurance) and "F" (flexibility). As you read, fill in your chart with information from the lesson.

	Define	Measure	Improve
C/E			
M/S			
M/E			
F			

New Vocabulary

▸ cardiorespiratory endurance
▸ muscular strength
▸ muscular endurance
▸ flexibility
▸ aerobic exercise
▸ anaerobic exercise

Real Life Issues

Building Fitness Levels. Mel is in the process of training for a 12-mile charity run, which will take place in a few months. His goal is to run the whole race, but he has only made it through ten miles during his practice runs. He knows that in order to reach his goal, he will have to improve his cardiorespiratory endurance.

Writing *Think of a physically challenging goal that you would like to work toward. Write a short paragraph describing which elements of fitness you would need to focus on in order to meet that goal.*

Elements of Fitness

Main Idea There are five elements of fitness that affect your health in different ways.

Are you physically fit if you can run five miles or do a dozen push-ups in a row? These are two of the five elements of health-related fitness that affect you in different ways.

- **Cardiorespiratory endurance** is *the ability of your heart, lungs, and blood vessels to send fuel and oxygen to your tissues during long periods of moderate to vigorous activity.* By maintaining good cardiorespiratory health, you can run a mile or go on a long hike without tiring. Good cardiorespiratory health lowers your risk of cardiovascular disease.

- **Muscular strength** is *the amount of force your muscles can exert.* You need muscular strength for all kinds of activities that put stress on your muscles, such as lifting, pushing, and jumping.

- **Muscular endurance** is *the ability of your muscles to perform physical tasks over a period of time without tiring.* Muscular endurance gives you the power to carry out daily tasks without fatigue, such as carrying boxes up and down a flight of stairs.

- **Flexibility** is *the ability to move your body parts through their full range of motion.* If you are flexible, you can touch your toes without bending your legs or put sunscreen on the center of your back. Flexibility can improve your athletic performance and reduce your risk of muscle strain and other injuries.

- **Body composition**—the ratio of fat to lean tissue in your body—is also an element of fitness. Having low overall body fat reduces your risk of cardiovascular disease and other health problems associated with being overweight.

Evaluating Your Fitness

Main Idea You can use different tests to evaluate each element of your fitness.

So how fit are you? Each of the tests described below measures a different element of fitness. By taking them all, you can figure out how you measure up in each area of fitness. Visit the online Fitness Zone to watch videos about the exercises in this section.

Measuring Cardiorespiratory Endurance

You can evaluate your cardiorespiratory endurance by doing a three-minute step test. You will need a sturdy bench or step about 12 inches high and a watch or clock with a second hand. Follow this procedure:

1. Step up onto the bench with your right foot. Bring up your left foot. Step back down, right foot first, then left foot.

2. Continue stepping up and back down for three minutes. Try to maintain a steady pace of about 24 steps per minute.

3. After three minutes, take your pulse. To do this, place two fingers of one hand on the wrist of your opposite hand. (Do not use your thumb, which has its own pulse.) Count the number of heartbeats you feel in 15 seconds. Then multiply that number by 4 to determine your pulse rate. Check your pulse rate against **Figure 12.6** on page 326 to see how you did on the test.

READING CHECK

Classify Which elements of fitness would help you run a marathon?

■ **Figure 12.5** The step test is one activity that requires cardiorespiratory endurance. *What are other activities that use this element of fitness?*

Figure 12.6 Fitness Test Scoring Chart

Each of the columns below provides scores for the fitness tests. If you scored at or above the number shown for each of these three tests, you are in good shape. If you scored below the number shown, you need to work on that element of fitness.

Step Test	Partial Curl-Ups	Right-Angle Push-Ups	Sit-and-Reach Test
Male teens: (heartbeats per minute) 85–95: Excellent 95–105: Good 105–126: Fair 126+: Needs improvement	Boys, ages 13–14: 21	Boys, age 14: 12 Boys, age 15: 14 Boys, age 16: 16 Boys, age 17: 18	Boys: 1 inch
Female teens: (heartbeats per minute) 85–95: Excellent 95–106: Good 106–126: Fair 126+: Needs improvement	Girls, ages 13–14: 18	Girls, ages 14–17: 7	Girls: 3 inches

READING CHECK

Compare and Contrast How are curl-ups and right-angle push-ups alike? How are they different?

■ **Figure 12.7** Curl-ups measure abdominal strength. *How might building abdominal strength improve your posture?*

Measuring Muscular Strength and Endurance

Different muscle groups require different exercises. The two exercises below will test the strength and endurance of your abdominal muscles and your upper body. After completing each exercise, check **Figure 12.6** to see how you did.

Partial Curl-Ups Use the following procedure to measure your abdominal strength:

1. Lie on your back with your knees bent and your feet about 12 inches from your backside. Extend your arms forward with your fingers pointing toward your knees.

2. Raise your head and upper body off the floor, sliding your hands forward. Touch your knees with your fingertips.

3. Slowly return to your original position.

4. Continue doing curl-ups at a rate of one every three seconds until you can no longer maintain this pace.

The number of curl-ups you can do without tiring is a measure of your abdominal strength and endurance.

Right-Angle Push-Ups The right-angle push-up is one test to gauge your upper body strength and endurance.

1. Lie facedown in the push-up position. Place your hands under your shoulders, with your legs parallel to each other and resting on your toes.

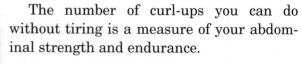

2. Straighten your arms and push up. Keep your back and knees straight. Bend your arms and lower your body until your elbows form a 90-degree angle, with your upper arms parallel to the floor.

3. Repeat this process, doing one push-up every three seconds until you can no longer maintain this pace.

Measuring Flexibility

The sit-and-reach test measures the flexibility of your lower back and hamstring muscles. To set up the test, tape a yardstick to the top of a box with 9 inches protruding over one end.

1. Remove your shoes. Place the box against a wall, or ask someone to hold the box in place, with the yardstick pointing out. Sit on the floor. Place the sole of one foot flat against the side of the box under the yardstick. Bend the other leg at the knee.

2. Extend your arms over the yardstick, with your hands placed one on top of the other, palms down.

3. Reach forward in this manner four times. The fourth time, hold this position for at least one second while a partner records how far you can reach.

4. Switch legs and repeat.

Getting Fit

Main Idea Use different forms of exercise to improve the various elements of your fitness.

To improve your overall fitness, you can choose from many different exercises and other activities. Most of these fit into two basic categories: aerobic and anaerobic.

Aerobic exercise includes *all rhythmic activities that use large muscle groups for an extended* **period** *of time.* Aerobic

■ **Figure 12.8** Right-angle push-ups are a way of measuring upper body strength and endurance. *What other activities require upper body strength and endurance?*

Academic Vocabulary

period *(noun):* the completion of a cycle

■ **Figure 12.9** The sit-and-reach test measures flexibility in your hips and legs. *What are some benefits of being flexible?*

exercise raises your heart rate and increases your body's use of oxygen. Jogging, swimming, and riding a bike are examples of aerobic exercise.

Anaerobic exercise involves *intense, short bursts of activity in which the muscles work so hard that they produce energy without using oxygen.* Sprinting and lifting weights are examples of this kind of exercise.

Improving Cardiorespiratory Endurance

Aerobic exercise is important for building cardiorespiratory endurance. Aerobic activities increase your heart rate and pump more blood throughout your body. Over time, your heart and lungs adapt to the demands made by aerobic activity by working more efficiently.

Regular aerobic exercise reduces your risk of cardiovascular disease. It also helps you manage your weight and lower your risk of type 2 diabetes, certain cancers, and other diseases associated with being overweight. The Real World Connection activity explains how to determine your target heart rate when doing aerobic exercise.

Improving Muscular Strength and Endurance

In contrast to aerobic activity, anaerobic exercises improve muscular strength and endurance. The more the muscles work, the stronger they will become. Exercises that strengthen the muscles are known as resistance or strength training. Free weights, exercise machines, or your own body weight provides resistance. There are three ways to use resistance to work your muscles:

(l to r)/McGraw-Hill Education

FITNESS ZONE

My goal this year is to get in great shape, so I started keeping a journal of what I eat and when I work out. With a journal it's easier to stick with my plan because I know exactly what I have to do. My success has kept me motivated. For more physical activity ideas, visit the Fitness Handbook and Fitness Zone sites in ConnectEd.

■ **Figure 12.10**
Lifting weights is one form of resistance or strength training. *What are the benefits of strength training?*

- **Isometric exercises** use muscle tension to improve strength with little or no movement of the body part. Pushing against a wall or other immovable object is an example of isometric exercise.

- **Isotonic exercises** combine movement of the joints with contraction of the muscles. Try lifting free weights or doing calisthenics, such as pull-ups, push-ups, and sit-ups. These exercises build flexibility as well as strength.

- **Isokinetic exercises** exert resistance against a muscle as it moves through a range of motion at a steady rate of speed. Various types of weight machines and other exercise equipment provide isokinetic exercise.

Increasing muscle mass boosts your metabolism so your body burns the energy you consume faster. That makes it easier to control your weight.

READING CHECK

Classify What are the three types of resistance exercise?

Real World CONNECTION

Targeting Cardiovascular Fitness

Your target heart range is the ideal range during aerobic activity. To calculate your target heart range:

1. Multiply your age by 0.7.

2. Subtract this number from 208 to get an estimate of your maximum heart rate. If you are 16 years old, your maximum heart rate will be 197 beats per minute.

3. Multiply this number by 50 percent to get your minimum heart rate for moderately intense activity.

4. Multiply the number in step 2 by 70 percent to get your maximum heart rate for moderately intense activity and the minimum for vigorous activity.

5. Multiply the number in step 2 by 85 percent to get your maximum target heart rate for any physical activity. Exercising above this rate is dangerous.

6. To figure out your heart rate during exercise, take your pulse for six seconds and multiply the result by 10.

Activity   Mathematics

Calculate your target heart range for moderate activity and for vigorous activity. Explain how you would adjust your activity level to stay within your target range for walking, running, and sprints.

Concept Operations and Algebraic Thinking
To solve this problem, change the percent to a fraction or to a decimal, and then multiply by the number.

Heartbeats per Minute

200 — ☀	○ **Maximum target rate**
190 —	
180 —	
170 —	
160 — ☀	○ **Target range for vigorous activity**
150 —	
140 —	
130 — ☀	○ **Target range for moderate-intensity activity**
120 —	
110 —	
100 — ☀	○ **Minimum target rate**
0 —	

Target Heart Range (16-year-old)

When lifting weights, begin each workout with one set of exercises using lighter weights. Gradually increase the amount of weight you use until you are lifting your maximum weight. Warm up before any kind of strength training with gentle aerobic activity, such as jogging or fast walking.

Improving Flexibility

Stretching exercises improve your flexibility, circulation, posture, and coordination, as well as ease stress. It may also reduce your risk of injury during other activities. Do the stretching exercises slowly, holding each stretch for 10 to 30 seconds. Don't bounce. If stretching causes pain you've pushed too far.

Exercise and Bone Strength

Exercise helps increase bone density and lowers the risk of osteoporosis. Weight-bearing exercises work with gravity, and are good for strengthening bones. Strength training, walking, aerobics, and dancing are all weight-bearing exercises.

■ **Figure 12.11** Weight-bearing exercises, which make your body work against gravity, are a way to build bone strength. *What weight-bearing activities do you do on a regular basis?*

LESSON 2 ASSESSMENT

 After You Read

Reviewing Facts and Vocabulary

1. What are the five elements of fitness?
2. Which element of fitness does the sit-and-reach test measure?
3. What kind of exercise would you do to improve your cardiorespiratory endurance?

Thinking Critically

4. **Analyze.** How will your target heart range for physical activity change as you grow older? Explain why.
5. **Evaluate.** Carmen wants to get in shape. She is planning to join a gym and use only the weight machines. Is this a good plan? Why or why not?

Applying Health Skills

6. **Goal Setting.** Create and perform a plan to improve your cardiorespiratory endurance.

Writing Critically

7. **Personal.** Create a fitness journal to track your food intake, calories consumed, and energy expended for one week. Analyze your results.

Real Life Issues ....

After completing the lesson, review and analyze your response to the Real Life Issues question on page 324.

Planning a Personal Activity Program

Real Life Issues ·······························

Getting Fit. Pete wants to get in better shape. He has decided to create a fitness plan, but he's not sure where to start. He's not even sure he knows how to determine what a good level of fitness is. He doesn't know which exercises to do, how often he should do them, or how long he should do them.

Writing *If you were Pete, what steps would you take to create an appropriate fitness plan? In a paragraph, describe the steps you would take.*

Your Fitness Plan

Main Idea **The physical activities you choose depend on factors such as your fitness goals and the activities you like.**

Identifying a specific fitness goal is a good way to get motivated to get in shape. You also need to consider your personal needs, such as your current level of fitness and the resources available to you. Turn to the Fitness Handbook to learn more about fitness goals.

Your Fitness Goals

Measuring your level of fitness can help you set fitness goals. If you found that you have good cardiorespiratory endurance but not much upper body strength, you might want to make building your upper body strength a goal of your activity plan. If your cardiorespiratory endurance needs to be strengthened, choose exercises that improve this aspect of fitness. Turn to the Health Skills Activity on page 334 for an example of how to set specific fitness goals.

GUIDE TO READING 📖

BIG Idea *Planning your physical activity can help you achieve specific fitness goals.*

Before You Read

Create Vocabulary Cards. Write each new vocabulary term on a separate note card. For each term, write a definition based on your current knowledge. As you read, fill in additional information related to each term.

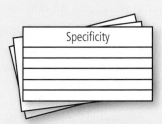

Specificity

New Vocabulary

▶ specificity
▶ overload
▶ progression
▶ warm-up
▶ workout
▶ cool-down
▶ resting heart rate

Personal Needs

When planning a personal activity program, choose activities that you enjoy and that you can realistically do. The following factors may affect your activity choices:

- **Cost.** Some activities require expensive equipment. Borrow or rent equipment to try a new sport.
- **Where you live.** Choose activities that you can do close to home, and that are best for your region. For example, is your local area flat or hilly? What is the climate like?
- **Your schedule.** Choose activities that fit into your schedule and habits. If you're not a morning person, a morning jog probably won't work for you.
- **Your fitness level.** Start slowly and choose activities that are right for your level of fitness.
- **Your overall health.** Do you have a health condition that may impact your exercise plan, such as asthma? Talk to a doctor before starting a new activity.
- **Personal safety.** When choosing activities, make sure that the environment where you perform the activity is safe.

Types of Activities

Teens should aim to get at least 60 minutes of physical activity most days. Choose different types of activity to meet specific fitness goals and to prevent boredom. An exercise plan can include activities such as walking, biking, or playing a pickup basketball or soccer game with friends, as well as school or community sports. Sedentary activities, or those activities that do not require physical activity, should be limited to a small part of your day. **Figure 12.13** shows a page from a fitness journal. The types of physical activities to be included in a fitness journal can include the following:

Moderate-Intensity Physical Activities These count toward your daily dose of physical activity. Examples include walking, climbing stairs, household chores, or yard work.

Aerobic Activities These raise your heart rate. Aim for at least three 20-minute sessions each week of vigorous aerobic activity. Examples include cycling, brisk walking, running, dancing, in-line skating, cross-country skiing, and most team sports.

Strength Training This develops muscle tone. Aim for at least two or three sessions per week of 20 to 30 minutes each, with at least one day off between sessions. Exercises that tone arm muscles include rowing, cross-country skiing, pull-ups, and push-ups. To tone legs, try cycling, running, and skating. Abdominal muscles can be toned by rowing or cycling, and doing abdominal crunches.

READING CHECK

Determine When might it be important to consult a doctor before trying a new physical activity?

■ **Figure 12.12** Measuring your resting heart rate is one way to track your fitness. *What is your resting pulse rate now?*

Figure 12.13 A Variety of Physical Activities

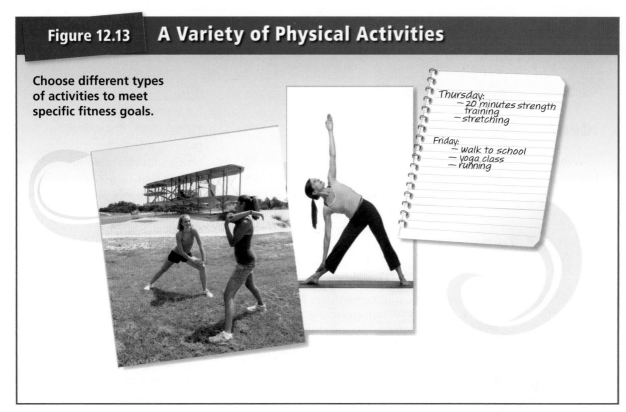

Choose different types of activities to meet specific fitness goals.

Thursday:
 – 20 minutes strength training
 – stretching

Friday:
 – walk to school
 – yoga class
 – running

Flexibility Exercises These include stretching for 10 to 12 minutes a day. Examples of flexibility exercises include gymnastics, martial arts, ballet, Pilates, yoga, or stretching.

Principles of Building Fitness

Main Idea Effective fitness plans focus on four principles: specificity, overload, progression, and regularity.

When designing your physical activity program, you will consider your needs and interests. In addition, you should focus on the four key principles of building a fitness plan: specificity, overload, progression, and regularity.

- **Specificity** means *choosing the right types of activities to improve a given element of fitness.* For example, strength-training activities will build muscular strength.

- **Overload** means *exercising at a level that's beyond your regular daily activities.* Increasing the demands on your body will make it adapt and grow stronger.

- **Progression** means *gradually increasing the demands on your body.* Try working a little harder or longer during each session, and more often during the week.

- **Regularity** means working out on a regular basis. You need at least three balanced workouts a week to maintain your fitness level. Include different activities to get the recommended hour of physical activity each day.

 **READING CHECK**

Make Inferences
Why do you need to increase the demands on your body over time to build fitness?

Health Skills > Activity

Goal Setting Skills

Identifying Fitness Goals

Wendy enjoys playing soccer and is interested in joining her school's team. She's not sure she's in good enough shape to try out though. She has pretty good muscle strength and endurance. But Wendy is concerned about her cardiorespiratory endurance. She always seems to run out of breath sooner than she should, and it sometimes slows her down. She's worried that she won't be able to keep up with the rest of the team or that she can't run around on the field for an hour at a time. She's also not sure she's flexible enough for all the dodging and maneuvering the game involves.

Writing Write a plan for Wendy to achieve her fitness goals. Remember to use the five-step process for goal setting:

1. Identify a specific goal. Write it down.
2. List the steps to reach your goal.
3. Identify potential problems and ways to get support.
4. Set up checkpoints to evaluate your progress.
5. Reward yourself with healthy rewards once you have achieved your goal.

Stages of a Workout

Main Idea An exercise session has three stages: warm-up, workout, and cool-down.

To get your body ready for physical activity and to avoid injuries, include three stages in every exercise session: the warm-up, the workout, and the cool-down. Turn to the Fitness Handbook to learn more about the stages of a workout and how to apply them to specific activities.

 **READING CHECK**

Explain What is the purpose of a warm-up?

Academic Vocabulary

instance *(verb):* to mention as a case or example

Warm-Up

A **warm-up** is *gentle cardiovascular activity that prepares the muscles for work.* Warming up before exercise increases blood flow, delivering needed oxygen and fuel to your muscles. It also gradually increases your pulse rate and body temperature. To warm up, choose an activity that will work the same muscles you're going to use during your workout. For **instance**, before a run, warm up by walking or jogging slowly.

Check your resting heart rate. After warming up your muscles, take a few minutes to stretch. Stretching can prepare your muscles for activity and increase your flexibility.

Workout

The **workout** is *the part of an exercise session when you are exercising at your highest peak.* Use the **F.I.T.T.** formula when planning your workouts:

- **F: Frequency of workouts.** Schedule at least three exercise sessions a week, but give your body time to rest between workouts. Include other types of physical activity during the week to get an hour of activity each day.

- **I: Intensity of workouts.** Push yourself hard enough to create overload. For aerobic activities, exercise within your target heart range. Check your heart rate during your workout. For strength training, you should feel strain on your muscles, but not pain.

- **T: Type of activity.** Vary your activities throughout the week to build different elements of fitness. If you jog Monday and Wednesday, try lifting weights on Tuesday and Thursday.

- **T: Time (duration) of workouts.** To build cardiovascular fitness, keep your heart rate within your target range for at least 20 minutes. Strength-training sessions should take 20 to 30 minutes, while flexibility can be increased in just 10 minutes of stretching.

Cool-Down

A **cool-down** is *low-level activity that prepares your body to return to a resting state.* The cool-down allows your heart rate, breathing, and body temperature to return to normal gradually. It also helps prevent muscle soreness. Cool-downs should include five to ten minutes of gentle activity. The cool-down stage is also a good time for stretching.

FITNESS ZONE

One of my friends showed me an exercise that she said would slim down just my thighs. Our coach, though, says there's no such thing as spot reducing. To get in shape, you have to change your eating habits by cutting down on fat and add a well-rounded exercise routine. For physical activity ideas, visit the Fitness Handbook and Fitness Zone sites in ConnectEd.

■ **Figure 12.14** Stretching your muscles helps prevent injuries. *What stage of a workout is the best time for stretching?*

Tracking Your Progress

Main Idea Track your progress to see how your fitness level increases over time.

One of the rewards of sticking to a physical activity program is seeing your level of fitness improve over time. You may notice that it takes you less time to walk to and from school, or you may not breathe as hard after climbing stairs.

Use a fitness journal, pedometer or other device to track your progress. List all of your activities, noting how long you work out, how often, and at what level. You'll see a noticeable difference in your fitness level if you stick with your plan for 12 weeks.

Another figure to list in your fitness journal is your **resting heart rate**—*the number of times your heart beats per minute when you are not active.* Before checking your resting heart rate, sit quietly for at least five minutes. Take your pulse for 15 seconds, then multiply the result by four. A typical pulse rate for teens and adults is between 60 and 100 beats per minute. As your fitness level increases, your resting heart rate will drop.

READING CHECK

Cause and Effect
How does regular exercise affect your resting heart rate?

LESSON 3 ASSESSMENT

After You Read

Reviewing Facts and Vocabulary

1. What personal factors can affect your choice of physical activities?

2. What are the four principles of building fitness?

3. What are the benefits of warming up before exercise and cooling down after exercise?

Thinking Critically

4. **Synthesize.** What activity might you choose if your fitness goals are to increase cardiorespiratory endurance and to strengthen your leg and abdominal muscles? How might increasing your flexibility help you achieve these fitness goals?

5. **Analyze.** How does where you live affect your choice of activities?

Applying Health Skills

6. **Analyzing Influences.** Draw five columns on a sheet of paper, labeled: "Cost," "Location," "Schedule," "Health," and "Safety." Add examples of how each influence might affect your physical activity choices.

Writing Critically

7. **Narrative.** Write a short story about a teen who designs and begins a fitness plan. List three fitness goals for this teen. Describe the types of activities the teen has chosen.

Real Life Issues ....

After completing the lesson, review and analyze your response to the Real Life Issues question on page 331.

Fitness Safety and Avoiding Injuries

BIG Idea *It is important to learn how to prevent injuries and respond to them when they occur.*

Before You Read

Create a T-Chart. Make a two-column chart on paper. Label the left column "Risks" and the right column "Prevention." As you read, fill in information about safety risks involved in different physical activities and prevention steps you can take to protect yourself.

Risks	Prevention

New Vocabulary

▶ frostbite
▶ hypothermia
▶ overexertion
▶ heat exhaustion
▶ heatstroke
▶ muscle cramps
▶ strains
▶ sprains

Real Life Issues

Preventing Injuries. In 2011, 48,000 bicyclists were injured in car crashes. Source: National Highway Traffic Safety Administration; Traffic Safety Facts; 2011.

> **19% of injured bicyclists were between 15–24 years old.**

> **20–25% of all bicyclists wear bicycle helmets.**

Writing *Write a paragraph encouraging a friend to use safety equipment when riding a bicycle.*

Safety First

Main Idea Safety precautions can help you avoid injuries during physical activity.

While getting regular physical activity, it is possible to injure yourself. A screening before starting a physical activity program can identify diseases and disorders that could make some activities unsafe. Other ways to protect yourself during exercise are to

- use the correct safety equipment for an activity;
- pay attention to other people, objects, and the weather;
- play or exercise at your skill level and know your limits;
- warm up before exercise and cool down afterward;
- stay within the areas designated for a given activity;
- obey all rules and restrictions; and
- practice good sportsmanship.

If you become ill or injured during a physical activity, get help immediately. Turn to the Fitness Handbook for more safety information.

The Right Equipment

Using the correct equipment can prevent injury. You might want to rent equipment when trying a new sport. Here are a few specific guidelines:

- Wear well-fitting athletic shoes that are designed for your sport or activity. Wear socks to cushion your feet and keep them dry. Choose comfortable, non-binding clothes that are appropriate for the weather.

- For cycling, always wear a helmet that fits you properly. Make sure the helmet is approved by Snell or ANSI. Use front and rear reflectors if you must ride at night. Wear light-colored clothing with reflective patches.

- For skating or skateboarding, wear a helmet, knee and elbow pads, gloves, and wrist guards.

- For contact sports, male players should wear a cup to protect the groin. For non-contact sports that involve running, they should wear an athletic supporter. Female players should wear sports bras.

- Special adaptive equipment helps those with disabilities take part in a variety of sports, from bowling to golf.

Figure 12.15 Using the right safety equipment can protect you from injury during physical activity. *What type of safety equipment is required for your favorite sport?*

Watching the Weather

Check the weather and avoid exercising outside during extreme weather, such as thunderstorms or blizzards.

Cold-Weather Risks Layers of clothing will keep you warm. You can remove layers as you warm up, or add more clothing if the temperature drops. **Figure 12.16** shows how to layer clothing. Follow these tips for cold-weather activity:

- Warm up and cool down, even in cold weather.
- Drink plenty of fluids. Cold air can lead to dehydration.
- Cover your nose and mouth to prevent breathing cold, dry air. If you have asthma, talk to your doctor before exercising outdoors in cold weather.

Two other health risks in cold weather are frostbite and hypothermia. **Frostbite** is *damage to the skin and tissues caused by extreme cold.* The skin becomes pale, hard, and numb. To treat frostbite, go to a warm place and thaw the affected areas with warm (not hot) water. As the skin thaws, it becomes red and painful. If the frostbite is severe or does not respond to treatment, seek medical help.

FITNESS ZONE

I learned how important it is to drink water when I started exercising. It helps prevent dehydration, cleans out the body, and promotes healing. For more physical activity ideas, visit the Fitness Handbook and Fitness Zone sites in ConnectEd.

Figure 12.16 Cold-Weather Layering

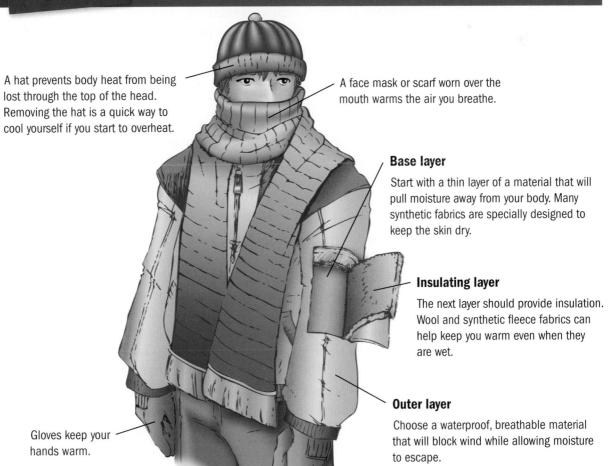

A hat prevents body heat from being lost through the top of the head. Removing the hat is a quick way to cool yourself if you start to overheat.

A face mask or scarf worn over the mouth warms the air you breathe.

Base layer

Start with a thin layer of a material that will pull moisture away from your body. Many synthetic fabrics are specially designed to keep the skin dry.

Insulating layer

The next layer should provide insulation. Wool and synthetic fleece fabrics can help keep you warm even when they are wet.

Outer layer

Choose a waterproof, breathable material that will block wind while allowing moisture to escape.

Gloves keep your hands warm.

Hypothermia, or *dangerously low body temperature,* occurs as a result of **exposure** to extreme cold, submersion in cold water, or wearing wet clothing in cold or windy weather. Hypothermia causes drowsiness, weakness, and confusion. Breathing and heart rate slow down, followed by shock and heart failure. Hypothermia requires emergency medical help. Try to warm the victim until help arrives.

Hot-Weather Risks Heavy sweating while exercising in hot weather can lead to dehydration, or excessive loss of water from the body. Drinking fluids before, during, and after physical activity can prevent dehydration. If you're exercising during hot weather, you may also need to replace sodium, chloride, and potassium. Sports drinks will replace these elements.

Hot-weather health problems may lead to **overexertion**, or *overworking the body.* This can cause **heat exhaustion**, *a form of physical stress on the body caused by overheating.* Symptoms include heavy sweating; cold, clammy skin; dizziness, confusion, or fainting; a weak, rapid pulse; cramps; shortness of breath; or nausea or vomiting. To recover, rest in a shady area, douse yourself with cold water, and fan your skin. If you don't feel better within half an hour, seek medical help.

Untreated heat exhaustion can lead to **heatstroke**, *a dangerous condition in which the body loses its ability to cool itself through perspiration.* Heatstroke can cause sudden death. If you recognize symptoms of heatstroke, call for medical help immediately and try to cool the person.

Sun and Wind Protection Sun and wind can pose a hazard in both hot and cold weather. Exposure to these elements can lead to the following:

- **Windburn,** or irritation of the skin caused by wind exposure. The skin's protective oil layer is stripped away, leaving it red, dry, and sore. Rubbing lotion into the skin can ease the pain. To reduce your risk of windburn, keep your skin covered and wear lip balm.

- **Sunburn,** a burning of the skin's outer layers. Mild sunburn makes the skin red and painful. Severe sunburn can cause blistering and swelling. Cool and moisturize the skin and take a mild analgesic pain reliever to ease the discomfort. Wear protective clothing when exercising in the sun. Use a sunscreen with a sun protection factor (SPF) of 15 or more, and reapply often. Avoid exercising outside when the sun's rays are most intense.

- **Skin cancer** can result from repeated or prolonged sun exposure. Sunscreens provide protection by blocking UVA, or ultraviolet A, rays, which lead to skin cancer.

Figure 12.17 This adaptive device was made to help people whose disabilities would make riding a bicycle difficult. *How do adaptive devices benefit people with disabilities?*

- **Eye damage** can be caused by exposure to ultraviolet (UV) rays. Wear sunglasses, a wide-brimmed hat in the summer, or UV-absorbing goggles during winter months.

Coping with Injuries

Main Idea You can treat minor sports injuries yourself, but major injuries require professional medical treatment.

You can identify and take action for both minor and major exercise-related injuries.

Minor Injuries

Muscles may become sore after exercise. Applying ice and taking pain relievers can help. Below are other minor injuries related to exercise:

- **Blisters,** fluid-filled bumps caused by friction. Well-fitting shoes and athletic socks can prevent blisters. Cover the blistered area, leave blisters intact, and let them heal.

- **Muscle cramps**, or *sudden and sometimes painful contractions of the muscles,* can occur when muscles are tired, overworked, or dehydrated. Stretching the affected muscle will usually relieve the cramps.

- **Strains** result from *overstretching and tearing a muscle.* Warm up before exercise to reduce the risk of strains. The symptoms are pain, swelling, and difficulty moving the affected muscle. Use the P.R.I.C.E. procedure, outlined in **Figure 12.18** on page 342 to treat strains.

 READING CHECK

Identify Problems and Solutions Name three health problems that can result from exercising in hot weather and explain how to prevent them.

- **Sprains** are *injuries to the ligaments around a joint* that produce pain, swelling, and stiffness. Use the P.R.I.C.E. procedure to treat minor sprains. If it hurts to move your joint, or you can't put weight on it, see your doctor.

- **Tendonitis** is inflammation and swelling in the tendons. Tendons are bands of fiber that connect muscles to bones. Treatment may include rest, medication, physical therapy, and in rare cases, surgery.

Major Injuries

While some minor injuries may be treated at home, major injuries require medical care. Here are some major injuries:

- **Fractures,** or broken bones, cause severe pain, swelling, bruising, or bleeding. If someone has broken a bone, get medical help immediately. Do not move the victim.

- **Dislocations** occur when a bone pops out of its normal position in a joint. The joint will be painful and may appear misshapen. Call for help immediately.

- **Concussion,** an injury to the brain can result in a severe headache, unconsciousness, or memory loss. A severe concussion can cause brain damage. Signs of brain damage include vomiting, confusion, seizures, or weakness on one side of the body. If any of these symptoms occur, seek medical help immediately.

READING CHECK

Explain What are the steps of the P.R.I.C.E. procedure?

Figure 12.18 **The P.R.I.C.E. Procedure**

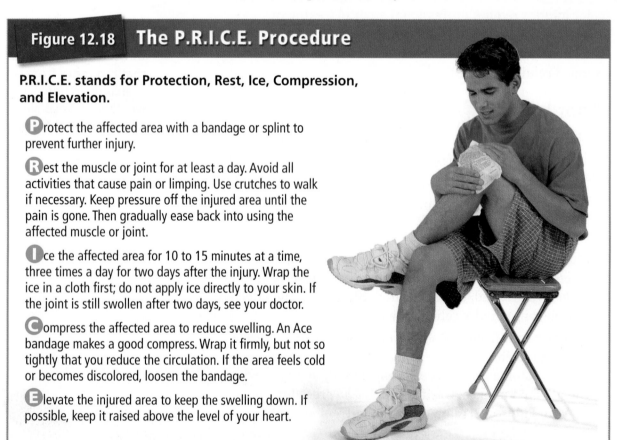

P.R.I.C.E. stands for Protection, Rest, Ice, Compression, and Elevation.

Protect the affected area with a bandage or splint to prevent further injury.

Rest the muscle or joint for at least a day. Avoid all activities that cause pain or limping. Use crutches to walk if necessary. Keep pressure off the injured area until the pain is gone. Then gradually ease back into using the affected muscle or joint.

Ice the affected area for 10 to 15 minutes at a time, three times a day for two days after the injury. Wrap the ice in a cloth first; do not apply ice directly to your skin. If the joint is still swollen after two days, see your doctor.

Compress the affected area to reduce swelling. An Ace bandage makes a good compress. Wrap it firmly, but not so tightly that you reduce the circulation. If the area feels cold or becomes discolored, loosen the bandage.

Elevate the injured area to keep the swelling down. If possible, keep it raised above the level of your heart.

Real World CONNECTION

Playing It Safe

Sports and other recreational activities are some of the most common causes of injury among teens. These injuries could be prevented if teens followed guidelines and safety precautions, and used the proper safety equipment for their sport.

In groups, choose a sport or recreational activity and conduct an online search. Use reliable online sources such as the Centers for Disease Control and Prevention (CDC) to find injury statistics, precautions for avoiding injuries, and types of protective equipment for this sport.

Use your research to create a blog or Web page that educates teens about how injuries occur in the sport that your group has chosen, and how to stay safe. Show

Activity Technology

examples of some protective equipment. Ask for permission to post the group's blog or Web page on the school's Web site.

LESSON 4 ASSESSMENT

 After You Read

Reviewing Facts and Vocabulary

1. What is the purpose of a health screening? How can it prevent injury during physical activity?

2. How should frostbite be treated? What can you do to prevent frostbite?

3. Name three symptoms of heat exhaustion.

Thinking Critically

4. **Analyze.** What distinguishes major injuries from minor injuries? How can you use the P.R.I.C.E. procedure to treat minor injuries?

5. **Synthesize.** Suppose you are playing Frisbee with some friends, and one of them falls and injures his ankle. How do you deal with the injury?

Applying Health Skills

6. **Practicing Healthful Behaviors.** Design a poster that illustrates the risks of sun and wind exposure. Include strategies for protecting yourself from these risks.

Writing Critically

7. **Expository.** Write a script for a one-minute public service announcement summarizing the importance of using the correct sports equipment. Your announcement should briefly describe the risks of injury.

Real Life Issues ....

After completing the lesson, review and analyze your response to the Real Life Issues question on page 337.

Hands-On HEALTH

 Activity ## Get Up and Get Fit

Now that you understand the benefits of fitness, use your knowledge to motivate others. Write a public service announcement (PSA) for a radio show. Conduct research to learn the physical, mental/emotional, and social health benefits of fitness. The PSA script should persuade others to get up and get fit.

What You'll Need

- computers with Internet access
- recording equipment (optional)

What You'll Do

Step 1

Work in groups of three or four. Identify at least five benefits of fitness, and five facts and examples demonstrating the benefits you selected.

Step 2

Write a script featuring at least three examples from your research. Support your position by citing at least one valid resource for each example.

Step 3

Present the PSA to the class as a role-play or a recording.

Apply and Conclude

Ask the entire class for feedback on each PSA. Discuss whether the message was clear, if valid examples were given, and whether the target audience was addressed.

Checklist: Advocacy

- ☑ Did I take a clear, health-enhancing stand?
- ☑ Can I support my position with reliable sources?
- ☑ Did I demonstrate an awareness of our target audience?
- ☑ Did I deliver the message with enough passion and conviction?

LESSON 1

Benefits of Physical Activity

Key Concepts

▸ Physical activity can benefit all sides of your health triangle.
▸ A sedentary lifestyle increases the risk of health problems.
▸ Several short periods of physical activity throughout the day can have the same benefits as one long workout.

Vocabulary

▸ physical activity (p. 318)
▸ physical fitness (p. 319)
▸ exercise (p. 319)
▸ sedentary (p. 321)

LESSON 2

Improving Your Fitness

Key Concepts

▸ The elements of fitness are five health-related components of fitness.
▸ Aerobic exercise improves cardiorespiratory endurance.
▸ Anaerobic exercises improve muscular strength and endurance.

Vocabulary

▸ cardiorespiratory endurance (p. 324)
▸ muscular strength (p. 324)
▸ muscular endurance (p. 325)
▸ flexibility (p. 325)
▸ aerobic exercise (p. 327)
▸ anaerobic exercise (p. 328)

LESSON 3

Planning a Personal Activity Program

Key Concepts

▸ Consider personal needs when planning a fitness program.
▸ Key fitness principles are specificity, overload, progression, and regularity.
▸ The F.I.T.T. formula will help you plan a successful workout.

Vocabulary

▸ specificity (p. 333)
▸ overload (p. 333)
▸ progression (p. 333)
▸ warm-up (p. 334)
▸ workout (p. 335)
▸ cool-down (p. 335)
▸ resting heart rate (p. 336)

LESSON 4

Fitness Safety and Avoiding Injuries

Key Concepts

▸ Wearing safety equipment will help protect you from injuries.
▸ The P.R.I.C.E. procedure can be used to treat minor injuries.
▸ Major injuries require medical care.

Vocabulary

▸ frostbite (p. 339)
▸ hypothermia (p. 340)
▸ overexertion (p. 340)
▸ heat exhaustion (p. 340)
▸ heatstroke (p. 340)
▸ muscle cramps (p. 341)
▸ strains (p. 341)
▸ sprains (p. 342)

LESSON 1

Vocabulary Review

Use the vocabulary terms listed on page 345 to complete the following statements.

1. _____ is the ability to carry out daily tasks easily.

2. To achieve specific fitness goals, use structured, purposeful physical activity, known as _____.

3. People whose lives include little physical activity can be described as _____.

Understanding Key Concepts

After reading the question or statement, select the correct answer.

4. Stronger muscles and bones, and greater energy, are examples of physical activity's
   a. physical benefits.
   b. mental benefits.
   c. emotional benefits.
   d. social benefits.

5. Which of the following is a mental/emotional benefit of physical activity?
   a. Lower blood pressure
   b. Better balance and coordination
   c. Reduced stress
   d. Forming new friendships

Thinking Critically

After reading the question or statement, write a short answer using complete sentences.

6. **Discuss.** Explain how physical activity can improve your social life.

7. **Identify.** Name two ways to fit physical activity into your daily life.

8. **Synthesize.** Give an example of how the physical, mental/emotional, and social benefits of physical activity are interrelated.

LESSON 2

Vocabulary Review

Choose the correct term in the sentences below.

9. Running a mile without stopping is a sign of good *cardiorespiratory endurance / muscular endurance.*

10. *Muscular strength / Flexibility* is the ability to move your body parts through their full range of motion.

11. Sprinting and lifting weights are examples of *aerobic exercise / anaerobic exercise.*

Understanding Key Concepts

After reading the question or statement, select the correct answer.

12. Which of the following is a good test of your cardiorespiratory fitness?
    a. The time it takes to run or walk a mile
    b. How many curl-ups you can do
    c. How heavy a weight you can lift
    d. Whether you can bend over and touch your toes

13. A healthy 30-year-old would have a target heart range between
    a. 60 and 120 beats per minute.
    b. 82 and 133 beats per minute.
    c. 94 and 159 beats per minute.
    d. 101 and 190 beats per minute.

Ken Karp/McGraw-Hill Education

14. Exercises to improve your flexibility are
    a. aerobic exercises.
    b. isometric exercises.
    c. isokinetic exercises.
    d. stretching exercises.

Thinking Critically

After reading the question or statement, write a short answer using complete sentences.

15. **Analyze.** Doing 50 curl-ups each day will improve what elements of fitness? What other activities can improve total fitness?

16. **Compare and Contrast.** Explain the different ways that aerobic and anaerobic exercise affect your body composition.

17. **Analyze.** Is swimming a good way to build bone mass? Why or why not?

LESSON 3

Vocabulary Review

Correct the sentences below by replacing the italicized term with the correct vocabulary term.

18. A *stretch* is gentle activity that prepares the muscles for work.

19. The part of an exercise session when you are exercising at your highest peak is called the *cool-down*.

20. Your *target heart rate* is the number of times your heart beats per minute when you are not active.

Understanding Key Concepts

After reading the question or statement, select the correct answer.

21. To build cardiovascular fitness, perform aerobic exercise at least
    a. twice a week for 20 minutes.
    b. three times a week for 20 minutes.
    c. five times a week for 10 minutes.
    d. one hour per day.

22. Which principle of building fitness involves gradually increasing the demands on your body?
    a. Specificity
    b. Overload
    c. Progression
    d. Regularity

23. If you have time to stretch only once during an exercise session, it's best to do it
    a. before warming up.
    b. after warming up.
    c. in the middle of your workout.
    d. while cooling down.

Thinking Critically

After reading the question or statement, write a short answer using complete sentences.

24. **Predict.** Explain what might happen if a teen builds a fitness plan around exercises that he or she strongly dislikes.

25. **Identify.** What are the four elements of the F.I.T.T. formula? How can the four elements help you become fit?

26. **Analyze.** How does your resting heart rate reflect your level of fitness? How does your active heart rate reflect your fitness level?

LESSON 4

Vocabulary Review

Choose the correct word in the sentences below.

27. *Overexertion/Heatstroke* is a dangerous condition in which the body loses its ability to cool itself through perspiration.

28. *Frostbite/Hypothermia* is damage to the skin and tissues caused by extreme cold.

29. Injuries to the ligaments around a joint are known as *strains/sprains*.

Assessment

Understanding Key Concepts

After reading the question or statement, select the correct answer.

30. Drowsiness, weakness, and slowed breathing and heart rate are symptoms of
    a. heat exhaustion.
    b. frostbite.
    c. hypothermia.
    d. concussion.

31. Stretching the affected muscle will usually relieve
    a. muscle cramps.
    b. strains.
    c. sprains.
    d. tendonitis.

32. Which of the following is *not* a major injury?
    a. Fracture
    b. Dislocation
    c. Concussion
    d. Sprain

Thinking Critically

After reading the question or statement, write a short answer using complete sentences.

33. **Describe.** What safety equipment is required for skating or skateboarding?

34. **Explain.** Why is it important to protect yourself from the sun during physical activity?

35. **Describe.** What are the steps in the P.R.I.C.E. procedure?

Technology PROJECT-BASED ASSESSMENT

Staying Informed About Physical Fitness

Background

Physical fitness is more than just doing exercise and maintaining a healthy, nutritious diet. Physical fitness requires being informed. Accurate information about the importance of physical activity helps individuals to make well-informed decisions about their health.

Task

Develop a Web site for your school showing different activities that students can do to stay fit. Include group, team, and individual activities.

Audience

Students in your school

Purpose

Provide information on physical fitness to your peers.

Procedure

1. Brainstorm and conduct an online search for activities that students can do to stay fit. Be sure to include information about clubs in school that may help with staying fit. You may also want to consider physical fitness for people with disabilities.

2. Divide the tasks among group members. Some members may want to do research, while others may want to help design and create the Web site.

3. Present the Web site to your class and ask students to complete a survey using an online survey tool to assess the effectiveness of the Web site. Also, ask students what other information they would like about physical fitness.

Math Practice

Calculating Distances. Huntsville High's school-wide olympics will promote physical activity. Races will be run in the gym. For one race, athletes will run one lap around the gym. That distance would be approximately the same as the perimeter of the gym.

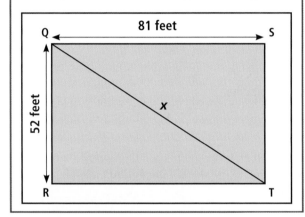

Diagram of Huntsville High School Gym

Q — 81 feet — S

52 feet

x

R — T

1. What is the perimeter of the gym?
   A. 1,112 feet
   B. 421 feet
   C. 266 feet
   D. 386 feet

2. One athlete covers about 5 feet every second. If she competed in the race today, approximately how many seconds would it take her to run the one-lap course?
   A. 53 seconds
   B. 55 seconds
   C. 532 seconds
   D. 260 seconds

3. A race across the gym diagonally is represented in the diagram by line x. Line x divides the gym into two congruent right triangles. What is the approximate length, in feet, of line x, the side the two triangles share?
   A. 421.2 feet
   B. 21.60 feet
   C. 133 feet
   D. 96.25 feet

Reading/Writing Practice

Understand and Apply. Read the passage below, and then answer the questions.

On Sunday, 17-year-old Rosa Martinez completed her first marathon. Her time of 2 hours and 45 minutes won't break any records, but she's proud to have finished the race—in her wheelchair.

"I lost the use of my legs in a car crash three years ago," says Rosa. "I was really depressed, but getting into wheelchair sports inspired me. I started focusing more on what I could do in my chair than on what I couldn't do."

To train for the marathon, Rosa says she did "a lot of aerobic exercises to strengthen my heart and lungs, and a lot of work on my upper body strength."

While she's proud of her achievement, Rosa isn't going to rest on her laurels. She's already looking ahead to next year's marathon, and she's determined to beat her time from this year.

1. In the final paragraph, the phrase "rest on her laurels" means
   A. take a break from exercising.
   B. go on to bigger challenges.
   C. keep doing the same activities.
   D. settle for what she's already achieved.

2. Which of the following would make the best title for this passage?
   A. The Winning Spirit
   B. How to Train for a Marathon
   C. Elements of a Fitness Program
   D. Better Wheelchair Designs

3. Describe the physical and mental qualities that make Rosa a successful athlete. How might she apply these qualities to other aspects of her life?

National Education Standards

Math: Measurement and Data
Language Arts: LACC.910.L.3.6, LACC.910.RL.2.4, LACC.910.W.3.8

TEENS *Speak Out*

©Erik Isakson/Blend Images LLC

Body Image and the Media

I t's no secret that the average person you see on the street doesn't look like a model in a magazine. The average fashion model is 7 inches taller than the average American woman, yet weighs 23 pounds less. Men in the media also have a typical look: broad-shouldered, narrow-waisted, muscle-clad, and free of body hair. That's far from the look of the average American male.

Some people think the bodies we see in the media are unrealistic and harmful. Others think they represent a healthful ideal. Take a look at what these teens have to say, and then decide how you feel.

Problems with Images in the Media

The "ideal body" presented in the media is exceptionally thin. A typical female fashion model has a body mass index of 16.8—thinner than 98 percent of all women in America, and thin enough to put her health at risk. These media images may be making teen girls dissatisfied with their own appearance. More than half of all teen girls and women say they are unhappy with their bodies. Increasing numbers of males are also unhappy with their body image.

> 66 Looking at magazines and seeing how perfect everyone's body is makes me feel like my body isn't good enough. When I look around, though, I see that I look pretty much like everyone else. "
>
> —Ned R., age 16

Benefits of Images in the Media

In the United States today, 66 percent of adults and 17 percent of teens are overweight. Over 30 percent of adults are considered obese. Health problems related to being overweight include type 2 diabetes and cardiovascular disease. Using models who look more like the typical American could give the impression that being overweight is normal and that it doesn't pose a health risk.

> 66 When I read magazines, I don't want to see people who look average. Media personalities are people to look up to—an ideal. With the obesity problem in America, media images that only show the average American may set an unhealthy example."
>
> —Joanna L., age 16

Activity Beyond the Classroom

1. **Investigate** images from a variety of different media, such as magazines, billboards, and television. Take notes on the type of males and females that are pictured.

2. **Survey** other teens to find out how they feel about this issue. Ask: Do you think the bodies you see in the media are healthy? Do they make you feel good or bad about yourself?

3. **Express** your views on this topic. Write a newspaper column summarizing what you've learned from other teens about the effect of media images on body image.

UNIT 5 Personal Care and Body Systems

Blend Images/Ariel Skelley/Getty Images

UNIT PROJECT

Hammering for a Good Cause

Using Visuals Habitat for Humanity is an international organization that recruits volunteers and raises funds to build and repair homes for low-income families in the United States and all over the world. Volunteers can raise money, gather building materials, or pick up a hammer and start building.

Get Involved. Locate a local organization that uses volunteers to build and repair homes for families in need. Find out how teens can participate. Share your findings with your classmates.

"The first wealth is health."
—Ralph Waldo Emerson, 19th-century writer and poet

CHAPTER 13

Personal Health Care

Ed Imaging

Lesson 1

Healthy Skin, Hair, and Nails

BIG Idea *Taking care of your skin, hair, and nails helps keep your whole body healthy.*

Lesson 2

Healthy Teeth and Mouth

BIG Idea *Your teeth and mouth need care to function well and keep you healthy.*

Lesson 3

Healthy Eyes and Ears

BIG Idea *Eyes and ears are sensitive organs that need protective care.*

Activating Prior Knowledge

Using Visuals The teen in this photo knows the importance of regular exams to keep her eyes healthy. In a few sentences, describe other ways you can take care of your eyes and protect your vision.

Chapter Launchers

Health in Action

Discuss the **BIG** Ideas

Think about how you would answer these questions:

▶ Why are personal hygiene habits so important?

▶ What problems can result from not taking care of your teeth and mouth?

▶ Could you be at risk for hearing proble

Assess Your Health

Read each statement. On a separate sheet of paper, write "yes," "sometimes," or "no" based on your typical behavior.

1. I recognize that using sunscreen will help keep my skin healthy.

2. I know where to seek help for acne and other skin disorders.

3. I brush my teeth at after every meal and floss daily.

4. I wear sunglasses in bright sunlight to protect my eyesight.

5. I avoid listening to loud music.

A "yes" response shows that you practice healthy behaviors. "Sometimes" indicates that you should analyze and possibly modify your behavior. A "no" response means that you should modify the behavior.

Healthy Skin, Hair, and Nails

BIG Idea *Taking care of your skin, hair, and nails helps keep your whole body healthy.*

Before You Read

Create a Table. Make a three-column table. Label the columns "Tissue," "Structure," and "Function." In the first column, list "Skin," "Hair," and "Nails." In the second column, describe the important structural features of each. In the third column, write the function of each.

Tissue	Structure	Function

New Vocabulary

▶ epidermis
▶ dermis
▶ melanin
▶ sebaceous glands
▶ hair follicles
▶ melanoma

Real Life Issues ·

Protecting the Skin You're In.

Source: Centers for Disease Control and Prevention, Youth Risk Behavior Surveillance Survey, 2011

> **13% of all high school students reported using indoor tanning devices.**

> **Almost 11% of high school students reported that they frequently used a sunscreen with an SPF of 15 or higher.**

Writing *Write a paragraph describing how sun exposure can harm your health.*

Your Skin

Main Idea Skin protects you from pathogens, regulates your body temperature, and helps you feel sensations.

What's the largest organ on the human body? You may be surprised to learn that the answer is the skin. The skin consists of two main layers, as shown in **Figure 13.1**. The **epidermis** is *the outer, thinner layer of the skin that is composed of living and dead cells.* Just underneath the **dermis** is *the thicker layer of the skin beneath the epidermis that is made up of connective tissue and contains blood vessels and nerves.* Cells in the epidermis make substances called *lipids,* which make your skin waterproof. This waterproofing helps the body maintain a proper balance of water and electrolytes. Other cells produce **melanin,** *a pigment that gives the skin, hair, and iris of the eyes their color*—the more melanin that your body produces, the darker the skin. The melanin in skin also helps protect the body from harmful ultraviolet (UV) radiation that causes skin cancer.

The skin performs three main functions to keep you healthy:

- **Protection.** The skin protects you from pathogens and internal damage. It acts as a barrier to prevent bacteria and viruses from entering your system. If this barrier is broken by a cut or other wound, the skin repairs itself to keep pathogens from entering the body.

- **Temperature control.** When your body temperature begins to rise, the blood vessels in the skin dilate, allowing heat to escape through the skin's surface. Sweat glands—structures within the dermis that release perspiration through ducts to pores on the skin's surface—cool the skin. If body temperature begins to drop, the blood vessels in the skin constrict, reducing the amount of heat lost and helping to maintain body heat.

- **Sensation.** Touch a hot stove, and your hand immediately pulls back. Why? The skin is a major sense organ. Nerve cells in the dermis act as receptors that are stimulated by changes in the outside environment. These receptors enable you to feel sensations such as pressure, pain, heat, and cold.

READING CHECK

Explain What are the dermis and epidermis?

Figure 13.1 **The Skin's Structure**

The skin is composed of two main layers, the epidermis and the dermis. These two layers are attached to bones and muscles by the subcutaneous layer, a layer of fat and connective tissue located beneath the dermis. Explain how the skin helps regulate body temperature.

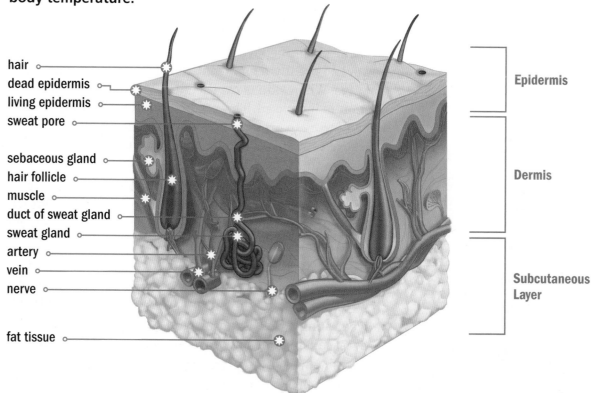

The dermis is a single thick layer composed of connective tissue, which gives the skin its elastic qualities. **Sebaceous glands**, *structures within the skin that produce an oily secretion called sebum,* are also found in the dermis. Sebum helps keep skin and hair from drying out. Blood vessels in the dermis supply cells with oxygenated blood and nutrients and help remove wastes from body cells.

Keeping Your Skin Healthy

Main Idea A daily routine will keep your skin healthy.

Keeping your skin healthy should be an important part of your daily routine. Some of the ways that you can keep your skin health, include:

- Wash your face every morning and evening with mild soap and water.
- Daily washing, bathing, or showering helps **remove** and slow the growth of bacteria that cause body odor.
- Avoid touching your face with your hands. This can introduce new bacteria to the skin's surface.
- Choose personal skin care products carefully to avoid irritation and the chance of allergic reaction.
- Follow a well-balanced eating plan that is rich in vitamins and minerals, especially vitamin A. Milk, green and yellow vegetables, and liver are good for healthy skin.

Academic Vocabulary

remove *(verb):* to get rid of

UV Protection

Some people believe that tanned skin looks good. A suntan, however, is really a sign that the skin has been damaged by UV rays. When skin is exposed to UV radiation, melanin production is increased. The production in melanin is the skin's way of trying to protecting itself from the UV rays. Prolonged exposure to UV rays can lead to skin cancer. To protect your skin from the sun's damaging rays,

- always wear sunscreen on exposed areas of skin. Use an SPF of 15 or higher that blocks both UVB and UVA (the more penetrating) rays. Apply it 15 to 30 minutes before going outside, even on cloudy days.
- wear protective clothing, including hats, long-sleeved shirts, and long pants. UV rays are most intense between 10:00 A.M. and 4:00 P.M., and stronger at higher altitudes.
- wear sunglasses. Exposure to UV rays can damage the eyes, causing burns, cataracts, and even blindness.
- avoid using tanning beds. Tanning beds are not safe, and prolonged exposure can lead to skin cancer.

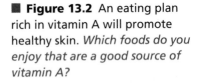

■ **Figure 13.2** An eating plan rich in vitamin A will promote healthy skin. *Which foods do you enjoy that are a good source of vitamin A?*

Health Skills Activity

Communication Skills

Is Tanning Worth the Risk?

It's a hot summer day as Shelley walks out of the locker room toward the pool. She has always taken pride in having a deep tan. Lately, however, she's been thinking about her aunt. After years of sun exposure, her aunt is now undergoing skin cancer treatments. Now Shelley doesn't think tanning is such a good idea.

"C'mon, Shelley!" says her friend Raye, motioning to Shelley. "The pool opens in ten minutes. Let's grab the best spots in the sun. School starts in a month, so we have to make the most of every sunny day!"

Shelley hangs back, biting her lip. She'd rather find a spot in the shade. She wants to convince Raye to do the same, but she doesn't want to hurt her friend's feelings.

Writing Compose your thoughts about what Shelley should say in a paragraph or two. Make your argument persuasive enough to change Raye's mind. Use the following tips as a guideline.

1. Use "I" messages.
2. Speak calmly and clearly.
3. Use a respectful tone.
4. Listen carefully and ask appropriate questions.
5. Use appropriate body language.

Body Piercing and Tattooing

Piercing and tattooing practices have been around for thousands of years. Unlike using makeup or changing hair color, however, piercings and tattoos are permanent. Both carry potential health risks because they break the physical barrier of the skin. This can result in infection from bacteria and the transfer of blood-borne pathogens from viruses such as hepatitis B, hepatitis C, and HIV through nonsterile needles. The American Dental Association also warns that oral piercing can damage your mouth and teeth. The decision to get a tattoo or piercing may also impact your social health, by limiting future job opportunities and relationships.

Skin Problems

Many skin problems can affect your self image, but are not life threatening. Check with a health care professional before purchasing any skin care product to make sure it's right for you. Common skin problems include

- **Acne.** When pores in the skin get clogged, bacteria causes inflammation, and pus forms. To treat acne, wash your face gently twice a day, apply over-the-counter treatments, and avoid using oily products or too much makeup. Touching and picking may cause scarring. Extreme cases may require prescription medication.

Tim Fuller Photography

National Cancer Institute (NCI)

- **Warts.** These are caused by a virus and are most commonly found on the hands, feet, and face. They can spread through direct contact with another person's wart.
- **Dermatitis, or eczema.** This is an inflamed or scaly patch of skin, usually from an allergic reaction. Keeping the area well moisturized can help reduce the irritation. A doctor may prescribe medications to treat dermatitis.
- **Fungal infections.** Ringworm and athlete's foot are infections that can be spread by contact with skin or infected clothing, or in public showers. Keep the infected area clean and dry, and treat with over-the-counter medicines.
- **Boils.** These form when **hair follicles**—*sacs or cavities that surround the roots of hairs*—become infected. The tissue becomes inflamed, and pus forms. Bursting or squeezing a boil can spread the infection. Treatment can include draining the pus and taking antibiotics.
- **Vitiligo.** A condition in which patches of skin lose melanin and have no pigment is called Vitiligo (vih-tuh-LY-go). These areas are extremely susceptible to burning when exposed to UV light, so they should always be covered.
- **Moles.** Though most moles are harmless, certain types may develop into **melanoma**, *the most serious form of skin cancer,* which can be deadly. Early detection and treatment are critical in controlling the spread of this cancer. See **Figure 13.3** on how to monitor the appearance of moles. Report any changes to a dermatologist.

READING CHECK

Describe What are two ways to protect your skin from UV rays?

| Figure 13.3 | **The ABCD's of Melanoma** |

Regularly checking the appearance of your moles is important for the early detection of melanoma.

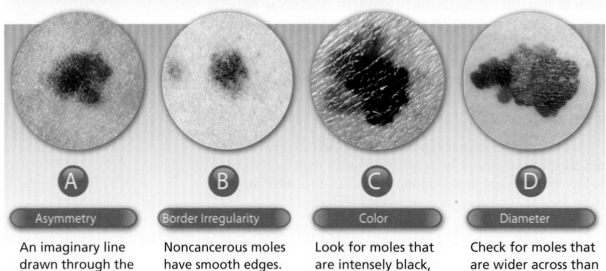

| A | B | C | D |
| Asymmetry | Border Irregularity | Color | Diameter |

An imaginary line drawn through the center of the mole does not produce matching halves.

Noncancerous moles have smooth edges. Suspect moles often have irregular edges.

Look for moles that are intensely black, possibly with a bluish tint, or that have an uneven color.

Check for moles that are wider across than the width of a pea.

Your Hair

Main Idea Your hair protects your skin from UV radiation and helps maintain body heat.

Hair grows on every surface of the skin, except for the palms of the hand and the soles of the feet. You have more than 100,000 hairs on your head alone. Although hair is composed of dead cells, living cells in the epidermis make new hairs and cause hair growth.

Hair helps protect the skin, especially the scalp, from exposure to UV radiation. The eyebrows and eyelashes protect the eyes from dust and other particles. Hair also reduces the amount of heat lost through the skin of the scalp.

Healthy hair begins with a well-balanced diet. Without proper nutrients, hair can become thin and dry. Daily brushing keeps dirt from building up and helps distribute the natural oils in your hair evenly. Regular shampooing will keep your hair healthy. It's best to limit the use of harsh chemical treatments such as dyes, bleach, or permanents. Also, avoid excessive use of heating irons or hot combs. Overexposure to these can cause hair to become dry and brittle.

Hair Problems

Normally, oil produced by sebaceous glands protects the skin from drying out and keeps hair soft and shiny. Dandruff—the dead skin cells that are shed as sticky white flakes when the scalp becomes too dry—usually can be treated with an over-the-counter dandruff shampoo. If itching or scaling persists, consult a health care professional who may prescribe another type of treatment.

Head lice are tiny parasitic insects that live in the scalp hair of humans. They feed on blood by biting through the skin of the scalp. Lice can infect anyone and are mainly transmitted by head-to-head contact or by using objects such as brushes, combs, or hats that have been used by an infected person. Using a medicated shampoo can kill the organisms. Washing sheets, pillowcases, combs, and hats with hot water and soap, as well as frequent vacuuming at home, can help prevent the spread of head lice or a repeat infection.

FITNESS ZONE

Want healthy hair? I guess we all do! My cousin is learning to be a hair stylist. She says that good nutrition and drinking lots of water helps keep your hair healthy. She says if your body is healthy and well nourished, your hair will be your shining glory. For more fitness tips, visit the Fitness Handbook and Fitness Zone sites in ConnecEd.

■ **Figure 13.4** Give your hair daily attention to keep it clean and healthy. *How do you choose hair care products that are right for your hair?*

Florian Franke/Purestock/SuperStock

Your Nails

Image Source/Alamy

Main Idea Nails help protect your fingers and toes.

Like your hair, your fingernails and toenails are made of closely packed dead cells that contain keratin. Cells beneath the root of the nail divide and multiply, causing the nail to grow. Nails protect and support tissues of fingers and toes.

Good care of the nails includes keeping them clean and evenly trimmed, which helps prevent split nails and hangnails. Use a nail file to shape and smooth nails, and keep cuticles pushed back. A cut, split, or break in the skin around the nail allows pathogens into the body and may lead to infection. Keep the area clean and apply an antibiotic ointment if necessary.

Trim toenails straight across and just slightly above the skin level to reduce the risk of infection and ingrown nails. Keeping nails short also reduces the risk of fungal infections under the nails. Fungal infections can be treated with antifungal medications.

■ **Figure 13.5** Keeping nails neatly clipped and filed improves your overall appearance. *List three other grooming habits that contribute to a healthy appearance.*

LESSON 1 ASSESSMENT

 After You Read

Reviewing Facts and Vocabulary

1. Define the terms *melanin* and *hair follicle*.

2. Explain the causes of acne. How is acne treated?

3. What viruses can you potentially contract through getting a tattoo?

Thinking Critically

4. **Apply.** Taking care to keep your nails clean and trimmed is important. Why might biting your nails be an unhealthy practice?

5. **Synthesize.** Explain how proper skin, hair, and nail care tells others that you care about your appearance.

Applying Health Skills

6. **Analyzing Influences.** Darla wants an eyebrow piercing because the lead singer in her favorite band has one. A friend offered to do the piercing for free. Write a letter to Darla and point out the influences on her choice. Remind her of the health risks.

Writing Critically

7. **Persuasive.** Write a brief dialogue between two teens. One wants to get a shoulder tattoo. The other explains the health and social risks.

Real Life Issues ....

After completing the lesson, review and analyze your response to the Real Life Issues question on page 356.

Healthy Teeth and Mouth

Real Life Issues

Preventive Health. Maria is afraid of going to the dentist, even though she knows that getting her teeth cleaned every six months is important to her health. She often makes an appointment, then gets nervous and cancels it. Maria's brother, Juan, overhears her canceling her latest appointment.

Writing *Write a dialogue between Maria and Juan. Juan should try to convince Maria that going to the dentist twice a year is important.*

Your Teeth

Main Idea **Every tooth has three main parts.**

Having healthy teeth is important for your appearance, but it's also important for your overall physical health. Your teeth break down foods into pieces that are small enough to easily digest. They also help form the shape and structure of your mouth. Your permanent teeth come in gradually, usually beginning when you are about five years old and continuing into young adulthood when your last permanent molars, or wisdom teeth, come in.

Parts of a Tooth

The **periodontium** (per-ee-oh-DAHN-tee-uhm) is *the area immediately around the tooth.* It is made up of the gum, periodontal ligaments, and the jawbone. The periodontium support the tooth and hold it in place. The tooth itself has three main parts: the crown, the neck, and the root, as shown in **Figure 13.6.** The crown is the visible portion of the tooth. It is protected with enamel, a hard substance made of calcium.

GUIDE TO READING

BIG Idea *Your teeth and mouth need care to function well and keep you healthy.*

Before You Read

Make an Outline. Use the headings and sub-headings in this lesson to make an outline of what you'll learn. Use this type of format to help you organize your notes.

> I.
>   A.
>   1.
>   2.
>   B.
> II.

New Vocabulary

- periodontium
- pulp
- plaque
- halitosis
- periodontal disease
- malocclusion

Figure 13.6   Cross Section of a Tooth

A protective layer of enamel covers the crown of a tooth. Inside the tooth, blood vessels supply the living tissue with oxygen and nutrients.

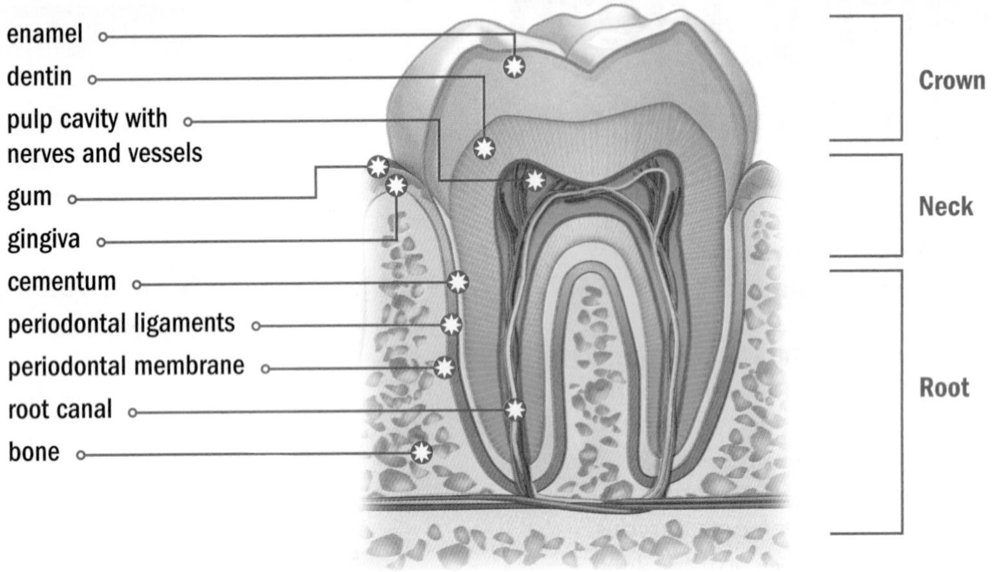

- enamel
- dentin
- pulp cavity with nerves and vessels
- gum
- gingiva
- cementum
- periodontal ligaments
- periodontal membrane
- root canal
- bone

Crown

Neck

Root

READING CHECK

Identify What are the three parts of the tooth?

■ **Figure 13.7** Healthy teeth are important to your overall health. *Explain how healthy teeth protect your health.*

Beneath the enamel is *dentin,* a layer of connective tissue that contributes to the shape and hardness of a tooth. The **pulp** is *the tissue that contains the blood vessels and nerves of a tooth.* Protected by the overlying layers of dentin and enamel, the pulp extends into the root canal. The neck of a tooth is between the crown and the root.

Keeping Your Teeth and Mouth Healthy

Main Idea You can make choices that help keep your teeth and mouth clean and healthy.

Oral hygiene, which includes brushing and flossing your teeth, is necessary for healthy, clean teeth. The bacteria that naturally inhabit your mouth metabolize the sugars in the foods you eat. They produce an acid that breaks down the protective layer of tooth enamel. Tooth decay occurs when the enamel is destroyed and bacteria penetrate the tooth.

Plaque is *a combination of bacteria and other particles, such as small bits of food, which adheres to the outside of a tooth.* Plaque damages the tooth by coating it, sealing out the saliva that normally protects the tooth from bacteria. If plaque builds up, the acids produced by bacteria break down the tooth enamel, resulting in a hole, or cavity. If decay spreads down to the pulp, the tooth may have to be removed.

Practicing good oral hygiene can prevent tooth decay and other diseases. Brushing your teeth after eating removes plaque from the surface of the teeth, before bacteria can produce the acid that harms teeth. Flossing between your teeth removes plaque in areas that cannot be reached with the bristles of a toothbrush.

To maintain your dental health, visit your dentist regularly. The dentist, or a dental hygienist, will clean your teeth and examine them for signs of decay. Dentists may use sealants to prevent tooth decay. In addition to visiting the dentist regularly, you can take the following steps to keep your teeth and gums healthy:

- Eat a well-balanced diet that includes foods containing phosphorus, calcium, and vitamin C.

- Reduce the number of sugary drinks and snacks you eat.

- Brush your teeth after every meal, and floss daily.

- Get regular dental checkups.

- Wear a mouth guard when you play contact sports or other activities to protect your mouth and teeth.

- Avoid all tobacco products. They stain teeth and cause gums to recede. They also increase the risk of oral cancer.

Tooth and Mouth Problems

Main Idea Neglecting your teeth can result in problems.

Some oral problems are caused by poor hygiene, others by poorly aligned teeth. Be alert to these common problems:

- **Halitosis**, or *bad breath,* can be caused by eating certain foods, poor oral hygiene, smoking, bacteria on the tongue, decayed teeth, and gum disease.

- Gum disease, or **periodontal disease**, *an inflammation of the periodontal structures,* is caused by bacterial infection. When plaque hardens, it builds up *tartar,* a hard, crustlike substance. This causes the gums to become irritated and swollen. This early stage is called *gingivitis* (jin-jih-VY-tis). If left untreated, the bone and tissue that support the teeth are destroyed, and teeth can be lost.

- **Malocclusion** (mal-uh-KLOO-zhun), *a misalignment of the upper and lower teeth,* or a "bad bite," can be caused by crowded or extra teeth, thumb sucking, injury, or heredity. If not treated, malocclusion can lead to decay, and affect a person's speech and ability to chew.

- **Impacted wisdom teeth** sometimes crowd and push on other teeth or become infected. They may need to be removed surgically.

READING CHECK

Cause and Effect Describe how plaque leads to tooth decay.

READING CHECK

Describe What are six possible causes of halitosis?

Examining Product Claims

The toothpaste aisle contains products that make many different claims. Some brands of toothpaste whiten teeth, others prevent cavities, some prevent bad breath, and others combine some or all of these claims. How can you tell if these claims are true?

Check for exaggerated or misleading claims on product labels. Does the label tell you how the product works?

Fresher Breath — Whiter Teeth, Too!

Determine whether the product is safe. Some tooth whiteners, for example, contain abrasives that may cause gum irritation.

Activity Technology

Investigate the claims made by various toothpaste manufacturers.

1. Begin by conducting an online search about product claims for toothpaste. A few good places to start are the Food and Drug Administration (FDA), professional dental associations, and nonprofit consumer protection organizations.

2. Find out what these organizations recommend that all toothpastes should do. Which features are important? If the information cannot be located on the Web sites, you may have to call or e-mail the organizations.

3. Create a multimedia slide presentation based on the information you were able to obtain. During your presentation, be able to cite sources and explain why you think the claims you support are reliable.

LESSON 2 ASSESSMENT

 After You Read

Reviewing Facts and Vocabulary

1. Define the terms *periodontal disease* and *plaque*.

2. What is the pulp of the tooth?

3. Explain how tooth decay happens.

Thinking Critically

4. **Infer.** Dentists may apply a sealant to children's teeth to protect them from decay. How do you think these sealants work?

5. **Compare and Contrast.** Which layers of the tooth are sensitive, and which are not? Explain.

Applying Health Skills

6. **Accessing Information.** Do research at the library or on the Internet to learn more about what an endodontist does.

Writing Critically

7. **Persuasive.** Write a short letter to a younger brother or sister describing the reasons why it's important to brush and floss teeth regularly.

Real Life Issues

After completing the lesson, review and analyze your response to the Real Life Issues question on page 363.

Healthy Eyes and Ears

How Loud is Too Loud? The graph reveals how a noise over 100 decibels can cause hearing damage with more than 15 minutes of unprotected exposure.

Source: National Institute on Deafness and Other Communication Disorders

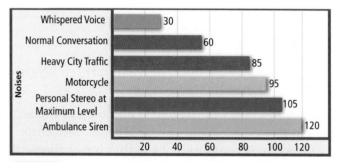

Noises:
- Whispered Voice — 30
- Normal Conversation — 60
- Heavy City Traffic — 85
- Motorcycle — 95
- Personal Stereo at Maximum Level — 105
- Ambulance Siren — 120

Writing *Write a letter to yourself describing ways that you can reduce your exposure to everyday noises.*

Your Eyes

Main Idea The eyes and their supporting structures are a complex of parts.

Most of the sensory information that travels to your brain comes from light signals received by your eyes. Structurally, your eyes sit in bony sockets, called orbits, at the front of your skull. A layer of fat cushions each eyeball inside its socket. Another structure is the lacrimal gland, which secretes tears into the eye through ducts. Tears are made of water, salts, mucus, and a substance that protects the eye from infection. As you blink, tears are moved across the surface of the eye. They keep the surface of the eyeball moist and clear of foreign particles.

Figure 13.8　The Eye

The eye collects light and sends signals to the brain, where images are formed.

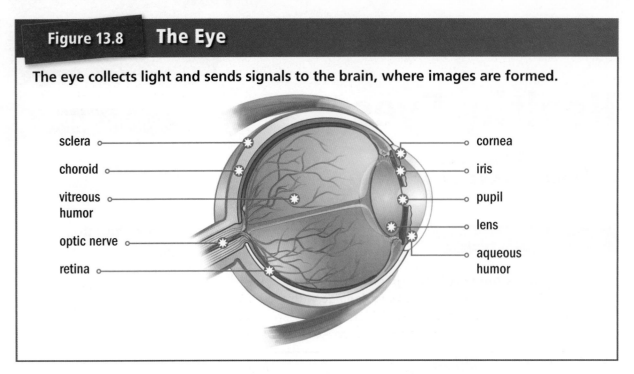

sclera
choroid
vitreous humor
optic nerve
retina

cornea
iris
pupil
lens
aqueous humor

Parts of the Eye

The eye consists of the optic nerve and three layers of the eyeball wall, as shown in **Figure 13.8.**

- The outermost layer of the eye is made up of the sclera and the cornea. The **sclera** (SKLEHR-uh), *the white part of the eye,* is composed of tough, fibrous tissue that protects the inner layers of the eye. The **cornea** at the front of the eye, is *a transparent tissue that bends and focuses light before it enters the lens.*

- Within the middle layer of the eye wall is the *choroid* (KOHR-oid), a thin structure that lines the inside of the sclera. Also within the middle layer is the *iris,* the colored **portion** of the eye that contains the pupil. The pupil is the hole through which light reaches the inner eye. In bright light, the pupil constricts; in dim light, it enlarges, or dilates, to let in more light.

- The **retina** is *the inner layer of the eye wall.* The retina contains millions of light-sensitive cells called rods and cones. Rods allow us to see in dim light. Cones function in bright light and allow us to see color. When light stimulates these cells, a nerve impulse travels to the brain through the *optic nerve,* which is located at the back of the eye.

Behind the iris and the pupil is the *lens* of the eye. The lens is transparent and helps refine the focus of images on the retina. The area between the cornea and the lens is filled with a watery fluid called *aqueous humor,* which provides nutrients to the eye. Between the lens and retina there is a cavity filled with a gelatin-like substance called *vitreous humor.* This helps the eyeball stay firm and keep its shape, and holds the retina against the choroid.

Academic Vocabulary

portion *(noun):* a part set off from the whole

Health Skills Activity

Decision Making

Fun in the Sun?

Clarissa's family is going on vacation to a sunny, warm climate, and she wants to buy a new pair of sunglasses. Clarissa and her friend Justine are trying on sunglasses at the mall. Justine encourages Clarissa to buy a particular pair.

"Those look *so* cool on you," says Justine. "You *have* to get them."

Clarissa reads the label to find out what kind of protection the sunglasses have. "I do like these, but there is no UV-protection label," she says. "We're going to be in the sun a lot, so I want to protect my eyes."

Justine shakes her head. "Those sunglasses look the best. C'mon, Clare, it's important to look good. You can worry about your health when you get old."

Writing Write an ending in which Clarissa decides between looking good and protecting her health. Include what she might do to have it both ways. Apply the six steps of the decision-making process to Clarissa's situation.

1. State the situation.
2. List the options.
3. Weigh the possible outcomes.
4. Consider values.
5. Make a decision and act.
6. Evaluate the decision.

Vision

When light passes through the cornea, pupil, and lens to reach the retina, an image forms. Light rays are first focused by the curved cornea, then later refined by the lens onto the retina. This light stimulates the rods and cones in the retina, sending nerve impulses to the brain through the optic nerve. The brain translates the nerve impulses into images that you recognize.

Your vision may be affected by the way images are produced on the retina. Having clear, or 20/20, vision means that the images are produced clearly and sharply on your retina. This means that you can stand 20 feet away from an eye chart and read the top eight lines. When the images produced on your retina are not clear, an eye doctor will perform tests to determine what type of disorder is affecting your vision. Two common vision disorders are nearsightedness, or *myopia*, and farsightedness, or *hyperopia*.

If you have 20/60 vision, you must be 20 feet from the chart in order to read it the way a person with normal vision can read it from 60 feet. A person with 20/60 vision is said to be nearsighted, which means being able to see close, but not far. Other components of vision include eye coordination, peripheral or side vision, and depth perception.

 READING CHECK

Explain How do light signals become images?

Figure 13.9 Eye Problems

Structural Problems	Cause	Treatment
Nearsightedness (myopia) The inability to see distant objects clearly	May occur naturally; cornea is misshaped, or eye is too long.	Contact lenses, eyeglasses, or laser surgery
Farsightedness (hyperopia) The inability see close objects clearly	May occur naturally; cornea is misshaped, or eye is too short.	Contact lenses, eyeglasses, or laser surgery
Astigmatism Blurred vision	May occur naturally; lens or cornea is misshaped.	Contact lenses, eyeglasses, or surgery
Strabismus Eyes off-center, turned inward or outward	Weak eye muscles	Vision therapy, contact lenses, eyeglasses, or surgery
Detached retina Blurred vision or bright flashes of light	Retina has become detached from the choroid due to injury or aging.	Laser surgery to reattach retina

Disease or Vision Problems	Cause	Treatment
Infections and Viruses (such as sties, pinkeye, hepatitis) Swelling, irritation, blurred vision, change in sclera color	Pathogens infect the eye or the tissue around the eye.	Medications such as antibiotics may cure infections. No treatment for viruses.
Glaucoma Cloudy, impaired vision, sometimes permanent eye damage	High pressure inside the eye damages the retina and optic nerve.	Laser treatment; early detection helps minimize damage.
Cataracts Foggy vision	Lens becomes cloudy and cannot focus light.	Surgical removal of old lens and replacement with an artificial lens
Macular degeneration Vision loss	Part of the retina opposite the lens deteriorates due to aging.	No cure exists. Treatment is limited.

 **READING CHECK**

Describe What are four things you can do to keep your eyes healthy?

Keeping Your Eyes Healthy

Main Idea Making healthy choices will keep your eyes healthy.

Eye problems are outlined in **Figure 13.9**. You can practice several healthful behaviors to help keep your eyes healthy.

- **Follow a well-balanced eating plan.** Include foods that contain vitamin A, such as carrots and sweet potatoes. (See **Figure 10.6**, p. 263, for other good sources of this vitamin.) Deficiency in vitamin A could result in night blindness—the inability to see well in dim light.

- **Protect your eyes.** Wear safety goggles when participating in activities in which your eyes could be injured. Keep dirty hands or other objects such as makeup applicators away from your eyes to reduce the risk of infection and injury. Wear sunglasses that block UV light, and never look directly into the sun or bright lights.

- **Rest your eyes regularly.** Take regular breaks when using the computer or reading. Looking up and away every 10 minutes reduces eyestrain.

- **Get regular eye exams.** Routine eye exams enable health care professionals to detect and treat eye disease in its early stages.

In some cases when a cornea is diseased, a corneal transplant may be recommended. Corneal transplants can restore vision and reduce pain. It is the most commonly performed transplant surgery in the United States.

Your Ears

Main Idea The inner, middle, and outer ear work together so you can hear.

The ear can be divided into three main sections, each with its own unique structures. The parts of the ear are shown in **Figure 13.10**, on page 372.

- **The Outer Ear.** The outer ear is the visible part of the ear, called the *auricle.* It channels sound waves into the *external auditory canal*. This canal leads to the remaining portion of the outer ear, called the eardrum. The skin of this canal is lined with tiny hairs and glands that produce wax that protect the ear from dust and foreign objects. The eardrum, also called the tympanic membrane, acts as a barrier between the outer and middle ear.

- **The Middle Ear.** Directly behind the eardrum are the **auditory ossicles**, *three small bones linked together that connect the eardrum to the inner ear.* The auditory ossicles are the smallest bones in the body. The middle ear is connected to the throat by the eustachian tube. When you swallow or yawn, this tube allows pressure to be equalized on each side of the eardrum.

- **The Inner Ear.** *The inner ear,* or **labyrinth**, consists of a network of curved and spiral passages that can be divided into three main parts. The *cochlea,* a spiral-shaped canal, is the area of hearing in the inner ear. The vestibule and the semicircular canals are where balance is controlled.

Figure 13.10 The Ear

The ear has two functions: hearing and balance. *Which parts of the ear are involved in hearing?*

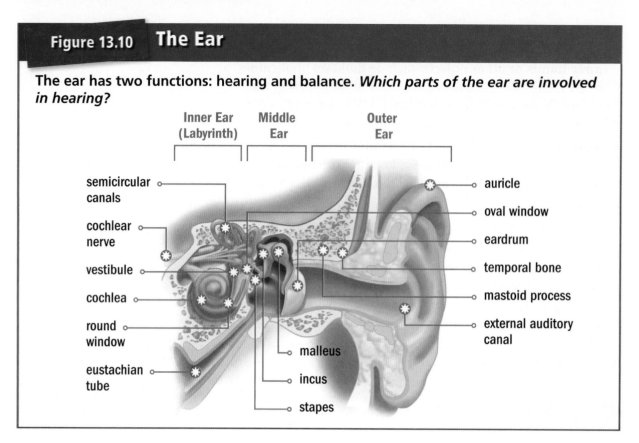

Inner Ear (Labyrinth)
Middle Ear
Outer Ear

semicircular canals
cochlear nerve
vestibule
cochlea
round window
eustachian tube
malleus
incus
stapes
auricle
oval window
eardrum
temporal bone
mastoid process
external auditory canal

Hearing and Balance

Receptors in your inner ear are stimulated by a sound wave. The impulse is then sent to your brain, where it is interpreted as a sound. These sound waves enter the external auditory canal, causing the eardrum to vibrate. The vibrations cause fluid in the cochlea to move, which stimulate receptor cells. These cells send a nerve impulse to the brain, where sound is interpreted. As this is occurring, receptor cells in the vestibule and the semicircular canals send messages to the brain about your sense of balance. Tiny hairs located in the ear sense movement and send nerve impulses to the brain. The brain makes adjustments to maintain balance.

Keeping Your Ears Healthy

Main Idea Caring for your ears helps prevent irritation, injury, infection, and damage to the ears, as well as hearing loss.

To protect your hearing, have your ears examined by a health care professional if you suspect an infection. Middle ear infections can damage the structure of the ear, but can be treated with antibiotics. Other ways to protect your hearing include wearing a hat that covers both the auricles and the earlobes in cold weather. Wear protective gear, such as a batting helmet, when playing sports. Keep foreign objects, including cotton-tipped swabs, out of the ear canal.

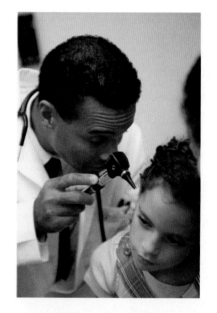

■ **Figure 13.11** A health care professional will check your ears during a routine physical examination. *What are other ways to keep your ears healthy?*

Preventing Hearing Loss

Exposure to loud noises can lead to temporary and sometimes permanent hearing loss, or deafness, over time. Hearing loss can be divided into two categories: conductive and sensorineural.

Conductive Hearing Loss In conductive hearing loss, sound waves are not passed from the outer ear to the inner ear, usually because of a blockage or injury to the inner ear. For example, middle-ear infections may cause fluid to build up within the middle ear.

Sensorineural Hearing Loss This problem may result from a birth defect, exposure to noise, growing older, and medication problems. One type of sensorineural hearing loss is **tinnitus**, *a condition in which a ringing, buzzing, whistling, roaring, hissing, or other sound is heard in the ear in the absence of external sound.* To prevent tinnitus, avoid loud music and wear earplugs in noisy environments and at loud concerts or sporting events. By limiting the length of time you are exposed to loud noise, you reduce the risk of permanent damage.

FITNESS ZONE

I love listening to music when I exercise, but I heard that wearing headphones with the volume turned up can cause permanent hearing loss over time. You might also miss warnings, like a car horn. I turn down the volume. I can still enjoy my music while protecting my hearing and my safety. For more fitness tips, visit the Fitness Handbook and Fitness Zone sites in ConnectEd.

LESSON 3 ASSESSMENT

 After You Read

Reviewing Vocabulary and Facts

1. What happens to eyes that have cataracts?

2. What is *astigmatism*?

3. Explain the function of the wax and tiny hairs in the ear canal.

Thinking Critically

4. **Infer.** What are four characteristics that you think good safety goggles should have?

5. **Identify.** What are two common activities that would require hearing protection?

Applying Health Skills

6. **Accessing Information.** Conduct research to learn about community health services for people with vision problems. Make a pamphlet that can be used as a reference on the availability of these community services.

Writing Critically

7. **Persuasive.** Write a script or skit featuring two teens. One teen is urging the other to avoid exposure to loud noises to reduce the risk of hearing impairment.

Real Life Issues . . . .

After completing the lesson, review and analyze your response to the Real Life Issues question on page 367.

Hands-On HEALTH

Activity Taking Care of the Skin You're In

This activity will use teamwork to help you identify the importance of personal health care and setting personal health goals.

What You'll Need

- one 6' × 3' piece of paper per group
- black, red and blue markers
- textbook
- paper and pencil/pen

What You'll Do

Step 1

Your team will draw a human silhouette using your black marker on your white paper.

Step 2

Your group will have five minutes to draw and label using the red marker all the body parts identified in Chapter 13 (without using your book). Next to that body part, write what a person needs to do to keep that body part healthy. (Example: Draw a picture of an eye, and add a pair of goggles, which would keep the eyes safe while swimming.)

Step 3

Using your textbook, use the blue marker to add any health behaviors you forgot to add in Step 2.

Apply and Conclude

Select one of the blue health behaviors you identified and create a plan for achieving that behavior.

Checklist: Goal Setting

- ☑ Clear goal statement
- ☑ Identifies a realistic goal
- ☑ Plan for reaching the goal
- ☑ Evaluation or reflection on the action
- ☑ Identifies and analyzes external and internal factor

LESSON **1**

Healthy Skin, Hair, and Nails

Key Concepts

▶ Skin has two main layers, the dermis and epidermis, over a layer of fat and connective tissue.

▶ A well-balanced diet rich in vitamin A is essential to your skin's health.

▶ Wear sunscreen with an SPF of 15 or higher to block UVB and UVA rays.

▶ Your hair protects your scalp from sun exposure and provides warmth.

Vocabulary

▶ epidermis (p. 356)
▶ dermis (p. 356)
▶ melanin (p. 356)
▶ sebaceous glands (p. 358)
▶ hair follicles (p. 360)
▶ melanoma (p. 360)

LESSON **2**

Healthy Teeth and Mouth

Key Concepts

▶ A tooth has three main parts: the crown, neck, and root.

▶ The buildup of plaque (bacteria) can lead to cavities.

▶ Periodontal disease begins with the buildup of plaque and tartar.

▶ Good dental care includes proper diet, brushing and flossing, dental checkups, protecting the mouth during sports, and avoiding use of tobacco products.

Vocabulary

▶ periodontium (p. 363)
▶ pulp (p. 364)
▶ plaque (p. 364)
▶ halitosis (p. 365)
▶ periodontal disease (p. 365)
▶ malocclusion (p. 365)

LESSON **3**

Healthy Eyes and Ears

Key Concepts

▶ Eyes gather and send light signals to the brain.

▶ Following a well-balanced eating plan, using eye goggles, and getting regular eye exams will protect your vision.

▶ Your ears change sound waves into electrical signals sent to the brain.

▶ To maintain your hearing, keep your ears clean and warm, avoid putting objects in them, protect them during sports, and avoid prolonged loud noise.

Vocabulary

▶ sclera (p. 367)
▶ cornea (p. 368)
▶ retina (p. 368)
▶ auditory ossicles (p. 371)
▶ labyrinth (p. 371)
▶ tinnitus (p. 373)

LESSON **1**

Vocabulary Review

Correct the sentences below by replacing the italicized term with the correct vocabulary term.

1. The *dermis* is the outer, thinner layer of the skin that is composed of living and dead cells.

2. Sweat is produced in the *sebaceous glands*.

3. *Melanoma* is a pigment in the skin.

4. An oily secretion called sebum is produced by the *dermis*.

Understanding Key Concepts

After reading the question or statement, select the correct answer.

5. Which substance cools the skin?
   a. Sebum
   b. Melanin
   c. Dandruff
   d. Sweat

6. What is the order of the structure of the skin, going from the inner body to the outer body?
   a. Epidermis, dermis, subcutaneous layer
   b. Subcutaneous layer, dermis, epidermis
   c. Dermis, subcutaneous layer, epidermis
   d. Subcutaneous layer, epidermis, dermis

7. Which of the following is a risk from overexposure to UV radiation?
   a. Hepatitis B
   b. Hepatitis C
   c. Skin cancer
   d. HIV

Thinking Critically

After reading the question or statement, write a short answer using complete sentences.

8. **Describe.** Why should you avoid exposure to the sun between 10:00 A.M. and 4:00 P.M.?

9. **Identify.** List some steps you can take to treat acne.

10. **Explain.** How does a balanced diet help you have healthy hair?

11. **Analyze.** What features make a mole on your skin suspicious? What should you do if you have a suspicious mole?

12. **Compare and Contrast.** What are the similarities and differences between moles and warts?

LESSON **2**

Vocabulary Review

Use the vocabulary terms listed on page 375 to complete the following statements.

13. The _____ is the area immediately around the teeth.

14. The part of the tooth that contains the blood vessels and nerves of a tooth is called the _____.

15. The combination of bacteria and other particles that adhere to the outside of a tooth is called _____.

16. When an inflammation of the periodontal structures occurs, you have (a)-_____.

Understanding Key Concepts

After reading the question or statement, select the correct answer.

17. What component of plaque works on sugars to create acids that cause cavities?
    a. Viruses
    b. Bacteria
    c. Moles
    d. Boils

18. Which minerals are most important for healthy teeth?
    a. Phosphorus and calcium
    b. Iron and magnesium
    c. Iron and phosphorus
    d. Magnesium and calcium

19. What is malocclusion?
    a. Bad breath
    b. A crust that forms from unremoved plaque
    c. A source of acids on teeth
    d. A misalignment of teeth

20. What part of the tooth is the connective layer between the enamel and the pulp?
    a. Dentin
    b. Root canal
    c. Periodontal ligaments
    d. Gum

Thinking Critically

After reading the question or statement, write a short answer using complete sentences.

21. **Describe.** What are two functions of the teeth?

22. **Identify.** Why is flossing as important as brushing?

23. **Infer.** Why should you brush your tongue when you brush your teeth?

24. **Predict.** If gingivitis is left untreated, it may destroy the bone tissue that supports the tooth. What may happen to the tooth that is supported by the affected bone?

25. **Analyze.** Why might malocclusion lead to tooth decay?

LESSON 3

Vocabulary Review

Choose the correct term in the sentences below.

26. The *cornea/sclera* is the tough white part of the eye.

27. The *retina/cornea* is the inner layer of the eye wall that contains millions of light-sensitive cells.

28. The *auditory ossicle/labyrinth* is made of three bones that connect the eardrum to the inner ear.

29. *Tinnitus/Auditory ossicle* is a ringing or other sound heard in the ear in the absence of external sound.

Understanding Key Concepts

After reading the question or statement, select the correct answer.

30. When you blink, tears move across the surface of the eyeball to
    a. moisten and clean the eye.
    b. reduce glare from sunlight.
    c. give the eye time to adjust to images.
    d. prevent eyestrain.

31. In which disease does abnormal pressure build up inside the eye?
    a. Cataracts
    b. Macular degeneration
    c. Strabismus
    d. Glaucoma

Assessment

32. What refines the focus of an image onto the retina?

   a. The lens

   b. The sclera

   c. The cornea

   d. The choroid

33. Which parts of the ear are involved in balance?

   a. Semicircular canals and vestibule

   b. Round window and eustachian tube

   c. Oval window and incus

   d. Temporal bone and cochlea

34. What object is safe to put in your ear?

   a. A cotton-tipped swab

   b. A pair of tweezers

   c. A small brush

   d. None of the above

Thinking Critically

After reading the question or statement, write a short answer using complete sentences.

35. Identify. What can you do now to protect your hearing in the future?

36. Describe. What is the effect of strabismus on the eyes?

37. Infer. Which important structure of the eye has no solid mass? Explain your answer.

38. Analyze. Which kind of hearing loss can result from a buildup of earwax? Explain your answer.

Mike Kemp/Rubberball/Getty Images

Technology PROJECT-BASED ASSESSMENT

Tattoos and Piercings—Risky Business?

Background

Tattoos and body piercing have become popular. Both tattoos and piercings, however, can cause serious health problems if unsanitary needles or methods are used.

Task

Create an animated video that can be shown on the Internet that illustrates the health risks of getting tattoos or body piercings.

Audience

Students in your class

Purpose

Point out specific health risks associated with either tattoos or body piercings. In your video, show how these health risks can affect one or more areas of the health triangle: physical, mental/emotional, or social health.

Procedure

1 Decide whether your video will be about body piercing or tattoos. Using the information in Chapter 13, review the health risks of the procedure you selected.

2 Use governmental Web sites to find more information on the health risks of body piercing or tattoos.

3 Conduct an online search of animated videos to see what themes are constant in every video that should be used in your group's video.

4 Collaborate as a group to create an animated video and decide on the story line and flow of the video.

5 Divide the tasks of creating the video among the members of the group. Every member should have a task to complete.

6 Present your video to your class. Ask for permission to upload your video to the school's Web site.

Math Practice

Interpret Graphs. Use the graph below of a UV Index for four weeks in Orlando and Jacksonville, Florida to answer questions 1–3.

Ultraviolet Index

The United States uses a UV Index (UVI) that goes from 1 to 11. The higher the number on the UVI, the more intense the exposure to UV rays will be on that day at noon. The bar graph below shows the average UVI for four different weeks for Orlando and Jacksonville.

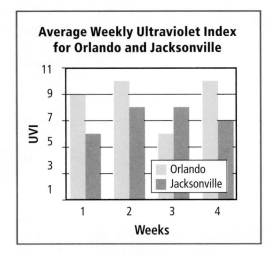

1. Which of the following statements is a correct interpretation of the bar graph above?
   A. The highest UVI for Jacksonville was 10.
   B. The lowest UVI for Orlando was 5.
   C. The mean UVI for Jacksonville was 7.25.
   D. The median UVI for Orlando was 5.50.

2. Which of these measures of central tendency would make Orlando seem to have as high a UVI as possible for the time span shown?
   A. Mean   C. Range
   B. Median   D. Mode

3. What is the mean UVI for Orlando during these four weeks?
   A. 8.50   C. 7.25
   B. 7.0   D. 7.50

Reading/Writing Practice

Understand and Apply. Read the passage below and then answer the questions.

(1) People start exercise programs for many reasons. (2) Some hope to lose weight, while others want to look better. (3) Many people are simply following doctor's orders. (4) But people often find that exercising helps them feel good as well as look good. (5) Here are two simple steps to help you get fit.

(6) Schedule your workout in the same way you schedule any important activity or part of your daily life. (7) Morning works best for most people. (8) Some people however, prefer to exercise later in the afternoon. (9) The important thing is to find the best time and stick to it.

1. Read sentence 6 in the essay. Which revision best supports the organization of the piece?
   A. There are three important things to remember in scheduling a workout.
   B. First, schedule your workout in the same way you schedule any important activity or part of your daily life.
   C. Second, schedule your workout in the same way you schedule any important activity or part of your daily life.
   D. First, it's hard to know where to begin to discuss scheduling workouts.

2. What change in *punctuation* should be made to sentence 8?
   A. Insert a colon after *people.*
   B. Insert a comma after *people.*
   C. Delete the comma after *however.*
   D. Delete the period after *afternoon.*

3. Write a newspaper column that asks and answers three questions from readers of different ages about exercise programs.

National Education Standards

Math: Measurement and Data, Represent and Interpret Data
Language Arts: LACC.910.L.3.6

CHAPTER 14

Skeletal, Muscular, and Nervous Systems

Ed-Imaging

Lesson 1

The Skeletal System

BIG Idea *The skeletal system provides a living structure for the body.*

Lesson 2

The Muscular System

BIG Idea *The muscular system enables the limbs and other parts of the body to move.*

Lesson 3

The Nervous System

BIG Idea *The nervous system sends messages through the nerves to coordinate all the body's activities.*

Ed-Imaging

Activating Prior Knowledge

Using Visuals As you look at this photo, consider how this teen is improving the health of his skeletal, muscular, and nervous systems. In a short paragraph, describe how physical activity and stretching can benefit your health.

Chapter Launchers

Health in Action

Discuss the BIG Ideas

Think about how you would answer these questions:

▶ What exercises strengthen muscles and bones?

▶ How does nutrition affect your muscles and bones?

▶ How can you prevent injuries to the muscular, skeletal, and nervous systems?

Assess Your Health

Read each statement. On a separate sheet of paper, write "yes," "sometimes," or "no" based on your typical behavior.

1. I protect my skeletal, muscular, and nervous systems by using the proper safety gear during physical activity.

2. I wear a safety belt when riding in a motor vehicle.

3. I wear a helmet while riding a bicycle, motorcycle, or when playing a contact sport.

4. I stretch my muscles before engaging in physical activities or sports.

5. I avoid using drugs and alcohol.

A "yes" response shows that you practice healthy behaviors. "Sometimes" indicates that you should analyze and possibly modify your behavior. A "no" response means that you should modify the behavior.

The Skeletal System

BIG Idea *The skeletal system provides a living structure for the body.*

Before You Read

Create a Table. Make a two-column table. Title the first column "Function" and the second column "Example." Write the functions of the skeletal system as listed in the chapter text. As you read, write in the names of specific bones that perform each function.

Function	Example

New Vocabulary

▶ cartilage
▶ ossification
▶ ligament
▶ tendon
▶ scoliosis
▶ osteoporosis

Real Life Issues ································

Speaking on Safety. At his school's health fair, David will give a presentation on the importance of safety gear—including wrist, elbow, and knee pads—when in-line skating or skateboarding. These activities are popular in his neighborhood, but many teens don't wear as much protective equipment as they should.

Writing *Write a paragraph listing the protective equipment needed for in-line skating and skateboarding, and the benefits of using this equipment.*

How the Skeletal System Works

Main Idea The skeletal system consists of bones and connective tissue.

Your skeletal system consists of 206 bones and the attached connective tissues. The bones of the skeleton range in size from the tiniest bone of the inner ear (about 0.25 cm long) to the longest bone of your thigh. The connective tissues cushion the bones, attach bone to bone, and attach bones to muscles.

Your skeletal system has many functions, including

• providing support for the body.

• protecting internal tissues and organs from damage.

• acting as a framework for attached muscles.

• allowing movement of limbs and digits.

• producing new red and white blood cells.

• storing fat and minerals, such as calcium and phosphorus.

Bones

Bones are made up of living tissues formed into different layers. The outer layer is hard, densely packed, compact bone. Beneath that is spongy bone, a less dense bone with a network of cavities filled with red bone marrow, where blood cells are produced. Some bones also contain yellow bone marrow, a type of connective tissue that stores fat. **Figure 14.1** shows the basic structure of a bone.

Bones are categorized by their shape, as shown in **Figure 14.2**. Shapes include long bones, short bones, flat bones, and irregular bones.

Connective Tissue

There are three types of connective tissue: cartilage, ligaments, and tendons. **Cartilage** is *a strong, flexible connective tissue* that can act as a cushion between two bones to reduce friction. It can also act as a flexible structure for soft parts of the body, such as the tip of the nose or the outer ear. All bones begin in the embryo as cartilage. Early in development, the cartilage hardens. This **ossification** (ah-sih-fih-KAY-shun) is *the process by which bone is formed, renewed, and repaired.*

Connective tissue can also hold parts of the body together. A **ligament** is *a band of fibrous, slightly elastic connective tissue that attaches one bone to another.* Ligaments attach to bones to create joints. For example, a ligament attaches the two bones of the forearm to each other, forming the pivot joint. A **tendon** is *a fibrous cord that attaches muscle to the bone.* Muscles contract to move parts of the body.

Joints

Joints, such as those shown in **Figure 14.3** on page 384, are points at which bones meet. Some joints, such as the ones between the bones of the skull, do not move. Flexible joints include ball-and-socket joints, hinge joints, pivot joints, and ellipsoidal joints.

Caring for the Skeletal System

Main Idea A healthy diet, exercise, protective gear, and regular checkups are ways to care for your skeletal system.

As you can see in **Figure 14.4** on page 385, your skeletal system supports your entire body. Your overall health depends on the health of your skeletal system. Eat a healthy diet, get regular physical activity, and have regular checkups to keep your skeletal system healthy. Foods high in calcium, vitamin D, and phosphorus help prevent skeletal disorders.

Figure 14.1

Bone Structure

Bone tissue is surrounded by calcium phosphate and other minerals. *Why is calcium important for bone health?*

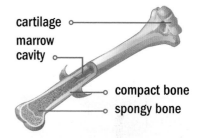

cartilage
marrow cavity
compact bone
spongy bone

Figure 14.2

Bone Shapes

A bone's shape is related to its function. *What are the functions of these bones?*

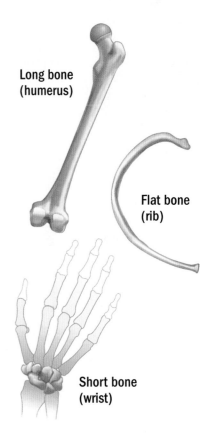

Long bone (humerus)

Flat bone (rib)

Short bone (wrist)

READING CHECK

Extend What kind of sports might require protective gear to protect your skeletal system?

During regular checkups, your doctor can screen you for skeletal disorders such as **scoliosis**, *a lateral or side-to-side curvature of the spine.* Weight-bearing exercise such as walking or weight training helps bones stay strong. Wearing protective gear during sports reduces the risk of bone fractures.

Understanding Skeletal Problems

Main Idea Injuries and disorders harm the skeletal system.

Poor nutrition, infections, sports injuries, and poor posture can lead to problems of the skeletal system. Degenerative disorders such as osteoporosis can also cause problems.

Fractures A fracture is any type of break in a bone. In some fractures, called compound fractures, the broken end of the bone breaks through the skin. In a simple fracture, the broken bone does not break through the skin. Fractures are also classified by the pattern of the break:

- **Hairline fractures:** if parts of the bone do not separate.
- **Transverse fractures:** when the fracture is completely across the bone.
- **Comminuted fractures:** pictured in **Figure 14.5** on page 386, when the bone shatters into more than two pieces.

Injuries to Joints Injuries to joints can occur from overuse, strain, or disease. The following are typical joint injuries:

- **Dislocation** results when a bone slips out of place, tearing the ligaments that attach the bone at the joint. A doctor may reset a joint and immobilize it until ligaments heal.
- **Torn cartilage** can result from a sharp blow to a joint or a severe twisting of a joint. Arthroscopic surgery can remove pieces of the damaged cartilage.
- **Bursitis** results from the painful inflammation of bursa, a fluid-filled sac that helps reduce friction in joints.
- **Bunions** are painful swellings of the bursae in the first joints of big toes. Wearing ill-fitting shoes can make bunions worse. Large bunions may require surgery.
- **Arthritis** is the inflammation of a joint, resulting from an injury, natural wear and tear, or autoimmune disease.

Repetitive Motion Injury Prolonged, repeated movements such as sewing or computer work can damage tissues. *Carpal tunnel syndrome* occurs when ligaments and tendons in the wrist swell, causing numbness, a tingling sensation in the thumb and forefinger, pain, and weakness in the hand.

Figure 14.3

Joints

The structure of a joint relates to the type of motion it can produce. *What kind of motion do these joints produce?*

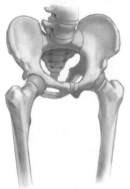

Hip (ball-and-socket joint)

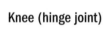

Knee (hinge joint)

Figure 14.4 | **The Skeletal System**

Your bones continue to grow, both in length and in thickness, until approximately age 25. At this age, bones usually stop growing, but may continue to thicken. *What is the process by which bones are formed?*

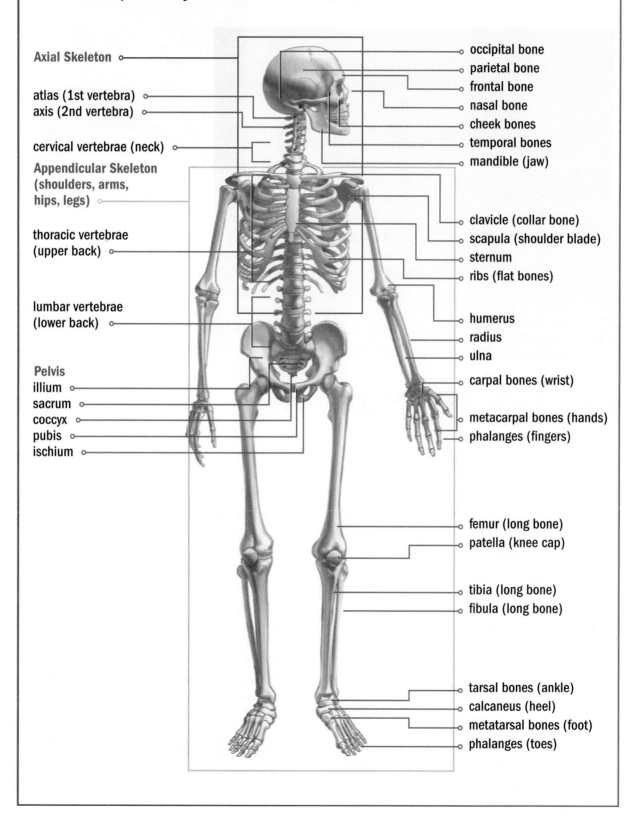

Axial Skeleton

atlas (1st vertebra)
axis (2nd vertebra)

cervical vertebrae (neck)

Appendicular Skeleton (shoulders, arms, hips, legs)

thoracic vertebrae (upper back)

lumbar vertebrae (lower back)

Pelvis
illium
sacrum
coccyx
pubis
ischium

occipital bone
parietal bone
frontal bone
nasal bone
cheek bones
temporal bones
mandible (jaw)

clavicle (collar bone)
scapula (shoulder blade)
sternum
ribs (flat bones)

humerus
radius
ulna

carpal bones (wrist)

metacarpal bones (hands)
phalanges (fingers)

femur (long bone)
patella (knee cap)

tibia (long bone)
fibula (long bone)

tarsal bones (ankle)
calcaneus (heel)
metatarsal bones (foot)
phalanges (toes)

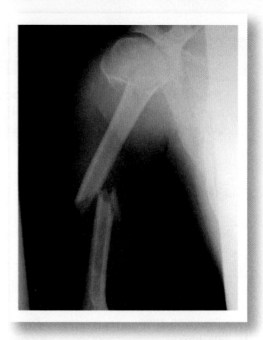

■ **Figure 14.5** Good nutrition and taking safety precautions can help avoid breaks like the fracture shown here. *How might poor nutrition lead to bone fracture?*

Osteoporosis is *a condition in which there is a progressive loss of bone tissue.* Bones weaken and become brittle. This disease affects millions of older Americans. Bone tissue loss is a natural part of aging, but healthful behaviors during your teen years can reduce your risk of developing osteoporosis later in life. A bone scan (in which X-rays measure bone density) can detect signs of osteoporosis. Eating foods containing calcium, vitamin D, and phosphorus will help bones remain strong and healthy. Regular weight-bearing physical activity, such as walking and weight training, stimulates bone cells to increase bone mass.

LESSON **1** ASSESSMENT

 After You Read

Reviewing Facts and Vocabulary

1. How does the skeletal system affect other body systems?

2. How do bones form?

3. How can you help avoid injury to your bones and joints?

Thinking Critically

4. **Analyze.** Some people are allergic to lactose, which is found in milk and many other dairy products. How can lactose-sensitive people get enough calcium for maintaining a healthy skeletal system?

5. **Evaluate.** How might behaviors that you practice as a teen affect your skeletal system later in life?

Applying Health Skills

6. **Accessing Information.** Conduct research to learn more about injury prevention related to activities that teens enjoy. Prepare a pamphlet or poster showing teens how they can avoid injuries.

Writing Critically

7. **Descriptive.** Write a paragraph describing three or more ways in which you have made healthy choices to protect your skeletal system this week.

Real Life Issues ....

After completing the lesson, review and analyze your response to the Real Life Issues question on page 382.

The Muscular System

Real Life Issues ··························

Friends Get Fit. Both Misaki and Cara are interested in getting fit. Misaki wants to join the track team eventually. Cara just wants to work on her muscle tone. They decide to go running before school three days a week. Each week, they plan to increase the distance they run to improve their fitness level.

Writing *Write a paragraph describing some of the benefits of running. Be sure to include benefits for other aspects of health besides muscular health.*

What Muscles Do

Main Idea The muscular system allows for voluntary and involuntary movements.

Like rubber bands, muscles are elastic; they stretch to allow a wide range of motion. This elasticity allows muscles to move the bones or organs to which they are attached. In this way, your muscular system allows you to move.

You might think that muscles work only when you do things such as pick up an object, catch a ball, or walk across a room. Some muscles in your body, however, are always at work. Even when you are sleeping, muscles help you breathe, make your heart beat, and move food through your digestive system. These involuntary processes occur without your knowing it. At other times, such as when you play the piano, make a dash toward first base, or shoot a basketball, you are using muscles that are under conscious, or voluntary, control. You are aware that you are controlling them. Without the use of both voluntary and involuntary muscles, you would not be able to perform these functions.

GUIDE TO READING

BIG Idea *The muscular system enables the limbs and other parts of the body to move.*

Before You Read

Create a K-W-L Chart. Make a three-column chart. In the first column, list what you **k**now about the muscular system. In the second column, list what you **w**ant to know about this topic. As you read, use the third column to summarize what you **l**earned about the topic.

K	W	L

New Vocabulary

▶ smooth muscles
▶ skeletal muscles
▶ flexor
▶ extensor
▶ cardiac muscle
▶ tendinitis
▶ hernia

How Muscles Work

Main Idea Muscles consist of long, fibrous cells that can shorten and stretch to make muscles move.

A muscle is made up of hundreds of long cells called muscle fibers. Major muscles in the body are made up of hundreds of bundles of these fibers. When these bundles are stimulated by nerve impulses, or signals, they contract, or shorten. When they relax, the bundles extend, or stretch. Some nerves stimulate many muscle fibers, especially large muscles such as your calf muscle or your biceps. In other areas, such as your eyes, a single nerve may provide impulses to only two or three muscle fibers.

Types of Muscles

The body contains three types of muscle tissue: smooth muscle, skeletal muscle, and cardiac muscle.

- **Smooth muscles** are *muscles that act on the lining of the body's passageways and hollow internal organs.* These muscles can be found in the digestive tract, the urinary bladder, the lining of the blood vessels, and the passageways that lead into the lungs. Smooth muscles are involuntary muscles.

- **Skeletal muscles** are *muscles attached to bone that cause body movements.* Skeletal muscle tissue has a *striated,* or striped, appearance under a microscope. Most of your muscle tissue is skeletal, and almost all skeletal muscles are under voluntary control. Skeletal muscles often work together and perform opposite actions to produce a movement. One muscle contracts while the other muscle relaxes. An example of this can be seen in **Figure 14.6**, which shows the biceps and triceps muscles of the upper arm. To bend and straighten your arm at the elbow, these muscles have opposite jobs. The **flexor** is *the muscle that closes a joint.* In this example, the biceps is the flexor. The **extensor** is *the muscle that opens a joint.* In this case, the triceps is the extensor. When the biceps contracts, the triceps extends and the joint closes. When the biceps extends, the triceps contracts and the joint opens. Identify other opposing skeletal muscles that appear in The Skeletal Muscles illustration, **Figure 14.8** on page 390.

- **Cardiac muscle** is *a type of striated muscle that forms the wall of the heart.* Cardiac muscle is involuntary and is responsible for the contraction of your heart. The heart contracts rhythmically about 100,000 times each day to pump blood throughout your body.

Figure 14.6

Muscle Movement

Skeletal muscles work in pairs to produce movement. *Which muscle in your arm is the flexor?*

Muscle Movement

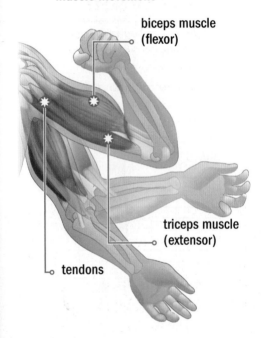

biceps muscle (flexor)

triceps muscle (extensor)

tendons

READING CHECK

Explain What are the two types of striated muscle, and how do they differ?

Caring for Your Muscles

Main Idea Eating a healthy diet and getting regular exercise will help you care for your muscular system.

Physical activity will keep your muscles strong and healthy. Muscles that remain unused for long periods of time will atrophy, or decrease in size and strength. Muscle tone is the natural tension in the fibers of a muscle. The following tips can help you care for your muscular system and maintain muscle tone:

- Get regular exercise.
- Eat high protein foods to build muscle.
- Practice good posture to strengthen back muscles.
- Use proper equipment and wear appropriate clothing to protect muscles during any physical activity.
- Warm up properly and stretch before exercising, and cool down after exercising to prevent injury.

READING CHECK

Identify What types of foods are good for the muscular system, and why?

Understanding Muscular Problems

Main Idea Caring for the muscular system can help prevent health problems and injuries.

Your muscles might be sore after strenuous activity, such as an all-day hike or bike ride. Although it can be painful, muscle soreness is usually temporary. However, other problems of the muscular system can be more serious. The recovery time varies according to the type and severity of the injury or disease.

- **Bruises** are areas of discolored skin that appear after an injury, usually a blow to the body. The injury causes the blood vessels beneath the skin to rupture and leak, resulting in a bruise. Large bruises can be treated with an ice pack to reduce initial swelling.

- **Muscle strains or sprains** result when muscles are stretched or partially torn from overexertion. Apply ice to strains to reduce swelling, and rest the affected area.

- **Tendinitis**, or *the inflammation of a tendon,* can be the consequence of injury, overuse, or natural aging. Treatment includes ultrasound or anti-inflammatory medication to reduce pain and swelling.

■ **Figure 14.7** Prepare your muscles by stretching before beginning a workout. *What injuries can be prevented by warming up before working out?*

- A **hernia** occurs when *an organ or tissue protrudes through an area of weak muscle.* Hernias commonly occur in the abdomen from straining to lift a heavy object. Surgery is usually recommended, and may be required, to repair hernias.

- **Muscular dystrophy** is an inherited disorder in which skeletal muscle fibers are progressively destroyed. There is no cure, but with early detection, muscle weakness can be delayed through exercise programs.

Figure 14.8 The Skeletal Muscles

The major muscle groups include the arms, legs, back, abdomen, shoulders, and chest. What functions do the facial muscles have?

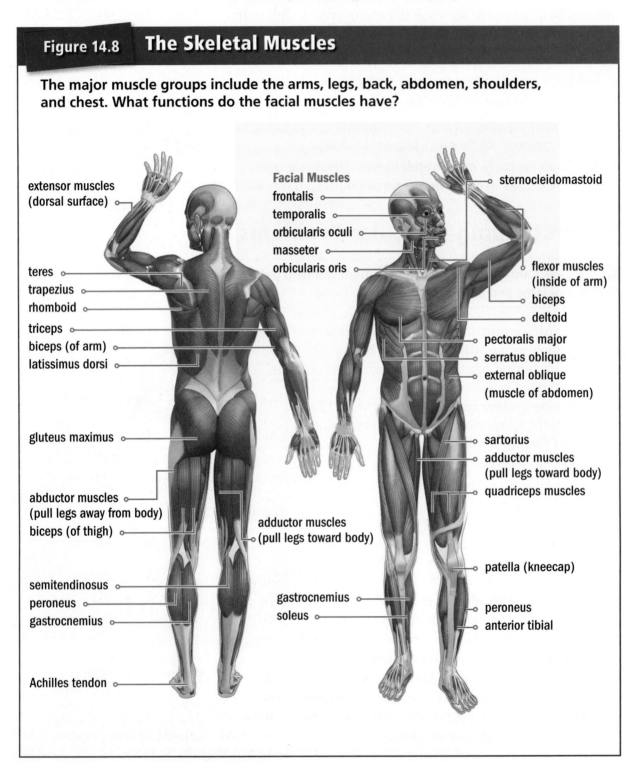

extensor muscles (dorsal surface)

teres
trapezius
rhomboid
triceps
biceps (of arm)
latissimus dorsi

gluteus maximus

abductor muscles (pull legs away from body)
biceps (of thigh)

semitendinosus
peroneus
gastrocnemius

Achilles tendon

Facial Muscles
frontalis
temporalis
orbicularis oculi
masseter
orbicularis oris

adductor muscles (pull legs toward body)

gastrocnemius
soleus

sternocleidomastoid

flexor muscles (inside of arm)
biceps
deltoid
pectoralis major
serratus oblique
external oblique (muscle of abdomen)

sartorius
adductor muscles (pull legs toward body)
quadriceps muscles

patella (kneecap)

peroneus
anterior tibial

Health Skills Activity

Practicing Healthful Behaviors

A Safety Check

One day after gym class, Quentin brought up the idea of weight training to his friend Tony.

"I read an article that said strength training builds muscle and helps burn fat," Quentin said.

"Well, I'm all for that," Tony replied. "I think we need to talk to a pro, though. You can get hurt if you lift weights the wrong way."

Quentin's father belongs to a fitness center close to the school. He arranges an appointment for the boys to talk to a professional strength coach. The strength coach shows them both how to use the equipment safely and effectively. As each piece of equipment is explained, Quentin and Tony decide to make a checklist to help them keep track of all the safety tips for each exercise.

Writing Write a checklist of safety tips for weight training:

1. Use a separate page for each exercise. Begin with general safety tips.

2. Include one weight-bearing exercise for every major muscle group.

3. Describe correct techniques, when and where to work out, and how to use equipment properly.

4. Make copies, stapling the pages together, to make a guide for the school gymnasium.

LESSON 2 ASSESSMENT

 After You Read

Reviewing Facts and Vocabulary

1. What are the functions of the muscular system?

2. Where is smooth muscle found?

3. What is a *hernia*? How can you get one?

Thinking Critically

4. **Apply.** What muscles are important for playing baseball? How can you protect these muscles from injury?

5. **Analyze.** How can you prevent muscle strains when you are participating in a new physical activity?

Applying Health Skills

6. **Decision Making.** Danielle strained her arm muscle. Her coach says it should heal by the gymnastics meet next week. Danielle wants to go kayaking this weekend. Describe Danielle's decision-making steps.

Writing Critically

7. **Narrative.** Write a story from the point of view of a muscle. Have the muscle describe itself, what it does in the body, how it works and the care it requires.

Real Life Issues

After completing the lesson, review and analyze your response to the Real Life Issues question on page 387.

The Nervous System

Real Life Issues

Feeling the Heat. It's Petra's turn to clean the stovetop. She remembers that her sister just heated up some milk to make hot chocolate. The electric burner does not appear to be hot. Still, Petra knows that a burner can still be warm enough to burn. She carefully holds her hand 2 inches over the burner to feel for heat. It feels cool enough, so she safely wipes down the stove.

Writing *Write a brief journal entry describing the ways you use your sense of touch. How does your sense of touch help you avoid danger?*

How the Nervous System Works

Main Idea The nervous system coordinates all of the activities in the body.

Your nervous system is a complex network that allows communication between the brain and parts of the body. It stores information and coordinates all activities, from breathing or digesting food to sensing pain and feeling fear. The brain, spinal cord, and nerves work together, transmitting messages between organs, tissues, and cells.

The nervous system has two main divisions. The *central nervous system* (CNS) consists of the brain and spinal cord. The *peripheral nervous system* (PNS) gathers information from inside and outside your body. It includes nerves that extend from the brain, spinal cord, and sensory receptors, such as those in the skin that sense pressure, temperature, or pain. The CNS receives messages from the nerves in the PNS, interprets them, and sends out a response.

Understanding Neurons

Main Idea Neurons transmit messages from the brain and spinal cord to the rest of the body.

Neurons, or *nerve cells,* transmit messages to and from the spinal cord and brain. The three types of neurons—sensory neurons, motor neurons, and interneurons—are classified by function. Sensory neurons carry messages from receptors in the body to the CNS. Motor neurons carry messages from the CNS back to muscles or glands in response to an impulse. Interneurons communicate with and connect other neurons. **Figure 14.9** illustrates the nerve impulse. A neuron consists of three main parts:

- The **cell body** of a neuron contains the nucleus, which regulates the production of proteins within the cell. Unlike other cells in the body, neurons have limited ability to repair damage or replace destroyed cells.

- **Dendrites** are branched structures that extend from the cell body in most neurons. Dendrites receive information and transmit impulses toward the cell body.

- **Axons** transmit impulses away from the cell body and toward another neuron, muscle cell, or gland.

 **READING CHECK**

Explain How is the cell body important to a nerve cell?

Figure 14.9 **The Nerve Impulse**

A nerve impulse begins when a sensory receptor is stimulated. The impulse travels to the CNS and is interpreted with the help of an interneuron. Then an impulse is sent to a muscle cell or gland in response to the stimulus.

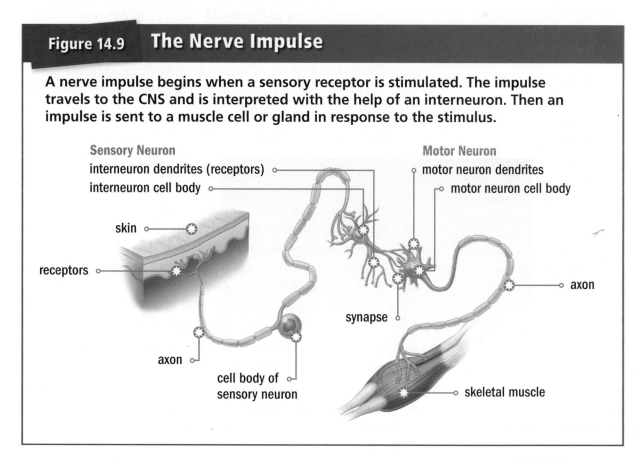

The Central Nervous System

Main Idea The central nervous system is made up of the brain and spinal cord.

The two organs that make up the CNS—the brain and spinal cord—send and receive impulses to and from nerves in the body. **Figure 14.10** shows how nerves extend to various parts of the body.

The spinal cord is a long column of nerve tissue about the thickness of your index finger. The tissue of the spinal cord is surrounded by several layers of connective tissue called the spinal meninges. The meninges, along with the vertebrae—the bones of the spine—help protect the spinal cord. The spinal cord is also bathed in cerebrospinal fluid that absorbs shock and nourishes the nerve tissue.

An adult human brain weighs up to 3 pounds and rests in the protective cavity formed by the bones of the skull. Like the spinal cord, the brain is protected from injury by layers of cranial meninges and cerebrospinal fluid. The brain depends on oxygen to survive. It can last for only four to five minutes without oxygen before suffering irreversible damage.

Sections of the Brain

The brain, shown in **Figure 14.11** on page 396, coordinates and controls the activities of the nervous system. Your brain helps you to receive and process messages; to think, remember, reason, and feel emotions; and to coordinate muscle movements. The brain has three main divisions: the cerebrum, the cerebellum, and the brain stem.

The Cerebrum The **cerebrum** (seh-REE-brum) is *the largest and most complex part of the brain.* Billions of neurons in the cerebrum are the center of conscious thought, learning, and memory. The cerebrum's right and left sides, or hemispheres, communicate with each other to coordinate movement. The right hemisphere controls the left side of the body, and the left hemisphere controls the right side of the body. The left hemisphere is the center of language, reasoning, and critical thinking skills. The right hemisphere is the center for processing music and art and comprehending spatial relationships. Each hemisphere has four lobes:

- **The frontal lobe** controls voluntary movements and has a role in the use of language. The prefrontal areas are thought to be involved with intellect and personality.

- **The parietal lobe** is involved with sensory information, including feelings of heat, cold, pain, touch, and body position in space.

FITNESS ZONE

Have you ever heard the expression "use it or lose it"? That's really true for keeping your brain healthy. Researchers at Harvard say that moving helps keep the brain healthy as well as the body. Now, when I feel like I've been sitting too long while studying, I get up and move around for a couple of minutes. It really helps when getting ready for a test. For more physical activity ideas, visit the Fitness Handbook and Fitness Zone sites in ConnectEd.

- **The occipital lobe** controls the sense of sight.
- **The temporal lobe** contains the sense of hearing and smell, as well as memory, thought, and judgment.

The Cerebellum The **cerebellum** (ser-eh-BEL-um) is *the second largest part of the brain.* It coordinates the movement of skeletal muscles. This area of the brain also continually receives messages from sensory neurons in the inner ear and muscles. It uses this information to maintain the body's posture and balance. Being able to carry out a complex series of muscle movements, such as serving a volleyball or playing the violin, is made possible by the cerebellum.

Figure 14.10 The Nervous System

Nerves extend to various parts of the body along the length of the spinal cord. *Which nerves are the longest?*

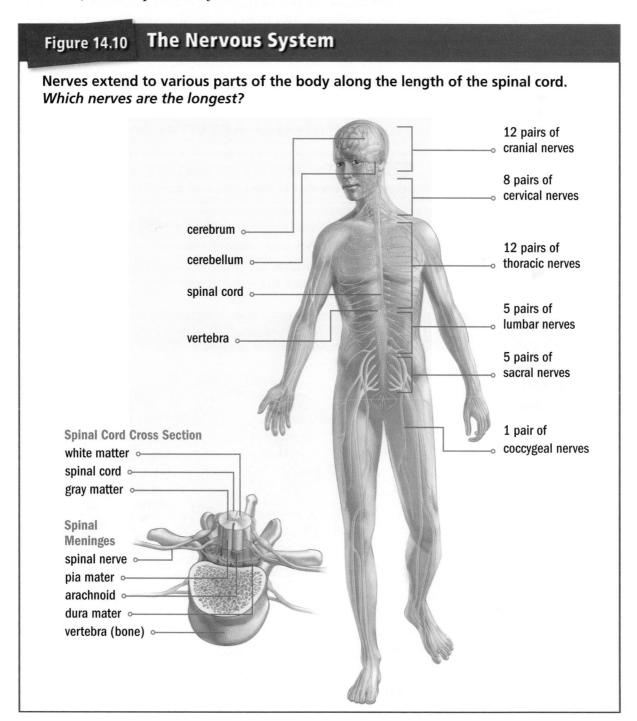

cerebrum
cerebellum
spinal cord
vertebra

12 pairs of cranial nerves

8 pairs of cervical nerves

12 pairs of thoracic nerves

5 pairs of lumbar nerves

5 pairs of sacral nerves

1 pair of coccygeal nerves

Spinal Cord Cross Section
white matter
spinal cord
gray matter

Spinal Meninges
spinal nerve
pia mater
arachnoid
dura mater
vertebra (bone)

The Brain Stem The **brain stem** is *a 3-inch-long stalk of nerve cells and fibers that connects the spinal cord to the rest of the brain.* Incoming sensory impulses and outgoing motor impulses pass through the brain stem. It has five parts:

- **The medulla oblongata** regulates heartbeat, respiratory rate, and reflexes such as coughing and sneezing.
- **The pons** helps regulate breathing and controls the muscles of the eyes and face.
- **The midbrain** controls eyeball movement, pupil size, and the reflexive response of turning your head.
- **The thalamus** relays incoming sensory impulses from the eyes, the ears, and from pressure receptors in the skin.
- **The hypothalamus** regulates body temperature, appetite, sleep, and controls secretions from the *pituitary gland,* affecting metabolism, sexual development, and emotions.

✓ READING CHECK

Describe What are the three sections of the brain, and what is the function of each?

The Peripheral Nervous System

Main Idea The peripheral nervous system is made up of the nerves that are not in the brain and spinal cord.

The peripheral nervous system (PNS) carries messages between the CNS and part of the body, signaling internal and external changes. The PNS is made up of the autonomic nervous system and the somatic nervous system.

Figure 14.11 The Brain

The brain coordinates all activities of the body. *What part of the brain coordinates muscle movements?*

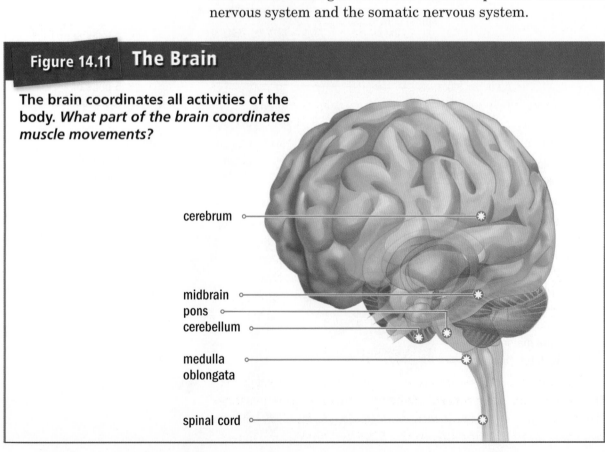

cerebrum

midbrain
pons
cerebellum

medulla
oblongata

spinal cord

Figure 14.12 **How Your Reflexes Work**

Reflexes can prevent injuries such as a burn from a hot stove.

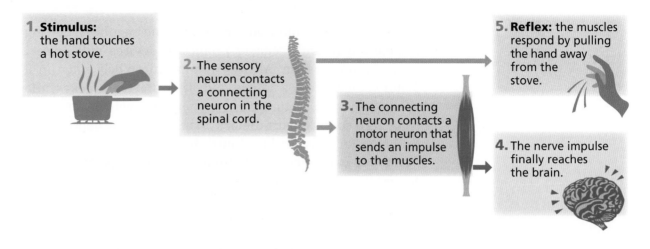

1. **Stimulus:** the hand touches a hot stove.

2. The sensory neuron contacts a connecting neuron in the spinal cord.

3. The connecting neuron contacts a motor neuron that sends an impulse to the muscles.

4. The nerve impulse finally reaches the brain.

5. **Reflex:** the muscles respond by pulling the hand away from the stove.

The Autonomic Nervous System

The *autonomic nervous system* contcs such involuntary functions as digestion and heart rate. It consists of a network of nerves divided into two smaller networks, the sympathetic nervous system and the parasympathetic nervous system.

- **The sympathetic nervous system** kicks in when you are startled, sending messages that cause your heart rate to increase. Blood vessels in your muscles dilate, allowing greater blood flow. This is the "fight-or-flight" response that prepares you to react in a dangerous situation. **Figure 14.12** illustrates this *reflex,* the body's spontaneous response to a stimulus, as when a doctor tests your knee-jerk reflex by tapping the ligament below your knee during a physical exam.

- **The parasympathetic nervous system** opposes the action of the sympathetic nervous system by slowing body functions. During periods of rest, it slows heartbeat, relaxes blood vessels, and lowers blood pressure to conserve energy. The parasympathetic nervous system stimulates production of saliva and stomach secretions to promote the digestion of food.

The Somatic Nervous System

The *somatic nervous system* involves voluntary responses that are under your control. Sensory neurons relay messages from the eyes, ears, nose, tongue, and skin to the CNS. Motor neurons carry impulses from the CNS to skeletal muscles.

 READING CHECK

Describe What happens during the "fight-or-flight" response?

Caring for Your Nervous System

Main Idea Making healthful choices can protect your nervous system from injury.

Eating a well-balanced diet, exercising regularly, getting enough sleep, and wearing protective devices will protect your nervous system. Always wear a safety belt when in a motor vehicle. Wear a helmet and other protective gear while riding a bicycle, motorcycle, or other open vehicle, or when enjoying a contact sport. Before diving, check the depth of the water. Never dive head first into shallow water or into water where you cannot see the bottom. Finally, drugs and alcohol can permanently damage nerve cells, so avoid using them.

Problems of the Nervous System

Injury to the nervous system affects the immediate tissues, and may lead to other problems, including the following:

- **Headaches.** Headaches can be caused by muscle tension, eyestrain, exposure to fumes, a sinus infection, dehydration, or food allergies. Migraines are recurrent headaches that may be accompanied by sensitivity to light.

- **Head injuries.** Each year, 473,000 American children and teens sustain brain injuries. Types of head injuries include concussion, a temporary loss of consciousness, contusion, a bruising of the brain tissues that causes swelling, and coma, caused by major trauma.

- **Spinal injuries.** Spinal cord injuries require medical care. Swelling of the spinal cord or the tissue around it can result in temporary loss of nerve function. Permanent nerve damage will result without treatment. If the spinal cord has been severed, paralysis results.

■ **Figure 14.13** Safe behaviors, such as diving only when you know the depth of the water, can reduce your risk of head and spinal injuries. *What other behaviors can protect you from head or spinal injury?*

- **Meningitis.** Meningitis is an inflammation of the spinal and cranial meninges caused by bacterial or viral infection. Meningitis is very serious and can result in death. Symptoms include fever, headache, light and sound sensitivity, and neck stiffness.

Some nervous system diseases are degenerative, which means they occur over time as cells break down. Multiple sclerosis, Parkinson's disease and Alzheimer's are examples of degenerative diseases. Other disorders result from injury or brain damage. **Epilepsy** is *a disorder of the nervous system that is characterized by recurrent seizures— sudden episodes of uncontrolled electrical activity in the brain.* Causes include brain damage at birth, infections, head injury, or exposure to toxins. Medications can help control seizures. **Cerebral palsy** refers to *a group of neurological disorders that are the result of damage to the brain before, during, or just after birth or in early childhood.* Physical therapy and medication help patients cope.

■ **Figure 14.14** Multiple sclerosis is an autoimmune disease, often resulting in impaired mobility. *What kinds of activities can you enjoy with someone who has limited voluntary muscle control?*

LESSON 3 ASSESSMENT

 After You Read

Reviewing Facts and Vocabulary

1. Where is the nucleus of a neuron located?
2. How can a reflex prevent injury?
3. What are some causes of nervous diseases and disorders?

Thinking Critically

4. **Analyze.** After sustaining a head injury, a patient is having trouble comprehending spatial relationships and controlling the left side of her body. What part of the brain might be damaged?
5. **Compare.** How are the functions of the autonomic nervous system and the somatic nervous system different?

Applying Health Skills

6. **Accessing Information.** Write a one-page report on current research into helping people with degenerative nervous system disorders, such as Parkinson's, multiple sclerosis, or Alzheimer's. Why are these diseases so difficult to treat?

Writing Critically

7. **Expository.** Write a paragraph describing what happens during a reflex action.

Real Life Issues ....

After completing the lesson, review and analyze your response to the Real Life Issues question on page 392.

Hands-On HEALTH

Activity Good "Housekeeping"

Imagine you were building a house that you plan to live in for 100 years. In this activity you will learn how your "house" is built, and how to care for it.

What You'll Need

- a computer with Internet access
- colored markers (black, red, blue, green)
- large sheet of paper

What You'll Do

Step 1

Work in groups to review the chapter. Conduct additional research to learn more about the skeletal, muscular, and nervous systems.

Step 2

Using reliable information that you obtained during your research, draw a life-size silhouette of the human body. Refer to your research, and label one major muscle, one major bone, and one part of the nervous system.

Step 3

Explain the components of each system, how it functions, and steps to keep the system healthy. Cite your sources.

Apply and Conclude

Identify three health-enhancing behaviors that will have a positive impact on your "house" for a lifetime.

Checklist: Accessing Information

- ✔ List all of your sources of information.
- ✔ Evaluate the sources to determine their reliability.
- ✔ Judge the appropriateness of your sources.

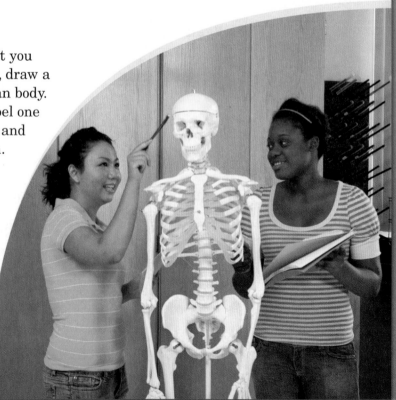

LESSON 1

The Skeletal System

Key Concepts

▶ Your skeletal system consists of bones, cartilage, ligaments, and tendons.

▶ Bones provide support and protection for the body, provide a framework for muscles, allow movement, produce blood cells, and store nutrients.

▶ Injuries to the skeletal system include fractures, dislocation of joints, and tears and inflammation of connective tissue.

Vocabulary

▶ cartilage (p. 383)
▶ ossification (p. 383)
▶ ligament (p. 383)
▶ tendon (p. 383)
▶ scoliosis (p. 384)
▶ osteoporosis (p. 386)

LESSON 2

The Muscular System

Key Concepts

▶ The muscular system includes smooth, skeletal, and cardiac muscle.

▶ Extensor and flexor muscles work together to perform opposite actions.

▶ Proper diet, regular exercise, stretching, and avoiding overexertion will strengthen muscles and prevent injury.

Vocabulary

▶ smooth muscles (p. 388)
▶ skeletal muscles (p. 388)
▶ flexor (p. 388)
▶ extensor (p. 388)
▶ cardiac muscle (p. 388)
▶ tendinitis (p. 389)
▶ hernia (p. 390)

LESSON 3

The Nervous System

Key Concepts

▶ The nervous system enables communication between the brain and all other areas of the body.

▶ Different parts of the brain are responsible for different functions of the body and different kinds of conscious thought.

▶ You can avoid injury to your nervous system by practicing healthful behaviors. Use a safety belt, wear a helmet, avoid drug and alcohol use, and check the depth of water before diving.

Vocabulary

▶ neurons (p. 393)
▶ cerebrum (p. 394)
▶ cerebellum (p. 395)
▶ brain stem (p. 396)
▶ epilepsy (p. 399)
▶ cerebral palsy (p. 399)

LESSON 1

Vocabulary Review

Use the vocabulary terms listed on page 401 to complete the following statements.

1. _____ is the connective tissue that can act as a cushion between two bones.

2. A fibrous cord called a(n) _____ attaches muscle to bone.

3. A lateral curvature of the spine is called _____.

4. The process by which bone is formed is called _____.

Understanding Key Concepts

After reading the question or statement, select the correct answer.

5. Which of the following is *not* a function of the skeletal system?
   a. Storing minerals and fats
   b. Producing red blood cells
   c. Responding to external stimuli
   d. Protecting internal tissues and organs

6. Which of the following is a condition that involves a progressive loss of bone tissue?
   a. Arthritis
   b. Osteoporosis
   c. Repetitive motion injury
   d. Scoliosis

7. In a hairline fracture,
   a. the break is completely across the bone.
   b. the two parts of the bone do not separate.
   c. the bone shatters into more than two pieces.
   d. one part of the bone protrudes through the skin.

Thinking Critically

After reading the question or statement, write a short answer using complete sentences.

8. **Contrast.** Contrast the shape of the skull bones with the shape of the arm bones, and relate these shapes to their different functions.

9. **Evaluate.** How are tendons and ligaments important for movement?

10. **Analyze.** How can nonrigorous activities, such as typing or sewing, lead to skeletal system injuries?

11. **Apply.** What kinds of protective gear can prevent injuries to the skeletal system?

12. **Analyze.** How will behaviors you practice during your teen years affect your chances of getting osteoporosis later in life?

LESSON 2

Vocabulary Review

Choose the correct word in the sentences below.

13. A(n) *flexor/extensor* is a muscle that closes a joint.

14. *Smooth/Cardiac* muscle is striated muscle in the heart.

15. All *smooth/skeletal* muscles are under involuntary control.

16. *Tendinitis/Hernia* results when an organ protrudes through an area of weak muscle.

17. A tendon problem resulting from overuse, injury, or natural aging is *tendinitis/hernia.*

Understanding Key Concepts

After reading the question or statement, select the correct answer.

18. Which type of muscle is *not* striated?
   a. Cardiac
   b. Extensor
   c. Flexor
   d. Smooth

19. Where is smooth muscle found?
   a. In the digestive tract
   b. In the walls of the heart
   c. In the triceps and biceps of the arms
   d. In the muscles controlling finger movement

20. How can you improve muscle tone?
   a. Practicing good posture
   b. Getting regular physical exercise
   c. Wearing safety equipment
   d. Eating fruits and vegetables

21. What problem of the muscular system is an inherited disorder?
   a. Hernia
   b. Muscle sprain
   c. Muscular dystrophy
   d. Tendinitis

Thinking Critically

After reading the question or statement, write a short answer using complete sentences.

22. **Connect.** What are some ways that the muscular system works together with other body systems?

23. **Infer.** Consider an injury to a muscle in the thigh. Why might it cause pain when straightening the leg, but not when bending the leg?

24. **Apply.** Why might it be important for the children of a person with muscular dystrophy to be screened for the disease?

25. **Apply.** How can strengthening muscles prevent injury?

LESSON 3

Vocabulary Review

Use the vocabulary terms listed on page 401 to complete the following statements.

26. The _____ coordinates the movement of skeletal muscles.

27. The _____ is the center of conscious thought.

28. Several important voluntary functions, such as breathing and heartbeat, are controlled by _____.

29. _____ is a disorder that is characterized by recurrent seizures.

30. A group of neurological disorders resulting from damage to the brain at birth is called _____.

Understanding Key Concepts

After reading the question or statement, select the correct answer.

31. Which of the following is *not* a part of the brain stem?
   a. Cerebellum
   b. Medulla oblongata
   c. Midbrain
   d. Thalamus

32. What can result from a spinal cord injury?
   a. Concussion
   b. Contusion
   c. Epilepsy
   d. Paralysis

33. Which disease or disorder is *not* degenerative?
   a. Alzheimer's disease
   b. Cerebral palsy
   c. Multiple sclerosis
   d. Parkinson's disease

34. What disease or disorder is characterized by sudden episodes of electrical activity in the brain?
  a. Alzheimer's disease
  b. Cerebral palsy
  c. Epilepsy
  d. Parkinson's disease

Thinking Critically

After reading the question or statement, write a short answer using complete sentences.

35. Analyze. How do the central nervous system and peripheral nervous system work together during the motor response to a stimulus?

36. Infer. How does the nervous system protect the brain and spinal cord?

37. Compare. Explain the difference between axons and dendrites.

38. Infer. Why might the symptoms of a degenerative disease of the nervous system be difficult to treat?

39. Describe. What might happen in the nervous system and muscles when a person is surprised by a sudden loud sound like a fire alarm going off nearby?

40. Explain. Why is it important to wear protective gear and to check the depth of water before diving?

Technology PROJECT-BASED ASSESSMENT

Sleep and Your Brain

Background

Getting enough sleep means more than resting your body. It's also a time when your brain prepares for the next day. To function at its best, your brain needs to go through five cycles while you sleep. If you don't get enough sleep, both your body and brain won't function at their best the next day.

Task

Create a podcast describing the importance of sleep to brain functioning.

Audience

Students in your class

Purpose

Help explain the importance of sleep in re-energizing the brain.

Procedure

1 Review the information on the nervous system in Chapter 14.

2 Conduct research online to learn about the importance of sleep to brain functions.

3 Create a podcast lasting two or three minutes on why the brain needs sleep. Also, using an online survey tool, develop a survey for classmates to fill out on the effectiveness of your podcast.

4 Play the podcast for your class.

5 Ask students to complete the survey to learn if they have any additional questions about why the brain needs sleep. The survey should also assess whether your podcast was effective.

Math Practice

Interpret Tables. A diet high in calcium and vitamin D can help prevent bone loss or osteoporosis. The table below shows the amount of calcium and vitamin D needed each day for different age groups. Use the table to answer Questions 1–3.

Daily Need for Calcium and Vitamin D		
Age Group	Calcium	Vitamin D
0 to 6 months	210 mg	200 IU
7 to 12 months	270 mg	200 IU
1 to 3 years	500 mg	200 IU
4 to 8 years	800 mg	200 IU
9 to 18 years	1,300 mg	200 IU
19 to 50 years	1,000 mg	200 IU
51 to 70 years	1,200 mg	400 IU
Over 70 years	1,200 mg	600 IU

Adapted from *"What Is Osteoporosis?"* National Institutes of Health Osteoporosis and Related Bone Diseases-National Resource Center, March 2006.

1. What is the difference between the Vitamin D daily need for babies, under 12 months, and the daily amount of vitamin D for people over 70?
   A. 200 IU   **C.** 600 IU
   B. 400 IU   **D.** 800 IU

2. Between which two age groups is there a decrease in the amount of calcium needed?
   A. 0 to 6 months and 7 to 12 months
   B. 4 to 8 years and 9 to 18 years
   C. 9 to 18 years and 19 to 50 years
   D. 19 to 50 years and 51 to 70 years

3. Which group needs the most calcium daily?
   A. 0 to 6 months   **C.** Over 50 years
   B. 9 to 18 years   **D.** Over 70 years

Reading/Writing Practice

Understand and Apply. Read the passage below, and then answer the questions.

> *Each year in the United States, 10,000 new cases of spinal cord injury are reported. These injuries may result from sports or recreational activities, motor vehicle crashes, falls, physical assaults, and gunshot wounds.*
>
> *Spinal injuries may result in paralysis, or the loss of muscle function and feeling in part of the body. An injury to the upper part of the spinal cord may result in* quadriplegia, *or paralysis of both upper and lower limbs.* Paraplegia, *paralysis of both lower limbs, is caused by an injury lower on the spinal column.*
>
> *Researchers are looking for ways to cure paralysis. Electrical sensors and stimulators can help quadriplegic victims flex their limbs. Possible cures include removal of scar tissue and transplantation of cells that promote nerve growth.*

1. What was the author's purpose?
   A. To explain how to cure paralysis
   B. To persuade people to wear helmets
   C. To describe the effects of spinal injuries
   D. To describe different types of paralysis

2. Which sentence best represents the main idea of the second paragraph?
   A. Paralysis can be cured.
   B. There are different degrees of paralysis.
   C. Paralysis can result from accidents.
   D. Paralysis cannot be cured.

3. Write a paragraph persuading a friend to wear a safety belt while riding in a motor vehicle. Provide details about spinal cord injuries to support your main points.

National Education Standards
Math: Measurement and Data
Language Arts: LACC.910.L.3.6, LACC.910.RL.2.4, LACC.910.W.3.8

CHAPTER 15

Cardiovascular, Respiratory, and Digestive Systems

Ed-Imaging

Lesson 1

The Cardiovascular and Lymphatic Systems

BIG Idea *The cardiovascular system moves blood through the body, while the lymphatic system circulates lymph throughout the body.*

Lesson 2

The Respiratory System

BIG Idea *The respiratory system provides oxygen to the blood and removes carbon dioxide from the body.*

Lesson 3

The Digestive System

BIG Idea *The digestive system provides nutrients and energy for your body through the digestion of food.*

Lesson 4

The Excretory System

BIG Idea *The excretory system removes wastes from the body.*

Activating Prior Knowledge

Using Visuals Regular aerobic activity will strengthen your heart muscle. Name some activities, other than bike riding, that you think might strengthen your heart.

Chapter Launchers

Health in Action

Discuss the **BIG** Ideas

Think about how you would answer these questions:

▶ When do you feel your heart rate change?

▶ What can cause your breathing to change?

▶ How does your stomach feel after you eat a large meal?

Assess Your Health

Read each statement. On a separate sheet of paper, write "yes," "sometimes," or "no" based on your typical behavior.

1. I care for my circulatory system by eating a well-balanced diet.

2. I recognize that regular physical activity will maintain my circulatory health.

3. I avoid tobacco use.

4. I understand why dental health is important to healthy digestion.

5. I practice good hygiene to prevent harmful bacteria from causing infection.

A "yes" response shows that you practice healthy behaviors. "Sometimes" indicates that you should analyze and possibly modify your behavior. A "no" response means that you should modify the behavior.

Sean Thompson/Photodisc/Getty Images

The Cardiovascular and Lymphatic Systems

Real Life Issues

The Beat Goes On. Marcos hears the announcement for the boys' 100-yard dash. His pulse quickens. This is it—the moment he's been training for all year. He gets into the starting position. At the sound of the starter pistol, he takes off. His legs are pumping, his lungs are burning, and his heart is racing as he crosses the finish line. He smiles, as the cheers of the crowd tell him he has won.

Writing *Write a persuasive paragraph that explains to a younger person why it's important to practice behaviors that keep your heart healthy.*

Why the Blood Circulates

Main Idea The cardiovascular system provides nutrients and oxygen, carries away wastes, and helps fight disease.

Your heart pumps blood to your body's cells 24 hours a day, even when you're asleep. Your heart accomplishes these important tasks:

• Carrying oxygen from the lungs to body cells

• Absorbing nutrients from food and delivering nutrients to body cells

• Carrying carbon dioxide, a waste gas, from your cells back to your lungs to be exhaled

• Delivering other waste products to the kidneys for removal from the body

• Helping the white blood cells fight disease by attacking infectious organisms

How Blood Circulation Works

Main Idea The cardiovascular system consists of the heart, blood, and blood vessels.

The cardiovascular system depends on the heart and its system of blood vessels to deliver blood throughout the body. If all of your blood vessels were laid end to end, they would stretch over 60,000 miles. That's enough to circle the earth almost two and a half times.

The Heart

Your heart is the muscle that makes the cardiovascular system work. Inside the heart are four chambers. The two top chambers are called the *atria*. The two lower chambers are called *ventricles*. A wall of tissue, the *septum,* separates the four chambers of the heart. Valves between the atria and ventricles allow blood to flow through the chambers.

At the top of the right atrium is an area of muscle that acts as a pacemaker for the heart. Electrical impulses stimulate the atria to **contract**, forcing blood into the ventricles. These electrical impulses travel through the heart to an area between the two ventricles. There they stimulate the muscles of the ventricles to contract, pumping blood out of the heart. Pumping the blood through the heart is only part of the process. **Figure 15.1** shows the pulmonary circulation.

Academic Vocabulary

contract *(verb):* to draw together

Figure 15.1 **Pulmonary Circulation**

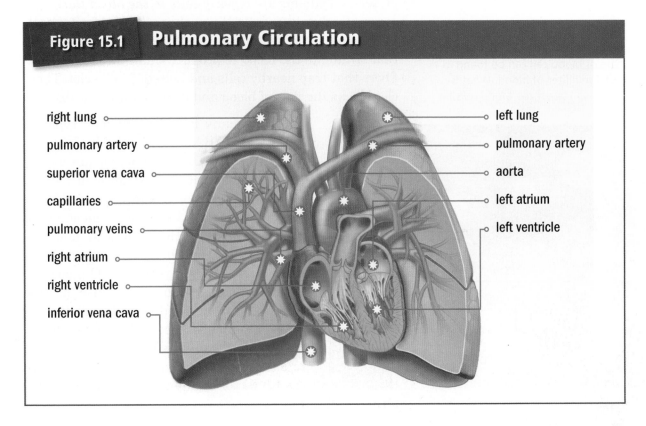

right lung
pulmonary artery
superior vena cava
capillaries
pulmonary veins
right atrium
right ventricle
inferior vena cava

left lung
pulmonary artery
aorta
left atrium
left ventricle

How Blood Circulates Pulmonary circulation is the process by which blood moves between the heart and the lungs. During this process, blood that has lost oxygen and picked up carbon dioxide and wastes receives fresh oxygen in the lungs. The oxygen-rich blood is circulated again through the body.

Blood

Blood is the fluid that delivers oxygen, hormones, and nutrients to the cells and carries away wastes. Blood is made up of the following components:

- **Plasma.** About 55 percent of total blood volume consists of **plasma**, *the fluid in which other parts of the blood are suspended.* Plasma is mainly water, but it also contains nutrients, proteins, salts, and hormones.

- **Red blood cells.** These cells make up about 40 percent of normal blood. They contain **hemoglobin**, *the oxygen-carrying protein in blood.* Hemoglobin contains iron that binds with oxygen in the lungs and releases the oxygen in the tissues. Hemoglobin also combines with carbon dioxide, which is carried from the cells to the lungs.

- **White blood cells.** These cells protect the body against infection. Some white blood cells surround and ingest the organisms that cause disease. Others form antibodies that provide immunity against a second attack from that specific disease. Still other types of white blood cells fight allergic reactions.

- **Platelets. Platelets** are *types of cells in the blood that cause blood clots to form.* When the wall of a blood vessel tears, platelets collect at the tear. They release chemicals that stimulate the blood to produce small thread-like fibers that trap nearby cells and help to form a clot. The clot blocks the flow of blood and dries to form a scab.

■ **Figure 15.2** Millions of each type of blood cell can be found in just 1 milliliter of blood. *What is the main role of red blood cells?*

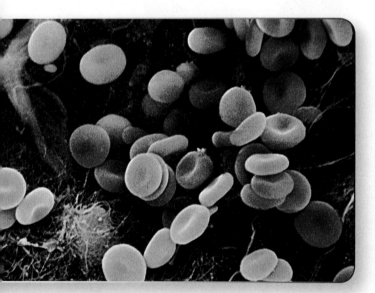

All humans have one of four types of blood: A, B, AB, and O. Each blood type is determined by the presence or absence of certain substances called antigens. Blood types A, B, or AB possess antigens, and a person must receive blood from someone with the same antigen. He or she can, however, receive type O blood, because it contains no antigens. People with type O blood are called universal donors, because anyone can receive their blood. Most blood also carries another substance called the Rh factor. If your blood contains Rh, you are referred to as *Rh positive.* Blood that doesn't have the Rh factor is called *Rh negative.*

Blood Vessels

The blood vessels that carry blood throughout the body are shown in **Figure 15.3**. There are three main types of blood vessels: arteries, capillaries, and veins.

- **Arteries** are *blood vessels that carry oxygenated blood away from the heart.* Arteries are vessels that branch into progressively smaller vessels called *arterioles.* The arterioles deliver blood to capillaries.

Figure 15.3 | The Cardiovascular System

A network of arteries, veins, and capillaries moves blood throughout the body, providing cells with oxygen and nutrients as well as removing wastes.

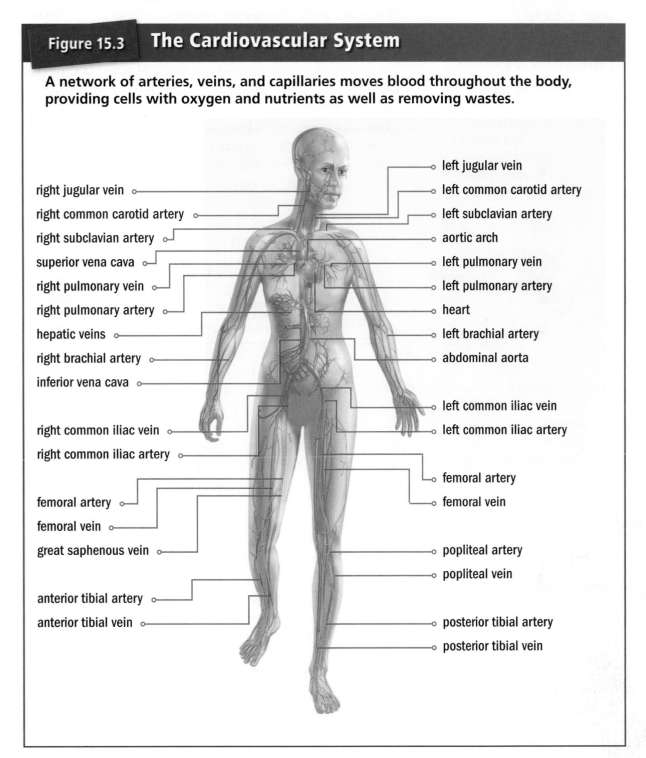

right jugular vein

right common carotid artery

right subclavian artery

superior vena cava

right pulmonary vein

right pulmonary artery

hepatic veins

right brachial artery

inferior vena cava

right common iliac vein

right common iliac artery

femoral artery

femoral vein

great saphenous vein

anterior tibial artery

anterior tibial vein

left jugular vein

left common carotid artery

left subclavian artery

aortic arch

left pulmonary vein

left pulmonary artery

heart

left brachial artery

abdominal aorta

left common iliac vein

left common iliac artery

femoral artery

femoral vein

popliteal artery

popliteal vein

posterior tibial artery

posterior tibial vein

- **Capillaries** are *small vessels that carry blood from arterioles and to small vessels called venules, which empty into veins.* Capillaries form a vast network throughout tissues and organs in the body, reaching almost all body cells. Capillaries near the skin's surface can also dilate, allowing heat to escape the body through the skin. They can also constrict to reduce heat loss if the body temperature drops below normal.

- **Veins** are *blood vessels that return blood to the heart.* While the walls of veins are thinner and less elastic than those of the arteries, they are still able to withstand the pressure exerted by blood flowing through them. The large veins, the *vena cava*, carry deoxygenated blood to the right atrium of the heart. Pulmonary veins carry oxygenated blood to the left atrium. Many veins throughout the body, especially those in the legs, have valves that help prevent the backflow of blood as it is pumped back to the heart. As surrounding muscles contract, they also exert pressure on vein walls, helping to move blood back through the veins.

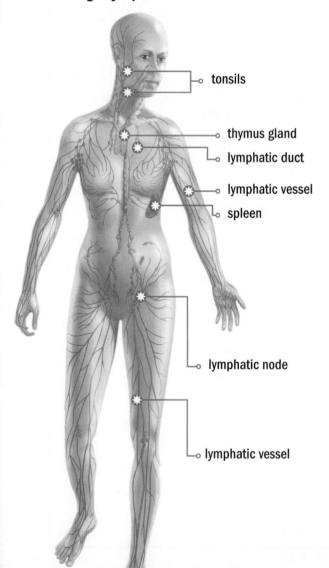

Figure 15.4

Lymphatic System

The lymphatic system is a system of vessels, much like the cardiovascular system, that helps protect the body against pathogens. *What moves lymph through lymph vessels?*

tonsils

thymus gland

lymphatic duct

lymphatic vessel

spleen

lymphatic node

lymphatic vessel

How Lymph Circulation Works

Main Idea The lymphatic system helps fight infection and provides immunity to disease.

The lymphatic system, shown in **Figure 15.4**, consists of a network of vessels and tissues that move and filter **lymph**, *the clear fluid that fills the spaces around body cells.* Like plasma, lymph contains water and proteins. It also contains fats and specialized white blood cells called lymphocytes. Like the white blood cells in the blood, these cells protect the body against pathogens. A **pathogen** is *a microorganism that causes disease.* There are two types of lymphocytes, B cells and T cells.

B cells multiply when they come in contact with a pathogen. Some of the new B cells produce antibodies that fight the pathogen. Other B cells create an immune response by preventing a second attack of the same disease. There are two main types of T cells, killer cells and helper cells. T cells multiply and enlarge when they come in contact with a pathogen.

One type of T cell, killer T cells, release toxins that prevent infections from spreading. Another type of T cell, the helper T cell, activates both the B cells and killer T cells. They also control the body's immune system.

Lymph is filtered by *lymph nodes,* small bean-shaped organs found in lymph vessels. White blood cells within lymph nodes trap and destroy pathogens.

Smooth muscles lining the walls of lymph vessels and surrounding skeletal muscles contract to move lymph toward the heart. Two large lymphatic ducts empty lymph into veins close to the heart, where the lymph is returned to the blood. The lymphatic system also includes certain organs and tissues—such as the spleen, thymus gland, tonsils, adenoids, and appendix—that help protect the body from infection.

Maintaining Your Circulatory Health

Main Idea Healthy habits can help protect the health of the cardiovascular and lymphatic systems.

Many problems with the cardiovascular and lymphatic systems first appear later in life. You can reduce your risk by making healthy decisions throughout your life. Here are some healthful behaviors that should become regular habits:

- Eat a well-balanced diet.
- Maintain a healthy weight.
- Participate in regular aerobic exercise for at least 30 minutes three or four times per week.
- Avoid secondhand smoke and using tobacco products.
- Avoid illegal drug use.
- Get regular medical checkups.

Blood Pressure

Maintaining pressure in the cardiovascular system is important for proper blood circulation. Pressure in arteries is created when the ventricles contract. As blood is forced into the arteries, arterial walls stretch under the increased pressure. When the ventricles relax and refill with blood, arterial pressure decreases. **Blood pressure** is *a measure of the amount of force that the blood places on the walls of blood vessels, particularly large arteries, as it is pumped through the body.*

A blood pressure reading includes two numbers. The first number measures your *systolic pressure*—the maximum pressure as your heart contracts to push blood into your arteries.

READING CHECK

Compare and Contrast How are the cardiovascular and lymphatic systems similar? How are they different?

FITNESS ZONE

I heard that antioxidants provide protection against conditions such as heart disease and cancer. They might even slow down the aging process. I read that researchers at Tufts University in Boston recommend these seven foods in your daily diet: prunes, raisins, blueberries, blackberries, kale, strawberries, and spinach. I should be able to eat at least one of those each day. For more fitness tips, visit the Fitness Handbook and Fitness Zone sites in ConnectEd.

The bottom, or second, number measures your *diastolic pressure*—the pressure at its lowest point when your ventricles relax. A healthy person's blood pressure will vary within a normal range of below 120/80. Exercise and stress will raise blood pressure. Blood pressure that is above 140/90 is considered high and places a strain on the heart.

Cardiovascular System Problems

Main Idea Some cardiovascular problems are inherited; others result from illness, diet, or aging.

These disorders of the cardiovascular system have wide-ranging effects and varying treatments:

- **Congenital heart defects** are conditions of the heart that are present at birth. A septal defect is a hole in the septum that allows oxygenated blood to mix with oxygen-depleted blood. Congenital heart defects may also result from poor health of a baby's mother during pregnancy. Medication and possibly surgery can sometimes repair the affected portion of the heart. In severe cases, a donor heart may be transplanted into a patient whose tissue and blood type matches the donor. The recipient must take anti-rejection drugs for the remainder of his or her life.

- **Heart murmurs** are caused by a hole in the heart, or a leaking or malfunctioning valve.

- **Varicose veins** are formed as result of the valves in veins not closing tightly enough to prevent backflow of blood.

- **Anemia** is a condition in which the ability of the blood to carry oxygen is reduced. The blood may contain low numbers of red blood cells or low concentrations of hemoglobin. The most common cause is iron deficiency.

- **Hemophilia** is an inherited disorder. The blood does not clot properly. Bruising and uncontrolled bleeding may occur spontaneously or due to injury. Treatment for hemophilia includes injections that introduce missing clotting proteins into the blood.

- **Leukemia** is a form of cancer in which white blood cells are produced excessively and abnormally. This causes the person to be susceptible to infection, severe anemia, and possibly uncontrolled bleeding. Chemotherapy, radiation, and bone marrow transplant are all treatment options.

 **READING CHECK**

Cause and Effect What cardiovascular disorder can be avoided through diet?

Lymphatic System Problems

Main Idea Problems of the lymphatic system can range from mild to life-threatening.

Disorders of the lymphatic system can range from mild to life-threatening. They may be caused by infection or heredity. Lymphatic system disorders include the following:

- **Tonsillitis.** Your tonsils help reduce the number of pathogens entering the body through the respiratory system. If the tonsils become infected, tonsillitis results. It is often treated with antibiotics, or surgery for chronic cases.

- **Immune deficiency.** Immune deficiency results if the immune system is weakened and can no longer protect the body against infection. It may be a congenital condition in which the body cannot make specialized white blood cells, limiting protection against infection. Other causes include HIV, chemotherapy, and sometimes aging.

- **Hodgkin's disease.** Also called Hodgkin's lymphoma, this type of cancer affects the lymph tissue found in lymph nodes and the spleen. Early detection and treatment is essential for recovery. Treatment may include removal of lymph nodes, radiation, and chemotherapy.

 **READING CHECK**

Apply What is the connection between the immune system and HIV?

LESSON 1 ASSESSMENT

 After You Read

Reviewing Facts and Vocabulary

1. Why is the cardiovascular system important to your overall health?
2. What behaviors will help you prevent high blood pressure?
3. What do the blood pressure numbers measure?

Thinking Critically

4. **Infer.** Why are people with type O blood called "universal donors"?
5. **Apply.** Why might lymph nodes become the main site of the body's response to an infection?

Applying Health Skills

6. **Advocacy.** Find out more about heart disease and how to prevent it. What foods and physical activities promote heart health? Design a Web site that promotes heart-healthy behaviors.

Writing Critically

7. Write a paragraph describing three healthy choices you have made to maintain your cardiovascular and lymphatic health.

Real Life Issues ....

After completing the lesson, review and analyze your response to the Real Life Issues question on page 408.

BIG Idea *The respiratory system provides oxygen to the blood and removes carbon dioxide from the body.*

Before You Read

Prepare Note Cards.
On separate index cards, list the various organs of the respiratory system. On the reverse side of each card, write the function of the organ.

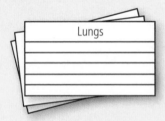
Lungs

New Vocabulary

▶ diaphragm
▶ trachea
▶ bronchi
▶ asthma
▶ tuberculosis
▶ emphysema

The Respiratory System

Real Life Issues

Respiratory Problems. Frequency of Asthma by Age Group
Source: Centers for Disease Control and Prevention, Vital and Health Statistics, National Surveillance of Asthma: United States, 2001–2010

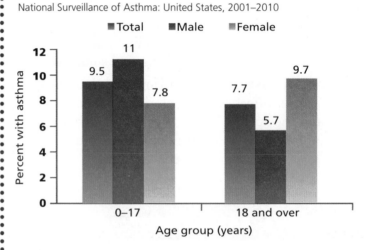

Writing *Write a paragraph describing what you think causes asthma and triggers asthma attacks.*

What Happens During Respiration

Main Idea The respiratory system provides oxygen to the blood and removes carbon dioxide from the body.

Your respiratory system removes carbon dioxide from the body and provides it with fresh oxygen. Inhaling and exhaling causes the lungs to expand and deflate slightly.

The process of respiration can be divided into two parts. In *external respiration,* oxygen moves from the lungs into the blood, and carbon dioxide moves from the blood into the lungs. In *internal respiration,* oxygen moves from the blood into the cells, and carbon dioxide moves from the cells into the blood. The continual exchange of gases in both external and internal respiration is essential for survival. Oxygen fuels the brain and allows your body to metabolize food for energy to move muscles.

How Respiration Works

Main Idea The respiratory system consists of the lungs, trachea, and diaphragm.

Your lungs automatically fill with air and are emptied in a rhythmic way. This rhythm changes with the level of your activity. You've probably noticed that when you do aerobic exercises, like running or fast walking, you tend to breathe harder than when you're sitting still. Breathing is regulated by the brain, which sends impulses to stimulate the muscles involved in respiration. This process provides your body with the oxygen it needs to keep going. It also removes carbon dioxide from the lungs. The lungs are found within the chest cavity and are protected by the ribs. In the base of the chest cavity is the **diaphragm** (DY-uh-fram), *a muscle that separates the chest from the abdominal cavity.*

As you inhale, the diaphragm and the muscles between your ribs contract. This contraction expands your chest cavity and your lungs. The pressure inside your lungs is lower than the pressure outside your body, so air naturally flows into your lungs to equalize the pressure. As you exhale, these same muscles relax and your chest cavity decreases. Pressure inside your lungs is higher, so air naturally flows out of your lungs to the outside, the area of lower pressure.

The Lungs

The structure of the lungs can be compared to the structure of a branching tree. Air moves into the lungs through the **trachea** (TRAY-kee-uh), or *the windpipe.* The trachea branches out into two **bronchi** (BRAHN-ky), *the main airways that reach into each lung.* The airways become smaller as they branch out deeper into the lungs. A network of tubes called *bronchioles* brings air closer to the site of external respiration. At the end of each bronchiole are groups of microscopic structures called *alveoli.* Shown in **Figure 15.5** on page 418, alveoli are thin-walled air sacs covered with capillaries. Gas exchange takes place as oxygen and carbon dioxide spread across the walls of the capillaries and alveoli.

Other Respiratory Structures

The respiratory system also includes structures in the upper airways. Air enters and exits your body through the nose and mouth. The membranes of the nose are lined with hairlike structures, called *cilia,* and with cells that produce mucus. The cilia and mucus work together to help prevent foreign particles such as dust, bacteria, and viruses from moving deeper into the respiratory system.

Academic Vocabulary

expand *(verb):* to open up

READING CHECK

Explain In what structures does gas exchange take place?

Figure 15.5 **The Respiratory System**

The lungs are the principal organs of the respiratory system. *Which structures in the diagram are also parts of the cardiovascular system?*

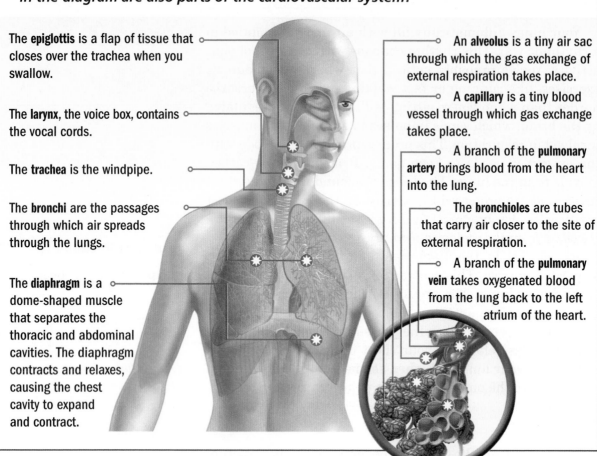

The **epiglottis** is a flap of tissue that closes over the trachea when you swallow.

The **larynx**, the voice box, contains the vocal cords.

The **trachea** is the windpipe.

The **bronchi** are the passages through which air spreads through the lungs.

The **diaphragm** is a dome-shaped muscle that separates the thoracic and abdominal cavities. The diaphragm contracts and relaxes, causing the chest cavity to expand and contract.

An **alveolus** is a tiny air sac through which the gas exchange of external respiration takes place.

A **capillary** is a tiny blood vessel through which gas exchange takes place.

A branch of the **pulmonary artery** brings blood from the heart into the lung.

The **bronchioles** are tubes that carry air closer to the site of external respiration.

A branch of the **pulmonary vein** takes oxygenated blood from the lung back to the left atrium of the heart.

The air that enters the respiratory system is filtered, warmed, and moistened. The air then moves into the *pharynx,* or throat, and then into the *trachea,* or windpipe. The tissue that lines the trachea is also lined with mucus and cilia to trap particles and prevent them from going deeper into the respiratory system.

Other structures that are not directly involved in respiration, but have important functions in the respiratory system, are the larynx and the epiglottis. The *larynx,* or voice box, connects the throat and the trachea. The larynx contains the vocal cords, two bands of tissue that produce sound when air forced between them causes them to vibrate.

The *epiglottis* is a flap of tissue located above the larynx. It folds down to close off the entrance to the larynx and trachea when you swallow. This is an involuntary action that keeps food or drink from entering the respiratory system. If you eat too quickly or talk or laugh while eating, your food may get past the epiglottis and "go down the wrong pipe." The piece of food stimulates the cough reflex to expel the material from your respiratory system.

Maintaining Your Respiratory Health

Main Idea Caring for your lungs can prevent many respiratory disorders.

Respiratory problems can affect the functioning of other body systems. Imagine not being able to climb a flight of stairs without running out of breath. The single most important decision you can make for your respiratory health is not to smoke. Smoking damages all parts of the respiratory system and is the main cause of lung cancer. Smoking can also cause bronchitis and emphysema, and increase the likelihood of asthma in children. Tobacco use also reduces the rate of lung growth in teens. Avoiding tobacco use and exposure to secondhand smoke will decrease your risk. Air pollution also increases the risk of respiratory health problems and certain types of cancers.

Regular physical activity is also important for a healthy respiratory system. Increased respiration during exercise improves the capacity of the lungs to pass oxygen into the blood.

Washing your hands regularly can help prevent infection. Bacteria and viruses can be easily transmitted to the respiratory system when contaminated hands touch the nose or mouth.

READING CHECK

Extend How might a friend's smoking habit affect your respiratory health?

Respiratory System Problems

Main Idea Problems of the respiratory system can be mild, such as a cold, or serious and even life threatening.

Problems of the respiratory system range from mild infections to disorders that can damage lung tissue and alveoli and prevent proper ventilation. When severe disease occurs a lung transplant may be recommended. A deceased donor may provide one or both lungs. Recent medical advances have enabled living donors to provide a portion of one lung to a recipient. Colds and influenza are common infections of the upper respiratory system. Other infections and disorders affect the lower respiratory tract.

- **Sinusitis** is an inflammation of the tissues that line the sinuses, air-filled cavities above the nasal passages and throat. The inflammation can result from allergies or an infection. Symptoms include nasal congestion, headache, and fever. Treatment includes nasal decongestant drops or sprays and antibiotics.

FITNESS ZONE

I like to work out with a friend. During our workouts, we talk. My PE teacher said that when we do aerobic exercises, we should be a little winded but still able to talk or sing. For more fitness tips, visit the Fitness Handbook and Fitness Zone sites in ConnectEd.

Real World CONNECTION

The Effects of Smoking

Ari's health teacher, Mrs. Gilcrest, held up a jar filled with a brown, gooey substance. Mrs. Gilcrest told the class that the jar represented the lungs of smokers. The brown sludge represented the amount of tar that gets into a smoker's lungs each year from smoking one pack of cigarettes a day. Ari thought about his Uncle Stan, who wears an oxygen tank because he has emphysema and has trouble breathing. "No wonder," thought Ari. "Uncle Stan smoked about a pack of cigarettes a day for as long as I can remember."

Activity Technology

Conduct an online search to learn the number of deaths each year that are caused by respiratory illnesses of tobacco users. The Centers for Disease Control and Prevention (CDC) is an example of a good Web site to use to find a wealth of information. Create a blog providing information and persuading teens to avoid tobacco use. Explain how avoiding tobacco use will reduce the risk of respiratory diseases.

- **Bronchitis** is an inflammation of the bronchi caused by infection or exposure to irritants such as tobacco smoke or air pollution. In bronchitis, the membranes that line the bronchi produce excessive amounts of mucus in the airways. This blocks the airways and leads to symptoms such as coughing, wheezing, and shortness of breath that worsen with physical activity. Treatment includes avoiding exposure to the irritant and taking antibiotics.

- **Asthma** (AZ-muh) is *an inflammatory condition in which the trachea, bronchi, and bronchioles become narrowed, causing difficulty breathing.* During an asthma attack, an involuntary contraction of smooth airway muscles leads to chest tightness and breathing difficulty. Acute asthma attacks can be relieved with an inhaler that dispenses medication to dilate, or widen, the airways.

■ **Figure 15.6** An inhaler can relieve an asthma attack. Long-term treatment of asthma includes using medication that reduces inflammation and avoiding substances that can trigger an attack, such as pollen, dust, animal dander, and tobacco smoke. *Why is it important for an asthmatic person to avoid air pollution?*

- **Pneumonia** is an inflammation of the lungs commonly caused by a bacterial or viral infection. In a common type of pneumonia, the alveoli swell and become clogged with mucus, decreasing the amount of gas exchange. Symptoms include cough, fever, chills, and chest pain. Bacterial pneumonia is treated with antibiotics.

- **Tuberculosis** is *a contagious bacterial infection that usually affects the lungs*. When a person is infected with tuberculosis, the immune system surrounds the infected area and isolates it. In this inactive stage, which can last for many years, a person doesn't show symptoms. However, if the immune system is weakened by illness or age, the infection can become active. During this active stage, symptoms include cough, fever, fatigue, and weight loss. Treatment includes antibiotics and hospitalization.

- **Emphysema** is *a disease that progressively destroys the walls of the alveoli*. Symptoms include breathing difficulty and chronic cough. Although the symptoms of emphysema can be treated, the tissue damage is permanent. Emphysema is almost always caused by smoking.

READING CHECK

Explain Why can you get tuberculosis from someone who doesn't show any symptoms of the disease?

LESSON 2 ASSESSMENT

 After You Read

Reviewing Facts and Vocabulary

1. What causes the lungs to fill with air?
2. Which problems with the respiratory system can be caused by smoking?
3. How can washing your hands protect your respiratory system?

Thinking Critically

4. **Compare.** How do internal respiration and external respiration differ?
5. **Apply.** A friend wants to quit smoking. You notice that just walking to school with you leaves her breathing hard. How can you encourage her to quit smoking?

Applying Health Skills

6. **Communication Skills.** Imagine you have a close family member who bicycles to work on major streets during rush hour. During this time, air pollution is at its worst, and a cyclist inhales a lot of it. Write a dialogue in which you encourage the family member to consider the negative effects of this practice. Explain the problems that can result.

Writing Critically

7. **Expository.** Write a paragraph explaining how oxygen and carbon dioxide are exchanged through the respiratory system.

Real Life Issues ....

After completing the lesson, review and analyze your response to the Real Life Issues question on page 416.

GUIDE TO READING

BIG Idea *The digestive system provides nutrients and energy for your body through the digestion of food.*

Before You Read

Create a Chart. Make a three-column chart like the one below. In the first column, list the organs of the digestive system. In the second, describe the function of each organ. In the third, list behaviors that contribute to the health of each organ.

Digestive organ	What it does	How to keep healthy

New Vocabulary

▶ mastication
▶ peristalsis
▶ gastric juices
▶ bile
▶ peptic ulcer
▶ appendicitis

The Digestive System

Real Life Issues ....

Fast-Food Folly. Joey has been looking forward to lunch with his uncle at Joey's favorite fast-food restaurant. He orders a double burger, a side of cheese fries, and a giant-size soda. They share a banana split for dessert. Soon after eating, Joey feels bloated and queasy. Later that afternoon, he feels tired, even though he hasn't done any physical activity. Even after a quick nap in front of the television, he still doesn't feel good. Joey wonders if it has something to do with his lunch.

Writing *Write a description of a time when the food you ate affected the way you felt afterward. Describe how your energy level was affected.*

What Happens During Digestion

Main Idea In digestion, foods are broken down and absorbed as nourishment or eliminated as waste.

The foods you eat provide nourishment. That food and drink, however, must be broken down into smaller nutrients to be absorbed into the blood and carried to the body's cells. The digestive system functions can be divided into three main processes:

- **Digestion** is the mechanical and chemical breakdown of foods within the stomach and intestines for use by the body's cells.

- **Absorption** is the passage of digested food from the digestive tract into the cardiovascular system.

- **Elimination** is the body's expulsion of undigested food or body wastes.

How Digestion Works

Main Idea The digestive system consists of the mouth, esophagus, stomach, and intestines.

Digestion includes two processes. The mechanical process involves chewing, mashing, and breaking food down. The chemical process involves secretions produced by digestive organs. **Figure 15.7** shows the organs involved in digestion.

- **Teeth.** The teeth break the food you eat into smaller pieces. **Mastication** (mas-tih-KAY-shun) is *the process of chewing,* which prepares food to be swallowed.

- **Salivary glands.** These glands produce digestive juices. Saliva contains an enzyme that begins to break down the starches and sugars in food into smaller particles.

- **Tongue.** The tongue prepares chewed food for swallowing by shaping it. The *uvula,* a small flap of tissue at the back of the mouth, prevents food from entering the nasal passages. The *epiglottis,* tissue covering the throat, prevents food from entering the respiratory system.

Academic Vocabulary

involve *(verb):* to require as a necessary accompaniment

Figure 15.7 The Digestive System

The organs of the digestive system break down and move food through the body, providing nutrients that are absorbed into the blood and transferred to cells.

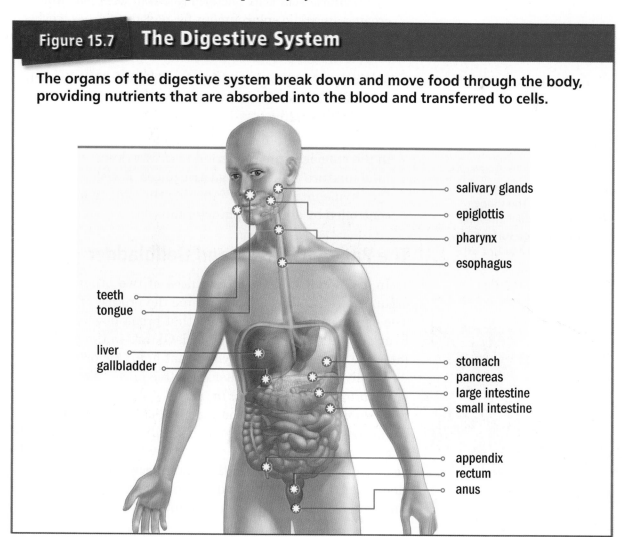

salivary glands
epiglottis
pharynx
esophagus

teeth
tongue

liver
gallbladder

stomach
pancreas
large intestine
small intestine

appendix
rectum
anus

The Esophagus

When food is swallowed, it enters the esophagus, the muscular tube about 10 inches long that connects the pharynx with the stomach. Food is moved through the esophagus, stomach, and intestine through **peristalsis** (pare-ih-STAWL-suhs), *a series of involuntary muscle contractions that moves food through the digestive tract*. The action of peristalsis begins as soon as food is swallowed. A sphincter muscle—a circular muscle at the entrance to the stomach—allows food to move from the esophagus into the stomach.

The Stomach

The stomach is a hollow, sac-like organ enclosed in a wall of muscles. These muscles are flexible and allow the stomach to expand when you eat. The stomach, shown in **Figure 15.9**, has three tasks:

- **Mixing foods with gastric juices. Gastric juices** are *secretions from the stomach lining that contain hydrochloric acid and pepsin, an enzyme that digests protein*. The hydrochloric acid kills bacteria taken in with food and creates an acidic environment for pepsin to do its work. Mucus produced by the stomach forms a protective lining so that the gastric juices do not harm the stomach.

- **Storing partially digested food and liquid.** The stomach holds the food for further digestion before it is moved into the small intestine.

- **Moving food into the small intestine.** As food is digested in the stomach, it is converted to *chyme* (kym), a creamy, fluid mixture of food and gastric juices. Peristalsis moves the chyme into the small intestine through an opening controlled by another sphincter muscle.

The Pancreas, Liver, and Gallbladder

In the small intestine, the juices of two other digestive organs mix with the food to continue the process of digestion. The pancreas produces enzymes that break down the carbohydrates, fats, and proteins in food. Glands in the wall of the intestine produce other enzymes that help this process.

The liver produces another digestive juice—**bile**, *a yellow-green, bitter fluid important in the breakdown and absorption of fats*. Bile is stored in the gallbladder between meals. At mealtimes, it is secreted from the gallbladder into the bile duct to reach the intestine and mix with fats in food. Bile acids dissolve the fats into the watery contents of the intestine. After the fat is dissolved, it is digested by enzymes from the pancreas and the lining of the intestine.

■ **Figure 15.8** The shape of the villi of the small intestine gives them a large surface area to maximize the amount of nutrients they can absorb. *Where do nutrients go once they are absorbed by the small intestine?*

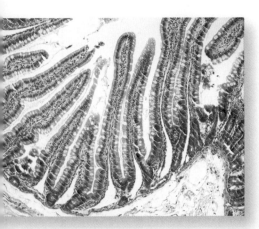

Figure 15.9   The Stomach

Digestion continues in the stomach. The three layers of stomach muscles each move in different directions to aid both mechanical and chemical digestion.

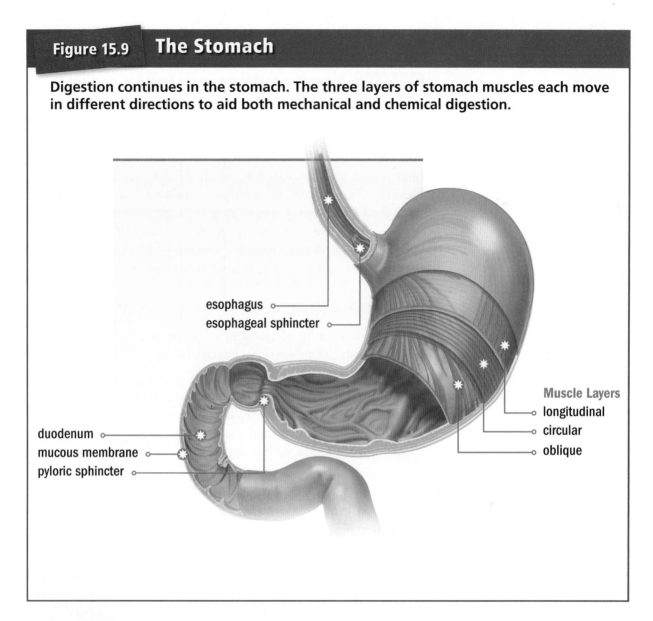

esophagus
esophageal sphincter

duodenum
mucous membrane
pyloric sphincter

Muscle Layers
longitudinal
circular
oblique

The Small and Large Intestines

The small intestine is 20 to 23 feet in length and 1 inch in diameter. It consists of three parts: the *duodenum,* the *jejunum,* and the *ileum.* As chyme enters the duodenum, it contains partially digested carbohydrates and proteins and undigested fats. This mixture is further dissolved by digestive juices secreted from the small intestine, liver, and pancreas.

About 90 percent of all nutrients are absorbed through the small intestine. The inner wall of the small intestine contains millions of fingerlike projections called *villi.* The villi are lined with capillaries that absorb the nutrients. Unabsorbed material leaves the small intestine in the form of liquid and fiber and moves by peristalsis into the large intestine.

The undigested parts of the food—fiber, or roughage—pass into the *colon,* or large intestine. The large intestine is about 2.5 inches in diameter and 5 to 6 feet in length. Its function is to absorb water, vitamins, and salts, and to eliminate waste.

READING CHECK

Interpret What are the differences between the small and large intestines?

Digestive System Problems

Main Idea Digestive problems range from indigestion to acute conditions that require immediate medical attention.

Taking care of your digestive system begins with the foods you eat and how you eat them. To maintain your digestive health, eat a variety of low-fat, high-fiber foods, wash your hands before preparing or eating meals, eat slowly and chew your food thoroughly, drink at least eight 8-ounce glasses of water a day, and avoid using food as a way of dealing with your emotions. Some digestive system problems may require medications and a visit to a health professional.

Functional Problems

The functioning of the digestive system may be affected by illness, stress, or eating a particular food. Functional problems of the digestive system include the following:

- **Indigestion** is a feeling of discomfort in the upper abdomen, sometimes with gas and nausea. It can be caused by eating too much food, eating too quickly, eating spicy or high-fat foods, or having a stomach disorder or stress.

- **Constipation** causes the feces to become dry and hard, making bowel movements difficult. It can be caused by not drinking enough water or not consuming enough fiber to move wastes through the digestive system.

- **Heartburn** is a burning sensation in the center of the chest that may rise up to the throat. It results from acid reflux, or the backflow of stomach acid into the esophagus. Using tobacco, alcohol, and aspirin, or eating spicy or greasy foods can cause heartburn.

■ **Figure 15.10** Eating plenty of fruits and vegetables can help prevent constipation. *What other health practices help you avoid constipation?*

- **Gas** produced from the breakdown of food is normal. Excessive gas can result in cramps or an uncomfortable feeling of fullness in the abdomen.
- **Nausea** is the feeling of discomfort that sometimes precedes vomiting. Motion sickness, pathogens, some medications, and dehydration can cause nausea.
- **Diarrhea** is the frequent passage of watery feces. It can be caused by bacterial or viral infections, some medications, a change in eating style, overeating, emotional turmoil, or nutritional deficiencies. Dehydration may result with each episode of diarrhea.

Structural Problems

The seriousness of structural problems of the digestive system can vary. Some problems are temporary or easily treated, others are serious and require immediate medical attention.

- **Tooth decay** may make it difficult to chew foods thoroughly. Brushing and flossing teeth daily can prevent tooth decay, along with regular dental checkups.
- **Gastritis** is an inflammation of the mucous membrane that lines the stomach. An increase in the production of stomach acid, use of tobacco or alcohol, bacterial or viral infections, and some medications can cause gastritis. Symptoms include pain, indigestion, decreased appetite, and nausea and vomiting.
- A **peptic ulcer** is *a sore in the lining of the digestive tract.* Peptic ulcers can be caused by a bacterial infection or the overuse of aspirin. Common symptoms include abdominal pain that worsens when the stomach is empty, nausea, and vomiting. Ulcers can cause stomach bleeding.
- **Gallstones** form when cholesterol in bile crystallizes. Gallstones can block the bile duct between the gallbladder and the small intestine. Symptoms of a blockage include pain in the upper right portion of the abdomen, nausea, vomiting, and fever.
- **Lactose intolerance** results from an inability to digest lactose, a type of sugar found in milk and other dairy products. Lactose is normally broken down by the enzyme *lactase.* People who are lactose intolerant do not produce enough lactase. Symptoms include abdominal cramps, bloating, gas, and diarrhea. Soy products are a good replacement for milk or dairy products.
- **Appendicitis** is *inflammation of the appendix,* the 3- to 4-inch tube at the tip of the large intestine. It can be caused by a blockage or bacterial infection. Symptoms include pain in the lower right abdomen and a fever.

FITNESS ZONE

I've heard all kinds of advice about how many meals a day you should eat. Everyone seems to have a different opinion about whether we should eat five or six smaller meals a day, or three large ones. It's really a personal decision based on what works best for you. Eating can help speed up your metabolism so you burn more calories. Eating smaller meals throughout the day keeps your body's engine revved up. For more fitness tips, visit the Fitness Handbook and Fitness Zone sites in ConnectEd.

READING CHECK

Compare Which structural problems of the digestive system can result from bacterial infection?

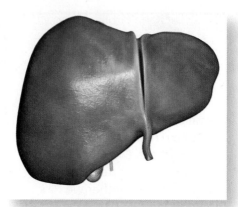

■ **Figure 15.11** Severe damage to the liver from cirrhosis may require a liver transplant. *What are the causes of cirrhosis?*

Decreased appetite, nausea, and vomiting will also occur. The appendix may burst, spreading infection throughout the abdomen, which can lead to death.

- **Colitis** is the inflammation of the large intestine, or colon. It may be caused by bacterial or viral infections. Symptoms can include fever, abdominal pain, and diarrhea that may contain blood.

- **Colon cancer** is the second leading cause of cancer death in the United States. It usually develops in the lowest part of the colon, near the rectum. A low-fat, high-fiber eating plan decreases the risk of colon cancer. Any rectal bleeding should be checked by a medical professional.

- **Hemorrhoids** are veins in the rectum and anus that may become swollen and inflamed. Hemorrhoids may occur with constipation, during pregnancy, and after childbirth. Signs of hemorrhoids include itching, pain, and bleeding.

- **Crohn's disease** causes inflammation of the lining of the digestive tract. Symptoms include diarrhea, weight loss, fever, and abdominal pain. The cause is not known, but seems to be associated with immune system problems.

- **Cirrhosis,** or scarring of the liver tissue, is caused by prolonged heavy alcohol use. Cirrhosis can lead to liver failure and may cause death.

(t)MedicalRF.com; (b) McGraw-Hill Education

LESSON 3 ASSESSMENT

 After You Read

Reviewing Facts and Vocabulary

1. What functions of the digestive system take place in the small intestine?

2. Describe the actions that cause food to move through the digestive tract.

3. What are three behaviors that help prevent indigestion?

Thinking Critically

4. **Evaluate.** What happens to the nutrients in food as it passes through the digestive system?

5. **Apply.** Create a menu with a full day of meals that you can serve to a friend who has lactose intolerance. Make sure that the menu you prepare contains foods high in calcium.

Applying Health Skills

6. **Advocacy.** Write a script for a play for elementary or middle school students on the importance of taking care of their teeth to protect their digestive systems.

Writing Critically

7. **Narrative.** Write a story from the point of view of a piece of food. Have the food describe its path through the digestive system, describing the function of each of the organs it meets.

Real Life Issues ....

After completing the lesson, review and analyze your response to the Real Life Issues question on page 422.

The Excretory System

The Artificial Kidney. Wendy is on her way to pick up her grandfather who has type 2 diabetes. Ever since his kidneys failed last year, Wendy's grandfather has been going to the clinic for dialysis. When Wendy arrives, her grandfather is not yet ready. At first glance, it looks to Wendy as if her grandfather is giving blood,  except the blood goes into a machine instead of a plastic bag. The machine acts like a real kidney, filtering wastes from the blood before returning the blood to her grandfather's body.

Writing *Write a letter to yourself describing ways you can reduce your risk for type 2 diabetes and prevent kidney failure.*

How Excretion Works

Main Idea The excretory system uses several organs to remove all types of wastes from the body.

Excretion is the process of removing wastes from the body. The body produces wastes in the form of solids, liquids, and gases. These wastes must be removed so that the body can function well.

The Lungs, Skin, and Large Intestine

The lungs expel carbon dioxide when you exhale. Sweating is another form of excretion. When sweat is produced, it removes excess water and salts through the pores. This excretion helps to regulate body temperature. As sweat evaporates on the surface of the skin, it cools the body. Sweating too much, however, can cause dehydration.

BIG Idea *The excretory system removes wastes from the body.*

Before You Read

Create an Outline. Preview this lesson by scanning the pages. Then organize the headings and subheadings into an outline. As you read, fill in the outline with important details.

| I. |
| A. |
| 1. |
| 2. |
| B. |
| II. |

New Vocabulary

▶ nephrons
▶ ureters
▶ urethra
▶ cystitis
▶ urethritis
▶ hemodialysis

Solid wastes produced by the digestive system are eliminated through the large intestine. Bacteria that live in the large intestine convert the undigested food materials into a semi-solid mass called *feces*.

The Liver

The liver plays an important role in the digestive system, and also removes certain toxins from the blood. It is the first organ to receive chemicals absorbed from the small intestine. The liver detoxifies the body by processing and excreting into bile such things as drugs, alcohol, and some cellular waste products.

The Urinary System

The urinary system consists of the kidneys, bladder, ureters, and urethra. The main function of the urinary system is to filter waste and extra fluid from the blood. Urine is liquid waste material excreted from the body. It consists of water and body wastes that contain nitrogen.

The Kidneys The kidneys, shown in **Figure 15.13**, are bean-shaped organs about the size of a fist. They are near the middle of the back, just below the rib cage, one on each side. The kidneys remove waste products from the blood through tiny filtering units called **nephrons** (NEH-frahnz), *the functional units of the kidneys*. Each kidney contains more than a million nephrons. Each nephron consists of a ball formed of small blood capillaries, called a *glomerulus,* which is attached to a small renal tubule that acts as a filtering funnel.

The kidneys adjust the amount of salts, water, and other materials excreted according to the body's needs. In this way, the kidneys **monitor** and maintain the body's acid-base and water balances. When the body becomes dehydrated, the pituitary gland releases *antidiuretic hormone* (ADH). This causes thirst and allows the kidneys to balance the fluid levels.

The Ureters From the kidneys, urine travels to the bladder through the ureters. The **ureters** (YUR-eh-terz) are *tubes that connect the kidneys to the bladder.* Each ureter is about 8 to

■ **Figure 15.12** The large surface area of your skin allows you to excrete water and salts when you sweat. *Why is it important to drink lots of water on a hot day?*

©liquidlibrary/PictureQuest

Academic Vocabulary

monitor *(verb):* to watch or keep track of

10 inches long. Muscles in the ureter walls tighten and relax to force urine down and away from the kidneys. Urine is passed from the ureters to the bladder about every 15 seconds.

The Bladder and Urethra The bladder is a hollow muscular organ located in the pelvic cavity. The bladder is held in place by ligaments attached to other organs and the pelvic bones. It stores about 2 cups of urine comfortably for two to five hours. Sphincter muscles help keep urine from leaking. The sphincter muscles close tightly like a rubber band around the opening of the bladder into the **urethra** (yur-EE-thruh), *the tube that leads from the bladder to the outside of the body.*

READING CHECK

Explain What vital body function do the kidneys provide?

Maintaining Your Excretory Health

Main Idea Healthful behaviors will help keep your excretory system healthy.

The excretory function removes wastes that can become toxic from the body. Healthy behaviors, like those listed below, will keep your excretory system healthy:

- Drink at least eight 8-ounce glasses of water each day.
- Limit your intake of caffeine and soft drinks, which can increase the amount of water lost through urination.

Figure 15.13 The Kidney

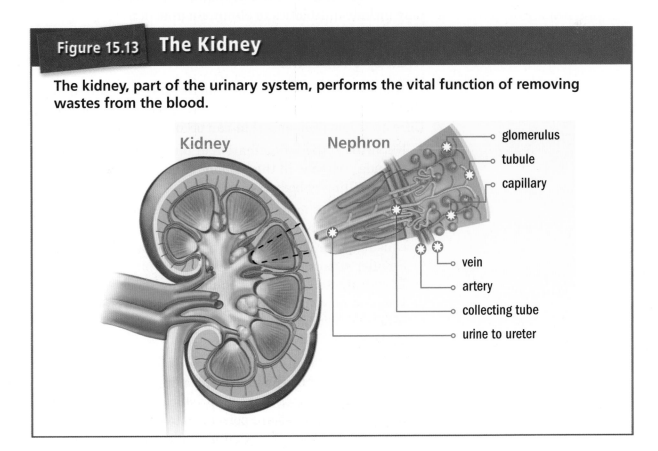

The kidney, part of the urinary system, performs the vital function of removing wastes from the blood.

Kidney   Nephron   glomerulus   tubule   capillary   vein   artery   collecting tube   urine to ureter

READING CHECK

Infer Why is it better to drink water rather than soft drinks when you are dehydrated?

- Follow a well-balanced eating plan.
- Practice good hygiene to prevent harmful bacteria from causing infection.
- Get regular medical checkups. Report changes in bowel habits and in the frequency, color, or odor of urine.

Excretory System Problems

Main Idea Excretory system problems commonly result from infection or blockage.

Disorders of the excretory system can have several different causes, including infection, blockage of urine, or natural aging. Two common disorders of the urinary system are cystitis and urethritis.

- **Cystitis** (sis-TY-tis) is *inflammation of the bladder,* most often caused by a bacterial infection. Left untreated, the infection can spread to the kidneys.
- **Urethritis** (yur-eh-THRY-tis) is *inflammation of the urethra.* It, too, can be caused by a bacterial infection.

Symptoms of both conditions include burning pain during urination, increased frequency of urination, fever, and possibly blood in urine. Treatment requires a visit to a doctor and may include antibiotics to eliminate infection.

Kidney Problems

Kidney disorders, some of which can be life threatening, should be treated and monitored by a medical professional. Here are some problems that can occur in the kidneys:

- **Nephritis** is the inflammation of the nephrons. Symptoms include a change in the amount of urine produced, fever, and swelling of body tissues.
- **Kidney stones** form when salts in the urine crystallize into solid stones. Kidney stones can move into the ureter, causing pain. They may also block the passage of urine. Smaller stones may be able to pass through naturally. Larger stones can be broken up using shock waves, so they can pass from the body through the ureters and urethra. In some cases, surgery is required to break up the stones.
- **Uremia** is a serious condition associated with decreased blood filtration by the kidneys, leading to abnormally high levels of nitrogen waste products remaining in the blood. These wastes are poisonous to body cells and can cause tissue damage, or death, if allowed to accumulate.

Kidney Failure

Kidney failure occurs when the kidneys lose their ability to function. It can be caused by infection, decreased blood flow, or diseases that damage kidney tissue. Here are treatments:

- **Hemodialysis** (HEE-moh-dy-AL-uh-sis) is *a technique in which an artificial kidney machine removes waste products from the blood.* A machine filters the blood. Hemodialysis takes three to five hours and is done three or four times per week, usually in a clinical setting.

- **Peritoneal dialysis** uses the *peritoneum,* a thin membrane that surrounds the digestive organs, to filter the blood. Substances that promote the removal of toxins enter into the abdomen through the catheter and are drained after filtration is complete.

- **Kidney transplant** is another treatment option for chronic kidney failure. This involves the replacement of a nonfunctioning kidney with a healthy kidney from an organ donor. An organ donor allows a healthy organ to be removed from his or her body and surgically placed into a patient who needs a healthy organ.

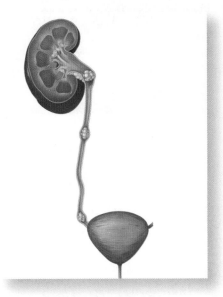

■ **Figure 15.14** Kidney stones larger than 1 cm must be broken up into smaller pieces to pass through the urethra. *How are larger kidney stones broken apart?*

LESSON 4 ASSESSMENT

 After You Read

Reviewing Facts and Vocabulary

1. What is the main function of the excretory system? What organs are part of it?

2. How are a *ureter,* a *urethra,* and *urethritis* different?

3. How can you prevent cystitis and urethritis?

Thinking Critically

4. **Evaluate.** What might pain during urination indicate? What should you recommend to a friend who experiences this?

5. **Analyze.** Why is it possible to donate a kidney and survive?

Applying Health Skills

6. **Analyzing Influences.** List the health behaviors that will help teens avoid problems that can affect the urinary system.

Writing Critically

7. **Comparative.** Write a brief paragraph comparing the way a kidney works and the way a hemodialysis machine works.

Real Life Issues ....

After completing the lesson, review and analyze your response to the Real Life Issues question on page 429.

Hands-On HEALTH

Activity
Look Inside the Body

This activity will help you understand how the body systems work, and what you can do to maintain and/or improve their amazing performance.

What You'll Need

- 20 or more index cards and a marker
- resources: your textbook, the Internet

What You'll Do

Step 1

Research a body system, writing the name of each organ from that system, one organ per index card. Place the index cards in order related to how the system works. For example, begin with the kidney (excretory system). Do the same for the digestive, lymphatic, respiratory, or circulatory systems.

Step 2

On the back of each index card, identify a health problem associated with this system and one lifestyle behavior to improve that problem.

Step 3

Mix the cards up, and pass them to the team on your left. When the teacher says "GO" place the cards in order. The first team done wins!

Apply and Conclude

Select one lifestyle behavior from the back of the cards you would be willing to do to improve your overall health.

Checklist: Self-Management Skills

✓ Identify healthy behaviors and habits

✓ Identify protective behaviors to avoid/manage unhealthy situations

✓ List organs within a body system in sequential order

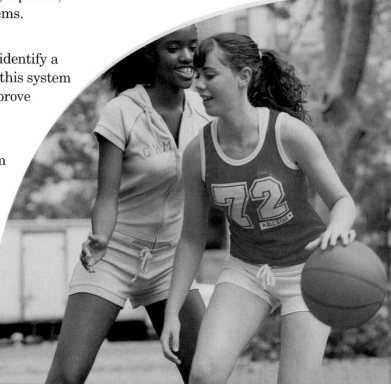

LESSON **1**

The Cardiovascular and Lymphatic Systems

Key Concepts

▶ The cardiovascular system includes the heart and blood vessels.

▶ The lymphatic system provides immunity against disease.

▶ Some problems of the cardiovascular system involve congenital heart defects or lifestyle factors.

Vocabulary

▶ plasma (p. 410)

▶ hemoglobin (p. 410)

▶ platelets (p. 410)

▶ capillaries (p. 412)

▶ lymph (p. 412)

▶ pathogen (p. 412)

▶ blood pressure (p. 413)

LESSON **2**

The Respiratory System

Key Concepts

▶ The lungs are the principal organs of the respiratory system.

▶ Avoiding tobacco smoke and other pollutants can keep your respiratory system healthy.

▶ Respiratory system problems include bronchitis, asthma, pneumonia, tuberculosis, and emphysema.

Vocabulary

▶ diaphragm (p. 417)

▶ trachea (p. 417)

▶ bronchi (p. 417)

▶ asthma (p. 420)

▶ tuberculosis (p. 421)

▶ emphysema (p. 421)

LESSON **3**

The Digestive System

Key Concepts

▶ The functions of the digestive system include digestion, absorption, and elimination.

▶ Digestion includes both mechanical and chemical processes.

▶ Digestive system problems might include indigestion, peptic ulcer, constipation, gallstones, cirrhosis, and colon cancer.

Vocabulary

▶ mastication (p. 423)

▶ peristalsis (p. 424)

▶ gastric juices (p. 424)

▶ bile (p. 424)

▶ peptic ulcer (p. 427)

▶ appendicitis (p. 427)

LESSON **4**

The Excretory System

Key Concepts

▶ The lungs, skin, liver, and large intestine remove wastes.

▶ The urinary system consists of the kidneys, bladder, ureters, and urethra, and filters waste and extra fluid from the blood.

▶ Problems of the excretory system include cystitis, urethritis, nephritis, kidney stones, uremia, and kidney failure.

Vocabulary

▶ nephrons (p. 430)

▶ ureters (p. 430)

▶ urethra (p. 431)

▶ cystitis (p. 432)

▶ urethritis (p. 432)

▶ hemodialysis (p. 433)

Vocabulary Review

Use the vocabulary terms listed on page 435 to complete the following statements.

1. _____ is the fluid in which other parts of the blood are suspended.

2. _____ are types of cells in the blood that cause blood clots to form.

3. A measure of the force that blood places on the walls of blood vessels as it is pumped through the body is called _____.

Understanding Key Concepts

After reading the question or statement, select the correct answer.

4. Which of the following is *not* a function of the cardiovascular system?
   a. Getting oxygen from air
   b. Producing red and white blood cells
   c. Removing carbon dioxide from the body
   d. Fighting disease by attacking infections

5. Congenital heart defects
   a. are present at birth.
   b. result from poor diet.
   c. affect mainly older people.
   d. can be prevented with regular exercise.

Thinking Critically

After reading the question or statement, write a short answer using complete sentences.

6. **Describe.** Describe and give examples of each type of blood vessel.

7. **Explain.** Explain how blood replaces oxygen with carbon dioxide.

8. **Analyze.** Why is early detection of high blood pressure important?

9. **Contrast.** Explain the difference between anemia and hemophilia.

Vocabulary Review

Choose the correct word in the sentences below.

10. The *trachea / bronchi* deliver air to and from the lungs.

11. The *diaphragm / trachea* is a muscle that changes the shape of the lungs.

12. In a(n) *bronchitis / asthma* attack, smooth muscles involuntarily contract and cause chest tightness.

Understanding Key Concepts

After reading the question or statement, select the correct answer.

13. Which of the following structures is the smallest?
    a. Bronchioles
    b. Diaphragm
    c. Lungs
    d. Trachea

14. Which behavior is *least* likely to prevent respiratory system problems?
    a. Smoking tobacco
    b. Washing your hands
    c. Getting regular exercise
    d. Eating fruits and vegetables

15. What problem of the respiratory system is almost always caused by smoking?
    a. Bronchitis   c. Pneumonia
    b. Emphysema   d. Tuberculosis

Thinking Critically

After reading the question or statement, write a short answer using complete sentences.

16. Describe. Describe the main function of the respiratory system.

17. Analyze. How might increased lung capacity benefit your health?

18. Describe. What is sinusitis? What causes it?

19. Apply. Why is it important for a person with asthma to avoid known allergens?

LESSON 3

Vocabulary Review

Use the vocabulary terms listed on page 435 to complete the following statements.

20. _____ is the series of muscle contractions that moves food through the digestive tract.

21. The stomach lining secretes _____, which contain hydrochloric acid and pepsin.

22. A(n) _____ is a sore in the lining of the digestive tract that can be caused by bacterial infection.

Understanding Key Concepts

After reading the question or statement, select the correct answer.

23. Which of the following is *not* one of the main functions of the digestive system?
   a. Absorption
   b. Digestion
   c. Elimination
   d. Circulation

24. Which substance is secreted by the liver?
   a. Bile
   b. Chyme
   c. Hydrochloric acid
   d. Mucus

25. Which of these tasks is *not* a function of the stomach?
   a. Storing food
   b. Moving food into the small intestine
   c. Absorbing nutrients from food
   d. Mixing food with gastric juices

26. Which of the following disorders involves a sensitivity to a sugar found in milk and other dairy products?
   a. Cirrhosis
   b. Gastritis
   c. Lactose intolerance
   d. Tooth decay

Thinking Critically

After reading the question or statement, write a short answer using complete sentences.

27. Describe. Describe how peristalsis moves food through the digestive tract.

28. Connect. What parts of foods do hydrochloric acid, pepsin, and bile work on?

29. Contrast. How are the roles of the small intestine and large intestine different?

30. Analyze. Why is it important to drink plenty of water when you have diarrhea or constipation?

LESSON 4

Vocabulary Review

Choose the correct word in the sentences below.

31. In *hemodialysis / urethritis,* a machine removes waste products from the blood.

32. The *ureters / nephrons* are the parts of the kidneys that filter blood.

33. The *ureter / urethra* carries urine from the bladder to the outside of the body.

34. *Cystitis / Urethritis* is an inflammation of the bladder caused by bacterial infection.

Kevin Peterson/Getty Images

Assessment

Understanding Key Concepts

After reading the question or statement, select the correct answer.

35. What role does skin play in excretion?
    a. Eliminating solid wastes
    b. Removing carbon dioxide
    c. Removing excess water and salts
    d. Breaking down toxic chemicals

36. Which of the following is *not* a recommended way to maintain the health of the excretory system?
    a. Having regular medical checkups, and reporting problems to your doctor
    b. Practicing good hygiene and personal health care
    c. Increasing your intake of caffeine and soft drinks
    d. Drinking eight 8-ounce glasses of milk each day

37. What problem of the urinary system could require hemodialysis?
    a. Cystitis
    b. Kidney failure
    c. Kidney stones
    d. Nephritis

Thinking Critically

After reading the question or statement, write a short answer using complete sentences.

38. **Infer.** Why might ingesting an unhealthful substance such as alcohol harm the liver first before any other organ?

39. **Analyze.** How does practicing good hygiene maintain the health of the urinary system?

40. **Apply.** Why is it important to address even mild cases of cystitis and urethritis?

41. **Infer.** Why might a patient choose a kidney transplant over hemodialysis?

Technology PROJECT-BASED ASSESSMENT

Create a True/False Test

Background

Everyone is familiar with tests. Tests measure your readiness to tackle a new topic or your mastery of a topic. One form of test question is true/false. These questions make a statement that must be judged to be either true or false, based on your knowledge of the topic.

Task

Using a free online survey tool, write a 15-question true/false survey to learn what your classmates know about organ transplants.

Audience

Students in your class

Purpose

Accurately and fairly test your classmates' knowledge of organ transplants.

Procedure

1. Review examples of true/false test questions provided by your teacher.

2. Conduct an Internet search to find information on organ transplants.

3. As a group, create the 15 questions for the online survey. Each member of the group should have input on the questions. Make sure you cover each section of the chapter. Consider creating questions that refer to diagrams or illustrations in the text.

4. Review your questions and prepare an answer key and scoring instructions for the survey.

5. Ask students to take the survey and tally the results. Present the results to your class for discussion.

Math Practice

Interpret Graphs. The bar graph below shows the percentages of high school students who were physically active for at least 60 minutes a day. Use the graph to answer Questions 1–3.

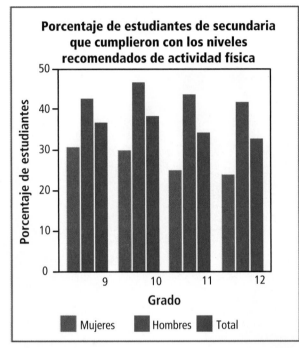

Porcentaje de estudiantes de secundaria que cumplieron con los niveles recomendados de actividad física

Adapted from: "Youth Risk Behavior Surveillance—United States, 2005"; Centers for Disease Control and Prevention, June 2006.

1. Which grade level had the lowest levels of physical activity for females?
   A. 9th grade   C. 11th grade
   B. 10th grade   D. 12th grade

2. At which grade level did at least half of the total students meet the currently recommended levels of physical activity?
   A. 9th grade   C. 12th grade
   B. 10th grade   D. None

3. Write a paragraph describing your general conclusion from the bar graph.

Reading/Writing Practice

Understand and Apply. Read the passage below, and then answer the questions.

(1) Have you ever heard of the influenza epidemic of 1918–1919? (2) Many people died worldwide. (3) In the United States, nearly 800,000 people died. (4) That's more than the number of Americans who died in World War I, World War II, and the Korea and Vietnam wars combined. (5) Influenza viruses still exist. (6) Why doesn't the flu kill as many people today?

(7) People in the health-care industry today know that they need to tell flu patients some things about how to feel better. (8) One of the most important treatments is simple—drink liquids. (9) People with the flu should drink lots of water, juice, and clear soups.

1. Which sentence below could be added after sentence 5 to support the first paragraph?
   A. Sick people should not drink liquids.
   B. Washing your hands often is important.
   C. Everyone can learn to wash their hands.
   D. You've probably had the flu yourself.

2. Which revision of sentence 7 is the most coherent and focused?
   A. Doctors and nurses need to know how to talk to sick people.
   B. Follow these logical and new rules of flu treatment to be safe.
   C. Health care professionals understand better how to treat the flu.
   D. To keep you safe from catching the flu, follow simple, new steps.

3. Create a poster using familiar children's book characters, words, and pictures to teach young flu patients to drink lots of liquids.

National Education Standards

Math: Measurement and Data, Data Analysis
Language Arts: LACC.910.L.3.6

Endocrine and Reproductive Health

Ed-Imaging

Activating Prior Knowledge

Using Visuals A teen with diabetes has a responsibility to manage his or her disease. What character traits do you think are important in someone who is managing a chronic disease?

Chapter Launchers

Health in Action

Discuss the BIG Ideas

Think about how you would answer these questions:

▶ What is the purpose of the endocrine system?

▶ Do you recognize the role of hormones during puberty?

▶ What behaviors are essential for the health of the reproductive systems?

Assess Your Health

Read each statement sheet. On a separate sheet of paper, write "yes," "sometimes," or "no" based on your typical behavior.

1. I get 8-10 hours of sleep each night.

2. I use stress management techniques to reduce stress.

3. I visit my health care provider for regular checkups.

4. I stretch my muscles before engaging in physical activities or sports.

5. I eat a healthful diet to lower my risk of diabetes and other diseases.

A "yes" response shows that you practice healthy behaviors. "Sometimes" indicates that you should analyze and possibly modify your behavior. A "no" response means that you should modify the behavior.

The Endocrine System

GUIDE TO READING

BIG Idea *Your body's endocrine system sends and receives chemical messages that control many body functions.*

Before You Read

Create Flashcards.
Write the name of each endocrine gland on the front side of a blank index card. Write the function of each on the reverse side. When you have completed the lesson, partner with a friend and use the cards to check your knowledge.

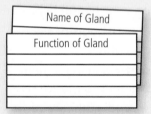

Name of Gland

Function of Gland

New Vocabulary

▶ endocrine glands
▶ thyroid gland
▶ parathyroid glands
▶ pancreas
▶ pituitary gland
▶ adrenal glands

Review Vocabulary

▶ hormones (Ch.3, L3)

Real Life Issues ....

A Close Call. Emily and Laura were walking home at dusk. Right after they stepped into the crosswalk, a car suddenly rounded the corner from behind them. Both girls dashed forward to get out of the way. Emily yelled at the driver to watch out. A few minutes later, Laura said her heart was still racing.

Writing *In a short essay, describe what your body feels like in a situation when you are suddenly startled or frightened.*

How the Endocrine System Works

Main Idea The endocrine system includes various organs that work together to regulate body functions.

Endocrine glands are *ductless or tubeless organs or groups of cells that secrete hormones directly into the bloodstream.* **Hormones** are *chemical substances that help regulate many of your body's functions.* Carried to their destination in the body through the blood, these chemical messengers influence physical and mental responses.

Hormones produced during puberty trigger physical and emotional changes in the body. Growth is controlled by certain hormones, and abnormally high or low amounts of these hormones may contribute to growth disorders. Factors such as stress, infection, and changes in the balance of fluids and minerals in the blood may affect hormone levels. Hormones work to maintain these balances in the body so that important processes and functions work more efficiently. **Figure 16.1** describes the major glands of the endocrine system and the body functions they regulate.

Figure 16.1 **The Endocrine System**

The glands of the endocrine system are located throughout the body.
Each gland has a particular function.

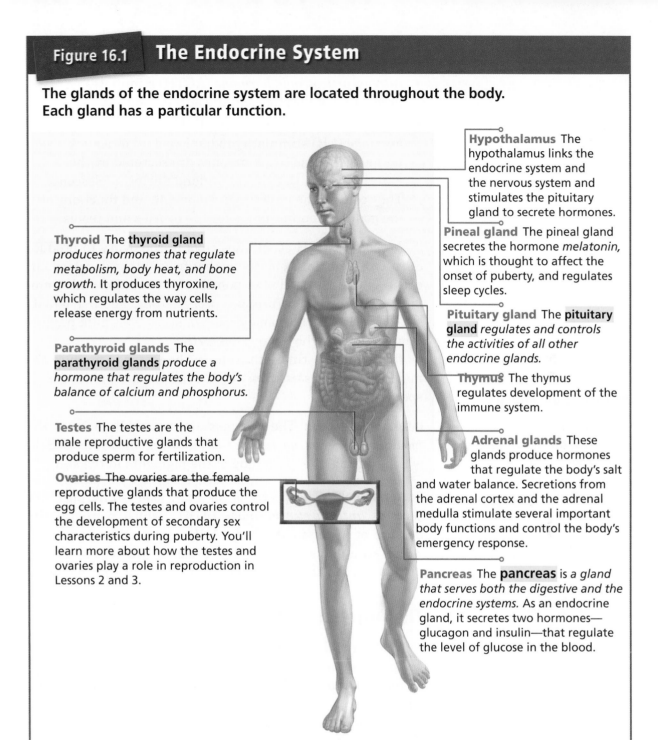

Hypothalamus The hypothalamus links the endocrine system and the nervous system and stimulates the pituitary gland to secrete hormones.

Pineal gland The pineal gland secretes the hormone *melatonin,* which is thought to affect the onset of puberty, and regulates sleep cycles.

Pituitary gland The **pituitary gland** *regulates and controls the activities of all other endocrine glands.*

Thymus The thymus regulates development of the immune system.

Adrenal glands These glands produce hormones that regulate the body's salt and water balance. Secretions from the adrenal cortex and the adrenal medulla stimulate several important body functions and control the body's emergency response.

Pancreas The **pancreas** is *a gland that serves both the digestive and the endocrine systems.* As an endocrine gland, it secretes two hormones—glucagon and insulin—that regulate the level of glucose in the blood.

Thyroid The **thyroid gland** *produces hormones that regulate metabolism, body heat, and bone growth.* It produces thyroxine, which regulates the way cells release energy from nutrients.

Parathyroid glands The **parathyroid glands** *produce a hormone that regulates the body's balance of calcium and phosphorus.*

Testes The testes are the male reproductive glands that produce sperm for fertilization.

Ovaries The ovaries are the female reproductive glands that produce the egg cells. The testes and ovaries control the development of secondary sex characteristics during puberty. You'll learn more about how the testes and ovaries play a role in reproduction in Lessons 2 and 3.

The Pituitary: The Master Gland

Known as the master gland, the pituitary gland has three sections, or lobes: anterior, intermediate, and posterior.

Anterior Lobe The anterior, or front, lobe of the pituitary gland produces these hormones:

- *Somatotropic,* or *growth hormone,* stimulates normal body growth and development by altering chemical activity in body cells.

- *Thyroid-stimulating hormone* (TSH) stimulates the thyroid gland to produce hormones.
- *Adrenocorticotropic hormone* (ACTH) stimulates production of hormones in the adrenal glands.
- *Follicle-stimulating hormone* (FSH) and *luteinizing hormone* (LH) stimulate production of all other sex hormones. These two hormones are secreted by the anterior lobe of the pituitary gland during adolescence. They control the growth, development, and functions of the gonads, another name for the ovaries and testes.

In females, FSH stimulates cells in the ovary to produce *estrogen,* a female sex hormone that triggers the development of ova, or egg cells. LH is responsible for ovulation and stimulates ovarian cells to produce *progesterone,* another female sex hormone. The hormone *prolactin* stimulates milk production in females who have given birth.

In males, LH stimulates cells in the testes to produce the male hormone *testosterone.* FSH controls the production of sperm.

Academic Vocabulary

intermediate *(adjective):* being at the middle place or stage

Intermediate Lobe The intermediate, or middle, lobe of the pituitary secretes *melanocyte-stimulating hormone* (MSH), which controls the darkening of the pigments in the skin.

Posterior Lobe The posterior, or rear, lobe of the pituitary secretes *antidiuretic hormone* (ADH), which regulates the balance of water in the body. ADH also produces oxytocin, which stimulates the smooth muscles in the uterus during pregnancy, causing contractions during the birth of a baby.

The Adrenal Glands

The **adrenal glands** *help the body deal with stress and respond to emergencies.* They each have two parts, the adrenal cortex and a smaller, inner region called the adrenal medulla, which controls a variety of body functions.

- The **adrenal cortex** secretes a hormone that inhibits the amount of sodium excreted in urine and maintains blood volume and blood pressure. It also secretes hormones that aid in the metabolism of fats, proteins, and carbohydrates. These hormones influence the body's response to stress and play a role in both the immune response and sexual function.

- The **adrenal medulla** is controlled by the hypothalamus and the autonomic nervous system. It secretes the hormones *epinephrine* (also called adrenaline) and *norepinephrine.* Epinephrine increases heartbeat and respiration, raises blood pressure, and suppresses the digestive process during periods of high emotion.

READING CHECK

Explain What are the functions of the thyroid gland and the parathyroid glands?

Maintaining Your Endocrine Health

Main Idea To keep your endocrine system working at its peak, you need to follow sound health practices.

Your endocrine health is directly related to your overall health. Remember to eat balanced meals to ensure that you get the nutrients you need, and use stress-management techniques. Sleep is also important to endocrine health. Teens need 8½ to 9 hours of sleep every night. Engage in regular physical activity to keep your body strong. Also, have regular medical checkups. Some hormonal disorders have symptoms you may not notice or recognize. A health care professional can perform tests to determine whether your endocrine function is normal.

Certain endocrine disorders can have lifelong effects on your health. Factors such as stress, infection, and changes in the balance of fluid and minerals in the blood can cause hormone levels to fluctuate. In many cases, these situations will correct themselves. Serious problems, including diabetes mellitus, hypothyroidisim, hyperthyroidism, goiter, or overproduction of adrenal hormones may require medication.

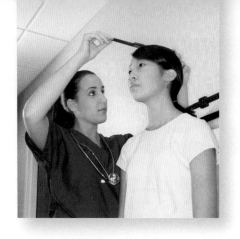

■ **Figure 16.2** Hormones produced by the pituitary gland play a significant role in determining height. *What may happen if the pituitary is damaged before adolescence?*

 READING CHECK

Identify What are three ways you can care for your endocrine system?

LESSON 1 ASSESSMENT

After You Read

Reviewing Facts and Vocabulary

1. What are *hormones*?
2. Name the hormone that stimulates normal growth and development.
3. What gland helps regulate the chemicals that control sleep?

Thinking Critically

4. **Infer.** Why do the hormones FSH and LH have different effects in men and women?
5. **Apply.** If the water in your body is not properly balanced, what endocrine gland (or part of a gland) may be malfunctioning?

Applying Health Skills

6. **Self-Management.** Sleep keeps the endocrine system healthy. For one week, log the number of hours you sleep each night. At the end of the week, calculate your average. Create a plan to get the appropriate amount of sleep each night.

Writing Critically

7. **Persuasive.** Write a script for a public service announcement reminding teens that everyone grows at a different rate. Include information about normal growth and genetic influences.

Real Life Issues ....

After completing the lesson, review and analyze your response to the Real Life Issues question on page 442.

The Male Reproductive System

GUIDE TO READING

BIG Idea *The male reproductive system is a series of organs involved in producing children.*

Before You Read

Create a Flow Chart. As you read, sketch the path that sperm take through each of the male reproductive organs.

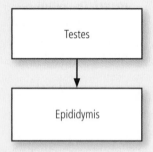

New Vocabulary

▶ sperm
▶ testosterone
▶ testes
▶ scrotum
▶ penis
▶ semen
▶ sterility

Real Life Issues

Getting Answers About Adolescence. Jackson has questions about the way his body is changing. His voice is changing and the hair above his lip is becoming thicker.
He often feels as though he is not in control of his emotions. He wants to know why he is experiencing all of these changes but is embarrassed to talk to anyone about it.

Writing *Write a letter to Jackson offering suggestions on what questions to ask, and advice on getting help from a trusted male adult or his doctor.*

How Male Reproduction Works

Main Idea The male reproductive system includes both external and internal organs that, with the help of hormones, allow physically mature males to produce children.

The two main functions of the male reproductive system are to produce and store **sperm**—*the male gametes*, or reproductive cells—and transfer them to the female's body during sexual intercourse. During the early teen years, usually between the ages of 12 and 15, the male reproductive system reaches maturity. At that time, hormones produced in the pituitary gland stimulate the production of **testosterone**, *the male sex hormone*. Testosterone initiates physical changes that signal maturity, including broadening of the shoulders, development of muscles and facial and other body hair, and deepening of the voice. Testosterone also controls the production of sperm. After puberty begins, a physically mature male is capable of producing sperm for the rest of his life.

External Reproductive Organs

A male's external reproductive organs include the testes, the penis, and the scrotum. The **testes**, also called *testicles,* are *two small glands that secrete testosterone and produce sperm.* They are located in the **scrotum**, *an external skin sac.* The **penis** is *a tube-shaped organ that extends from the trunk of the body just above the testes.* The penis is composed of spongy tissue that contains many blood vessels. When blood flow to the penis increases, it becomes enlarged and erect, causing an erection. Erections are normal body functions that occur more easily and more frequently during puberty. They can occur for no reason.

When the penis becomes erect, semen can be ejected from the body. **Semen** is *a thick fluid containing sperm and other secretions from the male reproductive system.* At the height of sexual arousal, a series of muscular contractions known as *ejaculation* may occur. *Fertilization*—the joining of a male sperm cell and a female egg cell—can result if ejaculation occurs during sexual intercourse.

At birth, the tip of the penis is covered by a thin, loose skin, called the *foreskin.* Some parents choose *circumcision*— surgical removal of the foreskin of the penis—for their male children. In general, circumcision is often chosen for cultural or religious reasons. Sperm cannot live in temperatures higher than the normal body temperature of 98.6 degrees F. The scrotum protects sperm by keeping the testes slightly below the normal body temperature. When body temperature rises, muscles attached to the scrotum relax, causing the testes to lower away from the body. If body temperature lowers, the muscles tighten, moving the testes closer to the body for warmth. Tight clothing that holds the testes too close to the body may interfere with sperm production.

READING CHECK

Explain How does fertilization occur?

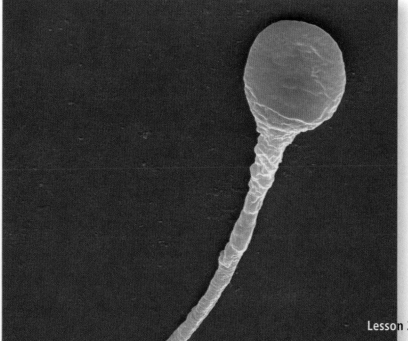

■ **Figure 16.3** Sperm are the male reproductive cells. Millions are produced each day. *What fluid contains sperm?*

Figure 16.4 Testosterone may spur the development of muscles in adolescence. *What else does testosterone influence?*

When a male begins to produce sperm, he may experience nocturnal emissions, an ejaculation that occurs when sperm are released during sleep. This is a normal occurrence to relieve the buildup of pressure as sperm begin to produce during puberty.

Internal Reproductive Organs

Although sperm are produced in the testes, which are suspended outside the body, they must travel through several structures inside the body before they are released. These structures include the vas deferens, the urethra, the seminal vesicles, and the prostate and Cowper's glands. **Figure 16.5** shows the path taken by sperm cells from the testes until they are released from the body.

Maintaining Reproductive Health

Main Idea Male reproductive health involves care and monitoring throughout a male's lifetime.

As with any other body system, the male reproductive system needs care. Ways of caring for the male reproductive system include practicing good personal hygiene, using adequate protection, practicing self-examination, and getting regular medical checkups.

- **Bathe regularly.** Males should shower or bathe daily, thoroughly cleansing the penis and scrotum. Uncircumcised males should take care to wash under the foreskin.

- **Wear protective equipment.** Use a protective cup or athletic supporter during physical activities to shield the external reproductive organs.

FITNESS ZONE

It's really important to me to succeed in whatever I start out to do. I know I need to take care of myself and keep in shape, even when I'm not feeling up to it. I am going out for track this year so I will be motivated to run every day. I know if I choose a sport I really like, I won't give up. For more physical activity ideas, visit the Fitness Handbook and Fitness Zone sites in ConnectEd.

Figure 16.5 Male Reproductive System

The internal structures of the male reproductive system work together to promote the delivery of sperm.

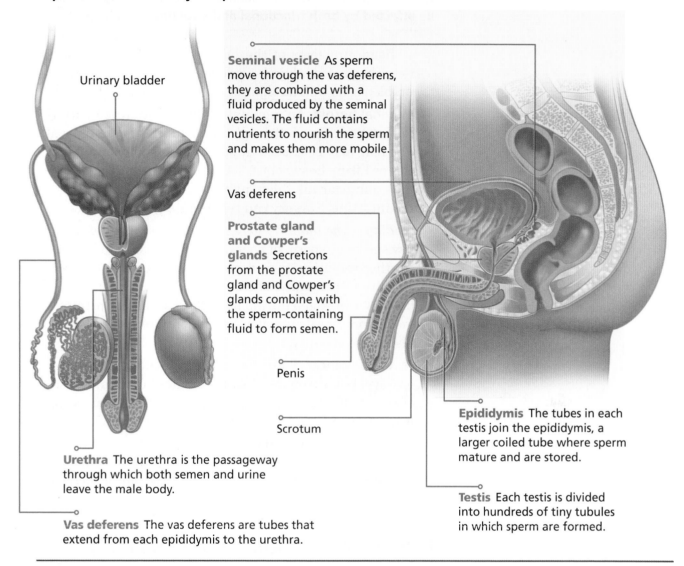

Urinary bladder

Seminal vesicle As sperm move through the vas deferens, they are combined with a fluid produced by the seminal vesicles. The fluid contains nutrients to nourish the sperm and makes them more mobile.

Vas deferens

Prostate gland and Cowper's glands Secretions from the prostate gland and Cowper's glands combine with the sperm-containing fluid to form semen.

Penis

Scrotum

Urethra The urethra is the passageway through which both semen and urine leave the male body.

Vas deferens The vas deferens are tubes that extend from each epididymis to the urethra.

Epididymis The tubes in each testis join the epididymis, a larger coiled tube where sperm mature and are stored.

Testis Each testis is divided into hundreds of tiny tubules in which sperm are formed.

- **Practice abstinence.** Abstain from sexual activity before marriage to avoid contracting sexually transmitted diseases (STDs).

- **Perform regular self-examinations.** Check the scrotum and testicles monthly for signs of cancer, as directed on page 450. Report any change to a physician. Even though lumps do not always mean cancer is present, early detection usually leads to successful treatment.

- **Get regular checkups.** All males should have regular checkups by a physician every 12 to 18 months. If an abnormality is found, the patient will be referred to a urologist, who specializes in care and problems of the male reproductive system.

READING CHECK

Identify What are the benefits of practicing abstinence to protect the reproductive system?

Male Reproductive System Problems

Main Idea The organs of the male reproductive system can be affected by both functional and structural problems.

Some problems of the male reproductive system are described below. Males should watch for the signs of these problems, as well as signs of infections from STDs.

- **Inguinal hernia.** An inguinal hernia occurs when part of the intestines push through a tear in the abdominal wall. The tear may be caused by straining the abdominal muscles or lifting heavy objects. Symptoms include a lump in the groin near the thigh, pain in the groin, or blockage of the intestine. Surgery can repair an inguinal hernia.

- **Sterility** is *the inability to reproduce*, as a result of too few sperm or sperm of poor quality. Exposure to X-rays or other radiation, toxic chemicals, and lead, can cause sterility. Other causes include hormonal imbalances, mumps contracted during adulthood, or using certain medications or drugs such as anabolic steroids. STDs can also cause sterility.

- **Testicular cancer.** Testicular cancer can affect males of any age, but occurs most often in males between the ages of 14 and 40. With early detection, most testicular cancer is treatable through surgery, radiation, or chemotherapy.

- **Prostate problems and prostate cancer.** The prostate gland can become enlarged as a result of an infection, a tumor, or age. Early detection of prostate cancer increases survival rates.

How to Do a Testicular Self-Exam (TSE)

The American Cancer Society recommends that males perform a self-exam for testicular cancer once a month.

1. Standing in front of a mirror, look for swelling. Examine each testicle with both hands. Roll the testicle gently between the thumbs and forefingers.

2. Cancerous lumps usually are found on the side of the testicle but can appear on the front. Find the epididymis, the soft tubelike structure behind each testicle, so that you won't mistake it for a lump.

3. Most lumps are not cancerous. If you do find a lump or experience pain or swelling, however, consult a health care professional.

 **READING CHECK**

Explain Why is it important to know where the epididymis is when doing a testicular exam?

Real World CONNECTION

TSE Awareness Campaign

To raise awareness of the importance of performing a monthly testicular self-exam (TSE), you will work with a group to develop an awareness program. Perform an online search to find facts such as risk factors and other facts about testicular cancer.

Activity Technology

Gather the information that was obtained from the online searches and what has been learned from the text and organize into groups of three or four students. As a group, create a podcast that has both audio and visual components for a public service announcement. Groups should work together to make sure that the campaign presents consistent information featuring the same key points.

1. Create a public service announcement script, blog, or podcast that raises awareness of TSE.

2. Create slides for a multimedia presentation about the warning signs of testicular cancer. Ask for permission to place the podcast and the multimedia presentation on the school's Web site.

3. Present the group's podcast and multimedia presentation to your class.

LESSON 2 ASSESSMENT

 After You Read

Reviewing Facts and Vocabulary

1. Sperm cannot survive at body temperature. How does the body protect sperm from heat?

2. What is *sterility*?

3. What are the vas deferens?

Thinking Critically

4. **Infer.** How might giving a mumps vaccine to a boy help protect his reproductive health later?

5. **Compare and Contrast.** How might a man's reproductive health concerns change at different periods of his life?

6. **Distinguish.** What is the difference between semen and sperm?

Applying Health Skills

7. **Practicing Healthful Behaviors.** Describe some behaviors that can help males maintain their reproductive health.

Writing Critically

8. **Persuasive.** Write an article persuading males of the need for protective equipment during football and other sports. Describe how shoulder pads, knee pads, and protective cups help prevent injury.

Real Life Issues ....

After completing the lesson, review and analyze your response to the Real Life Issues question on page 446.

GUIDE TO READING

BIG Idea *The female reproduction system matures at puberty and enables women to reproduce.*

Before You Read

Create a T-Chart. Set up a T-chart like the one pictured below to organize information about the parts of the female reproductive system and their functions.

Part	Function

New Vocabulary

▶ eggs
▶ ovaries
▶ uterus
▶ ovulation
▶ fallopian tubes
▶ vagina
▶ menstruation
▶ cervix

The Female Reproductive System

Real Life Issues

Being Teased. Jody and her friend Sandra decide to try out for the girls' basketball team at school. Later, however, Sandra tells Jody that she has changed her mind. Sandra is embarrassed to shower and change in the locker room because she's developing at a slower pace than most of the other girls. She's been teased by some of the girls and feels self-conscious in the locker room.

Writing *Write a supportive note to Sandra telling her that all girls mature at various rates. Be sure your note respects Sandra as she is and encourages her to understand she is not alone.*

Female Reproductive Organs

Main Idea The organs of the female reproductive system enable pregnancy to occur with the first monthly ovulation.

The female reproductive system has several functions, including producing female sex hormones and storing the **eggs**, *female gametes* or ova (singular: *ovum*). The **ovaries** are *the female sex glands that store the ova and produce female sex hormones.* They are located on each side of the **uterus**, *the hollow, muscular, pear-shaped organ that nourishes and protects a fertilized ovum until birth.* A female at birth has more than 400,000 immature ova. At puberty, the pituitary gland produces hormones that cause these ova to mature. **Ovulation** is *the process of releasing a mature ovum into the fallopian tube each month.*

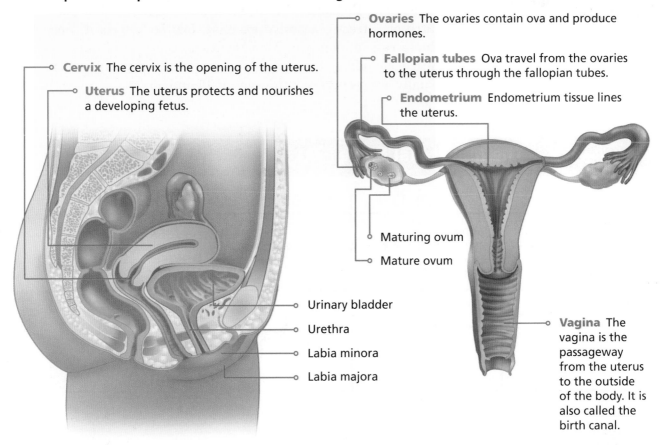

■ **Figure 16.7** Each month, an ovum is released that may unite with sperm, as shown here, in the process of fertilization. *If fertilization occurs, what type of cell is produced?*

Colin Anderson/Brand X Pictures

Female Reproductive Organs

The structures of the female reproductive system are shown in **Figure 16.8**. A mature ovum is released from an ovary and moves into one of the two **fallopian tubes**—*a pair of tubes with fingerlike projections that draw in the ovum.*

Figure 16.8 Female Reproductive System

The female reproductive system produces egg cells called ova, and each month provides a place for a fertilized ovum to grow.

Ovaries The ovaries contain ova and produce hormones.

Cervix The cervix is the opening of the uterus.

Fallopian tubes Ova travel from the ovaries to the uterus through the fallopian tubes.

Uterus The uterus protects and nourishes a developing fetus.

Endometrium Endometrium tissue lines the uterus.

Maturing ovum

Mature ovum

Urinary bladder

Urethra

Labia minora

Labia majora

Vagina The vagina is the passageway from the uterus to the outside of the body. It is also called the birth canal.

Tiny hairlike structures called *cilia* work to move the ovum with the help of muscular contractions in the fallopian tubes. Sperm from the male enter the female reproductive system through the **vagina**, *a muscular, elastic passageway that extends from the uterus to the outside of the body.*

If sperm are present in the fallopian tubes, the sperm cell and ovum may unite, resulting in fertilization. The fertilization of an egg by a sperm produces a cell called a *zygote.* When the zygote leaves the fallopian tube, it enters the uterus. The zygote attaches itself to the uterine wall. The uterine wall thickens with blood to nourish the zygote as it grows. The fetus remains in the uterus until birth.

Menstruation

After a female matures, the uterus prepares each month for possible pregnancy. If pregnancy doesn't occur, the thickened lining of the uterus, called the *endometrium,* breaks down into blood, tissue, and fluids. **Figure 16.9** shows the cycle of **menstruation**, *the shedding of the uterine lining.* The endometrium tissues pass through the **cervix**, *the opening to the uterus,* and into the vagina. Females wear sanitary pads or tampons to absorb the blood flow.

Most females begin their first menstrual cycle between the ages of 10 and 15. The menstrual cycle may be irregular at first. As a female matures, it usually becomes more predictable. Endocrine hormones control the cycle, but poor nutrition, stress, excessive exercise, low body weight, and illness can influence it. Menstruation occurs from puberty until *menopause,* the end of the reproductive years, which usually occurs between the ages of 45 and 55.

READING CHECK

Describe How does the uterine wall prepare for the zygote?

| Figure 16.9 | **The Menstrual Cycle** |

Days 1–8	Days 9–13	Day 14	Days 15–28
The cycle begins with the first day of mentruation.	The hormones FSH and LH cause an egg to mature in one of the ovaries.	Ovulation occurs and the mature egg is released into one of the fallopian tubes.	The egg travels through the fallopian tube to the uterus. If the egg is not fertilized the cycle starts again.

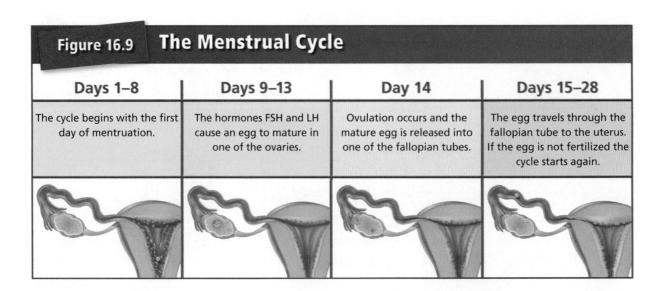

Maintaining Reproductive Health

Main Idea Good hygiene, breast self-exams, and abstinence from sexual activity help female reproductive system health.

Sound health practices, such as those described below, will help females care for their reproductive systems.

- **Bathe regularly.** It is especially important to shower or bathe daily, and change tampons or sanitary pads every few hours during the menstrual period.

- **Have regular medical exams.** Regular medical exams will include a test (Pap smear) for cancerous cells on the cervix, as well as a mammogram to test for breast cancer. Report any pain, discharge, or other signs of infection to your health care provider as soon as possible.

- **Practice abstinence.** Abstain from sexual activity to avoid unplanned pregnancy and STDs.

Breast Self-Exam

Breast cancer is the most common cancer and the second leading cause of death, after lung cancer, for women in the United States. The American Cancer Society recommends that females examine their breasts once a month, right after the menstrual period, when breasts are not tender. Early detection is critical for successful treatment of breast cancer. See **Figure 16.10** and follow these steps:

1. Lie down with a pillow under your right shoulder. Put your right arm behind your head. Place the three middle finger pads of your left hand on your right breast. Move your fingers in a circular motion, pressing first with light, then medium, then firm pressure. Feel for any lumps or thickening in the breast. Follow this process in an up-and-down path over the breast. Be sure to check all of the breast tissue, from the underarm edge to the middle of the chest bone, and from the collarbone to ribs. Repeat, using your right hand on your left breast.

2. Stand in front of a mirror with your hands pressed firmly on your hips. Inspect your breasts for *any* changes in size, shape, or appearance. Look for dimpling, rash, puckering or scaliness of the skin or nipple, or discharge.

3. Next, raise your arms over your head (palms pressed together), and look for changes.

4. Examine your underarms with your arms only slightly raised so you can more easily feel these areas.

READING CHECK

Describe What are three important steps for maintaining reproductive health in females?

Figure 16.10

Breast Self-Exam

In a vertical pattern, check from the underarm to the chest bone, and from the collarbone to the ribs.
How can the underarms be checked?

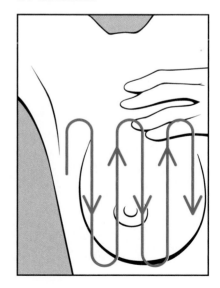

Female Reproductive System Problems

Main Idea Several disorders can affect the female reproductive system, and some can lead to infertility.

Both menstrual cramps and premenstrual syndrome are common in females. Toxic shock syndrome is uncommon.

- **Menstrual cramps** sometimes occur at the beginning of a menstrual period. Light exercise or applying a heating pad to the abdominal area may help relieve cramps. If cramps are severe, a health care professional may recommend an over-the-counter or prescription medicine.

- **Premenstrual syndrome (PMS)** is a disorder caused by hormonal changes. Symptoms include anxiety, irritability, bloating, weight gain, depression, mood swings, and fatigue. Regular physical activity and good nutrition may reduce the severity of symptoms.

- **Toxic shock syndrome (TSS)** is a rare but serious bacterial infection that affects the immune system and the liver. It can be fatal. To reduce TSS risk, use tampons with the lowest absorbency and change them often. If symptoms occur, such as fever, vomiting, diarrhea, rash, red eyes, dizziness, and muscle aches, see a doctor.

Infertility and Other Disorders

Infertility in females can have several causes.

- **Endometriosis** occurs when uterine tissue grows in the ovaries, fallopian tubes, or the lining of the pelvic cavity.

- **Sexually transmitted diseases** spread during sexual contact. Untreated STDs, such as gonorrhea and chlamydia, are associated with pelvic inflammatory disease (PID) and may cause infertility. Abstinence from sexual activity until marriage is the only way to avoid STDs.

- **Vaginitis** results in discharge, odor, pain, itching, or burning. *Candida* (yeast infection) and bacterial vaginosis are two common forms of vaginitis.

- **Ovarian cysts** are fluid-filled sacs on the ovary. Small, noncancerous cysts may disappear on their own. Larger cysts may have to be removed surgically.

- **Cervical, uterine, and ovarian cancers.** Early sexual activity and STDs such as human papillomavirus (HPV) increase the risk of cervical cancer. Regular exams are important for early detection and treatment. The Food and Drug Administration has approved a vaccine that prevents infection from four strains of the HPV virus.

Health Skills Activity

Communication Skills

Asking Difficult Questions

Jenny has just arrived for her yearly physical. She has been a patient of Dr. Silvio's since she was young and feels comfortable asking her female doctor questions about her health. During this exam, Jenny has a sensitive question to ask Dr. Silvio. She feels uncomfortable asking her father

the question, even though they've grown close since her mother died of cancer one year ago. Jenny feels awkward discussing a certain personal problem with him: her menstrual period. She wonders to herself: *Should I tell Dr. Silvio that I haven't started menstruating yet?*

Writing Write a dialogue in which Jenny talks about her concerns. Include a response from Dr. Silvio that is thoughtful and communicates that Jenny has nothing to worry about. Use these guidelines for effective communication.

1. Use "I" messages.
2. Speak calmly and clearly.
3. Be respectful.
4. Listen carefully and ask appropriate questions.
5. Use appropriate body language.

LESSON 3 ASSESSMENT

 After You Read

Reviewing Facts and Vocabulary

1. What is the function of the uterus?
2. Distinguish between *ova, ovaries,* and *ovulation.*
3. Identify a kind of cancer of the female reproductive system that is linked to a sexually transmitted disease (STD).

Critical Thinking

4. **Distinguish.** What is the difference between menstrual cramps and PMS?
5. **Infer.** Why do blocked fallopian tubes often result in infertility?

Applying Health Skills

6. **Advocacy.** Create a brochure that educates females about ways to promote reproductive health. Include preventive care such as hygiene, mammograms, and Pap smears.

Writing Critically

7. **Persuasive.** Some students in Mrs. Garcia's class are uncomfortable learning about the reproductive system of the opposite gender. Write a persuasive letter explaining why this education is important.

Real Life Issues ....

After completing the lesson, review and analyze your response to the Real Life Issues question on page 452.

Hands-On HEALTH

Activity Peer-to-Peer Education

Create a lesson for a class of middle school students to introduce structures, functions, and care of the male and female reproductive systems. Be prepared for "tough" questions that sixth graders might ask. Use communication skills appropriate for that age group.

What You'll Need

- computer with Internet access
- paper and pens or pencils
- poster paper, construction paper, markers, glue, tape, scissors
- props: paper cups, tennis balls, string

What You'll Do

Step 1

Work in groups to research information and outline your presentation. Include facts to support each point.

Step 2

Create visual aids and props, for example, to show how an egg cell passes through the female reproductive system.

Step 3

Include a clear, organized health message. Be sure the language is age appropriate.

Apply and Conclude

Discuss how your presentation will help younger students identify healthful lifestyle behaviors.

Checklist: Communication Skills

- ✔ Interaction between individuals
- ✔ Clear, organized message
- ✔ Respectful tone
- ✔ Listening skills
- ✔ Appropriate body language

LESSON **1**

The Endocrine System

Key Concepts

▶ The endocrine glands are organs or groups of cells that secrete hormones into the bloodstream.

▶ To care for your endocrine system, get plenty of rest, eat well, and avoid stress.

▶ Problems of the endocrine system include diabetes mellitus, hypothyroidism, goiter, and growth disorders.

Vocabulary

▶ endocrine glands (p. 442)
▶ hormones (p. 442)
▶ thyroid gland (p. 443)
▶ parathyroid glands (p. 443)
▶ pancreas (p. 443)
▶ pituitary gland (p. 443)
▶ adrenal glands (p. 444)

LESSON **2**

The Male Reproductive System

Key Concepts

▶ The external organs include the penis, testes, scrotum, seminal vesicles, vas deferens, epididymis, prostate and Cowper's glands, and urethra.

▶ Care of the male reproductive system involves getting medical checkups, bathing regularly, protecting against injury, performing self-exams, and abstaining from sexual activity.

▶ Problems include inguinal hernias, sterility, and testicular and prostate cancer.

Vocabulary

▶ sperm (p. 446)
▶ testosterone (p. 446)
▶ testes (p. 447)
▶ scrotum (p. 447)
▶ penis (p. 447)
▶ semen (p. 447)
▶ sterility (p. 450)

LESSON **3**

The Female Reproductive System

Key Concepts

▶ The major structures of the female reproductive system include the ovaries, fallopian tubes, uterus, cervix, and vagina.

▶ *Menstruation* is the shedding of the uterine lining.

▶ Breast self-exams should be done in two phases: lying down and standing in front of a mirror.

▶ Problems include menstrual cramps, PMS, vaginitis, infertility-related disorders such as endometriosis and PID, and cancer.

Vocabulary

▶ eggs (p. 452)
▶ ovaries (p. 452)
▶ uterus (p. 452)
▶ ovulation (p. 452)
▶ fallopian tubes (p. 453)
▶ vagina (p. 454)
▶ menstruation (p. 454)
▶ cervix (p. 454)

LESSON **1**

Vocabulary Review

Correct the sentences below by replacing the italicized term with the correct vocabulary term.

1. The *hypothalamus* produces hormones that regulate metabolism, body heat, and bone growth.

2. The *adrenal glands* regulate calcium.

3. The *thymus* regulates and controls the activities of all other endocrine glands.

Understanding Key Concepts

After reading the question or statement, select the correct answer.

4. What is the main role of the pituitary gland?
   a. Controls sleep
   b. Helps digestion
   c. Regulates other endocrine glands
   d. Adjusts water balance

5. Which gland is involved with the release of epinephrine?
   a. Thyroid
   b. Adrenal glands
   c. Pineal gland
   d. Thymus

Thinking Critically

After reading the question or statement, write a short answer using complete sentences.

6. **Infer.** What might happen if the adrenal glands stop regulating the body's salt and water balance?

7. **Analyze.** If you suddenly began having trouble sleeping, how might the pineal gland be involved?

8. **Infer.** Epinephrine helps you respond to dangerous situations. Why might it help to stop digesting food if you are in danger?

9. **Analyze.** What would happen if scientists applied luteinizing hormone to a female's ovary cells and to a male's testes cells?

10. **Compare and Contrast.** Discuss similarities and differences of the adrenal cortex and adrenal medulla.

LESSON **2**

Vocabulary Review

Use the vocabulary terms listed on page 459 to complete the following statements.

11. The _____ is the skin sac that holds the testes.

12. _____ is a thick fluid containing sperm from the male reproductive system.

13. The inability to produce children is called _____.

Understanding Key Concepts

After reading the question or statement, select the correct answer.

14. What is an inguinal hernia?
   a. A separation where part of the intestine pushes into the abdominal wall
   b. The inability to reproduce
   c. A type of cancer affecting the prostate
   d. An STD

15. How often should males conduct a testicular self-exam?
    a. Once a day
    b. Once a week
    c. Once a month
    d. Once a year

16. Where are sperm formed?
    a. In the urethra
    b. In the epididymis
    c. In the penis
    d. In the testes

17. What is the passageway through which both semen and urine leave the body?
    a. The urethra
    b. The testes
    c. The seminal vesicles
    d. The epididymis

Thinking Critically

After reading the question or statement, write a short answer using complete sentences.

18. **Analyze.** How does the scrotum respond to temperature, and for what purpose?

19. **Identify.** Where are you most likely to find a cancerous lump on a testicle?

20. **Infer.** What should a male infer if he begins to develop facial hair?

21. **Compare and Contrast.** How are testicular cancer and prostate cancer different?

22. **Infer.** Why is it important to see a health care provider right away if a testicular lump is discovered?

LESSON 3

Vocabulary Review

Choose the correct term in the sentences below.

23. The *ovaries / fallopian tubes* produce hormones.

24. *Menstruation / Ovulation* is the process of releasing a mature ovum into the fallopian tube each month.

25. The *cervix / uterus* is the hollow, muscular, pear-shaped organ inside a female's body.

26. *Toxic shock syndrome / Premenstrual syndrome* is a rare but serious bacterial infection.

Understanding Key Concepts

After reading the question or statement, select the correct answer.

27. What type of disease, if left untreated, is associated with pelvic inflammatory disease?
    a. STDs
    b. Vaginitis
    c. Ovarian cancer
    d. Ovarian cysts

28. What is the opening to the uterus called?
    a. Ovum
    b. Bladder
    c. Endometrium
    d. Cervix

29. What happens to the uterine wall as it prepares for a zygote?
    a. The wall shrinks.
    b. The wall dissolves.
    c. The wall thickens.
    d. The wall sheds skin cells.

30. How many ova mature each month in a female reproductive system?
    a. 1
    b. 100
    c. 400,000
    d. Millions

Thinking Critically

After reading the question or statement, write a short answer using complete sentences.

31. **Infer.** Charlotte and her friend Karen are the same age. Charlotte has started menstruating, but Karen has not. What could be the cause?

32. **Explain.** How can practicing healthful behaviors help a female maintain a healthy reproductive system and even prevent some infertility problems later in life?

33. **Analyze.** Why is it important for females to conduct a breast exam after a menstrual period ends?

34. **Synthesize.** How can good hygiene help prevent toxic shock syndrome?

Technology PROJECT-BASED ASSESSMENT

Care of the Reproductive System

Background

When it comes to serious health issues, teens often think, "It can't happen to me." Teens are not immune to developing serious health problems, such as those that can affect the male or female reproductive systems. For this project, small groups of students will create a multimedia presentation providing teens with information on how to prevent reproductive health problems.

Task

Create a multimedia presentation that illustrates a problem of the reproductive system. (If you are female, select a problem of the female reproductive system; if you are male, choose a problem of the male reproductive system.)

Audience

Students in your school and community

Purpose

Help teens become better informed about how to prevent problems related to the male and female reproductive systems.

Procedure

1. Review the student text and choose a particular problem related to the male or female reproductive system.

2. Research the problem online. Identify the cause, symptoms, and treatments (historical and present-day), as well as methods by which the occurrence of the problem could be reduced.

3. Search through approved Web sites to find suitable illustrations and video clips.

4. As a group, organize the relevant information and illustrations to create an informative and attractive multimedia presentation.

5. Present your presentation to your class. Ask for permission to possibly place the presentation on your school's Web site.

Math Practice

Reading Tables. The time between fertilization of an egg by a sperm and birth is a known as a *gestation period*. This amount of time varies from organism to organism. The table gives the average gestation periods (in days) for several different types of animals.

Animal	Gestation Period
Hamster	16.5 days
Ferret	42 days
Coyote	63 days
Lion	108 days
Human	267 days
Horse	337 days
Camel	406 days

1. What is the median of the gestation periods in the table?
   A. 108 days
   B. 177 days
   C. 389.5 days
   D. There is no median because all the numbers are different.

2. The actual gestation period can be stated as a range of days. For a human, that range is from 250 to 285 days. If the value in the table is the average gestation period, the value is the
   A. median.   **C.** mean.
   B. mode.   **D.** first quartile.

3. Compare the size of each animal listed to its gestation period. Predict the relative gestation period of a rhinoceros. Explain your prediction.

Reading/Writing Practice

Understand and Apply. Read the passage below, and then answer the questions.

Gigantism is a problem of the endocrine system caused when the pituitary gland secretes too much growth hormone during childhood before the bones have completed their growth cycle. As a result, the body's long bones become overdeveloped. The person grows to an abnormally tall height. This very rare disorder can be the result of a tumor in the pituitary gland.

A related disorder, acromegaly (ak-roh-MEG-uh-lee), occurs when the production of growth hormone continues after the normal growth cycle has ended. People with acromegaly experience abnormal growth of bones in the face, hands, feet, and skull.

1. Which outline best represents the passage?
   A. I. Gigantism
       A. Famous Giants
       B. Symptoms
   B. I. Endocrine Problems
       A. Gigantism
       B. Excessive Growth Hormone
   C. I. Growth Disorders
       A. Gigantism
       B. Acromegaly
   D. I. Functions of the Pituitary Gland
       A. Growth Hormone Production
       B. Sex Hormone Production

2. In which type of publication would this passage most likely appear?
   A. Encyclopedia   **C.** Letter from a doctor
   B. Fictional novel   **D.** A pamphlet

3. Write a short story about a person with gigantism. Give details about the disorder.

National Education Standards

Math: Measurement and Data
Language Arts: LACC.910.L.3.6, LACC.910.RL.2.4

CAREER CORNER Medical Support Careers

Dental Hygienist

Dental hygienists work in a dentist's office. They record your dental history, examine your mouth and teeth for obvious problems, take X-rays, and clean your teeth. A dental hygienist also teaches you how to take care of your teeth and mouth.

To learn more about becoming a dental hygienist, take science and communications classes in high school. After high school, a dental hygienist must complete a two-year course of study to earn an associate's degree.

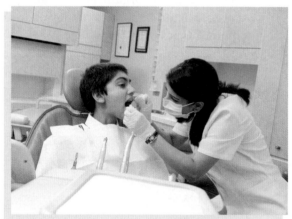

Physical Therapy Assistant

Physical therapy assistants work under the supervision of a physical therapist. They help patients practice movements, perform exercises, and learn to use mobility tools such as walkers. Physical therapy assistants must have patience and good communication skills. To learn more about becoming a physical therapy assistant, take human physiology and health classes in high school. Physical therapy assistants may need specialized training through an accredited program, and may need to pass a licensing exam.

Medical Laboratory Technician

A Medical Laboratory Technician (MLT) analyzes blood samples and often specializes in certain areas. In order to be a successful MLT, you must be precise, well organized, and attentive to detail.

To gain a greater understanding of the role of an MLT, take biology, chemistry, physics, and computer classes in high school. Most MLTs need a degree from a one- or two-year accredited program. After completing the required course work, you will need to pass a national certification test to qualify for employment.

CAREER SPOTLIGHT

Prosthetist

Paul Morton always knew what he wanted to do with his life. "I loved art, I liked to work with my hands, and I wanted to help people. And I grew up watching my dad." It's no wonder that Paul became a prosthetist, just like his dad.

Q. What do you do as a certified prosthetist?

A. I design, fabricate, and fit artificial limbs for patients. It might be for someone who lost a limb due to diabetes, someone who was born with a missing foot, or someone who was in a car accident.

Q. Do you need a college degree to do your job?

A. At my level, yes, you do. But you also can go for special training right after high school and be trained as a prosthetic technician, who focuses on building artificial limbs to precise specifications.

Q. How is your field changing?

A. Bionic technology has expanded. There are more opportunities for a person who wants to design and fabricate computer-controlled prosthetic devices.

Activity Beyond the Classroom

Writing Exploring Internet Resources. People often think that all medical professionals are doctors or nurses. On the contrary, many other kinds of medical careers exist. Use the Internet to research other careers in the medical or health field. What kinds of medical specialists are there? For example, who takes care of foot problems? Who operates scanning equipment, such as CAT, MRI, and ultrasound machines?

Create a brochure providing information on four careers you identified from your research. Include information on the type of education needed and how each profession helps promote physical health in the community.

McGraw-Hill Education

UNIT **6** Growth and Development

Chapter 17
The Beginning of the Life Cycle

Chapter 18
The Life Cycle Continues

UNIT PROJECT

Hunger Relief

Using Visuals **Project Bread** is dedicated to "alleviating, preventing, and ultimately ending hunger in Massachusetts." Project Bread provides funding for food for 400 emergency food programs, school breakfast and lunch programs, and 135 communities throughout Massachusetts. Project Bread's fundraising walk, the Walk for Hunger, attracts more than 50,000 walkers and 2,000 volunteers.

Get Involved. Learn about food banks and hunger relief agencies in your community. Find out how teens can volunteer. Share your findings with the class.

"All of us need to grow continuously in our lives."
— *Les Brown, author and motivational speaker*

CHAPTER 17

The Beginning of the Life Cycle

Ed-Imaging

Lesson 1
Prenatal Development and Care

BIG Idea *As a fetus develops during pregnancy, special care needs to be taken to ensure the fetus and mother remain healthy.*

Lesson 2
Heredity and Genetics

BIG Idea *Certain traits, such as eye and hair color, come from both of your parents.*

Lesson 3
Birth Through Childhood

BIG Idea *Infancy and childhood are times of great changes and growth.*

Activating Prior Knowledge

Using Visuals The mother in this photo is bonding with her baby daughter. In a few sentences, describe why it is important for parents to form strong bonds with their children at such a young age.

Chapter Launchers

Health in Action

Discuss the BIG Ideas

Think about how you would answer these questions:

▶ Why is the mother's health important to her fetus during pregnancy?

▶ What role did your parents have in your development?

▶ What changes do infants and children experience?

Assess Your Health

Read each statement. On a separate sheet of paper, write "yes," "sometimes," or "no" based on your typical behavior.

1. I understand why prenatal care is important to the health of the fetus.

2. I know why it is important for a pregnant female to remain active during pregnancy.

3. I recognize how heredity and genetic affects families.

4. I understand the development that occurs during the stages of childhood.

5. I recognize the need for childhood health screenings.

A "yes" response shows that you practice healthy behaviors. "Sometimes" indicates that you should analyze and possibly modify your behavior. A "no" response means that you should modify the behavior.

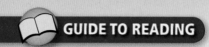

Prenatal Development and Care

GUIDE TO READING

BIG Idea *As a fetus develops during pregnancy, special care needs to be taken to ensure the fetus and mother remain healthy.*

Before You Read

Create a Table. Make a two-column table. Label the first column "Things to Avoid." Label the second column "Things to Do." Fill in the table as you read the lesson.

Things to Avoid	Things To Do

New Vocabulary

▶ fertilization
▶ implantation
▶ embryo
▶ fetus
▶ prenatal care
▶ fetal alcohol syndrome

Review Vocabulary

▶ sperm (Ch.16, L.2)
▶ egg (Ch.16, L.3)

Real Life Issues

Eating for Two. Amanda's older sister, Linda, is pregnant and has moved back home while her husband is overseas on a military assignment. Amanda's health class is starting a chapter on healthy pregnancies. Their mother says that good nutrition is very important for a healthy pregnancy, so they decide to plan menus that will be healthy for Linda and her growing baby.

Writing *Create a one-day menu and a shopping list that could help Amanda and her mother plan meals to keep Linda healthy during her pregnancy. Refer to the nutrition information you learned in Chapters 10 and 11.*

The Very Beginning

Main Idea A single cell, formed from one egg and one sperm, can grow into a complex human being.

The human body begins as one microscopic cell that is formed by *the union of a male sperm cell and a female egg cell,* called **fertilization**. This is also known as *conception.* The cell that results from fertilization is called a *zygote.*

The zygote begins to divide and travel through the fallopian tube, as shown in **Figure 17.1**. It divides many times, forming a cluster of cells by the time it reaches the uterus. Within a few days, **implantation**, *the process by which the zygote attaches to the uterine wall,* occurs. After about two weeks, the zygote becomes an **embryo** (EM-bree-oh), *a cluster of cells that develops between the third and eighth week of pregnancy.* This *group of developing cells* is called a **fetus** (FEE-tuhs) after about eight weeks.

Figure 17.1

Implantation

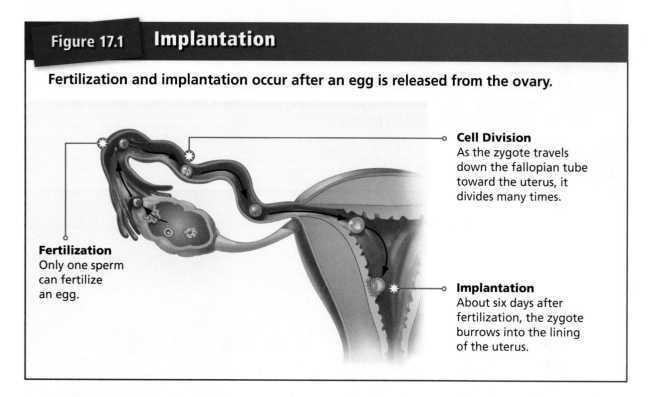

Fertilization and implantation occur after an egg is released from the ovary.

Cell Division
As the zygote travels down the fallopian tube toward the uterus, it divides many times.

Fertilization
Only one sperm can fertilize an egg.

Implantation
About six days after fertilization, the zygote burrows into the lining of the uterus.

The Growing Embryo

The cells of an embryo will continue to divide as it grows. **Eventually,** three layers of tissue are formed. Later, these layers develop into various body systems. One layer becomes the respiratory and digestive systems. A second layer develops into muscles, bones, blood vessels, and skin. The third layer forms the nervous system, sense organs, and mouth.

Two important structures form outside the embryo.

- The *amniotic sac* is a thin, fluid-filled membrane that surrounds and protects the developing embryo.

- The *umbilical cord* is a ropelike structure that connects the fetus with the mother's placenta. The *placenta* is thick, blood-rich tissue that lines the walls of the uterus during pregnancy and nourishes the embryo.

During pregnancy, the blood supply of the mother and the developing embryo are kept separate. Oxygen and nutrients are passed from the mother's blood to the embryo, and waste from the embryo is passed to the mother's blood. These wastes are excreted from the mother's body along with her own body wastes.

Substances that are harmful to a developing embryo can also pass through the umbilical cord. If a pregnant female uses tobacco, alcohol, or other drugs, those substances can cross the placenta and harm the developing embryo.

The time from conception to birth usually takes about 280 days, or nine months. The nine months are divided into three *trimesters* of three months each. **Figure 17.2** on page 472 shows the major changes that occur in each trimester.

Academic Vocabulary

eventually *(adverb):* at an unspecified later time

Figure 17.2   Stages of Embryonic and Fetal Development

Fetal development occurs over nine months. The nine months are divided into three trimesters, lasting three months each. An example of a single birth fetus during each trimester of development is shown below.

First Trimester (0 to 14 weeks)	Major Changes
0–2 weeks	A zygote may float freely in the uterus for 48 hours before implanting. The spinal cord grows. The brain, ears, and arms begin to form. The heart begins to beat. 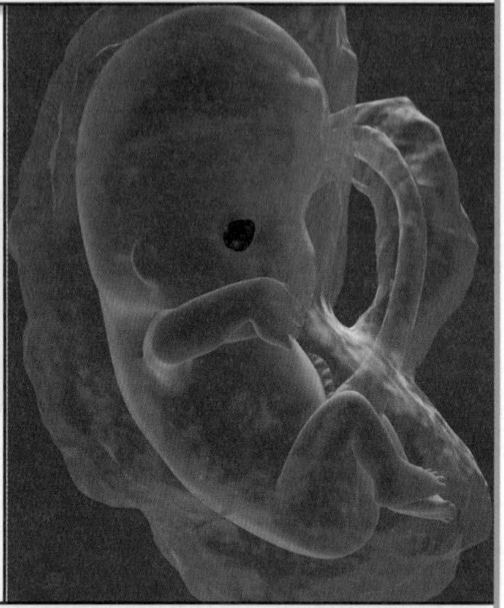
3–8 weeks	The embryo is about 1 inch long at 8 weeks. The mouth, nostrils, eyelids, hands, fingers, feet, and toes begin to form. The nervous system and cardiovascular system are functional.
9–14 weeks	The fetus develops a human profile. Sex organs, eyelids, fingernails, and toenails develop. By week 12 it can make crying motions and may suck its thumb.

Second Trimester (15 to 28 weeks)	Major Changes
15–20 weeks	The fetus can blink its eyes and becomes more active. The body begins to grow, growth of the head slows and the limbs reach full proportion. Eyebrows and eyelashes develop. 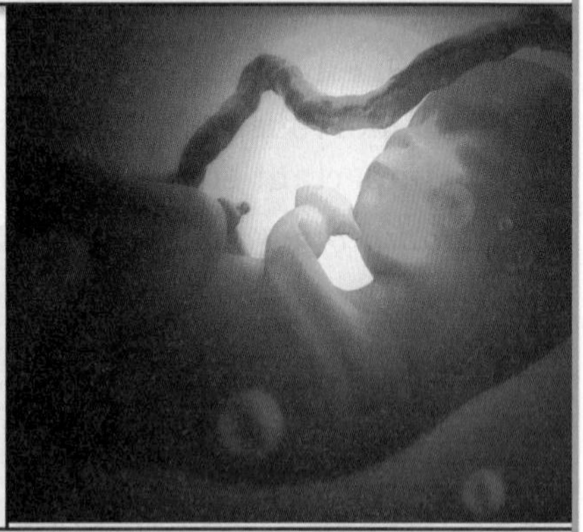
21–28 weeks	The fetus can hear conversations and has a regular cycle of waking and sleeping. Weight increases rapidly. The fetus is about 12 inches long and weighs a little more than 1 pound. The fetus may survive if born after 24 weeks, but will require special medical care.

Third Trimester (29 weeks to birth)	Major Changes
29–40 weeks	The fetus uses all five senses and begins to pass water from the bladder. Brain scans have shown that some fetuses dream during their periods of sleep in the eighth and ninth months of development. Approximately 266 days after conception, the baby weighs 6 to 9 pounds and is ready to be born. 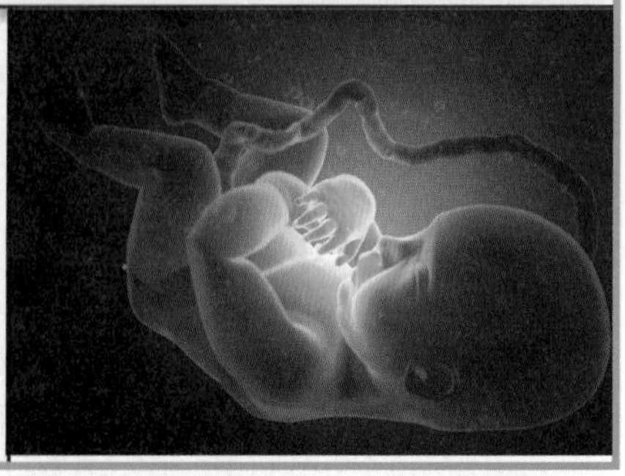

Multiple Births

In most cases, fertilization results in one embryo. Twins, triplets, and quadruplets, known as *multiple births,* can result when multiple embryos are formed. Identical twins result from a single zygote that splits into two separate embryos with identical traits and the same gender. Fraternal twins occur when two eggs are released and are fertilized by two different sperm. Fraternal twins can be different genders.

✔ READING CHECK

Describe What are some of the changes that occur during the second trimester?

A Healthy Pregnancy

Main Idea A pregnant female can maintain the health of her fetus in many different ways.

When a woman learns that she is pregnant, she should begin prenatal care to ensure her health and that of her growing baby. **Prenatal (pree-NAY-tuhl) care** refers to *the steps that a pregnant female can take to provide for her own health and the health of her baby.* Seeing a doctor regularly throughout the pregnancy will provide a new mother with the care and nutritional advice she needs.

What to Eat While Pregnant

An unborn baby receives nourishment from the mother. Pregnant females are encouraged to take prenatal vitamins to provide a balance of nutrients, such as:

- **Calcium** helps build strong bones and teeth, as well as healthy nerves, muscles, and developing heart rhythm.
- **Protein** helps form muscle and other tissue.
- **Iron** makes red blood cells and supplies oxygen to cells.
- **Vitamin A** helps in the growth of cells and bones and in eye development.
- **Vitamin B complex** aids in forming the nervous system.
- **Folic acid** is critical in development of the neural tube, which contains the central nervous system. It's recommended that all females of childbearing age consume 400 to 600 micrograms of folic acid daily.

Most pregnant females need to consume only an additional 300 calories per day to achieve a healthy weight gain during pregnancy. This is equivalent to drinking an extra 2½ cups of low-fat milk per day. Females at a healthy weight before becoming pregnant can gain between 25 and 35 pounds. Gaining too little weight can result in a small, undeveloped baby. Gaining too much weight can result in an early delivery. Extra weight also increases the mother's risk of high blood pressure, diabetes, and varicose veins.

■ **Figure 17.3** During pregnancy, good nutrition and rest keep both the mother and the developing child healthy. *Why are healthy snacks important to fetal development?*

Getty Images

Fitness During Pregnancy

Physical activity can help a pregnant female maintain a healthy weight during pregnancy. At the end of the pregnancy, it may become more difficult to maintain a fitness program. Before starting any exercise program, an expectant mother should discuss the importance of exercise and exercise programs with her health care provider.

■ **Figure 17.4** Regular exercise is an important part of a healthy pregnancy. *How can regular physical activity help the health of a pregnant female and the fetus?*

A Healthy Fetus

Main Idea Expectant mothers should avoid tobacco, alcohol, drugs, and environmental hazards.

An expectant mother should avoid substances that can harm her and her fetus. Tobacco, alcohol, and other drugs can be harmful to both mother and baby.

Avoid Tobacco Use

Smoking and using other tobacco products during pregnancy is harmful to the fetus. It is estimated that smoking accounts for up to 30 percent of low-birth-weight babies, 14 percent of premature births, and 10 percent of all infant deaths. Studies suggest that smoking may also affect growth, mental development, and behavior after a child is born. Research by the American Lung Association shows that pregnant females who are exposed repeatedly to secondhand smoke increase the risk of having a low-birth-weight baby.

Avoid Alcohol Use

When an expectant mother uses alcohol, so does her growing fetus. Alcohol passes through the umbilical cord to the fetus. The fetus, however, breaks down alcohol more slowly than the mother. This means that the alcohol level in the fetus's blood is higher, and it remains in the bloodstream for a longer period of time. A severe alcohol-related disorders is **fetal alcohol syndrome** (FAS), *a group of alcohol-related birth defects that includes both physical and mental problems.* Infants born with FAS may have learning, memory, and attention problems, as well as visual and hearing impairments.

Avoid Drug Use

Prescription or over-the-counter medications should be used only with the approval of a doctor or other health care professional. These substances can also harm the fetus.

Any use of illegal drugs poses a health risk to both the mother and the fetus. Drug abuse can harm the mother's health and make her less able to support the pregnancy. Drugs can also harm the development of the fetus. Infants born to mothers who use drugs may not grow at the same rate as infants born to mothers who do not use drugs. They may have respiratory or cardiovascular problems, mental impairments, or birth defects. In some cases, drug use may lead to the premature birth of the infant, or even a miscarriage. The baby may also be born addicted to the same drugs the mother used during pregnancy.

Avoid Hazards in the Environment

A pregnant female should avoid common hazardous substances in the environment. Family members can help by also being aware of these substances.

- **Lead.** Exposure to lead has been linked to miscarriage, low birth weight, mental disabilities, and behavior problems in children. Lead can be found in the paint of houses built before 1978, and in some glassware or dinnerware.

- **Mercury.** Pregnant females should avoid eating certain types of fish that are known to contain higher than average levels of mercury. These include shark, swordfish, and king mackerel.

- **Smog.** Medical studies have linked air pollution with birth defects, low birth weight, premature birth, stillbirth, and infant death. The greatest period of risk is the second month of pregnancy, when organs are developing.

- **Radiation.** Ionizing radiation, such as that found in X-rays, can affect growth and cause mental retardation.

Pregnant females should also use caution when using household chemicals. They should read all cleaning-product labels, wear gloves, and work in a well-ventilated area.

Complications of Pregnancy

Main Idea A pregnancy may have an unexpected outcome.

Most pregnancies result in the birth of a healthy baby. About 70 percent of all births occur through a vaginal delivery. Pregnancy complications, however, can result in a cesarean delivery, made through an incision in the mother's abdomen.

The complications of pregnancy can also result in a *premature birth*. This type of birth takes place at least three weeks before the due date. Serious complications may lead to *miscarriage,* the spontaneous expulsion of a fetus occurring before the twentieth week of pregnancy. The delivery of a fetus that has died after the twentieth week of pregnancy is called a *stillbirth.*

■ **Figure 17.5** Activities such as painting or using lead-based products should be avoided or done carefully during pregnancy. *Why is it important to keep a room well-ventilated while painting?*

Pregnancy Alert

LeAnn has just discovered she is pregnant with her first child. She and her husband, Walt, are both excited and scared. The thought of having a baby is overwhelming. LeAnn and Walt have begun listing everything they need to do. Both sets of parents are excited about the upcoming

birth. They've offered to help buy furniture and to help decorate the baby's room. Still, LeAnn and Walt have a lot to do before the baby is born. After her first prenatal visit to her doctor, LeAnn is more nervous than ever. There are so many things to think about while she is pregnant that she has started to become stressed. What can LeAnn and Walt do together to reduce her anxiety and help her focus on having a healthy pregnancy?

Writing You may know someone who has had an experience just like LeAnn's. Use the information you learned in this lesson to create a pamphlet on ways to plan for a healthy pregnancy. The following tips can be used as a guideline.

1. Identify substances to be avoided during pregnancy.
2. List ways to prepare for pregnancy and birth.
3. Develop strategies to incorporate physical activity.
4. Compare childbirth options.

READING CHECK

Explain Why is low birth weight a concern?

Sometimes, a medical reason causes a miscarriage or stillbirth. Using tobacco or drugs during pregnancy, however, can increase the risk of miscarriage or stillbirth. Receiving prenatal care during pregnancy can reduce the risk, or severity, of problems during pregnancy. During prenatal care, a doctor may be able to identify medical problems with the fetus before the birth. In some cases, medical care, and even surgery, may be performed on the baby before birth.

Other medical complications result from medical conditions affecting the pregnant female. *Gestational hypertension,* or high blood pressure during pregnancy, may occur after the twentieth week of pregnancy. A severe form of this is *preeclampsia.* Symptoms include high blood pressure, swelling, and large amounts of protein in the urine. Preeclampsia can prevent the placenta from getting enough blood to nourish the fetus. Treatment includes reducing blood pressure through bed rest or medication. Hospitalization may be necessary.

An *ectopic pregnancy* results when a zygote implants not in the uterus but in the fallopian tube, abdomen, ovary, or cervix. This makes it impossible for the fetus to receive nourishment and grow. An ectopic pregnancy cannot lead to the birth of a healthy fetus. It is also the number one cause of death in women in the first trimester of pregnancy.

Childbirth

Main Idea The birth of a baby takes place in three steps: labor, delivery, and afterbirth.

Expectant parents must decide where the birth will occur. Most births occur in hospital maternity wards staffed by nurses and doctors, with medical equipment to handle complications. Other options include birthing centers or home births, which offer a more comfortable environment. Midwives may attend home births and births at birthing centers.

As the birth approaches, the fetus becomes more crowded in the uterus. Birth occurs in three steps.

- **Step 1: Labor.** Muscle contractions of the uterus become regular, stronger, and closer together. This causes the cervix—the opening to the uterus—to dilate, or widen.

- **Step 2: Delivery.** Once the cervix is fully dilated, the baby passes through the birth canal and emerges from the mother's body. The baby takes its first breath and cries to clear its lungs of amniotic fluid.

- **Step 3: Afterbirth.** The placenta is still attached to the baby by the umbilical cord. Contractions, although weaker, will continue until the placenta (now called the *afterbirth*) is pushed from the mother's body.

■ **Figure 17.6** The birth of a healthy baby is a joyous event. *What steps can pregnant women take to help ensure a healthy pregnancy and delivery?*

 READING CHECK

Explain During delivery, what must happen before the baby can pass through the birth canal?

LESSON 1 ASSESSMENT

 After You Read

Reviewing Facts and Vocabulary

1. Describe an *embryo* and a *fetus.*
2. What is the difference between identical and fraternal twins?
3. What is the relationship between the placenta and the umbilical cord?

Thinking Critically

4. **Evaluate.** How is the diet of a pregnant female important to her growing fetus?
5. **Cause and Effect.** What are some of the risks that may occur if a pregnant female uses tobacco, alcohol, or drugs during her pregnancy?

Applying Health Skills

6. **Accessing Information.** Research library or Internet resources to learn more about gestational hypertension. Explain the warning signs and the need for proper treatment.

Writing Critically

7. **Persuasive.** Write a short essay from the point of view of a fetus persuading its mother to eat healthy foods during pregnancy.

Real Life Issues ....

After completing the lesson, review and analyze your response to the Real Life Issues question on page 470.

Heredity and Genetics

GUIDE TO READING

BIG Idea *Certain traits, such as eye and hair color, come from both of your parents.*

Before You Read

Make an Outline. Use the headings of this lesson to make an outline of what you'll read. Use a format like the one below to help you organize your notes.

I.
A.
1.
2.
B.
II.

New Vocabulary

▶ chromosomes
▶ genes
▶ DNA
▶ genetic disorders
▶ amniocentesis
▶ chorionic villi sampling
▶ gene therapy

Review Vocabulary

▶ heredity (Ch.1, L.2)

Real Life Issues

The Same Genes. Casey has never met Jason's brother, Sean, but has been asked to find him at the bus station and give him a ride back to school. Casey was glad to help out by picking up his friend's brother, but he isn't sure he will recognize Sean at the station. He is surprised when he spots a boy who looks almost identical to his friend.

Writing *Write a paragraph describing two related people you know, or have seen, that share common traits.*

Heredity

Main Idea Heredity is the passing of physical traits from parents to their children.

Each one of us inherits traits such as hair and eye color, as well as the shape of your earlobes, from your parents. Inherited traits, however, can also be influenced by the environment. For example, height is an inherited trait, but poor nutrition may limit growth during childhood. Other inherited traits that can be impacted by the environment include body size and the tendency for certain diseases, like diabetes.

Most of the cells in the human body contain a nucleus, or the control center of a cell. Inside each nucleus is a set of **chromosomes** (KROH-muh-sohmz), *thread-like structures found within the nucleus of a cell that carry the codes for inherited traits.* Most of the cells in the human body contain 46 chromosomes that are arranged in 23 pairs.

■ **Figure 17.7** Family members often share similar physical traits. *What determines your physical traits?*

Sections of chromosomes, called **genes**, are *the basic units of heredity.* Genes occur in pairs, just like chromosomes. One gene from each pair is inherited from each parent. You have thousands of genes in every cell of your body.

DNA

The chemical unit that makes up chromosomes is called **DNA**, or deoxyribonucleic (dee-AHK-si-ry-boh-noo-KLEE-ik) acid. All living things are made of DNA. DNA is made up of chemical building blocks arranged along a single molecule. Several of these molecules are linked together in a strand to form a DNA sequence, known as the genetic code. When a child is born, that child carries a combination of DNA sequencing from both parents. This DNA contains different proteins that result in individual traits. All the characteristics you have, such as your eye color, the amount of curl in your hair, and your height, are determined by your genetic code. This unique code is a combination of the DNA of both your parents. Only identical twins share the same DNA pattern.

Genetics and Fetal Development

Main Idea Chromosomes from a sperm and an egg unite to carry the hereditary traits from parents.

Passing on traits from parent to child involves genetics. Most human cells have 46 chromosomes, or 23 pairs. However, egg and sperm cells have half that number—23 chromosomes. When a sperm and egg unite during fertilization, the resulting zygote will have 46 chromosomes, 23 from each parent. These chromosomes carry the hereditary traits of the parents, which are passed on to their child. Heredity and the environment can affect human growth and development.

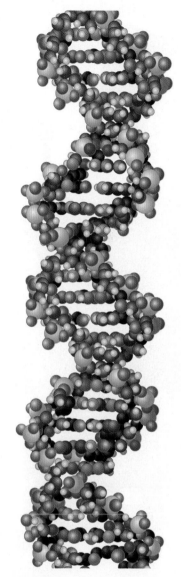

■ **Figure 17.8** DNA like this double helix, look like a long, twisted ladder. Nitrogen bases make up the rungs of this ladder. *What determines your own personal genetic code?*

A zygote divides many times, ultimately producing the trillions of cells that make up the human body. Between each cell division, each chromosome in the nucleus of the cell copies itself, producing two sets of the 46 chromosomes. The cell then divides, and the two sets of chromosomes separate. Each new cell then contains one complete set of the 46 chromosomes that are identical to the ones found in the first cell of the zygote.

Dominant and Recessive Genes

Each human trait is determined by at least one pair of genes. Some genes are *dominant,* while others are *recessive.* The traits of the dominant genes generally appear in the offspring when they are present. The traits of recessive genes usually appear only when the dominant genes are not present. For example, suppose an individual receives two genes for eye color, one for brown eyes and one for blue eyes. This individual will have brown eyes because the gene for brown eyes is dominant and the gene for blue eyes is recessive. If an individual has blue eyes, that means he or she has two recessive genes for blue eye color.

Genes and Gender

One pair of chromosomes determines gender. Females have two chromosomes that look exactly alike; these are called X chromosomes. Males have two different chromosomes, one shorter than the other. The shorter chromosome is the Y chromosome. The longer one is the X chromosome. See **Figure 17.9** to compare X and Y chromosomes.

Since sperm and egg cells contain only half the chromosomes of other cells, these cells have only one sex cell, not two. Because females have only X chromosomes, their egg cells contain only an X chromosome. Sperm, because they come from a male, contain either an X or a Y chromosome. Thus, the sperm from the male determines the gender of a child. If a sperm cell carries the X chromosome, the child will be a girl. If the sperm carries a Y chromosome, the child will be a boy.

■ **Figure 17.9** The body cells of a male have both an X chromosome and a Y chromosome. The body cells of a female have two X chromosomes. *Which parent determines the gender of a child, and why?*

Genetic Disorders

Main Idea Genetic disorders are caused by defects in genes.

FITNESS ZONE

In 1990, researchers working on the Human Genome Project began to identify all of the genes in human DNA. By 2003, they identified genes that are linked to more than 1,800 diseases. This knowledge can help diagnose, treat, and perhaps even prevent genetic disorders.

A person can inherit genes that contain a *mutation,* or abnormality. The mutation may have little or no effect on the person. These genetic mutations, however, may result in a birth defect or may increase the person's likelihood of developing a disease. These diseases, called **genetic disorders**, are *disorders caused partly or completely by a defect in genes.* Some genetic disorders are apparent at birth. Others may not show up until later in life. **Figure 17.10** lists some common genetic disorders.

Most genetic disorders cannot be cured, but some can be treated. Two technologies used to test for genetic disorders before birth are amniocentesis and chorionic villi sampling.

- **Amniocentesis** (am-nee-oh-sen-TEE-sis) is *a procedure in which a syringe is inserted through a pregnant female's abdominal wall to remove a sample of the amniotic fluid surrounding the developing fetus.* Doctors examine the chromosomes in fetal cells for genetic abnormalities. This test is performed 16 to 20 weeks after fertilization.

Figure 17.10	**Common Human Genetic Disorders**

Disorder	Characteristics
Sickle-cell anemia	Red blood cells have a sickle shape and clump together; may result in severe joint and abdominal pain, weakness, kidney disease, restricted blood flow
Tay-Sachs disease	Destruction of nervous system; blindness; paralysis; death during early childhood
Cystic fibrosis	Mucus clogs many organs, including lungs, liver, and pancreas; nutritional problems; serious respiratory infections and congestion
Down syndrome	Varying degrees of mental retardation, short stature, round face with upper eyelids that cover inner corners of the eyes
Hemophilia	Failure of blood to clot

- **Chorionic villi sampling** (kor-ee-ON-ik VIL-eye), or CVS, is *a procedure in which a small piece of membrane is removed from the chorion, a layer of tissue that develops into the placenta.* The tissue can be examined for genetic disorders or to determine the age and gender of the fetus. The procedure is done around the eighth week of fetal development.

It is also possible to test a child for genetic disorders after birth. For example, many states require that all newborns be tested for phenylketonuria (PKU). If PKU is diagnosed soon after birth, a baby's diet can be altered to stop possible mental retardation caused by this genetic disorder.

■ **Figure 17.11** Newborns can be checked for certain genetic disorders with the help of some very simple tests. *Why might it be difficult to diagnose a genetic disorder early in life?*

Battling Genetic Diseases

Main Idea ▸ Research is ongoing to correct genetic disorders.

The information obtained as a result of the Human Genome Project was an important first step in learning more about genetic diseases. With it, scientists have gained a greater understanding of how diseases progress. The information can also be used to identify people who may be susceptible to genetic diseases or disorders.

Genetic disorders occur when an individual is missing a functioning gene. Without the functioning gene, the body does not produce some of the substances it needs. One experimental treatment, **gene therapy**, is *the process of inserting normal genes into human cells to correct genetic disorders.* Scientists feel that once a defective gene is replaced with a normal gene, the cells can then begin producing the normal gene. At this time gene therapy is experimental. Placing parts of DNA from one organism into another is called *genetic engineering.*

Genetic Counseling

Genetic research has resulted in many ways to diagnose and treat genetic diseases. Genetic counselors can guide families of children with genetic disorders on treatment options. Genetic counseling offers options to people with a family history of some diseases. In Chapter 2, you learned about the importance of keeping a complete family medical history. With this information, genetic counselors can educate families about possible risks for certain diseases and guide families through their options. Having a faulty gene, however, does *not* guarantee the person will get the disease.

■ **Figure 17.12** Genetic research often takes place in a laboratory. *What sort of information might be learned in a genetics laboratory?*

 READING CHECK

Cause and Effect
What happens when a defective gene is replaced by a normal one?

Genetically Engineered Drugs

Genes used to treat diseases are not inserted directly into human beings. They are instead placed into other organisms, causing that organism to produce substances that can be used to treat human diseases and disorders. Through genetic engineering, some vaccines that can prevent disease have been produced.

LESSON 2 ASSESSMENT

After You Read

Reviewing Facts and Vocabulary

1. Define the terms *chromosomes* and *genes*.

2. How many chromosomes are found in most human cells? How many are found in egg and sperm cells?

3. Identify the difference between the chromosomes of a male and the chromosomes of a female.

Thinking Critically

4. **Evaluate.** When might a pregnant woman consider having CVS?

5. **Interpret.** The gene for brown hair is dominant, while the gene for blond hair is recessive. What genes does a brown-haired person have?

Applying Health Skills

6. **Accessing Information.** Use library or Internet resources to learn more about genetic research. Explain how this technology can prevent disease, and how it can impact personal, family, and community health.

Writing Critically

7. **Descriptive.** Write a short essay describing how genetic counseling and gene therapy might change someone's life.

Real Life Issues ....

After completing the lesson, review and analyze your response to the Real Life Issues question on page 478.

LESSON **3**

BIG Idea *Infancy and childhood are times of great changes and growth.*

Before You Read

Create a K-W-L Chart.
Make a three-column chart like the one below. In the first column, write what you **k**now about infancy and childhood. In the second column, write what you would **w**ant to know about this topic. As you read, fill in the third column describing what you have **l**earned.

K	W	L

New Vocabulary

▸ developmental tasks
▸ autonomy
▸ scoliosis

Birth Through Childhood

Real Life Issues ••••••••••••••••••••••••••

Trip Down Memory Lane. Henry and his parents are going to a family reunion. He's looking forward to seeing his grandparents, aunts, uncles, and cousins, all of whom live in other states. Earlier today he helped his mom put together some photo albums to take on the trip. At one point, she showed him a photo of a young child. It's Henry when he was five years old. He thought back to when he was younger and how he has changed since then.

Writing *Write three entries in Henry's diary as he remembers what he was like at the ages of 5, 8, and 10.*

Childhood

Main Idea Each child passes through four stages of development during infancy and childhood.

Our lives can be divided into eight developmental stages. The first four stages occur during infancy and childhood. Each of the eight stages is associated with certain **developmental tasks**, *events that need to happen in order for a person to continue growing toward becoming a healthy, mature adult.* The developmental tasks for each stage are summarized in **Figure 17.13**.

Infancy

Infancy is the time of fastest growth in a person's life. It is a time of learning: how to eat solid food, how to sit up, how to crawl, and how to walk. Infants also learn to trust others during this time.

Figure 17.13 Stages of Infancy and Childhood

Each stage of development is associated with a developmental task that involves a person's relationship with other people.

Infancy Birth to 12 months	Early Childhood Ages 1–3	Middle Childhood Ages 4–6	Late Childhood Ages 7–12
Opens and closes hands	Walks well	Dresses and undresses	Puberty may begin
May begin associating sounds with objects	Picks up objects without losing balance	Uses utensils to eat for most foods	Sensitivity about body image may begin
Imitates new word sounds	Throws balls overhead, but inaccurately	Becomes more independent	Develops sense of self
May walk a few steps	Draws recognizable pictures	Eager to explore the larger world	Recognizes unique personality traits
Experiences the five basic emotions	Begins showing defiance, disagreement	Craves praise and approval	Sense of competence develops
Forms strong attachment to parents	Behaves affectionately	Self-confidence grows	Becomes aware of dangers in the world
Begins to smile	May wish to help adults	Begins forming friendships	Deeper friendships develop
Wants companionship	Begins being bothered by fears	Becomes more outgoing and talkative	Relationships with parents change
Enjoys company of other children	Desires approval	Respects others' belongings	Begins facing moral decisions
Begins experiencing stranger anxiety	Bosses other children	May want to do things their own way	Peer pressure becomes stronger
Shows strong likes and dislikes	Takes part in brief group activities		

Early Childhood

During early childhood, children begin to feel proud of their accomplishments and are eager to try new tasks and to learn new things. During this stage, children also begin to learn to play as part of a group. Parents are encouraged to allow their children to try new things and to test their abilities. This helps a child in the early childhood stage develop a sense of **autonomy**, *the confidence that a person can control his or her own body, impulses, and environment.*

 READING CHECK

Infer How can a parent help a child develop a sense of autonomy?

Jill Braaten/McGraw-Hill Education

Middle Childhood

During middle childhood, children learn to initiate play rather than following the lead of others. Children at this stage must be taught to recognize emotions and practice expressing them in appropriate ways

Late Childhood

School becomes an important part of a child's life during late childhood. Children learn to get along with their peers, learn about different roles in society, and develop a conscience at this stage.

■ **Figure 17.14** Infants learn to trust and depend on others. *What do infants learn during the infancy stage of development?*

Childhood Health Screenings

Main Idea Many screening tests are performed in childhood to monitor the health and growth of a child.

Vision and hearing impairments are two problems that can affect a child's ability to learn and develop. Health screenings and immunizations can identify and prevent many problems that can affect development.

Vision and Hearing

Nearly one in every four school-aged children in the United States has a vision problem. The American Academy of Ophthalmology recommends that newborns receive a vision screening and that these screenings continue through childhood. Children may receive regular vision screenings at school. Oftentimes, school aged children do not receive a vision screening until age 18.

As well as vision problems, hearing impairment can also affect a childs ability to learn. Two or three in every 1,000 children in the United States are born with a hearing impairment severe enough to affect their language development. Some states require that infants are screened for hearing loss. Again, some school districts may provide children screenings to identify hearing impairments.

Scoliosis

Scoliosis, *an abnormal lateral, or side-to-side, curvature of the spine,* may begin in childhood and go unnoticed until a child is a teenager. The exact cause of scoliosis is unknown, but it is more common in girls. Many middle schools have developed screening methods to check students for scoliosis.

■ **Figure 17.15** Children may receive vision screenings through school, a pediatrician, or a health clinic. *Why is it important to get regular health screenings?*

Other Screenings

Children are tested for lead poisoning yearly until age four. Blood pressure screenings begin after age three. Children with a family history of cholesterol problems or anemia may also be screened for these conditions.

 READING CHECK

Explain Why are students screened for vision and hearing?

LESSON 3 ASSESSMENT

After You Read

Reviewing Facts and Vocabulary

1. What are some *developmental tasks* children learn in early childhood?

2. What is an important part of a child's life during late childhood?

3. What is *scoliosis*?

Thinking Critically

4. **Evaluate.** How can positive parenting affect the autonomy and independence of a child?

5. **Analyze.** What is the result when a parent allows a child autonomy?

Applying Health Skills

6. **Accessing Information.** Research library or Internet resources to learn more about vision and hearing screenings. Explain how these screenings could prevent problems later in life.

Writing Critically

7. **Descriptive.** Write a short essay about the changes a child will face from early childhood to late childhood.

Real Life Issues . . . .

After completing the lesson, review and analyze your response to the Real Life Issues question on page 484.

Hands-On HEALTH

Activity The Amazing Beginning of Human Life

This activity will help you understand vocabulary terms and identify valid resources that can be helpful for a new parent.

What You'll Need
- 40 or more index cards and a marker
- Internet and textbook

What You'll Do

Step 1
Review Chapter 17, identifying a minimum of 20 key vocabulary words. Write each vocabulary word on an index card and its definition on another card.

Step 2
Place all cards face down, mix them up, and on the teacher's command, turn over one card at a time, matching the word with its definition. The first team to match all the cards correctly wins.

Step 3
Your team will be assigned one lesson in Chapter 17. Create a resource list to learn more about the health topics. Access the resources by phone, or via the Internet and identify the type of help available from that resource.

Apply and Conclude
Share this list with your peers by creating a pamphlet listing valid resources.

Checklist: Accessing Information

- ✔ List all of your sources
- ✔ Evaluate the sources to determine their reliability
- ✔ Judge the appropriateness of your sources
- ✔ Describe and discuss the type of help available from your sources

Prenatal Development and Care

Key Concepts

▶ A human fetus begins with the joining of a female egg and a male sperm.

▶ A pregnant female must avoid tobacco, alcohol, drugs, and environmental hazards.

▶ Expectant parents have many options regarding the location and method of delivery of their baby.

▶ The birth of a baby takes place in three steps: labor, delivery, and afterbirth.

Vocabulary

▶ fertilization (p. 470)

▶ implantation (p. 470)

▶ embryo (p. 470)

▶ fetus (p. 470)

▶ prenatal care (p. 473)

▶ fetal alcohol syndrome (p. 474)

Heredity and Genetics

Key Concepts

▶ A person's DNA determines his or her individual characteristics.

▶ Children inherit genetic traits from their parents.

▶ Genetic disorders are caused by defects in genes.

▶ Genetic research provides an opportunity to correct some genetic disorders.

Vocabulary

▶ chromosomes (p. 478)

▶ genes (p. 479)

▶ DNA (p. 479)

▶ genetic disorders (p. 481)

▶ amniocentesis (p. 481)

▶ chorionic villi sampling (p. 482)

▶ gene therapy (p. 482)

Birth Through Childhood

Key Concepts

▶ Infants and children complete a variety of tasks during four stages of development.

▶ Health screenings during childhood can identify problems, and monitor growth and development.

▶ Keeping your health history and your immunizations current is important for your individual health.

Vocabulary

▶ developmental tasks (p. 484)

▶ autonomy (p. 485)

▶ scoliosis (p. 486)

LESSON 1

Vocabulary Review

Correct the sentences below by replacing the italicized term with the correct vocabulary term.

1. A(n) *fetus* is a cluster of cells that develops between the third and eighth week of pregnancy.

2. *Fetal alcohol syndrome* refers to the steps that a pregnant female can take to provide for her own health and the health of her baby.

3. The process by which a zygote attaches to the uterine wall is called *fertilization*.

Understanding Key Concepts

After reading the question or statement, select the correct answer.

4. Which of the following nutrients helps form the nervous system of an embryo?
   a. Calcium
   b. Vitamin A
   c. Folic acid
   d. Iron

5. Exposure to which of the following may cause birth defects in the second month of pregnancy?
   a. Lead
   b. Mercury
   c. Radiation
   d. Smog

6. What is the result when a zygote implants in the fallopian tube or ovary?
   a. Preeclampsia
   b. Ectopic pregnancy
   c. Miscarriage
   d. Stillbirth

Thinking Critically

After reading the question or statement, write a short answer using complete sentences.

7. **Analyze.** What is the role of prenatal care in protecting the health of the mother and the fetus?

8. **Evaluate.** What should parents consider when choosing a childbirth method?

9. **Explain.** How can drinking alcohol during pregnancy damage a fetus?

10. **Analyze.** What is the result if one healthy zygote splits into two?

11. **Compare and Contrast.** What are the similarities and differences between preeclampsia and an ectopic pregnancy?

LESSON 2

Vocabulary Review

Use the vocabulary terms listed on page 489 to complete the following statements.

12. _____ are the basic units of heredity.

13. The process of inserting normal genes into human cells to correct genetic disorders is called _____.

14. Disorders caused by a defect in genes are called _____.

Understanding Key Concepts

After reading the question or statement, select the correct answer.

15. How many chromosomes do most human cells have?
    a. 12
    b. 23
    c. 46
    d. 69

16. Where are chromosomes located?
    a. Within the DNA molecule
    b. In genes
    c. In the nucleus of a cell
    d. Outside a cell

17. What is the chemical compound that makes up genetic material?
    a. Genes
    b. DNA
    c. Chromosomes
    d. Genetic code

Thinking Critically

After reading the question or statement, write a short answer using complete sentences.

18. **Explain.** How does genetics play a role in fetal development?

19. **Synthesize.** How might a disorder like sickle-cell anemia be traced to its origin?

20. **Infer.** Why would a brown-eyed parent and a blue-eyed parent have a brown-eyed child?

21. **Predict.** What would a pregnant female expect to find out after having an amniocentesis?

LESSON 3

Vocabulary Review

Use the vocabulary terms listed on page 489 to complete the following statements.

22. The confidence that you can control your own body, impulses, and environment is called _____.

23. _____ are events that need to happen for a person to continue growing toward being a healthy, mature adult.

24. The abnormal lateral curvature of the spine is known as _____.

Understanding Key Concepts

After reading the question or statement, select the correct answer.

25. Which of the following is the time of fastest growth in a person's life?
    a. Infancy
    b. Early childhood
    c. Middle childhood
    d. Late childhood

26. Which of the following may lead to low self-esteem in children?
    a. Overprotective parents
    b. Parents who encourage questions
    c. Parents who encourage autonomy
    d. Attentive parents

27. Which group is more likely to be diagnosed with scoliosis?
    a. Boys
    b. Infants
    c. Girls
    d. Preschoolers

Ryan McVay/Getty Images

Assessment

28. Which of the following can impact a child's development?
    a. Vision impairments
    b. Hearing impairments
    c. Social factors
    d. All of the above

Thinking Critically

After reading the question or statement, write a short answer using complete sentences.

29. **Analyze.** What developmental tasks are involved as friendships and school become especially important during late childhood?

30. **Infer.** What might happen to a child who is constantly scolded for making a mess or for getting in the way?

31. **Infer.** If a parent is overprotective and does not let a child explore his surroundings, what may happen?

32. **Explain.** Why do most states require that students be screened for hearing and vision problems?

Technology PROJECT-BASED ASSESSMENT

Genes Count

Background

In 1962, James Watson, Francis Crick, and Maurice Wilkins received a Nobel Prize for their explanation of the chemical structure of DNA. Since then, our knowledge of genetics has increased rapidly. In 1992, research turned to human genome sequencing to identify the location of hundreds of thousands of human genes. (The term *genome* means the genetic material of an organism.)

Task

Conduct research online to learn the accomplishments of the Human Genome Project. Develop a multimedia slide presentation based on your findings.

Audience

Students in your class

Purpose

Understand the progress that has been made in recent years by the Human Genome Project.

Procedure

1. Use a variety of online resources to learn the progress of the Human Genome Project and to identify recent advances in genetic research.
2. Based on your findings, create the slides for your presentation.
3. Include information on genetic maps, summarize the progress that has been made to identify genes that cause specific genetic disorders, and describe how genetic counseling and genetic engineering are used.
4. Compile your notes and create your multimedia presentation. Assign part of the presentation to each member of your group.
5. Be sure to use several images for your presentation.
6. Present your presentation to your class.

Math Practice

Interpret Graphs. The pie chart below shows the number of weeks in a typical 40-week pregnancy designated for each trimester. Use the pie chart to answer questions 1–3.

Stages of Pregnancy

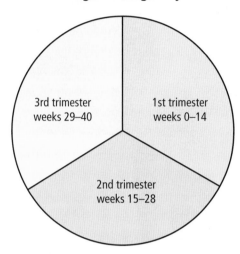

3rd trimester
weeks 29–40

1st trimester
weeks 0–14

2nd trimester
weeks 15–28

1. According to the information in the pie chart, what percentage of duration of the pregnancy makes up the first trimester?
   A. 14%   **C.** 35%
   B. 30%   **D.** 40%

2. The term *zygote* describes the fertilized egg in the first two weeks after conception. For what percentage of a typical 40-week pregnancy is the developing human called a zygote?
   A. 2%   **C.** 14%
   B. 5%   **D.** 35%

3. A baby is born early, after only 36 weeks of pregnancy. What percentage of the typical 40-week pregnancy did the baby complete?
   A. 10%   **C.** 90%
   B. 36%   **D.** 96%

Reading/Writing Practice

Understand and Apply. Read the passage below, and then answer the questions.

Jean Piaget (1896–1980) was the first theorist to study how children learn. Piaget analyzed facts about the way children develop cognitive abilities. He also conducted studies on the way children develop thinking skills.

One of Piaget's famous studies used pieces of candy to test children's discriminative abilities. Piaget placed equal numbers of the candy into two lines. He spread one line farther apart than the other line. The two- to three-year-olds tested saw that the rows had the same amount of candies, while the three- to four-year-olds tested believed the longer row had more candies.

Piaget's test showed that during this stage of development three to four year olds temporarily lose their ability to problem solve.

1. What is a *theorist*?
   A. A person who solves a problem
   B. A person who writes a story
   C. A person who analyzes a set of facts
   D. A person who uses a large vocabulary

2. The word *discriminative*, used in paragraph two, means which of the following?
   A. Objective
   B. Distinguish
   C. Prejudice
   D. Judicious

3. Considering Piaget's test, why do you think children between ages three to four temporarily lose the ability to solve?

National Education Standards

Math: Measurement and Data, Data Analysis
Language Arts: LACC.910.L.3.6

CHAPTER 18

The Life Cycle Continues

Lesson 1
Changes During Adolescence

BIG Idea *Adolescence begins with puberty as a person starts to mature physically, emotionally, and mentally.*

Lesson 2
Adulthood, Marriage, and Parenthood

BIG Idea *During adulthood, individuals may choose to get married and become parents.*

Lesson 3
Health Through the Life Cycle

BIG Idea *Middle and late adulthood are times of contribution and reflection.*

Activating Prior Knowledge

Using Visuals As you look at this photo, identify the different generations in the picture. What are some occasions that might bring generations together? How can maintaining relationships within different generations enhance the lives of each member of a family?

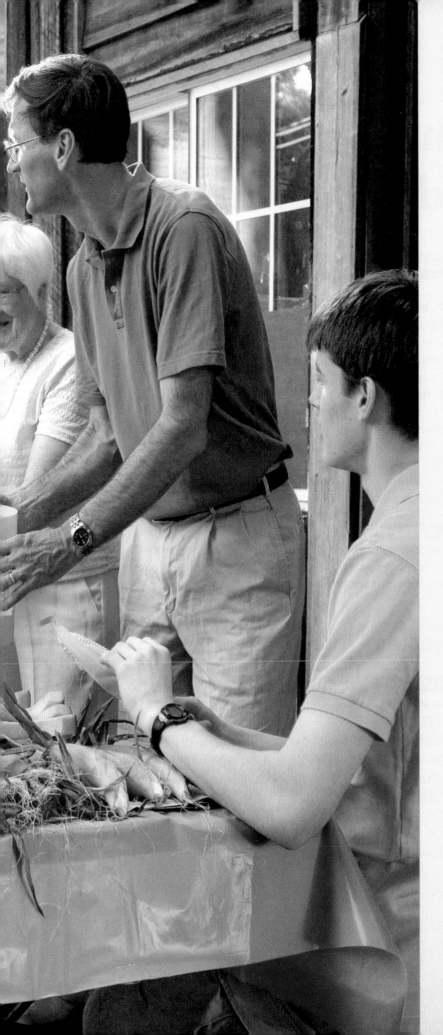

Chapter Launchers

Health in Action

Discuss the **BIG** Ideas

Think about how you would answer these questions:

▶ What sorts of changes occur during adolescence?

▶ How could marriage and parenthood affect your life?

▶ What challenges do adults face in their later years?

Assess Your Health

Read each statement. On a separate sheet of paper, write "yes," "sometimes," or "no" based on your typical behavior.

1. I avoid risky situations.

2. I have short-term and long-term goals and have identified the steps to reach them.

3. I use decision-making skills to make healthful decisions.

4. I use refusal skills to say "no" in difficult situations.

5. I talk with a parent or guardian about goal-setting and making healthful decisions.

A "yes" response shows that you practice healthy behaviors. "Sometimes" indicates that you should analyze and possibly modify your behavior. A "no" response means that you should modify the behavior.

Changes During Adolescence

GUIDE TO READING

BIG Idea *Adolescence begins with puberty as a person starts to mature physically, emotionally, and mentally.*

Before You Read

Create an Outline. Preview this lesson by scanning the pages. Then, organize the headings and subheadings into an outline. As you read, fill in the outline with important details.

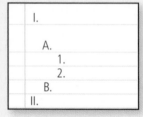

```
I.
   A.
      1.
      2.
   B.
II.
```

New Vocabulary

▶ adolescence
▶ puberty
▶ cognition

Review Vocabulary

▶ hormones (Ch.3, L.3)

Real Life Issues

Puberty Differs from Person to Person. Two fraternal twins, Seth and Claire, are not as close as they used to be. Lately, they have not been spending much time together. Claire is experiencing the physical, social, and mental/emotional changes that come with puberty, but Seth is not. Nothing has changed for him yet and he wonders if there is something wrong with him. Seth is feeling confused and scared.

Writing *Write a letter to Seth, explaining some of the reasons why you think Seth should not be worried about developing at a different pace than his sister. Reassure him that there is no need for concern.*

Puberty: A Time of Changes

Main Idea Adolescents begin moving toward adulthood during puberty.

Adolescence, *the period between childhood and adulthood,* is a time of many challenges and changes. Physical growth is one of the most noticeable changes during this period. Children grow, their voices change, and their bodies begin to fill out. After infancy, adolescence is the second fastest period of growth. During adolescence, changes occur in one's mental, emotional, and social life.

The beginning of adolescence is marked by the onset of puberty. **Puberty** is *the time when a person begins to develop certain traits of adults of his or her gender.* Puberty usually begins sometime between the ages of 12 and 18.

Hormones are chemical substances produced in glands that help regulate many of the body's functions. Testosterone, the male hormone, and female hormones estrogen and progesterone, create changes that affect teens during puberty.

Physical Changes

One of the most important and **significant** body changes that takes place during puberty is the development of *sex characteristics*. These are the traits related to a person's gender. Primary sex characteristics are directly related to the production of reproductive cells, or *gametes*. The male gametes are *sperm*. The production of sperm by the testes begins at puberty. The female gametes are the *eggs*, or *ova*. At birth, a female's body contains all the eggs she will ever produce. However, the eggs don't mature until puberty begins, with the onset of ovulation. The onset of these changes indicates sexual maturity and the ability to reproduce. You will read more about emotional maturity, which is required to parent a child, in Lesson 2.

Other changes occur during puberty and are associated with secondary sex characteristics. These are summarized in **Figure 18.1**.

Academic Vocabulary

significant *(adjective):* having meaning

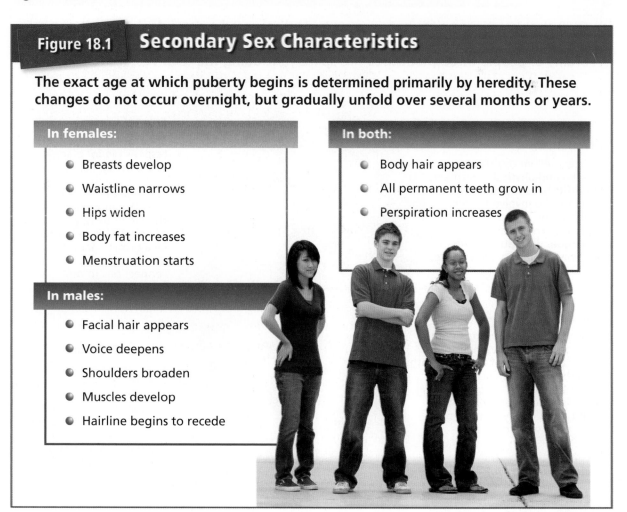

Figure 18.1 **Secondary Sex Characteristics**

The exact age at which puberty begins is determined primarily by heredity. These changes do not occur overnight, but gradually unfold over several months or years.

In females:
- Breasts develop
- Waistline narrows
- Hips widen
- Body fat increases
- Menstruation starts

In males:
- Facial hair appears
- Voice deepens
- Shoulders broaden
- Muscles develop
- Hairline begins to recede

In both:
- Body hair appears
- All permanent teeth grow in
- Perspiration increases

In any group of teens, there is a variety of body sizes and shapes. Growth rates vary from person to person. Each teen goes through puberty at his or her own pace. There is no real timetable for the physical changes that a teen experiences. Some teens develop the physical characteristics of adults before their friends do. You may experience changes sooner or later than your classmates do, and you may feel uncomfortable about these changes. Just remember that every teen experiences these changes, that they are normal, and that they will resolve themselves as time passes.

Mental and Emotional Changes

As your body goes through dramatic changes during adolescence, your brain is changing as well. By age six, a child's brain is about 95 percent of the size it will be when the child is an adult. The cerebrum, though, continues to develop during adolescence, increasing memory and cognition. **Cognition** is *the ability to reason and think out abstract solutions*. A child sees only a limited number of solutions to a problem.

Figure 18.2　Brain Development in Teens

Over the past 25 years, neuroscientists have discovered a great deal about the human brain. Recent imaging techniques have enabled scientists to examine the brains of people throughout their life spans—including the teen years.

Cerebellum

The cerebellum coordinates muscles and physical movement. Scientists have found evidence that it is also involved in the coordination of thinking processes. The cerebellum undergoes dramatic growth and change during adolescence.

Amygdala

The amygdala is associated with emotion. New studies indicate that teens use this part of the brain, rather than the more analytical frontal cortex that adults use in emotional responses. Scientists believe this might explain why teens sometimes react so emotionally.

Frontal Cortex

The frontal cortex is responsible for planning, strategizing, impulse control, and reasoning. The area undergoes a growth spurt when a child is 11 to 12 years of age. This is followed by a growth period, during which new nerve connections form.

Corpus Callosum

The corpus callosum connects the two sides of the brain. It is thought to be involved in creativity and problem solving. Research suggests that it grows and changes significantly during adolescence.

Adolescents become increasingly capable of solving problems in more complex ways. During this stage you will learn to

- anticipate the consequences of a particular action.
- think logically.
- understand different points of view.

Adolescence is a time of emotional change as well. You may begin to look outward to try to understand yourself and your place in society. Most adolescents begin to search for meaning, personal values, and a sense of self. These new mental and emotional developments can be tied to physical changes and growth in adolescent brains. The brain development of teens is explored in **Figure 18.2**.

Social Changes

You may notice that you are also experiencing social changes during adolescence. Friends become a major part of a teen's social experience. As you explore new interests through a variety of classes and extracurricular activities, you'll make new friends and meet people from many cultural and social backgrounds. This is a time when you'll begin to appreciate how ethnic and cultural diversity can enrich your life. Likewise, your peers may challenge what you believe in and what you think is right and wrong. You can practice responsible decision making skills when a friend asks you to do something that goes against your own personal values. In general, strong friendships begin when people realize they share similar goals, experiences, and values.

Accomplishments in Adolescence

(**Main Idea**) Adolescents will develop independence, find their identity, and establish their personal values.

You are familiar now with some of the physical, mental, emotional, and social changes that teens experience. These changes are not experienced separately. A teen may be dealing with several issues at the same time. This makes adolescence a wonderful, but often difficult, time of life.

READING CHECK

Describe How does a person change emotionally during adolescence?

■ **Figure 18.3** Friends from different cultural backgrounds share their traditions and cultural interests. *What traditions or interests would you share?*

During adolescence, you will be maturing, and you will accomplish a series of specific developmental tasks that you will find helpful as you make the transition from adolescence to adulthood. Some of these tasks include achieving emotional independence from your parents, developing an identity, and adopting a system of personal values. You will begin to establish career goals, and you will find that practicing healthful, responsible behaviors will help you achieve these goals.

As you recognize the adjustments you are going through, evaluate your progress by asking yourself the questions at the end of each developmental task description below.

- **Emotional and psychological independence.** During adolescence, you may find yourself moving back and forth between wanting independence and wanting the security of your family. Teens who have ongoing, open communication with their parents or guardian have the advantage of seeking advice and feedback about the decisions they need to make. A parent can help teens learn problem-solving skills. When parents or guardians discuss and explain situations, rules, and reasons in the decision-making process, teens learn from their modeling. Your family's support and guidance can help you become more emotionally and socially independent. During this time, you will develop confidence and build your self-esteem as you become more independent. *In what ways are you a different person than you were two years ago?*

■ **Figure 18.4** Teens begin to make decisions about their future and goals at this stage of their lives. *What are your vocational goals?*

- **Personal value system.** Parents and guardians provide a set of rules for appropriate and inappropriate behavior for younger children. This helped lay the foundation for your own values. You will now begin to assess your values when they differ from the values expressed by your peers and others. *Have you begun to establish personal beliefs and values that enhance your health and well-being? Are you acting in ways that support those standards?*

- **Vocational goals.** The teen years are a time to begin identifying your vocational goals for the future. As you explore the possibilities open to you and develop new interests, you may discover that some of these interests can lead to a career. *Have you set long-term goals and identified steps to reach those goals?*

- **Control over behaviors.** Adolescents make decisions every day about whether to participate in risky behaviors that may harm their health. Considering your values and your short-term and long-term goals will give you a firm basis for making healthful decisions and avoiding risky situations. *Identify two recent events that challenged you to show emotional maturity and avoid a risky behavior.*

 **READING CHECK**

Identify What are your vocational goals for the future?

 LESSON 1 ASSESSMENT

After You Read

Reviewing Facts and Vocabulary

1. Define the terms *adolescence* and *puberty*.

2. What are the reproductive cells of females called? What are those of males called?

3. What are some secondary sex characteristics of males that develop during puberty?

Thinking Critically

4. **Infer.** How does the fact that the cerebrum is still developing during adolescence explain some teenage behavior?

5. **Explain.** During which time frame does adolescent development typically occur?

Applying Health Skills

6. **Refusal Skills.** At times, your peers might encourage you to do something that you know is wrong. Write a scenario describing how you would handle such a situation.

Writing Critically

7. **Persuasive.** Your friend is being asked to participate in an activity that you think is unsafe and unwise. Write a short letter suggesting why you think this is a mistake and how to get out of it. Be encouraging and supportive, but suggest alternatives.

Real Life Issues ....

After completing the lesson, review and analyze your response to the Real Life Issues question on page 496.

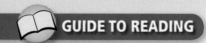

📖 GUIDE TO READING

BIG Idea *During adulthood, individuals may choose to get married and become parents.*

Before You Read

Create a K-W-L Chart. Make a three-column chart like the one below. In the first column, write what you **k**now about parenthood. In the second column, write what you **w**ant to know about this topic. As you read, fill in the third column describing what you have **l**earned.

K	W	L

New Vocabulary

▶ physical maturity
▶ emotional maturity
▶ commitment
▶ adoption
▶ self-directed
▶ unconditional love

Review Vocabulary

▶ extended family (Ch.7, L.1)

Adulthood, Marriage, and Parenthood

Real Life Issues ....

Wedding Bells? Lily's older sister, Maya, will graduate from college at the end of the year. Maya confides in Lily that she and her boyfriend are thinking about getting married as soon as they graduate. Lily likes Maya's boyfriend, but has learned in her health class that marriage is a very big step.

Writing *Write a letter from Lily to Maya, explaining what she thinks marriage involves and why it's an important step that requires a lot of thought.*

Maturing Physically and Emotionally

Main Idea Adulthood is reached when both physical maturity and emotional maturity are achieved.

Most people reach **physical maturity**, *the state at which the physical body and all its organs are fully developed,* in late adolescence or their early twenties. Being physically mature does not make you an adult, however. To be an adult, you need to develop emotionally as well.

Emotional maturity is *the state at which the mental and emotional capabilities of an individual are fully developed.* Emotionally healthy individuals have positive values and goals. They are able to give and receive love, have the ability to face reality and deal with it, and have the capacity to learn from life experiences. Relationships with peers, family, and friends can have a positive effect on a person's physical and emotional health.

Stages of Adulthood

Main Idea Three major stages make up the adult years.

The adult years are made of three different stages. Each stage can be characterized by its own accomplishments and is associated with a goal that involves a person's relationship with other people.

The three stages are as follows:

- **Young adulthood.** This stage lasts from 19 to 40 years of age. The goal of this stage is to develop intimacy. A person in young adulthood tries to develop close personal relationships. Many people will decide to get married and start a family during this stage.

- **Middle adulthood.** This stage occurs between the ages of 40 and 65. The goal here is to develop a sense of having contributed to society. A person in this stage looks outside themselves and cares for others through such activities as grandparenting or volunteering.

- **Late adulthood.** This stage lasts from the age of 65 to death. The goal of a person in this stage is to feel satisfied with his life. A person in this stage tries to understand the meaning and purpose of her life.

READING CHECK

Infer What positive events can happen during middle adulthood?

■ **Figure 18.5** Your extended family may include aunts, uncles, and grandparents, as well as your parents and brothers or sisters. *Which members of your extended family do you enjoy a close relationship with?*

Todd Wright/Blend Images LLC

Marriage

Main Idea Marriage is a commitment to share your life with another person.

Most people marry because they fall in love and are ready to enter into a lasting, intimate relationship. Married couples share togetherness and support each other in hard times as well as in good times.

Deciding to Marry

There are differences between a dating relationship and marriage. When a couple agree that marriage may be in their future, their relationship becomes more serious, and they make a deeper **commitment**—*a promise or a pledge*—to each other. They also consider the long-term consequences when making decisions. In this way, the couple show they understand the relationship between mental, emotional, and physical health. If one partner has any doubts or questions about the other partner's reasons for marrying, these questions should be explored and resolved before the marriage actually takes place.

Successful Marriages

The decision to make the commitment to marry is only the first step in a successful marriage. *Marital adjustment*—how well a person adjusts to marriage and to a spouse—depends on the following factors:

- **Communication.** Couples need to be able to share their feelings and express their needs and concerns to each other. Demonstrating communication skills helps couples build and maintain a healthy marriage relationship.

- **Emotional maturity.** Emotionally healthy people try to understand their partner's needs and are willing to compromise. They don't always think of themselves first; they consider what is best for the relationship as a whole.

- **Values and interests.** When couples share similar attitudes about the importance of good health, spirituality, ethical standards, morality, family, and friendships, they spend more time together, which strengthens a marriage.

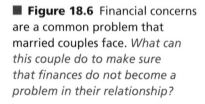

■ **Figure 18.6** Financial concerns are a common problem that married couples face. *What can this couple do to make sure that finances do not become a problem in their relationship?*

Resolving Conflict in Marriage

Conflict arises occasionally, even in the strongest marriages, because people can't completely agree on everything all the time. Learning how to get along involves how to recognize causes of these conflicts and finding ways to resolve them so the relationship can be strengthened.

Issues that can cause problems in marriages include the following:

- Differences in spending and saving habits
- Conflicting loyalties involving family and friends
- Lack of communication
- Lack of intimacy
- Jealousy, infidelity, or lack of attention
- Decisions about having children and arranging child care
- Abusive tendencies or attitudes

In a successful marriage, partners respect, trust, and care for each other. **Conflicts** are resolved fairly without damaging the self-esteem of either partner. The development of good communication and conflict-resolution skills can help reduce the impact of conflict on a marriage. Sometimes couples need counseling to resolve marital conflict.

Teen Marriage

Some people begin talking about marriage at a young age. In most states it's illegal for a couple to marry before the age of 18 without parental permission. As well as meeting legal requirements, marriage is a step that requires emotional maturity. Emotional maturity enables partners to deal with the problems and decisions of marriage. Most teens are still struggling to figure out their own identity and set goals for the future. It is unlikely that they have had a chance to determine their life path, or what they want in a marriage partner. That's one reason about 60 percent of marriages involving teens end in divorce. Statistics show a high probability that a teen marriage will end in the first few years.

Teens who do get married may soon begin to realize that they have increased responsibilities that interfere with their personal freedoms and their educational or career goals. They may find they do not have enough life experience to make this important and lasting decision. Financial pressures can add stress to the marriage. Marriage difficulties may arise as the newness wears off and teens recognize the responsibilities required to make a successful marriage.

Parenthood

Main Idea Parenting demands many added responsibilities.

Many married couples decide to start a family. Some couples choose to adopt a child. **Adoption** is *the legal process of taking a child of other parents as one's own.* Couples may choose to raise foster children by becoming a legal guardian.

Academic Vocabulary

conflict *(noun):* a competitive or opposing act, or incompatibles

READING CHECK

Identify What are three issues that often cause problems in a marriage?

FITNESSZONE

Every summer I visit my grandparents. They have a habit of having a mid-afternoon snack. My grandmother calls it "sweets for my sweetie." Their afternoon snack is always something healthy. Gran may cut up an apple and drizzle some chocolate or butterscotch syrup over it. It tastes just like a candied apple, and my grandparents are both in good shape, so I guess it works. For more physical activity ideas, visit the Fitness Handbook and Fitness Zone sites in ConnectEd.

■ Figure 18.7 Setting limits and curfews is one of the responsibilities of being a parent. *What limits do your parents or guardians place on you?*

Most parents find raising a child to be both challenging and rewarding, and they take great joy in loving and caring for their children. Prospective parents need to understand the changes in lifestyle and responsibility that will take place before and after the birth of their baby. These responsibilities will continue for many years after birth. Parents must provide protection, food, clothing, shelter, education, and medical care. Parenting also involves providing guidance, instilling values, setting limits, and giving unconditional love.

Providing Guidance

Involved parents will teach children that each individual is responsible for one's own successes and failures. Parents should encourage children and help them develop a sense of pride in their accomplishments. Parents also need to guide and protect their children while teaching them to make their own decisions. Watching children learn to get along with others and solve their own problems is a satisfying experience for a parent. The role of the extended family—grandparents and other family members—is important in promoting a healthy family. When children see family members interacting in a mature, loving, and caring manner, they are more likely to grow up to be healthy and productive.

Instilling Values

Parents with a strong value system may pass these aspects of life on to their children. Children develop a set of *values* as they grow. Values make up the system of beliefs and standards of conduct that people find important and that will guide the way they live. Values help children develop strong character and the ability to resist negative influences they may encounter. Teaching positive values, ethics, and respect for cultural diversity can help children grow to become happy, productive, and mature adults.

Setting Limits

When children learn limits, they become **self-directed**, or *able to make correct decisions about behavior when adults are not present to enforce rules.* The limits parents set for their children during childhood are very different, but just as important, as the limits they set for adolescence. Limits on bedtime, eating habits, and television exposure will change as children get older. Limits on adolescents can include curfews and use of the family car. It is a parent's responsibility to set limits. During the teen years, children will be able to handle more responsibility.

Giving Unconditional Love

Another responsibility of parenthood is providing children with **unconditional love**, *love without limitations or quantifications.* Parents need to show their children love at all times. Receiving unconditional love helps a child thrive by meeting a child's basic needs. Meeting basic needs is part of Maslow's hierarchy of needs, which you learned about in Chapter 3. Before having children, couples should carefully examine the lifelong responsibilities and requirements of being a parent.

Teen Parenting

Becoming a parent is challenging and difficult at any stage in life. This is particularly true when the parents are teenagers. Teens who are pregnant may choose to put their baby up for adoption or raise the child themselves. Consequences of teen parenthood are many, but may include:

- Financial difficulties
- Restrictions on educational and career plans
- Emotional stress
- Limitations on social and personal life

Although parenthood is very rewarding, it also requires maturity. Teens need to consider the responsibilities of parenthood as they make decisions that could alter their lives.

■ **Figure 18.8** Teen parents may feel they are missing part of their own childhood. *What other stresses may impact teen parents?*

 READING CHECK

Explain How do limits change as a child grows older?

LESSON 2 ASSESSMENT

After You Read

Reviewing Facts and Vocabulary

1. Distinguish between *physical maturity* and *emotional maturity.*
2. How do the goals of young adulthood differ from the goals of middle adulthood?
3. Which factors determine how well a person will adjust to marriage?

Thinking Critically

4. **Predict.** What are some factors that may cause conflict, even in a good marriage?
5. **Identify.** What is a parent giving a child by offering unconditional love?

Applying Health Skills

6. **Accessing Information.** Use print or online resources to research the legal rights and responsibilities of teen parents. Compare and contrast these rights with the rights of adult parents.

Writing Critically

7. **Expository.** Write a short essay explaining the disadvantages of marrying during the teen years.

Real Life Issues ....

After completing the lesson, review and analyze your response to the Real Life Issues question on page 502.

GUIDE TO READING

BIG Idea *Middle and late adulthood are times of contribution and reflection.*

Before You Read

Create a Table. Make a two-column table. Label the first column "Middle Adulthood." Label the second column "Late Adulthood." Fill in the major milestones for each stage as you read the lesson.

Middle Adulthood	Late Adulthood

New Vocabulary

▶ transitions
▶ empty-nest syndrome
▶ integrity

Health Through the Life Cycle

Real Life Issues ....

Where Did They Go? Anna's grandmother lives less than a mile from Anna and her family. Anna visits her grandmother twice a week on her way home from school. Her grandmother lives alone, and Anna worries that she is lonely. Anna's grandfather died two years ago. Anna has learned in school that writing down your feelings is a good way to begin dealing with them.

Writing *Write a scenario between Anna and her grandmother in which Anna encourages her grandmother to write about her feelings.*

Middle Adulthood

Main Idea Many changes occur during middle adulthood.

Middle adulthood is the stage of development that spans from age 40 to age 65. These years are full of **transitions**, *critical changes that occur at all stages of life*. These changes can include family and individual accomplishments, including children graduating from college, the arrival of the first grandchild, achievement of a satisfying career goal, or recognition of an individual's contribution to the community. These accomplishments can enhance a person's physical, mental, emotional, and social health throughout adulthood.

Health Concerns

Each phase of adulthood has its own unique concerns that affect one's health and well-being. Scientific research continues to bring new discoveries that are advancing disease prevention and improving nutrition for better health.

These advances have made a significant difference in the lives of older adults. In general, people nowadays are experiencing a healthier older adulthood. Older adults, however, still have to be aware of the conditions that result from aging.

- **Eyesight** changes with age. The eyes of an adult in this stage may have difficulty bringing images into focus.

- **Hearing** may decrease, particularly if a person was exposed to loud sounds constantly throughout one's life.

- **Muscles and joints** may be affected by arthritis. Arthritis affects nearly half the people over the age of 65.

- **Bones** may become brittle and more likely to break. Osteoporosis most commonly affects older women, but anyone can get osteoporosis.

- **Teeth and gums** can become decayed and diseased without proper care.

- **Heart disease** may occur due to heredity or lifestyle factors, such as a lifetime of too little physical exercise and a diet too high in saturated fat.

- **Cancer** may occur, so it's recommended that adults in this stage get regular screenings.

■ **Figure 18.9** Routine eye exams are part of staying healthy at any age. *How are the lives of middle adults changing?*

Physical Transitions

As you have learned, physical change doesn't stop when adolescence ends; it continues throughout the life cycle. The rate of change may slow down, however. People in middle adulthood experience physical changes as their bodies begin to age. Females enter *menopause,* or the end of ovulation and menstruation, between the ages of 45 and 55. This means that a woman can no longer become pregnant.

Research indicates that most people who practiced healthful behaviors as teens and young adults, such as weight management, nutritious eating, and regular physical activity, experience better physical health for a longer period. Adults who have developed lifelong healthful habits and continue to be active stay healthy by eating low-fat, high-fiber diets and avoiding tobacco, alcohol, and other drugs. Strength training has been proven to offer significant benefits to most adults. Benefits include increasing muscle mass, preserving bone density, and protecting major joints from injury.

Mental Transitions

Just as physical exercise strengthens the body, mental activities strengthen the brain. Solving puzzles, reading, and playing strategy games provide mental stimulation. An adult who exercises his or her brain remains mentally active. Learning should be a lifelong pursuit. At midlife, many adults begin new careers, return to school, and learn new hobbies.

■ Figure 18.10 Middle adults still play an active role in the workplace. *Why is it healthful to make learning a lifelong pursuit?*

✓ **READING CHECK**

Explain How can skills developed in adolescence help you in middle adulthood?

The use of computers gives older adults opportunities to broaden their access to information. People reach middle adulthood with a great deal of knowledge and experience, which they can use to pursue new interests.

Emotional Transitions

The emotional transitions people experience during middle adulthood can be similar to those experienced during puberty. By this time in life, a person may take pride in personal accomplishments. A middle adult may also experience some disappointments. Adults at this stage are said to be having a midlife crisis when they have questions and concerns about whether they have met their goals, feel loved and valued, and have made a positive difference in the lives of others.

Social Transitions

Most social transitions during middle adulthood focus on family. Often people are faced with the death of a parent or need to adjust to children growing up and leaving home. *The feelings of sadness or loneliness that accompany seeing children leave home and enter adulthood* are called **empty-nest syndrome**. Those who maintain healthy relationships with family and friends have less difficulty adjusting to these changes. For many, this is a time to apply their talents and life experiences to community programs. They may pursue new interests and make new friends. Developing good social skills earlier in life can help ease these transitions.

Late Adulthood

Main Idea People in late adulthood may reflect on their lives and accomplishments.

Late adulthood begins after age 65. One of the goals at this stage is to look back with satisfaction and a sense of fulfillment. Older adults may review the events of their lives and their achievements. If they have lived their lives with **integrity**, meaning they have made decisions with *a firm adherence to a moral code,* then they may feel fulfilled. A person who considers family a high priority may have succeeded in a career while providing for the family. People who remain committed to a system of values throughout life will have a sense of satisfaction. Many older adults are able to look back without regret and feel proud of their accomplishments.

Public Health Policies and Programs

The Social Security system, created in 1935, provides benefits to older adults as well as to people with disabilities. To assist with health care needs, the government offers Medicare to those over 65 and Medicaid to those with low incomes and limited resources. Advances in disease detection, prevention, and treatment have allowed older adults to maintain independent and satisfying lives.

Because better health care is available today, people can expect to live longer after retirement. For this reason, financial planning is essential. Some companies provide retirement benefits, but many workers still must plan ahead with personal or company-provided long-term savings plans. With the addition of these savings funds, in conjunction with Social Security benefits, the poverty rate for older adults has been reduced. Many are finding that, because of a lifetime of practicing healthful behaviors, the years after retirement are fulfilling and rewarding.

■ **Figure 18.12** Many older adults enjoy active lives. *Why is older adulthood such a rewarding time for many?*

 READING CHECK

Identify What programs help people in late adulthood?

LESSON 3 ASSESSMENT

 After You Read

Reviewing Facts and Vocabulary

1. What is *empty-nest syndrome*?
2. What are some examples of activities older adults can participate in to remain mentally active?
3. What sort of physical changes do females in middle adulthood experience?

Thinking Critically

4. **Infer.** Why is it so important today for adults to plan financially for retirement?
5. **Identify.** How have changes in nutrition and health care changed the lives of older adults?

Applying Health Skills

6. **Accessing Information.** Research library or Internet resources to learn about the Social Security system. Identify problems that may arise with this system in the not-so-distant future. Share what you learn in the form of an informational poster or pamphlet.

Writing Critically

7. **Descriptive.** Write a short essay explaining the emotional transitions one may have in middle adulthood.

Real Life Issues ....

After completing the lesson, review and analyze your response to the Real Life Issues question on page 508.

Hands-On HEALTH

Activity · **Skills for a Happy Marriage**

Many couples find marriage to be an exciting and challenging experience. Your task is to identify the skills needed for a successful partnership. You will develop a list of ten questions couples might have about how to communicate in a marriage and resolve conflicts in a healthful way. Then you will create a resource packet of information couples might find helpful.

What You'll Need

- paper and pencils
- computer with Internet access

What You'll Do

Step 1

Work in groups of two or three to review Chapter 18. Conduct research on the Internet for additional skills needed for a healthy marriage.

Step 2

List ten questions covering communication and conflict-resolution skills for a successful partnership.

Step 3

Review the chapter again and write answers to each of the questions you listed in step 2.

Apply and Conclude

Develop a resource packet or pamphlet to help couples identify strategies to resolve conflicts in healthy ways.

Checklist: Conflict-Resolution Skills

- ✓ Taking turns explaining each side without interruptions
- ✓ "I" messages
- ✓ Listening skills
- ✓ Brainstorming solutions
- ✓ Solutions that benefit both sides

LESSON 1

Changes During Adolescence

Key Concepts

▶ Adolescents begin moving toward adulthood during puberty.

▶ During adolescence, the brain develops important pathways.

▶ Adolescents will develop independence, find their identity, and establish their personal values.

Vocabulary

▶ adolescence (p. 496)

▶ hormones (p. 497)

▶ puberty (p. 496)

▶ cognition (p. 498)

LESSON 2

Adulthood, Marriage, and Parenthood

Key Concepts

▶ Adulthood is reached when physical and emotional maturity are achieved.

▶ There are three major stages during the adult years.

▶ Marriage is a commitment to share your life with another person.

▶ A married couple may have several different options when they decide to start a family.

Vocabulary

▶ physical maturity (p. 502)

▶ emotional maturity (p. 502)

▶ extended family (p. 503)

▶ commitment (p. 504)

▶ adoption (p. 505)

▶ self-directed (p. 506)

▶ unconditional love (p. 507)

LESSON 3

Health Through the Life Cycle

Key Concepts

▶ Physical, mental/emotional, and social changes occur during middle adulthood.

▶ Developing lifelong healthful habits that include healthy nutrition and physical activity help promote health throughout the life span.

▶ People in late adulthood have the opportunity to reflect on their lives and accomplishments.

Vocabulary

▶ transitions (p. 508)

▶ empty-nest syndrome (p. 510)

▶ integrity (p. 510)

LESSON **1**

Vocabulary Review

Use the vocabulary terms listed on page 513 to complete the following statements.

1. The period of time between childhood and adulthood is called _____.

2. The time when a person begins to develop certain traits of adults is called _____.

3. The ability to reason and think out abstract solutions is called _____.

4. Chemical substances called _____ help regulate the body's many functions.

Understanding Key Concepts

After reading the question or statement, select the correct answer.

5. Assessing your own values when they might be different than the values of your peers shows that you are developing which of the following?
   a. Personal value system
   b. Emotional independence
   c. Vocational goals
   d. Self-control

6. Which is *not* an example of a vocational goal?
   a. Entering trade school
   b. Entering college
   c. Learning to repair automobiles
   d. Passing the science test tomorrow

7. Which is the term for the development of certain traits of adults of your gender?
   a. Adolescence
   b. Emotional maturity
   c. Puberty
   d. Physical maturity

Thinking Critically

After reading the question or statement, write a short answer using complete sentences.

8. **Describe.** Describe the characteristics of a good friend.

9. **Extend.** Not everyone develops at the same rate. What does this mean?

10. **Analyze.** How can parents or guardians make the transition to emotional independence easier for teens?

11. **Infer.** What sort of decisions might you make that could be potentially risky?

12. **Synthesize.** How will the values you developed earlier in life impact you as you go through the stages of development?

LESSON **2**

Vocabulary Review

Correct the sentences below by replacing the italicized term with the correct vocabulary term.

13. *Emotional maturity* is the point at which the body and its organs are developed.

14. Children learn to make decisions about behavior when adults aren't present as they become more *physically mature.*

15. Giving love to a child without question in all situations is called *commitment.*

16. Legally taking someone else's child to raise as your own is known as *integrity*.

Understanding Key Concepts

After reading the question or statement, select the correct answer.

17. Which is a responsibility of parenthood?
    a. Instilling values
    b. Setting limits
    c. Giving unconditional love
    d. All of the above

18. In a marriage, the partners don't always think of themselves first. Rather, they consider what is best for the relationship. What is this known as?
    a. Communication
    b. Emotional maturity
    c. Physical maturity
    d. Values and interests

19. Setting limits can help children become which of the following?
    a. Self-directed
    b. Physically mature
    c. Emotionally mature
    d. Committed

20. Which is the stage at which the physical body and all its organs are fully developed?
    a. Physical maturity
    b. Emotional maturity
    c. Marital adjustment
    d. Commitment

Thinking Critically

After reading the question or statement, write a short answer using complete sentences.

21. **Describe.** What happens in the three stages of development during the adult years?

22. **Analyze.** What should a couple consider before entering into marriage?

23. **Describe.** What particular challenges do teens who get married face?

24. **Analyze.** What are the different ways a couple may start a family?

LESSON 3

Vocabulary Review

Use the vocabulary terms listed on page 513 to complete the following statements.

25. The middle adult years are often full of _____, critical changes that occur at all stages of life.

26. When their children grow up and leave home, middle adults may suffer from _____.

27. People in late adulthood may reflect on whether they lived their lives according to their moral code, or with _____.

Understanding Key Concepts

After reading the question or statement, select the correct answer.

28. Which provides financial assistance to older adults, as well as disabled individuals?
    a. Social Security
    b. Medicare
    c. Medicaid
    d. None of the above

29. What is a social and mental/emotional change that middle and older adults experience?
    a. Sadness and loneliness when their children move away
    b. Pride in having met personal goals
    c. The opportunity to pursue talents and interests
    d. All of the above

Assessment

30. Arthritis may affect which of the following during late adulthood?
   a. Heart
   b. Eyesight and hearing
   c. Teeth and gums
   d. Joints

31. What is a potential benefit of strength training in later life?
   a. Preserving bone density
   b. Improving sleep habits
   c. Improving mental sharpness
   d. Maintaining a healthy weight

Thinking Critically

After reading the question or statement, write a short answer using complete sentences.

32. Assess. What role have computers played in changing the lives of people in late adulthood?

33. Synthesize. What is a midlife crisis?

34. Predict. How might a person deal with empty-nest syndrome?

Technology — PROJECT-BASED ASSESSMENT

The Journey Ahead

Background

During adolescence, many changes and transitions occur. Many teens feel the need to begin expressing more independence from their parents. They also begin showing more interest in developing close relationships with other teens. The teen years can be difficult because of these changes and uncertainties.

Task

Collaborate in groups to develop a podcast aimed at middle school students. Each student will play the role of an expert and discuss one developmental task. The goal is to reassure younger students about the changes they will soon experience.

Audience

Middle school students

Purpose

Analyze the changes that teens experience. Reassure younger students that these changes are normal.

Procedure

1 Review the sections in the chapter describing developmental changes experienced by teens.

2 Select one developmental change that teens commonly experience.

3 Conduct research online on that developmental task using reliable sources.

4 Discuss your findings with the group. Decide on a theme for the podcast. One student will play the role of moderator, and the other students will provide information on a developmental task.

5 Practice presenting the podcast. If possible, make an MP3 of your podcast to play for middle school students. Play the podcast, or perform it live for a group of middle school students.

Math Practice

Interpret Graphs. From information she gathered at the public library, Marsha made the following bar graph that shows the number of people of each age group in her town last year. Use the graph to answer Questions 1–3.

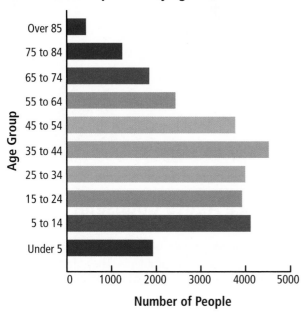

Town Population by Age Last Year

1. Which age group made up the greatest percentage of Marsha's town population last year?
   A. Under 5   C. 35 to 44
   B. 5 to 14   D. Over 85

2. Approximately how many people were below the age of 35?
   A. 4,000   C. 14,000
   B. 4,500   D. 18,500

3. If the town's total population was approximately 28,100, what percentage of the population was 15 to 24 years of age?
   A. 14%   C. 40%
   B. 24%   D. 50%

Reading/Writing Practice

Understand and Apply. Read the passage below, and then answer the questions.

(1) Teenagers are working in greater numbers than ever before. (2) Some choose jobs in fields they feel strongly about, like conservation or recycling. (3) They might join the parks department to plant trees. (4) Teenagers interested in working with younger children can work in community centers, or tutor students after school. (5) Many teens work in their own neighborhoods, doing yard work.

(6) Studies show that teens are successful workers. (7) They relate well to other teens, as well as to older adults. (8) Most teenage workers find a pleasant surprise—they like their jobs more than they expected to. (9) Many teens make lasting friendships on the job. (10) They also learn new skills related to their future career and educational choices.

1. Which detail below supports the idea that teenagers are successful in the workforce?
   A. Many teen have entered the workforce.
   B. Many join the parks department.
   C. Some teens work in conservation jobs.
   D. They relate well to other teens and to older adults.

2. The writer wants to add the following detail to this passage:
   Some teenagers—girls and boys—babysit neighborhood children.

   Based on the organization of this piece, after which number should this detail be added?
   A. Sentence 1   C. Sentence 6
   B. Sentence 4   D. Sentence 7

3. Write an essay giving examples of why you think teens can contribute in the workplace.

National Education Standards

Math: Measurement and Data
Language Arts: LACC.910.L.3.6, LACC.910.RL.2.4, LACC.910.W.3.8

CAREER CORNER Medical Specialists

Respiratory Therapist

Respiratory therapists help people who have trouble breathing due to a number of lung conditions, such as emphysema or asthma. They usually work as part of a team with physicians, nurses, and other medical professionals.

If a career as a respiratory therapist interests you, take communications, biology, and other science classes in high school. Educational requirements for respiratory therapists vary from state to state. Some states require a two-year college degree, and others require a four-year college degree.

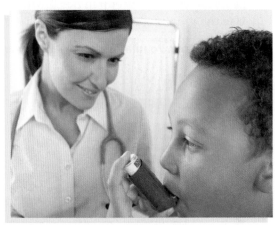

Substance Abuse Counselor

Substance abuse counselors assess and treat people who have substance abuse problems. They often work closely with doctors, social workers, psychologists, and other health professionals to address the physical, social, and psychological needs of the user and the user's friends and family. If you're interested in a career as a substance abuse counselor, take communications, psychology, and sociology classes in high school. Substance abuse counselors need an undergraduate degree and a master's degree in psychology or counseling. Most states also require certification and licensing.

Medical Records Technician

Medical records technicians maintain patients' health information. Medical records management is a rapidly changing field. Until recently, medical information was recorded on paper and then filed. Medical records are now often kept on computers.

Medical records technicians need to be detail oriented and well organized. Taking computer classes in high school can help you decide if a career as a medical records technician is right for you. After high school, you may need an associate's degree in information management from a community college or vocational school.

CAREER SPOTLIGHT

HIV/AIDS Educator

Livia Phillips was a freshman in college when she became an HIV/AIDS educator for Students Teaching AIDS to Youth (STAY). She volunteered thousands of hours of her time and received scholarships that paid for her undergraduate studies. Livia is now pursuing a nursing degree.

Q. Why did you want to become a health educator?

A. I wanted to be a resource person for health information. In school, I was the girl that anyone could talk to about their health problems and I loved helping my friends.

Q. What is the highlight of your career?

A. The last two years I worked as an educator in the HIV division of a substance abuse treatment facility. I educated residents about HIV/AIDS and offered testing and counseling services. It was very satisfying work and I hope to do it again.

Q. What makes you successful at your career?

A. I don't judge. I help people realize the risks they take with certain behaviors. I try to empower them to make good choices.

Activity Beyond the Classroom

Writing Drug Prevention Careers. Use print and online resources to learn more about substance prevention organizations. Many of these groups are nonprofit organizations that obtain funding from government and other sources. What types of careers might be available through one of these organizations? Research one nonprofit organization that aims to prevent substance abuse.

Make a list of at least five different careers within this organization that teens might consider. Develop a one-page informational flyer for two of the careers. Provide facts on the job responsibilities and education needed for these careers.

UNIT 7 Drugs

UNIT PROJECT

Improving Your Community

Using Visuals Students Against Destructive Decisions, also known as SADD, is a national organization that helps teens be positive role models. Their goal is to empower teens to abstain from destructive behaviors, such as drinking, drug use, and driving while impaired. Members educate their peers about the dangers of alcohol and drug use and provide prevention strategies.

Get Involved. Locate other organizations that discourage alcohol and drug use. Find out how teens in your community can volunteer to help their peers.

"It is our choices . . . that show what we truly are, far more than our abilities."
—J. K. Rowling, author

CHAPTER 19

Medicines and Drugs

Lesson 1
The Role of Medicines

BIG Idea *Medicines are divided into classes and have different effects on different people.*

Lesson 2
Using Medicines Safely

BIG Idea *Medicines are safe only if they are used for the intended purpose and according to the directions on the label.*

Activating Prior Knowledge

Using Visuals The teen and the pharmacist in this photo are discussing the appropriate way to take a medication. Write a short plan that describes ways that teens can find out how to use medicine properly. Be sure to discuss which sources of information are reliable.

Chapter Launchers

Health in Action

Discuss the **BIG Ideas**

Think about how you would answer these questions:

▶ What are reasons people take medicines?

▶ What are possible consequences of not following the instructions on a medicine label?

Assess Your Health

Read each statement. On a separate sheet of paper, write "yes," "sometimes," or "no" based on your typical behavior.

1. I understand the role of taking medicines and drugs in maintaining health.

2. I recognize that vaccines prevent disease.

3. I only use medicines according to directions from my doctor.

4. I avoid overusing medicines such as antibiotics.

5. I know the difference between over-the-counter and prescription medicines.

A "yes" response shows that you practice healthy behaviors. "Sometimes" indicates that you should analyze and possibly modify your behavior. A "no" response means that you should modify the behavior.

GUIDE TO READING

BIG Idea *Medicines are divided into classes and have different effects on different people.*

Before You Read

Create a Cluster Chart. Draw a circle and label it "Medicines." Create four surrounding circles labeled "Prevent Disease," "Fight Pathogens," "Relieve Pain," and "Promote Health." As you read, fill in the chart with more circles and details about the kinds of medicines discussed in the lesson.

New Vocabulary

▶ medicines
▶ drugs
▶ vaccine
▶ side effects
▶ additive interaction
▶ synergistic effect
▶ antagonistic interaction

The Role of Medicines

Real Life Issues

Choosing Medicines Wisely. Grant has a cold with a cough and runny nose. He checks the medicine cabinet for any cold medications that will help him feel better. He finds more than one type of cold medicine in the cabinet. Grant is not sure which one he should take.

Writing *Write a paragraph that explains what Grant should look for in a cold medicine. For example, what symptoms he wants to relieve, how much he should take, and how many hours a dose will last.*

Types of Medicines

Main Idea Medicines are classified based on how they work in your body.

People use medicines to help restore their health when they are ill. **Medicines** are *drugs that are used to treat or prevent diseases or other conditions.* **Drugs** are *substances other than food that change the structure or function of the body or mind.* All medicines are drugs, but not all drugs are medicines. Drugs are effective in treating illness when taken as directed by a physician or according to the label instructions. Medicines that treat or prevent illness can be classified into four broad categories:

• Medicines that help prevent disease
• Medicines that fight pathogens
• Medicines that relieve pain and other symptoms
• Medicines that manage chronic conditions, help maintain or restore health, and regulate body's systems

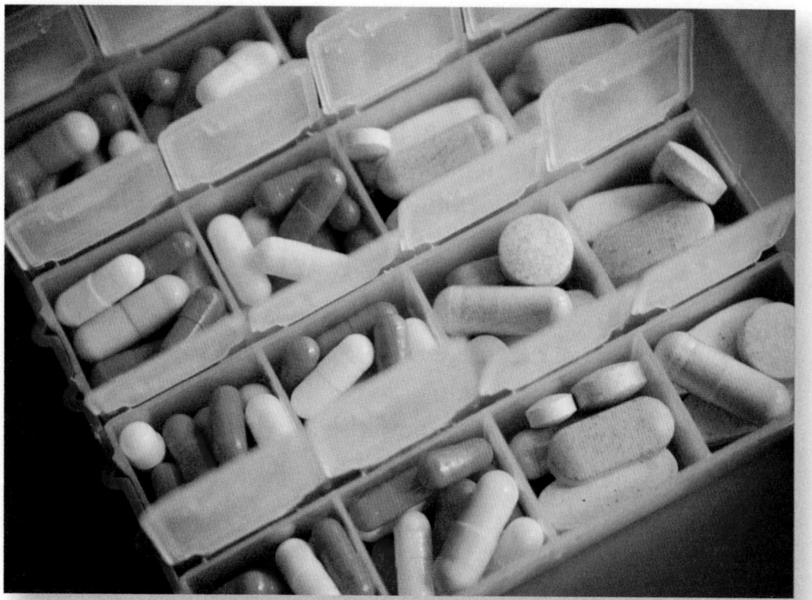

Design Pics/Kristy-Anne Glubish

Figure 19.1 Many types of medications are available. *What was the last medication that you used, and for what purpose did you use it?*

Preventing Disease

Today, we have medicines that prevent disease. About 95 percent of children receive **vaccines**, *a preparation that prevents a person from contracting a specific disease.*

Vaccines Vaccines contain weakened or dead *pathogens* that cause the disease. When injected into your body, the vaccine produces antibodies that fight those pathogens. Your body also produces memory cells that recall how to make these antibodies. This provides you with long-lasting protection against these specific pathogens.

The protection from some vaccines, however, fades over time. The vaccines for tetanus must be given periodically. For other vaccines, like those that prevent the flu, a new vaccine is required every year.

Antitoxins Antitoxins, like vaccines, prevent disease. They can also help **neutralize** the effects of toxins. Antitoxins fight the bacteria that produce substances toxic to the body. Antitoxins are usually produced by injecting animals with safe amounts of a specific toxin. This stimulates the animal's immune system to produce antibodies. These antibodies are then used to make an antitoxin.

Fighting Pathogens

Medicines can also help your body fight the pathogens that cause illness.

Antibiotics *Antibiotics* are a class of drug that destroy disease-causing microorganisms, called *bacteria.* Antibiotics such as penicillin work either by killing harmful bacteria in the body or by preventing bacteria from reproducing.

Academic Vocabulary

neutralize *(verb):* to counteract the effect of

When antibiotics were first introduced, they were considered a miracle drug because they saved so many lives. Some antibiotics, however, can cause nausea or stomach pain. Allergies are another side effect of antibiotic use. Tell your doctor if you experience any negative side effects of antibiotics, or if you know you are allergic to an antibiotic. Antibiotics can also lose their effectiveness. The bacteria that antibiotics kill have adapted to the drug over time.

Bacteria can develop a resistance in two ways: when antibiotics are overused, and when the patient does not finish taking the full prescription. If you do not finish taking all of a prescription, you may not kill all of the bacteria. The remaining bacteria may develop a resistance, or immunity, to treatment.

Antivirals and Antifungals Antibiotics are effective only against bacteria. They do not cure illnesses caused by viruses. Antiviral drugs are available to treat some viral illnesses, such as the flu. These medicines suppress the virus, but do not kill it. A person who takes antiviral medication for cold sores or fever blisters, which are caused by viruses, will still have the virus in his or her body. As a result, the person often has symptom-free periods followed by flare-ups when symptoms reappear. Like bacteria, viruses can develop a resistance to medications. Fungi are another type of pathogen that can infect the body. Antifungals can suppress or kill fungus cells, such as athlete's foot and ringworm.

Relieving Pain

The most commonly used medicines are *analgesics,* or pain relievers. Analgesics range from relatively mild medicines, such as aspirin, to strong narcotics, such as opium-based morphine and codeine. Aspirin is used to relieve pain and reduce fever. Other analgesics fight *inflammation,* or redness, swelling, and pain.

Even though aspirin is a widely used drug, it can cause stomach upset, dizziness, and ringing in the ears. Children who take aspirin when they have a fever are at risk of developing Reye's syndrome, a potentially life-threatening illness of the brain and liver. For that reason, aspirin should not be given to anyone under the age of 20 unless directed by a health care professional. Some people who are sensitive to aspirin take acetaminophen or ibuprofen. Acetaminophen is the recommended analgesic for children.

Pain Reliever Dependence Certain types of medicines that relieve pain can be addictive. These medicines, usually called narcotics, require a doctor's prescription. Patients who use these drugs can become physically or psychologically dependent on them.

READING CHECK

Describe Explain how vaccines prevent a person from getting a disease.

Managing Chronic Conditions

Some medicines are used to treat chronic conditions. These medicines maintain or restore health, and offer people with chronic diseases a higher level of wellness.

Allergy Medicines Antihistamines reduce allergy symptoms such as sneezing, itchy or watery eyes, and a runny nose. They block the chemicals released by the immune system that cause an allergic response. For people with allergies such as those to peanuts or bee stings, severe symptoms can appear suddenly. An allergic reaction can lead to death. Individuals who know they are allergic to substances that cause severe reactions can ask a doctor to prescribe a single-dose shot of epinephrine. The medication is designed to slow down or stop an allergic reaction. The patient is taught to self-administer a shot with a single-dose injector.

Body-Regulating Medicines Some medicines regulate body chemistry. Insulin used by people with diabetes regulates the amount of sugar in their blood. Asthma sufferers may take medicines every day to control symptoms and prevent attacks. They may also use inhalers during an asthma attack. Cardiovascular medicines are taken to regulate blood pressure, normalize irregular heartbeats, or regulate other functions of the cardiovascular system.

Antidepressant and Antipsychotic Medicines Medications can also help people suffering from mental illnesses. These medicines can help regulate brain chemistry, or stabilize moods. For example, mood stabilizers are often used in the treatment of mood disorders, depression, and schizophrenia. Proper medication can help people with these diseases live healthy lives. As with other prescribed medications, it is important to talk to your doctor before you stop taking the medication, even if you feel better.

Cancer Treatment Medicines Some cancers can be treated and even cured. Some medicines can be used to treat cancer. These medicines can reduce rapid cell growth and help stop the spread of cancer cells. One drug, chemotherapy, uses chemicals to kill fast-growing cancer cells. Immunotherapy, or biological therapy, uses the body's immune system to fight the cancer cells. Because these medications can also destroy healthy cells, serious side effects may occur as part of the treatment. Other medications can help treat the side effects.

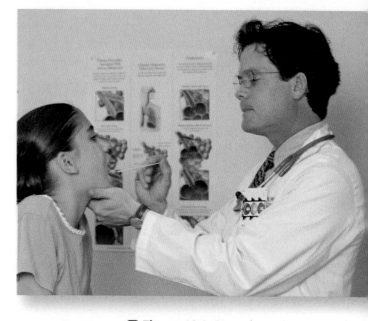

■ **Figure 19.2** Strep throat is a bacterial infection that is treated with antibiotics prescribed by a doctor. *Why is it important to take all of the antibiotics a doctor prescribes, even if you are feeling better?*

 READING CHECK

Describe Give two reasons that a person would take a body-regulating medicine.

Figure 19.3 Medications help many people with conditions such as asthma and diabetes live active, normal lives. *How do medicines work to control these diseases?*

Louis-Paul St-Onge/Getty Images

Taking Medications

Main Idea Medicines enter the body in a variety of ways.

Medicines can be delivered to the body in many ways. Factors that determine how a medicine is taken include what the medicine is used for, and how it will most quickly and effectively help a person.

- **Oral medicines** are taken by mouth in the form of tablets, capsules, or liquids. These medicines pass from the digestive system into the bloodstream.
- **Topical medicines** are applied to the skin. Transdermal skin patches also deliver a medicine through the skin.
- **Inhaled medicines,** such as asthma medicines, are delivered in a fine mist or powder.
- **Injected medicines** are delivered through a shot, and go directly into the bloodstream.

However you take a medicine, it is always important to follow the directions on the medicine label.

Reactions to Medications

Main Idea The effect of medicine depends on many factors.

Medicines can have a variety of effects. They can cause **side effects**, *reactions to medicine other than the one intended.* Some side effects may be mild, such as drowsiness, but others may be more severe, and can even cause death.

Medicine Interactions

When two or more medicines are taken together, or when a medication is taken with certain foods, the combination may have a different effect than when the medicine is taken alone. Types of medicine interactions include the following:

- **Additive interaction** occurs when *medicines work together in a positive way.* For example, an anti-inflammatory and a muscle relaxant may be prescribed to treat joint pain.
- **Synergistic effect**—*the interaction of two or more medicines that results in a greater effect than when each medicine is taken alone*—occurs when one medicine increases the strength of another.

- **Antagonistic interaction** occurs when *the effect of one medicine is canceled or reduced when taken with another medicine.* For example, someone who receives an organ transplant must take anti-rejection medicines. If the person is diabetic and takes insulin, the anti-rejection medicine may decrease the effectiveness of the insulin.

Tolerance and Withdrawal

When a person takes a medication for a long period of time, the body can become used to the medication. Problems that may occur include:

- **Tolerance** is a condition in which the body becomes used to the effect of a medicine. The body requires increasingly larger doses to produce the same effect. Sometimes a person will experience "reverse tolerance." In this condition, the body requires less medicine.

- **Withdrawal** occurs when a person stops using a medicine on which he or she has become physiologically dependent. Symptoms of withdrawal can include nervousness, insomnia, severe headaches, vomiting, chills, and cramps which gradually ease in time. Talk to your health care provider if you experience withdrawal.

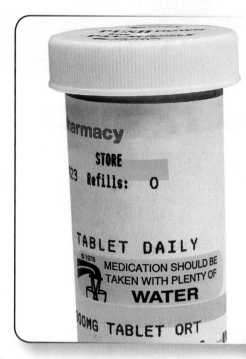

■ **Figure 19.4** Medicine labels include important information about possible side effects and interactions. *Why is it important to read this information before you take the medicine?*

LESSON 1 ASSESSMENT

 After You Read

Reviewing Facts and Vocabulary

1. Define the term *medicine* and the term *drugs*.

2. What types of medicines fight pathogens? What types of medicines prevent disease?

3. Compare a *synergistic effect* with an *antagonistic interaction.*

Thinking Critically

4. **Analyze.** Why are vaccines given to children at a young age?

5. **Evaluate.** Explain why people should not stop taking prescribed medications without talking to their doctor.

Applying Health Skills

6. **Accessing Information.** Use reliable online resources to find information on new and experimental drugs. Write a paragraph evaluating one of the drugs, and list the reasons why you think the information is reliable.

Writing Critically

7. **Descriptive.** Write a paragraph describing why it's important to take a medicine as your doctor prescribed.

Real Life Issues ....

After completing the lesson, review and analyze your response to the Real Life Issues question on page 524.

LESSON 2

Ingram Publishing

GUIDE TO READING

BIG Idea *Medicines are safe only if they are used for the intended purpose and according to the directions on the label.*

Before You Read

Make a T-Chart. Make a two-column chart like the one below. Label one column "Prescriptions" and the other column "OTCs." As you read, fill in the first column with information about prescription medicines. Fill in the second column with information about over-the-counter (OTC) medicines.

Prescriptions	OTCs

New Vocabulary

▸ prescription medicines
▸ over-the-counter (OTC) medicines
▸ medicine misuse
▸ medicine abuse
▸ drug overdose

Using Medicines Safely

Real Life Issues

Safety First. Monica is on the swim team and has an earache. She visits her doctor, who prescribes an anti-biotic. Monica is supposed to take the medicine for ten days. Her friend Amy, who is also on the swim team, thinks she may have an ear infection, too. Amy doesn't want to go to the doctor, though, so she asks Monica if she can share her medicine.

Writing *Write a dialogue in which Monica explains to Amy why she doesn't think she should share her medication.*

Standards for Medicines

Main Idea Medicines are regulated to make them safe.

All new medicines in the United States must meet standards set by the Food and Drug Administration (FDA). Before approving a drug for use, the FDA receives information about a medicine's chemical composition, intended use, effects, and possible side effects. Drug manufacturers test new medicines according to FDA guidelines. That includes completing at least three clinical trials for a drug. During a clinical trial, the drug is tested on human volunteers. They are monitored to determine the drug's effectiveness and to identify any harmful side effects.

Sometimes, if a drug hasn't yet completed clinical trials but is thought to be effective, people with life-threatening illnesses are allowed to use the drug. This usage is referred to as *experimental*. Patients are given experimental drugs only after clinical trials show that the drugs are safe and may be effective in treating their illness.

The FDA does not regulate herbal and dietary supplements. These supplements do not go through the same testing procedures or meet the same strict requirements for safety and proven effectiveness. Many people believe that herbal supplements are safe because some are advertised as "natural." Even supplements made from natural compounds can have harmful side effects or interactions. Never take any supplement without telling your health care provider first.

Prescription Medicines

Prescription medicines are *medicines that are dispensed only with the written approval of a licensed physician or nurse-practitioner.* A licensed pharmacist dispenses these medicines. Prescription medicines provide only the amount of medicine that is needed to treat your condition. If more medicine is needed, your health care provider must approve a refill. A prescription medicine should be taken *only* by the person whose name appears on the label.

■ **Figure 19.6** Medicines are regulated by the FDA, but herbal supplements are not. *Explain whether herbal supplements are safer than medicines.*

Over-the-Counter (OTC) Medicines

Over-the-counter (OTC) medicines, or *medicines you can buy without a doctor's prescription* are available without a prescription. The FDA considers these medicines to be safe if they are used as the label directs. However, all medicines can harm you if not used according to the directions.

While all OTC medicines are available without a prescription, the distribution of some OTC medicines is controlled. For example, cold medications that contain pseudoephedrine must be kept behind the pharmacy counter. These medications can be used to make highly addictive, illegal drugs.

Medicine Labels

When the FDA approves a medicine, it is considered safe when used as directed. The FDA requires that all prescription and OTC medicine labels contain information telling consumers how to use the medicine safely and effectively. The requirements for prescription and OTC medicine labels differ. Prescription medicine labels must also include any special instructions for taking the medicine, the prescribing doctor's name, the patient's name, the pharmacy's name and address, the date the prescription was filled, the prescription number, and whether refills are allowed. **Figure 19.7** on page 532 shows the information that must appear on all OTC medicine labels.

READING CHECK

Describe How would you obtain a prescription medicine?

Figure 19.7 Over-The-Counter Medicine

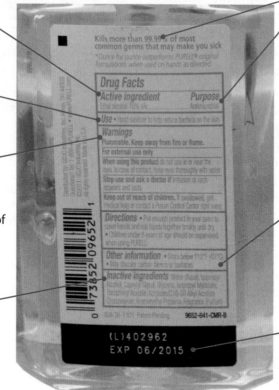

Active Ingredient: Ingredient that treats condition, including amount per unit

Uses: Conditions or symptoms treated by the product

Warnings: Side effects, interactions, when to talk to a doctor, when not to take the product, keep out of reach of children

Inactive Ingredients: Substances added to the product that do not help treat the condition, such as flavor and color

Purpose: Product category and what the product is supposed to do, such as antacid

Other Information and Directions: Some information may be printed on the opposite side of the label. This information may include how to take the medicine, how to store the product, and required information about certain ingredients, such as sodium

Expiration Date: The date you should no longer use the medicine

Medicine Misuse

> **Main Idea** Taking medicines unnecessarily or without following the label instructions is dangerous.

Medicine misuse can prevent the user from getting the full benefit of the medicine and can have serious health consequences. **Medicine misuse** involves *using a medicine in ways other than the intended use.* Examples of medicine misuse are:

- Failing to follow the instructions on or in the package
- Giving a prescription medicine to a person for whom it was not prescribed, or taking another person's medicine
- Taking too much or too little of a medicine
- Taking a medicine for a longer or shorter period than prescribed or recommended
- Discontinuing use of a medicine without informing your health care provider
- Mixing medicines without the knowledge or approval of your health care provider

Medicine Abuse

Intentionally taking medications for nonmedical reasons is **medicine abuse**. Most teens—96 percent—use medicines correctly. Some, however, think that medicines requiring a prescription and OTC medicines are safer than illegal drugs. Abusing any medicine is dangerous and illegal. Teens should avoid using drugs to:

- To lose weight or stay awake while studying. A healthy diet and exercise are the safest way to maintain a healthy weight. Getting plenty of sleep and managing your time wisely will help you study effectively.

- To fit in with peers. A dangerous trend is the emergence of "pill parties," where teens mix whatever OTC and prescription medicines are available. Mixing medicines, drugs, or alcohol is extremely dangerous.

- Avoid taking any medicine that was prescribed to someone else. Medicines are prescribed to treat a specific illness. It's illegal and unsafe to use a drug not prescribed to you.

One danger of medicine misuse is **drug overdose**—*a strong, sometimes fatal reaction to taking a large amount of a drug.* Misusing medicines can also lead to addiction. Never use a medicine other than how it is prescribed or intended.

READING CHECK

List What are some ways that teens might abuse medicines?

LESSON 2 ASSESSMENT

 After You Read

Reviewing Facts and Vocabulary

1. How do *prescription medicines* differ from *OTC medicines*?

2. List four pieces of information that must be on an OTC medicine label. Describe the purpose of each piece of information.

3. What is *medicine misuse*? How does it differ from *medicine abuse*?

Thinking Critically

4. Analyze. Why does the FDA regulate medicines and the information on medicine labels?

5. Evaluate. What are three ways you can avoid medicine abuse?

Applying Health Skills

6. Advocacy. Create a bookmark that gives information on the importance of correct medicine use.

Writing Critically

7. Expository. Create a script for a commercial or PSA that explains how people can use their health care providers, pharmacists, and medicine labels to ensure that they are using their medicines properly.

Real Life Issues ....

After completing the lesson, review and analyze your response to the Real Life Issues question on page 530.

Hands-On HEALTH

Activity — Stay Informed

Medicines can treat many health problems. If they're taken improperly, however, the same medicines can cause health problems. Working in teams of four or five, compete to determine which team can better define medicine terms. Then, create an informative pamphlet or poster on how to use medicines wisely.

What You'll Need

- textbook for each student
- six index cards per student
- pens or pencils
- small paper bag, one per team
- markers, paints, and poster board

What You'll Do

Step 1

Form teams of four or five. Write each vocabulary term on an index card and the definition of each term on a different index card.

Step 2

Place all the cards into a bag.

Step 3

At your teacher's signal, exchange bags with another team and match each term with the correct definition.

Apply and Conclude

Create a poster or pamphlet listing medicine facts. Persuade others to make positive health choices regarding the use of medicines.

Checklist: Advocacy

☑ Clear, health-enhancing message

☑ Support for the position with relevant information

☑ Awareness of the audience

☑ Encouragement of others to make healthful choices

LESSON 1

The Role of Medicines

Key Concepts

▶ Medicines can help manage chronic conditions, including allergies, diabetes, asthma, and depression, and can treat cancer.

▶ Medicines can be taken orally or topically. They can be inhaled or taken by injection.

▶ When taking medicines, some people may experience side effects or allergies.

▶ When medicines interact, they may have a different effect than intended.

Vocabulary

▶ medicines (p. 524)
▶ drugs (p. 524)
▶ vaccine (p. 525)
▶ side effects (p. 528)
▶ additive interaction (p. 528)
▶ synergistic effect (p. 528)
▶ antagonistic interaction (p. 529)

LESSON 2

Using Medicines Safely

Key Concepts

▶ The FDA regulates medicines and their distribution to make sure that medicines are safe and effective.

▶ Written approval in the form of a prescription is needed for some medicines, but OTC medicines can be bought without a prescription.

▶ Medicines can be dangerous if they are not used as directed.

Vocabulary

▶ prescription medicines (p. 531)
▶ over-the-counter (OTC) medicines (p. 531)
▶ medicine misuse (p. 532)
▶ medicine abuse (p. 533)
▶ drug overdose (p. 533)

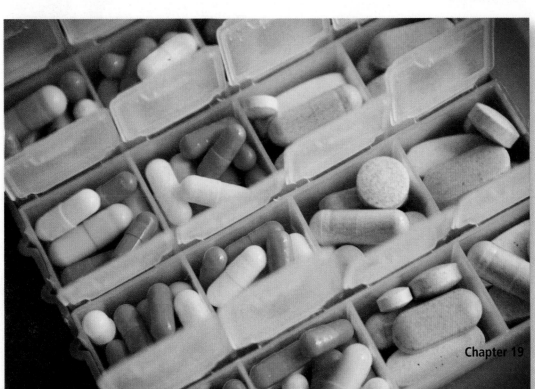

Ingram Publishing

Vocabulary Review

Correct the sentences below by replacing the italicized term with the correct vocabulary term.

1. A(n) *drug* is the interaction of two or more medications that results in a greater effect than when each medicine is taken alone.

2. People may experience *synergistic effects* while taking a medicine, which are effects that are not intended.

3. *Antagonistic interactions* are used to treat or prevent disease, and are dangerous when mixed with alcohol.

4. A(n) *side effect* occurs when medicines work together in a positive way.

5. A(n) *additive interaction* is a substance other than food that changes the structure or function of the body or mind.

Understanding Key Concepts

After reading the question or statement, select the correct answer.

6. A vaccine for polio will do which of the following?
   a. Causes the body to make antibodies to fight polio
   b. Causes people to develop the polio disease
   c. Protects the body against the measles virus
   d. Protects people against polio for a short time

7. What type of medicine might your doctor prescribe if you have an ear infection caused by bacteria?
   a. Antifungal
   b. Antitoxin
   c. Antiviral
   d. Antibiotic

8. What type of medicine might your doctor prescribe if you have the flu?
   a. Antitoxin
   b. Antibiotic
   c. Antiviral
   d. Antifungal

9. Why would you take an antihistamine?
   a. To cure an allergy
   b. To relieve allergy symptoms
   c. To slow an allergic reaction
   d. To build immunity

10. If a person takes two medicines at the same time and one is less effective than when taken alone, what is this called?
    a. Additive interaction
    b. Antagonistic interaction
    c. Side effect
    d. Synergistic effect

11. What type of medication is aspirin?
    a. Analgesic
    b. Antiviral
    c. Antihistamine
    d. Antibiotic

Thinking Critically

After reading the question or statement, write a short answer using complete sentences.

12. **Analyze.** How do bacteria become resistant to certain types of antibiotics?

13. **Evaluate.** Why do narcotic pain relievers require a doctor's prescription?

14. **Describe.** How can medication help people with mental illness?

15. **Compare.** Why might taking a medicine orally rather than taking a medicine topically help a person?

16. **Explain.** What can result if a person takes a medication for a long period of time?

17. **Evaluate.** Anne does not like the side effects of the prescription medicine she is taking. Why should she talk to her doctor before she stops taking the medicine?

18. **Describe.** How does a vaccine work? Can a vaccine help cure disease? If so, how?

19. **Explain.** What is the difference between a synergistic effect and an antagonistic effect when taking medicines?

LESSON 2

Vocabulary Review

Use the vocabulary terms listed on page 535 to complete the following statements.

20. Medicines that you can buy without a doctor's prescription are called _____.

21. If you take a large amount of a medicine, you could have a life-threatening reaction known as a(n) _____.

22. Intentionally taking medicines for nonmedical reasons is known as _____.

23. Medicines that are available only with the recommendation of a doctor, and are dispensed only by a licensed pharmacist are called _____.

24. Failing to follow the instructions on or included with a medicine package is an example of _____.

Understanding Key Concepts

After reading the question or statement, select the correct answer.

25. Which does *not* describe a prescription medication?
    a. Only a specified amount is distributed.
    b. Written approval is required.
    c. It should be taken only by the person it is prescribed to.
    d. It can be purchased without a doctor's recommendation.

26. Which must have proven safety and effectiveness before being sold?
    a. Prescription medicines
    b. Herb-based diet pills
    c. Protein shake drink mix
    d. Vitamins

27. Which of the following information is *not* on an OTC medicine label?
    a. The directions for taking the medicine
    b. The expiration date
    c. The inactive ingredients in the medicine
    d. The name of the pharmacy

28. Which of the following is a way a person could misuse a medicine?
    a. Taking only half of the prescription with your doctor's approval
    b. Saving half of an antibiotic prescription in case you get sick later
    c. Following the instructions on the medicine label
    d. Taking two medicines at the same time, as directed by a doctor

29. Why might teens abuse medicines?
    a. They take medicine only as prescribed by a doctor.
    b. They do not like the side effects of a certain medicine.
    c. They believe the medicine may help them study longer.
    d. They think the medicine is hard to obtain.

30. Which is *not* a risk of abusing medicines?
    a. Addiction
    b. Death from heart failure
    c. Paranoia
    d. Taking too little medicine

31. Which of the following organizations tests and approves all prescription medicines before they are sold to the public?
    a. the Centers for Disease Control and Prevention
    b. the U.S. Department of Agriculture
    c. the Food and Drug Administration
    d. the Consumer Product Safety Commission

Thinking Critically

After reading the question or statement, write a short answer using complete sentences.

32. **Explain.** Why should you talk to your doctor before taking herbal supplements?

33. **Analyze.** Why does the FDA limit the distribution of certain OTC medicines?

34. **Explain.** Why should you always read the label before taking a medicine?

35. **Evaluate.** How do the FDA guidelines for approving medicines protect the health of the public?

36. **Discuss.** How can taking medicines not prescribed to you, or mixing medicines, harm your health?

Technology PROJECT-BASED ASSESSMENT

Explaining Vaccines

Background

Your body's immune system has the ability to recognize and destroy bacteria and viruses. It remembers features of these pathogens to fight them. By exposing your immune system to parts of a pathogen, or a pathogen that has been altered so it cannot hurt you, a vaccine allows your body to prepare for that pathogen if you become infected.

Task

Create an online cartoon that explains to young children how vaccines work.

Audience

Students in grades 1 through 3

Purpose

Explain why vaccines work, and how important they are in maintaining your health.

Procedure

1. Review the text and write notes describing some basic facts about vaccines.

2. Conduct an online search to learn how the body's immune system works. Find answers to these questions: How do vaccines prepare the body to fight pathogens? How do vaccines and antitoxins fight disease?

3. Search for online cartoons that are geared to children in grades 1 through 3 to see the different styles and formats that you might use.

4. Create and organize your cartoon. Use words and pictures that clearly show how vaccines work.

5. Present your cartoon to your class or, if possible, to students in grades 1 through 3.

Math Practice

Interpret Graphs. The table below shows the percentages of nonmedical use of psychotherapeutics among 12- to 17-year-olds. Use the table to answer Questions 1–3.

Non-Medical Use of Psychotherapeutics Among 12- to 17-Year-Olds

	Past Year		Past Month	
	1999	2000	1999	2000
Any Psycho-therapeutic*	7.1%	7.1%	2.9%	3.0%
Pain Relievers	5.5%	5.4%	2.1%	2.3%
Tranquilizers	1.6%	1.6%	0.5%	0.5%
Sedatives	0.5%	0.5%	0.2%	0.2%
Stimulants**	2.1%	2.4%	0.7%	0.8%

* Denotes the non-medical use of any prescription-type pain reliever, tranquilizer, stimulant, or sedative; does not include over-the-counter drugs.
** Includes methamphetamine

Source: U.S. Department of Health and Human Services, Substance Abuse and Mental Health Services Administration, NHSDA

1. Of the 12,000 students surveyed in 1999, how many of them have never used any psychotherapeutic drugs in the past year?
   A. 10,000
   B. 11,148
   C. 852
   D. 967

2. After reviewing the chart above, determine the mean, median, and mode of the 2000 data.

3. On one line graph, show the trend, both annually and monthly, of prescription drug abuse in 1999.

Reading/Writing Practice

Understand and Apply. Read the passage below, and then answer the questions.

> Aspirin was first introduced by a German company in 1899. Centuries earlier, people used similar chemicals to ease pain and reduce fever. The ancient Greeks used a bitter powder extracted from the bark of willow trees to treat pain. In the 1700s, physicians treated patients with another willow-derived substance that was later discovered to be the chemical salicin.
>
> By the mid-1800s, European pharmacists used an acid form of salicin to treat arthritis. However, patients who took salicylic acid would often suffer from a painful side effect, a severe upset stomach. With the introduction of the milder aspirin, many patients experienced pain relief with fewer stomach problems.

1. What was the author's purpose in writing this piece?
   A. To list the dosages of aspirin
   B. To describe the types of pain relievers
   C. To explain how aspirin works
   D. To tell how aspirin products are used

2. Which sentence best represents the main idea of the second paragraph?
   A. Arthritis causes pain and swelling.
   B. Salicylic acid relieves pain, but can irritate the digestive tract.
   C. European pharmacists tend to prescribe painkillers other than aspirin.
   D. Older aspirin-like products caused stomach irritation.

3. Write a pamphlet that outlines the importance of following a doctor's instructions when taking medicines.

National Education Standards

Math: Measurement and Data, Data Analysis
Language Arts: LACC.910.L.3.6

Lesson 1
The Health Risks of Tobacco Use

BIG Idea *The chemicals in all tobacco products harm your body.*

Lesson 2
Choosing to Live Tobacco-Free

BIG Idea *Avoiding tobacco use will bring lifelong health benefits.*

Lesson 3
Promoting a Smoke-Free Environment

BIG Idea *Secondhand smoke is harmful, but there are ways you can reduce your exposure.*

Activating Prior Knowledge

Using Visuals As this billboard shows, people are getting the message out about the dangers of tobacco use. Write a short paragraph comparing and contrasting the anti-tobacco and pro-tobacco messages you have seen in advertisements.

Chapter Launchers

Health in Action

Discuss the **BIG** Ideas

Think about how you would answer these questions:

▶ In what ways does tobacco harm your body?

▶ What are the consequences of tobacco use?

▶ Why should you avoid secondhand smoke?

Assess Your Health

Read each statement. On a separate sheet of paper, write "yes," "sometimes," or "no" based on your typical behavior.

1. I avoid tobacco products.

2. I stay away from situations where tobacco products may be used.

3. I practice refusal skills in advance so that I know what I would say if someone offers me tobacco.

4. I surround myself with people who do not use tobacco.

5. I know how to find help if someone I know wants to quit smoking.

A "yes" response shows that you practice healthy behaviors. "Sometimes" indicates that you should analyze and possibly modify your behavior. A "no" response means that you should modify the behavior.

 **GUIDE TO READING**

BIG Idea *The chemicals in all tobacco products harm your body.*

Before You Read

Create a Venn Diagram. Draw a Venn diagram that has two overlapping circles. Label one circle "Tobacco Smoke" and the other circle "Smokeless Tobacco." Write the risks of each in the circles. Put the risks shared by both in the area where the circles overlap.

Tobacco Smoke

Smokeless Tobacco

New Vocabulary

▶ addictive drug
▶ nicotine
▶ stimulant
▶ carcinogen
▶ tar
▶ carbon monoxide
▶ smokeless tobacco
▶ leukoplakia

The Health Risks of Tobacco Use

Real Life Issues •

Dangers of Tobacco Use. A teen who decides to start using tobacco can affect their health now and in the future. Source: U.S. Department of Health and Human Services, *National Women's Health Information Center.*

> In 2011, 18% of high school students smoked cigarettes.

> 13% of high school students smoked cigars

Writing *Considering what you already know about tobacco use, write a letter expressing your feelings to a close friend or relative who has begun using tobacco.*

Health Risks of Tobacco Use

Main Idea **All forms of tobacco contain chemicals that are dangerous to your health.**

Advertisements for tobacco products often feature healthy, attractive people, sending the message that using tobacco has no health consequences. So what's the truth about tobacco? All tobacco products display warning labels stating that using tobacco products can be harmful to an individual's health.

Medical studies have shown that tobacco use is the leading cause of preventable death and disability in the United States. Any form of tobacco use, such as smoking, chewing, or dipping tobacco, can cause health problems. Smoking has been linked to lung disease, cancers, and heart disease. About 90 percent of adult smokers began the habit as teenagers. Most teens think that they can just quit whenever they choose. The reality is that quitting is difficult. It's easier to avoid tobacco use rather than quit later.

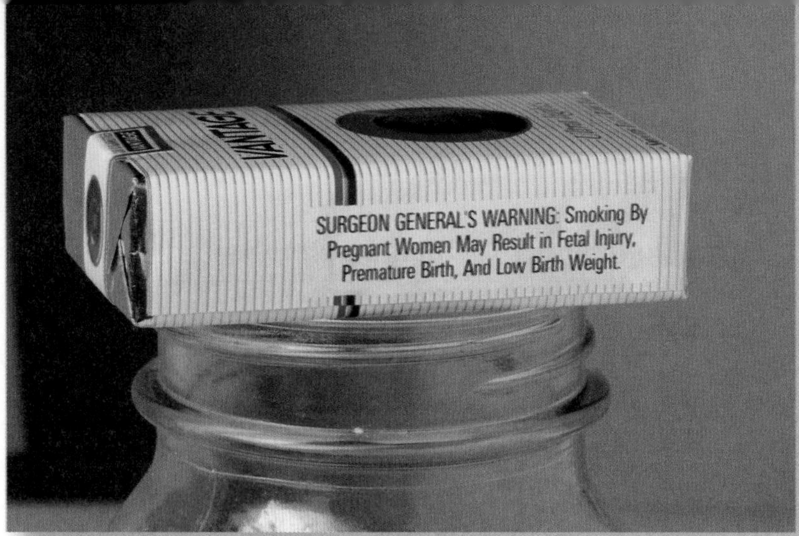

■ **Figure 20.1** All tobacco products must carry warning labels showing they are harmful to your health. *Why do people who use tobacco ignore these warnings?*

Nicotine

Tobacco users have difficulty quitting because tobacco contains an **addictive drug**, *a substance that causes physiological or psychological dependence*. All tobacco products contain **nicotine**, *the addictive drug found in tobacco leaves*. Nicotine is a **stimulant**, *a drug that increases the action of the central nervous system, the heart, and other organs*. Using nicotine raises blood pressure, and increases the heart rate.

Poisonous Substances in Tobacco Smoke

Tobacco is an addictive and toxic drug. It's a **carcinogen**, *a cancer-causing substance*. Tobacco smoke contains tar and carbon monoxide. It also contains the same poisonous compounds found in products such as paint, rat poison, and toilet cleaner.

Tar Cigarette smoke contains **tar**, *a thick, sticky, dark fluid produced when tobacco burns*. The tar damages a smoker's respiratory system by paralyzing and destroying cilia, the tiny hairlike structures that line the upper airways and protect the body against infection. Tar also destroys the alveoli, or air sacs, which absorb oxygen and rid the body of carbon dioxide. Lung tissue is also damaged, reducing lung function. Smokers are susceptible to diseases such as bronchitis, pneumonia, emphysema, heart disease, and cancer. According to the CDC, more people in the United State die from lung cancer than any other type of cancer.

Carbon Monoxide **Carbon monoxide**, *a colorless, odorless, and poisonous gas,* is another compound found in cigarette smoke. It is absorbed more easily than oxygen. Carbon monoxide deprives the body's tissues and cells of oxygen. It also increases the risk of high blood pressure, heart disease, hardening of the arteries, and other circulatory problems.

■ **Figure 20.2** Cigarette filters do not protect smokers from the more than 50 carcinogens, including cyanide and arsenic, which are in tobacco products. The filters themselves contain poisonous chemicals such as those used in insecticides, paint, toilet cleaner, antifreeze, and explosives. *How can you warn others about the risks of using tobacco products?*

Academic Vocabulary

decade *(noun):* a group or set of ten

Pipes, Cigars, and Smokeless Tobacco

Main Idea No tobacco product is safe to use.

The dangers of tobacco use are not limited to smoking cigarettes. The smoke from pipes and cigars also causes serious health consequences. Cigars contain significantly more nicotine and produce more tar and carbon monoxide than cigarettes. One cigar can contain as much nicotine as an entire pack of 20 cigarettes. Pipe and cigar smokers also increase the risk of developing cancers of the lips, mouth, throat, larynx, lungs, and esophagus.

Another form of tobacco that some believe to be safer to use than cigarettes is **smokeless tobacco**, *tobacco that is sniffed through the nose, held in the mouth, or chewed.* Smokeless (sometimes called "spit") tobacco products are *not* a safe alternative to smoking. The nicotine and carcinogens in these products are absorbed into the blood through the mucous membranes in the mouth or the digestive tract.

The harmful chemicals of smokeless tobacco are absorbed into the body at levels up to three times the amount of a single cigarette. That's because the exposure to harmful chemicals in smokeless tobacco is often three times longer than that of a smoked cigarette. Using smokeless tobacco also irritates the sensitive tissues of the mouth, causing **leukoplakia** (loo-koh-PLAY-kee-uh), or *thickened, white, leathery-looking spots on the inside of the mouth that can develop into oral cancer.* Smokeless tobacco causes cancers of the mouth, throat, larynx, esophagus, stomach, and pancreas. People who chew eight to ten plugs of tobacco each day take in the same amount of nicotine as a smoker who smokes two packs of cigarettes a day. Smokeless tobacco is as addictive as smoked tobacco, making quitting just as difficult as it is for someone who uses smoked tobacco.

Harmful Effects of Tobacco Use

Main Idea Tobacco use causes both short-term and long-term damage to your body.

Health officials have warned the public about the dangers of tobacco use for several **decades**. **Figure 20.3** shows how smoking can damage one body system, the lungs. If a pregnant female smokes, she risks the health of her fetus, as well as her own health. Pregnant women who smoke during pregnancy risk giving birth to an infant with a low birth weight and other health problems.

Short-Term Effects

Some effects of tobacco use occur immediately. These short-term effects include the following:

- **Brain chemistry changes.** The addictive properties of nicotine cause the body to crave more of the drug. The user may experience withdrawal symptoms, such as headaches, nervousness, and trembling as soon as 30 minutes after the last tobacco use.

- **Respiration and heart rate increase.** Breathing during physical activity becomes difficult and endurance is decreased. Nicotine may cause an irregular heart rate.

- **Taste buds are dulled and appetite is reduced.** Tobacco users often lose much of their ability to enjoy food.

- **Users have bad breath, yellowed teeth, and smelly hair, skin, and clothes.** If tobacco use continues for any length of time, these unattractive effects can become permanent.

Long-Term Effects

Over time, tobacco use can cause damage to many body systems, as shown in **Figure 20.4** on page 546. People who are exposed to others who smoke can also suffer many health problems. Here are some of those health problems:

- **Chronic bronchitis** can occur when the cilia in the bronchi become so damaged that they are useless. This leads to a buildup of tar in the lungs, causing chronic coughing and excessive mucus secretion.

- **Emphysema** is a disease that destroys the tiny air sacs in the lungs. The air sacs become less elastic, making it more difficult for the lungs to absorb oxygen. A person with advanced emphysema uses up to 80 percent of his or her energy just to breathe.

- **Lung cancer** can develop when the cilia in the bronchi are destroyed, and extra mucus cannot be expelled. Cancerous cells can multiply, block the bronchi, and move to the lungs. Nearly 90 percent of lung cancer deaths are caused by smoking.

- **Coronary heart disease and stroke** can be caused by nicotine. Nicotine constricts blood vessels, which cuts down blood flow to the body's limbs. Nicotine also contributes to plaque buildup in the blood vessels, which can lead to hardened arteries, a condition called *arteriosclerosis*. Arteries may become clogged, increasing the risk of heart attack and stroke. The risk of developing heart disease is greater for smokers than for nonsmokers.

- **A weakened immune system** from long-term tobacco use makes the body more vulnerable to disease.

■ **Figure 20.3** Smokers cause severe damage to their lungs. Compare the healthy lung (top) with the one damaged by tobacco smoke (bottom). *How do tar and the other substances in tobacco smoke affect the respiratory system and its ability to function?*

 READING CHECK

Describe Name two ways in which the health of long-term tobacco users will suffer.

Figure 20.4

Long-Term Health Risks of Tobacco

Tobacco use increases your risk of developing many severe health problems.

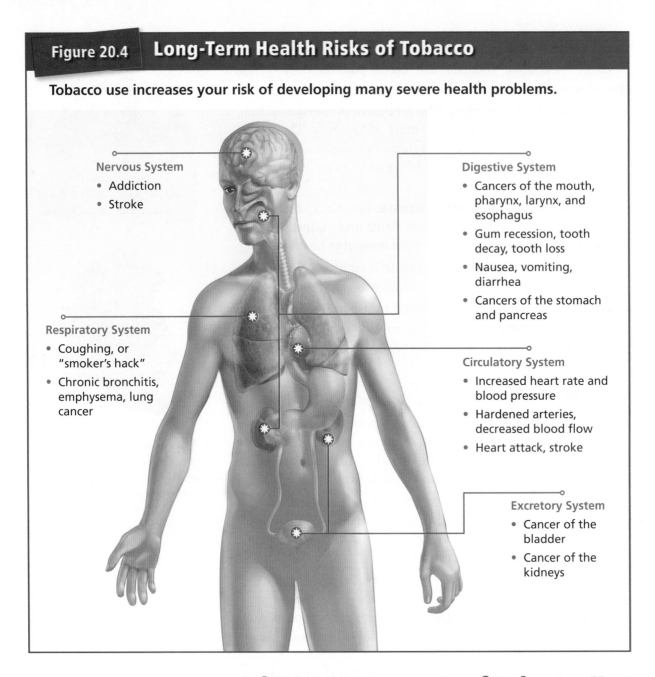

Nervous System
- Addiction
- Stroke

Digestive System
- Cancers of the mouth, pharynx, larynx, and esophagus
- Gum recession, tooth decay, tooth loss
- Nausea, vomiting, diarrhea
- Cancers of the stomach and pancreas

Respiratory System
- Coughing, or "smoker's hack"
- Chronic bronchitis, emphysema, lung cancer

Circulatory System
- Increased heart rate and blood pressure
- Hardened arteries, decreased blood flow
- Heart attack, stroke

Excretory System
- Cancer of the bladder
- Cancer of the kidneys

Other Consequences of Tobacco Use

Main Idea As well as health risks, tobacco use is costly.

Making the decision to avoid the use of tobacco products will safeguard your health.

- **Costs to society.** Tobacco-related illnesses cost the United States about $167 billion each year. Productivity suffers when smokers call in sick due to tobacco-related illnesses.
- **Cost to individuals.** A person smoking one pack of cigarettes a day will spend about $3,561 a year on the habit.
- **Legal consequences.** Selling tobacco products to individuals under the age of 18 is illegal. Using tobacco products on school property may lead to suspension or expulsion.

Real World CONNECTION

Health Risks of Tobacco

Latoya knows that tobacco use causes serious health problems. She wants to encourage her friends to avoid tobacco use. She does an online search to learn more about the health effects of tobacco use. Latoya decides to search Web sites such as the CDC and the National Cancer Institute (NCI) to find statistics about tobacco-related deaths.

Activity Technology

Using the CDC and NCI Web sites, along with other reliable and safe Web sites, conduct an Internet search to learn more about tobacco use among teens. Search for the following information:

1. How many teens begin smoking each year?

2. How can tobacco use affect a teen's physical health?

3. What impact can tobacco use have on a teen's mental/emotional and social health?

Once your research is complete, create a Web page urging teens who use tobacco to quit. Include information urging teens who have never used tobacco not to start the habit.

LESSON 1 ASSESSMENT

 After You Read

Reviewing Facts and Vocabulary

1. What is an *addictive drug*? What is the addictive drug in tobacco?

2. List three types of toxic substances found in cigarette smoke. Why are these substances harmful?

3. Explain four ways using tobacco immediately affects your body.

Thinking Critically

4. **Identify.** What are three ways in which tobacco use affects the respiratory system?

5. **Analyze.** In addition to protecting your health, explain reasons you should not use any form of tobacco.

Applying Health Skills

6. **Advocacy.** Write an editorial for a newspaper that encourages people to quit using tobacco products, and explain the long-term effects of tobacco use on the body.

Writing Critically

7. **Persuasive.** Create a pamphlet raising awareness of the health risks of tobacco use. Include information on the long-term effects of tobacco use.

Real Life Issues ....

After completing the lesson, review and analyze your response to the Real Life Issues question on page 542.

GUIDE TO READING

BIG Idea *Avoiding tobacco use will bring lifelong health benefits.*

Before You Read

Make a T-Chart. Make a two-column chart like the one below. Label one column "Start" and the other column "Quit." Fill in the first column with reasons why teens start using tobacco. Fill in the second column with reasons why tobacco users want to quit using tobacco.

Start	Quit

New Vocabulary

▶ nicotine withdrawal
▶ nicotine substitutes
▶ tobacco cessation program

Choosing to Live Tobacco-Free

Real Life Issues ....

Quitting Smoking. Juan started smoking a year ago. Juan does not like the hold that tobacco has on him, so he has decided to quit. It's been harder than he thought it would be. His friends Joe and Pamela want to help Juan become tobacco-free, but they aren't sure how.

Writing *Write a short essay explaining how Juan's friends can encourage him to quit and support his efforts.*

Teens and Tobacco

Main Idea Fewer teens are starting to use tobacco.

The number of nonsmokers in the United States, including teens, is on the rise. Knowing the health risks of tobacco use helps teens make the healthful decision to stay tobacco-free. However, some teens are influenced by tobacco company advertisements and other pressures. These teens may begin to use tobacco.

Why Some Teens Use Tobacco

Teens start smoking for many reasons. Some teens falsely believe that smoking will help control their weight or cope with stress. Others believe that smoking will make them seem mature and independent. The truth is that smoking reduces the body's capacity for physical activity, so it actually may lead to weight gain. Health problems caused by tobacco use and nicotine dependency may increase the tobacco user's stress level. Many times, teens are influenced to try tobacco products by movies, TV, and advertisements. Media images may convince teens that tobacco use is glamorous.

Figure 20.5 Teens who choose a tobacco-free lifestyle will feel mentally and physically better than teens who use tobacco. *What are some healthful ways to control weight and relieve stress?*

Reduced Tobacco Use Among Teens

More teens recognize the health risks of tobacco use and are avoiding the use of tobacco products. The CDC reports that 82 percent of high school students nationwide do not smoke. Several factors contribute to this trend:

- **Tobacco legislation.** In 1998, tobacco companies and 46 states reached a legal settlement that restricts tobacco advertising aimed at young people. Tobacco companies are required to fund ads that discourage young people from smoking. It is illegal for anyone under the age of 18 to purchase tobacco products in the United States.

- **No-smoking policies.** Legislation has limited smoking in public places and businesses.

- **Family values.** Teens whose parents avoid tobacco use are more likely to avoid tobacco use themselves.

- **Positive peer pressure.** Teens who do not smoke act as healthy role models for other teens.

- **Health risks.** More teens understand that tobacco use can lead to diseases, such as health disease, cancer, and respiratory problems.

 READING CHECK

Explain Why has tobacco use among teens decreased?

| Figure 20.6 | **Teens Smoking Less** |

High school students who reported smoking a cigarette in the last 30 days

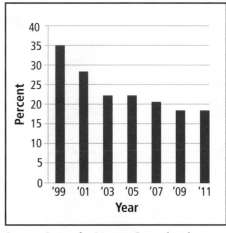

Source: Center for Disease Control and Prevention, 2012

Benefits of Living Tobacco-Free

Main Idea A tobacco-free lifestyle has many benefits.

If you do not use tobacco, you have better cardiovascular endurance and lung function. You can improve your fitness level and athletic performance. When you avoid tobacco, you reduce your risk of lung cancer, heart disease, and stroke.

Living tobacco-free has mental/emotional and social benefits, too. You will have a sense of freedom because you know that you are not dependent on an addictive substance. You will have less stress because you do not have to worry about tobacco-related health problems. You will have more confidence in social situations because you look and feel better.

Strategies for Avoiding Tobacco

The best way to avoid the negative consequences of tobacco use is never to start using tobacco products. With these strategies, you can stick to your decision to live tobacco-free:

- **Surround yourself with positive influences.** Being around people who share your healthy-living values and beliefs will strengthen your commitment to lead a tobacco-free life. Choose friends who do not use tobacco.

- **Reduce peer pressure.** By staying away from situations where tobacco products may be used, you reduce the chance of being pressured to use tobacco.

- **Be prepared with refusal skills.** Practice in advance what you will say if someone offers you tobacco. Be assertive, and leave the situation if the pressure continues. Be confident and stand up for your healthy choices.

Quitting Tobacco Use

Main Idea There are good reasons for quitting tobacco use.

Teens who use tobacco give these reasons for quitting:

- They begin to have health problems, such as asthma, coughing, or respiratory infections.

- They realize the high cost of tobacco or find it difficult to purchase tobacco products if they are under 18.

- They realize that using tobacco can lead to other risky behaviors, such as the use of alcohol and other drugs.

- They understand the damaging effects of secondhand smoke and do not want to harm others.

- They feel more powerful because they are not controlled by an addiction to nicotine.

■ **Figure 20.7** Health care professionals can help tobacco users find the resources they need to successfully quit using tobacco. *Why is it difficult for people to quit?*

Alistair Berg/Getty Images

Figure 20.8 Choosing friends who support a tobacco-free lifestyle will help you stay tobacco-free. *What are some health benefits of living tobacco-free?*

Ending the Addiction Cycle

Overcoming nicotine addiction can be difficult, but millions of people have succeeded. It is not impossible. It is common to experience symptoms of **nicotine withdrawal**, *the process that occurs in the body when nicotine, an addictive drug, is no longer used.* Symptoms can include irritability, difficulty concentrating, anxiety, sleep disturbances, and cravings for tobacco. To relieve the symptoms, some people use **nicotine substitutes**, *products that deliver small amounts of nicotine into the user's system while he or she is trying to give up the tobacco habit.* These include gum, patches, nasal sprays, and inhalers. Some are over-the-counter products; others require a doctor's prescription. Smoking while using nicotine substitutes is dangerous due to increased nicotine exposure.

Getting Help to Quit Tobacco Use

People who want to quit can try the following strategies:

- **Prepare for the quit day.** Set a target date, and stick to it. Prepare your environment and avoid tobacco triggers.

- **Get support and encouragement.** Tell everyone you know about your plan to quit. Support from family and friends will increase a person's chance of success.

- **Access professional health services.** Seek advice from a doctor, enroll in a **tobacco cessation program**—*a course that provides information and help to people who want to stop using tobacco*—or join a support group. Other helpful resources include the American Lung Association, the American Cancer Society, the Centers for Disease Control and Prevention (CDC), and local hospitals. Many high schools also sponsor tobacco cessation programs.

- **Replace tobacco use with healthy behaviors.** Try sugarless gum or carrots until cravings pass. Physical activity, good nutrition, avoiding drugs and alcohol, and using stress-management techniques can help you succeed.

READING CHECK

Describe What are the symptoms of nicotine withdrawal?

Health Skills Activity

Advocacy

Helping Teens Stay Tobacco-Free

Isabella recently quit smoking with the help of her friend, Anthony. Isabella told Anthony that she started smoking after giving in to the peer pressure. She also believed that tobacco products were not harmful to her health. She felt that tobacco products in a locked case at her local supermarket looked appealing. She realized how tobacco products were harming her health when she couldn't run a mile without stopping to catch her breath. Anthony and Isabella want to discourage other teens from using tobacco by asking store managers to use less prominent product displays.

Writing Write a persuasive letter to a store manager asking that tobacco products be removed from open display areas. Include the following information in your letter:

1. Health problems caused by tobacco use.
2. The number of deaths each year from tobacco use.
3. The financial cost of tobacco-related health issues.

LESSON 2 ASSESSMENT

 After You Read

Reviewing Facts and Vocabulary

1. What are four reasons that smoking among teens is on a downward trend?
2. List three reasons that you might use to convince a friend to quit using tobacco products.
3. Why might some people use nicotine substitutes when quitting smoking?

Thinking Critically

4. **Evaluate.** How will staying tobacco-free benefit your physical, mental/emotional, and social health?
5. **Synthesize.** Explain how the media influences teens to use and not to use tobacco products.

Applying Health Skills

6. **Refusal Skills.** Write a scenario describing a teen being pressured to use tobacco. Develop three refusal statements that the teen can use to avoid tobacco use.

Writing Critically

7. **Narrative.** Write a short story from the point of view of someone who is trying to quit smoking. Include at least three reasons for quitting and why it might be difficult to quit.

Real Life Issues ....

After completing the lesson, review and analyze your response to the Real Life Issues question on page 548.

Promoting a Smoke-Free Environment

Photodisc/Getty Images

Real Life Issues ·····························

Avoiding Secondhand Smoke.
Ken visits his aunt and uncle once a week. Ken's uncle smokes and often lights up a cigarette while Ken is in the room. Ken doesn't like breathing secondhand smoke, but he doesn't want to offend his uncle. He is not sure what he can do to convince his uncle not to smoke.

Writing *Write a paragraph about what you would do in this situation if you were Ken.*

Health Risks of Tobacco Smoke

Main Idea Tobacco smoke can harm nonsmokers.

The health effects of tobacco smoke affect smokers and nonsmokers alike. Nonsmokers who breathe air containing tobacco smoke are also at risk for health problems. **Environmental tobacco smoke (ETS)**, or secondhand smoke, is *air that has been contaminated by tobacco smoke*. ETS is composed of **mainstream smoke**, *the smoke exhaled from the lungs of a smoker*, and **sidestream smoke**, *the smoke from the burning end of a cigarette, pipe, or cigar*. Because mainstream smoke has been exhaled by a smoker, it contains lower concentrations of carcinogens, nicotine, and tar. For this reason, sidestream smoke is more dangerous than mainstream smoke. ETS from cigarettes, cigars, and pipes contains more than 4,000 chemical compounds. More than 50 of those chemicals are cancer-causing carcinogens. Some studies show that infants and young children who are exposed to ETS are more likely to develop asthma than their peers who are not exposed to ETS. Inhaling ETS is a serious health risk.

GUIDE TO READING

BIG Idea *Secondhand smoke is harmful, but there are ways you can reduce your exposure.*

Before You Read

Make an Outline. Use the headings of this lesson to make an outline of what you will learn about the risks of smoking. Use a format like this to help you organize your notes.

    I.
       A.
          1.
          2.
       B.
    II.

New Vocabulary

• environmental tobacco smoke (ETS)
• mainstream smoke
• sidestream smoke

Review Vocabulary

• *Healthy People* (Ch.1, L.4)

■ **Figure 20.9** Smoking is prohibited in many restaurants, and some restaurants are required to have a nonsmoking section. *How does this rule protect the health of restaurant customers and employees?*

READING CHECK

Analyze How do mainstream smoke and sidestream smoke differ?

Health Risks to Nonsmokers

Secondhand smoke causes about 3,400 deaths from lung cancer every year. ETS causes eye irritation, headaches, ear infections, and coughing in people of all ages. It worsens asthma and other respiratory problems, and it increases the risk of coronary heart disease.

Health Risks to Unborn Children and Infants

Choosing to live tobacco-free is one of the healthiest choices a pregnant female can make for her baby. Smoking during pregnancy can seriously harm the developing fetus. Nicotine passes through the placenta, constricting the blood vessels of the fetus. Carbon monoxide reduces the oxygen levels in the blood of the mother and fetus. This increases the risk of impaired fetal growth, spontaneous miscarriage and prenatal death, premature delivery, low birth weight, deformities, and stillbirths. The infant may also suffer from growth and developmental problems during early childhood.

Babies of mothers who smoked during pregnancy or who are exposed to ETS are more likely to die of sudden infant death syndrome (SIDS). Infants exposed to ETS after birth are twice as likely to die of SIDS. They may have severe asthma attacks, ear infections, or respiratory tract infections.

■ **Figure 20.10** Parents protect the health and development of their children by staying tobacco-free. *How can tobacco use harm young children?*

Health Risks to Young Children

Young children are particularly sensitive to ETS. Children of smokers are more likely to be in poor health than children of nonsmokers. Consider these facts:

- Children of smokers tend to have a higher incidence of sore throats, ear infections, and upper respiratory problems than children of nonsmokers.

- Secondhand smoke can slow lung development. Children who live with smokers are more likely to have weaker lungs than children of nonsmokers.

Also, children learn by example. The children of smokers are more than twice as likely to smoke themselves.

Reducing Your Risks

Main Idea You can take action to reduce the effects of ETS.

Since you spend so much time in your home, you can make an effort to minimize the health effects of ETS. If a family member smokes, encourage that person to quit by telling him or her the health effects of tobacco smoke. Try to establish smoke-free areas in the house, or make a rule that smokers go outside. If a smoker cannot always smoke outside, air cleaners can help remove some contaminants from the air. Open windows to allow fresh air in.

If you have a visitor who smokes, politely request that he or she does not smoke inside your home. If you are visiting a home in which someone smokes, try to stay outside or in a different room as much as possible. Ask to open the windows to provide fresh air. Suggest meeting elsewhere, such as in your home or at a library. In restaurants and other public places, request seating in a nonsmoking area. Express your preference wherever you can for a smoke-free environment.

I hate the smell of cigarette smoke. It reminds me that my grandfather has emphysema. He coughs and struggles to breathe all the time. When he was my age, he ran track and played football. Smoking made it hard for him to breathe, and so he stopped being active. I don't ever plan to smoke because I've seen what it did to my grandfather's health. For more physical activity ideas, visit the Fitness Handbook and Fitness Zone sites in ConnectEd.

Image Source/Getty Images

TEENS Making a Difference

"Take a stand against tobacco."

A Voice for Change

Diamond J., of South Carolina, is a member of Rage Against the Haze. This youth-run anti-tobacco movement is for teens ages 13 to 17. Its sole purpose is to spread the anti-tobacco message to as many people in as many places possible.

During the fall, RAGE volunteers are busy traveling to Friday night football games at area high schools. "When teens come to our table, we share information about RAGE. We often get students asking to join because they want to take a stand against tobacco in their school and in their community."

For Diamond, volunteering with RAGE is a lot of fun because she gets to travel and meet new people. But it's more than just having fun. "I have lost loved ones to tobacco use. I'm volunteering my time for something that is very worthwhile."

Activity Write your answers to the following questions in your personal health journal.

1 Does your school or community have a youth-run program similar to Rage Against the Haze? If so, what specific activities does it do?

2 Why does RAGE target children and teens with an anti-tobacco message?

3 What can you do to make a difference in the tobacco habits of your peers?

Creating a Smoke-Free Society

READING CHECK

List What are three ways you can reduce your exposure to ETS?

Main Idea In most states, it is illegal to sell tobacco to teens under the age of 18, and it is illegal to smoke in public places.

Medical research shows that any exposure to secondhand smoke can cause health problems. When a smoker chooses to smoke, that person makes a decision that affects his or her health, and the health of others. In the United States, efforts to create a smoke-free society continue to grow. The health effects of tobacco use, and the cost of health-related illnesses are just some of the reasons for this movement.

According to the U.S. Surgeon General, the only way to fully protect people from the damaging health effects of ETS is to prohibit smoking in public places. Many states now prohibit smoking in any workplace. Advertisements aimed at young people encourage them not to smoke, and public service announcements encourage parents not to smoke near their children.

Supporting National Health Goals

One of the goals of *Healthy People 2020* is to reduce tobacco use and the number of tobacco-related deaths. States and local communities are also supporting the efforts to create a smoke-free society. Laws prohibiting the sale of tobacco to minors have been enacted. Some states have successfully sued tobacco companies to recover the costs of treating tobacco-related illnesses. The money awarded in these cases may be used to fund anti-smoking campaigns or to offset the medical costs related to tobacco use. Community activities that promote a healthy lifestyle provide everyone with the opportunity to practice healthful behaviors. These activities allow anyone to become a role model encouraging others to avoid tobacco use.

■ **Figure 20.11** Laws restrict where people may smoke, as well as who can buy tobacco products. *What are the benefits of having smoke-free public places?*

LESSON 3 ASSESSMENT

 After You Read

Reviewing Vocabulary and Facts

1. What is *environmental tobacco smoke,* and what chemical does it contain?

2. List three ways that ETS affects children.

3. What are two public policies aimed at reducing ETS?

Thinking Critically

4. **Analyze.** How can smoking during pregnancy have long-term effects on the child?

5. **Explain.** Why should you try to avoid ETS, and how can you reduce your exposure to ETS?

Applying Health Skills

6. **Analyzing Influences.** Keep a log of how many tobacco ads you see in one week. Note what type of media was used (print, audio, video), where you saw the ad, and who was targeted. List steps you could take to eliminate these influences in your community.

Writing Critically

7. **Personal.** Describe a situation in which you were exposed to ETS. Include how you felt afterward physically and mentally. Then write about how you could have prevented being exposed to ETS in that situation.

Real Life Issues ....

After completing the lesson, review and analyze your response to the Real Life Issues question on page 553.

Hands-On
HEALTH

Activity **Creating a Smoke-Free Society for the Future**

You have been asked to represent the smoke-free youth of America. Prepare a speech to persuade government officials to create laws prohibiting tobacco use in public buildings and eliminating external factors that influence people to use tobacco.

What You'll Need

- paper and pencil
- Internet and valid health resources

What You'll Do

Step 1

For two days, record and analyze the external factors that you observe in the media that influence the use of tobacco products (for example, TV and movies). Discuss with your class.

Step 2

Research the impact of advertising on tobacco use and what you can do to advocate for a smoke-free America.

Step 3

Research the impact of tobacco and environmental tobacco smoke (ETS) on a person's health.

Apply and Conclude

Present a speech using your observations and research that will persuade government officials to change existing tobacco laws.

Checklist: Advocacy, Analyzing Influences

☑ Shows a clear, health-enhancing message

☑ Supports the position with relevant information

☑ Encourages others to make healthful choices

☑ Shows passion and conviction

☑ Identifies and analyzes external/internal factors

LESSON 1

The Health Risks of Tobacco Use

Key Concepts

▶ All tobacco products contain harmful chemicals, including carcinogens and nicotine, an addictive drug.

▶ In addition to carcinogens and other toxic substances, tobacco smoke contains tar and carbon monoxide.

▶ Tobacco use causes both short-term and long-term damage to your body, such as lung cancer and heart disease.

▶ Tobacco use can also cause legal, social, and financial problems.

Vocabulary

▶ addictive drug (p. 543)

▶ nicotine (p. 543)

▶ stimulant (p. 543)

▶ carcinogen (p. 543)

▶ tar (p. 543)

▶ carbon monoxide (p. 543)

▶ smokeless tobacco (p. 544)

▶ leukoplakia (p. 544)

LESSON 2

Choosing to Live Tobacco-Free

Key Concepts

▶ Some teens are influenced by peers or the media to use tobacco, but many who start want to quit.

▶ The number of tobacco-free teens is steadily increasing.

▶ You can avoid using tobacco by carefully choosing your friends, avoiding places where tobacco is present, and having a refusal plan.

▶ Tobacco users often find it difficult to quit using tobacco because they experience nicotine withdrawal.

Vocabulary

▶ nicotine withdrawal (p. 551)

▶ nicotine substitutes (p. 551)

▶ tobacco cessation program (p. 551)

LESSON 3

Promoting a Smoke-Free Environment

Key Concepts

▶ Environmental tobacco smoke (ETS) is harmful to nonsmokers and smokers because it contains toxic substances.

▶ A pregnant mother should avoid tobacco to protect the fetus.

▶ ETS is especially harmful to infants and young children.

▶ Laws and public policies are reducing ETS in public places.

Vocabulary

▶ environmental tobacco smoke (ETS) (p. 553)

▶ mainstream smoke (p. 553)

▶ sidestream smoke (p. 553)

▶ *Healthy People* (p. 557)

LESSON **1**

Vocabulary Review

Correct the sentences below by replacing the italicized term with the correct vocabulary term.

1. A(n) *stimulant* is a cancer-causing substance.

2. Tobacco users can become addicted to the *carbon monoxide* in tobacco.

3. When tobacco burns, it produces a thick, sticky, dark fluid known as *leukoplakia*.

Understanding Key Concepts

After reading the question or statement, select the correct answer.

4. Leukoplakia can develop into which condition?
   a. Emphysema   c. Heart disease
   b. Oral cancer   d. Bad breath

5. Which of the following are ways that tobacco harms the cardiovascular system?
   a. Increased heart rate, hardened arteries, chronic bronchitis
   b. Increased heart rate, hardened arteries, increased risk of heart attack
   c. Increased heart rate, chronic bronchitis, emphysema
   d. Chronic bronchitis, emphysema, lung cancer

6. Which of the following is a way that tobacco use immediately affects the body?
   a. Chronic bronchitis
   b. Increased risk of cancer
   c. Leukoplakia
   d. Increased heart rate

Thinking Critically

After reading the question or statement, write a short answer using complete sentences.

7. **Explain.** Is smokeless tobacco less harmful than cigarettes? Why or why not?

8. **Analyze.** If you started smoking today and continued to smoke until you are 30 years old, would you have a higher risk of developing cancer? If so, why?

9. **Explain.** How does nicotine cause an increased risk of stroke?

10. **Discuss.** Kate says that it is her choice to smoke, and that she is the only one who has to worry about her health. How does her decision to smoke affect other people?

11. **Identify.** How can using tobacco impact the social life of a tobacco user?

LESSON **2**

Vocabulary Review

Use the vocabulary terms listed on page 559 to complete the following statements.

12. When a tobacco user no longer uses tobacco, the body no longer gets nicotine, and the user experiences _____.

13. A(n) _____ can be used to deliver small amounts of nicotine to the body while a tobacco user is quitting tobacco.

14. A person who wants to successfully quit tobacco could join a(n) _____ that will help the person learn how to quit.

Understanding Key Concepts

After reading the question or statement, select the correct answer.

15. Which is a strategy to keep you from becoming a tobacco user?
    a. Practicing refusal statements
    b. Saving money
    c. Trying a cigarette .
    d. Using nicotine gum

16. Which of the following is a reason that a teen tobacco user should quit the habit?
    a. Tobacco use looks sophisticated.
    b. Their friends also smoke.
    c. They can quit as adults.
    d. They will experience health problems.

17. Why do people who are trying to quit tobacco experience physical symptoms such as irritability and anxiety?
    a. They are not committed to quitting.
    b. They are experiencing nicotine withdrawal.
    c. They are upset that other people want them to quit.
    d. They are not engaged in healthier behaviors.

18. Which is *not* a strategy that can help people give up tobacco?
    a. Develop a new daily routine.
    b. Take up a physical activity.
    c. Model adults who smoke.
    d. Join a support group.

19. People gain which health benefit by *not* using tobacco?
    a. Good refusal skills
    b. Friends who do not use tobacco
    c. Lower risk of many diseases
    d. Less money to spend on other interests

Thinking Critically

After reading the question or statement, write a short answer using complete sentences.

20. **Describe.** Why do tobacco users try to quit the habit?

21. **Infer.** Why did tobacco use among teens begin to decrease after 1998?

22. **Identify.** Why do some teens choose to use tobacco and other teens do not?

23. **Predict.** What are two reasons that teen tobacco users find quitting tobacco difficult?

24. **Describe.** What can a person do to quit using tobacco successfully?

LESSON 3

Vocabulary Review

Choose the correct term in the sentences below.

25. Environmental tobacco smoke is another name for *secondhand smoke / sidestream smoke.*

26. *Mainstream smoke / Sidestream smoke* comes from a smoker's lungs.

27. Higher concentrations of toxic substances are in *mainstream smoke / sidestream smoke.*

Understanding Key Concepts

After reading the question or statement, select the correct answer.

28. Which is *not* a way that smoking during pregnancy affects the fetus?
    a. Prenatal death
    b. High birth weight
    c. Premature delivery
    d. Developmental problems

29. How can you reduce your risk from ETS?
    a. Become a smoker.
    b. Take vitamins.
    c. Allow visitors to smoke in your house.
    d. Visit places that are smoke-free.

30. Which is a way that the government is reducing ETS exposure?
    a. Banning smoking in public places
    b. Distributing more tobacco licenses
    c. Forming youth antismoking groups
    d. Giving away air cleaners

31. ETS can cause which of these conditions?
    a. Headache
    b. Lung cancer
    c. Sudden infant death syndrome
    d. All of the above

Thinking Critically

After reading the question or statement, write a short answer using complete sentences.

32. **Identify.** If parents stop smoking, how will that decision help their children's health?

33. **Apply.** Your friend invites you over for dinner, but you know that your friend's parent smokes. What can you do?

34. **Infer.** What can you do to promote public policies that support a smoke-free environment?

35. **Explain.** How does the *Healthy People 2010* program promote health?

36. **Analyze.** Jim smokes around his friends, who do not smoke. Jim's friends say that they're not worried about the health effects of breathing the ETS created by Jim's tobacco smoke. Are they right *not* to be worried about their health? Explain your answer.

Technology PROJECT-BASED ASSESSMENT

Smoking in the Movies

Background

Smoking tobacco is discouraged or forbidden in many places. Restaurants, public buildings, schools, offices, airplanes, and even some outdoor parks are smoke-free zones. Yet, many movies continue to show people using tobacco products. In this activity, you will encourage filmmakers to stop showing people smoking in film.

Task

Create a blog opposing smoking in movies, especially in movies aimed at teens.

Audience

Students in your school

Purpose

Take a public position on the dangers of media that influence teens to smoke tobacco.

Procedure

1. Review your text for information on the health effects of tobacco use.

2. Conduct an Internet search to determine why people start smoking and what influences teens to smoke. Visit government Web sites and health advocacy organizations to learn their positions on smoking in movies and on television.

3. Search for blogs to study the language, style, and format to use in your own blog.

4. Create a blog making a clear argument opposing smoking in movies.

5. Speak with your principal to see if the blog can be placed on your school's Web site and updated periodically.

Math Practice

Interpret Graphs. Fred has decided to conduct a survey over the school year to find out what percentage of students in 8th, 10th, and 12th grades at his school reported cigarette use each month. Use the information Fred gathered in the graph to answer Questions 1–3.

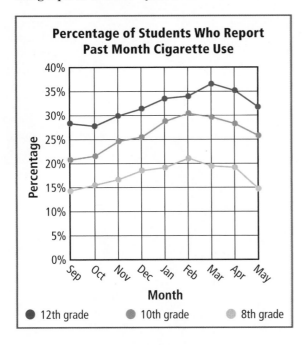

Percentage of Students Who Report Past Month Cigarette Use

● 12th grade ● 10th grade ● 8th grade

1. Which grade level had the highest percentage of smokers from September to May?
   A. 8th grade
   B. 10th grade
   C. 12th grade
   D. All grade levels were the same.

2. During which time span was smoking the most common for 8th, 10th, and 12th graders?
   A. Sep.–Oct.   C. Jan.–Mar.
   B. Nov.–Jan.   D. Mar.–May

3. In general, what trend is common to all three grade levels from September to May, according to the chart shown here? What influences do you think contributed to this trend?

Reading/Writing Practice

Understand and Apply. Read the passage below, and then answer the questions.

Today you will make a number of decisions. These decisions may or may not have a lasting effect on your life. One decision that will have a lasting effect is the decision not to start smoking. As you know, smoking is dangerous to your health. Smokers have a greater chance than nonsmokers of dying of lung and heart diseases. Cigarette smoke is also dangerous to others: secondhand smoke harms the health of nonsmokers, including the smoker's friends and family.

Smoking is also addictive. Tobacco poisons the user, but the nicotine in tobacco smoke makes the user want—and need—more. Many smokers admit that they would like to quit, but they think they can't. Starting is easy; quitting is hard. If you start smoking now, you may be starting a habit that is dangerous to you, your friends, and your family.

1. Which sentence best summarizes the writer's view of smoking?
   A. Many smokers have difficulty quitting.
   B. Smoking is a harmful habit.
   C. Starting is easy, quitting is hard.
   D. Smokers will get lung cancer.

2. How does the writer support the statement that nicotine is addictive?
   A. By explaining that nicotine makes users want more
   B. By stating how easy it is to quit smoking
   C. By telling how easy it is to start smoking
   D. By describing the dangers to nonsmokers

3. Create a podcast to encourage teens not to start smoking.

National Education Standards

Math: Measurement and Data
Language Arts: LACC.910.L.3.6, LACC.910.W.3.8, LACC.910.WHST.2.6

CHAPTER 21

Alcohol

Ed-Imaging

Lesson 1

The Health Risks of Alcohol Use

BIG Idea *Alcohol use can harm your body and your brain and cause you to make poor decisions.*

Lesson 2

Choosing to Live Alcohol-Free

BIG Idea *Choosing not to use alcohol protects you from dangerous health consequences.*

Lesson 3

The Impact of Alcohol Abuse

BIG Idea *Problem drinking and alcoholism harm both the drinkers and the people around them.*

Activating Prior Knowledge

Using Visuals The teens in this photo are looking at a car that was crashed as a result of a drunk-driving accident. Summarize how choosing to stay alcohol-free can help teens avoid risky behaviors such as drunk driving.

Chapter Launchers

Health in Action

Discuss the **BIG Ideas**

Think about how you would answer these questions:

▶ How does alcohol use contribute to risky behaviors?

▶ In what ways does problem drinking harm people?

▶ How can you encourage others to stay alcohol-free?

Assess Your Health

Read each statement. On a separate sheet of paper, write "yes," "sometimes," or "no" based on your typical behavior.

1. I avoid getting into a vehicle driven by someone who has used alcohol.

2. I recognize the short-term and long-term health effects of using alcohol.

3. I choose to live alcohol-free

4. I recognize that teen use of alcohol is of illegal.

5. I understand the effects of living with an alcoholic and know where to seek help.

A "yes" response shows that you practice healthy behaviors. "sometimes " indicates that you should analyze and possibly modify your behavior. A "no" response means that you should modify the behavior.

The Health Risks of Alcohol Use

GUIDE TO READING

BIG Idea *Alcohol use can harm your body and your brain and cause you to make poor decisions.*

Before You Read

Make an Outline.
Use the headings of this lesson to make an outline of what you'll learn about the harmful effects of alcohol use. Use a format like this to help you organize your notes.

I.
A.
1.
2.
B.
II.

New Vocabulary

▶ ethanol
▶ fermentation
▶ depressant
▶ intoxication
▶ binge drinking
▶ alcohol poisoning

Real Life Issues

Knowing the Risks. Sarah is worried because she has seen her younger sister, Jamie, experiment with alcohol. Jamie is only in high school, and Sarah feels that her sister does not fully understand the risks of her behavior. Jamie doesn't think there is a problem because she has had alcohol only a few times with her friends. Sarah is concerned that her sister is putting herself at risk, and she knows it's against the law to drink if you're under 21.

Writing *Write a dialogue in which Sarah tries to convince Jamie to change her behavior. Sarah should communicate her concern to Jamie.*

Alcohol

Main Idea Alcohol is an addictive drug.

Alcohol advertisements show images of happy, healthy-looking young adults to promote and sell a drug that is addictive, physically damaging, and often an entry into other drug use. Alcohol, or more accurately, **ethanol**—*the type of alcohol in alcoholic beverages*—is a powerful and addictive drug. It's a drug that can have serious consequences for teens who use alcohol. Using alcohol during the teen years can affect brain development. Ethanol can be produced synthetically, or naturally by fermenting fruits, vegetables, and grains. **Fermentation** is *the chemical action of yeast on sugars.* Water, flavoring, and minerals are mixed with ethanol to produce beverages such as beer, wine, and flavored malt-liquor drinks. Alcohol also can be processed to create spirits, or liquors, such as whiskey and vodka.

Short-Term Effects of Alcohol

Main Idea Alcohol impairs the central nervous system.

Alcohol is a **depressant**, *a drug that slows the central nervous system.* Using alcohol slows reaction time, impairs vision, and diminishes judgment. If a person consumes too much alcohol, he or she will become intoxicated. **Intoxication** is *the state in which the body is poisoned by alcohol or another substance, and the person's physical and mental control is significantly reduced.* Alcohol stays in a person's system until the liver can metabolize it, or break it down. The amount of alcohol that can cause intoxication varies from person to person. **Figure 21.2** on page 568 describes many of alcohol's effects.

Factors that Influence Alcohol's Effects

Some factors that influence the onset and intensity of alcohol's effects include:

- **Body size.** A smaller person feels the effect of the same amount of alcohol faster than a larger person does.
- **Gender.** Alcohol generally moves into the bloodstream faster in females than in males, because females tend to have smaller bodies than males.
- **Food.** Food in the stomach slows down the passage of alcohol into the bloodstream.
- **Rate of intake.** If a person drinks alcohol faster than the liver can break it down, the person becomes intoxicated.
- **Amount.** As the amount of alcohol consumed increases, the level of alcohol in the bloodstream rises.
- **Medicine.** Alcohol can interfere with the effects of medicines, and medicines can heighten the effects of alcohol.

■ **Figure 21.1** Alcohol impairs both physical and mental abilities. *How can alcohol use decrease your performance in activities that you enjoy?*

Figure 21.2

Short-Term Effects of Alcohol

Physical and mental impairment begin with the first drink of alcohol and increase as more alcohol is consumed.

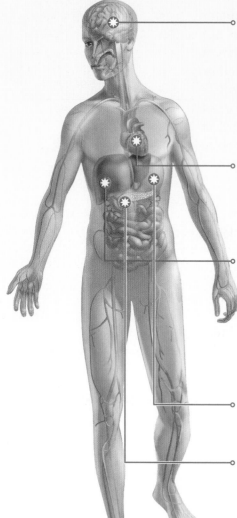

Changes to the Brain

- **Development.** Pathways and connections necessary for learning may be permanently damaged.
- **Memory.** Thought processes are disorganized, and memory and concentration are dulled.
- **Judgment and control.** Judgment is altered and coordination is impaired. Movement, speech, and vision may be affected.
- **Risk of stroke.** Alcohol use may increase risk of stroke in young people.

Cardiovascular Changes

- **Heart.** Small amounts of alcohol can increase the heart rate and blood pressure. High levels of alcohol have the opposite effect, decreasing heart rate and blood pressure. Heart rhythm becomes irregular. Body temperature drops.

Liver and Kidney Problems

- **Liver.** Toxic chemicals are released as the liver metabolizes alcohol. These chemicals cause inflammation and scarring of the liver tissue.
- **Kidneys.** Alcohol causes the kidneys to increase urine output, which can lead to dehydration.

Digestive System Problems

- **Stomach.** Alcohol increases stomach acid production and can cause nausea and vomiting.

Pancreas Problems

- **Pancreas.** Consuming large amounts of alcohol quickly can cause pancreatitis, which is accompanied by acute, severe pain. The pancreas produces enzymes that break down nutrients in foods. Alcohol use can disrupt the absorption of these nutrients.

Alcohol and Drug Interactions

Alcohol can change the effect of medicines. These interactions can lead to illness or death. Medicines that may cause reactions have warning labels that advise people not to use alcohol. Note these typical alcohol-drug interactions:

READING CHECK

Describe How can alcohol change the way a medicine affects your body?

- The body may absorb the drug or alcohol more slowly, increasing the length of time that alcohol or the drug is in the body.

- Alcohol use can decrease the effectiveness of some medications, and increase the effectiveness of others.

- Enzymes in the body can change some medications into chemicals that can damage the liver or other organs.

Long-Term Effects of Alcohol

Main Idea Alcohol use can have negative effects on a person's health.

Alcohol use can have long-term effects on a user's physical, mental/emotional, and social health. The effects of alcohol use may also be felt by the people who are close to someone who uses alcohol. Excessive alcohol use over a long period of time can damage many of the body systems, such as:

- damage to brain cells and a reduction in brain size,
- increase in blood pressure, which may lead to a heart attack or stroke,
- buildup of fat cells in the liver, which can lead to cell death,
- damage to the digestive lining of the stomach causing ulcers and cancer of the stomach, and
- destruction of the pancreas.

If a person stops using alcohol, some of the physical effects of long-term alcohol use can be reversed over time. The negative effects of alcohol use can also include damage to relationships with family, friends, and others. Excessive alcohol use over a prolonged period of time can damage most body systems. **Figure 21.4** on page 570 shows more of the long-term effects of alcohol abuse.

Binge Drinking and Alcohol Poisoning

Main Idea Consuming a large amount of alcohol over a short period of time can be fatal.

Some people choose to drink large amounts of alcohol during one session. **Binge drinking**, *drinking five or more alcoholic drinks at one sitting,* is a serious problem. Rapid binge drinking is sometimes done on a bet or a dare. Whatever the reason for binge drinking, it can have dangerous consequences. Drinking any alcohol can impair a drinker's physical and mental abilities.

Binge drinking can severely impair the drinker's body systems. It can lead to **alcohol poisoning**—*a severe and potentially fatal physical reaction to an alcohol overdose.* Alcohol acts as a depressant on body organs. Involuntary actions, such as breathing and the gag reflex that prevents choking may be impaired. Alcohol is also a stomach irritant.

READING CHECK

Explain How does long-term alcohol use affect the liver?

■ **Figure 21.3** The effects of alcohol depend on many factors, including gender and body size. *Why might women be affected more by alcohol use than men?*

Tim Fuller Photography

Figure 21.4

Long-Term Effects of Alcohol

Alcohol has a negative effect on many of the body organs, and excessive long-term alcohol use can cause death.

The Brain	The Cardiovascular System	The Digestive System	The Pancreas
Addiction Physical dependence can lead to the inability to control the frequency and amount of drinking. **Loss of brain functions** Loss of verbal skills, visual and spatial skills, and memory. **Brain damage** Excessive use of alcohol can lead to brain damage and to a reduction of brain size. The learning ability and memory of adolescents who drink even small amounts can be impaired.	**Heart damage** The heart muscles become weakened and the heart becomes enlarged, reducing its ability to pump blood. This damage can lead to heart failure. Reduced blood flow can also damage other body systems. **High blood pressure** Damages the heart and can cause heart attack and stroke.	**Irritation of digestive lining** Can lead to stomach ulcers and cancer of the stomach and esophagus. **Fatty liver** Fats build up in the liver and cannot be broken down, leading to cell death. **Alcoholic hepatitis** Inflammation or infection of the liver. **Cirrhosis of the liver** Liver tissue is replaced with useless scar tissue. Cirrhosis can lead to liver failure and death.	**Swelling of the pancreas lining** The passageway from the pancreas to the small intestine can become blocked, and chemicals needed for digestion cannot pass to the small intestine. The chemicals begin to destroy the pancreas itself, causing pain and vomiting. A severe case of pancreatic swelling can lead to death.

Effects of Alcohol Poisoning

A person who drinks too much alcohol may eventually pass out. Even though the person is unconscious, alcohol that is in the stomach continues to enter the bloodstream. So, even if someone is unconscious, that person's blood alcohol level will continue to rise. This increases the risk of alcohol poisoning. It is dangerous to assume that a person who has passed out after consuming a lot of alcohol will be fine if left to "sleep it off." Some of the symptoms of alcohol poisoning include:

- mental confusion and stupor.
- coma and an inability to be roused.
- vomiting and seizures.
- slow respiration—ten seconds between breaths or fewer than eight breaths per minute.
- irregular heartbeat.
- hypothermia or low body temperature—pale or bluish skin color.

If you suspect that a person has alcohol poisoning, call 911 immediately.

Health Skills Activity

Refusal Skills

A Drink at Home

Zach is going to Karen's house on Saturday to watch a couple of movies with friends. When Zach arrives at Karen's house, he is surprised to find that many of the teens there are drinking beer.

Karen sees him. "Hi, Zach. Do you want a beer?"

"Where did you get those?" asks Zach. "Aren't your parents home?"

"No, they're not home. What they don't know won't hurt them," says Karen. "Come on, I'll get you a beer. It's no big deal."

Zach has an agreement with his parents that he won't drink alcohol. He knows his parents would not approve of him drinking. He doesn't want his friends to think he's not part of the crowd, but he knows he needs to make it clear that he is not going to drink.

Writing Write a script completing the dialogue between Zach and Karen. How can Zach refuse alcohol and still remain friends with Karen? When writing your dialogue, use the following refusal skills:

1. Say no in a firm voice.
2. Explain why you are refusing.
3. Suggest alternatives to the proposed activity.
4. Back up your words with body language.
5. Leave the situation if necessary.

LESSON 1 ASSESSMENT

 After You Read

Reviewing Facts and Vocabulary

1. What is *intoxication*? What influences how fast a person becomes intoxicated?

2. Explain how alcohol acts as a *depressant* on the central nervous system.

3. What is *binge drinking*? What can happen as a result of binge drinking?

Thinking Critically

4. **Analyze.** Is it safe to take an over-the-counter medicine after drinking alcohol? Explain your reasoning.

5. **Describe.** How can drinking even moderate amounts of alcohol permanently affect teens?

Applying Health Skills

6. **Advocacy.** Write a PSA script that advises other teens of the dangers of binge drinking. Include the health risks and what to do if someone drinks too much alcohol.

Writing Critically

7. **Persuasive.** Write a letter to your city council member asking that police prosecute stores that sell alcohol to anyone who is underage.

Real Life Issues ....

After completing the lesson, review and analyze your response to the Real Life Issues question on page 566.

BIG Idea *Choosing not to use alcohol protects you from dangerous health consequences.*

Before You Read

Make a Cause-and-Effect Concept Map. Draw a box around the phrase "Teen Alcohol Use," as shown here. Write at least three consequences of alcohol use, each in its own box. Connect the consequence boxes to the Teen Alcohol Use box as shown.

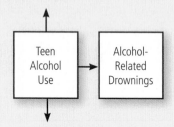

New Vocabulary

▶ psychological dependence
▶ physiological dependence
▶ alcohol abuse
▶ alcoholism

Choosing to Live Alcohol-Free

Real Life Issues •

Teens and Alcohol. According to a national survey, 45 percent of teens had at least one drink of alcohol in a 30-day period. Other statistics show that:

> **20 percent of teens had consumed alcohol before the age of 13.**

Source: Centers for Disease Control and Prevention, *Youth Risk Behavior Surveillance System, 2011*

Writing *Write a paragraph describing some consequences of teen alcohol use.*

Alcohol Use

Main Idea Several factors influence teen alcohol use.

Alcohol use by responsible adults is legal. Anyone who drinks alcohol, however, increases the risk of the negative consequences. One consequence of alcohol use is **psychological dependence**, *a condition in which a person believes that a drug is needed in order to feel good or to function normally.* Overuse of alcohol can lead to a **physiological dependence**, *a condition in which the user has a chemical need for a drug.*

Factors that Influence Alcohol Use

A teen's choices about alcohol use are influenced by:

- **peer pressure.** When alcohol use is not an accepted activity in a group, a teen will not feel pressure to drink.

- **family.** When a teen's parents discourage and avoid the use of alcohol, the teen is more likely to do the same.

- **media messages.** Media messages may make alcohol use seem glamorous and fun.

■ **Figure 21.5** Exercise, eating a balanced diet, and avoiding alcohol can help teens stay healthy. *Why should teens avoid alcohol?*

Advertising Techniques

Companies that produce alcohol spend billions of dollars each year to associate their product with youthful, healthy people who seem to be having fun. These companies advertise in ways that are visible to teens and children, on billboards, TV, and radio, and in magazines and newspapers. Alcohol companies also sponsor sporting events, music concerts, art festivals, exhibits, and college events. Manufacturers and advertisers, however, never show consumers the negative side of alcohol use.

Alcohol companies target teens and young adults by marketing beverages that appear safer than other alcoholic beverages. These drinks are sweet and look similar to non-alcoholic carbonated beverages. In reality, these beverages contain alcohol.

Health Risks of Alcohol Use

Main Idea Alcohol can harm more than just your health.

In the United States, nearly 30 people die each day as a result of alcohol-related traffic collisions. Alcohol use is linked to deaths from drowning, fire, suicide, and homicide. A nondrinker's risk of being injured increases if the friends that person is with are drinking.

Alcohol and the Law

It is illegal for anyone under the age of 21 to buy, possess, or consume alcohol. For teens who break the law, the consequences can be very serious. Teens who use alcohol can be arrested and sentenced to a youth detention center.

 READING CHECK

Explain How do advertisements try to encourage teens to start drinking?

■ Figure 21.6 Your decision to avoid alcohol is influenced by the people around you. *How can having friends who do not use alcohol help you to stay alcohol-free?*

We learned about alcohol use at a school assembly. The speaker said that people who start drinking before the age of 15 are five times more likely to have alcohol-related problems later in life. They also said that new research shows that alcohol use by teens can harm our brains and that our brains are still developing. After hearing that, I have even more reason to refuse alcohol. For more fitness tips, visit the Fitness Handbook and Fitness Zone sites in ConnectEd.

Any arrest and conviction can affect a teen's future. An arrest can limit college and employment options. Breaking the law can also damage a teen's reputation and cause that teen to lose the trust of friends and family members.

Alcohol and Violence

Teens can protect their health by avoiding situations where alcohol is present. Fights are more likely to break out at parties where alcohol is used. Teens who are involved in fights face school or police disciplinary action. Teens who drink are also more likely to be victims or perpetrators of violent crimes, such as rape, aggravated assault, and robbery. It is estimated that alcohol use is a factor in one-third to two-thirds of sexual assaults or date-rape cases.

Alcohol and Sexual Activity

Alcohol impairs judgment and lowers inhibitions, and can cause a person to compromise his or her values. Teens who use alcohol are more likely to become sexually active at an earlier age, and to engage in unprotected sexual activity. Approximately 22 percent of sexually active teens use alcohol or drugs before engaging in sexual activity. Teens who drink often are twice as likely to contract an STD as teens who do not drink.

Real World CONNECTION

Analyzing the Media

What media images come to mind when you think of advertisements for alcohol? Many ads feature attractive young people whose message seems to be, "You can be like us if you use this product." What are other reasons the people in the ad are enjoying themselves? Conduct an online search using government Web sites and other trusted sources to examine the guidelines and perceived messages for advertisements for alcohol.

Activity Technology

Use a critical eye when examining alcohol ads in magazines, television, and billboards. Select three alcohol ads and ask yourself the following questions:

1. What is really being advertised?
2. What is the hidden message?
3. What is the truth?
4. How do the advertisers distort the truth about alcohol?

After analyzing the ads, select one and create a multimedia slide presentation showing how the ad distorts the truth. Use your creativity to show how the ad, or even just elements of the ad, can be misleading.

Alcohol and the Family

Young people who live in a household in which **alcohol abuse** occurs—*the excessive use of alcohol*—are four times more likely to abuse alcohol themselves. Other effects on the family of alcohol abuse include

- neglect, abuse, or social **isolation**.
- economic hardship.
- personal use of alcohol themselves.
- mental illness or physical problems.

Studies show that a person who begins drinking alcohol as a teen is four times more likely to develop alcohol dependence than someone who waits until adulthood to use alcohol. This dependence, called **alcoholism**, is *a disease in which a person has a physical or psychological dependence on drinks that contain alcohol*.

Academic Vocabulary

isolation *(noun):* the state of being withdrawn or separated

Alcohol and School

Most schools have adopted a zero-tolerance policy for students found using alcohol on school property. Students who use alcohol may become ineligible for or be suspended from school activities or graduation, or expelled from school. These students could also be placed in an alternate education program. Ultimately, a student may find that his or her options for choosing a college or job may be limited.

READING CHECK

List Name three risky behaviors that can be caused by alcohol use.

Avoiding Alcohol

George Doyle/Stockbyte/Getty Images

Main Idea You will experience many benefits if you choose to live alcohol-free.

Living alcohol-free is a choice that some adults make. Many other adults choose to drink alcohol occasionally and responsibly. Alcohol is addictive, and once you start drinking, it may be difficult to stop. Teens who start drinking by age 15 are five times more likely to become dependent on alcohol than people who do not start drinking until age 21.

Benefits of Living Alcohol-Free

Many teens make the commitment to stay alcohol-free. Avoiding alcohol will help you with the following:

- **Maintaining a healthy body.** You will avoid the damage alcohol can do to the brain and body organs and decrease the likelihood of being injured in an accident.
- **Establishing healthy relationships.** You can be open and honest with your family about your activities and habits.
- **Making healthy decisions.** Avoiding intoxication will allow you to make decisions that protect your health.
- **Avoiding risky behaviors.** You will reduce the risk of making unhealthy choices, such as drinking and driving.
- **Avoiding illegal activities.** You can avoid arrest and legal problems by being alcohol-free. Purchasing or possessing alcohol is against the law for anyone under 21.
- **Avoiding violence.** Avoiding alcohol reduces your risk of being a victim of or participating in a violent crime.
- **Achieving your goals.** Being alcohol-free allows you to stay focused on your short-term and long-term goals.

■ **Figure 21.7** Avoiding alcohol use can help you avoid risky behaviors. *How can teen alcohol use put your future at risk?*

Refusing Alcohol

At times, it may be difficult to avoid situations in which you are pressured to use alcohol. Saying no is much easier when you know how you will respond *before* you are faced with the situation. If you find yourself in a situation where alcohol is present, be assertive and use refusal skills. Call your parents or another trusted adult for a ride home, if needed. Here are some examples of refusal statements:

- "I don't like the taste."
- "No, thanks. I need to be in top shape for the game this week."
- "I don't drink alcohol—besides, I'm heading home."
- "I really can't, my parents would be angry. We have an agreement."
- "I don't want to risk getting kicked off the team."

Another strategy to avoid alcohol use is to plan alcohol-free activities with friends. Avoid parties or social gatherings where alcohol is served. Practice your refusal skills to build confidence when you are with peers who may use alcohol.

■ **Figure 21.8** Having a strategy to stay alcohol-free will help you avoid the risks of alcohol use. *What are some ways to avoid using alcohol?*

 READING CHECK

Explain How can staying alcohol-free help you stay physically and mentally healthy?

LESSON 2 ASSESSMENT

 After You Read

Reviewing Facts and Vocabulary

1. What is *alcohol abuse,* and how are teens likely to be affected if it occurs in their family?

2. How does *alcohol abuse* differ from *alcoholism*?

3. Explain whether teens are at risk of alcohol dependence.

Thinking Critically

4. **Evaluate.** What are four possible consequences of poor decisions made while under the influence of alcohol?

5. **Analyze.** Explain how teens can stay alcohol-free.

Applying Health Skills

6. **Refusal Skills.** You arrive at a party and see that other teens are drinking alcohol. Write a dialogue describing how you use refusal skills effectively to stay alcohol-free.

Writing Critically

7. **Descriptive.** Write a scenario between a teen and a parent. The teen is providing reasons for remaining alcohol-free.

Real Life Issues ....

After completing the lesson, review and analyze your response to the Real Life Issues question on page 572.

The Impact of Alcohol Abuse

Real Life Issues

Living with Alcohol Abuse. Lily's mother drinks alcohol every day. She is always drunk and out of control. Lily thinks her mother may be an alcoholic. Her mother says that drinking helps her unwind at the end of the day. Recently, Lily's mother lost her job because she was drunk at work. Lily often finds herself taking care of her mother whenever she is drunk.

Writing *Write a letter to Lily's mother as if you were Lily. In your letter, explain how her mother's drinking is affecting Lily's life.*

Alcohol and Driving

Main Idea Drinking and driving is very dangerous.

Driving after drinking can have disastrous and even deadly results. One-fifth of all teen drivers involved in fatal car accidents have a blood alcohol concentration of 0.01 percent. A person's **blood alcohol concentration (BAC)** is *the amount of alcohol in a person's blood, expressed as a percentage.* BAC depends on the quantity and type of alcohol that was consumed, the rate of consumption, and body size and gender. **Figure 21.9** illustrates how the alcohol content varies in common alcoholic beverages. Any amount of alcohol in the blood can cause the following:

• Slow reflexes

• Reduced ability to judge distances and speeds

• Increase in risk-taking behaviors

• Reduced concentration and increased forgetfulness

Figure 21.9 Comparing Beer, Wine, and Spirits

Each of these beverages contains the same amount of pure alcohol.

Drink	Alcohol by Volume	Alcohol Content
Beer *(12 oz.)*	4%	0.5 oz.
Wine *(5 oz.)*	10%	0.5 oz.
Vodka or Whiskey *(1.25 oz.)*	40%	0.5 oz.

Driving While Intoxicated

Driving while intoxicated (DWI), or driving under the influence (DUI), is illegal. Adult drivers who have a BAC of 0.08 percent can be charged with drunk driving. For those under 21, there is no acceptable BAC, since it's illegal to use alcohol. The consequences for DWI or DUI include

- injuries to or death of the driver and others.

- arrest, jail time, court appearance and fine or bail, a police record, and possible lawsuits.

- severely restricted driving privileges and/or immediate confiscation of driver's license.

- higher auto insurance rates or a canceled insurance policy.

Riding in a vehicle with a driver who has been drinking is just as dangerous as if you were the one drinking and driving. If someone you're with has been drinking, find a ride with someone who has not been drinking, or call home for a ride.

READING CHECK

Explain Why is it dangerous to drive after drinking alcohol?

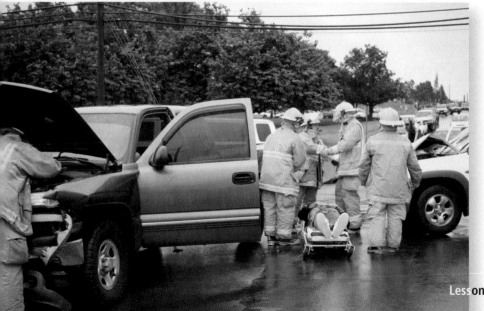

■ **Figure 21.10** More than 10,000 people died in 2010 in alcohol-related crashes. *What are other consequences of drinking and driving?*

Alcohol and Pregnancy

Main Idea A female who drinks during pregnancy can harm her fetus.

When a pregnant female drinks, alcohol passes directly from her body into the bloodstream of the fetus. A fetus processes alcohol much more slowly than the mother does. As a result, there is more alcohol in the fetus's system for a longer period of time. Infants born to mothers who drink during pregnancy are at risk of **fetal alcohol syndrome (FAS)**, *a group of alcohol-related birth defects that include physical and mental problems.*

Effects of Fetal Alcohol Syndrome (FAS)

The effects of FAS are both severe and lasting. Infants born with FAS may have the following problems:

- Small head and deformities of face, hands, or feet
- Heart, liver, and kidney defects
- Vision and hearing problems
- Central nervous system problems, developmental disabilities, and poor coordination
- Difficulties learning and short attention span
- Hyperactivity, anxiety, and social withdrawal

FAS is one of the leading preventable causes of mental retardation. Females who are trying to become pregnant or may be pregnant should not drink *any* alcohol.

Alcoholism

Main Idea Alcoholism is a disease that affects the person who drinks and others around him or her.

Alcoholics are physically or psychologically dependent on alcohol. The symptoms of alcoholism include the following:

- **Craving**—Feeling a strong need for alcohol to manage tension or stress, and a preoccupation with alcohol
- **Loss of control**—Inability to limit alcohol consumption
- **Physical dependence**—Withdrawal symptoms, such as nausea, sweating, shakiness, and anxiety
- **Tolerance**—A need to drink increasingly more alcohol in order to feel its effects

■ **Figure 21.11** When a pregnant female drinks, so does her fetus. *What effect can alcohol have on a fetus?*

READING CHECK

List Name four problems that a baby born to a mother who drank alcohol during pregnancy may have.

TEENS Making a Difference

"A young person can make a difference."

Advocating for Change

Elianna Y., of California, joined the substance-free group Friday Night Live as a freshman. Three years later, Elianna is one of the leaders of the group.

Under Elianna's leadership, Friday Night Live is targeting alcopops, sweet, malt-flavored beverages that contain distilled spirits. "They're popular with teens, even though it's illegal for teens to drink alcohol. Currently alcopops are taxed the same as beer," says Elianna. Friday Night Live has petitioned California legislators to raise the tax.

"I feel like I am making a difference in people's lives not only with the policy changes, but also as a role model," said Elianna. "Being young doesn't mean you can't do anything. A young person can make a difference in the community."

Activity Write your answers to the following questions in your personal health journal:

1 What issues in your community would motivate you to become a leader?

2 Describe two benefits that might result from the change in this tax law.

3 How can you make a difference in your community regarding alcohol and teens?

Alcoholics

An **alcoholic** is *an addict who is dependent on alcohol.* The behavior of alcoholics varies—some are aggressive and violent, while others may become withdrawn. Alcoholism is not limited to any age, race, or ethnic or socioeconomic group.

Growing scientific evidence suggests that alcoholism is partially due to genetics. One study shows that children of alcoholics are four times more likely to become alcoholics. Environmental factors such as family, friends, culture, peer pressure, availability of alcohol, and stress also contribute to alcoholism. The age at which a person starts drinking also influences the risk of alcoholism. Teens who start drinking are at a higher risk of becoming alcoholics during their lifetime than people who begin drinking as adults.

READING CHECK

Describe What are the symptoms of alcoholism?

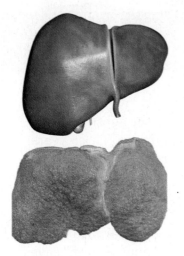

■ **Figure 21.12** Alcohol causes serious damage to the liver. Compare the healthy liver (top) with the liver that has been damaged by alcohol abuse. *What effect does alcohol have on the liver?*

Figure 21.13

Steps to Recovery

Step 1—Admission

• The person admits to having a drinking problem and asks for help.

Step 2—Detoxification

• The person goes through detoxification, a process in which the body adjusts to functioning without alcohol.

Step 3—Counseling

• The person receives counseling to help him or her learn to change behaviors and live without alcohol.

Step 4—Recovery

• The person takes responsibility for his or her own life.

Stages of Alcoholism

Alcoholism develops in three stages: abuse, dependence, and addiction. All alcoholics do not, however, experience each stage equally.

- **Stage 1—Abuse.** It may begin with social drinking. A physical and psychological dependence develops. The person may experience memory loss and blackouts, and may begin to lie or make excuses to justify his or her drinking.
- **Stage 2—Dependence.** The person becomes physically dependent on alcohol. The drinker tries to hide the problem, but performance on the job, at school, and at home suffer.
- **Stage 3—Addiction.** In the final stage of alcoholism, the person is addicted. At this stage, the liver may be already damaged, and so less alcohol may be required to cause drunkenness. If the alcoholic stopped drinking, he or she would experience severe withdrawal symptoms.

Effects on Family and Society

Main Idea Alcohol abuse plays a role in crimes and has negative effects on people who are around problem drinkers.

In the United States about 88,000 people die each year from excessive alcohol use. Alcohol abuse affects more than just the drinker. It is a major factor in the four leading causes of accidental death: car accidents, falls, drowning, and house fires. Alcohol also plays a major role in violent crimes, such as homicide, forcible rape, and robbery.

Often, people close to alcoholics develop mentally unhealthy behaviors, such as *codependency*. Codependents ignore their own emotional and physical needs and instead focus their energy and emotions on the needs of the alcoholic. In the process, codependents lose their self-esteem and their trust in others, and their own physical health suffers.

Treatment for Alcohol Abuse

Main Idea Alcoholics can recover if they get treatment.

Alcoholism cannot be cured, but it *can* be treated. **Recovery** is *the process of learning to live an alcohol-free life*. Recovering alcoholics must make a lifelong commitment to **sobriety**, which is *living without alcohol*.

The steps to recovery outlined in **Figure 21.13**, include admitting that alcohol use is a problem; detoxification, or adjusting to functioning without alcohol; receiving counseling to change behaviors; and recovery, or taking responsibility for one's own life. Here are a few of the resources and programs available to help alcoholics and problem drinkers, as well as their families and friends:

- **Al-Anon/Alateen** helps families and friends learn to deal with the effects of living with an alcoholic.

- **Alcoholics Anonymous** provides help for alcoholics.

- **Mothers Against Drunk Driving (MADD)** provides education to prevent underage drinking.

- **National Association for Children of Alcoholics** provides help for children of alcoholics.

- **National Drug and Alcohol Treatment Referral Routing Service** provides treatment referral and information about treatment facilities.

- **Students Against Destructive Decisions (SADD)** provides peer-led education about avoiding alcohol use.

- **SAMSHA's National Clearinghouse for Alcohol and Drug Information** provides information about alcohol and other drugs.

READING CHECK

Explain Describe what an alcoholic must do in order to recover.

LESSON **3** ASSESSMENT

After You Read

Reviewing Vocabulary and Facts

1. What is *blood alcohol concentration (BAC),* and what is the legal BAC for teen drivers?

2. What is *fetal alcohol syndrome (FAS),* and what causes it?

3. Why does an alcoholic go through detoxification when trying to become sober?

Thinking Critically

4. Analyze. How does a moderate amount of alcohol, which otherwise might not be harmful to a female, have the potential to harm her fetus?

5. Discuss. What are two possible outcomes of drinking and driving?

Applying Health Skills

6. Accessing Information. Use reliable sources to identify community and other resources that help alcoholics and their families. Then, design a webpage that lists these resources.

Writing Critically

7. Narrative. Write a one-page story showing how a teen can use refusal skills to avoid getting into a car with a driver who has been drinking.

Real Life Issues ·

After completing the lesson, review and analyze your response to the Real Life Issues question on page 578.

Hands-On HEALTH

Activity ## You Booze, You Cruise, You Lose

According to Students Against Destructive Decisions (SADD), more than 16,000 people die every year in alcohol- and drug-related car accidents. That's one person killed every 33 minutes and one person injured every two minutes. This activity encourages you to prevent driving under the influence of drugs and alcohol in your school and community.

What You'll Need

- five sheets of paper
- access to the Internet

What You'll Do

Step 1

Make a list of five reasons to avoid alcohol use and driving. Write each reason on each sheet of paper.

Step 2

On the sheets of paper create realistic scenarios in which a person would refuse a ride from an impaired driver.

Step 3

Role-play each scenario, supporting your refusal skills with the top five reasons you wrote earlier.

Apply and Conclude

Prior to a school activity, create a campaign to advocate for sober driving, using a variety of media such as newspaper ads, red ribbons on car antennas, or posters.

Checklist: Refusal Skills, Advocacy

- ✓ Say no, and explain why you are refusing
- ✓ Propose alternatives to the activity
- ✓ Use body language to back up your words
- ✓ Walk away from the situation if necessary
- ✓ Show support for the position with relevant information

The Health Risks of Alcohol Use

Key Concepts

▶ Alcohol is an addictive drug that slows the central nervous system (CNS) and impairs physical abilities and judgment.

▶ Mixing alcohol with medicines or other drugs is extremely dangerous.

▶ Long-term excessive alcohol use harms many of the body systems, and can possibly damage adolescent brain development processes.

▶ People who binge drink put themselves at serious risk of alcohol poisoning.

Vocabulary

▶ ethanol (p. 566)
▶ fermentation (p. 566)
▶ depressant (p. 567)
▶ intoxication (p. 567)
▶ binge drinking (p. 569)
▶ alcohol poisoning (p. 569)

Choosing to Live Alcohol-Free

Key Concepts

▶ Peer pressure, advertising, and family can influence a teen's choice to use or not to use alcohol.

▶ Alcohol use leads to risky behaviors that have serious consequences.

▶ Teens who do not use alcohol are more likely to make healthy decisions that maintain their health.

▶ In order to stay alcohol-free, teens should try to avoid situations where alcohol will be present, and they should have a refusal plan.

Vocabulary

▶ psychological dependence (p. 572)
▶ physiological dependence (p. 572)
▶ alcohol abuse (p. 575)
▶ alcoholism (p. 575)

The Impact of Alcohol Abuse

Key Concepts

▶ No one should drive after drinking alcohol because it impairs mental and physical abilities.

▶ If a female drinks while she is pregnant, her baby may be born with mental and physical birth defects.

▶ Dependence on alcohol causes alcoholics to harm themselves and the people around them.

▶ Alcoholics can recover if they seek help.

Vocabulary

▶ blood alcohol concentration (BAC) (p. 578)
▶ fetal alcohol syndrome (FAS) (p. 580)
▶ alcoholic (p. 581)
▶ recovery (p. 582)
▶ sobriety (p. 582)

Vocabulary Review

Correct the sentences below by replacing the italicized term with the correct vocabulary term.

1. Drinking alcohol can lead to *fermentation,* a state of reduced physical and mental control.

2. *Intoxication* is when a person drinks five or more alcoholic drinks at one sitting.

3. *Binge drinking* is a potentially fatal reaction to an alcohol overdose.

Understanding Key Concepts

After reading the question or statement, select the correct answer.

4. Which is *not* a short-term effect of alcohol?
   a. Coordination is impaired.
   b. Vision is impaired.
   c. More stomach acid is produced.
   d. Judgment is altered.

5. Which type of person is most likely to be quickly affected by alcohol?
   a. A small female who has not eaten
   b. A large female who just ate dinner
   c. A small male who just ate dinner
   d. A large male who has not eaten

6. Which is a potential consequence of long-term excessive alcohol use?
   a. A heart attack
   b. The need for a liver transplant
   c. Swelling of the brain
   d. An increased ability to control drinking

Thinking Critically

After reading the question or statement, write a short answer using complete sentences.

7. **Describe.** Why are teens who drink more likely to put themselves in risky situations?

8. **Evaluate.** Terry takes a 12-hour allergy medication, and then drinks alcohol after waiting an hour. Explain why this is not a safe behavior.

9. **Explain.** How does long-term excessive drinking affect the brain?

10. **Analyze.** Dana drinks only occasionally, but when he drinks he does things that he later regrets. His friends assure him that he is fun when he drinks, and that he should not worry about his actions. Does Dana have a drinking problem? Explain.

11. **Evaluate.** A person passes out after drinking. Explain the dangers of leaving the person alone, and what action should be taken.

Vocabulary Review

Use the vocabulary terms listed on page 585 to complete the following statements.

12. Teens are likely to experience neglect or abuse if there is _____ in their family.

13. People who have a chemical need for alcohol have a _____.

14. A dependence on drinks with alcohol is called _____.

15. When a person believes that alcohol use is needed to feel good or function normally, that person has a _____.

Understanding Key Concepts

After reading the question or statement, select the correct answer.

16. Which of the following influences teens to stay alcohol-free?
    a. Having peers who drink alcohol
    b. Seeing alcohol ads on TV
    c. Having parents who disapprove of alcohol use
    d. Attending alcohol-sponsored sporting events

17. Which is *not* a result of the high-risk behaviors associated with alcohol use?
    a. Increased deaths in traffic accidents
    b. Increased frequency of date rape
    c. Suspension from sports teams
    d. Improved job prospects

18. If you stay alcohol-free, which is a likely benefit?
    a. Decreased likelihood of getting a sexually transmitted disease
    b. Increased likelihood of being injured in an accident
    c. Decreased likelihood of making responsible decisions
    d. Increased likelihood of being a victim of violent crime

19. Which is *not* a strategy to remain alcohol-free?
    a. Planning an alcohol-free party
    b. Practicing refusal statements
    c. Attending parties with people who use alcohol
    d. Calling for a ride home if alcohol is present

Thinking Critically

After reading the question or statement, write a short answer using complete sentences.

20. **Explain.** What are the dangers of adult alcohol use? What are the dangers for teens?

21. **Analyze.** How does sponsoring community events and music concerts help alcohol companies sell their products?

22. **Discuss.** How does teen alcohol use impact the community?

23. **Evaluate.** Why are alcohol-free teens more likely to achieve their long-term goals than teens who use alcohol?

24. **Examine.** Why is avoiding gatherings where alcohol is present the best way to stay alcohol-free?

LESSON 3

Vocabulary Review

Choose the correct term in the sentences below.

25. *Blood alcohol concentration (BAC)/Fetal alcohol syndrome (FAS)* is a condition that babies can be born with if a female drinks while pregnant.

26. The amount of alcohol in a person's blood that is expressed as a percentage is *blood alcohol concentration/fetal alcohol syndrome.*

27. An addict who is dependent on alcohol is in/an *sobriety/alcoholic.*

Understanding Key Concepts

After reading the question or statement, select the correct answer.

28. It is illegal for adults to drive when they have what BAC level?
    a. 0.01   c. 0.05
    b. 0.02   d. 0.08

29. Which is *not* likely to occur if a person is caught drinking and driving?
    a. A field sobriety test will be conducted.
    b. The person loses driving privileges.
    c. The person loses insurance coverage.
    d. There will be no penalty.

30. A baby born with FAS may have which of the following?
    a. Anxiety
    b. Facial deformities
    c. Kidney problems
    d. All of the above

31. Which of the following is *not* part of the stages of alcoholism?
    a. Becoming intoxicated regularly
    b. Going through detoxification
    c. Making excuses for alcohol-related problems
    d. Alcohol taking control of the drinker's life

Thinking Critically

After reading the question or statement, write a short answer using complete sentences.

32. **Identify.** What effects of alcohol make it risky to drive after drinking?

33. **Apply.** Jacob's mother is seven months pregnant. She thinks it is alright to start having an occasional glass of wine with dinner because the fetus is mostly developed. What should Jacob tell her?

34. **Analyze.** What factors increase one's likelihood of becoming an alcoholic?

35. **Synthesize.** Jesse knows she has a drinking problem and wants help. Explain what she can do to find help.

Technology PROJECT-BASED ASSESSMENT

The Power of Persuasion

Background

Companies that sell alcohol often use television advertising to convince people to buy and use their products. Alcohol commercials use powerful combinations of video and audio to sell their product. Anti-alcohol ads may also appear on television. These ads aim to reduce alcohol use among teens.

Task

Create a Web page about how saying no to alcohol is an attractive and practical option for teens.

Audience

Students in your class

Purpose

Persuade teens to say no to alcohol use.

Procedure

1. Working with a small group, create a list of situations in which teens may be pressured by other teens to use alcohol.

2. Select two possible situations that teens in your school may encounter. Conduct research using reliable sources to learn ways that advertisers might try to persuade teens to use their products.

3. Create a Web page discussing the situations and the possible outcomes, both good and bad. Be sure to include a list of images and possibly videos of commercials for your Web page.

4. Make any necessary revisions to your Web page.

5. Present your final product to your class for input.

Math Practice

Interpreting Graphs. The graph below shows the relationship between blood alcohol level and breath alcohol level. After reviewing the graph, answer the questions that follow.

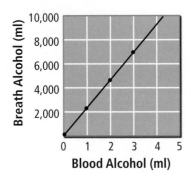

1. Based on the graph, how many milliliters of blood are equivalent to 3150 ml of air?
   A. 1 ml
   B. 1.5 ml
   C. 3 ml
   D. 2100 ml

2. What type of function is shown on the graph?
   A. Exponential
   B. Higher degree
   C. Linear
   D. Quadratic

3. An increase in body temperature of 1.8 degrees F results in an increase of 7 percent in the test results. If a graph similar to the one shown were drawn for a person with a temperature of 97 degrees F, would the slope be greater than or less than the slope of this graph?

Reading/Writing Practice

Understand and Apply. Read the passage below, and then answer the questions.

> *Having a natural mentor can help teens make positive choices, according to a study by Students Against Destructive Decisions (SADD). Natural mentors can include parents, other family members, teachers, coaches, members of the clergy, and other trusted adults.*
>
> *In the study, 46% of teens who have a natural mentor reported having a higher sense of self. Only 25% of teens who did not have a natural mentor agreed that they have a higher sense of self.*
>
> *More than half of the students surveyed also said that having a natural mentor impacted them in a positive way. Teens are willing to talk to their natural mentors about avoiding alcohol, drugs, and sexual activity.*

1. The study's findings show that
   A. a mentor has no influence on teens.
   B. a mentor can help teens make positive choices.
   C. a natural mentor is always a parent.
   D. teens with a natural mentor have a higher sense of self.

2. Based on this article, a natural mentor is
   A. another teen who may be one or two years older.
   B. a parent or other trusted adult.
   C. an adult the teen is paired with through a matching program.
   D. a teacher or school counselor who has had training as a mentor.

3. Write a one-page, persuasive essay describing the importance of having a natural mentor. Describe how mentors can help teens.

National Education Standards

Math: Measurement and Data, Represent and Interpret Data
Language Arts: LACC.910. RL.2.4

CHAPTER 22 Illegal Drugs

Activating Prior Knowledge

Using Visuals While looking at this photo, think of reasons why someone might take illegal drugs. List healthful activities that could take the place of drug use. Explain why these kinds of activities are safer than using illegal drugs.

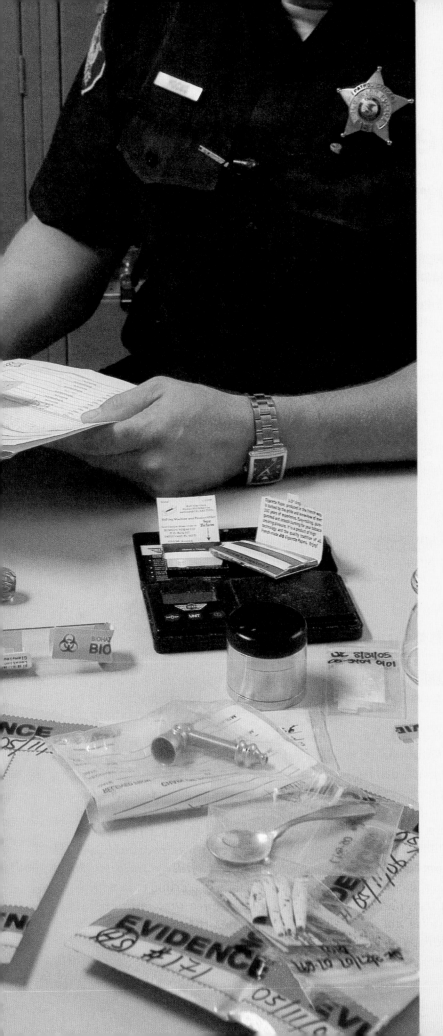

Chapter Launchers

Health in Action

Discuss the BIG Ideas

Think about how you would answer these questions:

▶ Why do some people use illegal drugs?

▶ How can you respond to peer pressure to use drugs?

▶ How has peer pressure influenced you to do something safe and healthy?

Assess Your Health

Read each statement. On a separate sheet of paper, write "yes," "sometimes," or "no" based on your typical behavior.

1. I understand the health effects of using illegal substances.

2. I recognize that peers may try to influence me to use drugs, and I know how to resist that peer pressure.

3. I choose to live drug-free.

4. I recognize that dangers of driving with someone who has used drugs.

5. I choose to advocate for a drug-free life.

A "yes" response shows that you practice healthy behaviors. "Sometimes" indicates that you should analyze and possibly modify your behavior. A "no" response means that you should modify the behavior.

📖 **GUIDE TO READING**

BIG Idea *Drug misuse and substance abuse are life-threatening behaviors.*

Before You Read

Create a K-W-L Chart.
Make a three-column chart. In the first column, list what you **k**now about the negative effects of illegal drugs. In the second column, list what you **w**ant to know about this topic. As you read, use the third column to summarize what you **l**earned.

K	W	L

New Vocabulary

▶ substance abuse
▶ illegal drugs
▶ illicit drug use
▶ overdose
▶ addiction

Review Vocabulary

▶ psychological dependence (Ch.21, L.2)
▶ physiological dependence (Ch.21, L.2)

Health Risks of Drug Use

Real Life Issues •••••••••••••••••••••••••

Percent of emergency room visits involving selected illicit drugs: Persons Aged 20 or younger, 2010.
Source: Office of Applied Studies, SAMHSA, Drug Abuse Warning Network, 2006.

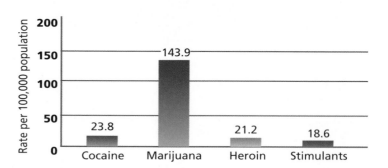

Writing *Consider what you already know about illicit drug use. Write a paragraph describing some of the potential dangers associated with drug use.*

Substance Abuse

Main Idea Substance abuse includes the use of illegal substances, as well as the misuse of legal substances.

Medicines cure and prevent disease. Sometimes medicines are accidentally used in an improper way. At other times, medicines are intentionally abused. **Substance abuse** is *any unnecessary or improper use of chemical substances for non-medical purposes.* It includes the overuse, or multiple use of a drug, use of an illegal drug, or use of a drug with alcohol.

Some abused substances are **illegal drugs**, *chemical substances that people of any age may not lawfully manufacture, possess, buy, or sell.* Using illegal drugs is a crime called **illicit drug use**, *the use or sale of any substance that is illegal or otherwise not permitted.* This includes the sale of prescription drugs to those for whom the drugs are not intended.

Factors That Influence Teens

Teens are faced with many choices, including the use of drugs. Many factors influence the choices a teen makes about drug use. Some influences can include:

- **Peer pressure**, or the influence of your friends or social group. Peers can influence teens to avoid illegal drug use. Teens whose friends avoid drug use are more likely to say no to drugs themselves.

- **Family members** can help teens resist drug use. Parents and other family members can encourage teens to abstain from drug use.

- **Role models** such as coaches, athletes, actors, and professionals who speak about the benefits of being drug-free.

- **Media messages** on TV, radio, Web sites, movies, and music can influence how you feel about drug use.

- **Perceptions of drug behavior** that may lead teens to believe that drug use is higher than it is in reality. According to the CDC, more than 90 percent of ninth-graders have never used marijuana.

- **Misleading information** about some drugs can lead teens to think that certain drug use can be beneficial. Some teens believe that steroid use boosts sports performance.

READING CHECK

Describe How could having a prescription for a legal drug lead to illicit drug use?

■ **Figure 22.1** Role models can influence you to avoid drugs. *How would having drug-free role models give you an advantage in resisting drugs?*

How Drugs Affect Your Health

Main Idea Illegal drug use can lead to death.

Unlike medicines, illegal drugs are not monitored for quality, purity, or strength. They don't come with labels that list safety guidelines or suggested dosage. Drug abuse affects your physical, mental/emotional, and social health.

- **Physical health.** A serious danger of drug abuse is the risk of an **overdose**, or *a strong, sometimes fatal reaction to taking a large amount of a drug*. For some illegal drugs, users inject the substances with a needle. This increases the risk of contracting diseases such as hepatitis B and HIV.

- **Mental health.** Drug use may impair a teen's ability to reason and think. The illegal drug Ecstasy alters the brain's structure and function. The influence of illegal drug use may cause teens to behave in ways that go against their values. **Figure 22.3**, on page 595, shows how drug use affects the brain.

Academic Vocabulary

alter *(verb):* to make different

- **Social health.** Teens who use drugs may lose friendships with teens who choose to live drug-free. Relationships with family members may also suffer. Some teens may have to accept the legal consequences of drug use. Substance abuse is a leading cause of crime, suicide, and unintentional injuries.

READING CHECK

Explain How does tolerance affect a drug user?

Other Effects of Drug Use

Teens who use illegal drugs may also experience unwanted physical reactions that may result in death. These reactions can occur with a teen's first drug use or even if a teen has used a drug in the past and believes that he or she can tolerate the drug. The manufacture of some illegal drugs is not regulated, so the compounds in each batch of a drug may be different. Other consequences are:

- **Tolerance.** This is a condition in which the body becomes accustomed to the drug and causes the user to experience a need for more and more of the drug to achieve the desired effect.

- **Psychological dependence.** *Psychological dependence* is a condition that develops over time and causes a person to believe that a drug is needed in order to feel good or to function normally.

- **Physiological dependence.** A user develops a chemical need for a drug. Symptoms of withdrawal occur when the effects of the drug wear off. Symptoms can include nervousness, insomnia, headaches, vomiting, chills, and cramps. In some cases, withdrawal symptoms are severe and can result in death.

- **Addiction.** Another serious consequence is **addiction**, *a physiological or psychological dependence on a drug.* An addict is someone who requires persistent, compulsive use of a substance known by the user to be harmful. Because addiction always involves both psychological and physiological dependence, people who are addicted to a substance have great difficulty in stopping its use on their own. Professional intervention to stop using illegal drugs is often necessary.

Trying a drug just once or using a drug only a few times can quickly lead to a serious cycle of addiction. In the addiction cycle, a user takes a drug to experience short-term pleasure. As the effects of the drug wear off, the user then experiences the physical and psychological consequences of withdrawal. In order to relieve these symptoms, the user takes the drug again to to relieve the pain and to repeat the feelings of short-term pleasure once again. The cycle continues until a person gets medical help to stop using the drug.

■ **Figure 22.2** Use of the illegal drug Ecstasy, a stimulant, causes structural and functional changes in the brain. *How can drugs affect your mental health?*

Figure 22.3   **Drug Use and the Brain**

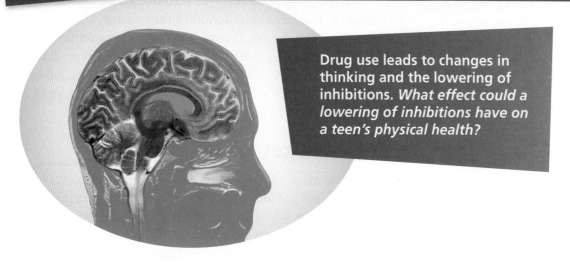

Drug use leads to changes in thinking and the lowering of inhibitions. *What effect could a lowering of inhibitions have on a teen's physical health?*

Drugs Take a Heavy Toll

Main Idea In addition to the physical risks to a person's health, substance abuse can damage all aspects of your life.

Some people believe that drugs can help them escape from their problems. Drug use, however, can actually create problems that affect a user's physical, mental/emotional, and social health.

Consequences for the Individual

Teens who use illegal drugs may stop pursuing their interests and goals, and the goals their parents, teachers, and other adults set for them. Taking drugs lowers inhibitions, which may lead teens to engage in behaviors that can harm their physical health. These behaviors may include engaging in sexual activity, which increases the risk of acquiring sexually transmitted infections, or acting recklessly.

Drug use is also a leading factor in teen depression and suicide. Teens who are involved in drug use are more likely to be arrested. Teens who are convicted of a drug offense can be sent to jail. Teen drug use can also lead to increased violence, crime, and accidental death.

Consequences for Friends and Family

When a teen abuses drugs, it affects everyone in his or her life. Teens who use illegal drugs may lose interest in healthy activities. They may stop spending time with friends who value a drug-free lifestyle. Family members who feel responsible for their loved ones feel the burden of the emotional and financial costs of drug abuse.

Consequences for Others

A developing fetus receives nutrients through the mother's placenta. If a pregnant female takes drugs, those drugs are passed to the fetus. These drugs have a greater effect on the fetus than the mother. The fetus may be born with birth defects, behavioral problems, or a drug addiction. After birth, a nursing mother's breast milk may contain traces of drugs that are passed to the baby.

Consequences for Society

People who abuse drugs cause harm to society. Illegal drug use can result in a rise in drug-related crime and violence. Driving while intoxicated (DWI) or driving under the influence (DUI) can result in collisions that cause injuries and deaths. Drug abuse also affects our nation's economy. Research by the Office of National Drug Control Policy shows that drug abuse costs the U.S. economy $180 billion per year. These costs result from

- lost work hours and productivity due to drug-related illnesses, jail time, accidents, and deaths.
- health care costs and legal fees.
- law enforcement costs and insurance costs due to drug-related damages, injuries, and deaths.

The consequences of drug abuse—mental, emotional, physical, legal, and social—are 100 percent preventable. By choosing a drug-free lifestyle, you avoid these consequences.

READING CHECK

Infer How can illicit drug use by one person affect people who do not use drugs?

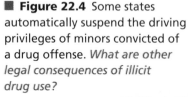

■ **Figure 22.4** Some states automatically suspend the driving privileges of minors convicted of a drug offense. *What are other legal consequences of illicit drug use?*

Real World CONNECTION

Trends in Illegal Drug Use

Conduct an online search to determine whether the number of teens who have never used drugs is increasing or decreasing.

A good place to start your search is with the Centers for Disease Control and Prevention's Web site and searching for the most recent National Youth Risk Behavior Survey. Use Web sites that end with ".gov" to ensure that they are safe and remember to keep a list of the sources of your research.

Activity Technology

Once you have found your information, complete the following activity:

1. Write a blog or create a podcast describing the factors that are influencing teens' choices not to use drugs.

2. Include information on the impact of drug use on the individual, family, friends, and the community.

3. Include information on the physical, mental/emotional, social, and legal consequences of drug use.

4. Present your blog or podcast to the class and possibly submit it to the school Web site.

LESSON 1 ASSESSMENT

 After You Read

Reviewing Facts and Vocabulary

1. What is an *overdose*?
2. How can an addiction affect your health?
3. How does drug abuse affect society?

Thinking Critically

4. **Infer.** Why might an addiction to a drug become more expensive as the body develops a tolerance to the drug?
5. **Analyze.** Distinguish between *substance abuse* and *illicit drug use*. How are these terms similar? How are they different?

Applying Health Skills

6. **Accessing Information.** Conduct a survey of teens. Ask: What percentage of teens do you think use drugs? Compare your information to statistics from reliable sources. Create a poster presenting your information.

Writing Critically

7. **Persuasive.** Write a dialogue between you and a friend who is thinking about trying an illegal drug. Tell your friend the consequences of drug use.

Real Life Issues ....

After completing the lesson, review and analyze your response to the Real Life Issues question on page 592.

Marijuana, Inhalants, and Steroids

GUIDE TO READING

BIG Idea Three often-abused drugs that can have serious physical and mental side effects are marijuana, inhalants, and anabolic steroids.

Before You Read

Create a Chart. Create a chart with three columns. Label the columns "Marijuana," "Inhalants," and "Steroids." As you read, list the physical, mental, and legal consequences of each.

Marijuana	Inhalants	Steroids

New Vocabulary

▶ marijuana
▶ paranoia
▶ inhalants
▶ anabolic-androgenic steroids

Real Life Issues

Driving While on Drugs

> Studies have shown that 18% of fatally injured drivers tested positive for an illegal substance.

> In 2011, 8.2% of teens who were 16 years old drove under the influence.

> Between 2001 and 2006, 14.1 percent of teens reported driving under the influence of marijuana.

National Institute of Drug Abuse, Drug Facts: Drugged Driving, Revised October 2013

Writing *Write a paragraph that describes what can happen when driving under the influence of illegal drugs.*

Marijuana

Main Idea Using marijuana has serious physical, mental, social, and legal consequences.

Every day, you make choices based on information that's available to you. Before deciding to see a particular movie, you may read a review. Before deciding what to eat, you might browse the list of ingredients. Drugs like marijuana can be mixed with unknown chemicals and have unexpected effects on your health. Even when you are certain of the source of a drug, using it may cause serious harm to your health. Misusing any drugs can have serious consequences to your health.

In 2013, two U.S. states decriminalized the use of ***marijuana***, *a plant whose leaves, buds, and flowers are usually smoked for their intoxicating effects*. In these states, a person will not be charged with a crime if he or she is found carrying less than one ounce of marijuana. Other states have passed laws legalizing the use of marijuana for medical purposes. However, a prescription from a doctor is required to purchase marijuana from an authorized medical marijuana retailer. Despite these changes to laws affecting the use of marijuana, it is still illegal for anyone under the age of 21 to use marijuana or carry any amount of the substance. It is also illegal for anyone to use marijuana in public or to drive while under the intoxicating effects of marijuana.

Marijuana is considered a possible *gateway drug,* a drug that may lead the user to try other drugs. All forms of marijuana are mind-altering and can damage the user's health. This is particularly true if marijuana use begins during the teen years. A New Zealand study shows that teens who are heavy users of marijuana lost an average of 8 IQ points between ages 13 and 38.

Figure 22.5   ## Health Risks of Marijuana

The effects of marijuana use varies from person to person, and can be influenced by a person's mood and surroundings.

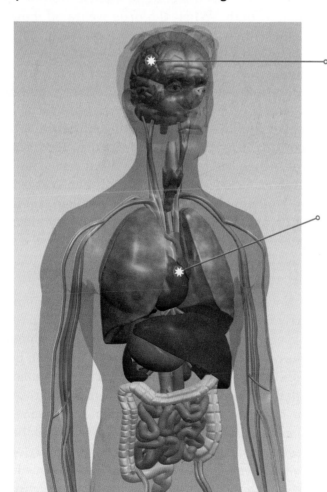

- Hallucinations and paranoia
- Impaired short-term memory, reaction time, concentration, and coordination

- Lung irritation, coughing
- Heart and lung damage
- Increased risk of lung cancer
- Weakened immunity to infection

- Increased appetite
- Increased risk of stillbirth and birth defects
- Changed hormone levels
- In females, risk of infertility
- In males, lowered sperm count and testosterone levels

Physical Consequences of Marijuana Use

Because marijuana is often smoked, users face the same health risks as tobacco smokers. Marijuana smoke contains more cancer-causing chemicals than tobacco smoke. Medical evidence shows that a compound in marijuana named delta-9-tetrahydrocannabinol, or THC, does reduce the pain and nausea caused by illnesses such as cancer and HIV/AIDS. However, the FDA has approved drugs that deliver THC without the other health risks associated with marijuana use. Smoking marijuana may also damage the immune system, making the user more susceptible to infections. Many of the physical effects of marijuana use are summarized in **Figure 22.5** on page 599.

Marijuana also poses risks to the reproductive system. In males, it interferes with sperm production and *lowers* levels of the male hormone testosterone. In females, marijuana *raises* testosterone levels which may lead to *infertility*, or the inability to bear children.

Mental and Emotional Consequences

Academic Vocabulary

intense *(adjective):* existing in an extreme degree

Marijuana raises levels of a brain chemical called *dopamine*. This chemical produces a pleasurable feeling. In some users, marijuana triggers the release of so much dopamine that the user reaches a feeling of **intense** well-being or elation, called a "high." When the drug wears off, however, the pleasure sensation stops, often dramatically. This abrupt letdown is called a "crash."

Marijuana users can experience slow mental reflexes and may suffer from sudden feelings of anxiety and **paranoia**, *an irrational suspiciousness or distrust of others*. The user might feel dizzy, have trouble walking, and have a hard time remembering things that just happened. Short-term memory is affected, which can lead to problems at school and at work. Users often experience distorted perception, loss of coordination, and trouble with thinking and problem solving. A few hours after use, the person can become very sleepy. These consequences can be deadly if the user is driving a vehicle.

Driving and Marijuana Use

The National Highway Traffic Safety Administration (NHTSA) estimates 10.3 million drivers involved in car crashes were on drugs. Driving under the influence of marijuana can be dangerous because marijuana interferes with depth perception, increases reaction time, causes sleepiness, impairs judgment, and slows reflexes. The penalties and legal consequences of driving under the influence of any drug—including marijuana—include suspension of a driver's license, fines, loss of eligibility for federal college loans, and possibly a jail term. If injury or death of another person results, the driver may face serious legal prosecution as well as devastating emotional consequences.

Inhalants

Main Idea Inhalants can cause the death of brain cells.

Inhalants are *substances whose fumes are sniffed or inhaled to give effect.* Some inhalants are prescribed by doctors to treat allergies, asthma, and other medical conditions. However, some substances are inhaled to achieve a high. Solvents, aerosols, glues, paints, varnishes, and gasoline can cause brain damage.

Most inhalants depress the central nervous system. Immediate effects include a glassy stare, slurred speech, impaired judgment, nausea, coughing, nosebleeds, fatigue, and lack of coordination. Using inhalants can lead to permanent loss of brain cells. Long-term use can cause liver and kidney damage, blindness, brain damage, paralysis, cardiac arrest, and death.

All inhalants are extremely dangerous, and many are labeled as poisons. Inhalants can be harmful even if you are not trying to abuse them. They can be accidentally inhaled when doing household chores. When using inhalants, work in a well-ventilated room and wear a mask if a project requires long exposure to the fumes.

Consequences of Steroid Use

Main Idea Steroids can cause severe health problems.

Anabolic-androgenic steroids are *synthetic substances similar to male sex hormones. Anabolic* refers to muscle building, and *androgenic* refers to increased male characteristics. Steroids may be prescribed for some medical conditions, but using steroids without medical supervision is dangerous.

Steroid use can result in unnatural muscle growth. When combined with physical conditioning, steroids can increase muscle strength, but the tendons and ligaments do not get stronger which can lead to injury. Other side effects include weight gain, acne, high blood pressure, and liver and kidney tumors. Steroid users who inject the drug may contract HIV or hepatitis B. These drugs may also cause violent behavior, extreme mood swings, depression, and paranoia. The effects on males include shrinking testicles, reduced sperm count, baldness, development of breasts, and an increased risk for prostate cancer. The effects on females include facial hair, baldness, menstrual cycle changes, and a deepened voice.

Any nonmedical use of steroids is illegal. Athletes who use steroids can face expulsion from a team or event, monetary fines, tarnished reputation, and jail time.

■ **Figure 22.6** Marijuana contains 421 different chemicals. The main psychoactive ingredient, THC (delta-9-tetrahydrocannabinol), is stored in body fat, and traces of it can be present in the blood for as long as a month. *Why could a marijuana user fail a drug test weeks after using the drug?*

READING CHECK

Explain Where does the term *anabolic-androgenic steroids* come from?

Health Skills Activity

Decision Making

Making Choices About Drugs

Naomi and her best friend, Jill, and two new friends, Gwen and Miriam, are driving together to a concert by their favorite band. Gwen, stops at a nearby park.

"My brother told me about this," says Gwen as she pulls out a paper bag. "You can get high by sniffing this stuff."

"I heard the same thing!" says Miriam. "Do you guys want to try?"

Jill turns to Naomi. "I'm not so sure. I don't want to do drugs," whispers Jill.

Naomi learned in her health class how dangerous inhalants can be. Gwen and Miriam are fun to hang out with, though, and Naomi doesn't want to disappoint her two new friends.

Writing Apply the six steps of the decision-making model to Naomi's situation.
1. State the situation.
2. List the options.
3. Weigh the possible outcomes.
4. Consider your values.
5. Make a decision and act.
6. Evaluate the decision.

LESSON 2 ASSESSMENT

After You Read

Reviewing Facts and Vocabulary

1. What body systems are harmed by smoking marijuana?

2. What are *inhalants*?

3. Why does using steroids for increasing muscle strength often result in injury?

Thinking Critically

4. **Infer.** Marijuana users often inhale the smoke very deeply and hold it in their lungs longer than cigarette smokers do. How might this practice make marijuana more dangerous than smoking tobacco?

5. **Compare.** How do the effects of steroids differ in males and females?

Applying Health Skills

6. **Accessing Information.** Research reliable sources to learn more about the dangers of accidentally or purposefully inhaling chemicals. Create a poster showing how inhalants can affect your physical health.

Writing Critically

7. **Persuasive.** Write a public service announcement describing the dangers of driving while under the influence of marijuana.

Real Life Issues..................................

After completing the lesson, review and analyze your response to the Real Life Issues question on page 598.

Psychoactive Drugs

Real Life Issues •••••••••••••••••••••••••••••

Skip this "Trip." It's early fall, and school has just started. Alex and his friends are at a party hosted by some college kids down the street. The party is a little wilder than they anticipated. At one point, a guy offers Alex a "hit" of LSD. "It's awesome," says the young man, who looks a little out of it. "No thanks," says Alex. "I'm not into drugs."

Writing *Write a paragraph that describes at least three safe activities that can give your body a physical adrenaline rush or stimulate the senses without the use of illegal drugs.*

Effects of Psychoactive Drugs

Main Idea Psychoactive drugs change the functioning of the central nervous system.

The central nervous system (CNS) is amazingly complex. Every human activity, from bending a finger to solving complicated problems, involves the CNS. **Psychoactive drugs,** *chemicals that affect the central nervous system and alter activity in the brain,* change the functioning of the CNS. The four main groups of psychoactive drugs are stimulants, depressants, opiates, and hallucinogens.

Some psychoactive drugs have medicinal value. When psychoactive drugs are misused or abused, a person's health and the functioning of all the body systems are seriously affected. The effects on a teen's developing brain and body can be especially damaging.

GUIDE TO READING

BIG Idea *Psychoactive drugs affect the central nervous system and can be especially damaging to the developing brain and body of a teen.*

Before You Read

Make Flash Cards. As you read the lesson, write each vocabulary term on the front of an index card. Write the definition on the back of each card. Use the cards to quiz a partner on the terms and their meanings.

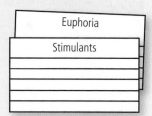

New Vocabulary

▶ psychoactive drugs
▶ designer drugs
▶ hallucinogens
▶ euphoria
▶ depressants
▶ stimulants
▶ opiates

Figure 22.7 Health Risks of Psychoactive Drugs

Type of Drug	Consequences to Your Health
Stimulants	
Amphetamines	• Decreased appetite, weight loss, malnutrition • High blood pressure, rapid heartbeat, heart failure, death • Aggressiveness, increased tolerance, addiction
Cocaine	• Nausea, abdominal pain, malnutrition, headache, stroke, seizure, heart attack, death • Exposure to HIV through contaminated needles, addiction
Crack	• Extreme addiction with the same consequences as cocaine • Rapid increase in heart rate and blood pressure, death
Methamphetamine (Meth)	• Memory loss, heart and nerve damage • Increased tolerance, addiction
Depressants	
Barbiturates	• Fatigue, confusion, impaired muscle coordination • Reduced heart rate, blood pressure, respiratory function, death
GHB	• Drowsiness, nausea, vomiting, loss of consciousness • Impaired breathing, coma, death
Rohypnol (roofies)	• Decreased blood pressure, drowsiness, memory loss, gastrointestinal disturbances
Tranquilizers	• Depression, fever, irritability, loss of judgment, dizziness
Opiates	
Codeine	• Reduced respiratory function, respiratory arrest, death • Exposure to HIV through contaminated needles, addiction
Heroin	• Confusion, sedation, unconsciousness, coma, addiction
Morphine	• Rapid onset of tolerance, addiction
Opium	• Nausea, constipation, addiction
Oxycodone (OxyContin®)	• Drowsiness, nausea, constipation, addiction • Reduced respiratory function, respiratory arrest, death
Hallucinogens	
DXM (tussin)	• Nausea, dizziness, lack of coordination, rashes • Hallucinations, disorientation, paranoia, panic attacks, seizures
Ecstasy (MDMA)	• Confusion, depression, paranoia, muscle breakdown
Ketamine	• Kidney and cardiovascular system failure, death • Memory loss, numbness, impaired motor function
LSD	• Delusions, illusions, hallucinations, flashbacks, numbness, tremors
Mescaline (peyote)	• Delusions, illusions, hallucinations, flashbacks, numbness, tremors
PCP	• Loss of appetite, depression, panic, aggression, violent actions
Psilocybin (mushrooms)	• Delusions, illusions, hallucinations, paranoia, extreme anxiety, nausea

Consequences of Psychoactive Drug Use

Psychoactive drug use can result in health problems and addiction. Using psychoactive drugs often leads to poor judgment and behaviors, which may put teens at risk for unintentional injuries, violence, STDs, unintended pregnancy, and suicide. Choosing a drug-free life can protect your health.

Club Drugs, Stimulants, and Depressants

Main Idea Club drugs, stimulants, and depressants can cause irreversible health damage.

Certain drugs are classified by their effects. They may speed up or slow down the senses, or affect judgment.

Club Drugs

The term *club drug* describes drugs found at concerts and dance clubs. These drugs are sometimes disguised in foods, or slipped into drinks and taken without a person's knowledge. Many club drugs are **designer drugs**, *synthetic drugs that are made to imitate the effects of other drugs.* Designer drugs can be several hundred times stronger than the drugs they imitate.

Ecstasy (MDMA) Ecstasy, or MDMA, has both stimulant and hallucinogenic effects. **Hallucinogens** are *drugs that alter moods, thoughts, and sense perceptions, including vision, hearing, smell, and touch.* Ecstasy may cause short-term **euphoria**, *a feeling of intense well-being or elation.*

Rohypnol Rohypnol, or "roofies," are **depressants**, or sedatives that are colorless, odorless, and tasteless. These are *drugs that tend to slow the central nervous system.* It is called the "date-rape" drug. Unwanted physical contact, unplanned pregnancies, and exposure to HIV and STDs can result. It's illegal to give someone a drug without his or her knowledge. Engaging in sexual activity with a person under the influence of a date-rape drug is a criminal offense.

GHB GHB, or gamma hydroxybutyric acid, is another CNS depressant. It is available as a clear liquid, a white powder, and in a variety of tablets and capsules. Like Rohypnol, it can be used as a date-rape drug.

Purple Drank or Sizzurp This drug is another depressant. Prescription cough syrup is an ingredient. One of the main ingredients is codeine. Users can become addicted and can overdose, possibly leading to death.

READING CHECK

Apply What should you do if you are at a party where people are taking psychoactive drugs?

■ **Figure 22.8** Never allow a stranger to handle your drink at a social event. *Why are Rohypnol and GHB often known as date-rape drugs?*

Academic Vocabulary

available *(adjective):* present or ready for immediate use

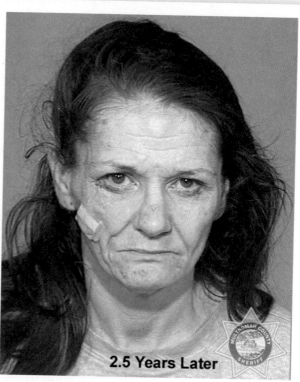

2.5 Years Later

■ **Figure 22.9** One common effect of meth use is delusions of bugs crawling on the user's skin. As a result, many meth users scratch and pick at their skin until they develop unsightly sores. *What other physical effects can meth use cause?*

Courtesy of Multnomah County Sheriff, Tim Fuller Photography

Ketamine Ketamine is an anesthetic used to treat animals. It causes hallucinations and may result in respiratory failure.

Meth Methamphetamine, or meth, is a stimulant. **Stimulants** are *drugs that speed up the central nervous system.* Meth is a white, odorless powder that easily dissolves in alcohol or water. Meth may provide a short-term feeling of euphoria, but often its use also causes depression, paranoia, and delusions. Meth use can cause death.

LSD (Acid) Acid, or lysergic acid diethylamide (LSD), can cause hallucinations and severely distorted perceptions of sound and color. Flashbacks—states in which a drug user experiences the emotional effects of a drug long after its actual use—can also occur. Users may experience emotions ranging from extreme euphoria, to panic, to terror or deep depression. The resulting behaviors can lead to serious injury or death.

■ **Figure 22.10** "Energy drinks" often have so much caffeine that their labels must warn pregnant women and people with high blood pressure not to drink them. *Why is the small serving size of a typical energy drink misleading?*

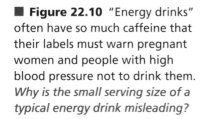

Other Stimulants

Stimulants speed up the CNS. The nicotine in tobacco products is a highly effective stimulant. The caffeine in coffee, tea, cola, and power drinks are all stimulants. "Energy" or "power" drinks often contain four to ten times the amount of caffeine as a regular-sized cola. Other dangerous stimulants include cocaine, amphetamines, and methamphetamines.

Cocaine Cocaine is a rapidly acting, powerful and highly addictive stimulant. Cocaine is a white powder extracted from the leaves of the coca plant. Using cocaine is illegal. Users may experience a surge of self-confidence and euphoria. The feelings of confidence induced by cocaine are followed by an emotional letdown. Regular use can lead to depression, fatigue, paranoia, and physiological dependence. Cocaine use can cause malnutrition and, especially among teens, may result in cardiac problems. When cocaine is injected, users risk contracting HIV or hepatitis B from infected needles. Overdosing can result in death.

Crack An even more dangerous form of cocaine is crack, also called rock or freebase rock. Crack reaches the brain seconds after being smoked or injected. Once in the blood, it causes the heart rate and blood pressure to soar to dangerous levels. Death may result from cardiac or respiratory failure. Mixing crack (or any drug) with alcohol can be fatal. Both substances combine in the liver, increasing the risk of death from liver failure.

Amphetamines Amphetamines are highly addictive. Some people use amphetamines to stay alert, to improve athletic performance, or to lose weight. It is easy to develop a tolerance to amphetamines, causing the user to ingest more and more of the substance. Regular use can result in an irregular heartbeat, paranoia, aggressive behavior, and heart failure.

Other Depressants

Depressants are drugs that tend to slow the central nervous system and can have negative, sometimes deadly effects on your health. Depressants are dangerous because they can slow heart and respiration rates and lower blood pressure. Alcohol is a commonly used depressant. Combining small amounts of depressants can cause shallow breathing, weak or rapid pulse, coma, and death.

Barbiturates Barbiturates are sedatives that are rarely used for medical purposes. Using barbiturates can cause mood changes, excessive sleepiness, and coma. Users may feel intoxicated. Combining barbiturates with alcohol can be fatal.

Tranquilizers Tranquilizers are depressants that relieve anxiety, muscle spasms, sleeplessness, and nervousness. When tranquilizers are overused, they can cause physiological and psychological dependence, coma, and death.

■ **Figure 22.11** Both cocaine and crack are dangerous illegal stimulants. *How are cocaine and crack related?*

 READING CHECK

Compare Name two legal stimulants that are commonly used.

 **READING CHECK**

Explain Why can combining depressants be dangerous?

Design Pics/Leah Warkentin

Hallucinogens and Opiates

Main Idea Hallucinogens and opiates seriously alter the sensory controls in the brain.

The sections of the brain that interpret sensory input can be permanently damaged by the effects of psychoactive drugs. Hallucinogens overload the brain's sensory controls. Opiates, which are highly addictive, cause confusion and dull the senses.

Hallucinogens

Hallucinogens can cause serious mental/emotional and physical consequences for users. These drugs alter mood, and impair judgement, thoughts, and sense perception. Users may behave in ways that they normally would not. A person who uses hallucinogens may believe that he or she is invincible. Using these drugs can also cause increases in heart and respiratory rates, which can lead to heart and respiratory failure. Using these drugs can cause coma. Hallucinogens have no medical use.

Ecstasy, ketamine, acid (LSD), phencyclidine (PCP or angel dust), dextromethorphan (DXM), psilocybin (mushrooms), and mescaline (peyote) are examples of powerful and dangerous hallucinogens. Hallucinogens overload the sensory controls in the brain, causing confusion, intensified sensations, and hallucinations. The altered mental states caused by hallucinogens can last for several hours or several days. Users can also experience flashbacks, or states in which they feel emotional effects of a drug long after its actual use. The effects of using these drugs are extremely unpredictable. Users sometimes harm themselves physically, or behave violently and harm others.

PCP PCP, or angel dust, is one of the most dangerous of all drugs, and its effects vary greatly from user to user. The drug creates a distorted sense of time, increased muscle strength, increased feelings of violence, and the inability to feel pain. Overdoses can cause death, but most PCP-related deaths are caused by the destructive behavior or disorientation that the drug produces. For example, PCP users have died in fires because they became disoriented and had no sensitivity to the pain of burning. Flashbacks can occur at any time, causing pain, confusion, and lack of control.

DXM DXM, or "tussin," is a cough suppressant sold as an over-the-counter medicine. When used in the recommended dosage, DXM is not dangerous. When misused, it can cause hallucinations, paranoia, panic attacks, nausea, increased heart rate and blood pressure, seizures, and addiction.

☑ READING CHECK

Apply Why are hallucinogens sometimes fatal?

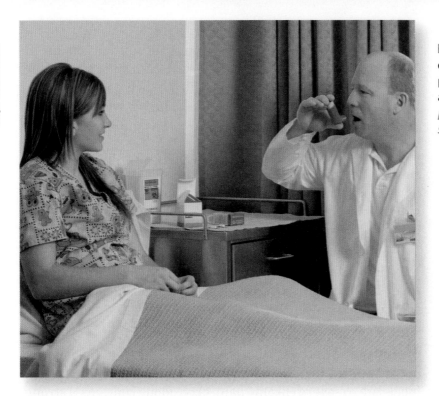

■ **Figure 22.12** Talk to your doctor or pharmacist about any prescription medications that you are given. *Why is it important for pharmacists to keep records of all sales of opiates?*

Mushrooms and Peyote Psilocybin (mushrooms) and mescaline (peyote cactus) are hallucinogens found in nature as a fungus and a plant. When eaten, they cause hallucinations, nausea, and flashbacks. Use of these drugs can also lead to poisoning and death when dealers harvest toxic species.

Opiates

Opiates, or narcotics, are *drugs such as those derived from the opium plant that are obtainable only by prescription and are used to relieve pain.* Morphine, oxycodone, and codeine are common examples of opiates. When opiates are used according to the directions provided by a health care professional, they are an effective pain reliever. Abusing opiates dulls the senses, causes drowsiness, constipation, slow and shallow breathing, convulsions, coma, and death. Pharmacists record all sales of opiates because the drug is addictive.

Codeine Codeine is a highly addictive ingredient in some prescription cough medicines. Even if a user takes codeine as prescribed, drowsiness can occur. The drug should not be used before driving a vehicle. Codeine use can cause dizziness, labored breathing, low blood pressure, seizures, and respiratory arrest.

Some people may be allergic to codeine. These people may experience difficulty breathing or mood changes. Codeine use has also been linked to death in infants. The CDC has issued a warning against giving any medications containing codeine to infants or small children. If you or someone you know experiences a problem after taking codeine, call 911 immediately.

Morphine Morphine is a much stronger drug than codeine. It is sometimes prescribed to treat severe pain, but is generally used for only a short time. Side effects include fast or slow heartbeat, seizures, hallucinations, blurred vision, rashes, and difficulty swallowing.

READING CHECK

Explain How are heroin and morphine related?

Heroin Heroin is a processed form of morphine that is injected, snorted, or smoked. Heroin comes in many forms, including a white or brownish powder and a black, sticky tar. Dealers may mix heroin with medicines or household substances to create other forms, such as "cheese" or "cheese heroin."

Heroin slows breathing and pulse rate. It can also cause infection of the heart lining and valves, as well as liver disease. Infectious diseases such as HIV and hepatitis B can also result from the use of infected needles. Large doses can cause coma or death, and fetal death if the user is pregnant.

Oxycodone When used properly under the supervision of a doctor, oxycodone is a prescription drug that helps to relieve moderate to severe chronic pain. Oxycodone contains a strong opiate. It is often referred to by the brand name OxyContin®. A side effect of this drug is suppression of the respiratory system, which cause death from respiratory failure.

LESSON 3 ASSESSMENT

 After You Read

Reviewing Facts and Vocabulary

1. On what body system do *psychoactive drugs* act?
2. Name the four types of drugs described in this lesson and give an example of each.
3. What are *opiates*?

Thinking Critically

4. **Evaluate.** An acquaintance offers you a drug that she says is natural. Does this mean it is safe to take? Why or why not?
5. **Apply.** Why is it important to follow directions from your doctor or pharmacist when taking a prescription drug such as codeine?

Applying Health Skills

6. **Advocacy.** Research the different types of designer drugs, the forms they take, and how they affect health. Use what you have learned to design a Web site that warns about the dangers of these drugs.

Writing Critically

7. **Narrative.** Write a script convincing a friend not to try drugs. Include data on the harmful effects and health consequences of drug use.

Real Life Issues ....

After completing the lesson, review and analyze your response to the Real Life Issues question on page 603.

Living Drug-Free

Real Life Issues •••••••••••••••••••••••••••

A Sister's Advice. When Penny arrives home, her younger sister is on her bed, crying. "Lisa, what happened?" asks Penny. "A girl at the recreation center offered me drugs," Lisa says. "I didn't know what to do, so I just ran off." Penny gives Lisa a comforting hug. "Sounds to me like you did the right thing," she assures her. Lisa looks up, her eyes red. "But they're going to make fun of me at school on Monday," she says. "Maybe so," replies Penny, "but would they make fun of you for eating right and working out? I always tell people who offer me drugs that I'm an athlete, and I'm not into that."

Writing *Make a list of reasons to say no to drug use.*

Resisting Pressure to Use Drugs

Main Idea Most teens never experiment with illegal drugs.

By deciding not to use drugs, you protect your health, and become a role model to others. Peer pressure can be intense during the teen years. When the subject of drug use comes up, you may be told that "everybody's doing it." This claim is not true. Most teens never experiment with illegal drugs. Almost 62 percent of high school students have never tried marijuana, and more than 90 percent have never tried cocaine.

Committing to Be Drug-Free

It is sometimes difficult to make decisions quickly. You may not have time to consider all of the consequences until after a dangerous situation has passed. You may feel unsure about saying no to drug use. If your friends put pressure on you to use drugs, you may need to decide whether you want to continue to remain friends with someone who uses drugs.

***BIG* Idea** By deciding not to use drugs, you promote your own health and influence others to do the same.

Before You Read

Create an Outline. Preview this lesson by scanning the pages. Then, organize the headings and subheadings into an outline. As you read, fill in the outline with important details.

```
I.
    A.
        1.
        2.
    B.
II.
```

New Vocabulary

▶ drug-free school zones
▶ drug watches
▶ rehabilitation

■ **Figure 22.13** Activities you enjoy can help you avoid situations where drugs may be available. *What types of activities would you choose?*

Alain Shroder/Getty Images

To commit to remaining drug-free, choose friends who share your attitude about drug use, and avoid places where drugs may be available.

Even a teen who has used drugs in the past can choose not to use drugs in the future. Refusal skills can help you say no to drugs. Thinking of and practicing refusal statements ahead of time will help you feel comfortable using them. Examples of refusal statements include:

- "No thanks, I don't do drugs."
- "I can't. I'm on medication."
- "I'm not interested. That stuff makes me sick."
- "No. I have to be in great shape for tomorrow's game."

Healthy Alternatives

Choosing friends who value a drug-free lifestyle and participating in activities that do not involve drugs can help you avoid drug use. It can also help to build self-esteem, provide role models, reinforce values, and help you make new friends. The following activities are just a few healthy alternatives to drug use.

- **Hobbies.** Enjoy hobbies such as photography, cooking, art, or music.
- **Sports.** Get physical activity through outdoor recreation, team, and individual sports.
- **Community activities.** Participate in neighborhood events, political movements, community service, religious activities, and local clubs.
- **School organizations.** Get involved in service groups, honor societies, and advocacy groups at school.

READING CHECK

Apply What can you say to someone who pressures you to use drugs by telling you that "everybody's doing it"?

Drug Prevention Efforts

Main Idea Schools and communities are working together to support students in their efforts to be drug-free.

Everyone can help reduce substance abuse by committing to remain drug-free. Schools and communities also provide ways to help young people avoid drugs.

School Efforts

Near schools, **drug-free school zones**, *areas within 1,000 to 1,500 feet of schools and designated by signs, within which people caught selling drugs receive especially severe penalties,* have been established. Penalties are often double what they might be for the same drug offense committed elsewhere. Other efforts at schools to eliminate drug use include drug education classes, zero-tolerance policies, and the expulsion of students found using drugs. Some schools conduct locker searches and maintain police patrol on campus.

Community Efforts

Communities across the nation are taking action to prevent drug abuse. **Drug watches** are *organized community efforts by neighborhood residents to patrol, monitor, report, and otherwise try to stop drug deals and drug abuse.* Anti-drug programs in your community can help protect your family and friends from drug abuse.

Becoming Drug-Free

Main Idea Many types of counseling are available for those who want to become drug-free.

Once someone begins using drugs, addiction can occur rapidly. **Figure 22.15** on page 614 lists some warning signs of drug abuse. The following steps can guide you in helping a friend or family member.

- Identify sources of help in your community.
- Talk to the person when he or she is sober. Express your affection and concern, and describe the person's behavior without being judgmental.
- Listen to the person's response. Be prepared for anger and denial.
- Offer to go with your friend or family member to a counselor or support group.

READING CHECK

Contrast How are some of the penalties for illicit drug use in a drug-free school zone more severe than in other areas?

■ **Figure 22.14** The penalties for using, selling, or possessing drugs in a drug-free school zone are more severe than in other areas. *How far from a school is the border of a typical drug-free school zone?*

613

Figure 22.15 Warning Signs of Drug Use

The following behaviors may indicate that a person has a drug problem.

- Lies about the drugs he or she is using, constantly talks about drugs
- Stops participating in activities that once were an important part of his or her life
- Changes eating or sleeping habits, shows rapid weight loss
- Takes unnecessary risks, participates in unsafe behaviors
- Gets in trouble with authorities, such as school administrators or police
- Seems withdrawn, depressed, tired, and cares less about appearance
- Has red-rimmed eyes and runny nose not related to colds or allergies
- Has blackouts and forgets what he or she did under the influence
- Has difficulty concentrating

Getting Help

Drug abuse is a treatable condition. **Rehabilitation** is *the process of medical and psychological treatment for physiological or psychological dependence on a drug or alcohol.* Most drug users need the help family, friends, and counseling to end their addiction.

Drug treatment centers offer a safe place to withdraw from drug use. Many of these centers provide medications to help with the physical and psychological effects of withdrawal. Types of drug treatment centers include:

- **Outpatient drug-free treatment.** These programs usually do not include medications and often use individual or group counseling.
- **Short-term treatment.** These centers can include residential therapy, medication therapy, and outpatient therapy.
- **Maintenance therapy.** Intended for heroin addicts, this treatment usually includes medication therapy.
- **Therapeutic communities.** These are residences for drug abusers. The centers include highly structured programs that may last from six to 12 months.

Drug counselors can also help people **adjust** to a life without drugs. Some counselors use behavioral change strategies to help a person become drug-free. These strategies include avoiding the people who supply or do the drug, recalling negative consequences of doing the drug, practicing refusal skills, and filling free time with planned, healthy activities.

Former drug users may also attend support groups. These meetings are gatherings of people who share a common problem. Support groups provide the long-term support that the recovering user needs to remain drug-free, in addition to family support.

Academic Vocabulary

adjust *(verb):* to bring to a more satisfactory state

TEENS Making a Difference

"We don't have time to consider using drugs."

Staying Drug-Free

Sandra L., of Florida, is president of her school's chapter of Drug Free Youth in Town (DFYIT). "Every weekend thirty to fifty of our 300 members do community service. The purpose is to have fun and not have time to think about drugs," says Sandra.

DFYIT teens pledge to be drug-free and agree to random drug tests. "We wear T-shirts with the words 'Drug-free and willing to prove it,'" Sandra says.

Through Sandra's leadership, DYFIT members have volunteered at a number of events, including an arts festival, an AIDS walk, a food and toy drive, and haunted houses on Halloween.

The 300-plus members of DFYIT rely on Sandra to organize their volunteer work. "Knowing that I am helping them stay away from drugs is rewarding."

Activity Write your answers to the following questions in your personal health journal:

1. What fun activities could teens volunteer for in your community?
2. How would you organize a similar club at your school?
3. List at least six reasons why it's important to stay drug-free.

LESSON 4 ASSESSMENT

After You Read

Reviewing Facts and Vocabulary

1. What are three healthy alternatives to using drugs?
2. What is a *drug-free school zone*?
3. Describe *rehabilitation*.

Thinking Critically

4. **Analyze.** Why is it important to commit to being drug-free before drugs are offered to you?
5. **Apply.** Former drug users try to fill their free time with healthy activities. What kinds of healthy activities could a former user try in order to remain drug-free?

Applying Health Skills

6. **Practicing Healthful Behaviors.** List five healthy alternatives to drug use, and share your ideas with the class.

Writing Critically

7. **Narrative.** Write a short story about someone who has stopped using drugs. Describe the types of community resources that provide help.

Real Life Issues ....

After completing the lesson, review and analyze your response to the Real Life Issues question on page 611.

Hands-On HEALTH

Activity Saying No

Illegal drug use can harm your physical, mental/emotional, and social health. You will create a presentation for younger students, providing information about illegal drugs and advice on how to use refusal skills to avoid drug use.

What You'll Need

- a computer with Internet access
- one large sheet of white paper and colored markers

What You'll Do

Step 1

Working with a partner, review the text for information on the risks of drug use. On the large sheet of paper, list the names of ten illegal drugs, and the effects of each drug.

Step 2

Create a computer presentation that helps younger students understand the risks of using illegal drugs. Show them how to use refusal skills.

Step 3

Show your presentation to a class of middle school students, and demonstrate refusal skills.

Apply and Conclude

Ask the middle school students for feedback. Write a short reflective paper on the effectiveness of your presentation, including what you would do differently.

Checklist: Refusal Skills

✓ Say no, and explain why you are refusing

✓ Propose alternative activities to the group

✓ Use body language to back up your words

✓ Walk away from the situation if necessary

Anti-Drug Rally!
- Practice Refusal Skills
- Learn how to get help
Today at 4:00 pm
Main Auditorium

The Health Risks of Drug Use

Key Concepts

▶ It is illegal to use, sell, or possess illegal drugs, or to sell or use prescription medications for nonmedical purposes.

▶ Misusing medicines and drugs can lead to addiction.

▶ Drug abuse can have negative consequences for the user, friends, family, and society.

Vocabulary

▶ substance abuse (p. 592)
▶ illegal drugs (p. 592)
▶ illicit drug use (p. 592)
▶ overdose (p. 593)
▶ addiction (p. 594)

Marijuana, Inhalants, and Steroids

Key Concepts

▶ Using marijuana can damage a user's health.

▶ Sniffing inhalants can permanently damage or kill brain cells.

▶ The nonmedical use of anabolic-androgenic steroids is illegal and can pose serious health risks.

Vocabulary

▶ marijuana (p. 599)
▶ paranoia (p. 600)
▶ inhalants (p. 601)
▶ anabolic-androgenic steroids (p. 601)

Psychoactive Drugs

Key Concepts

▶ Many illegal drugs often contain unknown ingredients.

▶ Stimulants include legal drugs, such as nicotine, caffeine, and many medicines, and illegal drugs, such as cocaine and crack.

▶ Hallucinogens can result in flashbacks that often cause mental and emotional problems that can occur long after use.

Vocabulary

▶ psychoactive drugs (p. 603)
▶ designer drugs (p. 605)
▶ hallucinogens (p. 605)
▶ euphoria (p. 605)
▶ depressants (p. 605)
▶ stimulants (p. 606)
▶ opiates (p. 609)

Living Drug-Free

Key Concepts

▶ Choosing to remain drug-free will protect your health.

▶ You can use refusal skills to maintain your commitment to remaining drug-free.

▶ Drug-free school zones and drug watches are two ways that communities help young people avoid drugs.

Vocabulary

▶ drug-free school zones (p. 613)
▶ drug watches (p. 613)
▶ rehabilitation (p. 614)

Thinking Critically

After reading the question or statement, write a short answer using complete sentences.

6. Infer. How might the legal consequences of drug use interfere with a teen's future educational and career goals?

7. Analyze. How can illicit drug use affect you if you and your friends do not use drugs?

LESSON 2

Vocabulary Review

Choose the correct term in the sentences below.

8. *Paranoia/Marijuana* is an irrational suspiciousness or distrust of others.

9. Using *anabolic-androgenic steroids/inhalants* leads to loss of brain cells.

10. *Anabolic-androgenic steroids/Inhalants* can increase muscle strength but not tendon and ligament strength, and causes injuries.

Understanding Key Concepts

After reading the question or statement, select the correct answer.

11. Which consequence of using marijuana can lead to reproductive system problems?
 a. Increased appetite
 b. Feelings of paranoia
 c. Heart and lung damage
 d. Changes in testosterone level

12. Which substance has a legal medical use when used as an inhalant?
 a. Gasoline   **c.** Solvent
 b. Nitrous oxide   **d.** Varnish

13. Which drugs are *not* usually taken in through the respiratory system?
 a. Aerosols
 b. Anabolic-androgenic steroids
 c. Marijuana
 d. Nitrous oxide

LESSON 1

Vocabulary Review

Use the vocabulary terms listed on page 617 to complete the following statements.

1. Drug users often find it difficult to stop using drugs without help because _____ involves both psychological and physiological dependence.

2. _____ is any unnecessary or improper use of chemical substances for nonmedical purposes.

3. Taking more than the recommended amount of a prescription drug can lead to serious health problems or even death from a(n) _____.

Understanding Key Concepts

After reading the question or statement, select the correct answer.

4. Which of the following is *not* usually a factor in deciding to use an illegal drug?
 a. Peer pressure at school
 b. The original source of the drug
 c. How role models live their lives
 d. Messages on television and in movies

5. Which of the following can be a negative consequence of drug use?
 a. Temporary euphoria
 b. Decrease in tolerance
 c. Contraction of an STD
 d. Strengthened refusal skills

Thinking Critically

After reading the question or statement, write a short answer using complete sentences.

14. Analyze. How could using marijuana harm your social interactions with friends?

15. Infer. Why might it be difficult for law enforcement officials to discover and prevent illegal inhalant use?

16. Extend. If an athlete chooses to use steroids to increase muscle mass, how does this perceived benefit actually turn out to be a negative consequence?

LESSON 3

Vocabulary Review

Choose the correct word in the sentences below.

17. *Stimulants / Depressants* speed up the central nervous system.

18. Drugs that cause *euphoria / hallucinations* give users a temporary feeling of intense well-being.

19. *Opiates / Hallucinogens* are often obtainable by prescription, but are heavily monitored because they can cause serious addiction.

Understanding Key Concepts

After reading the question or statement, select the correct answer.

20. Which type of psychoactive drug is best known for altering sense perceptions?
  a. Depressants
  b. Hallucinogens
  c. Opiates
  d. Stimulants

21. Which type of psychoactive drug is used medically to block pain messages to the brain?
  a. Depressants   **c.** Opiates
  b. Hallucinogens   **d.** Stimulants

22. Which hallucinogen is classified as a designer drug?
  a. DXM   **c.** MDMA
  b. LSD   **d.** PCP

23. Why are Rohypnol and GHB linked to exposure to STDs?
  a. They are highly addictive.
  b. They can be used as date-rape drugs.
  c. They are usually taken intravenously.
  d. They affect the body's immune response.

Thinking Critically

After reading the question or statement, write a short answer using complete sentences.

24. Infer. Dangerous drugs are often even more dangerous when mixed together. Why might a drug user take a depressant after taking a stimulant?

25. Evaluate. Why could driving under the influence of a psychoactive drug contribute to an accident?

26. Analyze. Why is it important for doctors and pharmacists to monitor the legal medical use of opiates?

LESSON 4

Vocabulary Review

Use the vocabulary terms listed on page 617 to complete the following statements.

27. _____ is a way to help drug users fight addiction.

28. A community effort to monitor and report illicit drug use is called a(n) _____.

DreamPictures/Blend Images LLC

Assessment

29. Penalties for drug use are often double what they might be for the same drug offense committed outside of a(n) _____.

Understanding Key Concepts

After reading the question or statement, select the correct answer.

30. Which is *not* a way that communities and schools are helping to prevent drug use?
   a. Organizing drug watches
   b. Providing drug treatment centers
   c. Establishing drug-free school zones
   d. Making it illegal to prescribe addictive medicines

31. Which is *not* a warning sign of drug use?
   a. Allergic reactions
   b. Regular hangovers
   c. Difficulty concentrating
   d. Change in sleeping habits

32. Which type of drug treatment strategy involves a meeting of people who share a common problem?
   a. Support group
   b. Outpatient therapy
   c. Medication therapy
   d. Individual counseling

Thinking Critically

After reading the question or statement, write a short answer using complete sentences.

33. **Connect.** How is peer pressure related to a teen's decision to use or avoid drugs?

34. **Analyze.** Why might it be difficult to determine whether a person has a problem with illegal drugs?

35. **Evaluate.** Why is it important to recognize when someone has an addiction to drugs and to discuss the problem with him or her?

Technology PROJECT-BASED ASSESSMENT

Drugs: Truth and Consequences

Background

It is your choice to abstain from illegal drug use. Knowing about the types of illegal drugs and the dangers associated with using them will help reinforce your decision to lead a drug-free life.

Task

Create a multimedia slide presentation that provides information about drugs and the consequences of drug abuse among teens.

Audience

Students and adults in your community.

Purpose

Educate others about the kinds of illegal drugs that are available. Present information about the effects of illegal drugs and the problems related to teen substance abuse.

Procedure

1 Conduct an online search about the types of illegal drugs and the health consequences of drug use.

2 Use statistics from reliable sources to prove the information you find.

3 Collaborate as a group to create slides for your presentation. Make sure to include any audio or video clips to illustrate your points. Make any necessary revisions.

4 Create the final version of your multimedia presentation.

5 Show your presentation to the class. Ask permission to make the presentation available on your school's Web site.

Math Practice

Reading Tables. Nonmedical use of substances known as *anabolic steroids* is considered substance abuse. The consequences of misusing steroids involves more than health risks. There are also legal consequences. The table shows the abuse of anabolic steroids in a 2004 study that involved students from both public and private schools.

Percentage of Teens in Grades 8, 10, and 12 Who Use or Have Used Steroids			
Grade	8th	10th	12th
Ever used	1.9%	2.4%	3.4%
Used in past year	1.1%	1.5%	2.5%
Used in past month	0.5%	0.8%	1.6%

1. If 12,000 of the students studied were tenth graders, how many of them have not used anabolic steroids in the past month?
   A. 96
   B. 180
   C. 11,820
   D. 11,904

2. If 20,000 of the students were eighth graders, how many of them have not ever used anabolic steroids?
   A. 380
   B. 3800
   C. 16,200
   D. 19,620

3. Examine the values in the table. Provide a logical explanation as to why the percentages are higher for older students.

Reading/Writing Practice

Understand and Apply. Read the passage below, and then answer the questions.

It was New Year's Eve, and Carrie was anxious to go to a friend's party. Carrie and her friend, Camille, decided to drive over together. They picked up Camille's new boyfriend, Carl, on the way.

"Hey, let's have some fun tonight," said Carl. "I bought this new drug that everyone's talking about. It'll make the party more fun. What do you say?"

Carrie put her hand up. "I don't want to take drugs. I can have fun without them."

"Aw, come on, everyone's trying it," said Carl. Camille looked at Carrie, then shrugged and looked only slightly apologetic. She really liked this new guy.

"No. It's not for me," Carrie said. "I don't want to use drugs."

1. What is the author's purpose in this piece?
   A. To show how refusal skills work
   B. To illustrate the dangers of drug use
   C. To illustrate the dangers of driving while under the influence of drugs
   D. To tell about the importance of supporting peers through a difficult time

2. What else can Carrie do to avoid drug use?
   A. Describe the physical effects of drug use.
   B. Leave the party and go home.
   C. Tell Camille that Carl's a bad influence.
   D. All of the above

3. Create a pamphlet with pictures or graphics and strong refusal statements showing middle school students how to use refusal skills when offered illegal drugs.

National Education Standards

Mathematics: Measurement and Data, Data Analysis
Language Arts: LACC. 910L.3.6, LACC. 910.RL. 2.4

TEENS *Speak Out*

©Kristy-Anne Glubish/Design Pics/Corbis

Why Limit Where People Can Smoke?

Most people know that smoking is hazardous to your health. Smoking causes cancer, heart disease, high blood pressure, and many other health problems. However, people still debate the hazards of secondhand smoke. People also debate whether the rights of nonsmokers should be considered over the rights of smokers. There are laws in place to limit where people can smoke. For example, federal law prohibits smoking on airplanes, and many cities and towns do not allow smoking in restaurants and public buildings. Is it fair to restrict smoking?

There are two sides to this issue. Read the viewpoints and consider both positions. How do they compare to your own thoughts on this topic?

Benefits of Restricting Smoking

There is no debate. The National Institutes of Health reports that secondhand smoke contains more than 50 chemicals that can cause cancer. Restricting where people smoke limits the exposure of nonsmokers to these harmful substances. Nonsmokers should not have to breathe secondhand smoke. It's a health risk that should not be forced on others. Smokers may decide to expose themselves to the dangers of smoking, but they should not have the right to decide to expose nonsmokers to those health dangers too.

> **❝** I don't want to smell someone else's smoke, and that's my choice. Besides, secondhand smoke kills people."
>
> —Tashauna J., age 15

Benefits of Unrestricted Smoking

Many people dislike the smell of tobacco smoke, but the rights of smokers should be balanced against the rights of nonsmokers. Some studies show there are dangers linked to secondhand smoke, but other studies have shown no danger. If a smoker is not hurting others, he should be able to smoke wherever he wants. Smokers who are trying to quit need time to adjust to a nonsmoking lifestyle. In several states, restaurants restrict smoking. So do airports, and many businesses. As long as smokers are polite and don't blow their smoke toward nonsmokers, they should be able to smoke wherever they want.

> **❝** If the jury is still out on the harm caused by secondhand smoke, why should I or anyone else care where people smoke?"
>
> —Kevin C., age 16

Activity Beyond the Classroom

1. **Summarize** what you learned about the impact of secondhand smoke after reading Chapter 20. Write a one-page letter to the editor persuading others of your point of view. What other issues besides health risks should be considered when determining where people should be able to smoke?

2. **Analyze** laws in your city and state that restrict smoking. Where is smoking prohibited? What laws govern the size and location of areas where smoking is permitted? What age limits apply to people in your state who want to purchase cigarettes or other tobacco products?

UNIT 8 Diseases and Disorders

Chapter 23
Communicable Diseases

Chapter 24
Sexually Transmitted Diseases
and HIV/AIDS

Chapter 25
Noncommunicable Diseases
and Disabilities

UNIT PROJECT

Building Healthy Communities

Using Visuals Special Olympics provides year-round sports training and athletic competition to over 2.25 million people with intellectual disabilities. The nonprofit organization relies on the time, energy, and enthusiasm of its volunteers.

Get Involved. Conduct research on other organizations in your community that help people affected by disease or disability. Contact one organization and find out how teens can volunteer to help.

624

"The future depends on what we do in the present."
— *Mahatma Gandhi, early 20th-century philosopher and peace activist*

CHAPTER 23

Communicable Diseases

Ed-Imaging

Lesson 1

Understanding Communicable Diseases

BIG Idea *Learning about communicable diseases and how they spread can help you prevent them.*

Lesson 2

Common Communicable Diseases

BIG Idea *You can lower your chances of catching a communicable disease by learning about the causes and symptoms of these diseases, and how to avoid them.*

Lesson 3

Fighting Communicable Diseases

BIG Idea *By learning about and practicing prevention strategies, you can help your body stay healthy.*

Lesson 4

Emerging Diseases and Pandemics

BIG Idea *Today, infectious diseases have the potential to spread quickly throughout the world.*

Activating Prior Knowledge

Using Visuals Take a look at the photo on this page. Why do you think this teen is at a doctor's office? Have you ever felt the way she does? What did you do to get better? Explain your thoughts in a short paragraph.

Chapter Launchers

Discuss the **BIG** Ideas

Think about how you would answer these questions:

▶ How often do you get a cold or the flu?

▶ How do you think you get these illnesses?

▶ What do you do to recover from them?

Assess Your Health

Read each statement. On a separate sheet of paper, write "yes," "sometimes," or "no" based on your typical behavior.

1. I know that communicable diseases are spread from one person to another.

2. I know how to reduce my risk of illness from bacteria.

3. I stay home when I have a cold or the flu to avoid infecting others.

4. I know how getting vaccines reduced my risk of communicable disease.

5. I understand how pathogens invade the body and cause illness.

A "yes" response shows that you practice healthy behaviors. "Sometimes" indicates that you should analyze and possibly modify your behavior. A "no" response means that you should modify the behavior.

LESSON 1

Ken Karp/McGraw-Hill Education

BIG Idea *Learning about communicable diseases and how they spread can help you prevent them.*

Before You Read

Create a T-Chart.
Make a T-chart and label the columns "How communicable diseases are caused" and "How communicable diseases are spread." As you read, fill in the chart with information about both topics.

Causes	Ways Spread

New Vocabulary

- communicable disease
- infection
- virus
- bacteria
- toxins
- vector

Review Vocabulary

- pathogens (Ch.15, L.1)

Understanding Communicable Diseases

Real Life Issues ••••••••••••••••••••••••••••

Taking Precautions. Nolan is very excited about his family's upcoming vacation in Central America because it will be his first time outside the United States. His friends warn him to be careful about drinking unbottled water while in Central America. They say the water might not be safe to drink. Nolan wonders what precautions he can take.

Writing *Write a paragraph explaining how Nolan might prepare for the trip. Suggest places where he could find information about potential health risks and how to avoid them.*

Understanding the Causes of Communicable Diseases

Main Idea Communicable diseases are caused by several kinds of microorganisms.

You've probably "caught" an illness from someone before. The illness you contracted was a **communicable disease**, *a disease that is spread from one living organism to another or through the environment*. Such illnesses are also known as *contagious* and *infectious* diseases.

Communicable diseases can occur when pathogens, microorganisms that cause disease, enter your body. If your body does not fight off the invaders quickly and successfully, you develop an **infection**, *a condition that occurs when pathogens in the body multiply and damage body cells*. **Figure 23.1** lists diseases caused by common pathogens, which include viruses, bacteria, fungi, protozoa, and rickettsias.

Viruses

Two of the most common communicable diseases—the cold and the flu—are caused by viruses. A **virus** is *a piece of genetic material surrounded by a protein coat*. In order to reproduce, viruses invade the cells of living organisms.

Once a virus has penetrated a cell, it begins to multiply. The new viruses burst out of the cell and start taking over other cells. As the virus multiplies and spreads, disease sets in, and the body's immune system jumps into action. Usually, the virus runs its course and is killed by the immune system. Antibiotics do not work against viruses, but can sometimes treat the symptoms of a virus.

Bacteria

Bacteria are *single-celled microorganisms* that live almost everywhere on earth. Most bacteria are harmless. Some are even helpful, like the ones that help you digest food. Unfortunately, some bacteria do cause diseases.

Disease-causing bacteria can produce **toxins**, *substances that kill cells or interfere with their functions*. Unlike diseases caused by viruses, a bacterial disease can often be treated with antibiotics. However, because of the overuse of these drugs, some bacteria have become resistant to antibiotics as they have evolved.

READING CHECK

Describe How does a virus affect the body?

Figure 23.1	Diseases by Type of Pathogen

Every common communicable disease can be traced to a particular type of pathogen. *Which of the diseases listed here have you experienced?*

Viruses	Bacteria	Fungi	Protozoa	Rickettsias
• common cold • influenza (flu) • viral pneumonia • viral hepatitis • polio • mononucleosis • measles • AIDS • viral meningitis • chicken pox • herpes • rabies • smallpox • West Nile virus	• bacterial foodborne illness • strep throat • tuberculosis • diphtheria • gonorrhea • Lyme disease • bacterial pinkeye • bacterial pneumonia • bacterial meningitis	• athlete's foot • ringworm • vaginal yeast infection	• malaria • amoebic dysentery • sleeping sickness	• typhus • Rocky Mountain spotted fever

Other Pathogens

Other types of organisms that can cause communicable diseases include the following:

- **Fungi** are plantlike organisms that can cause diseases of the lungs, the mucous membranes, and the skin. Athlete's foot is a common fungal disease.

- **Protozoa** are single-celled microorganisms that are larger and more complex than bacteria. Malaria is an example of a disease caused by protozoa.

- **Rickettsias**, which resemble bacteria, often enter the body through insect bites. Typhus is caused by rickettsias.

How Diseases Spread

Main Idea Diseases can be transmitted in a variety of ways.

Pathogens infect humans and other living things in a variety of ways. Knowing how diseases are transmitted is your first line of defense against them.

Direct Contact

Many pathogens are transmitted through direct contact with an infected person. This includes touching, biting, kissing, and sexual contact. Other transmission methods include

- **puncture wounds.** A person can get tetanus from stepping on a rusty nail.

- **childbirth.** A pregnant woman may transmit an infection to her unborn child through the placenta.

- **contact with infected animals.** Animal bites and scratches can sometimes transmit disease.

Indirect Contact

Academic Vocabulary

contact *(noun):* union or touching of surfaces

You don't have to be in direct contact with a person to become infected. Indirect contact can be just as dangerous.

Contaminated Objects If you touch a contaminated object (for example, a doorknob), you could pick up pathogens. The pathogens can enter your body if you then rub your eyes. To protect yourself, keep your hands away from your mouth, nose, and eyes, and wash your hands regularly.

Vectors Pathogens are often spread by a **vector**, *an organism that carries and transmits pathogens to humans or other animals.* Common vectors include flies, mosquitoes, and ticks. Diseases that spread this way, such as malaria, West Nile virus, and Lyme disease, are called *vector-borne* diseases.

Contaminated Food and Water When food is improperly handled or stored, harmful bacteria can develop. This is true not only for meat and fish but for fruits and vegetables as well. Water supplies that become contaminated with human or animal feces can also cause illnesses such as hepatitis A.

Airborne Transmission

When an infected person sneezes or coughs, pathogens are released into the air as tiny droplets that can travel as far as 10 feet. Even when the droplets evaporate, the pathogens may float on dust particles until they are inhaled. Other pathogens such as fungal spores are also small enough to spread this way. Diseases spread by airborne transmission include chicken pox, tuberculosis, influenza, and inhalation anthrax.

Taking Precautions

Main Idea You can take steps to prevent infection.

There is no guaranteed way to avoid communicable diseases completely, but a few simple practices can dramatically reduce your risk. As you learn about these practices, think about how you can include them in your daily life.

Wash Your Hands

Washing your hands regularly with soap and warm water is the single most effective way to protect yourself from catching or spreading disease. Always wash your hands

- before you eat.
- after you use the bathroom.
- after handling pets.
- before and after inserting contact lenses or applying makeup.
- after touching an object handled by an infected person.

READING CHECK

Identify List three ways communicable diseases can be spread.

■ **Figure 23.2** Japan is known for being a very polite society. People who have colds or the flu often wear masks when they go outdoors. *Why do you think people in some cultures wear masks when they are ill?*

631

Protect Yourself from Vectors

Some vector-borne diseases, such as West Nile virus and bird flu, are on the rise. To protect yourself, follow these steps:

- Limit the time you spend outdoors at dawn and dusk, when mosquitoes are most active.
- Wear pants and long-sleeved shirts to avoid insect bites.
- Use insect repellent, and avoid contact with dead birds.

Other Prevention Strategies

These additional strategies will also help reduce your risk of getting or spreading communicable diseases:

- Avoid sharing personal items, such as eating utensils.
- Handle food properly. (See Chapter 10, Lesson 4.)
- Eat well and exercise. Getting the nutrients your body needs and staying fit will help you fight against infection.
- Avoid tobacco, alcohol, and other drugs.
- Abstain from sexual contact.
- Cover your mouth when you cough or sneeze, and wash your hands after using a tissue.

READING CHECK

Explain How do a healthful diet and regular physical activity help you avoid communicable diseases?

LESSON 1 ASSESSMENT

After You Read

Reviewing Facts and Vocabulary

1. Define the word *communicable*.
2. List three ways that communicable diseases are spread through indirect contact.
3. How is a virus different from bacteria?

Thinking Critically

4. **Analyze.** The fungus that causes athlete's foot lives in warm, moist places. What can you do to reduce your risk of infection when you are in gym locker rooms or other public places?
5. **Synthesize.** If you had a cold, what actions would you take to prevent spreading the illness to other people?

Applying Health Skills

6. **Practicing Healthful Behaviors.** Create an e-mail announcement that your school could send to parents at the beginning of the school year. In your e-mail, give strategies for avoiding communicable diseases such as the flu or the common cold.

Writing Critically

7. **Narrative.** Write a short story from the point of view of bacteria or a virus. Describe how the bacteria or virus finds its way into someone's body, and what happens when it gets there.

Real Life Issues ....

After completing the lesson, review and analyze your response to the Real Life Issues question on page 628.

Common Communicable Diseases

Real Life Issues

Passing It On.

> **Students miss nearly 22 million school days each year due to the common cold.**

> **More than 200,000 people in the U.S. are hospitalized for the flu each year.**

Source: Centers for Disease Control and Prevention, Seasonal Influenza-Associated Hospitalizations in the United States

Writing *Write a paragraph explaining how you can protect yourself and others from the spread of illnesses.*

Respiratory Infections

Main Idea Many diseases begin as respiratory infections.

Many communicable diseases occur in the **respiratory tract**, *the passageway that makes breathing possible.* This passageway includes the nose, throat, and lungs. The respiratory tract connects the outside world to the inside of your body. However, a few habits can help you avoid getting sick:

- Avoid close contact with sick people. If you're ill, stay home.
- Wash your hands often.
- Avoid touching your mouth, eyes, and nose.
- Eat right and get physical activity to strengthen your immune system.
- Abstain from smoking.

Colds, influenza, pneumonia, strep throat, and tuberculosis are the most common respiratory infections.

Common Cold

The common cold is a viral infection that causes inflammation of the **mucous membrane**, *the lining of various body cavities, including the nose, ears, and mouth*. Sneezing, a sore throat, and a runny nose are the most common symptoms. Cold germs spread through direct contact with an infected person, indirect contact with contaminated objects, or airborne transmission.

Because a cold is caused by a virus, there is no cure. Your body has to fight off the infection. The best treatment is to get plenty of rest and drink liquids.

Influenza

Influenza, or the flu, is a viral infection of the respiratory tract. Symptoms include high fever, fatigue, headache, muscle aches, and coughing. Like the common cold, the flu can spread through the air or through direct or indirect contact.

Because the flu is a viral infection, antibiotics can't cure it. Antiviral drugs may be effective in treating flu symptoms if taken early enough. Usually, though, most people treat the flu with proper nutrition, plenty of rest, and lots of liquids. Many people choose to get a flu vaccination once a year. This shot protects you from one type of flu virus that may be common that year. The CDC recommends that everyone age 6 months or older get a yearly flu vaccine.

Pneumonia

In severe cases, the flu can lead to **pneumonia**, *an infection of the lungs in which the air sacs fill with pus and other liquids*. Its symptoms are similar to those of the flu, which means that sometimes people can have pneumonia without realizing it. People who are vulnerable to pneumonia include older adults and those who already have the flu.

Pneumonia can be caused by a virus or by bacteria. Viral pneumonia is sometimes treated with antiviral drugs. Bacterial pneumonia, if diagnosed early enough, can be treated with antibiotics. Pneumonia can be fatal, especially when it strikes older adults and people with lung or heart problems.

Strep Throat

Strep throat is a bacterial infection spread by direct contact with an infected person or through airborne transmission. Symptoms include sore throat, fever, and enlarged lymph nodes in the neck. Left untreated, strep throat can lead to serious conditions, including heart damage. Strep throat can be treated with antibiotics.

 **READING CHECK**

Identify Name at least three respiratory infections.

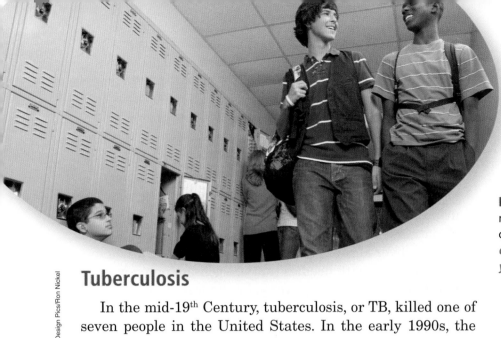

■ **Figure 23.3** There are many methods of spreading communicable diseases. *What can you do every day to reduce your risk of infection?*

Tuberculosis

In the mid-19th Century, tuberculosis, or TB, killed one of seven people in the United States. In the early 1990s, the rates of TB in the U.S. began to rise again. However, recent statistics indicate that TB is again decreasing in the U.S.

TB is a bacterial disease that usually attacks the lungs. It spreads through the air. TB is not spread by shaking another person's hand, sharing food or drink, touching bed linens or toilet seats, sharing toothbrushes, or kissing. Symptoms of TB include fatigue, coughing, fever, weight loss, and night sweats. The disease affects people with weakened immune systems due to other diseases.

Most people who are infected with TB bacteria never develop the disease because the immune system prevents the bacteria from multiplying and spreading. For those people with TB disease, antibiotics are prescribed. In some cases, the disease has become resistant to some antibiotics. In these cases, a doctor must prescribe several antibiotics at one time to treat the disease.

Hepatitis

Main Idea There are three common types of hepatitis.

Hepatitis is a viral infection that causes inflammation of the liver. There are at least five different kinds of hepatitis, but the most common are types A, B, and C. Symptoms include **jaundice**, *a yellowing of the skin and eyes*. Some people also develop **cirrhosis**, or *scarring of the liver.* Vaccines are available for hepatitis A and B, but because the disease comes from a virus, there is no cure.

- **Hepatitis A** usually attacks the digestive system through contact with the feces of an infected person. Common symptoms include fever, vomiting, fatigue, abdominal pain, and jaundice. The best ways to avoid hepatitis are to stay away from people already infected and to wash your hands thoroughly after using any public restroom.

READING CHECK

Identify Which of the body's organs is affected by hepatitis?

- **Hepatitis B** has symptoms similar to those of hepatitis A, but it can cause liver failure and cirrhosis. This virus can be spread through sexual contact or contact with an infected person's blood. You can avoid getting hepatitis B by not sharing personal care items such as razors and toothbrushes, by abstaining from sexual activity and use of illegal drugs, and by not getting tattoos and body piercings.

- **Hepatitis C** is the most common blood-borne infection in the United States. Symptoms include jaundice, dark urine, fatigue, abdominal pain, and loss of appetite. Hepatitis C can lead to chronic liver disease, liver cancer, and liver failure. The disease is most often spread by direct contact with needles that are contaminated with infected blood. You can lower your chances of infection by not sharing personal care items and abstaining from illegal drug use and sexual activity.

Other Communicable Diseases

Main Idea Stay informed about communicable diseases.

Respiratory infections and hepatitis are the most common communicable diseases, but there are many more. The more you know about these diseases and how they are transmitted, the better your chances of not getting them. **Figure 23.4** provides information about additional communicable diseases.

Figure 23.4 Common Communicable Diseases

	Mononucleosis	Measles	Encephalitis	Meningitis	Chicken pox
Type/ Transmission	Virus; spread by direct contact, including sharing eating utensils and kissing	Virus; spread by coughs, sneezes, or a person talking	Virus; carried by mosquitoes	Virus or bacteria; spread by direct or indirect contact	Virus; spread through air or contact with fluid from blisters
Symptoms	Chills, fever, sore throat, fatigue, swollen lymph nodes	High fever, red eyes, runny nose, cough, bumpy red rash	Headache; fever; hallucinations; confusion; paralysis; disturbances of speech, memory, behavior, and eye movement	Fever, severe headache, nausea, vomiting, sensitivity to light, stiff neck	Rash, itching, fever, fatigue
Treatment/ Prevention	Rest if tired	No definite treatment; vaccine for prevention	If caused by herpes simplex virus, antiviral medicine; if caused by another virus, no known treatment	Viral meningitis: antiviral medicine if severe; bacterial meningitis: antibiotics; vaccine available	Rest, stay home so others aren't infected; vaccine available

Health Skills Activity

Decision Making

Caring for Your Immune System

It's Friday night, and Zoë is at home trying to rest. She feels feverish and fatigued. Her phone rings. She blows her nose and answers it.

"Hey, Zoë! It's Elisabeth. Your birthday's tomorrow—are we all still going to the Ice Cream Company for one of those 12-person sundaes? That was so fun last year!"

"I don't know, Elisabeth," Zoë says. "I feel terrible, and I don't want to spread my germs around. I should stay home."

Elisabeth doesn't give up. "But it's your birthday! Nobody's going to get sick. Come on, we're counting on you!"

"I'll think about it," Zoë replies. "Let me call you back later."

Writing Write a paragraph in which Zoë explains her decision to Elisabeth. Use the six steps of decision making as a guideline:

1. State the situation.
2. List the available options.
3. Weigh the possible outcomes of each option.
4. Consider your values.
5. Make a decision, and act on it.
6. Evaluate the decision.

LESSON 2 ASSESSMENT

After You Read

Reviewing Facts and Vocabulary

1. How is a common cold different from the flu?

2. What are three ways to prevent a respiratory tract infection?

3. Can hepatitis be treated successfully with antibiotics? Explain.

Thinking Critically

4. **Explain.** Why do you think the respiratory tract is where most infections from communicable diseases occur?

5. **Cause and Effect.** How does hepatitis spread from one illegal drug user to another?

Applying Health Skills

6. **Accessing Information.** Research the Web site for your state's Health Department. Write a brief summary of the information you find about communicable diseases in your state.

Writing Critically

7. **Persuasive.** Create a handout for elementary school students about the importance of washing your hands regularly. The handout should convince young people that hand washing is one of the best ways to avoid catching communicable diseases.

Real Life Issues ....

After completing the lesson, review and analyze your response to the Real Life Issues question on page 633.

Fighting Communicable Diseases

BIG Idea *By learning about and practicing prevention strategies, you can help your body stay healthy.*

Before You Read

Create Vocabulary Cards. Write each new vocabulary term on a separate note card. For each term, write a definition based on your current knowledge. As you read, fill in additional information related to each term.

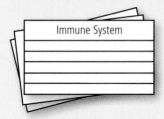

Immune System

New Vocabulary

▸ immune system
▸ inflammatory response
▸ phagocytes
▸ antigens
▸ immunity
▸ lymphocyte
▸ antibody
▸ vaccine

Real Life Issues ●

Too Busy to Stay Healthy. Sang's friend Ashley always seems to have a cold. Ashley complains that she's tired all the time because her schedule is so busy. In addition to school, Ashley has a part-time job and also volunteers at a local animal shelter. Sang has also noticed that Ashley frequently skips lunch.

Writing *Write a dialogue between Sang and Ashley in which they discuss how Ashley's behavior may be contributing to her colds. The girls should come up with ideas for Ashley that will help her avoid catching colds so often.*

Physical and Chemical Barriers

Main Idea Physical and chemical barriers make up your body's first line of defense against pathogens.

You wear a coat or sweater to stay warm, a hat to keep the sun off your head and face, and a helmet during many sports activities. Your coat, sweater, hat, and helmet are all barriers that protect your body, but have you ever stopped to think about how your body deals with invasion from microscopic pathogens? Your body has its own built-in barriers to handle these tiny invaders.

There are two kinds of barriers that help protect you: physical and chemical. Physical barriers, such as the skin, block pathogens from invading your body. Chemical barriers, such as the enzymes in tears, destroy those invaders. See **Figure 23.5** for more examples of physical and chemical barriers that defend you against pathogens.

Figure 23.5 Physical and Chemical Barriers

Your body uses physical and chemical barriers to fight pathogens. *Which barriers are physical? Which are chemical?*

Tears and saliva contain enzymes that disable and even destroy pathogens.

Mucous membranes form a protective lining for your mouth, nose, and many other parts of your body. These membranes produce mucus, a sticky substance that traps pathogens before they can cause infection, then carries the trapped pathogens to other parts of the body for disposal.

Skin is like a personal coat of armor, stopping most pathogens in their tracks as they try to enter the body.

Cilia are small hairs that line parts of your respiratory system. Cilia sweep mucus and pathogens to the throat, where they can be swallowed or coughed out.

Gastric juice in the stomach destroys many pathogens that enter your body through the nose or mouth.

The Immune System

Main Idea Your body's immune system is your best ally in the fight against communicable diseases.

Although your body's physical and chemical barriers stop many pathogens before they can cause infection or disease, pathogens can—and do—sneak past these defenses. That's when your **immune system**, *a network of cells, tissues, organs, and chemicals that fights off pathogens,* goes to work. The immune system fights pathogens using two major strategies: the inflammatory response and specific defenses.

The Inflammatory Response

Have you ever gotten a splinter or a cut? If so, you probably remember that the affected area became red and swollen. These are symptoms of the **inflammatory response**, *a reaction to tissue damage caused by injury or infection.* This response prevents further injury to the tissue and stops the invading pathogens. Your immune system knows a foreign object such as a wood splinter might have pathogens on it. It also knows that a cut could allow pathogens to get into your body. That's why it triggers the inflammatory response.

The inflammatory response, which works against all types of pathogens, includes the following actions.

1. In response to tissue damage and invading microorganisms, blood vessels near the injury expand. This allows more blood to flow to the area and begin fighting the invading pathogens.

2. Fluid and cells from the bloodstream cause swelling and pain because of pressure on the nerve endings.

3. **Phagocytes**, *white blood cells that attack invading pathogens,* surround the pathogens and destroy them with special chemicals. Pus, a mass of dead white blood cells and damaged tissue, may build up at the site of inflammation as a response to bacteria.

4. With the pathogens killed and tissue damage under control, the body begins to repair the tissue.

Specific Defenses

Although the inflammatory response kills many pathogens, some may survive. So, in addition to the inflammatory response, the immune system triggers specific defenses in reaction to certain pathogens. This process is called the *immune response.* When the immune system recognizes a particular pathogen, it activates specific defenses in an attempt to prevent this type of infection from occurring again.

Figure 23.6 describes the immune response. During this process, your immune system reacts quickly to **antigens**, *substances that can trigger an immune response.* Antigens are found in toxins and on the surfaces of pathogens. Macrophages, a type of phagocyte, make antigens recognizable to white blood cells. This **enables** the white blood cells to destroy the pathogens. The result of this type of immune response is known as **immunity**, *the state of being protected against a particular disease.*

Academic Vocabulary

enable *(verb):* to make possible

Lymphocytes

The **lymphocyte**, a *specialized white blood cell that coordinates and performs many functions of specific immunity,* plays an important role in the immune response. There are two types of lymphocytes: T cells and B cells.

T Cells This type of lymphocyte has a variety of functions:

- **Helper T cells** trigger the production of B cells and killer T cells.

- **Killer T cells** attack and destroy infected body cells. These cells don't attack the pathogens, only the infected cells.

- **Suppressor T cells** coordinate the actions of other T cells. They suppress, or "turn off," helper T cells when the infection has been cleared.

Figure 23.6   The Immune Response

1. Pathogens invade the body.
2. Macrophages engulf the pathogen.
3. Macrophages digest the pathogen, and T cells recognize antigens of the pathogen as invaders.
4. T cells bind to the antigens.
5. B cells bind to antigens and helper T cells.
6. B cells divide to produce plasma cells.
7. Plasma cells release antibodies into the bloodstream.
8. Antibodies bind to antigens to help other cells identify and destroy the pathogens.

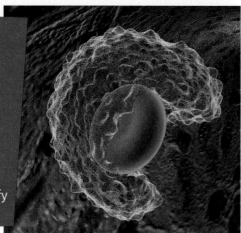

B Cells These lymphocytes have just one job: producing antibodies. An **antibody** is *a protein that acts against a specific antigen*. Each B cell is programmed to make one type of antibody that is specific to a certain pathogen. The different purposes of antibodies include

- attaching to antigens to mark them for destruction.
- destroying invading pathogens.
- blocking viruses from entering body cells.

Immune System Memory

Your immune system also "remembers" the antigens it has dealt with in the past. When antigens activate certain T cells and B cells, the cells become *memory lymphocytes.* These special memory cells circulate in your bloodstream and through the lymphatic system, shown in **Figure 23.7** on page 642. When memory cells recognize a former invader, the immune system sends antibodies and killer T cells to stop the invasion. For example, if you've had measles or been vaccinated against it, your immune system remembers and will attack the antigens for the measles virus.

Your immune system's memory not only identifies invading pathogens. It also helps you develop immunity from certain diseases. There are two types of immunity: active and passive.

Active Immunity This type of immunity develops from natural or artificial processes. Your body develops naturally acquired active immunity when it is exposed to antigens from invading pathogens. Artificially acquired active immunity is developed from a **vaccine**, *a preparation of dead or weakened pathogens that are introduced into the body to stimulate an immune response.*

 READING CHECK

Identify Name three ways that your immune system helps protect you against pathogens.

Figure 23.7

Immunity and the Lymphatic System

The lymphatic system circulates antibodies to give you protection against many diseases. This protection can last throughout your life. *What role do lymphocytes play in fighting disease?*

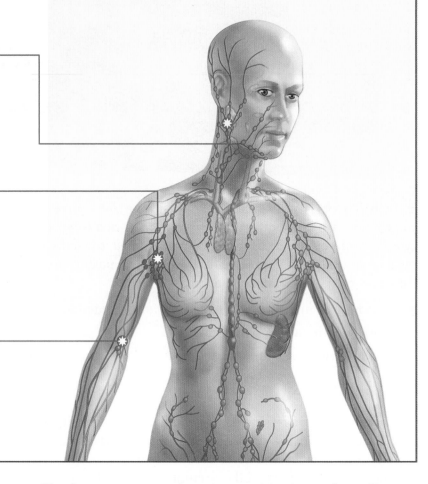

The lymphatic system is part of your immune system. It includes your tonsils, lymph nodes, and a network of vessels, similar to blood vessels, that transport lymph, or tissue fluid.

Lymph nodes can become enlarged when your body is fighting an infection because of the increased number of lymphocytes. If swelling lasts for three days, see your health care professional.

Lymphocytes are produced by lymph nodes. These nodes occur in groups and are concentrated in the head and neck, armpits, chest, abdomen, and groin.

Vaccines cause your immune system to produce disease-fighting antibodies without causing the disease itself. Today, more than 20 serious human diseases can be prevented by vaccination. For some diseases, you need to be vaccinated only once in your life. For other diseases, such as measles, tetanus, and influenza, you may need to be vaccinated at regular intervals.

Passive Immunity You acquire passive immunity when your body receives antibodies from another person or an animal. This type of immunity is temporary, usually lasting only a few weeks or months. Like active immunity, passive immunity can be either natural or artificial.

Natural passive immunity occurs when antibodies pass from mother to child during pregnancy or while nursing. Artificial passive immunity happens when you receive an injection prepared with antibodies that are produced by an animal or a human immune to the disease.

Prevention Strategies

Main Idea Strategies for preventing the spread of disease include practicing healthful behaviors, tracking diseases, and getting vaccinations.

The immune system is a powerful fighter against infection, and you can keep it tuned up and in good working order by eating a nutritious, well-balanced diet and getting regular physical activity. In addition, you can take preventive measures to avoid disease and stay healthy. These include frequent hand washing, handling food properly, avoiding insect bites, and abstaining from sexual contact.

Tracking Reportable Diseases

Community, national, and global efforts also play a crucial role in fighting communicable diseases. Agencies such as the Centers for Disease Control and Prevention (CDC) and the World Health Organization (WHO) keep a constant watch on the spread of diseases around the world. By tracking infections such as hepatitis, influenza, and yellow fever, they can often predict where the diseases might strike next. This information helps countries prepare and develop their own prevention strategies.

READING CHECK

Explain Why is it important to track communicable diseases?

Vaccinations

In the past, smallpox killed hundreds of millions of people. Today, thanks to the smallpox vaccine, the disease has been essentially wiped out. Scientists and health care workers are always trying to stay one step ahead of communicable diseases and develop new vaccines. Vaccines fall into four categories:

- **Live-virus vaccines** are made from pathogens grown in laboratories. This process removes most of the pathogens' disease-causing characteristics. The pathogens are weak, but they can still stimulate the immune system to produce antibodies. The vaccine for measles, mumps, and rubella (MMR) and the vaccine for chicken pox are produced this way.
- **Killed-virus vaccines** use dead pathogens. Even though the pathogens are no longer active, they still stimulate an immune response. Flu shots, the Salk vaccine for polio, and the vaccines for hepatitis A, rabies, cholera, and plague are all killed-virus vaccines.
- **Toxoids** are inactivated toxins from pathogens. They are used to stimulate the production of antibodies. Tetanus and diphtheria immunizations use toxoids.

- **New and second-generation viruses** are on the cutting edge of disease-fighting technology. One example is the vaccine for hepatitis B, which is made from genetically altered yeast cells.

Immunization for All When you receive a vaccine, you are not only keeping yourself healthy, but you are also helping to protect everyone around you. Vaccination reduces the number of people who are at risk for a communicable disease. That's why it's important to keep your immunizations up-to-date.

To find out which immunizations you need, ask your family physician or local health department. Maintaining a record of your vaccinations will help you keep track of when you need "booster" shots.

Most schools and preschools require students to show proof of current immunizations before admission. Each state also has its own laws about immunization and school attendance. Make sure you know and follow the public health policies and government regulations in your community. Remember, everyone can play an active role in preventing the spread of communicable diseases.

LESSON 3 ASSESSMENT

 After You Read

Reviewing Facts and Vocabulary

1. What is the purpose of the inflammatory response?
2. What is the difference between *active immunity* and *passive immunity*?
3. What is a *phagocyte*?

Thinking Critically

4. **Analyze.** Discuss the meaning of *memory* as it applies to the immune system. How is it similar to your brain's memory?
5. **Synthesize.** You could say that your good health is the result of a successful partnership between you and your body. Support this statement using facts from the lesson.

Applying Health Skills

6. **Analyzing Influences.** A healthy immune system depends on a healthful diet and regular physical activity. Consider the influences that might affect your ability to practice these habits. In what ways do these influences make it easier for you to stay healthy? In what ways do they make it more difficult?

Writing Critically

7. **Expository.** Write a paragraph explaining why keeping your own vaccinations up to date is a duty not only to yourself but also to the people around you.

Real Life Issues ....

After completing the lesson, review and analyze your response to the Real Life Issues question on page 638.

Emerging Diseases and Pandemics

Real Life Issues............................

Bacteria in Your Food.

> In 2011, more than one million cases of Salmonella were reported in the U.S.

> About 400 people die each year from severe cases of Salmonella contamination.

Source: Centers for Disease Control and Prevention, Division of Foodborne, Bacterial and Mycotic Diseases.

> **Writing** *Think about foods that you or your family have purchased or prepared. Write a paragraph describing how bacteria can be found in foods, which can lead to a foodborne illness.*

Emerging Infections

Main Idea Some diseases are becoming more dangerous and widespread.

Vaccines and modern technology have saved millions of lives, but communicable diseases continue to be the top cause of deaths worldwide. Health experts label some communicable diseases as **emerging infections**, *communicable diseases whose occurrence in humans has increased within the past two decades or threatens to increase in the near future.*

Scientists now believe that some diseases once thought to be noncommunicable may, in fact, be caused by infectious pathogens. Such diseases include Alzheimer's, diabetes, and coronary artery disease. Many factors are involved in the development and spread of these diseases. See **Figure 23.8** on page 646 to learn more about how emerging infections spread.

GUIDE TO READING

BIG Idea *Today, infectious diseases have the potential to spread quickly throughout the world.*

Before You Read

Organize Information. Make a table and label the columns "Disease," "How It's Spread," and "Prevention Strategies." As you read, fill in the chart with information about the emerging infections discussed in this lesson.

Disease	How It's Spread	Prevention Strategies

New Vocabulary

▶ emerging infections
▶ giardia
▶ epidemic
▶ pandemic

Figure 23.8 **Factors Behind Emerging Infections**

Emerging infections spread in several ways. *Why is Lyme disease increasing today?*

The Factor	How It Happens	Examples
Transport across borders	Infected people and animals carry pathogens from one area to another; sometimes spread by insect carriers such as mosquitoes.	Dengue fever, found mostly in South and Central America and Asia, has now appeared in the southwestern United States. West Nile encephalitis has spread from Asia and Africa to Europe and the Americas. Both diseases are carried by mosquitoes.
Population movement	As residential areas expand, people move closer to wooded areas.	Lyme disease in the United States
Resistance to antibiotics	Widespread use of antibiotics gives rise to drug-resistant pathogens.	The pathogens that cause tuberculosis, gonorrhea, and a type of pneumonia are resistant to one or more antibiotics.
Changes in food technology	Mass production and distribution of food mean that a small amount of pathogens can infect a great number of people.	*E. coli* and *Salmonella* have been responsible for widespread outbreaks of illness.
Agents of bioterrorism	Some pathogens are deadly even in tiny amounts, and they can be dispersed over a large area.	In 2001, envelopes containing anthrax spores were sent to government and media figures in the United States.

Avian Influenza

Avian influenza is caused by a virus that occurs naturally among birds. Wild birds carry the virus in their intestines and usually do not get sick from it. However, the virus has spread to domesticated birds, such as chickens, ducks, and turkeys, through contact with water, feed, cages, or dirt infected by wild birds. Avian flu is passed to humans if there is direct contact with infected birds or contaminated surfaces. In rare cases, mostly in Asia, people have died from avian influenza. Because there is no vaccine and no cure, health authorities are watching this disease very carefully.

READING CHECK

Explain Why are health organizations so worried about avian influenza?

H1N1 Virus

The H1N1 virus is a respiratory virus normally found in pigs. It is a combination of human, pig, and avian flu viruses, and can spread from human to human. Symptoms include fever, sore throat, runny nose, body aches, and fatigue. More than 70 countries, including the United States, have reported human cases of the H1N1 virus. In late 2013, the H1N1 virus again emerged in the U.S., primarily causing illness among young and middle-aged adults.

Salmonella and *E. coli*

Salmonella and *E. coli* are bacteria that sometimes live in animals' intestinal tracts. If people come in contact with these bacteria by eating contaminated food produced by these animals, they may become ill. Illnesses can spread quickly to large areas if contamination occurs in central agricultural or food-processing facilities and contaminated food products are distributed to cities and towns all over the world. Storing foods carefully and cooking meat to proper temperatures will kill *Salmonella* and *E. coli* bacteria. For more tips on avoiding foodborne illness, see Chapter 10, Lesson 4.

Recreational Water Illnesses

Swimming is a fun activity, but if the water is not regularly treated with disinfectants, chlorine, or other chemicals, you run the risk of getting a *recreational water illness,* or RWI. RWIs can occur when water is contaminated by harmful strains of bacteria such as *E. coli* or by **giardia**, *a microorganism that infects the digestive system.*

RWIs are most commonly spread through swallowing or having contact with water contaminated with untreated sewage or feces from humans or animals. RWIs are on the rise throughout the world, particularly in areas where raw sewage is dumped in untreated waterways. To help prevent RWIs, don't swim when you have diarrhea. Try not to let water in your mouth, and definitely try not to swallow it. Also, remember to practice good hygiene: take a shower before swimming, and wash your hands after using the bathroom.

Other Emerging Infections

Other emerging infections with serious health concerns include HIV/AIDS, Lyme disease, West Nile Virus, SARS, and mad cow disease. As with other highly communicable diseases, awareness is the first step toward prevention.

HIV/AIDS is not a new disease, but it is spreading quickly and has become a global health threat. You will learn more about HIV and AIDS in Chapter 24.

■ **Figure 23.9** Swimming is a fun way to stay fit, but it can pose a risk of getting an RWI. *What actions can you take to avoid RWIs?*

Lyme Disease This disease is transmitted to humans through tick bites. Lyme disease is on the rise because, as suburban **communities** grow, people build their homes ever closer to heavily wooded areas, where ticks thrive. To protect yourself, avoid bushy areas with high grass. When you go hiking, use insect repellent and cover skin.

West Nile Virus Mosquitoes sometimes feed on birds carrying the West Nile virus, a pathogen commonly found in Africa, the Middle East, and West Asia. When infected mosquitoes bite humans, they often transfer the virus. About 20 percent of those bitten will develop West Nile fever, a potentially severe illness.

SARS Severe Acute Respiratory Syndrome, or SARS, is a viral illness first reported in Asia in 2003. The illness spread to more than two dozen countries, killing almost 800 people. Health and government agencies were able to contain the virus and stop the spread of illness.

Mad Cow Disease This is also known as *bovine spongiform encephalopathy,* or BSE. This disease, which affects the brain functions of cattle, has reached epidemic proportions in Great Britain. An **epidemic** is *a disease outbreak that affects many people in the same place and at the same time*. Scientists are worried that BSE could spread to humans.

How Diseases Affect the World

Main Idea Diseases can spread with amazing speed.

The world's countries are connected through trade and travel. These connections make it easy for infectious diseases to spread. Sometimes a disease becomes a **pandemic**—*a global outbreak of an infectious disease*. An outbreak of avian flu or *E. coli* in a small area of the globe can quickly spread and threaten the health of entire countries, even continents.

Medical treatment and prevention requires constant research to find the causes and the cures for emerging diseases. The U.S. government has launched programs that will educate the public about flu pandemics. Health agencies plan for pandemics and develop rapid-response strategies to reduce their impact.

The Impact of Travel

The mobility of people in our globalized world contributes to the spread of disease. For example, an American tourist can pick up an infection in another country, return home, and spread it to his family, friends, and coworkers.

READING CHECK

Explain Why can pandemics spread so quickly throughout the world?

■ **Figure 23.10** Travel is exciting, but it can also pose health risks. *How does air travel contribute to the spread of infection?*

Mutation of Pathogens

The increased development of antibiotics has saved countless lives. However, because antibiotics are so widely used, some pathogens have mutated into new forms that are resistant to antibiotics. Pathogens become drug-resistant in a three-step process:

- Pathogens invade the body and cause illness.
- Antibiotics attack the pathogens.
- The pathogens that survive the antibiotics reproduce, creating a new generation of drug-resistant pathogens.

LESSON 4 ASSESSMENT

 After You Read

Reviewing Facts and Vocabulary

1. What is an *emerging infection*?
2. How are recreational water illnesses most commonly spread?
3. How is a *pandemic* different from an *epidemic*?

Thinking Critically

4. **Evaluate.** If a friend told you that you don't need to worry about infectious diseases because you can always take antibiotics, what would you say?
5. **Analyze.** In a Colorado meatpacking plant, a vat of hamburger meat has been infected with *E. coli* bacteria. Weeks later, people in a dozen American states get sick. How might the contamination have occurred over such a large area?

Applying Health Skills

6. **Accessing Information.** Choose one emerging disease from this lesson that you want to know more about. Research how the disease spreads, and find as many tips for avoiding the disease as you can.

Writing Critically

7. **Expository.** You have been asked to write a column for an airline magazine that explains emerging diseases to travelers. Think about what air travelers in particular need to know about how diseases spread, and what they can do to stop a disease from becoming a pandemic.

Real Life Issues ....

After completing the lesson, review and analyze your response to the Real Life Issues question on page 645.

Hands-On HEALTH

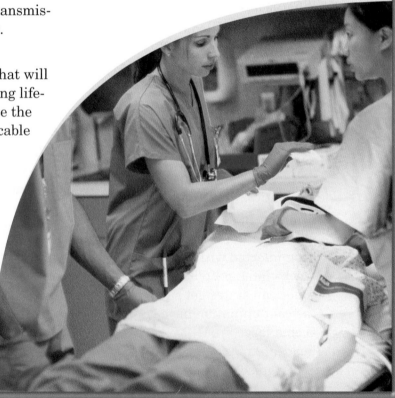

Activity Keeping My Community Disease-Free and Healthy

Imagine that the people in a community next to yours have an undiagnosed communicable disease. Your goal is to create and present an emergency plan that will reduce the risk of contracting this disease in your community.

What You'll Need

- textbook
- markers and one poster board per group
- Internet access

What You'll Do

Step 1

Using the textbook and the Internet, research how to prevent the transmission of communicable diseases.

Step 2

Create a poster presentation that will educate your audience regarding lifestyle behaviors that will reduce the risk of contracting a communicable disease.

Step 3

Identify protective behaviors, listing the steps a person can take to reduce their risk of contracting the disease.

Apply and Conclude

Present your poster which encourages your audience to make healthful choices.

Checklist: Self-Management Skills

✓ Demonstrate healthful behaviors, habits, and/or techniques

✓ Identify protective behaviors (e.g., first-aid techniques, safety steps, strategies) to avoid/manage unhealthy or dangerous situations

✓ Listing steps in correct order

©Corbis

Understanding Communicable Diseases

Key Concepts

▶ Communicable diseases are caused by pathogens.

▶ Communicable diseases can spread through direct contact, indirect contact, and airborne transmissions.

▶ To reduce your risk of disease, wash your hands regularly, eat properly, exercise, and avoid contact with vectors.

Vocabulary

▶ communicable disease (p. 628)
▶ pathogens (p. 628)
▶ infection (p. 628)
▶ virus (p. 629)
▶ bacteria (p. 629)
▶ toxins (p. 629)
▶ vector (p. 630)

Common Communicable Diseases

Key Concepts

▶ Many communicable diseases start in the respiratory tract.

▶ Hepatitis, a widespread viral disease, attacks the liver.

▶ Practicing good hygiene and avoiding risk behaviors can help protect you from some communicable diseases.

Vocabulary

▶ respiratory tract (p. 633)
▶ mucous membrane (p. 634)
▶ pneumonia (p. 634)
▶ jaundice (p. 635)
▶ cirrhosis (p. 635)

Fighting Communicable Diseases

Key Concepts

▶ Physical and chemical barriers stop or destroy many pathogens before they can cause disease.

▶ Your body's immune system fights pathogens with the inflammatory response and specific defenses.

▶ Vaccinations protect you from specific communicable diseases.

Vocabulary

▶ immune system (p. 639)
▶ inflammatory response (p. 639)
▶ phagocytes (p. 640)
▶ antigens (p. 640)
▶ immunity (p. 640)
▶ lymphocyte (p. 640)
▶ antibody (p. 641)
▶ vaccine (p. 641)

Emerging Diseases and Pandemics

Key Concepts

▶ Emerging infections, such as avian influenza, are on the rise.

▶ Infections spread more easily and quickly than in the past.

▶ Pathogens can mutate and become resistant to antibiotics.

Vocabulary

▶ emerging infections (p. 645)
▶ giardia (p. 647)
▶ epidemic (p. 648)
▶ pandemic (p. 648)

LESSON 1

Vocabulary Review

Correct the sentences below by replacing the italicized term with the correct vocabulary term.

1. A(n) *infection* is an organism that causes disease.

2. A substance that kills cells or interferes with their functions is called a(n) *vector*.

3. When pathogens in the body multiply and damage body cells, a(n) *virus* results.

Understanding Key Concepts

After reading the question or statement, select the correct answer.

4. The common cold and influenza are caused by
   a. overeating.
   b. viruses.
   c. bacterial infection.
   d. exposure to toxins.

5. Malaria, West Nile virus, and Lyme disease are examples of diseases that are spread by
   a. vectors.
   b. contaminated utensils.
   c. sexual contact.
   d. contaminated water.

Thinking Critically

After reading the question or statement, write a short answer using complete sentences.

6. **Explain.** If the body's immune system cannot fight off an infection, what happens?

7. **Identify.** Name the process by which bacteria multiply themselves.

8. **Synthesize.** Describe at least three strategies for reducing your risk of getting or spreading communicable diseases.

9. **Evaluate.** Consider your role in preventing disease. How do your behaviors affect the health of your community as well as your own health?

LESSON 2

Vocabulary Review

Use the vocabulary terms listed on page 651 to complete the following statements.

10. The lining of body cavities (such as the mouth) is made of _____.

11. The passageway that makes breathing possible is the _____.

12. Influenza can lead to _____, a potentially fatal infection of the lungs.

Understanding Key Concepts

After reading the question or statement, select the correct answer.

13. Which of the following habits probably will *not* help you avoid respiratory tract infections?
    a. Rinsing with mouthwash
    b. Frequent hand washing
    c. Avoiding close contact with ill people
    d. Abstaining from smoking

14. Some strains of tuberculosis have become resistant to which form of treatment?
    a. Bed rest   c. Dietary changes
    b. Surgery   d. Antibiotics

15. What is the most common blood-borne infection in the United States?
    a. Hepatitis A   c. Hepatitis C
    b. Hepatitis B   d. None of the above

Thinking Critically

After reading the question or statement, write a short answer using complete sentences.

16. Evaluate. What is the best treatment for the common cold?

17. Identify. Receiving a flu vaccine once a year is especially important for which groups of people?

18. Explain. Why do doctors sometimes have to prescribe several antibiotics for a person in order to treat one disease?

19. Synthesize. Explain how peer pressure might contribute to the spread of hepatitis B.

LESSON 3

Vocabulary Review

Choose the correct word in the sentences below.

20. *Antigens/Lymphocytes* are substances that are capable of triggering an immune response.

21. *Inflammation/Immunity* is the state of being protected against a particular disease.

22. A preparation of dead or weakened pathogens used to stimulate an immune response is called a(n) *vaccine/antibody*.

Understanding Key Concepts

After reading the question or statement, select the correct answer.

23. What is the role of phagocytes in the inflammatory response?
  a. They prevent pus from building up.
  b. They surround and destroy pathogens.
  c. They trigger the production of T cells.
  d. They cause blood vessels to expand.

24. If you receive antibodies from another person or an animal instead of producing them in your own body, it is called
  a. communicable disease.
  b. specific defense.
  c. active immunity.
  d. passive immunity.

25. Live-virus, killed-virus, toxoid, and second-generation virus are all categories of
  a. vaccines.
  b. antigens.
  c. preventive strategies.
  d. antibiotics.

26. To remain effective, some vaccinations
  a. must have passive immunity.
  b. contain amateur pathogens.
  c. must be repeated at regular intervals.
  d. are most successful if given when a person is young.

Thinking Critically

After reading the question or statement, write a short answer using complete sentences.

27. Identify. What two major strategies does the immune system use to fight pathogens?

28. Explain. Why do health agencies like the CDC and WHO track and monitor the spread of diseases?

29. Synthesize. If you do not receive up-to-date immunizations, how might your future be affected?

LESSON 4

Vocabulary Review

Correct the sentences below by replacing the italicized term with the correct vocabulary term.

30. West Nile encephalitis is an example of a(n) *acute infection*.

Assessment

31. A global outbreak of an infectious disease is called a(n) *mutation*.

32. *Antibody* is a microorganism that infects the digestive system.

Understanding Key Concepts

After reading the question or statement, select the correct answer.

33. The incidence of emerging infections is
  a. decreasing.
  b. increasing.
  c. holding steady.
  d. virtually nonexistent, thanks to modern medicine.

34. The most effective way to prevent infection from *Salmonella* and *E. coli* is to
  a. visually inspect food before eating.
  b. avoid eating salmon.
  c. cook meat thoroughly.
  d. wash your hands after you eat.

35. Which of the following is *not* a strategy for preventing the spread of RWI?
  a. Relying on chlorine treatments
  b. Staying out of the water when you have diarrhea
  c. Keeping water from entering your mouth when you are swimming
  d. Taking a shower before swimming

Thinking Critically

After reading the question or statement, write a short answer using complete sentences.

36. **Explain.** How do emerging infections happen?

37. **Describe.** What are the three steps of pathogen mutation?

38. **Evaluate.** What is the impact of travel on the spread of diseases?

Technology PROJECT-BASED ASSESSMENT

Victory for Vaccines

Background

Polio is a communicable disease caused by a virus. The disease can affect the brain and spinal cord and cause paralysis. In the early 1950s, a polio epidemic in the United States killed many people, mostly children. By 1975 the disease was almost completely eliminated in the United States.

Task

Conduct research on Dr. Jonas Salk and Dr. Albert Sabin, including their role in the near eradication of polio. Present your findings to the class by developing a Web site or video.

Audience

Students in your class

Purpose

Make people aware of the importance of vaccinations in controlling disease.

Procedure

1 Use online resources to find articles about polio.

2 Learn how the vaccines for polio were discovered.

3 Tell how the two vaccines differ from each other.

4 Explain why there are still some cases of polio in the United States.

5 Describe the efforts that are being made to eliminate polio in the rest of the world. What health groups are involved in the effort?

6 Prepare your Web site or video based on your research and present to your class.

Math Practice

Solve Problems. Use the passage below to answer Questions 1–3.

> *If you have ever had a bacterial infection, you have seen how quickly bacteria can multiply in your body. Bacteria reproduce by dividing in two in a process known as* binary fission. *Under ideal conditions, binary fission takes about 15 minutes. However, this time can vary from 10 minutes to 24 hours.*
>
> *Starting with a single bacterium, how can you find out how many bacteria exist after a certain length of time? After one reproductive cycle, you have two bacteria, or 2^1. After two cycles, you have four, or 2^2. You can summarize this pattern with the formula $B = 2^n$, where B is the number of bacteria, and n is the number of reproductive cycles.*

1. One bacterium has a reproductive cycle of 30 minutes. How many bacteria will there be at the end of four hours?
   A. 16
   B. 120
   C. 256
   D. 512

2. How many bacteria exist after seven reproductive cycles?
   A. 14
   B. 64
   C. 128
   D. It depends on the length of the reproductive cycle.

3. What would be the shape of a graph on which time is plotted on the *x*-axis and number of bacteria is plotted on the *y*-axis? Where is the slope of the line the steepest?

Reading/Writing Practice

Understand and Apply. Read the passage below, and then answer the questions.

> *I can't wait to go camping again with my family this summer. We always have a great time. Last year my best friend, Randy, came with us. I'm hoping he wants to go again, even after the argument we had last time.*
>
> *It was late in the afternoon at the campsite, and we were walking along the river. I knew the mosquitoes would be coming out soon, so I took a bottle of insect repellent out of my backpack and sprayed it on my exposed skin. I told Randy he should do the same, but he just laughed. "You worry too much," he said.*
>
> *I told him mosquitoes carry diseases that can spread to people, and that it's important to prevent insect bites. He put on the insect repellent, but he was annoyed. Things were tense between us for a while, but we got over it. Still, I wonder if it will happen again this year.*

1. When did the author decide it was time to apply insect repellent?
   A. Noon   C. Sunset
   B. Late afternoon   D. Before going to bed

2. What reason did the author give Randy as to why it's important to put on insect repellent?
   A. Insect bites can be painful.
   B. Mosquitoes are annoying.
   C. Mosquitoes carry diseases that can spread to people.
   D. Insect bites are the leading cause of infection among teens.

3. How do you think the author felt during this encounter? Do you think he handled the situation appropriately? Explain.

National Education Standards

Math: Arithmetic with Polynomials and Rational Expression, Perform arithmetic operations on polynomials.
Language Arts: LACC.910.RL.2.4

CHAPTER 24

Sexually Transmitted Diseases and HIV/AIDS

Lesson 1
Sexually Transmitted Diseases

BIG Idea *Sexually transmitted diseases (STDs) are highly communicable infections that are contracted through sexual contact.*

Lesson 2
Preventing and Treating STDs

BIG Idea *All STDs are preventable and most can be treated, but some are incurable.*

Lesson 3
HIV/AIDS

BIG Idea *HIV is the virus that causes AIDS, a disease that weakens the body's immune system and may have fatal consequences.*

Lesson 4
Preventing and Treating HIV/AIDS

BIG Idea *HIV/AIDS is preventable and treatable, but it is incurable.*

Activating Prior Knowledge

Using Visuals Take a look at the photo on this page. What effect can the work of this scientist and others have on the knowledge and behaviors of teens? Explain your thoughts in a short paragraph.

Chapter Launchers

Health in Action

Discuss the **BIG Ideas**

Think about how you would answer these questions:

▶ What do you know about infections that are spread through sexual contact?

▶ Why is it important for you to know about these infections?

Assess Your Health

Read each statement. On a separate sheet of paper, write "yes," "sometimes," or "no" based on your typical behavior.

1. I am aware of STDs, how they are transmitted, and their symptoms.

2. I abstain from behaviors that may expose me to STDs.

3. I can say no to something I do not want to do, or do not feel ready to do.

4. I know what behaviors can lead to the transmission of HIV.

5. I avoid all behaviors that can transmit HIV.

A "yes" response shows that you practice healthy behaviors. "Sometimes" indicates that you should analyze and possibly modify your behavior. A "no" response means that you should modify the behavior.

BIG Idea *Sexually transmitted diseases (STDs) are highly communicable infections that are contracted through sexual contact.*

Before You Read

Create a Cluster Chart. Draw a circle and label it "STDs." Use surrounding circles to identify common STDs. As you read, continue filling the chart with more details about each type of infection.

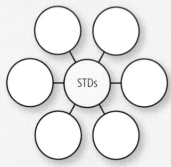

STDs

New Vocabulary

▶ sexually transmitted diseases (STDs)
▶ sexually transmitted infections (STIs)
▶ asymptomatic

Review Vocabulary

▶ communicable disease (Ch.23, L.1)
▶ epidemic (Ch.23, L.4)

Sexually Transmitted Diseases

Real Life Issues ....

Developing Awareness. Today is the third session of the juniors' human sexuality class, and the topic is diseases and infections that are spread through sexual contact. Tricia and her friends do not want to go to class. They say they've heard it all before, and besides, it's embarrassing to talk about. Joe and his friends think they should attend because the more they know, the safer they'll be.

Writing *Write reasons for going to the class that Joe's group might suggest to Tricia's group. Include reasons why teens are at risk for STDs.*

What Are STDs?

Main Idea **Anyone who has sexual contact with another person risks contracting a sexually transmitted disease.**

Sexually transmitted diseases (STDs) are *infections spread from person to person through sexual contact.* Also known as **sexually transmitted infections (STIs)**, STDs are communicable diseases that can be easily transmitted from one person to another. For an infection to occur, a person must engage in sexual activity that involves direct genital contact or the exchange of semen or other body fluids with someone infected with an STD.

Some STDs are caused by a bacterial infection and can be cured with medication. Other STDs are caused by viruses and are incurable. Early diagnosis and treatment are crucial to controlling or curing an STD.

However, several of the most common STDs are often **asymptomatic**, meaning that *individuals show no symptoms, or the symptoms are mild and disappear after the onset of the infection.* This lack of symptoms makes STDs particularly dangerous. A person may not realize that he or she is infected. Therefore, he or she may not seek treatment. An individual with an undiagnosed STD may unknowingly pass the infection on to future sexual partners.

Any person who has sexual contact with another person risks contracting an STD. The risk of contracting an STD also increases as the number of sexual partners increases. It is estimated that more than 9 million young people between the ages of 15 and 24 will become infected with an STD each year. As **Figure 24.1** shows, many of these cases will not be diagnosed, treated, or reported, creating a serious health crisis.

Females are more likely to suffer complications from STDs, and the effects are more serious in females than in males. Both the physical and psychological effects on people infected with STDs are significant, and so the consequences for health care in the United States are serious as well. The Centers for Disease Control and Prevention (CDC) estimates that direct medical costs associated to STDs are now at more than $15.3 billion a year.

READING CHECK

Explain What makes STDs particularly dangerous for teens?

Figure 24.1	**STDs in the United States**

This chart shows the discrepancy between the estimated number of new STD cases in the United States and the number of reported cases. *Why do you think such a large percentage of STDs are undiagnosed and unreported?*

STD	Estimated Number of New Cases Each Year	Reported Cases (2012)
Chlamydia	2.86 million*	1,422,976
Genital Herpes	776,000	300,000
Gonorrhea	820,000	334,826
Trichomoniasis	1 million*	219,000
Syphilis	55,400	49,903

2008*

Source: National Center for HIV/AIDS, Viral Hepatitis, STD, and TB Prevention, 2013.

Common STDs

Main Idea There are approximately 25 different STDs, six of which are considered the most common.

Of the approximately 25 STDs worldwide, the following six are considered the most common: genital HPV infections, chlamydia, genital herpes, gonorrhea, trichomoniasis, and syphilis. **Figure 24.3** shows the symptoms and possible long-term effects of these STDs.

Genital HPV Infections

Genital HPV infections are caused by human papillomavirus (HPV), a group of more than 100 kinds of viruses. More than 30 of these viruses are **transmitted** through sexual contact. Close to 20 million people in the United States are infected with HPV each year.

HPV infections can cause genital warts, which appear as bumps or growths near or on the genitals. Most genital HPV infections do not have symptoms and will disappear without medical treatment. However, some HPV infections, if not diagnosed and treated, may cause abnormal Pap tests or, more seriously, may result in certain types of cervical cancer. A vaccine treatment is now available for protection against HPV. It is not a cure, but is recommended to reduce the number of cases of cervical cancer.

Chlamydia

Chlamydia is a bacterial infection that affects the reproductive organs of both males and females. In 2012 almost 1.5 million new cases of chlamydia were reported. As with genital HPV, chlamydia often produces no symptoms. Thus, sexually active teens may not know they are infected, do not seek testing, and go untreated. Chlamydia is still the most common STD among teens.

If left untreated, chlamydia can cause serious complications. Females can develop pelvic inflammatory disease (PID) and suffer chronic pelvic pain or infertility. Untreated chlamydia can also lead to infertility in males. Pregnant females with chlamydia can deliver prematurely, and the infants born to infected mothers may develop eye disease or pneumonia, as well as fatal complications. Females with chlamydia are up to five times more likely to become infected with HIV if exposed to the virus. You will learn more about HIV and AIDS in Lesson 3.

Academic Vocabulary

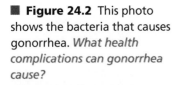

transmit *(verb):* to send from one person or place to another

■ **Figure 24.2** This photo shows the bacteria that causes gonorrhea. *What health complications can gonorrhea cause?*

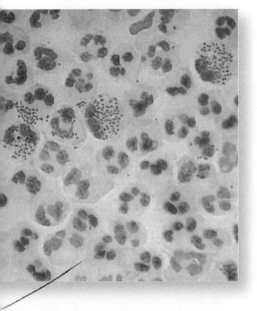

Figure 24.3 | **STD Symptoms**

This chart shows the symptoms and possible long-term effects of common STDs. *Why is delayed treatment for STDs never a healthy choice?*

STD	Symptoms in Males	Symptoms in Females	Possible Long-Term Effects
Genital HPV Infection	Genital warts on the penis, scrotum, groin, anus, or thigh	Genital warts in or around the vagina, vulva, cervix, or anus	Development of cervical cancer in females
Chlamydia	Penis discharge; burning during urination; itching or burning sensations around penis	Lower abdominal or back pain; nausea; fever; bleeding between periods; pain during intercourse; muscle ache; headache; abnormal vaginal discharge; burning sensation when urinating	In males, inflammation of urethra In females, inflammation of cervix, damage to fallopian tubes, chronic pelvic pain, infertility
Genital Herpes	Blisters on or around genitals or rectum; sores that can take weeks to heal; flu-like symptoms, including fever and swollen glands	Blisters on or near vagina or rectum; sores that can take weeks to heal; flu-like symptoms, including fever and swollen glands	Psychological distress; can cause life-threatening infection in baby born to mother with the disease
Gonorrhea	Burning sensation when urinating; green, yellow, or white discharge from penis; painful, swollen testicles	Pain or burning when urinating; increased vaginal discharge; vaginal bleeding between periods	In males, painful condition of testicles leading to infertility if untreated (epididymitis) In females, chronic pelvic pain and infertility
Trichomoniasis	Temporary irritation inside penis; mild burning after urination or ejaculation	Thick, gray or yellowish green vaginal discharge with strong odor; painful urination; vaginal itching	Discomfort; higher susceptibility to other STDs; premature or low-birth-weight babies born to infected pregnant females
Syphilis	Single sore on the genitals (sores disappear but infection remains); skin rash	Single sore on the vagina (sores disappear but infection remains); skin rash	Serious damage to internal organs, including brain, heart, and nerves

Genital Herpes

Genital herpes is caused by the herpes simplex virus. Herpes simplex 1 usually causes cold sores in or near the mouth. Herpes simplex 2 typically causes genital sores. Both types, however, can infect the mouth and the genitals. In the United States it is estimated that one of every six people has contracted genital herpes.

Many people infected with genital herpes are asymptomatic and are not aware they have the infection. If symptoms do occur, the first outbreak will usually appear as blisters on the genitals or rectum within two weeks of the virus being transmitted. The blisters break, leaving sores that can take several weeks to heal. Usually the first sores are followed by shorter, less severe outbreaks that can occur on and off for years. Antiviral treatments can lessen the frequency of outbreaks, but there is no cure for genital herpes.

READING CHECK

Explain Why is it important that STDs are diagnosed as soon as possible?

Gonorrhea

Gonorrhea is a bacterial STD that usually affects mucous membranes. Gonorrhea is the second most commonly reported infectious disease in the United States. The CDC estimates that more than 820,000 Americans are infected with gonorrhea each year, but only half of these are reported.

Many males with gonorrhea are asymptomatic, and infected females show only mild symptoms. Left untreated, gonorrhea can cause severe health problems, such as infertility. The bacteria can also spread to the bloodstream and cause permanent damage to the body's joints. Females can pass the infection to their babies during childbirth. These babies may contract eye infections that cause blindness.

Trichomoniasis

Trichomoniasis is caused by a microscopic protozoan that results in infections of the vagina, urethra, and bladder. About 3.7 million people in the U.S. have the disease. However, only 30 percent of these people develop symptoms.

Although the disease may not produce symptoms, some males have a temporary irritation inside the penis, mild discharge, or slight burning during and after urination or ejaculation. Many infected females often experience *vaginitis,* an inflammation of the vagina characterized by discharge, odor, irritation, and itching. Females with trichomoniasis are also more likely to contract HIV if they are exposed to it. Babies born to females with trichomoniasis are often premature and have low birth weights.

Syphilis

Syphilis, an infection caused by a small bacterium called a spirochete, attacks many parts of the body. People with syphilis develop sores in the genital area lasting a couple of weeks. The disease is passed from one person to another by direct contact with the sores during sexual activity.

Syphilis progresses through three stages. During the primary stage, a sore appears on the external genitals or the vagina. At this stage, the disease can be easily treated. If the infection goes untreated, the sore heals, but the infection remains. In the second stage, the infection produces a skin rash. As in the first stage, the untreated rash will disappear, but the infection remains and progresses to the third stage. During this stage, syphilis can damage internal organs, cause brain dementia, and may cause death.

The STD Epidemic

Main Idea Accurate health information and responsible behavior will help fight the STD epidemic.

The United States currently faces an STD epidemic. The CDC estimates that each year, 20 million people are infected with an STD. Almost half are under the age of 24. Many STD cases go undiagnosed and untreated because of

- **embarrassment or fear.** Some people are too embarrassed or afraid to seek medical help.

- **lack of symptoms.** Many people infected with STDs are asymptomatic, and are unaware they have a disease.

- **misinformation.** If STD symptoms disappear without treatment, the infected person may mistakenly believe the disease has been cured. People may not have all the facts and may receive wrong information from friends.

- **notification policies.** State laws require health care providers to report certain but not all STDs. People who have contracted HPV infections or genital herpes are not required to report their infections or inform any partners of their condition. Infected individuals may unknowingly transmit the disease to others.

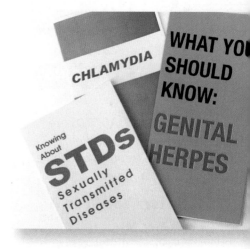

■ **Figure 24.4** The CDC provides important updated information about STDs at their Web site. *Where else might you find reliable information about STDs?*

LESSON 1 ASSESSMENT

 After You Read

Reviewing Facts and Vocabulary

1. What is a *sexually transmitted disease*?

2. Name four common STDs.

3. In the United States, approximately how many people are infected with an STD each year?

Thinking Critically

4. **Synthesize.** How can you communicate the danger of STDs to other teens?

5. **Analyze.** Why is it important to learn about the reasons STDs go undiagnosed and untreated?

Applying Health Skills

6. **Accessing Information.** Create a directory identifying resources in your community where teens can find accurate information about the diagnosis and treatment of STDs.

Writing Critically

7. **Expository.** Describe the cause-and-effect relationship between the reasons many STDs go unreported and undiagnosed and the current STD epidemic in the United States.

Real Life Issues ....

After completing the lesson, review and analyze your response to the Real Life Issues question on page 658.

LESSON 2

GUIDE TO READING

BIG Idea *All STDs are preventable and most can be treated, but some are incurable.*

Before You Read

Create an Outline. Preview this lesson by scanning the pages. Then organize the headings and subheadings into an outline. As you read, fill in your outline with important details.

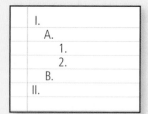

New Vocabulary

▶ antibiotics
▶ HPV vaccine

Review Vocabulary

▶ abstinence (Ch.1, L.3)
▶ refusal skills (Ch.2, L.1)

Preventing and Treating STDs

S. Olsson/PhotoAlto

Real Life Issues

Making Personal Decisions. Maria and Jake have been dating for six months. Together, they work on the school newspaper, play tennis, and enjoy eating out. Lately, though, Jake has been more and more insistent that they take their relationship "to the next level." Maria has had two previous sexual relationships, but she has now decided to practice abstinence. She made her decision based partly on what she learned about STDs. "It's not worth all the risks," Maria tells Jake. She is committed to her decision to practice abstinence.

Writing *Write a dialogue between Maria and Jake. Include an explanation of the risks that you think Maria is talking about, and show how Jake might respond in a healthful way.*

Prevention Through Abstinence

Main Idea The most successful method to prevent the spread of STDs is abstinence.

About 9 million American teens contract STDs annually. Some are bacterial infections, such as chlamydia or gonorrhea, that can be treated and cured with antibiotics. **Antibiotics** are *a class of chemical agents that destroy disease-causing microorganisms while leaving the patient unharmed.* Others, such as genital herpes and HPV, are incurable viral infections. Any STDs that are not diagnosed early and treated can result in serious permanent or long-term health consequences.

The only method that is 100 percent successful in preventing the contraction and spread of STDs is abstinence. Abstinence before marriage is the best way to avoid STDs.

664 Chapter 24 Sexually Transmitted Diseases and HIV/AIDS

■ **Figure 24.5** Group outings are one way to enjoy the company of friends and avoid pressure to engage in sexual activity. *Why might you want to discuss your commitment to abstinence with your friends?*

Fuse/Getty Images

To help you protect your health and stay committed to abstinence, follow these **guidelines**:

- Set personal limits on physical affection.
- Avoid dating someone who is sexually active or who pressures you to go beyond your limits.
- Avoid situations where you may feel pressured to engage in sexual activity.
- Avoid people who make fun of your decisions or urge high-risk behaviors, including use of alcohol or drugs.
- Choose group outings where you can enjoy the company of friends and avoid pressure to engage in sexual activity.
- Be clear about your decision to practice abstinence, and discuss it with others who are close to you.
- Practice refusal skills. Use words and body language to resist the pressure to engage in sexual activity.

Academic Vocabulary

guideline *(noun):* an outline of conduct

Understanding the Risks

Each month, about 1.5 million teens are diagnosed with an STD. This age group is at high risk partly because many teens are unaware of a partner's past behavior. It's impossible to look at someone and tell if that person has an STD. Because many STDs go undiagnosed, it is not enough for a partner simply to say that he or she is uninfected. Abstinence is the only sure method of preventing STDs.

Health Skills Activity

Refusal Skills

The Lines of Defense

Mark and Alyssa have been dating for a year. Both are committed to practicing abstinence. It hasn't always been easy, but they believe firmly that it is the safest and most caring method of avoiding STDs and unintended pregnancy.

One afternoon, Alyssa's friend Emma says, "Hey, guys, I'm having a party tonight. For once, my parents won't be there. The new guy I'm dating is bringing his band. It's going to be great. "

Mark and Alyssa glance at each other, and then Alyssa shakes her head. "Sounds like fun," she replies, "but things could get out of hand."

Mark agrees, "Yeah, thanks for the invitation, but I think we'll pass."

"Oh, give me a break!" says Emma, annoyed. "What are you afraid of?"

Writing Write a dialogue in which Mark and Alyssa use refusal skills to respond to Emma. Follow these steps:

1. Say no in a firm voice.
2. Explain why you are refusing.
3. Suggest alternatives to the proposed activity.
4. Back up your words with body language.
5. Leave if necessary.

Avoiding High-Risk Behaviors and STDs

Avoiding high-risk behaviors can help prevent people from contracting STDs. High-risk behaviors include

- **being sexually active with more than one person.** This includes having a series of sexual relationships with one person at a time. However, being sexually active with even one partner puts a person at risk.

- **engaging in unprotected sex.** Even protected sex, or barrier protection, is not 100 percent effective in preventing the transmission of STDs. Abstaining from sexual activity is the only method that is 100 percent effective in avoiding STDs.

- **engaging in sexual activity with high-risk partners.** Such partners include those with a history of being sexually active with more than one person and those who have injected illegal drugs. Taking a person's word about past behaviors is not wise. Sexual activity with just one infected person puts you at risk.

- **using alcohol and other drugs.** Alcohol can lower inhibitions and cause teens to engage in sexual activity when they might ordinarily choose not to. To safeguard your health, it's important to be in control of your decisions.

 READING CHECK

Explain How can the use of alcohol or other drugs increase a person's risk of contracting an STD?

Figure 24.6 | **Diagnosis Methods and Treatments for STDs**

Diagnosis methods and successful treatments for common STDs vary.

STD	Diagnosis Method	Treatment/Cure
Genital HPV Infection	Pap test in females; genital warts diagnosed by a physical examination	No cure; warts may clear up without medication or by using medications applied by patient; or may clear up with treatments performed by a health care provider
Chlamydia	Urine tests; tests on specimen collected from the infected site	Treated and cured with antibiotics
Genital Herpes	Visual inspection by a health care professional; testing of infected sore; blood tests	No cure; antiviral medication can shorten and prevent outbreaks
Gonorrhea	Laboratory test (Gram's stain); urine test	Treated and cured with antibiotics; successful treatment becoming difficult due to increase of drug-resistant strains; medication stops infection but cannot repair damage done by disease
Trichomoniasis	Physical examination and laboratory test	Prescription drug, metronidazole, given by mouth in a single dose; both partners should receive treatment at same time
Syphilis	Physical examination; blood test	Curable with penicillin or other antibiotics; treatment will not repair damage already done

HPV Vaccine

The Food and Drug Administration (FDA) has approved the **HPV vaccine**, *a vaccine that can prevent cervical cancer, pre-cancerous genital lesions (or sores), and genital warts caused by genital HPV infection.* This vaccine protects against four types of HPV infections. Health officials recommend the vaccine for females 9 to 26 years old. In 2011, the CDC also recommended that males between the ages of 11–26 receive the vaccine. Studies are under way to learn if the vaccine has health benefits for males. At this time, no vaccines are available for any other types of STDs.

Diagnosing and Treating STDs

Main Idea Only a health care professional can accurately diagnose and treat an STD.

If STDs are not diagnosed and treated early, serious long-term consequences can result. Teens who believe they might be infected with an STD should talk to a health care professional. Many public health clinics provide information and treatment free of charge. **Figure 24.6** summarizes diagnosis and treatment methods for common STDs.

Keep in mind that not all genital infections are STDs; some are localized skin infections or rashes. Only a trained professional can determine which test will most effectively screen for a particular STD. When an STD has been diagnosed, a health care professional will prescribe the most effective medication and monitor the patient's treatment. STDs cannot be cured using common household products, homemade remedies, or over-the-counter treatments. Also, remember that taking medicines prescribed to others is risky.

Antibiotics can effectively treat bacterial STDs, but viral STDs are incurable. However, medications can lessen the discomfort from sores and skin irritations caused by STDs.

READING CHECK

Explain Which types of STDs can be treated and possibly cured? Which cannot?

Act Responsibly

Everyone has an obligation to prevent the spread of STDs. One way to help control this epidemic is to practice abstinence. A second way is to report any known infections. Public health clinics can sometimes help locate past partners to make sure they get medical treatment. Ultimately, however, it is the responsibility of any person infected with an STD to notify everyone with whom he or she has had sexual contact. Informing someone else about a possible STD infection could save a life.

LESSON 2 ASSESSMENT

After You Read

Reviewing Facts and Vocabulary

1. What is the only 100 percent effective method for preventing the spread of STDs?

2. Identify two high-risk behaviors that can lead to contracting an STD.

3. What is the *HPV vaccine*?

Thinking Critically

4. **Synthesize.** Predict situations that could lead to pressures to engage in sexual activity, and identify ways to avoid these situations.

5. **Analyze.** Explain the causes and consequences of teen health risk behaviors that could result in STD infection, and describe prevention strategies.

Applying Health Skills

6. **Refusal Skills.** Write a scenario in which one teen is pressuring another to engage in behavior that puts both at high risk for contracting an STD. The second teen should use refusal skills to respond to the pressure.

Writing Critically

7. **Persuasive.** Write a public service announcement urging teens to get medical help for all health problems, including suspected STDs. Include local resources for medical care.

Real Life Issues ....

After completing the lesson, review and analyze your response to the Real Life Issues question on page 664.

HIV/AIDS

Real Life Issues

Finding Out the Facts. Mariano, Cal, and Janine heard a rumor at school that one of the seniors tested positive for HIV. "I'm not going to play ball with the seniors. I don't want to come into contact with their sweat," says Janine.

Cal laughs. "Get real. You can't get HIV from sweat!"

"Well, sweat's a bodily fluid, isn't it?" Janine shoots back.

"I'm not sure exactly how HIV/AIDS is passed around," says Mariano slowly. "Let's figure out where we can go or who we can ask to find out."

Writing *Write a list of true/false questions about HIV/AIDS that Mariano and his friends might have about how HIV is transmitted. After reading the lesson, return to your list and answer each question.*

What Is HIV/AIDS?

Main Idea HIV/AIDS weakens the body's immune system.

Human immunodeficiency virus (HIV) is *a virus that attacks the immune system.* Once HIV enters the body, it finds and destroys the white blood cells that fight disease. The final stage of an HIV infection is **acquired immunodeficiency syndrome (AIDS)**, *a disease in which the immune system is weakened.*

AIDS has become one of the deadliest diseases in human history. More than 36 million people around the world have died of this disease, including more than 636,000 Americans. Health care officials estimate that currently 35 million people are living with HIV/AIDS. The statistics are alarming:

- Approximately 39 percent of newly reported HIV/AIDS infections in 2009 were reported among people aged 15 to 24.

- Half of all new HIV infections are among young people. Every day, about 2,000 young people become infected.

©KidStock/Blend Images/Corbis

GUIDE TO READING

BIG Idea *HIV is the virus that causes AIDS, a disease that weakens the body's immune system and may have fatal consequences.*

Before You Read

Create a K-W-L Chart. Make a three-column chart. In the first column, list what you **k**now about HIV/AIDS. In the second column, list what you **w**ant to know about this topic. As you read, use the third column to summarize what you **l**earned.

K	W	L

New Vocabulary

▶ human immunodeficiency virus (HIV)
▶ acquired immunodeficiency syndrome (AIDS)

Review Vocabulary

▶ pandemic (Ch.23, L.4)
▶ mucous membranes (Ch.23, L.2)
▶ lymphocytes (Ch.23, L.3)
▶ antibodies (Ch.23, L.3)

Figure 24.7

Worldwide HIV Infection Rates for Adults

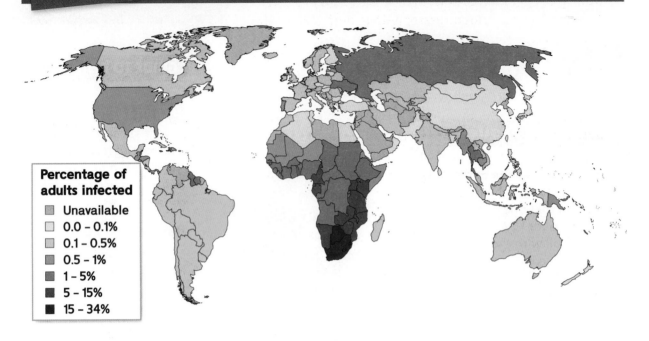

Percentage of adults infected

- Unavailable
- 0.0 – 0.1%
- 0.1 – 0.5%
- 0.5 – 1%
- 1 – 5%
- 5 – 15%
- 15 – 34%

READING CHECK

Explain How are HIV and AIDS related?

As **Figure 24.7** shows, HIV infection is a worldwide concern. Health care officials consider HIV/AIDS a *pandemic,* a global outbreak of infectious disease. Many experts and scientists consider HIV/AIDS to be the most serious public health problem facing the world. The seriousness of the HIV/AIDS pandemic is greatly increased because many of the young people who are infected do not know it.

Understanding HIV/AIDS

Main Idea HIV/AIDS is transmitted in a variety of ways.

HIV is a fragile virus and cannot live outside the human body. Exposure to air at room temperature kills the virus. HIV cannot be spread through airborne transmission, through casual contact such as shaking hands or hugging, or from insect bites. Although the virus has been found in sweat, tears, and saliva of infected persons, the amount is too small to be considered dangerous.

HIV is transmitted among humans only when one person's infected blood, semen, or vaginal secretions comes in contact with another person's broken skin or mucous membranes. Mucous membranes can be found in the mouth, eyes, nose, vagina, rectum, and the opening in the penis.

HIV is spread in three ways:

- **During sexual intercourse.** HIV can enter the blood-stream through microscopic openings in tissues of the vagina, anus, mouth, or the opening in the penis. People with STDs are more vulnerable to HIV infection because STDs cause changes in the body's membranes that increase the likelihood of HIV transmission.

- **By sharing needles.** Anyone who uses needles contaminated with HIV allows the virus to enter directly into his or her bloodstream. Needles used for body piercings and tattoos also can come in contact with contaminated blood.

- **From mother to baby.** A pregnant female infected with HIV can pass the virus to her unborn baby through the umbilical cord, during childbirth, or through breastfeeding. If an expectant mother knows she's infected, she can take medication that might prevent her child from contracting HIV. During childbirth, the doctor and nurses will work to prevent the newborn from coming into contact with the mother's blood, and the mother will be asked not to breastfeed her infant. In the United States pregnant females are routinely tested for HIV.

READING CHECK

List What are the three ways that HIV is spread?

How HIV/AIDS Affects the Immune System

HIV attacks the body's immune system by destroying *lymphocytes*. These are specialized white blood cells that perform many immune functions, such as fighting pathogens. Helper T cells stimulate B cells to produce antibodies, which help destroy pathogens that enter the body. When HIV enters certain cells, including lymphocytes, it reproduces itself and eventually destroys the cell. **Figure 24.8** shows how HIV attacks cells. As more cells are destroyed, the immune system becomes weaker and weaker. The body then becomes vulnerable to *AIDS-opportunistic illnesses,* infections the body could fight off if the immune system were healthy.

HIV infection moves through identifiable stages before progressing to AIDS:

- **Acute Infection Stage.** This stage begins within two to four weeks of infection. A person may become sick with a severe flu-like illness. Symptoms include fever, swollen glands, sore throat, rash, muscle and joint aches, fatigue, and headache. During this stage, the virus is reproducing rapidly in the body. During the acute infection stage, the levels of HIV in the bloodstream are very high. During this stage, the person with HIV can easily transmit the virus to others through sexual activity or drug use, as well as other methods of transmission.

Academic Vocabulary

confine *(verb):* to keep within limits

Figure 24.8 How HIV Attacks Cells

Once inside the cell, HIV is safe from attack by the immune system's antibodies.
How does this make HIV particulary dangerous?

1. HIV attaches to cell surface.
2. Virus core enters cell and goes to nucleus.
3. Virus makes a copy of its genetic material.
4. New virus assembles at cell surface.
5. New virus breaks away from host cell.

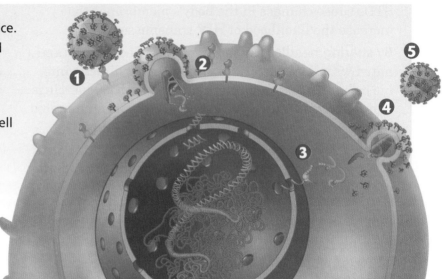

- **Clinical Latency Stage.** During this stage, the HIV virus continues to reproduce in the body. This stage is sometimes referred to as the asymptomatic stage because the person may not feel ill. The clinical latency stage can last for decades if a person is being treated for HIV.

- **AIDS.** When the helper T-cells drop to less than 200, or one of several AIDS-opportunistic infections are present, the disease has progressed to the AIDS stage.

While the three stages are identifiable, no timetable exists to determine how long each individual will remain in a stage, or how healthy a person with HIV will remain. Several factors determine each individual's health status. These include how healthy the person was when infected, and whether the person maintains good nutrition, exercises regularly, and avoids tobacco use after infection.

Giving or Receiving Blood: Is It Safe?

Some people fear that they might be infected with HIV when donating or receiving blood. In the United States, health care professionals always use sterile needles to draw blood. In addition, all donated blood has been tested for HIV since 1985. According to the CDC, "The U.S. blood supply is among the safest in the world."

TEENS Making a Difference

Promoting AIDS Education

Hoonie K., of Pennsylvania, decided to take a stand by becoming a Student AIDS Educator (SAE). "There's a lot of misinformation and prejudices about HIV/AIDS. My goal is to educate my peers so they take action to prevent it."

SAEs meet with an adult adviser each week to obtain the latest information about HIV. They also help students who need more information or want to be tested. "We direct them to the appropriate places," Hoonie explains. "Everything is anonymous."

Hoonie hopes to expand the group's efforts by hosting a concert to raise money for HIV/AIDS research. "We want more opportunities to educate our peers."

Activity Write your answers to the following questions in your personal health journal.

1. Write three questions you would ask a Student AIDS Educator.
2. Where can you find accurate HIV/AIDS information in your community?
3. Why is it important to educate people about HIV/AIDS?

LESSON 3 ASSESSMENT

 ## After You Read

Reviewing Facts and Vocabulary

1. How does HIV affect the human immune system?

2. How can you protect yourself from contracting HIV/AIDS?

3. Why do the body's antibodies fail to protect people from HIV?

Thinking Critically

4. **Analyze.** Why has the CDC implemented mandatory testing for all donated blood?

5. **Synthesize.** How does the immune system respond to the presence of HIV in the body?

Applying Health Skills

6. **Advocacy.** Create a poster or public service announcement that warns teens about the risks of contracting HIV/AIDS.

Writing Critically

7. **Expository.** Write an essay that explains the relationship between HIV and AIDS. Discuss why the infection is considered one of the world's deadliest diseases.

Real Life Issues ....

After completing the lesson, review and analyze your response to the Real Life Issues question on page 669.

PhotoAlto/Eric Audras/Getty Images

Preventing and Treating HIV/AIDS

GUIDE TO READING

BIG Idea *HIV/AIDS is preventable and treatable, but it is incurable.*

Before You Read

Organize Information. Make a three-column chart. Label the columns "Prevention," "Diagnosis," and "Treatment." As you read, fill in the chart with information about how HIV/AIDS can be prevented, diagnosed, and treated.

Prevention	Diagnosis	Treatment

New Vocabulary

▶ Antibody screening test
▶ Western blot
▶ rapid test

Academic Vocabulary

estimate *(verb):* to determine roughly the size or extent of

Real Life Issues

Worried About HIV. Tony is concerned about his older sister, Kari. She confided that she enjoys college life, but she and her friends are under a lot of pressure to have sex and to experiment with alcohol and other drugs. Kari is committed to abstinence, so she's chosen to hang out with people who respect her decision. Yesterday, however, one of her closest friends called to tell her that he tested positive for HIV. Kari was shocked and upset by the news that someone she cares about deeply is infected with HIV.

Writing *Write a dialogue between Tony and Kari. How might Tony express his concern and support for his sister?*

Preventing HIV/AIDS

Main Idea There are many actions you can take to avoid contracting HIV/AIDS.

The Centers for Disease Control and Prevention (CDC) estimates that more than one million Americans live with HIV, and 50,000 are infected each year. About 8,300 of those will be young people between the ages of 13 and 24. Teens who are sexually active or who use intravenous drugs have a particularly high risk for contracting HIV/AIDS. Take a look at **Figure 24.10**. The graph shows the number of HIV/AIDS cases reported among teens between 1999 and 2011.

There is no way to tell just by looking whether a person is infected with HIV. The CDC **estimates** that about 20 percent of the people in the United States who are infected with HIV do not know they are infected. Because they are unaware

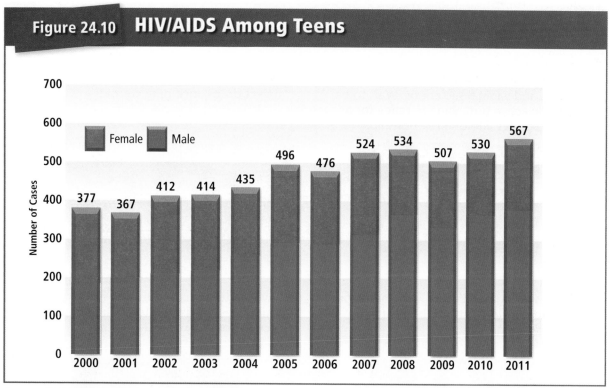

Figure 24.10   HIV/AIDS Among Teens

Number of Cases

Female   Male

Year	Cases
2000	377
2001	367
2002	412
2003	414
2004	435
2005	496
2006	476
2007	524
2008	534
2009	507
2010	530
2011	567

Source: CDC, HIV Surveillance in Adolescents and Young Adults, Stage 3 (AIDS) Classifications among Adolescents Aged 13-19 Years, by Sex, 2000–2011.

that they are HIV-positive, they may unknowingly spread the virus to others.Fortunately, you can take action to prevent the spread of HIV/AIDS. The following healthful behaviors will help protect you from infection:

- Practice abstinence.
- Do not share needles.
- Avoid situations where drug and alcohol use might compromise your decision making.
- Use refusal skills when you feel pressured to engage in risky behaviors.

For more information on HIV prevention strategies, see **Figure 24.11** on page 676. Notice that the listed behaviors all involve relationships. For example, sharing a needle always involves another person. Ask yourself:

- What do I know about the people in my life and their behaviors?
- Will they put me at risk for getting HIV/AIDS?
- How can I be sure another person is not HIV-positive?

Knowing as much as you can about the people around you and their behaviors can help you make responsible and informed decisions. It's also a good idea to practice refusal skills so you are prepared when pressured to engage in high-risk behaviors.

READING CHECK

Identify What are successful methods to avoid contracting HIV/AIDS?

Figure 24.11 **Ways to Prevent HIV and AIDS**

Practice abstinence. HIV is spread through semen and vaginal secretions. Wait to be sexually active until you are ready for a monogamous, lifelong relationship.

Avoid sharing needles or syringes used to inject drugs, including steroids.

Avoid sharing needles, knives, and razors used for cutting, tattoos, body art, or body piercing.

Avoid situations and events where you might feel pressured to engage in sexual activity or drug and alcohol use.

(br)Ingram Publishing, (tl)Nancy Ney/Alamy, (tr)Ingram Publishing, (bl)Glow Images

Diagnosing HIV/AIDS

Main Idea Several tests are used to diagnose HIV/AIDS.

Because 1 in 5 people with HIV are unaware that they have the infection the CDC recommends that everyone aged 13 to 64 be tested at least once. People in high-risk groups should be tested more frequently.

Typically, a blood sample or an oral specimen from between the inside of the cheek and the gum is collected and sent to a laboratory for analysis. Results are usually available within two weeks. At most testing sites, qualified personnel are available to answer questions, make referrals, and explain results.

Types of Laboratory HIV Tests

After collected samples are sent to a laboratory, technicians screen them for HIV antibodies. A person's body does not naturally have HIV antibodies: they are produced only in the presence of an infection. The most common laboratory tests used to screen, diagnose, and confirm HIV antibodies are the antibody screening test and the rapid test.

Antibody Screening Test The first test technicians run on a sample is an antibody screening test, *a test that screens for the presence of HIV antibodies in the blood*. If the results are positive, that means HIV antibodies are present, and the test is repeated. If the second test is also positive, then the Western blot test is run. Both tests are more than 98 percent accurate.

Follow-up Tests If a test for HIV is returned showing a positive result, an additional test is done to confirm the first test. Further testing is also done to distinguish the type of HIV antibodies, and to look directly at the virus. One of the tests that may be used is the Western Blot test, a test that detects HIV antibodies and confirms the results of earlier tests.

Cost of Testing The tests to determine the presence of HIV can be costly. The tests are typically offered in a doctor's office, local health department, hospital, and sites that specialize in HIV testing. Public health departments will offer the test free-of-charge. The local public health department can provide information on free testing, and may offer locations for free tests on the agency's website.

Additional Tests Two other tests—the RNA and the CD4—may be run when a Western blot test comes back positive. The RNA, or viral load test, shows how many copies of the virus are circulating in the blood. The CD4 test looks at the number of white blood cells in a sample of blood. These two tests give a more complete picture of an HIV-infected person's condition. They can also help doctors monitor the disease and determine how much medicine, if any, a patient needs.

A **rapid test** is *an HIV test that produces results in only 20 minutes*. It can be used in situations where the infected person might not come back to learn the results of the test. A blood sample is collected and analyzed immediately on site. Although the results are fairly accurate, they are considered preliminary.

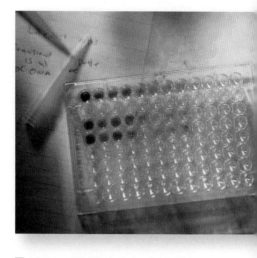

■ **Figure 24.12** In a positive EIA test, HIV antibodies bind to the HIV antigens on a plastic bead coated with HIV proteins. *Under what circumstances would HIV antibodies appear in a tissue or blood sample?*

■ **Figure 24.13** A viral load test measures the amount of HIV in a person's blood. *What's the difference between a viral load test and a CD4 test?*

READING CHECK

Explain. Why is testing necessary for those who suspect they have contracted HIV/AIDS?

Home Testing Kits One FDA-approved test uses a cheek swab to test the presence of the HIV virus in saliva. Results are available in 20 minutes. However, because the levels of HIV antibodies in saliva might be low, this test can provide false-negative results. A false-negative means that the test shows a person as HIV-negative, although the person is HIV-positive. Another home testing kit requires that the person provide a spot of dried blood on a test strip. The kit is then sent to a lab where a test is completed and the person can call in for results in a few days. This type of test is more accurate than a test using a cheek swab. The manufacturers of these tests provide confidential counseling and referrals to health care providers for treatment.

Benefits of Early Diagnosis

There are several benefits to early testing and diagnosis of HIV/AIDS. Early detection allows a person to

- begin proper medical care early to slow the progress of the virus.
- avoid behaviors that could spread HIV to others.
- gain peace of mind when the results are negative.

Treating HIV/AIDS

Main Idea Medications can slow the growth of HIV/AIDS, but there is no cure.

Since the early 1980s, drugs have been developed that slow the growth of the human immunodeficiency virus and treat some of the symptoms. No drug yet exists to cure HIV/AIDS. Many of the drugs available are also used to treat opportunistic infections. As you learned in Lesson 3, these infections occur in individuals who do not have healthy immune systems. AIDS-opportunistic illnesses include pneumonia and some types of cancers that can ultimately cause death.

To slow the growth of the AIDS virus, people take a combination of drugs, a treatment known as antiretroviral therapy.

■ **Figure 24.14** Medical research has developed drugs that protect the immune systems of people with HIV. *How can early diagnosis aid the treatment of HIV?*

Real World CONNECTION

AIDS Awareness Campaign

HIV/AIDS affects everyone. Conduct an Internet search to find the number of people, along with age ranges, who have been diagnosed with AIDS from 2007 up to the present. The CDC and the National Institutes of Health (NIH) are good Web sites to begin your search for this information.

Activity Technology

Work in groups to create a campaign to raise awareness of HIV/AIDS prevention. Use the information that was gathered from the Internet search. Each member of your team will be responsible for completing one item for the campaign. Campaign materials can include the following:

- A blog, Web page, or wiki
- Public service announcement script for a podcast
- Opinion article for an e-newsletter or the school's Web site

The materials should encourage teens to avoid behaviors that can put them at risk for HIV/AIDS.

LESSON 4 ASSESSMENT

After You Read

Reviewing Facts and Vocabulary

1. How is HIV detected?
2. What test is used to confirm a diagnosis of HIV?
3. HIV/AIDS home testing kits may not be trustworthy. What should you do to make sure the results are accurate?

Thinking Critically

4. **Synthesize.** What are the benefits of getting tested for HIV when an infection is suspected?
5. **Analyze.** When and for what reasons are blood or tissue samples tested more than once for HIV?

Applying Health Skills

6. **Advocacy.** Working in small groups, plan a classroom, school, or community project to help support AIDS research.

Writing Critically

7. **Expository.** Write an essay discussing how a teen's health and social life might be affected if the teen tested positive for HIV.

Real Life Issues

After completing the lesson, review and analyze your response to the Real Life Issues question on page 674.

Hands-On HEALTH

Checklist: Practicing Healthful Behaviors

- ✓ Demonstration of health knowledge about STDs
- ✓ Identification of protective behaviors to avoid STDs
- ✓ Step-by-step plan to avoid risks of contracting STDs
- ✓ Implementation of plan

 Activity ## STDs: A Game of Risks

You'll play a game called "STDs: A Game of Risks" and create a plan for reducing your risk of contracting an STD.

What You'll Need

- 32 index cards
- black marker

What You'll Do

Step 1

Work in teams of five or six. Write "STD," "Symptoms," "Long-Term Effects," and "Treatment" on four index cards. Spread these category cards in a row on a table or desk.

Step 2

Take five more index cards and write the name of one STD on each. For each STD you choose, write the symptoms, long-term effects, and treatment on three more cards. When finished, shuffle the cards and place them upside down in a stack.

Step 3

At your teacher's signal, turn over one card at a time and place the card under the correct category card. Arrange the cards to align them with the correct STD.

Apply and Conclude

Create a plan to avoid STDs. Include specific steps you can take.

LESSON **1**

Sexually Transmitted Diseases

Key Concepts
▶ STDS are transmitted through sexual contact.
▶ The risk of being infected with an STD increases as the number of sexual partners increases.
▶ STDs sometimes produce no symptoms. People may not realize they are infected, and therefore do not seek treatment.

Vocabulary
▶ sexually transmitted diseases (p. 658)
▶ sexually transmitted infections (p. 658)
▶ asymptomatic (p. 659)
▶ epidemic (p. 663)

LESSON **2**

Preventing and Treating STDs

Key Concepts
▶ Practicing abstinence is the only 100 percent successful method for preventing the transmission of STDs.
▶ It is crucial for people to seek diagnosis and treatment if they think they are infected with an STD.
▶ Bacterial STDs can be treated and cured with antibiotics, but viral STDs have no cure.

Vocabulary
▶ antibiotics (p. 664)
▶ abstinence (p. 664)
▶ refusal skills (p. 665)
▶ HPV vaccine (p. 667)

LESSON **3**

HIV/AIDS

Key Concepts
▶ HIV is the virus that causes AIDS.
▶ HIV destroys white blood cells, weakening the body's immune system.
▶ HIV is transmitted from one person to another through sexual intercourse, by sharing contaminated needles, or from mother to infant.

Vocabulary
▶ human immunodeficiency virus (HIV) (p. 669)
▶ acquired immunodeficiency syndrome (AIDS) (p. 669)
▶ pandemic (p. 670)
▶ mucous membranes (p. 670)
▶ lymphocytes (p. 671)
▶ antibodies (p. 671)

LESSON **4**

Preventing and Treating HIV/AIDS

Key Concepts
▶ HIV/AIDS has no cure at present.
▶ Medication can slow the progression of HIV, but cannot completely stop it.

Vocabulary
▶ Antibody Screening Test (p. 677)
▶ Western blot (p. 677)
▶ rapid test (p. 677)

LESSON **1**

Vocabulary Review

Use the vocabulary terms listed on page 681 to complete the following statements.

1. People infected with STDs often do not realize they have an infection because many STDs can be _____.

2. Health experts say that the United States currently faces an STD _____.

Understanding Key Concepts

After reading the question or statement, select the correct answer.

3. STDs can be passed from person to person through
   a. casual contact such as shaking hands.
   b. the air by coughing or sneezing.
   c. sexual contact.
   d. all of the above.

4. If left untreated, all STDs
   a. can lead to serious health problems.
   b. will eventually cure themselves.
   c. will become asymptomatic.
   d. lead to infection by HIV/AIDS.

Thinking Critically

After reading the question or statement, write a short answer using complete sentences.

5. **Describe.** Give one reason why STDs go undiagnosed and untreated.

6. **Compare and Contrast.** What are the differences in the ways that STDs affect males and females?

LESSON **2**

Vocabulary Review

Correct the sentences below by replacing the italicized term with the correct vocabulary term.

7. Many STDs can be treated and some cured with medications called *HPV vaccines*.

8. *A refusal skill* is the deliberate decision to avoid sexual activity.

Understanding Key Concepts

After reading the question or statement, select the correct answer.

9. Which is *not* a high-risk behavior?
   a. Engaging in sexual activity with multiple partners
   b. Engaging in unprotected sexual activity
   c. Using alcohol and other drugs
   d. Abstaining from sexual activity

10. Getting a diagnosis and treatment is
    a. acting responsibly.
    b. crucial for those infected with STDs.
    c. a healthful behavior.
    d. all of the above.

11. Treatment of an STD
    a. does not prevent reinfection.
    b. isn't always necessary.
    c. can be postponed.
    d. always cures the infection.

Thinking Critically

After reading the question or statement, write a short answer using complete sentences.

12. **Compare and Contrast.** Identify the differences and similarities between viral and bacterial STDs.

13. **Discuss.** Which STD can be prevented by a vaccine? What are its limitations? Who is eligible to receive this vaccination?

14. **Evaluate.** Why is preventing STD transmission more effective than treating STDs?

15. **Explain.** What are antibiotics? How are they used to treat STDs?

LESSON 3

Vocabulary Review

Choose the correct term in the sentences below.

16. *HIV / Mucous membrane* is transmitted through the bloodstream.

17. *Lymphocyte / AIDS* is the final stage of HIV infection.

18. Health care officials consider AIDS to be a(n) *antibody / pandemic*.

Understanding Key Concepts

After reading the question or statement, select the correct answer.

19. During the course of HIV/AIDS, the infected person
    a. gets stronger.
    b. should not hug anyone or shake hands.
    c. needs less and less medication.
    d. becomes vulnerable to opportunistic illnesses.

20. Which of the following is *not* a way that HIV attacks cells?
    a. The virus attaches itself to the cell's surface.
    b. The virus makes a copy of its genetic material.
    c. The virus shrinks cells.
    d. The new virus assembles at cell surface.

21. It is difficult for antibodies to fight AIDS because
    a. HIV weakens antibodies.
    b. HIV destroys white blood cells.
    c. HIV is protected once it enters cells.
    d. HIV mutates rapidly.

Thinking Critically

After reading the question or statement, write a short answer using complete sentences.

22. **Explain.** Describe how HIV infection progresses in the body.

23. **Identify.** Name three ways HIV is transmitted.

24. **Evaluate.** What misinformation causes some people to stay away from those infected with HIV?
    Why is this information wrong?

LESSON 4

Vocabulary Review

Use the vocabulary terms listed on page 681 to complete the following statements.

25. The _____ test is the first test that technicians use to screen for HIV.

26. If the initial test produces positive results twice, a(n) _____ test is run.

27. The _____ allows samples to be tested on site rather than sending them to labs.

Understanding Key Concepts

After reading the question or statement, select the correct answer.

28. A person who thinks he or she is infected with HIV/AIDS should
    a. use a home testing kit.
    b. hide the condition from others.
    c. get a medical diagnosis right away.
    d. hope that symptoms do not appear.

S. Olsson/PhotoAlto

Assessment

29. About one-fourth of the people infected with HIV/AIDS
    a. are males.
    b. are females.
    c. don't know they are infected.
    d. will never develop symptoms.

30. People who are infected with HIV/AIDS, but don't know it,
    a. won't become as ill as those who know they have the virus.
    b. don't need to change their high-risk behaviors.
    c. don't need to practice abstinence.
    d. can unknowingly spread the virus to others.

Thinking Critically

After reading the question or statement, write a short answer using complete sentences.

31. **Explain.** The number of HIV/AIDS infections is much higher in developing nations than in developed countries. Why might this number be higher in some countries?

32. **Identify.** What are some benefits of early diagnosis of HIV/AIDS?

Technology PROJECT-BASED ASSESSMENT

Knowledge is Power

Background
Access to accurate information can prevent the spread of STDs. How much do the teens in your school know about STDs?

Task
Use a free online survey tool to create a survey to determine areas in which students in your school are uninformed or misinformed about STDs. Create a Web page that provides students with accurate information about STDs, including the fact that abstinence is the only 100 percent effective way to prevent infection.

Audience
Teens in your school

Purpose
Educate students about STDs by providing accurate information.

Procedure
1. Collaborate as a group to use the information in the chapter to make up the questions for the online survey.
2. Select 20 or more students in your school and ask them to answer the questions.
3. Analyze the answers to determine areas in which students are uninformed or misinformed.
4. Create a Web page that provides the needed information.
5. Present your group's Web page to your class.

Math Practice

Interpret Graphs. The bar graph below shows the number of new cases of different STDs reported in the United States. Study the bar graph, and then answer the questions.

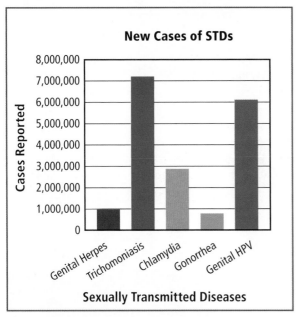

New Cases of STDs

Source: Centers for Disease Control and Prevention, 2004

1. If the population of the United States is about 296 million, what is the ratio of infection for trichomoniasis this year?
   - **A.** 1 in 4
   - **B.** 1 in 40
   - **C.** 1 in 400
   - **D.** 1 in 4,000

2. How much more common is genital HPV infection than genital herpes?
   - **A.** Twice as common
   - **B.** Three times more common
   - **C.** Four times more common
   - **D.** Six times more common

3. Using the bar graph, explain how you would predict the rates of STD infection for people in your state.

Reading/Writing Practice

Understand and Apply. Read the passage below, and then answer the questions.

During the late 1980s, Ryan White was the face of AIDS for many Americans. Ryan contracted AIDS through a blood transfusion. Many members of his community mistakenly believed that AIDS could spread through casual contact. They pressured the school board to ban Ryan from attending school. Ryan's family took his case to court, and he was eventually allowed to return to school.

Ryan became an AIDS educator. He spoke of the need for everyone to learn about AIDS and to treat affected people with compassion and dignity. Ryan lived for six years following his AIDS diagnosis. He died in 1990 at age 18. Later that year, Congress passed the Ryan White Comprehensive AIDS Resources Emergency (CARE) Act. Today, the act provides about $1.5 billion annually to care for people living with HIV/AIDS.

1. What is the purpose of this passage?
   - **A.** To describe an early case of AIDS
   - **B.** To describe the fear of HIV/AIDS
   - **C.** To show how Ryan fought AIDS
   - **D.** To blame public officials

2. What was the result of the publicity surrounding Ryan's case?
   - **A.** It helped Ryan live longer.
   - **B.** It allowed Ryan to return to school.
   - **C.** It provided the public with factual information about HIV/AIDS.
   - **D.** It increased the hostility against Ryan.

3. Write a paragraph explaining how HIV can and cannot be spread. Explain why the virus can be spread only in certain ways.

National Education Standards

Math: Measurement and Data, Data Analysis
Language Arts: LACC.910. RI.1, LACC.910. RL.2.4

CHAPTER **25**

Noncommunicable Diseases and Disabilities

Activating Prior Knowledge

Using Visuals Look at the photo on this page. Write a short paragraph describing how these teens are actively promoting their health. What technologies and devices can help people with disabilities lead full and active lives?

Chapter Launchers

Health in Action

Discuss the **BIG** Ideas

Think about how you would answer these questions:

▶ Why should you learn about heart disease and cancer?

▶ What do you know about asthma or allergies?

▶ Why is it important to be educated about physical and mental challenges?

Assess Your Health

Read each statement. On a separate sheet of paper, write "yes," "sometimes," or "no" based on your typical behavior.

1. I understanding the role of regular physical activity in reducing the risk of cardio-vascular disease.

2. I recognize the risk factors for heart disease and stroke.

3. I avoid tobacco use to main-tain cardiovascular health and reduce the risk of cancer.

4. I recognize that healthful behaviors will reduce my risk of cancer.

5. I recognize that advocacy can enact laws to help people with physical and mental challenges.

A "yes" response shows that you practice healthy behaviors. "Sometimes" indicates that you should analyze and possibly modify your behavior. A "no" response means that you should modify the behavior.

GUIDE TO READING

BIG Idea *Preventive behaviors can reduce your risk for cardiovascular disease and stroke.*

Before You Read

Create a Cluster Chart. Draw a circle and label it "Cardiovascular Disease," or CVD. Use surrounding circles to identify factors that contribute to this disease. As you read, continue filling in the chart with more details.

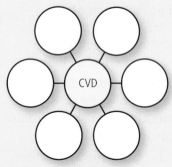

New Vocabulary

▶ noncommunicable disease
▶ cardiovascular disease
▶ hypertension
▶ atherosclerosis
▶ arteriosclerosis
▶ angina pectoris
▶ arrhythmias
▶ stroke

Cardiovascular Disease

Real Life Issues

Neglected Hearts.

> About 600,000 people die of heart disease in the United States every year—that's 1 in every 4 deaths.

> Coronary heart disease alone costs the United States $108.9 billion each year. This total includes the cost of health care services, medications, and lost productivity.

Source: Centers for Disease Control and Prevention; Heart Disease, Facts and Statistics

Writing *Think about what you already know about heart disease. Write a paragraph explaining why you think it is important to fight heart disease before it even develops.*

Cardiovascular Disease

Main Idea The heart, blood, and blood vessels are at risk for a number of potentially serious diseases.

You've learned about communicable diseases and how to prevent them. Some illnesses, however, are not infectious. A **noncommunicable disease** is *a disease that is not transmitted by another person, a vector, or the environment.* One of the most common noncommunicable diseases is **cardiovascular disease**, or CVD, *a disease that affects the heart or blood vessels.* CVD is responsible for about 25% of all U.S. deaths, killing more than a million Americans every year.

Cardio refers to the heart, and *vascular* refers to the blood vessels. As you learned in Chapter 15, the cardiovascular system works tirelessly to keep you alive and well. A problem in just one part of the system can jeopardize your health. That's why it's important to know about the variety of cardiovascular diseases and how to prevent them.

Types of Cardiovascular Disease

Main Idea There are many different types of CVDs.

There are quite a few different cardiovascular diseases. As you read about these diseases, think about how each one is caused and what you can do to reduce your risk.

Hypertension

High blood pressure, or **hypertension**, can damage the heart, blood vessels, and other body organs if it continues over a long period of time. It is also a major risk factor for other types of CVDs. Because hypertension often has no symptoms in its early stages, it is sometimes called a "silent killer."

Hypertension can occur at any age, but it is more common among people over the age of 35. It is estimated that about one-third of American adults have hypertension. To treat hypertension, patients should manage their weight, get adequate physical activity, and eat a nutritious diet. Medication for hypertension is also available.

Atherosclerosis

When you were born, the lining of your blood vessels were smooth and elastic. What is the condition of your blood vessels today? If you smoke, have high blood pressure, or have high cholesterol levels, fatty substances called *plaques* can build up on your artery walls. This condition is known as **atherosclerosis**, *a disease characterized by the accumulation of plaque on artery walls.* People with atherosclerosis have a condition called **arteriosclerosis**, *hardened arteries with reduced elasticity.* **Figure 25.1** compares a healthy artery with unhealthy ones.

Go to **glencoe.com** and use this code to complete the Student Web Activity on heart disease and maintaining healthy blood cholesterol levels.

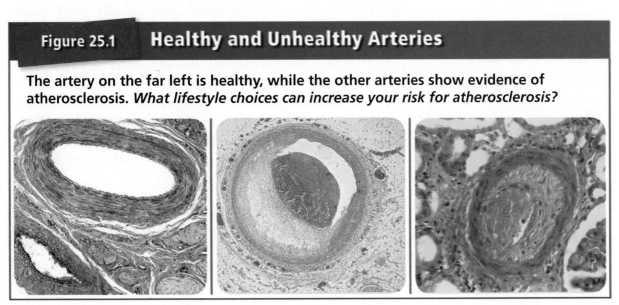

Figure 25.1 **Healthy and Unhealthy Arteries**

The artery on the far left is healthy, while the other arteries show evidence of atherosclerosis. *What lifestyle choices can increase your risk for atherosclerosis?*

READING CHECK

Identify What are some healthful behaviors that can help prevent atherosclerosis?

The main cause of atherosclerosis is making unhealthful food choices—specifically, foods that have large amounts of saturated fat and cholesterol. Sometimes a blood clot forms near plaque buildup and blocks the artery. If this artery supplies blood to the heart or the brain, a heart attack or stroke may result.

Diseases of the Heart

Every day your heart pumps about 100,000 times, moving blood to all parts of your body. Like every other organ in your body, it needs oxygen from the blood in order to function. When the blood supply to the heart is restricted, the heart does not get the oxygen it needs. Under these conditions, a heart attack can occur. The result can be heart muscle damage or even sudden death due to cardiac arrest.

Heart attack and cardiac arrest are not quite the same thing. *Heart attack* occurs due to insufficient blood supply to the heart. *Cardiac arrest,* in which the heart stops beating in a rhythmic way, occurs due to an electrical problem with the heart. To diagnose and treat heart disease, doctors use several techniques, shown in **Figure 25.2**. Commonly diagnosed heart diseases include angina pectoris, arrhythmias, heart attack, and congestive heart failure.

Angina Pectoris *Chest pain that results when the heart does not get enough oxygen* is called **angina pectoris** (an-JY-nuh PEK-tuh-ruhs). This pain, which usually lasts from a few seconds to minutes, is a warning sign that the heart is temporarily not getting enough blood. Angina is usually caused by atherosclerosis and should be taken seriously. It sometimes can be treated with medication.

Arrhythmias *Irregular heartbeats,* or **arrhythmias**, happen when the heart skips a beat or beats very fast or very slowly. Arrhythmias are quite common and usually don't cause problems. However, some types are serious, and should always be checked by a doctor. In one type of arrhythmia, called *ventricular fibrillation,* the electrical impulses that **regulate** heart rhythm become rapid or irregular. This is the most common cause of cardiac arrest. Cardiopulmonary resuscitation (CPR) and using an automated external defibrillator are the only ways to help someone suffering a cardiac arrest. You will learn more about these first-aid methods in Chapter 27.

Heart Attack A heart attack occurs when a reduced or blocked blood supply damages the heart muscle. Many heart attacks cause intense chest pain, but about 25 percent produce no symptoms or unusual symptoms such as shortness of breath. Milder symptoms may be common in women.

Figure 25.2 Diagnosing and Treating Heart Disease

As medical technology advances, more diagnostic tools and treatment options become available. *Which treatment option uses a small balloon to clear a blocked artery?*

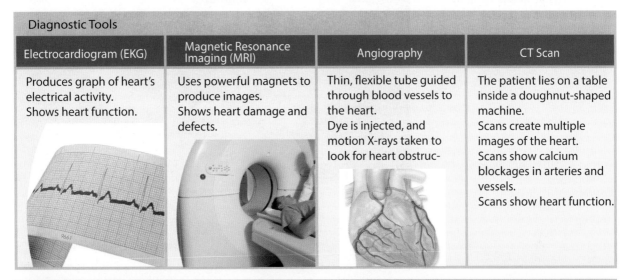

Diagnostic Tools

Electrocardiogram (EKG)	Magnetic Resonance Imaging (MRI)	Angiography	CT Scan
Produces graph of heart's electrical activity. Shows heart function.	Uses powerful magnets to produce images. Shows heart damage and defects.	Thin, flexible tube guided through blood vessels to the heart. Dye is injected, and motion X-rays taken to look for heart obstruc-	The patient lies on a table inside a doughnut-shaped machine. Scans create multiple images of the heart. Scans show calcium blockages in arteries and vessels. Scans show heart function.

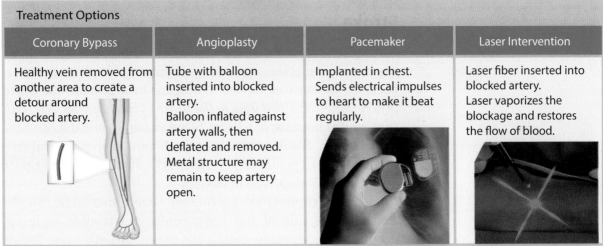

Treatment Options

Coronary Bypass	Angioplasty	Pacemaker	Laser Intervention
Healthy vein removed from another area to create a detour around blocked artery.	Tube with balloon inserted into blocked artery. Balloon inflated against artery walls, then deflated and removed. Metal structure may remain to keep artery open.	Implanted in chest. Sends electrical impulses to heart to make it beat regularly.	Laser fiber inserted into blocked artery. Laser vaporizes the blockage and restores the flow of blood.

Anyone who experiences the following warning signs of heart attack should call 911 immediately:

- Pressure, fullness, squeezing, or aching in chest area
- Pain spreading to arms, neck, jaw, abdomen, or back
- Chest discomfort, with shortness of breath, lightheaded feeling, sweating, nausea, or vomiting

Congestive Heart Failure This occurs when the heart gradually weakens and can no longer maintain its regular pumping rate and force. Congestive heart failure cannot be cured, but it can improve through continuous treatment, such as medication and practicing healthy behaviors. A heart transplant may also be recommended. A transplant center considers each case individually based on factors such as age, other health concerns, and willingness to make lifestyle changes.

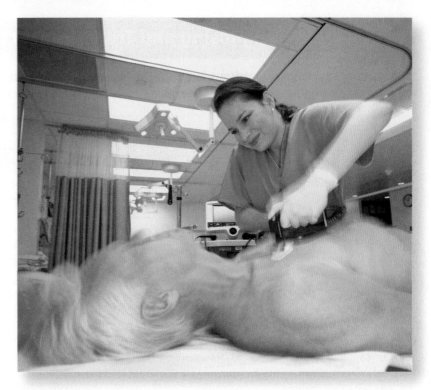

Figure 25.3 In many cases sudden cardiac arrest can be reversed if CPR or electric shock using a defibrillator is applied. *Why is it important to have defibrillators available in many different public places?*

Stroke

Cardiovascular disease can affect the brain as well as the heart. Sometimes an artery supplying blood to the brain becomes blocked or bursts, resulting in a **stroke**—*an acute injury in which blood flow to the brain is interrupted.* A stroke that occurs because of a burst blood vessel is called a *cerebral hemorrhage.* Stroke can cause problems such as paralysis. The damage depends on the size of the stroke and what part of the brain is deprived of oxygen.

Warning signs of stroke include severe headache, numbness on one side of the body, confusion, trouble walking, dizziness, and trouble seeing out of one or both eyes. Today, treatments exist that can stop a stroke as it is occurring. Drugs known as clot busters can break up a clot and restore the normal flow of blood to the brain.

What Teens Need to Know

Main Idea CVD can begin during the teen years.

Did you know that CVD can start to develop during adolescence or even childhood? Autopsy results on adolescents who died from causes other than CVD have found that one in six already had evidence of CVD. Those who had a history of known risk factors, such as smoking or diabetes, were more likely to have blood vessel damage. A teen with damaged blood vessels may not experience any symptoms until later in adulthood, but the danger is already there.

Figure 25.4

CVD Risk Factors You Can Control

Risk Factor	Preventive Measure	Why It's Important
Tobacco Use	Avoid using tobacco.	About 20 percent of deaths from CVDs are smoking related. For teens, tobacco use is the biggest risk factor.
	Avoid secondhand smoke.	About 46,000 nonsmokers who are exposed to secondhand smoke die from CVDs each year.
High Blood Pressure	Have your blood pressure checked regularly.	High blood pressure strains your cardiovascular system.
	Eat healthfully, exercise regularly, and manage your weight.	
High Cholesterol	Eat fewer high-fat and high-cholesterol foods, and get regular physical activity.	High cholesterol can cause plaque to form in your arteries.
Physical Inactivity	Be sure you get at least 30 to 60 minutes of physical activity every day.	Physical activity strengthens your heart and helps you maintain a healthy weight.
Excess Weight	Maintain a healthy weight.	Excess weight puts a strain on the heart and raises blood pressure and blood cholesterol levels. It also increases your risk for type 2 diabetes (a risk factor for heart disease).
Stress	Use stress-management techniques.	Constant stress raises blood pressure.
Alcohol and Drug Use	Abstain from alcohol and other drugs.	Too much alcohol raises blood pressure and can cause irregular heartbeat or heart failure. Some illegal drugs increase heart rate and blood pressure and can result in heart failure.

Risk Factors

The American Heart Association has identified several factors that increase the risk of heart attack and stroke. **Figure 25.4** lists some actions you can take to reduce your risk of CVD. Unfortunately, there are also some risk factors that are unavoidable:

- **Heredity.** Children whose parents have CVD are more likely to develop CVD themselves.

- **Gender.** Men have a greater risk than women of developing CVD and having heart attacks. However, research shows that older women are less likely than men of the same age to survive a heart attack.

- **Age.** The risk of CVD increases with age. Approximately 80 percent of people who die of CVD are 65 or older.

Knowing about these risk factors can help you make healthful decisions to reduce your risk. For example, if CVD runs in your family, you can make a strong commitment to control your weight, exercise regularly, avoid tobacco use, and eat foods low in fat and cholesterol.

READING CHECK

Identify What are three decisions you could make today to reduce your risk of CVD?

Real World CONNECTION

An Exercise Campaign

Moderate exercise helps keep people healthy. Exercise burns calories, builds muscle, and helps the heart stay strong. Some people feel that exercise must be done for at least an hour and must be strenuous. Conduct an Internet search for examples of forms of exercise that are not strenuous and time-consuming.

Activity Technology

Exercise can be as simple as a brisk walk, and most of the benefits are gained in the first half hour. More people should know that moderate exercise for just 30 minutes per day is an effective way to maintain or improve cardiovascular health.

Based on your Internet research, create a campaign to encourage teens to exercise for 30 minutes per day. The campaign should help teens understand why exercise is important, what kinds of exercises help their cardiovascular systems, and how much exercise people need. Create a Web page for the school's Web site that will inspire teens to begin a moderate exercise program.

LESSON 1 ASSESSMENT

 After You Read

Reviewing Facts and Vocabulary

1. Define *cardiovascular disease*.
2. What can happen if hypertension continues over a long period?
3. What is a *stroke*?

Thinking Critically

4. **Compare and Contrast.** How is stroke similar to heart attack? How is it different?
5. **Synthesize.** How can practicing healthy lifestyle behaviors today help lower your risk for cardiovascular disease in the future?

Applying Health Skills

6. **Practicing Healthful Behaviors.** Evaluate your daily habits. What decisions can you make today to replace unhealthful choices with healthful ones?

Writing Critically

7. **Persuasive.** Imagine you have a friend who says that you don't need to worry about CVD until you are older. Write a letter convincing this friend that it's important to start taking preventive measures now.

Real Life Issues

After completing the lesson, review and analyze your response to the Real Life Issues question on page 688.

Cancer

Real Life Issues ....

Making a Healthful Choice. Amy wants her granddad to quit smoking. She learned in school that smoking can lead to health problems, such as heart disease and cancer. Amy's mom wants him to quit too. Amy and her mom have decided to write letters to Granddad telling him how they feel when they see him smoking. They want him to know how important he is to them, and why they want him to stay healthy.

Writing *What should Amy write to encourage her granddad to stop smoking? How can she use the letter to express her concern and support? Summarize your thoughts in a paragraph.*

What Is Cancer?

Main Idea Cancer has a variety of forms and affects different areas of the body.

Cells are the building blocks of your body. Approximately 100 trillion of these tiny structures make up who you are. The cells in your body are constantly growing, dividing, dying, and replacing themselves. Although most new cells are normal, some are not. When abnormal cells reproduce rapidly and uncontrollably, they can build up inside otherwise normal tissue. This *uncontrollable growth of abnormal cells* is called **cancer**.

How Cancer Harms the Body

When abnormal cells build up in the body, they can form a **tumor**. This is *an abnormal mass of tissue that has no natural role in the body.* Many people equate tumors with cancer. However, the presence of a tumor does not necessarily mean that a person has cancer. In fact, there are two kinds of tumors: benign and malignant.

Figure 25.5

Types of Cancer

Organ Affected (new cases/year)	Some Risk Factors	Symptoms	Screening and Early Detection Methods
Skin (60,000) Most common type of cancer in the United States	Exposure to ultra-violet (UV) radiation from the sun, tanning beds, sunlamps, and other sources	Change on the skin, especially a new growth, a mole or freckle that changes, or a sore that won't heal	Physical exam, biopsies
Breast (209,005) Second leading cause of cancer death for women	Genetic factors, obesity, alcohol use, physical inactivity	Unusual lump; nipple that thickens, changes shape, dimples, or has discharge	Self-exam, mammogram
Prostate (196,038) Found mostly in men over 55	Possible hereditary link, possible link to high-fat diet	Frequent or painful urination; inability to urinate; weak or interrupted flow of urine; blood in urine or semen; pain in lower back, hips, or upper thighs	Blood test
Lung (201,144) Leading cause of cancer deaths in the United States	Exposure to cigarette smoke, radon, or asbestos	No initial symptoms; later symptoms include cough, shortness of breath, wheezing, coughing up blood, hoarseness	Chest X-ray
Colon/Rectum (131,607) Second leading cause of cancer deaths in the United States	Risk increases with age	Often no initial symptoms; later, blood in feces; frequent pain, aches, or cramps in stomach; change in bowel habits; weight loss	Test for blood in the stool, rectal exam, colonoscopy
Mouth (41,380*) Occurs mostly in people over 40	Use of tobacco, chewing tobacco, or alcohol	Sore or lump on mouth that doesn't heal; unusual bleeding; pain or numbness on lip, mouth, tongue, or throat; feeling that something is caught in the throat; pain with chewing or swallowing; change in voice	Dental/oral exam
Cervix (11,818)	History of infection with human papillomavirus (HPV)	Usually no symptoms in early stages; later, abnormal vaginal bleeding, increased vaginal discharge	Pap test
Testicle (7,920*) Most common cancer in men ages 15 to 34	Undescended testicle; family history of testicular cancer	Small, hard, painless lump on testicle; sudden accumulation of fluid in scrotum; pain in region between scrotum and anus	Self-exam

CDC, Cancer Statistics by Cancer Type, 2101.

*National Cancer Institute, Surveillance, Epidemiology, and End Results Program, Cancer Statistics, 2011

A **benign**, or *noncancerous*, tumor grows slowly. It is surrounded by membranes that prevent it from spreading. Does this mean it is harmless? No. Even if a benign tumor does not spread, it could still interfere with normal body functions. For example, a benign tumor in the brain could block the brain's blood supply.

A **malignant**, or *cancerous*, tumor does not stay in one place. It spreads to neighboring tissues and enters the blood or lymph to travel to other parts of the body. This process, *the spread of cancer from the point where it originated to other parts of the body*, is called **metastasis**. As cancer cells spread throughout the body, they divide and form new tumors.

Many cancers kill normal cells as they compete with them for nutrients in the body. Whether a tumor is benign or malignant, it can put pressure on your organs and tissues and interfere with body functions. It can also block arteries, veins, and other passages that work best when they are unobstructed.

Types of Cancer

Cancers can develop in almost any part of the body and are classified according to the tissues they affect:

- **Lymphomas** are cancers of the immune system.
- **Leukemias** are cancers of the blood-forming organs.
- **Carcinomas** are cancers of the glands and body linings, including the skin and the linings of the digestive tract and lungs.
- **Sarcomas** are cancers of connective tissue, such as bones, ligaments, and muscles.

Figure 25.5 lists common types of cancers, grouped according to the body organ in which they first develop.

Risk Factors for Cancer

Main Idea Risk factors for cancer include lifestyle behaviors.

Every day, your body produces countless numbers of healthy, normal cells—but it also produces some abnormal ones. Your immune system usually kills these abnormal cells before they become cancerous. However, when the immune system is weak or the abnormal cells multiply faster than the immune system can destroy them, cancer may develop.

Carcinogens

Many cancers develop because of exposure to a **carcinogen**, or a *cancer-causing substance*. Tobacco and UV light are two of the most common carcinogens that cause cancer.

Tobacco Use The number one cause of cancer deaths in the United States is tobacco use. About 70 different carcinogens have been identified in tobacco and tobacco smoke. Consider these numbers:

- In 2010, 201,144 new cases of lung cancer related to smoking diagnosed were.
- About 90 percent of lung cancers are caused by smoking.
- Smokers live 10 fewer years than nonsmokers, on average.

FITNESS ZONE

When I told my doctor that some of my older relatives have gotten cancer, she told me that eating a healthy diet and making healthy lifestyle choices can reduce my risks. Making healthier choices could mean 375,000 fewer cancer diagnoses in the U.S. every year. That convinced me to make better choices to protect my health. For more fitness tips, visit the Fitness Handbook and Fitness Zone sites in ConnectEd.

Academic Vocabulary

link *(verb):* to connect

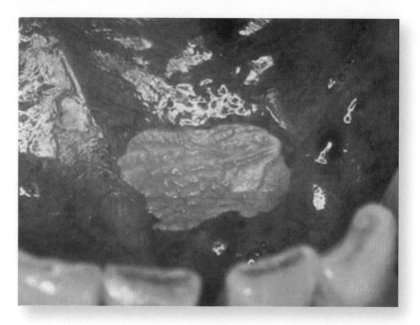

■ **Figure 25.6** This person's cancer may have been caused by using smokeless tobacco. *Smokeless tobacco is a major risk factor for what kind of cancer?*

The tobacco risk factor is not limited to smoking. Smokeless tobacco is a major risk factor for oral cancer, which affects the lips, mouth, and throat. Nonsmokers who are exposed to secondhand smoke are also at risk because they breathe in nicotine and other toxic chemicals.

Radiation Another carcinogen that commonly causes cancer is radiation. The glow of a suntan might look attractive, but a suntan is actually your skin's reaction to damage from the sun. UV radiation from the sun is the main cause of skin cancer. Tanning beds and sunlamps also emit UV radiation, which is just as damaging as the sun's rays.

Sexually Transmitted Diseases

Some sexually transmitted viruses have the ability to cause cancer. For example, certain forms of human papillomavirus, or HPV, can cause cervical cancer. The hepatitis B virus, another sexually transmitted virus, can cause cancer in the liver.

Dietary Factors

Being obese increases a person's risk for cancer. A diet that is high in fat and low in fiber is often linked with colon, breast, and prostate cancers. Here's why:

- **Fats** make colon cells more vulnerable to carcinogens. Colon cells divide faster if the diet is high in fat, increasing the risk that abnormal cells will form.
- **Dietary fiber** speeds the movement of waste through the intestines and out of the body. If a person's diet is low in fiber, the waste moves more slowly, giving carcinogens in the waste more time to act on the body's cells.

READING CHECK

Explain Why is it important to eat high-fiber foods?

Reducing Your Risk

(**Main Idea**) You can reduce your risk of cancer by practicing healthful behaviors.

Sometimes cancer seems to strike people at random. That is one of the most frightening aspects of the disease. Although some factors may be beyond your control, you can dramatically reduce your risk for cancer by practicing healthful behaviors. See **Figure 25.7** for steps you can start taking today.

READING CHECK

Identify What are three actions you could take to reduce your risk for cancer?

Figure 25.7 **How You Can Reduce Your Cancer Risk**

There are many healthful behaviors you can practice to reduce your risk for cancer. *How many of these behaviors do you already practice?*

Protect your skin from UV radiation.

Avoid tobacco and alcohol. Tobacco is the single major cause of cancer death in the United States. Excess alcohol increases the risk of several types of cancer, including mouth and throat cancer.

Practice abstinence from sexual activity to reduce the risk of sexually transmitted diseases. Hepatitis B can cause liver cancer, and HPV can cause cancers of the reproductive organs.

Be physically active.

Maintain a healthy weight.

Eat nutritious foods. Include 2–4 servings of fruits and 3–5 servings of vegetables every day. These foods are good sources of fiber, and some contain compounds that act against carcinogens.

Follow an eating plan that is low in saturated fat and high in fiber.

Recognize the warning signs of cancer. Do regular self-exams to detect cancer early.

Detecting and Treating Cancer

Main Idea Successful cancer treatment depends on early detection and the right kind of treatment.

As medical technology continues to advance, doctors are able to detect cancer earlier than in past years. **Figure 25.8** shows the many warning signs for cancer. There have also been many advances in treatment options. The survival rate for people with cancer depends on two main factors: early detection and the type of cancer.

Early detection, the most critical factor in successful cancer treatment, depends on both self-examination and medical examination.

- **Self-examination** involves checking your own body for possible signs of cancer. Many types of cancer, including those of the breasts, testicles, and skin, are discovered early through self-examination.

- **Medical examination**, or medical screening, involves testing by a doctor for early signs of cancer. About half of all new cancer cases each year are detected during a routine medical screening.

If a doctor thinks cancer is a possibility, a **biopsy**—*the removal of a small piece of tissue for examination*—may be ordered. A biopsy is usually necessary to determine whether cancer is present. To help determine a tumor's location and size, doctors use X-rays and other imaging techniques.

Figure 25.8 Warning Signs of Cancer

The warning signs listed below do not necessarily indicate cancer, as there may be other causes. However, all are serious enough to bring to a doctor's attention right away. A person with any of these warning signs should see a doctor as soon as possible. *What is a warning sign of skin cancer?*

- **Fever, fatigue, pain, and discoloration of the skin.** These general signs can sometimes indicate cancer.

- **Change in bowel habits or bladder function.** This may suggest colon, bladder, or prostate cancer.

- **Sores that will not heal.** Persistent sores on skin, mouth, or genitalia should be examined promptly.

- **Unusual bleeding or discharge.** This could be present in phlegm, stool, urine, or discharge from vagina or nipples.

- **Thickening or lump in breast or other body part.** Many cancers can be felt through the skin.

- **Indigestion or trouble swallowing.** Though usually harmless, these symptoms can sometimes indicate cancer of the stomach, esophagus, or throat.

- **Change in wart or mole.** Change in color or size might indicate skin cancer.

- **Nagging cough or hoarseness.** This could indicate cancer of the lungs, larynx, or thyroid.

Treatment Options

The methods used to treat cancer depend on several factors, such as the type of cancer and whether a tumor has spread from its original location. Treatment might include one or more of the methods listed below.

- **Surgery** removes some or all of the cancerous masses from the body.
- **Radiation therapy** uses radioactive substances to kill cancer cells and shrink cancerous masses.
- **Chemotherapy** uses chemicals to destroy cancer cells.
- **Immunotherapy** activates a person's immune system to recognize specific cancers and destroy them.
- **Hormone therapy** uses medicines to interfere with the production of certain hormones, such as estrogen, that help cancer cells grow. These treatments kill cancer cells or slow their growth.

When treatment works and the cancer is either gone or under control, the cancer is said to be in **remission**. This is *a period of time when symptoms disappear.* Today, more and more cancer survivors are able to lead full, active lives.

READING CHECK

Identify Which cancer treatment option uses chemicals to destroy cancer cells?

LESSON 2 ASSESSMENT

After You Read

Reviewing Facts and Vocabulary

1. What is *metastasis*?
2. What are the two important methods for early cancer detection?
3. Identify three cancer treatment options.

Thinking Critically

4. **Synthesize.** Based on what you know about your own lifestyle and what you now know about the risk factors for cancer, do you need to change any of your behaviors? Explain.
5. **Evaluate.** How does technology help in detecting and treating cancer?

Applying Health Skills

6. **Refusal Skills.** Based on what you have learned in this lesson, write down what you might say to someone trying to pressure you into using tobacco.

Writing Critically

7. **Expository.** Research the procedures used for early detection of cancer. Then write a short essay analyzing the benefits of health screenings, checkups, and early detection. Include information from your research to support your analysis.

Real Life Issues ....

After completing the lesson, review and analyze your response to the Real Life Issues question on page 695.

LESSON 3

Photodisc/Getty Images

GUIDE TO READING

BIG Idea *Practicing self-management strategies can help reduce the severity of allergies, asthma, diabetes, and arthritis.*

Before You Read

Create Vocabulary Cards. Write each new vocabulary term on a separate index card. For each term, write a definition based on your current knowledge. As you read, fill in additional information related to each term.

New Vocabulary

▶ allergy
▶ histamines
▶ asthma
▶ diabetes
▶ autoimmune disease
▶ arthritis
▶ osteoarthritis
▶ rheumatoid arthritis

Allergies, Asthma, Diabetes, and Arthritis

Real Life Issues

Using Good Judgment. Eliza feels good about volunteering at the local animal shelter on weekends. She loves taking care of the animals but she always ends up leaving with a runny nose, an itchy throat, irritated eyes, and sneezing attacks. Eliza thinks she may have allergies but she has not seen a doctor. She is worried that she won't be able to continue working with animals if she can't find a way to control her allergy symptoms.

Writing *Write a letter offering advice on how you think Eliza should handle her allergy symptoms.*

Allergies

Main Idea Allergies are caused by a variety of substances.

If you are sneezing and have a runny nose, you might not have a cold. Rather, you might have an **allergy**—*a specific reaction of the immune system to a foreign and frequently harmless substance.* These substances are present in your environment and make their way into your body. Have you noticed how many advertisements for allergy medications are on TV and in magazines? That's because allergies are a very common noncommunicable illness.

The substances that cause allergies are called *allergens.* Allergens include pollen, certain foods, dust, mold spores, chemicals, insect venom, dander from animals, and certain medicines.

Allergens produce the allergic reactions of sneezing and a runny nose in a four-step process.

1. The allergen enters the body, which treats the allergen as a foreign invader.

2. Antigens on the surface of allergens attach to special immune cells in the linings of the nasal passage.

3. These immune cells release **histamines**, *chemicals that can stimulate mucus and fluid production.*

4. Histamines cause sneezing, itchy eyes, runny nose, and other allergy symptoms.

There are many kinds of allergic reactions. Some allergies produce hives—itchy raised bumps on the skin. More serious reactions that can be life threatening include the following:

- Severe hives
- Itching or swelling of an area stung by an insect
- Difficulty breathing or swallowing
- Swelling of the tongue, mouth, or eyes
- Sharp drop in blood pressure, which can cause dizziness

Diagnosing Allergies

Sometimes you can diagnose an allergy yourself. For example, you may notice that you break out in a rash after eating certain foods. In many cases, though, tests are needed to identify the source of an allergic reaction. Blood tests and skin tests are common methods. During a skin test, small amounts of possible allergens are applied to a scratched area of the skin. If a person is allergic to any of the allergens, the skin will swell and turn red.

Treating Allergies

The simplest way to treat an allergy is to avoid the allergen that causes it. When avoidance is not possible, people with allergies can take *antihistamines*. These medicines help control allergy symptoms. Talk to your doctor or pharmacist about which medication may be most helpful for you. Some antihistamines may aggravate other medical conditions, such as heart conditions or lung problems.

Allergies range from mild to life threatening. People with long-lasting or severe allergies should seek medical attention. If you suffer from severe allergies, your doctor may prescribe antihistamines or a single, injectable dose of medicine that you carry with you at all times. If someone you know experiences a severe allergic reaction, call 911 immediately.

READING CHECK

Describe How does the production of histamines affect the body?

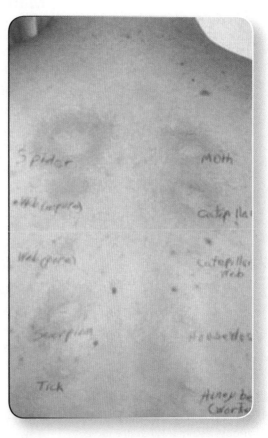

■ **Figure 25.9** Skin tests can determine which substances cause an individual to have allergic reactions. *Why are several different substances used when doing skin patch tests?*

CDC/Dr. Frank Perlman, M.A. Parsons

Asthma

Main Idea Asthma has no cure, but it can be managed.

More than 18 million adults and 7 million children in the United States have **asthma**—*an inflammatory condition in which the small airways in the lungs become narrowed, causing difficulty in breathing.* This disease can develop at any age, but about one-third of asthma sufferers are under the age of 18. Asthma can be life threatening, so those who have it must learn to manage it.

READING CHECK

Identify Name three triggers that can cause an asthma attack.

The bronchial tubes of people with asthma are highly sensitive to certain substances called *triggers*. Common asthma triggers include air pollution, pet dander, tobacco smoke, microscopic mold, pollen, and dust mites. Sometimes an asthma attack may be triggered by exercise. During an asthma attack, the muscles of the bronchial walls tighten and produce extra mucus. Symptoms may range from minor wheezing to severe difficulty in breathing.

Managing Asthma

People with asthma are usually under a doctor's care and take prescribed medications. They also help themselves with these self-management strategies:

- **Monitor the condition.** Learn to recognize the warning signs of an attack: shortness of breath, chest tightness or pain, coughing, or sneezing. Responding quickly can help prevent attacks or keep them from getting worse.

- **Manage your environment.** Avoid exposure to tobacco smoke, wash bedding frequently, and be aware of the air quality in your area.

- **Manage stress.** Stress can trigger an asthma attack. Learn relaxation and stress-management techniques to reduce your risk.

- **Take medication properly.** Medications help relieve symptoms, prevent flare-ups, and make air passages less sensitive to triggers. Many people with asthma use *bronchodilators,* or inhalers. These devices deliver medicine that relaxes and widens respiratory passages.

■ **Figure 25.10** These are some environmental conditions that trigger asthma. *What are some ways that people with asthma can manage their condition?*

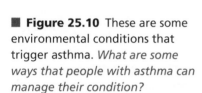

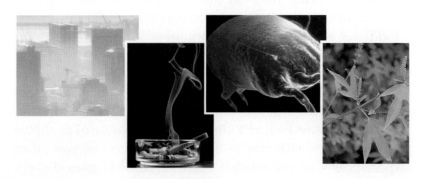

(cr)MedicalRF.com, (l)Don Bayley/Getty Images, (c)JC Zachariasen/PhotoAlto, (r)©Steven P. Lynch

Diabetes

Main Idea Type 2 diabetes is on the rise.

It's likely you know someone with **diabetes**—*a chronic disease that affects the way body cells convert sugar into energy.* It is one of the fastest-growing diseases in the United States, with almost 2 million new cases diagnosed in 2010. Young people are especially at risk today. Symptoms of diabetes include the following:

- Frequent urination
- Excessive thirst
- Unexplained weight loss
- Sudden changes in vision
- Tingling in hands or feet
- Frequent fatigue
- Sores that are slow to heal
- More infections than usual

In a person with diabetes, the pancreas produces too little or no insulin, a hormone that helps glucose from food enter body cells and provide them with energy. Some diabetics do produce enough insulin, but their cells don't **respond** normally to it. As a result, glucose builds up in the blood instead of being delivered to cells.

The only way to diagnose diabetes is through a blood test. Diabetes can be successfully managed with medication, a healthful eating plan, and regular moderate exercise. If the disease is not treated, the long-term effects include blindness, kidney failure, limb amputations, heart disease, and stroke.

Academic Vocabulary

respond *(verb):* to react in response

Type 1 Diabetes

Type 1 diabetes accounts for 5 percent of all diabetes cases in young people. It appears suddenly and progresses quickly. The body fails to produce insulin, glucose builds up in the blood, and cells don't get the energy they need. Over time, the high blood sugar level can cause damage to the eyes, kidneys, nerves, and heart.

Scientists have not yet been able to determine what causes type 1 diabetes. Some suspect an environmental trigger—for example, an unidentified virus—that stimulates an immune response. The body begins attacking itself and destroys the cells of the pancreas that produce insulin. Type 1 diabetes is thus known as an **autoimmune disease**, *a condition in which the immune system mistakenly attacks itself, targeting the cells, tissues, and organs of a person's own body.* People with type 1 diabetes must take daily doses of insulin, either through injections or through a specially attached pump.

Campaigning for Health

Fiona and Bernard's health class just learned about the increase of type 2 diabetes among teens. "I don't want our generation to be known as the diabetes generation," Fiona says, concerned.

"What can we do?" asks Bernard.

"Let's start by asking that the sodas and candy in vending machines be replaced with healthier choices," Fiona suggests.

"Sounds great, but we also need to raise awareness of the issue," says Bernard. "We need to create some posters and information sheets, and maybe a petition."

Writing Write an ending to the scenario in which Fiona and Bernard organize their campaign. Use these guidelines:

1. List the reasons why healthier food items should be offered.

2. Create a brochure explaining the risk of diabetes and the importance of healthy foods.

3. Draft a petition to be signed by other students.

Type 2 Diabetes

Type 2 diabetes accounts for 90 to 95 percent of all cases of diabetes. It typically appeared after age 40, but growing numbers of younger people—even children and teens—are developing this disease. In this form of diabetes, the body is unable to make enough insulin or to use insulin properly. Some scientists fear that type 2 diabetes will become an epidemic for two reasons: there are more older people in the population, and there are more obese and inactive young people. The increase in childhood obesity is directly linked to the increase in type 2 diabetes among children.

To help prevent type 2 diabetes, include these healthful behaviors in your life:

- **Choose low-fat, low-calorie foods.** People whose eating plans are high in fat, calories, and cholesterol have an increased risk of diabetes.

- **Participate in regular physical activity.** Being active helps control weight and lower blood cholesterol levels.

People with diabetes can live full, normal lives if they manage their condition. This includes monitoring their blood sugar levels, making healthful eating decisions, engaging in regular physical activity, and taking prescribed medications.

READING CHECK

Explain What is one reason that type 2 diabetes is increasing among young people?

Tim Fuller Photography

Arthritis

Main Idea Arthritis is a major cause of disability.

Arthritis is *a group of more than 100 different diseases that cause pain and loss of movement in the joints.* Arthritis affects people of all ages, though it is more common in older adults. The two main forms of arthritis are osteoarthritis and rheumatoid arthritis. Both can be debilitating, limiting movement in the affected joints. There is currently no cure for either type, but self-management techniques can reduce pain and improve movement.

Osteoarthritis

Half of all arthritis cases involve **osteoarthritis**—*a disease of the joints in which cartilage breaks down.* Cartilage is the strong, flexible tissue that cushions your joints. Osteoarthritis causes the cartilage to become pitted and frayed. In time, it may wear away completely, causing the bones to rub painfully against each other.

People with osteoarthritis experience aches and soreness, especially when moving. Osteoarthritis mainly affects the large, weight-bearing joints, such as the knees and hips. However, the fingers, feet, lower back, and other joints are also at risk. Several strategies can reduce your risk:

- **Control your weight.** Maintaining a healthy weight reduces stress on your joints.

- **Stay active.** Physical activity strengthens your joints.

- **Prevent sports injuries.** Warm up before exercising, participate in strength training, and use protective equipment to avoid joint injuries.

- **Protect against Lyme disease.** If left untreated, Lyme disease can result in a rare form of osteoarthritis. When walking in wooded areas, use insect repellent and wear long-sleeved shirts and pants.

■ **Figure 25.11** Staying active will help keep your joints strong. *What other healthful behaviors can help prevent arthritis?*

Rheumatoid Arthritis

Rheumatoid arthritis is *a disease characterized by the debilitating destruction of the joints due to inflammation.* It is three times more common in women than in men. Symptoms usually first appear between the ages of 20 and 50, but the disease can also affect young children. Some of the symptoms and side effects include

- joint pain, inflammation, swelling, and stiffness.
- deformed joints that can't function normally.
- possible fever, fatigue, and swollen lymph glands.

Rheumatoid arthritis is caused by an autoimmune disorder. It affects mainly the joints in the hand, foot, elbow, shoulder, neck, knee, hip, and ankle. The effects are usually *symmetrical,* meaning that both sides of the body develop the same symptoms at the same time. Treatments focus on relieving pain, reducing inflammation, and keeping the joints flexible. Treatment methods include medication, exercise, rest, joint protection, and physical and occupational therapy.

LESSON 3 ASSESSMENT

After You Read

Reviewing Facts and Vocabulary

1. What are *histamines*? What role do they play in allergies?

2. Name three strategies for managing asthma.

3. What are the two main forms of arthritis?

Thinking Critically

4. **Synthesize.** If someone has allergies, is it safer to stay indoors or to get as much fresh air as possible? Explain.

5. **Evaluate.** Many people have diabetes but are not aware of it. What makes this lack of awareness dangerous?

Applying Health Skills

6. **Practicing Healthful Behaviors.** Make a three-column chart. In the first column, list the four diseases described in this lesson. In the second column, identify risk factors for each disease. In the third column, write down actions you can take to reduce your risk for each disease.

Writing Critically

7. **Narrative.** Write a story about a teen who has one of the diseases covered in this lesson. Describe how the condition affects the teen's daily life and how he or she manages the disease.

Real Life Issues . . . .

After completing the lesson, review and analyze your response to the Real Life Issues question on page 702.

Physical and Mental Challenges

Real Life Issues ....

Dealing with a Disability. Peter was born with a physical disability that affects the way he walks. He doesn't need a wheelchair or a cane, but when he walks, he looks very different from most people. It also takes him longer to get from one place to another. Because he moves more slowly, Peter is always the last one picked for team sports. He sometimes hears people laughing at him.

Writing *What would you say to someone who laughs at Peter? In a paragraph, explain why this behavior is wrong.*

Physical Challenges

Main Idea Most physical challenges affect sight, hearing, and motor ability.

About 37 to 56 million American adults have some type of **disability**—*any physical or mental impairment that limits normal activities, including seeing, hearing, walking, or speaking.* The range of physical challenges is quite broad. However, as **Figure 25.12** on page 710 shows, most physical challenges fall into one of three categories: sight impairment, hearing impairment, or motor impairment.

Sight Impairment

More than 3 million Americans are either blind or have low vision. In the United States, about 1.3 million people are legally blind, and at least 5 million more have some degree of sight impairment that cannot be corrected with glasses or contact lenses. Sight impairment is more common among older adults, but it can affect people of all ages.

GUIDE TO READING

BIG Idea *People with physical and mental challenges deserve to be treated with dignity and respect.*

Before You Read

Create a T-Chart.
Make a two-column chart. Label one column "Physical Challenges" and the other "Mental Challenges." As you read, fill in the columns with examples and descriptions of each.

Physical Challenges	Mental Challenges

New Vocabulary

▸ disability
▸ profound deafness
▸ mental retardation
▸ Americans with Disabilities Act

Figure 25.12

Dealing with Physical Challenges

Sight, hearing, and motor impairment are examples of physical disabilities. *How has technology affected people with disabilities?*

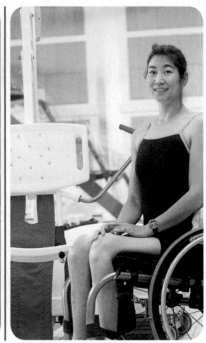

The common causes of blindness include

- **complications from diabetes,** in which high blood sugar levels lead to damage of the retina. Complications from diabetes are the leading cause of blindness.

- **macular degeneration,** a disease in which the retina degenerates. This is the main cause of blindness in people over 55.

- **glaucoma,** a disease that damages the eye's optic nerve.

- **cataracts,** a condition in which the eye's lens becomes clouded.

Blindness can also result from an injury, but disease is a much more common cause. Regular eye exams can lead to early diagnosis of many conditions and help prevent blindness or slow its progress.

READING CHECK

Identify What is the number one cause of blindness?

Hearing Impairment

Almost 30 million Americans have disabilities that affect their ability to hear. Hearing problems range from mild to severe. The most acute form is **profound deafness,** *hearing loss so severe that a person affected cannot benefit from mechanical amplification, such as a hearing aid.*

A variety of factors cause hearing impairment, including the following:

- **Heredity.** If one or both parents have hearing impairment, their child is more likely to develop it as well.

- **Injury.** An injury to the ears or head, such as a skull fracture, can cause hearing loss.

- **Disease.** Ear infections, brain tumors, measles, and other conditions can lead to hearing loss.

- **Obstruction.** Hearing loss is sometimes caused by a buildup of wax or a bone blockage in the ear.

- **Nerve damage.** Nerve damage often occurs with age, but it can also be the result of repeated exposure to loud noises, such as stereos, traffic, video games, and concerts.

Hearing loss due to loud noises is easy to prevent: wear earplugs if you're exposed to loud noise, and turn down the **volume** if you're wearing earphones while listening to music. Anyone who works around loud machinery, airplanes, or other sources of high decibel levels should wear earplugs to protect their hearing. To learn more about how your ears work and ways you can protect your hearing, see Chapter 13, Lesson 3.

We live in a noisy world, and some experts think the increase in environmental noise is why hearing loss may be occurring earlier in people's lives than it did a few decades ago. Hearing impairment can be a gradual process. If you ever notice any change in your hearing, it may be time to visit an *audiologist,* a specialist in hearing problems.

> **Academic Vocabulary**
>
> **volume** *(noun):* the degree of loudness

Motor Impairment

Tasks that are simple for most people—tying a shoe, climbing the stairs, opening a jar, lifting a glass—can be a challenge for people with a motor impairment. Motor impairments result when the body's range of motion and coordination are affected by a brain injury or a nervous system disorder.

People with motor impairments cope with physical challenges in different ways, depending on their situation. The following treatments and devices have helped many people with motor impairment adapt to their situation and lead full, active lives:

- **Physical therapy** helps people keep their joints flexible and their muscles stretched, improving their ability to move around.

- **Occupational therapy** helps people learn how to perform everyday functions so that they can lead independent lives.

- **Assistive devices** include motorized wheelchairs and special computers, as well as artificial limbs for people with limb amputations. These devices help people perform everyday tasks. People who cannot use their hands and arms can also use mouth sticks or head sticks to operate a wheelchair or send instructions to a computer.

Mental Challenges

Image Source/Getty Images

Main Idea Mental disabilities have been linked to several different causes.

READING CHECK

List What are three factors that may cause mental retardation?

One challenge that affects a person's ability to live independently is **mental retardation**. This is *a below-average intellectual ability present from birth or early childhood and associated with difficulties in learning and social adaptation.* Several factors have been found to cause mental retardation, including injury, disease, and brain abnormality. Additional factors include the following:

- **Genetic disorders** such as Down syndrome, phenylketonuria (PKU), Tay-Sachs, and Fragile X syndrome result in babies born with mental retardation.

- **Behaviors during pregnancy** can have a serious impact. Pregnant women who use alcohol or other drugs greatly increase the risk that their babies will be born with mental retardation, low birth weight, or fetal alcohol syndrome.

- **Rubella infection** during pregnancy puts the baby at risk. Immunization against rubella either during childhood or in the first three months of pregnancy reduces this risk.

- **Restricted oxygen supply** can cause mental retardation during birth. Head injury, stroke, and certain infections such as meningitis can also limit oxygen supply, causing mental retardation in older individuals.

■ **Figure 25.13** Guide dogs are trained to assist the visually and hearing impaired. *How do you think guide dogs help people with visual and hearing impairments?*

Accommodating Differences

Main Idea It is important to provide equal treatment and opportunities for people with physical and mental challenges.

In recent decades, the federal government has begun to address the difficulties of living in a society that may not meet the needs of people with disabilities. Advocacy efforts have resulted in laws and policies that address discrimination. These policies are based on the following principles:

- Public transportation vehicles and building entrances must be wheelchair accessible, so that people with motor impairments can readily participate in business and social activities.

- People should be evaluated on the basis of individual merit, not on assumptions about disabilities.

- People with disabilities, to the extent they are able, should have the same opportunities as everyone else.

In 1990, the U.S. government passed the **Americans with Disabilities Act** (ADA), *a law prohibiting discrimination against people with physical or mental disabilities in the workplace, transportation, public accommodations, and telecommunications.* The ADA includes the following provisions:

- **Employers** with 15 or more employees must give qualified individuals with disabilities an equal opportunity to benefit from employment-related opportunities.

- **State and local governments** must provide accessible entrances in buildings and communicate effectively with people who have hearing, vision, or speech disabilities.

- **Telephone companies** must set up telecommunications relay services (TRS) that allow callers with hearing and speech challenges to communicate through an assistant.

In 1998, the government passed another law, the Workforce Investment Act. This law ensures that any information posted to a Web site by a government agency must be accessible by those who are disabled.

■ **Figure 25.14** Federal law requires that accommodations be made for people with disabilities. *Can you name other ways that our society has helped people with disabilities?*

LESSON 4 ASSESSMENT

After You Read

Reviewing Facts and Vocabulary

1. What are three common causes of blindness?

2. What is an assistive device?

3. Is mental retardation preventable? Explain.

Thinking Critically

4. **Analyze.** What are some challenges that someone with a sight or hearing impairment might have commuting to work each day?

5. **Evaluate.** Why is it important to make buildings and services accessible to people with physical and mental challenges?

Applying Health Skills

6. **Advocacy.** Create a flyer that promotes better understanding of physical and mental challenges and empathy for people with these disabilities. Include appropriate information and statistics.

Writing Critically

7. **Expository.** Write about the accommodations your school has made to assist people with physical or mental challenges. Describe these accommodations and explain whether your school needs to make any additional accommodations.

Real Life Issues ....

After completing the lesson, review and analyze your response to the Real Life Issues question on page 709.

Hands-On HEALTH

Activity A Family Letter

Your teacher will present a set of index cards with the names of noncommunicable diseases. You will conduct research on one of these diseases and develop a plan for reducing risks of getting the disease. Then, you will write a letter to your family persuading them to make healthy choices to prevent getting the disease.

What You'll Need
- print and online resources
- paper and pens or pencils
- envelopes (optional)

What You'll Do

Step 1

Select one card from your teacher and conduct research on the signs, symptoms, risk factors, and treatment for the disease printed on the card.

Step 2

Based on your findings, list at least four healthful behaviors that can reduce the risk of getting this disease.

Step 3

Create a health-enhancing action plan to reduce your risks for this disease.

Apply and Conclude

Write a letter to your family suggesting specific healthful behaviors all of you can adopt to reduce the risks associated with this disease.

Checklist: Practicing Healthful Behaviors

- ☑ Identification of protective behaviors
- ☑ Steps demonstrating healthful habits
- ☑ Knowledge of healthful behaviors, habits, and techniques

714 Chapter 25 Hands-On Health

Cardiovascular Disease

Key Concepts

▶ Cardiovascular disease can begin developing in your teens.
▶ You can reduce your risk for CVD by avoiding tobacco, alcohol, and other drugs; maintaining a healthy weight; and getting regular physical activity.
▶ Some risk factors for CVD, such as heredity, cannot be avoided.

Vocabulary

▶ noncommunicable disease, cardiovascular disease (p. 688)
▶ hypertension, atherosclerosis, arteriosclerosis (p. 689)
▶ angina pectoris (p. 690)
▶ arrhythmias (p. 690)
▶ stroke (p. 692)

Cancer

Key Concepts

▶ Cancer is the uncontrollable growth of abnormal cells.
▶ Avoiding carcinogens like tobacco and ultraviolet radiation can reduce your risk for some kinds of cancers.
▶ Many cancers can be treated successfully if detected early.

Vocabulary

▶ cancer, tumor (p. 695)
▶ benign, malignant, metastasis (p. 696)
▶ carcinogen (p. 697)
▶ biopsy (p. 700)
▶ remission (p. 701)

Allergies, Asthma, Diabetes, and Arthritis

Key Concepts

▶ Allergic reactions are caused by allergens.
▶ Taking proper medication and practicing management techniques can reduce the number and severity of asthma attacks.
▶ Type 2 diabetes is strongly associated with obesity.
▶ Arthritis causes pain and loss of movement in the joints.

Vocabulary

▶ allergy (p. 702)
▶ histamines (p. 703)
▶ asthma (p. 704)
▶ diabetes, autoimmune disease (p. 705)
▶ arthritis, osteoarthritis (p. 707)
▶ rheumatoid arthritis (p. 708)

Physical and Mental Challenges

Key Concepts

▶ Sight, hearing, and motor impairments are common physical disabilities.
▶ The Americans with Disabilities Act provides accommodation for people with physical and mental disabilities.

Vocabulary

▶ disability (p. 709)
▶ profound deafness (p. 710)
▶ mental retardation (p. 712)
▶ Americans with Disabilities Act (p. 713)

LESSON 1

Vocabulary Review

Correct the sentences below by replacing the italicized term with the correct vocabulary term.

1. *Heart attack* is an acute injury in which blood flow to the brain is interrupted.

2. High blood pressure is also known as *atherosclerosis*.

3. A disease that affects the heart or blood vessels is called a *noncommunicable disease*.

Understanding Key Concepts

After reading the question or statement, select the correct answer.

4. Which of the following statements is true about stroke?
   a. A stroke can cause paralysis.
   b. A stroke is an acute injury that affects the liver.
   c. During a stroke, blood flow to the brain increases.
   d. During a stroke, the brain gets too much oxygen.

5. Which of the following statements is *not* true about tobacco use?
   a. About 20 percent of deaths from cardio-vascular disease are smoking related.
   b. People who smoke less than a pack a day are generally safe from cardiovascular disease.
   c. Cardiovascular disease can be caused by exposure to secondhand smoke.
   d. For teens, tobacco use is the number one risk factor for cardiovascular disease.

Thinking Critically

After reading the question or statement, write a short answer using complete sentences.

6. **Explain.** What is the difference between a communicable disease and a noncommunicable disease?

7. **Describe.** How can a high cholesterol level cause atherosclerosis?

8. **Analyze.** What happens during congestive heart failure?

9. **Explain.** Why is it important to learn about cardiovascular disease as a teen, rather than waiting until you are older?

LESSON 2

Vocabulary Review

Use the vocabulary terms listed on page 715 to complete the following statements.

10. A(n) _____ is an abnormal mass of tissue that has no natural role in the body.

11. Cancer-causing substances are called _____.

12. During a(n) _____, a doctor removes a small piece of tissue for examination.

Understanding Key Concepts

After reading the question or statement, select the correct answer.

13. Which of the following is true about malignant tumors?
    a. They are inconvenient but harmless.
    b. They stay in their original location.
    c. They travel to other parts of the body via the blood or lymph.
    d. They occur only in older adults.

14. Which of the following statements is true about cancer?
    a. Smoking is the leading cause of cancer deaths in the United States.
    b. Cancer is a hereditary disease.
    c. People who live in moderate or cool climates have a low risk for cancer.
    d. Metastasis can be stopped with a healthful diet.

15. What percentage of all cancer deaths are caused by dietary risk factors?
    a. 10   c. 30
    b. 20   d. 40

Thinking Critically

After reading the question or statement, write a short answer using complete sentences.

16. **Describe.** What happens during metastasis?

17. **Explain.** Why is it important to pay attention to the moles on your skin?

18. **Identify.** What are three cancers that can be detected through self-examination?

19. **Evaluate.** What is the connection between abstaining from sexual activity and reducing cancer risk?

LESSON 3

Vocabulary Review

Correct the sentences below by replacing the italicized term with the correct vocabulary term.

20. Chemicals that can stimulate mucus and fluid production are called *allergens*.

21. *Arthritis* affects the way body cells convert sugar into energy.

22. *Allergy* is a condition in which the airways in the lungs become narrowed.

Understanding Key Concepts

After reading the question or statement, select the correct answer.

23. Severe hives and difficulty swallowing are symptoms of a serious
    a. asthma attack.
    b. diabetic seizure.
    c. allergic reaction.
    d. arthritic condition.

24. The only way to diagnose diabetes is by
    a. watching for the key symptoms.
    b. undergoing a biopsy procedure.
    c. receiving an eye exam.
    d. getting a blood test.

25. The main areas affected by osteoarthritis are
    a. internal organs, such as the liver.
    b. weight-bearing joints, such as the knees.
    c. the neck and shoulders.
    d. the sinuses.

Thinking Critically

After reading the question or statement, write a short answer using complete sentences.

26. **Identify.** What are four strategies for managing asthma?

27. **Explain.** Why are some scientists concerned that type 2 diabetes will become an epidemic?

28. **Synthesize.** How can your family reduce asthma triggers in your home?

LESSON 4

Vocabulary Review

Use the vocabulary terms listed on page 715 to complete the following statements.

29. _____ is hearing loss so severe that hearing aids have no effect.

30. The _____ is a law that prohibits discrimination against people with disabilities.

Assessment

Understanding Key Concepts

After reading the question or statement, select the correct answer.

31. Glaucoma and diabetes complications are two common causes of
    a. deafness.
    b. mental illness.
    c. blindness.
    d. paralysis.

32. What percentage of Americans have some type of disability?
    a. 5
    b. 10
    c. 20
    d. 40

33. Advocates for people with physical and mental challenges believe that
    a. people are defined by their disabilities.
    b. people with disabilities should have different opportunities.
    c. everyone must learn to read braille.
    d. buses and building entrances should be wheelchair accessible.

Thinking Critically

After reading the question or statement, write a short answer using complete sentences.

34. **Identify.** What are the three main categories of physical challenges?

35. **Analyze.** What is the role of heredity in hearing impairment?

36. **Explain.** What are three ways that assistive devices help people with motor impairments?

37. **Evaluate.** Discuss the impact of the Americans with Disabilities Act. How does it affect the lives of people with physical and mental challenges?

Technology · PROJECT-BASED ASSESSMENT

Reducing Risk

Background

Scientists have identified behaviors and treatments that decrease the risk of noncommunicable diseases. While some risk factors for these diseases are related to heredity, gender, and age, many other factors can be modified to reduce disease risk.

Task

Choose one of the diseases discussed in the chapter, research it, and develop a multimedia presentation illustrating the nature of the disease.

Audience

Students in your class and adults in the community

Purpose

Inform people about the nature, risk factors, and treatment of a particular noncommunicable disease.

Procedure

1. Choose a noncommunicable disease discussed in the chapter.

2. Conduct an Internet search to learn more about the disease.

3. Find illustrations and video clips showing the effect of the disease.

4. Include information on positive, preventive measures that lower risks related to the disease. Be sure to also include recent medical advances in the diagnosis and treatment of the disease.

5. Collaborate as a group to create a multimedia presentation incorporating all these aspects.

6. Present your presentation to your class.

Math Practice

Interpret Graphs. Frequent sunburns can lead to melanoma, a deadly type of skin cancer. The bar graph below shows the percentage of young people ages 11 to 18 who reported getting sunburned. Use the graph to answer the

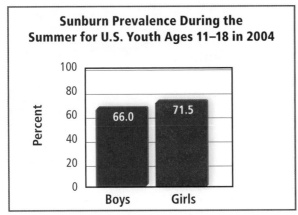

Sunburn Prevalence During the Summer for U.S. Youth Ages 11–18 in 2004

Adapted from "Cancer Statistics Presentation 2007," American Cancer Society, Inc., 2007.

questions that follow.

1. Which group makes up the greatest percentage of all youth surveyed?
   A. Girls who got a sunburn that summer
   B. Girls who did not get a sunburn that summer
   C. Boys who got a sunburn that summer
   D. Boys who did not get a sunburn that summer

2. What percentage of boys did *not* get a sunburn that summer?
   A. 28.5%
   B. 34.0%
   C. 66.0%
   D. 71.5%

3. If you examined a representative sample that consisted of 500 boys, how many would have gotten a sunburn that summer?

Reading/Writing Practice

Understand and Apply. Read the passage below and then answer the questions.

> It's the last home game of the season, and Lincoln High School's basketball team is headed for another victory. The team has compiled its best-ever record this season. Some people attribute this success to Shawn, the assistant manager.
>
> Shawn is mentally retarded. His impairment prevents him from being a regular player, but he loves helping manage the team and practicing with the players. He's so devoted to the team that players say he inspires them to play harder.
>
> As the clock runs down, cheers fill the gymnasium. Lincoln wins! At the team's annual banquet, the coach presents Shawn with a special Most Valuable Player award for his contribution to the team.

1. How would you best describe Shawn's role on the basketball team?
   A. He helps organize the equipment.
   B. He is the team's point guard.
   C. His enthusiasm inspires the players to give their best effort on the court.
   D. He practices but doesn't play.

2. What message does Shawn's MVP award send to other students and faculty?
   A. Mental retardation is a barrier to athletic achievement.
   B. Teens with mental disabilities can make valuable contributions.
   C. It is important to treat Shawn differently.
   D. Other teams should ask Shawn to be their assistant manager.

3. Describe the effect that Shawn's success might have on other students with physical or mental challenges.

National Education Standards

Math: Measurement and Data, Data Analysis
Language Arts: LACC.910.RT.1, LACC.910.RL.2.4

Public Health Specialist

Public health specialists help educate communities and organizations on how to reduce communicable diseases, occupational diseases, and foodborne illnesses. They also run wellness programs. Other responsibilities of a public health specialist include speaking to groups about disease prevention, monitoring disease trends, and attending professional meetings and conferences.

Most public health specialists need a master's degree to stay competitive in the field. By taking classes in biology, statistics, and economics, you can start acquiring the background and skills necessary for this career.

Dietitian and Nutritionist

Dietitians and nutritionists promote healthful eating habits and suggest dietary changes in order to prevent and treat illness. They supervise the preparation of meals and coordinate programs to educate people about nutrition. Dietitians and nutritionists need at least a bachelor's degree. Some states require additional licenses or certifications. A variety of classes, including nutrition, biology, and biochemistry, will help you prepare for this field.

Nurse

Nurses help treat ill or injured patients in a wide variety of settings. Although many nurses work in hospitals, many others work in clinics, private homes, and branches of the U.S. military. In the United States and many other countries, there is a rapidly growing demand for qualified nurses. More and more men are joining the nursing profession.

To legally practice nursing, an individual must become a registered nurse (RN). This process includes earning a two- or four-year college degree and then passing a licensing exam. To help prepare for a career in nursing, take classes in biology, psychology, algebra, and anatomy.

CAREER SPOTLIGHT

Allergist

Pratibha Vakharia always knew that she was destined to be a doctor. In high school, she took biology and other science classes. During her medical training, Dr. Vakharia worked with children who had allergies and asthma—an experience that led her to open a private practice as an allergist.

Q. What are the things you love about your job?

A. *I get to treat patients of all ages. I treat everything from food allergies to drug reactions to insect stings. I love the process of finding the culprit behind an allergic reaction.*

Q. What surprised you the most about being an allergist?

A. *Many days, I'm treating things other than the allergy. An allergist develops a bond with the patients where they can discuss other concerns, like stress or family issues.*

Q. What are some advantages of being an allergist?

A. *The field has a lot of variety and flexibility compared to surgical branches. It is easier to set up a balance between family and professional life.*

Activity Beyond the Classroom

Writing Communicate with Health Professionals. Visit or call a health professional in your school or community. This professional could be a school nurse, a family doctor, an employee of your local health department, or another expert. Ask this person to describe the top three concerns about the health of teens in your area. Ask for specific details and examples. Based on what you learn, create a brochure, poster, or blog to share this information with your classmates. Feel free to do your own research on these health concerns, and add your findings to your report.

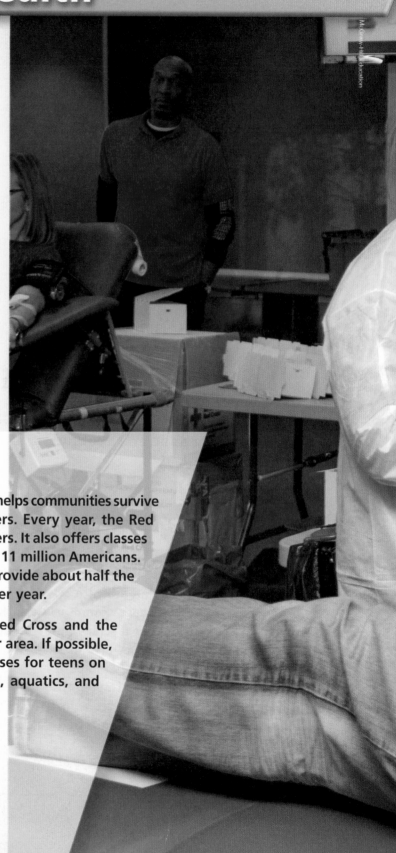

UNIT 9 Safety and Environmental Health

UNIT PROJECT

Lifesaving Services

Using Visuals The American Red Cross helps communities survive weather emergencies and natural disasters. Every year, the Red Cross responds to more than 70,000 disasters. It also offers classes in first aid, CPR, and water safety to about 11 million Americans. Blood drives conducted by the Red Cross provide about half the nation's blood supply, or 6 million pints per year.

Get Involved. Learn more about the Red Cross and the services this organization provides in your area. If possible, arrange to take one of the Red Cross classes for teens on subjects such as child care, first aid, CPR, aquatics, and water safety.

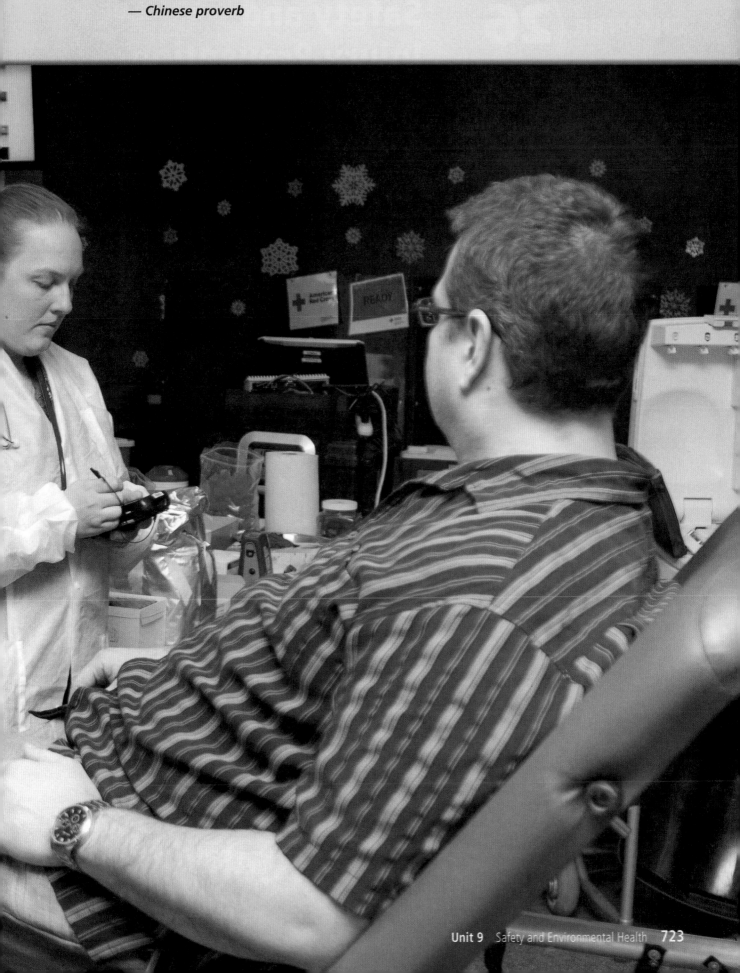

"Diligence is a priceless treasure; prudence a protective charm."
— Chinese proverb

CHAPTER 26

Safety and Injury Prevention

Lesson 1

Personal Safety and Protection

BIG Idea *Learning basic safety precautions can help you avoid threatening or harmful situations.*

Lesson 2

Safety at Home and in Your Community

BIG Idea *Reducing the potential for accidents can help you stay safe at home and at work.*

Lesson 3

Outdoor Safety

BIG Idea *Common sense and caution can minimize the risk of accidental injuries during outdoor activities.*

Lesson 4

Safety on the Road

BIG Idea *Drivers, pedestrians, and others on the road need to follow rules to stay safe.*

Activating Prior Knowledge

Using Visuals Look at the photo on this page. What action do you see these teens taking to protect their health? On a sheet of paper, brainstorm a list of other actions you could take to protect your health and safety while participating in recreational activities.

Chapter Launchers

Health in Action

Discuss the **BIG** Ideas

Think about how you would answer these questions:

▶ What basic precautions do you take in your daily life?

▶ How do you protect your safety at home?

▶ How do you stay safe outdoors and on the road?

Assess Your Health

Read each statement. On a separate sheet of paper, write "yes," "sometimes," or "no" based on your typical behavior.

1. I avoid walking alone at night or in isolated areas.

2. I always let my family know where I am going, when I will be back, and I call them if my plans change.

3. I keep personal information private when I am online.

4. I tell my parents about friends I meet online, the same way I talk about friends at school.

5. I avoid telling others when I am home alone.

A "yes" response shows that you practice healthy behaviors. "Sometimes" indicates that you should analyze and possibly modify your behavior. A "no" response means that you should modify the behavior.

GUIDE TO READING

BIG Idea *Learning basic safety precautions can help you avoid threatening or harmful situations.*

Before You Read

Create a T-Chart. Make a chart with two columns labeled "Personal Safety" and "Internet Safety." As you read, fill in each column with information about types of risks and how to avoid them.

Personal Safety	Internet Safety

New Vocabulary

▶ personal safety
▶ self-defense
▶ cyberbullying

Personal Safety and Protection

Real Life Issues ....

Safety First. In 2011, 20 percent of students in grades 9–12 were bullied and 16 percent of students were electronically bullied.

> **70% of students say they have seen bullying in their schools.**

> **Only about 20 to 30% of students who are bullied report it to an adult.**

Source: U.S. Dept. of Health and Human Services, www.stopbullying.gov

Writing *Write a paragraph describing safety strategies that could reduce your chances of becoming a victim of a crime.*

Safety Strategies

Main Idea The key to personal safety is learning how to recognize and avoid dangerous situations.

Did you know that teens are the victims of more violent crimes than any other age group? Teens are more likely than children to go out at night, but they are less likely than adults to protect their **personal safety**—*the steps you take to prevent yourself from becoming the victim of crime.*

People living in urban areas report the highest rates of violent crime. However, crime can occur in any neighborhood and among any ethnic or socioeconomic group. About half of all violent crime occurs within one mile of a victim's home, and many victims know their attackers.

To reduce your risk of becoming a crime victim, always be aware of your surroundings and take precautions to protect yourself and your belongings. Whenever you leave your home, keep the following tips in mind.

- If you carry a cell phone, make sure it's easy to get to. Remember that 911 will connect you with emergency services anywhere in the United States.
- Avoid walking alone at night or in isolated areas, such as alleys or parks. Stick to brightly lit, well-traveled streets.
- Walk briskly and confidently. Wear comfortable shoes so that you can move quickly.
- Carry your wallet or purse in a place that makes it difficult to grab. Avoid openly displaying expensive jewelry, electronics, or anything that would attract a thief.
- If you drive, park your car in a well-lit area and lock it. Before getting in, check to make sure no one is inside, and lock the doors as soon as you get in.
- Never hitchhike or give a ride to anyone you do not know well. Keep in mind that even someone you've met before could be dangerous.
- Get on and off public transportation in busy, well-lit areas. Sit near the driver or with a group of people.
- Know the locations of nearby public places where you can seek help if you need it.
- Let your family know where you're going and when you'll be back. Call them if your plans change.

Learning to Protect Yourself

One way that you can protect yourself from crime is to avoid the places where it is likely to occur. Be aware of what's happening around you, even when you are in familiar places. If you cannot avoid a dangerous situation, you can do the next best thing: know how to protect yourself. **Self-defense** includes *any strategy for protecting yourself from harm.* One self-defense strategy is to project a strong, confident image. Criminals are more likely to attack those who look vulnerable, confused, or inattentive. Show confidence by holding your head high and walking with a deliberate stride.

FITNESS ZONE

I'm really careful to avoid exercising outside after dark. When I go for a walk with my mom in the evenings, we wear reflective clothing so other people can see us, and we use flashlights to see where we're going. We also walk against traffic so we can see what is coming toward us. For more physical activity ideas, visit the Fitness Handbook and Fitness Zone sites in ConnectEd.

■ **Figure 26.1** Walking with groups can help protect you from being a victim of crime. *Why does being in a group offer protection?*

If you think you are being followed in a public place, let the stalker know that you are aware of his or her presence. Try changing directions or crossing the street. If necessary, seek help from someone nearby or enter a business that's open. If you are attacked or about to be attacked, do whatever is necessary to escape, such as running, yelling, or kicking. Shout "fire" instead of "help"—it's more likely to get a response.

Self-defense classes can teach you additional strategies for protecting yourself. When you hear "self-defense," you may think of martial arts fighting, and some classes do teach these skills. However, self-defense classes can also teach you how to size up a situation, figure out what to do, and catch your attacker off-guard. Most important, these classes can give you the confidence you need to defeat an attacker.

Staying Safe Online

Main Idea Teens need to protect themselves online.

The Internet is a useful resource, but it can also be a dangerous place. The hazards you can **encounter** range from upsetting situations, like being insulted in a chat room, to physical threats, such as Internet predators.

When you're online, you need to know how to protect yourself. Here are a few precautions to take when you're online:

- **Keep your identity private.** Avoid posting personal information in any public space. This includes your full name, address, phone number, financial information, passwords, the name of your school, and anything else a stranger could use to track you down in the real world.

- **Keep online relationships online.** Agreeing to meet in person with someone you've met online can be risky.

READING CHECK

Cause and Effect
Give two examples of behaviors that can help you avoid a dangerous situation.

Academic Vocabulary

encounter *(verb):* to experience

■ **Figure 26.2** Self-defense classes can boost your confidence and help you take charge of your own safety. *What other strategies can you use to protect yourself?*

Health Skills — Activity

Decision Making

Meeting a Friend Online

Lately, Marisa has been spending time in a chat room for teens who share her hobby, photography. She's also posted some of her nature photos online. Marisa has met some interesting people in the chat room. One is a guy named Craig, who loves her photos and says she's a talented photographer. He's asked her to post photos of herself and any

she's taken at her school and in her neighborhood. Marisa enjoys Craig's compliments, but she's not sure if she should share photos that show personal details, such as what school she goes to or the area she lives in. What should she do?

Writing Use the decision-making process to help Marisa decide how to respond to Craig in a way that protects her safety.

1. State the situation.
2. List the options.
3. Weigh the possible outcomes.
4. Consider values.
5. Make a decision and act.
6. Evaluate the decision.

- **Don't respond to inappropriate messages.** If anyone sends you a message that makes you feel uncomfortable for any reason, tell a parent or other trusted adult.

- **Let your parents or guardians know what you're doing online.** Tell them about the people you meet online, the same way you'd talk to them about your friends in the real world.

Coping with Cyberbullies

About 16 percent of teens say they have experienced *cruel or hurtful online contact*, or **cyberbullying**. Such contact can come from people you know or from strangers. It can range from immature and annoying to threatening and scary.

To avoid becoming a target of cyberbullying, be careful how you communicate online. When you use a Web site, learn and follow its rules for postings. Also, be careful how you word your messages. What may be a joke to you may come across as an attack or insult to someone else. Avoid getting into "flame wars," trading insults back and forth.

If you receive hurtful messages, don't respond to them. Cyberbullies are often looking for attention, and if you don't react, they'll go find someone who will. If the bullying continues, however, seek help from a trusted adult. Save the messages as evidence and contact your Internet service provider (ISP). It may be possible to block all future communications from the cyberbully. If any actual crime has been committed, such as making violent threats, contact the police.

READING CHECK

List What are three types of information you should keep private while online?

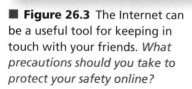

■ **Figure 26.3** The Internet can be a useful tool for keeping in touch with your friends. *What precautions should you take to protect your safety online?*

Avoiding Internet Predators

Internet predators use online contact to build up trust so they can lure victims into a face-to-face meeting. To avoid falling victim to Internet predators, follow the general guidelines for online safety. Keep your identity private, and don't agree to meet in person with someone you've met online. If you ever find yourself in an online conversation that makes you feel uncomfortable or threatened—for any reason—log off and let a trusted adult know about the incident.

LESSON **1** ASSESSMENT

 After You Read

Reviewing Facts and Vocabulary

1. What steps can you take to protect yourself from an attack when entering or leaving a car?

2. Name two threats you may encounter on the Internet.

3. How can you avoid becoming the target of a cyberbully?

Thinking Critically

4. **Evaluate.** Why is it important to avoid dangerous situations, even if you know how to defend yourself?

5. **Synthesize.** Gina is walking home from school when she notices someone is following her. What could she do to protect herself?

Applying Health Skills

6. **Communication Skills.** Suppose you have been posting on a message board about current events. The group is debating a political issue that you have strong opinions about. Write a message you could post that expresses your opinions in a way that is respectful toward those who disagree with you.

Writing Critically

7. **Creative.** Write lyrics for a pop song or rap about personal safety. Choose a topic in this lesson as the basis for your lyrics.

Real Life Issues ....

After completing the lesson, review and analyze your response to the Real Life Issues question on page 726.

Safety at Home and in Your Community

Real Life Issues ··························

Fire Safety. Lucius and his family are moving into a new house. As he's examining his bedroom on the second floor, his dad comes in and looks out the window. "We'll need to find a place to store a ladder," he says. "This window's your

emergency exit in case of fire. Come to think of it, we need to develop a fire safety plan for the whole house."

> **Writing** *What do you think is involved in developing a fire safety plan? Why is it important to have such a plan?*

The Accident Chain

> **Main Idea** Many accidental injuries are preventable.

Every year, more than 20 million children and teens require medical attention or face restricted activity due to **unintentional injuries**. These are *injuries resulting from an unexpected event*. You can prevent these injuries by breaking the **accident chain**, *a sequence of events that leads to an unintentional injury*. **Figure 26.4** on page 732 shows how stopping just one of the events in the chain can prevent the injury.

Keeping Your Home Safe

> **Main Idea** Safety precautions can prevent injuries at home.

Accidents in the home are one of the top causes of injury and death in the United States. Common types of household accidents include fires, falls, and poisonings. You can reduce the risk of these accidents by taking safety precautions.

GUIDE TO READING 📖

BIG Idea *Reducing the potential for accidents can help you stay safe at home and at work.*

Before You Read

Organize Information. Use a T-chart to organize the information in this lesson. On one side, list causes of accidental injuries. On the other side, list safety precautions that can prevent them.

Causes	Safety Precautions

New Vocabulary

▶ unintentional injuries
▶ accident chain
▶ fire extinguisher
▶ smoke alarm
▶ Occupational Safety and Health Administration (OSHA)

Review Vocabulary

▶ carbon monoxide (Ch.20, L.1)
▶ peer mediation (Ch.9, L.2)

Figure 26.4 The Accident Chain

Breaking any of the links in this chain can prevent the accident and the resulting injury.

An Unsafe Situation Mark's alarm clock didn't go off this morning. As a result, he overslept and has to rush to get ready for school.

An Unsafe Habit Mark often leaves his books on the stairs.

An Unsafe Action Mark hurries down the stairs without watching where he's going.

The Accident Mark trips over his books and falls down the stairs.

The Consequences Mark lands on his wrist and sprains it. He's also late for school.

Preventing Fires

Common causes of household fires include burning candles and incense, smoking, kitchen fires, and faulty electrical wiring. To prevent fires in your home, follow these precautions:

- Keep matches, lighters, and candles away from children. Don't leave burning candles unattended.
- Make sure that smokers extinguish cigarettes completely, and that no one smokes in bed.
- Don't leave cooking food unattended. Clean stoves and ovens to prevent grease buildup, which can catch fire.
- Follow the operating instructions for using space heaters.

If a fire does occur, two lifesaving devices can help you escape without harm:

- **Fire extinguisher**, *a portable device for putting out small fires.* Keep an all-purpose fire extinguisher in your kitchen—one that is approved for flammable materials, flammable liquids, and electrical fires. Make sure that everyone in the house knows how to use it.
- **Smoke alarm**, *a device that produces a loud warning noise in the presence of smoke.* Having working smoke alarms in your home more than doubles your chances of surviving a house fire. Every home should have a smoke alarm on each floor, near the kitchen and bedrooms.

It's also imortant to plan an escape route ahead of time. Identify an escape path from every room of your home and a designated spot to meet up with your family after you get out.

■ Figure 26.5 Test your smoke alarms once a month, and change the batteries twice a year. *How do smoke alarms protect your safety?*

When you are escaping from a fire, stay close to the ground so that you can crawl under the smoke. If your clothes catch fire, stop, drop, and roll to put out the flames.

Staying Safe with Electricity

Because wiring problems are a common source of house fires, knowing about electrical safety can help prevent electrical fires as well as electric shock. Here are some safety tips to follow:

- Avoid overloading your electrical system.
- Inspect electrical cords regularly. If you find any worn or exposed wiring, unplug the appliance *immediately* and don't use it anymore.
- Make sure extension cords are properly rated for their intended use and have polarized (three-prong) plugs.
- Do not run electrical cords under rugs or behind baseboards. Don't let furniture sit on cords, and don't attach cords to walls using nails or staples.
- Avoid using an electrical appliance near water, and *never* reach into water to retrieve a dropped appliance without first unplugging it.
- In homes with small children, cover unused outlets with safety caps.

Preventing Falls

Falls are responsible for about half of all accidental deaths in the home. To reduce the risk of injury from falls, take precautions in these areas of the home:

- **Stairs.** Keep stairways well lit, in good repair, and free of clutter. Staircases should have sturdy handrails, and all stair coverings should be securely fastened down. Never put small rugs at the foot of a staircase.

- **Bathrooms.** Put nonskid mats or strips in the tub or shower. Keep a night-light in the bathroom.
- **Windows.** If there are small children in the home, install window guards on the upper floors. However, make sure the windows can be opened completely in case of a fire.
- **Kitchens.** Keep the floor clean, and mop up spills promptly. Use a step stool to get things down from high places.
- **Living areas.** Keep the floor clear of clutter. Use nonskid rugs or place nonskid mats under rugs. Keep phone and electrical cords out of the flow of traffic.

Preventing Poisonings

Many common household items can be harmful or even fatal if swallowed. **Figure 26.6** shows some poisonous products that might be found in different parts of the home. To prevent poisonings, follow these tips:

- **Store products safely.** Store all medications and other hazardous substances in childproof containers, and keep them out of the reach of children. Put locks or safety latches on cabinets where dangerous chemicals are stored. Discard medicines that are past their expiration date. Don't store household chemicals near pet food or water dishes, and clean up spills promptly.

Figure 26.6 **Common Household Poisons**

Room	Poisons that might be found there
Bathroom	• Medications of any kind • Mouthwash • Hair spray • Toilet bowl cleaner • Astringents (such as rubbing alcohol) • Antiseptics
Bedroom	• Mothballs and crystals • Perfumes and colognes • Nail polish remover and nail glue remover
Kitchen	• Cleaning products • Rust removers • Drain cleaners • Furniture polish or floor wax • Metal polishes
Living room	• Lead paint (especially if chipped or peeling) • Poisonous houseplants
Garage or shed	• Pesticides • Fertilizer • Pool cleaners

- **Pay attention to labels.** Unless directed by a doctor, never take more of a drug than the label recommends. Check with your doctor if you are taking two or more drugs. Follow instructions for using household chemicals. Mixing chemicals can result in dangerous fumes, explosions, home fires, and burns. Also, fuel-burning appliances, such as barbecue grills or kerosene lamps, must be properly vented to prevent carbon monoxide poisoning.

Using Computers Safely

Using a computer for a long period of time can lead to eyestrain and sore muscles. It can also cause injuries to the wrists, hands, or arms. To reduce these problems

- adjust your position from time to time.
- stretch your hands, arms, and body.
- stand up and walk around for a few minutes every hour.
- sit in a "neutral body position," a comfortable posture in which your joints are naturally aligned.
- Blink your eyes to moisten them and reduce eyestrain.

Handling Firearms Safely

Nearly half the households in the United States contain one or more guns. Gun accidents result in an estimated 650 deaths and 15,000 injuries per year. Most gunshot injuries in the home occur when a child finds a loaded gun. Children need to know that guns are dangerous and can kill people. Instruct them never to touch a gun and to leave the area and tell an adult if they find one.

Adults should also take precautions with firearms. When handling a gun, always assume that it is loaded. Never point a gun at anyone. Add a trigger lock, and keep your finger off the trigger except when firing. Store guns unloaded and in a locked cabinet, lock ammunition away separately, and keep the keys where children can't find or get to them.

Guarding Against Intruders

Accidents aren't the only threat to the safety of your home. There is also the risk that an intruder could break into your home. To keep intruders out, follow these guidelines:

- Keep your doors and windows locked. Deadbolt locks are the most secure kind. If doors or windows are damaged, repair them promptly. Don't hide a spare key outside the house. Instead, give a key to a neighbor you trust.
- Use a peephole to identify people who come to the door. Don't open the door to a stranger. Never tell people that you're home alone.

■ **Figure 26.7** Setting up your computer workstation correctly will reduce eyestrain, fatigue, headaches, and injury. *What other precautions can you take when working on your home computer?*

READING CHECK

Describe What are three ways that you can stay safe at home?

Real World CONNECTION

Accidents and Unintentional Injuries

The graph below compares the top five causes of nonfatal unintentional injuries to Americans between the ages of 15 and 19. Study the graph, then answer the questions that follow.

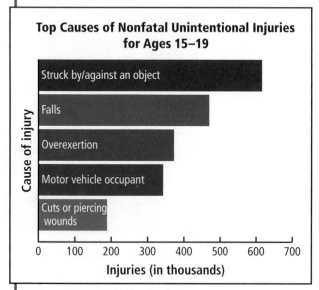

Top Causes of Nonfatal Unintentional Injuries for Ages 15–19

Source: National Electronic Injury Surveillance System-All Injury Program (NEISS-AIP) from the Consumer Product Safety Commission, 2009

Activity Mathematics

Assume the total number of nonfatal unintentional injuries for this age group was 1.9 million.

1. What percentage of these injuries resulted from falls?

2. What percentage of all injuries do the five causes listed here account for?

3. **Writing** Identify at least three steps you could take to reduce your risk of accidents and unintentional injuries.

Concept Numbers and Operations: Percents A percent can be used to express the relationship between two numbers (A and B). To calcuate what percent of B is represented by A, use this formula: $(A \div B) \times 100$. For example, $(2 \div 8) \times 100 = 25$. Therefore, 2 is 25 percent of 8.

- Make sure your answering machine does not tell callers you are away from home.
- If you come home and see something suspicious, such as an unfamiliar car parked in your driveway or a window that's been forced open, don't go inside. Instead, call the police from a neighbor's house.

Keeping Your Community Safe

Main Idea You can work with others to protect your safety at school, at work, and in your community.

You have a right to be safe everywhere you go—at school, at work, and in your community. Many communities are taking the following steps to make neighborhoods safer:

- **Increased police presence.** Putting more police officers on the streets can reduce crime by as much as 15 percent.

- **Neighborhood Watch programs.** Through these programs, citizens watch for suspicious activity and report it to the police.

- **After-school programs.** These programs give students a place to go during after-school hours, when many crimes are committed. Keeping students at school or at a community center makes them less likely to commit crimes and less likely to become victims.

- **Improved lighting in public areas.** Better lighting can discourage crime by making it harder to commit crimes under cover of darkness.

Safety at School

Violence in schools can include fights between students, bullying, gang activity, and the presence of weapons. Other problems that can make school an unsafe environment include vandalism and alcohol and drug use. Eliminating these problems takes a joint effort by school staff, students, and parents.

- **School staff** can develop security procedures, such as hiring security guards, working cooperatively with the police, or using metal detectors to keep weapons out. Schools can also put disciplinary policies in place to deal with offenders. Some schools have adopted "zero-tolerance" policies, which means that a student can be expelled or **suspended** for a single offense.

- **Students** can develop peer mediation programs to help settle conflicts. They can report crimes or other suspicious activities to school staff. They can also clean up graffiti, lead anti-violence groups, and get others involved in community service.

- **Parents** can play a role by being aware of the conditions at the school. They can become involved in school affairs by joining parent-teacher groups, chaperoning field trips, and helping out in the classroom.

Academic Vocabulary

suspend *(verb):* to bar temporarily

Safety on the Job

Millions of teens in the United States hold full-time or part-time jobs. Part-time or summer jobs offer a way to earn extra cash, build responsibility, and learn useful skills. However, work also has its risks. Each year, about 52,600 teen workers suffer injuries or illnesses serious enough to send them to a hospital emergency room.

 **READING CHECK**

Cause and Effect List three problems that can make school less safe. Identify three strategies for dealing with these problems.

The federal government has enacted laws to protect the health of young workers. First, all employers must meet standards set by the **Occupational Safety and Health Administration (OSHA)**. OSHA is *the agency within the federal government that is responsible for promoting safe and healthful conditions in the workplace.* Other laws place limits on the kinds of jobs that teens can do. For example, workers under 18 years old are not allowed to drive forklifts, work as miners or loggers, operate certain types of power-driven equipment, or work with explosives or radioactive materials.

Teen employees and their employers can take additional steps to prevent work-related injuries. Young workers can be aware of the risks of their jobs, follow safe work practices, and refuse to work in unsafe conditions. Employers can provide adequate training and supervision.

■ **Figure 26.8** The law places restrictions on the types of work that teens can do. *Why might it be unsafe for teens to do certain jobs?*

LESSON 2 ASSESSMENT

 After You Read

Reviewing Facts and Vocabulary

1. Define *unintentional injuries*.
2. Identify two important pieces of fire safety equipment.
3. What are two steps you can take to prevent poisonings in your home?

Thinking Critically

4. **Synthesize.** Seventeen-year-old Claude finds his father's shotgun on the kitchen table. It looks like his dad was interrupted in the middle of cleaning it. What should Claude do?
5. **Analyze.** What factors may make teens especially vulnerable to being injured on the job?

Applying Health Skills

6. **Practicing Healthful Behaviors.** Think of a specific job that a teen might have. Develop a list of strategies for preventing injuries on that job.

Writing Critically

7. **Narrative.** Write a short story about an accident involving a teen. Your story should clearly show each of the steps in the accident chain and how all of them work together to result in the accident.

Real Life Issues ....

After completing the lesson, review and analyze your response to the Real Life Issues question on page 731.

Purestock/SuperStock

Outdoor Safety

Real Life Issues

Playing It Safe. Outdoor activities can be fun, but can also pose a risk for injury. This chart shows the percentage of accidents caused by various sports.

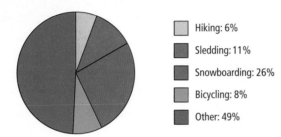

- Hiking: 6%
- Sledding: 11%
- Snowboarding: 26%
- Bicycling: 8%
- Other: 49%

Source: U.S. Dept of Health and Human Services

Writing *Write a paragraph describing the safety equipment that is needed to participate in various outdoor activities.*

Outdoor Recreation

Main Idea Planning ahead can protect you from injury during outdoor activities such as camping, hiking, and winter sports.

The most important general rule for all outdoor activities is to plan ahead. Here are some specific ways to do just that:

- **Know your limits.** Stick with tasks that match your level of ability. Brush up on necessary skills ahead of time.
- **Bring supplies.** Take plenty of safe drinking water. Never drink the water from lakes, rivers, or streams. Plan simple meals and store the food safely. Don't forget first-aid supplies and any medications you normally take.
- **Plan for the weather.** Check the local weather forecast and plan for expected conditions. See Chapter 12 for tips on hot-weather and cold-weather activities.
- **Wear appropriate clothing.** Choose clothes that are right for the weather and will protect you from poisonous plants and insects. Dress in layers.
- **Tell people your plans.** Let your family know where you're going and when you'll be back. If possible, carry a cell phone for emergencies. A sports whistle can also be useful as a way to signal for help.

GUIDE TO READING

BIG Idea *Common sense and caution can minimize the risk of accidental injuries during outdoor activities.*

Before You Read

Create a Cluster Chart. In the center of a sheet of paper, write "Outdoor Activities" and circle it. Surround it with circles labeled "Camping and Hiking," "Winter Sports," "Swimming and Diving," and "Boating." As you read, add information about staying safe during each type of activity.

New Vocabulary

▶ personal flotation device (PFD)

Review Vocabulary

▶ frostbite (Ch.12, L.4)
▶ hypothermia (Ch.12, L.4)

■ **Figure 26.9** Packing the right supplies will help guarantee that outdoor activities are safe as well as fun. *What supplies would you bring on a camping trip?*

Tim Fuller Photography

Camping and Hiking

There's nothing like a day out on the trails or a night sleeping under the stars. Just make sure you steer clear of bears, poison ivy, and sprained ankles! To enjoy your stay in the woods and reduce your risk of injury, follow these tips:

- **Camp with a group.** Having at least one other person with you means one person can go get help if the other is ill or injured.

- **Stick to well-marked trails.** In case you do get lost, bring a map and a compass, and know how to use them.

- **Be cautious around wildlife.** Don't feed wild animals. Avoid keeping food in or near sleeping areas, where wild animals may come looking for it.

- **Take care with fires.** Before starting a campfire, make sure it's **legal**. Keep fires at least 10 feet away from the tent. Put your fire out completely before going to bed.

- **Respect the environment.** If there aren't any trash bins at your campsite, pack your waste with you when you go.

Winter Sports

When you take part in cold-weather activities, wear warm, layered clothing to protect you from frostbite (skin and tissue damage) and hypothermia (dangerously low body temperature). To avoid sunburn, you should also apply sunscreen to all exposed skin. The sunscreen should have an SPF of at least 15. Make sure you have a buddy with you to help out in an emergency.

Academic Vocabulary

legal *(adjective):* permitted by law

READING CHECK

Predict Give two examples of problems that might occur while engaging in outdoor recreation.

Tips for specific winter activities include the following:

- **Sledding.** Make sure your equipment is in good condition. Choose safe spots to sled in: gently sloped hills with plenty of space and a level area to come to a stop at the bottom. Don't sled on or near frozen lakes, because the ice may not be solid.

- **Ice skating.** Skate only in designated areas. Never skate where you don't know the thickness of the ice. Wear skates that fit comfortably and support your ankles.

- **Skiing, snowboarding, and snowmobiling.** Wear an approved, properly fitting ski helmet. Make sure that your other equipment, such as your snowmobile, boots, and bindings, are in good condition. Stick to marked trails that are appropriate for your level of ability. Remember to look both ways and uphill before crossing or merging onto a trail. When heading downhill, give the people ahead of you the right of way, since they may not be able to see you coming from behind. If you need to stop, get to the side of the trail, out of the path of others.

Water Safety

Main Idea Following safety precautions can prevent drowning and other water-related injuries.

Swimming, boating, and other water sports are great ways to beat the summer heat. However, it's important not to lose sight of water safety. Every year, nearly 3,500 people die from drowning. Although most drowning incidents involve young children, people of all age groups need to pay attention to water safety guidelines.

Swimming and Diving

The most important rule for safety in the water is to know how to swim. Know your limits as a swimmer. If you're just learning, don't try to keep up with skilled swimmers. Instead, stick to shallow areas where your feet can touch the bottom. If you are a strong swimmer, keep an eye on friends who aren't as skilled as you are. No matter how good a swimmer you are, never swim alone. Even experienced swimmers can suffer a muscle cramp or other medical emergency.

Here are additional rules for safe swimming and diving:

- Swim only in designated areas where a lifeguard is present. Obey "No Swimming" and "No Diving" signs.

- Dive only into water that you know is deep enough. Diving into shallow water could result in permanent spinal cord damage or death.

■ **Figure 26.10** Proper clothing and equipment are two of the keys to outdoor winter safety. *What kind of clothing should you choose for cold-weather activities?*

- When swimming, always enter the water feet first. Check for hidden rocks and other hazards.
- Avoid swimming near piers and reefs. These areas are subject to rip currents that can drag you into open water.
- If you get caught in a current, swim with the current until it releases you, then swim back to the shore.
- Pay attention to the weather. When it's hot, drink plenty of fluids and reapply sunscreen frequently. If you start to shiver, it's time to get out of the water.
- Be prepared for emergencies. Knowing first aid can help you save a life.

Boating

Every year, more people die in boating accidents than in airplane crashes or train wrecks. Following a few common-sense guidelines can help you stay safe while boating:

- Make sure the person handling the boat is experienced. Never get into a boat with an operator who has been using alcohol or other drugs.
- Always wear a **personal flotation device (PFD)**, better known as a *life jacket,* when you go out in a boat. PFDs come in a wide variety of types and styles for boaters of different ages and levels of swimming ability. Inflatable toys or "water wings" are *not* a substitute for an approved PFD.
- Plan ahead and check weather reports. If a storm is predicted, do not go out onto the water. If you are already on the boat, head back to shore immediately.
- Make sure someone on land knows where you are and when you expect to be back.

When canoeing or kayaking, be prepared to fall into the water. Because the water is likely to be cold, dress in layers and choose synthetic fabrics that will wick moisture away from your body. **Figure 26.11** shows some survival techniques you can use if you fall into deep water. Know your limits when canoeing or kayaking, and don't attempt rivers or rapids that are beyond your abilities. Make sure you know how to handle a boat properly and recognize river hazards before heading out on the water.

The same safety rules that apply to boating also apply to personal watercraft. According to the U.S. Coast Guard, 60 percent of all accidents involving personal watercraft occur because of a lack of experience and speeding. Some states have additional laws governing the use of personal watercraft devices. For example, there may be an age limit for operating one or a test you have to pass before you can use one.

READING CHECK

Classify List two safety tips you should follow when swimming and two tips for safe boating.

Figure 26.11 Preventing Drowning

If you fall into cold water while wearing a PFD, assume one of these positions. If you are not wearing a PFD, tread water gently while keeping your head out of the water.

A. The Heat Escape Lessening Posture (H.E.L.P.) involves drawing your knees up and holding your arms tight across your chest, elbows bent. Keep your head out of the water to avoid losing body heat.

B. If you are with a group of people, huddle in a circle with your chests pressed together to hold in body heat. Small children should be sandwiched between adults or larger people.

LESSON 3 ASSESSMENT

 After You Read

Reviewing Facts and Vocabulary

1. Identify three strategies for preventing accidental injuries while hiking or camping.
2. List three general safety guidelines for participating in winter sports.
3. What is the main safety rule for diving?

Thinking Critically

4. **Analyze.** You and your friend Jake are skiing. Jake suggests trying the advanced slope, even though you're both beginners. What are the possible consequences of going along with this idea?
5. **Synthesize.** You and your family are taking a boat out on the lake for the afternoon. What supplies and safety equipment should you bring with you?

Applying Health Skills

6. **Decision Making.** Some friends invite you to go on a canoe ride. You've never canoed before and don't know how to handle the boat. On a sheet of paper, outline a response to this situation, using the six steps of the decision-making process.

Writing Critically

7. **Personal.** Write a journal entry about a day spent doing some kind of outdoor activity. You may describe an activity you have actually done or a fictitious one. In your journal, discuss the steps you took to protect your health and safety while outdoors.

Real Life Issues ....

After completing the lesson, review and analyze your response to the Real Life Issues question on page 739.

Safety on the Road

BIG Idea *Drivers, pedestrians, and others on the road need to fol-low rules to stay safe.*

Before You Read

Organize Information. Draw a chart with three columns. In the first column, list facts you already know about traf-fic safety. In the second, list questions about this topic you would like to have answered. As you read, fill in the third col-umn with the answers.

Facts	Questions	Answers

New Vocabulary

▶ vehicular safety
▶ graduated licensing
▶ road rage
▶ defensive driving

Real Life Issues

Limits on Driving. Shang was excited when he passed his driver's test. The first thing he did was to ask his dad if he could borrow the car that night to take a friend to a movie. Shang was surprised and disappointed when his dad said, "I don't think that's a good idea. You just got your license, and driving at night is a lot trickier. You should wait until you've been driving for a while."

Writing *Write a dialogue in which Shang and his dad use good communication and conflict resolution skills to reach an acceptable solution.*

Auto Safety

Main Idea Paying attention and following the rules of the road are the keys to safe driving.

Motor vehicle crashes are the leading cause of death for people between the ages of 15 and 20. Young drivers are more than twice as likely to be involved in a crash as the rest of the population. This is why **vehicular safety**—*obeying the rules of the road and exercising common sense and good judgment while driving*—is such an important issue for teens.

The most important rule of driving safety is: Pay attention. According to the National Highway Traffic Safety Adminis-tration (NHTSA), at least 80 percent of car crashes happen when a driver is distracted. The driver may be talking on a cell phone, drowsy, or lost in thought. Reduce distractions when you drive by positioning the seat and mirrors and fas-tening your safety belt before starting the engine. Adjust the radio and temperature controls before moving.

Here are some examples of things you need to pay attention to when you're in the driver's seat:

- **Other drivers.** Be aware of the cars around you and how they're moving. Make sure other drivers can see you by switching on your headlights at night and in bad weather.

- **Road conditions.** Reduce your speed if the road is icy or wet, if heavy snow or rain is limiting your vision, if a lane narrows, if there are sharp curves ahead, or if there is construction or heavy traffic.

- **Your physical state.** Drowsiness can impair your reaction time and your judgment. If you feel tired, try to wake yourself up by stopping for a snack or exercise. If you're still drowsy, pull over at the nearest safe, well-lit area and call home.

- **Your emotional state.** Being angry or upset can affect your driving. Ask someone else to drive, or if you're alone, pull over to a safe spot until you calm down.

Passengers riding with teen drivers are responsible for reducing the risk of accidents too. Avoid distracting the driver so they can pay attention to the road. Passengers are also responsible for their own safety. Avoid getting into a vehicle with an impaired driver. Ask the person for the keys, and if he or she refuses, call a parent or other trusted adult for help.

Teen Drivers

Young drivers may be more likely to get into an accident because they lack the experience and skills needed to drive safely. They are more likely to underestimate the hazards of the road. They may also take more risks such as speeding, running red lights, or driving after using alcohol or drugs.

To protect inexperienced drivers and others on the road, many states have graduated driver's licensing programs. **Graduated licensing** is *a system that gradually increases driving privileges over time.* Many programs have three stages: learner, provisional, and full driver's license. Each stage has a different set of driving restrictions. Another leading cause of accidents is distracted driving. Some states have instituted laws requiring that all drivers use only hands-free cell phone devices, and have banned sending text messages while driving. Other states have put further restrictions on teen drivers, prohibiting teen drivers from using cell phones at all while driving.

■ **Figure 26.12** Getting lessons from an experienced driver will help you improve your driving skills. *Why might young or inexperienced drivers be more likely to get into accidents?*

Figure 26.13　Driving Do's and Don'ts

Do:	Don't:
● Maintain a safe speed—not too fast, not too slow.	● Drive after using alcohol or any other depressant.
● Maintain a safe distance from other cars. Follow the three-second rule; when the car in front of you passes an object, you should pass it at least three seconds later.	● Drive while drowsy.
	● Use a cell phone while driving.
	● Be distracted by adjusting the radio or other controls.
● Signal all turns.	● Eat food while driving.
● Obey traffic signals.	● Drive with someone who has been drinking alcohol or using illegal drugs.
● Let other drivers merge safely.	
● Wear your safety belt, and make sure your passengers wear theirs.	● Use your horn inappropriately. It's meant to be a warning signal; save it for that.

READING CHECK

Identify Problems and Solutions Name two actions you can take to stay safe while driving.

Avoiding Road Rage

You're driving along when another driver suddenly swerves into your lane without signaling, forcing you to slam on your brakes. Some drivers respond to this type of situation with **road rage**. This means *responding to a driving incident with violence.* Examples of road rage behaviors can include

- honking, shouting, gesturing, or flashing lights.
- chasing or tailgating another vehicle.
- cutting off another car or forcing it off the road.
- deliberately hitting or bumping another car.
- threatening or physically attacking another driver.

If you see these kinds of behaviors, stay a safe distance away. If you're threatened, lock your doors and drive to the nearest police station. Never try to retaliate, or the conflict could turn deadly.

Being a Responsible Driver

Unfortunately, you can't always trust other drivers to drive safely. To protect yourself, you need to drive defensively. **Defensive driving** means *being aware of potential hazards on the road and taking action to avoid them.*

When you drive defensively, you stay alert and take responsibility for your behavior. You also watch out for potential dangers. A car that is weaving, crossing the center line, making wide turns, or braking without warning may have an impaired driver. If you spot such a vehicle, keep your distance, or pull over and notify the police. **Figure 26.13** lists other ways to protect yourself while driving.

Sharing the Road

Main Idea Everyone on the road shares a responsibility to follow traffic laws.

You share the road with other motorists, pedestrians, cyclists, and people on skates, scooters, or small motor vehicles. When you're driving, watch for other vehicles and pedestrians. When you're on foot, on a bike, or skating, be aware of vehicles and follow the rules of the road.

Pedestrian Safety

Always use the sidewalk. If there is no sidewalk, walk on the left side of the road, facing oncoming traffic. This will make it easier for cars to see you. It also makes it easier for you to see them, and get out of the way if a driver comes too close. Before you cross a street, look left, then right, then left again. Cross only at marked crosswalks, or at a corner. Make sure the cars have seen you and stopped before you step into the street.

Bicycle Safety

Riding a bike is a great way to travel around and get exercise at the same time. Here are tips for safe cycling:

- Always wear a safety-approved helmet that fits properly.
- Follow the rules of the road, and obey traffic laws.
- Signal turns about half a block before reaching the intersection. Extend your left arm straight out to the side to signal a left turn. Bend your left arm upward at the elbow to signal a right turn.
- Ride single file, and keep to the far right side of the road. Watch out for obstacles such as opening car doors, sewer gratings, soft shoulders, and cars pulling into traffic.
- Do not tailgate motor vehicles or ride closely behind a moving vehicle.
- Look left, right, and left again before riding into the stream of traffic.
- Wear bright colors in the daytime and reflective clothing at night. Place reflectors on the front and rear of your bike, on both wheels, and on both pedals.

■ **Figure 26.14** Cyclists ride with the flow of traffic and obey the same traffic signs and signals as cars. *What are some safety measures you can take when riding a bike?*

Skating Safety

To protect yourself while skating, wear the proper equipment, including: helmet, knee and elbow pads, wrist guards, and gloves. If you're a beginner, avoid skating in high traffic areas. Watch out for pedestrians, cyclists, and others on the road. Avoid skating in the street, and cross streets safely when you come to them. If you start to lose your balance, crouch down so that you won't have as far to fall. Try to keep your body loose and roll, rather than absorbing the force of the fall with your arms, which can cause wrist injuries.

Small Motor Vehicle Safety

Small motor vehicles include motorcycles, mopeds, and all-terrain vehicles. Motorcycles and mopeds are motor vehicles, just like cars, and are subject to the same traffic laws. Motorcyclists must have a special license in addition to their driver's license.

According to the NHTSA, motorcyclists and passengers are 35 times more likely to die in a crash than automobile drivers and passengers. Head injuries cause the most deaths in motorcycle accidents. In 20 states, all motorcyclists and passengers must wear protective helmets. In another 27 states, motorcyclists and passengers under the age of 18 are required to wear a helmet.

Helmets should meet the standards set by the U.S. Department of Transportation (DOT). Wearing sturdy clothing that covers the arms and legs also provides some protection. Passengers should avoid riding with a motorcyclist who is impaired by drug or alcohol use.

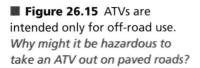
■ **Figure 26.15** ATVs are intended only for off-road use. *Why might it be hazardous to take an ATV out on paved roads?*

Another type of small vehicle is the all-terrain vehicle (ATV). ATVs have either three or four wheels. These off-road vehicles are used for recreation, as well as for work on farms and ranches. It's important to take safety precautions when operating ATVs. About 46 percent of all injuries and deaths from ATV use occur among children and teens under age 16.

In 2008, the Consumer Product Safety Commission (CPSC) banned ATVs with three wheels. The CPSC is also proposing other rules for safe ATV use. Those proposals include licensing ATV users, restricting people under age 16 from using ATVs, and requiring all ATV users to complete safety classes. To operate ATVs safely, keep these guidelines in mind:

- Only one person should ride on an ATV at a time.
- Avoid using attachments that will reduce the stability and braking of the ATV.
- Wear appropriate gear when riding an ATV. In addition to a DOT-approved helmet, you should wear eye protection, a long-sleeved shirt, long pants, gloves, and boots that cover your ankles.
- Avoid taking an ATV out on paved roads.
- Avoid ATV drivers who have been using alcohol or drugs.

 READING CHECK

Classify Identify two safety rules that apply to pedestrians, cyclists, and skaters.

LESSON 4 ASSESSMENT

 After You Read

Reviewing Facts and Vocabulary

1. What is the most important rule of driving safety?
2. Identify three behaviors associated with road rage.
3. What piece of safety equipment is required for both cycling and in-line skating?

Thinking Critically

4. **Evaluate.** According to an old saying, "It's better to be alive than right." How could this saying be applied to vehicular safety?
5. **Evaluate.** What are some of the risks associated with operating motorcycles, mopeds, and ATVs?

Applying Health Skills

6. **Advocacy.** Work with a small group to produce a safety guide that educates teens and others on how to stay safe while operating a motorcycle, moped, or ATV. Produce your guide as a video, public service announcement, brochure, or comic book.

Writing Critically

7. **Expository.** List three risks you might face while driving, skating, or riding a bicycle. Then write a paragraph explaining how your behavior can increase or reduce these risks.

Real Life Issues ....

After completing the lesson, review and analyze your response to the Real Life Issues question on page 744.

Hands-On
HEALTH

Activity ## Planning Ahead for Fun

Your class has been invited to travel to an outdoor resort for a weekend of fun activities. Your task is to think ahead and create a safety plan so that the weekend is free of accidents, injuries, and illnesses.

What You'll Need

- 12 index cards
- one red marker and one green marker per group

What You'll Do

Step 1

In your group, identify 12 accidents, illnesses, or injuries. Write them in red marker on the front of each index card.

Step 2

On the back of each card, use a green marker to write what a person can do to prevent that accident, illness, or injury from happening.

Step 3

Pass your cards to the next group. Read the safety issue on the front and, without looking at the back, identify what can be done to prevent the accident. Flip over the card to check your answer.

Apply and Conclude

Create an individual safety plan that will enable you to be safe and have fun during your weekend trip.

Checklist: Self-Management Skills

✔ Demonstrate healthful behaviors, habits, and techniques

✔ Identify protective behaviors (such as first-aid techniques, safety steps, or strategies) to help you avoid and manage unhealthy or dangerous situations

✔ List steps in correct order

Personal Safety and Protection

Key Concepts
▶ Personal safety precautions can help you avoid becoming a victim of crime.
▶ Self-defense strategies include being able to spot dangerous situations and knowing how to evade or fight off an attacker.
▶ Internet safety strategies include keeping your identity private and not responding to inappropriate messages.

Vocabulary
▶ personal safety (p. 726)
▶ self-defense (p. 727)
▶ cyberbullying (p. 729)

Safety at Home and in Your Community

Key Concepts
▶ The accident chain includes an unsafe situation, an unsafe habit, and an unsafe action.
▶ Home safety hazards include fires, electric shock, falls, poisonings, firearms, and intruders.
▶ Neighborhood Watch programs, after-school programs, and improved lighting in public areas can reduce crime.

Vocabulary
▶ unintentional injuries (p. 731)
▶ accident chain (p. 731)
▶ fire extinguisher (p. 732)
▶ smoke alarm (p. 732)
▶ carbon monoxide (p. 735)
▶ peer mediation (p. 737)
▶ OSHA (p. 738)

Outdoor Safety

Key Concepts
▶ It's important to plan ahead for all kinds of outdoor activities.
▶ To prevent drowning and other water-related injuries, know how to swim and know your limits as a swimmer.

Vocabulary
▶ frostbite (p. 740)
▶ hypothermia (p. 740)
▶ personal flotation device (PFD) (p. 742)

Safety on the Road

Key Concepts
▶ To drive safely, pay attention and follow traffic laws.
▶ Graduated licensing programs help young drivers develop the experience and skills they need to drive safely.
▶ Motorists, pedestrians, cyclists, and others can share the road safely by watching out for each other and obeying traffic laws.

Vocabulary
▶ vehicular safety (p. 744)
▶ graduated licensing (p. 745)
▶ road rage (p. 746)
▶ defensive driving (p. 746)

LESSON 1

Vocabulary Review

Correct the sentences below by replacing the italicized term with the correct vocabulary term.

1. Recognizing and avoiding dangerous situations is a part of *everyday precautions*.

2. Learning how to size up a situation, figure out what to do, and catch your attacker off-guard are examples of *martial arts*.

3. Cruel or hurtful online contact is called *harassment*.

Understanding Key Concepts

After reading the question or statement, select the correct answer.

4. If you think you are being followed in a public place, you should
   a. pretend you aren't aware of the stalker.
   b. go into a business that's open.
   c. challenge your attacker.
   d. avoid making a scene.

Thinking Critically

After reading the question or statement, write a short answer using complete sentences.

5. **Analyze.** How does letting your family know your plans protect your personal safety when you go out?

6. **Make Inferences.** Why might a person trying to escape from an attacker be more likely to get a response by shouting "fire" instead of "help"?

7. **Compare and Contrast.** How do the tactics used by cyberbullies differ from those used by Internet predators?

LESSON 2

Vocabulary Review

Use the vocabulary terms listed on page 751 to complete the following statements.

8. The kinds of accidents that pose a real danger are the ones that result in a(n) _____.

9. The _____ is a sequence of events that leads to an unintentional injury.

10. A(n) _____ is a portable device for putting out small fires.

11. A(n) _____ is a device that produces a loud warning noise in the presence of smoke.

12. The agency within the federal government that is responsible for promoting safe and healthful conditions in the workplace is called _____.

Understanding Key Concepts

After reading the question or statement, select the correct answer.

13. How often should smoke alarms be tested to make sure they are working?
   a. Every week
   b. Every month
   c. Twice a year
   d. Once a year

14. What is responsible for approximately half of all accidental deaths in the home?
   a. Poisonings
   b. Fire
   c. Electrical shock
   d. Falls

15. Which of the following steps can students take to improve the safety of their schools?
  a. Hire security guards.
  b. Put metal detectors at school entrances.
  c. Adopt zero-tolerance policies for offenses.
  d. Develop peer mediation programs.

Thinking Critically

After reading the question or statement, write a short answer using complete sentences.

16. Identify. What are the five steps in the accident chain?

17. Explain. How does following rules for electrical safety help prevent home fires?

18. Analyze. Why is it dangerous to mix household chemicals, such as cleaning fluids?

LESSON 3

Vocabulary Review

Use the vocabulary terms listed on page 751 to complete the following statement.

19. Wearing warm, layered clothing will protect you from _____, or a dangerously low body temperature.

20. Another name for a(n) _____ is a life jacket.

21. Damage to the skin and tissue caused by the cold is _____.

Understanding Key Concepts

After reading the question or statement, select the correct answer.

22. When skiing, snowboarding, or snowmobiling, you should give the right of way to
  a. the people ahead of you.
  b. the people coming from behind you.
  c. the people to your left.
  d. the people to your right.

23. The only safe place to swim is
  a. in a swimming pool.
  b. in a lake or river.
  c. in a designated area with a lifeguard present.
  d. near piers and reefs.

24. Which of the following water safety rules applies *only* to boating?
  a. Know how to swim.
  b. Don't go out alone.
  c. Pay attention to the weather.
  d. Always wear a life jacket.

Thinking Critically

After reading the question or statement, write a short answer using complete sentences.

25. Explain. Why should you avoid keeping food in or near sleeping areas while camping?

26. Evaluate. What is the advantage of having a buddy with you for all types of outdoor activity?

27. Apply. What should you do if you get caught in a current while swimming?

LESSON 4

Vocabulary Review

Correct the sentences below by replacing the italicized term with the correct vocabulary term.

28. *Traffic law* is a system that gradually increases driving privileges over time.

29. Responding to a driving incident with violence is called *highway anger*.

30. *Responsiveness* means being aware of potential hazards on the road and taking action to avoid them.

Assessment

Understanding Key Concepts

After reading the question or statement, select the correct answer.

31. The National Highway Traffic Safety Administration (NHTSA) estimates that at least 25 percent of car crashes happen when a driver
    a. is not paying attention.
    b. is not wearing a safety belt.
    c. is angry or upset.
    d. is on wet or icy roads.

32. The proper place to ride a bicycle is
    a. on the left side of the road, facing oncoming traffic.
    b. on the far right side of the road.
    c. as close to the middle of the road as possible.
    d. on the sidewalk.

33. All-terrain vehicles (ATVs) should be ridden only
    a. on paved roads.
    b. by licensed drivers.
    c. for recreation.
    d. by one person at a time.

Thinking Critically

After reading the question or statement, write a short answer using complete sentences.

34. **Analyze.** What factors make teen drivers more likely to be involved in accidents?

35. **Evaluate.** What are the advantages of graduated licensing?

36. **Apply.** What should you do if you start to lose your balance while on a skateboard?

Technology PROJECT-BASED ASSESSMENT

Preventing Poisonings

Background

Households contain a surprising number of toxic substances. Cleaners, paints, medicines, and even some houseplants can be poisonous. Many home poisonings involve small children. A tiny amount of a toxic substance can be deadly. Treatment often depends on the type of poison. The good news is that most poisonings can be prevented.

Task

Create a public service announcement (PSA) that can be played over the school intercom.

Audience

Families in your community

Purpose

Provide families with information to prevent household poisonings.

Procedure

1. Review the poison prevention information presented in this chapter.

2. Conduct an Internet search to find additional information on poison prevention.

3. Collaborate as a group to compile a list of steps people can take to prevent household poisonings. Add emergency measures that can be taken if someone is poisoned.

4. Find the 24-hour toll-free number for the Centers for Disease Control and Prevention's poison control hotline.

5. Present the PSA to your class. Get permission from your principal to play the PSA over the school intercom.

Math Practice

Interpret Graphs. The bar graph below shows the percentage of total hospitalized injuries based on a sample size of 650,000 people. Use the graph to answer Questions 1–3.

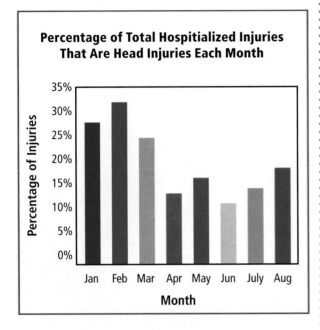

Percentage of Total Hospitialized Injuries That Are Head Injuries Each Month

1. If 32% of the sample size were injured in February, how many people were injured?
   - **A.** 175,000
   - **B.** 208,000
   - **C.** 324,000
   - **D.** 475,000

2. Which month had the fewest number of injuries?
   - **A.** January
   - **B.** April
   - **C.** June
   - **D.** August

3. Approximately how many injuries occurred during the month with the fewest injuries?
   - **A.** 117,000
   - **B.** 91,000
   - **C.** 71,500
   - **D.** 65,000

Reading/Writing Practice

Understand and Apply. Read the passage below, and then answer the questions.

A fire broke out Thursday night in the home of the Levin family in Deep Valley. Mr. and Mrs. Levin and their two children, Sam and Jamie, escaped unhurt.
   "Our smoke detector saved our lives," reported Debbie Levin. "It woke us all up out of a sound sleep. Dave and I went to check on the kids, but they were already on their way out—crawling under the smoke just the way we taught them."
   "It's really worth the effort to make a fire safety plan and have drills with your kids," added Dave Levin.
   The Levins escaped to the home of their neighbors, the Johnsons, and called the fire department. Firefighters were able to extinguish the blaze before it caused significant damage.

1. Which of the following sentences would best complete the third paragraph?
   - **A.** The fire was caused by bad wiring.
   - **B.** We have a fire extinguisher on hand.
   - **C.** The Johnsons let us use their phone to call the fire department.
   - **D.** Our kids knew exactly what to do in this situation.

2. What is the purpose of this passage?
   - **A.** To urge people to buy fire extinguishers
   - **B.** To report a neighborhood fire
   - **C.** To explain how smoke detectors work
   - **D.** To generate sympathy for the Levins

3. Write a conclusion for this article that describes how the fire in the Levin home started. Include advice on how readers can protect themselves from fires in their own homes.

National Education Standards
Math: Measurement and Data
Language Arts: LACC. 910. L.3.6, LACC.910.RL.2.4

Ed-Imaging

CHAPTER 27

First Aid and Emergencies

Lesson 1

Providing First Aid

BIG Idea *Knowing how to perform first aid can save a life in an emergency.*

Lesson 2

CPR and First Aid for Shock and Choking

BIG Idea *Medical emergencies that are life threatening include loss of breathing, shock, and choking.*

Lesson 3

Responding to Other Common Emergencies

BIG Idea *You can use first aid to deal with common emergencies such as muscle and bone injuries, impaired consciousness, animal bites, nosebleeds, and poisoning.*

Lesson 4

Emergency Preparedness

BIG Idea *Planning ahead and knowing what to expect can help you survive severe weather and natural disasters.*

Activating Prior Knowledge

Using Visuals Look at the picture on this page. What skills is this teen learning? How could these skills help protect the teen's health and the health of others? Write a paragraph explaining why this teen decided to take this class.

Chapter Launchers

Health in Action

Discuss the BIG Ideas

Think about how you would answer these questions:

▶ When was the last time you or someone who was with you suffered an injury?

▶ How did you respond to the injury when it happened?

Assess Your Health

Read each statement. On a separate sheet of paper, write "yes," "sometimes," or "no" based on your typical behavior.

1. I know the three C's of first aid.

2. I understand how CPR helps a victim of cardiac arrest.

3. I know what to do to help someone who is choking.

4. I know how the P.R.I.C.E. procedure is used.

5. I understand how to prepare for emergencies in my area.

A "yes" response shows that you practice healthy behaviors. "Sometimes" indicates that you should analyze and possibly modify your behavior. A "no" response means that you should modify the behavior.

Rick Brady/McGraw-Hill Education

Providing First Aid

📖 GUIDE TO READING

BIG Idea *Knowing how to perform first aid can save a life in an emergency.*

Before You Read

Create a Comparison Chart. Divide a sheet of paper into three columns. Label the columns "First Steps," "Bleeding," and "Burns." As you read, fill in information about each topic.

First Steps	Bleeding	Burns

New Vocabulary

▶ first aid
▶ Good Samaritan laws
▶ universal precautions

Real Life Issues

Helping Out a Stranger. Eva was driving in her neighborhood when she saw someone lying by the side of the road. She pulled over and got out of her car. The person was a woman wearing a bicycle helmet. An overturned bike was lying nearby. Cautiously, Eva touched her shoulder. "Hey, are you okay?" she asked. "Can you move?" The woman responded with a muffled groan.

Writing *Write a conclusion to this story that shows how Eva responded to this emergency and what effect her actions had on herself and on the stranger she helped.*

First Steps in an Emergency

Main Idea The three steps for responding to an emergency are *check*, *call*, and *care*.

If you ever find yourself in an emergency—like a car crash, a hurricane, or even a terrorist attack—would you know what to do? In a situation like this, knowing first aid could save someone's life. **First aid** is *the immediate, temporary care given to an ill or injured person until professional medical care can be provided*. In the seconds and minutes right after an emergency strikes, first aid can mean the difference between life and death. By learning and using proper first-aid procedures, you can help prevent victims from suffering further injury and reduce the number of victims who die.

Recognizing an emergency is the first step in responding to it. The next step is to check the scene to make sure it's safe for you to respond. Look out for hazards such as downed electrical lines or oncoming traffic that might put your own life at risk if you approach. Remember, you can't help the other person if you become injured yourself. Once you've determined that the scene is safe, you can follow the three Cs of emergency care.

The three Cs include these steps:

- **Check** the victim. A victim who is unconscious or has a life-threatening condition (for example, someone who is not breathing) needs immediate care. Only move the victim if he or she is in direct physical danger or if you must move the victim in order to provide lifesaving care.

- **Call** 911 or your local emergency number. If the victim is in need of immediate care, get someone else at the scene to call 911 while you provide first aid. If no one else is present, make the call yourself. Emergency operators may be able to talk you through the steps of helping the victim. Stay on the line until help arrives.

- **Care** for the victim. If possible, get the victim's permission before giving first aid. If the victim refuses help, respect this decision. However, if the victim can't speak to give permission, don't hesitate to provide care. Most states have **Good Samaritan laws**—*statutes that protect rescuers from being sued for giving emergency care.*

Universal Precautions

One risk of giving first aid is that blood and other body fluids can carry pathogens, including the viruses that cause AIDS and hepatitis B. Health care workers follow **universal precautions**—*steps taken to prevent the spread of disease through blood and other body fluids when providing first aid or health care.* These steps require people who provide first aid or medical care to treat all body fluids as if they could carry disease. Universal precautions include

- wearing sterile gloves whenever you could come into contact with someone's blood or body fluids.
- washing hands immediately after providing first aid.
- using a mouthpiece, if one is available, when providing rescue breathing (see Lesson 2).

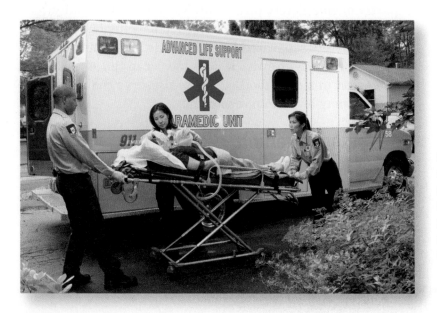

■ **Figure 27.1** Although first aid can help keep a victim alive, it is still important to call emergency services and get the victim professional medical care. *What is the number for emergency services in your area?*

First Aid for Bleeding

Main Idea The steps for treating bleeding depend on the type of injury and how severe it is.

Academic Vocabulary

minor *(adjective):* not serious or involving risk to life

Minor injuries that cause bleeding, such as small cuts and scrapes, can usually be treated at home. Severe bleeding, however, can be a life-threatening emergency. The appropriate first aid for bleeding depends on what type of wound you are dealing with and how severe the bleeding is.

Types of Open Wounds

Open wounds are injuries in which the skin is broken. Types of open wounds include the following:

- **Abrasions,** also known as scrapes. These occur when the skin is scraped against a hard surface, bursting the tiny blood vessels in the outer layer of skin. The chief danger with this type of wound is that dirt and bacteria can penetrate the skin. It's important to clean the wound well to prevent infection and speed healing.

- **Lacerations.** These are cuts caused by a sharp object slicing through layers of skin. Minor lacerations can be treated at home, but medical care is needed for deep cuts, cuts that won't stop bleeding, and cuts on the face and neck. These wounds may require stitches. A tetanus booster may also be needed.

- **Punctures.** A puncture wound is a small but deep hole caused by a sharp and narrow object (such as a nail) that pierces the skin. Puncture wounds do not usually cause heavy bleeding, but they do carry a high risk of infection, including tetanus infection. If a puncture wound is deep, dirty, or in the foot, see a doctor. The victim may need a tetanus shot or booster.

- **Avulsions.** An avulsion occurs when skin or tissue is partly or completely torn away. Such wounds usually require stitches. If a body part, such as a finger, is partly or completely separated from the body, seek emergency medical care right away. If possible, wrap the severed body part in a cold, moist towel to preserve the tissue; doctors may be able to reattach it.

Controlling Bleeding

When treating an open wound on someone other than yourself, wear clean protective gloves. If medical help is needed, call 911 before taking any other steps. Next, wash the wound thoroughly with mild soap and running water to remove dirt and debris.

Follow these steps to control the bleeding:

- If possible, raise the wounded body part above the level of the heart.

- Cover the wound with sterile gauze or a clean cloth.

- Press the palm of your hand firmly against the gauze. Apply steady pressure to the wound for five minutes or until help arrives. Do not stop to check the wound; you may interrupt the clotting of the blood.

- If blood soaks through the gauze, add another gauze pad on top of the first and continue to apply pressure.

- Once the bleeding slows or stops, **secure** the pad firmly in place with a bandage, strips of gauze, or other material. The pad should be snug, but not so tight that you can't feel the victim's pulse.

- If you can't stop the bleeding after five minutes, or if the wound starts bleeding again, call for medical help (if you have not done so already). Continue to apply pressure to the wound until help arrives.

Academic Vocabulary

secure *(verb):* to make firm or to fasten

TEENS Making a Difference

"You can put a smile on someone's face"

Helping in Times of Need

Ashley S., of West Virginia, followed in her mother's footsteps when she decided to volunteer. "My mom worked with Red Cross so I wanted to get involved, too." So far, Ashley has helped with six blood drives. "After the person has donated blood, I take them to a bed and give them a snack and a drink to get energy into them. It's a lot of responsibility because you need to be careful how you help them or they can get hurt."

She also helps teach first aid for pets. "We demonstrate how to wrap gauze around a cut, or if the leg is broken, how to put splints on to immobilize it." Ashley has taken first aid classes to learn how to work with babies and toddlers. In the event of a local disaster, she'll be able to go into shelters and help with the kids while the parents are doing other tasks. "Volunteering is fun. You can put a smile on someone's face and help them during times of tragedy."

 Activity

❶ What are three things you could do if there was a disaster in your community?

❷ What are some of the activities that you might like to do if you volunteered with Red Cross?

❸ What are some of the benefits of helping others?

READING CHECK

Explain What is internal bleeding?

Certain types of injuries can cause internal bleeding—blood from a damaged blood vessel entering one of the body's cavities. Internal bleeding is difficult to detect. However, bleeding from the eyes, nose, mouth, or ears may be a sign that internal bleeding is occurring. Internal bleeding requires emergency care, so call for help right away. While waiting for help to arrive, you can take steps to prevent the victim from going into shock (see Lesson 2).

First Aid for Burns

Main Idea Treatment for burns depends on the severity of the burn.

Burns can occur in a variety of ways. Burns caused by heat are the most common type. They may occur as a result of exposure to flame, touching a hot object such as a stove, scalding with hot water or steam, or overexposure to the sun. Burns can also result from exposure to electricity and to certain chemicals, such as bleach. Electrical and chemical burns require special first-aid procedures.

Figure 27.2 Types of Burns

Burns are classified according to the amount of damage they cause.

First-degree burns involve only the outer layer of skin. This outermost layer is called the epidermis. In a first-degree burn, the skin becomes red, and the burned area may become swollen and painful. First-degree burns are considered minor burns unless they involve a major joint or cover large areas of the hands, feet, face, groin, or buttocks.

Second-degree burns involve the epidermis and the underlying layers of skin (the dermis). The skin becomes very red and develops blisters. There is severe pain and swelling. A second-degree burn no larger than 2 to 3 inches in diameter can be treated as a minor burn. Larger burns, or burns that affect the hands, feet, face, groin, buttocks, or a major joint, require professional medical care.

Third-degree burns, the most serious kind, involve all layers of the skin and may penetrate the underlying tissues. The skin may be charred black or may appear white and dry. It may also be possible to see muscle and even bone. These burns can destroy nerve endings, so victims may not experience pain. Third-degree burns require immediate medical attention.

Figure 27.2 shows how to distinguish the different types of burns. First-degree burns and small second-degree burns are considered minor and can be treated with these steps:

1. Cool the burned area by holding it under cold, running water for at least five minutes. If this isn't possible, immerse the burned area in cool water or wrap it in cold, wet cloths. Do not use ice, which may cause frostbite and further damage the skin.

2. Cover the burn loosely with a sterile gauze bandage.

3. The victim may take an over-the-counter pain reliever. Make sure the victim isn't allergic to the medication.

4. Minor burns usually heal without further treatment, though the skin may be discolored. If signs of infection develop—including increased pain, redness, fever, swelling, or oozing—seek medical help.

Some second-degree burns and all third-degree burns require immediate medical care. Call 911 and provide first aid until help arrives. Cover the burned area with a clean, moist cloth, but do not remove burned clothing unless it is still smoldering. Do not immerse a large burned area in cold water; the victim could go into shock. Be prepared to give first aid for shock or loss of circulation (see Lesson 2).

READING CHECK

Classify Which kind of burn always requires professional medical care?

LESSON 1 ASSESSMENT

 After You Read

Reviewing Facts and Vocabulary

1. What are the three first steps for responding to an emergency?

2. Identify the four types of open wounds.

3. Describe the procedure for treating a minor burn.

Thinking Critically

4. **Synthesize.** Suppose that you are looking after your seven-year-old neighbor. The boy steps on a tack and gets a puncture wound in his foot. How would you respond?

5. **Evaluate.** Which types of open wounds are most likely to require professional medical care? Why?

Applying Health Skills

6. **Advocacy.** Write a persuasive flyer designed to encourage other teens to learn first aid. Your flyer should explain the value of knowing first aid and the situations in which it can be useful.

Writing Critically

7. **Narrative.** Write a short story in which a teen responds to a medical emergency and provides appropriate first aid.

Real Life Issues ....

After completing the lesson, review and analyze your response to the Real Life Issues question on page 758.

LESSON 2

📖 GUIDE TO READING

BIG Idea *Medical emergencies that are life threatening include loss of breathing, shock, and choking.*

Before You Read

Organize Information.
Make a three-column chart. Label the columns "CPR," "First Aid for Shock," and "First Aid for Choking." As you read, fill in the appropriate columns with the steps in each first-aid procedure.

CPR	Shock	Choking

New Vocabulary

▶ chain of survival
▶ defibrillator
▶ cardiopulmonary resuscitation (CPR)
▶ rescue breathing
▶ shock

CPR and First Aid for Shock and Choking

Real Life Issues ························

Learning CPR. Lauren babysits her nephew on Tuesday nights while her sister attends class. Her sister wants Lauren to take CPR and child safety classes at a local community college so that she will know how to respond to an emergency. Lauren isn't sure that she wants to spend her weekend learning first aid and CPR since she feels she may never have to use these life-saving techniques.

Writing *Write a letter to Lauren, explaining why it could be important for Lauren to learn CPR and first aid.*

The Chain of Survival

Main Idea In a medical emergency, a victim's life depends on a specific series of actions called the *chain of survival.*

The most urgent medical emergencies are often those in which the victim is unresponsive, or unable to speak or react to his or her surroundings. This condition can result from a heart attack, a stroke, or cardiac arrest. In this type of emergency, you need to act quickly, because the first few minutes after a medical crisis are usually the most critical. The key is to know what to do, remain calm, and take action.

An unresponsive victim is in immediate danger. Her or his best hope lies in the **chain of survival**, *a sequence of actions that maximize the victim's chances of survival.*

©RubberBall

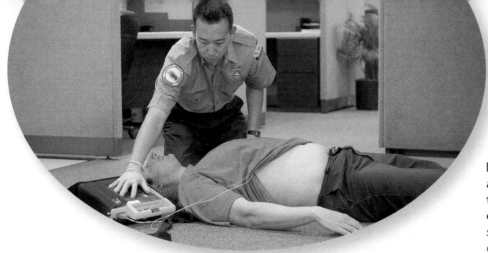

■ **Figure 27.3** The increased availability of AEDs has improved the survival rate for victims of cardiac arrest. *What is the next step after defibrillation in the chain of survival?*

The links in the chain of survival include the following:

- **A call to emergency medical services.** This first step is important for all victims. The 911 operator will ask you questions about the victim's condition and instruct you on what to do next. If the victim's heart has stopped, you will be instructed to move on to the next link in the chain of survival.

- **CPR,** or cardiopulmonary resuscitation. CPR gives the victim a chance to survive until medical help arrives.

- **Defibrillation.** A **defibrillator** is *a device that delivers an electric shock to the heart to restore its normal rhythm.* An increasing number of public places now provide automated external defibrillators (AEDs).

- **Advanced care.** Paramedics and other trained medical personnel can provide the care needed to keep the victim alive on the way to the hospital.

CPR

Main Idea CPR can save the life of a person whose heartbeat or breathing has stopped.

The second link in the chain of survival is to perform CPR on the victim. CPR, or **cardiopulmonary resuscitation**, is *a first-aid procedure that combines rescue breathing and chest compressions to supply oxygen to the body until normal body functions can resume.* It takes training from a certified professional to learn how to perform CPR correctly. However, if no trained person is present, it's better to have CPR done by an untrained person than to wait for paramedics to arrive. A person who is untrained in giving CPR can perform an updated form of the procedure called Hands-Only™ CPR. Hands-Only™ CPR eliminates the need for rescue breathing on chest compressions and focuses on chest compressions. In some parts of the country, 911 dispatchers are taught how to talk an untrained person through the steps of CPR.

READING CHECK

Make Inferences
What is the purpose of the chain of survival?

Health Skills Activity

Communication Skills

Calling Emergency Services

Kenji is at the mall when he sees another shopper suddenly clutch his chest and collapse. Kenji checks the victim and realizes that he has lost consciousness. Grabbing his cell phone, he dials 911 as a crowd gathers. The 911 dispatcher asks, "What is your emergency?" and "Where are you calling from?" Kenji struggles to stay calm as he answers the questions. He knows that the victim's life depends on it.

Tim Fuller Photography

Writing Write a dialogue in which Kenji responds to the 911 dispatcher's questions. Follow these guidelines:

1. Listen carefully to the questions.
2. Respond clearly and concisely, providing only the requested information.
3. Follow the dispatcher's instructions, or repeat them exactly to other rescuers. Confirm that other rescuers are following the instructions.
4. Stay on the line until instructed to hang up.

CPR for Adults

Before performing CPR on an adult, check to see if the person is conscious. Tap the victim on the shoulder while shouting, "Are you okay?" If the victim doesn't respond, start the chain of survival by calling 911. Then begin performing the steps for CPR shown in **Figure 27.4**.

CPR for Infants and Children

If the victim is an infant or a young child (under eight years old), the cycle of CPR is still 30 chest compressions for every two rescue breaths. However, the procedure is different in several ways:

- Check to see if the child is breathing *before* calling 911. If the child is not breathing, give five cycles of CPR—about two minutes' worth—before making the call.
- When performing rescue breathing on infant or young child, place your mouth over the child's nose and mouth at the same time—not the mouth only, as for an adult.
- Do not use a face mask designed for adult CPR when performing CPR on an infant or young child.

READING CHECK

Compare and Contrast List three ways in which CPR for infants differs from CPR for adults.

Figure 27.4 Adult CPR

The basic process of Hands-Only™ CPR for adults focuses on chest compressions.

Check to see if the victim is breathing. Look, listen, and feel for normal breathing for five to ten seconds. Signs of normal breathing include

▶ seeing the person's chest rise and fall.

▶ hearing breathing sounds, including wheezing, gurgling, or snoring.

▶ feeling air moving out of the person's mouth or nose.

Begin chest compressions. To position your hands correctly, follow these steps:

1. Use your fingers to find the end of the victim's sternum (breastbone), where the ribs come together.
2. Place two fingers over the end of the sternum.
3. Place the heel of your other hand against the sternum, directly above your fingers (on the side closest to the victim's face).
4. Place your other hand on top of the one you just put in position. Interlock the fingers of your hands and raise your fingers so they do not touch the person's chest.

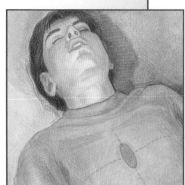

Perform chest compressions using the following procedure:

1. Straighten your arms, lock your elbows, and line your shoulders up so they are directly above your hands.
2. Press downward on the person's chest, forcing the breastbone down by 1.5 to 2 inches (3.8 to 5 cm).
3. Begin compressions at a steady pace. You can maintain a rhythm by counting, "One and two and three and . . ." Press down each time you say a number.
4. Try to perform 100 chest compressions per minute.

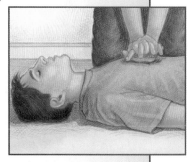

Figure 27.5 **Infant and Child CPR**

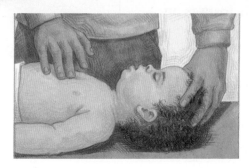

▶ This image shows how to position your fingers to perform chest compressions on an infant.

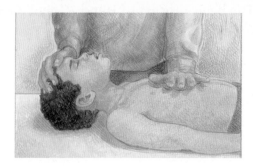

▶ This image shows how to position your hands for chest compressions on a child between one and eight years old.

- To perform chest compressions on an infant, position your fingers on the baby's sternum. Press the sternum down about one-third to one-half the depth of the infant's chest.

- To perform chest compressions on a child between one and eight years old, you can either use the heel of one hand or use both hands as in adult CPR. Position your hands about two finger widths above the end of the sternum, and press the sternum down about one-third to one-half the depth of the child's chest.

Figure 27.5 illustrates how to position your hands when performing CPR on an infant or a young child.

Other Emergencies

Main Idea Choking and shock are life-threatening medical emergencies that require immediate attention.

Academic Vocabulary

survival *(noun):* the continuation of life or existence

The chain of **survival** does not apply to every medical emergency. If a person is choking, rescue breathing will not help because the airway is blocked. Knowing the specific first-aid procedures for choking and shock can save lives.

First Aid for Choking

Choking occurs when an object, such as a piece of food, becomes stuck in a person's windpipe, cutting off the flow of air. Clutching the throat is the universal sign for choking. Other signs of choking include an inability to speak, difficulty breathing, an inability to cough forcefully, turning blue in the face or lips, and loss of consciousness. Figure 27.6 shows how to help someone who is choking.

If you see these signs in an adult, help the person immediately by performing abdominal thrusts. If someone else is nearby, ask that person to call 911 while you help the victim. For a choking infant, perform back blows and chest thrusts to dislodge the object. **Figure 27.6** illustrates these procedures.

If the choking victim is unconscious, lower the person to the floor and try to clear the airway. Reach into the mouth and sweep the object out with one finger. Be careful not to push the obstruction deeper into the throat. If the obstruction cannot be dislodged, begin performing CPR. The chest compressions may dislodge the object.

If you begin to choke when you're alone, you can perform abdominal thrusts on yourself by covering your fist with the other hand and pushing upward and inward. Another method is to bend over and position your abdomen over a rigid structure, such as a countertop or the back of a chair. Press against it to thrust your abdomen upward and inward.

READING CHECK

Explain How can you help a choking adult?

Figure 27.6 | **Treatment for Choking**

Use abdominal thrusts on a choking adult. For an infant, alternate back blows with chest thrusts. *Why do you think different methods are used for adults and infants?*

If an adult is choking:

1. Stand behind the victim and wrap your arms around his or her waist. (For a pregnant or obese victim, wrap your arms around the rib cage.)
2. Make a fist with one hand and grasp it with your other hand.
3. Pull your hands into the abdomen with a quick, upward thrust.
4. Repeat the abdominal thrusts until the object is dislodged.

If an infant is choking:

1. Sit down and hold the baby facedown over your forearm, which should be resting on your thigh.
2. With the heel of your hand, give the infant five gentle but firm blows between the shoulder blades.
3. If this doesn't dislodge the object, turn the infant faceup, with the head lower than the body. Perform five chest compressions as you would when performing infant CPR.
4. If the baby still isn't breathing, have someone call emergency services immediately while you repeat the back blows and chest thrusts. If breathing doesn't resume, begin infant CPR.

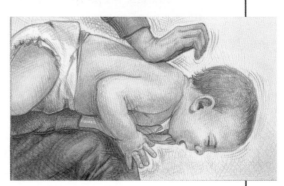

First Aid for Shock

Shock is *a life-threatening condition in which the heart is not delivering an adequate supply of blood to the body.* Symptoms of shock include

- cold, clammy skin, which may appear pale or grayish.
- weak, rapid pulse and altered breathing.
- dull, staring eyes, which may have dilated pupils.
- faintness, weakness, confusion, or loss of consciousness.

■ **Figure 27.7** A person suffering from shock should lie down with the legs elevated, unless the person has an injury to the head, neck, legs, or spine. *What purpose might raising the victim's legs serve?*

If someone displays these symptoms, call 911 right away. Get the victim to lie down and raise his legs about 12 inches if he is conscious and doesn't have an injury to the head, neck, legs, or spine. If the victim has any wounds or other injuries, give first aid for these while you wait for help. Some shock victims become anxious or agitated, so try to keep the person calm. Monitor the victim's breathing, and be prepared to start CPR immediately if breathing stops. Loosen the victim's clothing and try to keep him warm and comfortable. Don't give the victim anything to eat or drink. If the victim vomits, drools, or starts bleeding from the mouth, roll him into the recovery position (see Lesson 3).

LESSON 2 ASSESSMENT

 After You Read

Reviewing Facts and Vocabulary

1. Identify the steps in the chain of survival.
2. What is the basic cycle of CPR?
3. What is the universal sign for choking?

Thinking Critically

4. **Evaluate.** Why is calling emergency services the first step in the chain of survival?
5. **Analyze.** Explain how the strategy for responding to choking differs depending on whether the victim is an adult or an infant.

Applying Health Skills

6. **Accessing Information.** Use community or Internet resources to find out where and when CPR classes are offered in your area. If possible, arrange to take one of these classes.

Writing Critically

7. **Creative.** Write a jingle that uses rhyme and rhythm to help people remember the steps for a first-aid procedure discussed in this lesson.

Real Life Issues ·······················

After completing the lesson, review and analyze your response to the Real Life Issues question on page 764.

Responding to Other Common Emergencies

Real Life Issues

Feeling Faint. Kim has been looking forward to the school dance all month. She's so excited on the day of the dance that she forgets to eat lunch. At the dance, she is having a great time out on the hot, crowded dance floor. Then she starts to feel dizzy, and the next thing she knows, she's lying outside on the ground. Looking up, she sees a teacher and a couple of her friends. "You fainted," the teacher explains. "You need to lie still for a while." Kim is embarrassed. She wants to reassure the teacher that she's fine and go back to the dance.

Writing *Write a dialogue between Kim and the teacher. Show how the two of them deal with Kim's situation in a way that protects her health.*

Muscle, Joint, and Bone Injuries

Main Idea Muscle and joint injuries can be minor or severe, but bone injuries are always medical emergencies.

As you learned in Chapter 12, sports and other physical activities can cause injuries to your muscles, joints, and bones. These kinds of injuries can occur in other situations as well. For example, you could sprain your ankle by tripping over a branch on the sidewalk, break your arm in a car crash, or dislocate your shoulder falling from a ladder.

You can take safety precautions to help avoid injuries such as these. However, you still need to be prepared in case accidents happen. That's why you should know the proper first-aid procedures for treating injuries such as strains, sprains, fractures, and dislocations.

GUIDE TO READING

BIG Idea *You can use first aid to deal with common emergencies such as muscle and bone injuries, impaired consciousness, animal bites, nosebleeds, and poisoning.*

Before You Read

Make a T-Chart. On one side of the chart, list common medical emergencies. On the other side, list strategies to deal with each type of emergency.

Emergencies	Strategies

New Vocabulary

▶ fracture
▶ dislocation
▶ unconsciousness
▶ concussion
▶ poison
▶ poison control center
▶ venom

Review Vocabulary

▶ strain (Ch.12, L.4)
▶ sprain (Ch.12, L.4)

Muscle and Joint Injuries

Two common and fairly minor injuries are strains and sprains. A strain is a tear in a muscle, while a sprain is an injury to the ligaments around a joint. These injuries produce similar symptoms, including pain, stiffness, swelling, difficulty moving the affected body part, and discoloration or bruising of the surrounding skin. Strains and sprains vary in severity. Severe strains and sprains will require medical care. Call 911 for emergency medical help if

- the victim is unable to move the affected muscle or joint.
- the pain is severe.
- the injury is bleeding.
- the joint appears deformed.
- you hear a popping sound coming from the joint.

You can treat minor strains and sprains with the P.R.I.C.E. procedure, which includes these steps:

- **Protect** the affected area by wrapping it in a bandage or splint.
- **Rest** the injured body part for at least a day.
- **Ice** the area to reduce swelling and pain. Wrap ice cubes in a cloth or towel and hold it against the affected area for 10 to 15 minutes at a time, three times a day.
- **Compress** the affected area by wrapping it firmly, but not too tightly, in a bandage.
- **Elevate** the injured body part above the level of the heart, if possible.

You can gradually begin to use the affected body part again as the pain and swelling subside. If the swelling lasts more than two days, see a doctor.

Fractures and Dislocations

Injuries to bones include fractures and dislocations. A **fracture** is *a break in a bone;* a **dislocation** is *a separation of a bone from its normal position in a joint.* Symptoms for fractures and dislocations include severe pain, swelling, bruising, and inability to move the affected body part. The limb or joint may be visibly misshapen, discolored, or out of place.

Fractures and dislocations are emergencies that require immediate medical care. The first-aid procedures for both conditions are the same:

1. Call 911 or your local emergency medical service.
2. Do your best to keep the victim still and calm.
3. If the skin is broken, rinse it carefully to prevent infection, taking care not to disturb the bone. Cover the wound with a sterile dressing, if available.

FITNESS ZONE

A lot of people like the saying, "No pain, no gain," but my coach says that feeling pain during exercise means something is wrong. Coach says that if you feel pain, you should stop exercising right away. Listen to your body. It knows the difference between real pain and the mild discomfort of a muscle working. For more fitness tips, visit the Fitness Handbook and Fitness Zone sites in ConnectEd.

4. If necessary, apply a splint. A splint will immobilize the injured body part to prevent further injury. Attach any kind of rigid support—such as a board or stick—to the injured body part with strips of cloth, immobilizing the area extending above and below the injured bone.

5. Apply an ice pack to reduce pain and swelling.

6. If the injury does not affect the head, neck, legs, or spine, have the victim lie down and raise his or her legs about 12 inches to prevent shock.

READING CHECK

Compare and Contrast Name one way in which fractures and dislocations are similar and one way in which they are different.

Unconsciousness

Main Idea A victim who loses consciousness for any amount of time requires medical care.

Unconsciousness is *the condition of not being alert or aware of your surroundings.* Victims who are unconscious are not able to respond to simple commands. They also cannot cough or clear their throats, putting them at risk of choking. Nearly any major injury or illness can cause unconsciousness. Alcohol and drug abuse can also cause a person to lose consciousness.

If you encounter someone who has lost consciousness, call 911, check the victim's breathing, and be prepared to perform CPR if necessary. If the victim is breathing and does not seem to have an injury to the spine, lay the victim down on his or her side. Bend the top leg so that the hip and knee joints form right angles. Gently tilt the victim's head back to open the airway. This position, known as the recovery position, will help the victim breathe. Keep the victim warm until help arrives.

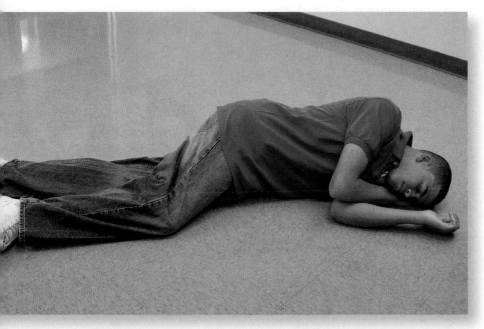

■ **Figure 27.8** The recovery position is the safest position for an unconscious person because the airway is protected. *Why is it important to keep the airway open?*

READING CHECK

Explain What can you do to help someone who is unconscious, has fainted, or has a concussion?

Fainting

Fainting is a temporary loss of consciousness that occurs when not enough blood is flowing to the brain. If you see someone faint, try to catch the person to stop him or her from falling. Lay the victim on the floor or ground and elevate the legs. Loosen any tight clothing around the victim's neck. If the person vomits, quickly roll him or her into the recovery position. If the victim does not regain consciousness within a couple of minutes, call 911. If the victim regains consciousness, keep the person lying still for at least 10 to 15 minutes.

A single episode of fainting may not be serious, but it is a warning that medical attention is needed. Victims of fainting should see a doctor as soon as possible if they have never fainted before or they are fainting frequently.

Concussion

A **concussion** is *a jarring injury to the brain that can cause unconsciousness.* Anyone who loses consciousness or experiences memory loss or confusion because of a head injury might have a concussion. Call 911 for all cases of suspected concussion. If the victim is conscious, have him or her lie down. Use first aid to treat any bleeding while you wait for help to arrive. If the victim is unconscious, avoid moving him or her if there is reason to suspect a head or neck injury. Otherwise, you can place the victim in the recovery position.

Other Common Emergencies

Main Idea It's important to learn first-aid procedures for emergencies such as animal bites, nosebleeds, and poisoning.

Other common medical emergencies include animal bites, nosebleeds, and poisoning. Learning proper first-aid procedures will help you stay calm and respond appropriately in the event of an emergency.

Animal Bites

Animal bites can transmit serious diseases such as rabies, a viral infection that can be deadly if not treated immediately. Once a person develops symptoms of rabies, the disease cannot be cured. However, a vaccine can prevent the disease if it is given within two days of exposure to the virus. Anyone who is bitten by an unknown or wild animal should **seek** emergency medical care immediately.

In general, animal bites should be treated like any other open wound. If you're providing first aid to a bite victim, wash your hands thoroughly and put on protective gloves.

Then wash the bite area thoroughly with mild soap and water. Apply pressure as needed to stop any bleeding. Apply antibiotic ointment and a sterile dressing. If the wound swells, apply ice wrapped in a towel for ten minutes. A tetanus booster shot may be required for any bite that has broken the skin. If the bite develops signs of infection (such as redness, pain, or swelling), seek emergency medical care.

Nosebleeds

Nosebleeds can occur after an injury to the nose or when very dry air causes the lining of the nose to become irritated. An occasional nosebleed isn't a cause for concern. If your nose starts bleeding, sit down and squeeze the soft part of the nose between your thumb and finger, holding the nostrils closed, for five to ten minutes. Breathe through your mouth and lean forward to avoid swallowing the blood. An ice pack or cold compress applied to the bridge of the nose may also help. If the bleeding doesn't stop after 20 minutes, seek emergency medical help.

Poisoning

A **poison** is *any substance that causes injury, illness, or death when it enters the body*. The substance can be a solid, liquid, or gas. Almost 2.5 million cases of poisoning occur in the United States each year, resulting in nearly 1,000 deaths. **Figure 27.10** on page 776 shows some ways poisons can enter the body.

The first step in any case of suspected poisoning is to call a **poison control center**, *a round-the-clock service that provides emergency medical advice on how to treat victims of poisoning*. You can reach the National Poison Control Hotline at 1-800-222-1222. Keep this number near your phone, and dial it at once in any case of suspected poisoning. Even if you aren't sure the victim has been poisoned, call right away, rather than wait for symptoms to develop. Some poisons require quick action to minimize damage or prevent death. When you call, be prepared to provide

- your name, location, and telephone number.
- the victim's condition, age, and weight.
- the name of the poison, when it was taken, and the amount of poison that was involved. If you do not have this information, tell as much as you know.

The poison control expert will provide you with step-by-step instructions on how to treat the victim. Do not give the victim any medication unless the expert tells you to do so.

■ **Figure 27.9** Pinching the nostrils closed will stop almost all nosebleeds. *What factors can trigger a nosebleed?*

Figure 27.10 | **Forms of Poisoning**

Poisons can enter the body in several ways.

How Poison Enters the Body	Examples	What Action to Take
Swallowing	Household cleaners, medicines	Call poison control and follow instructions. You may be instructed to give the victim a small amount of milk or water or to induce vomiting. Do not take these actions unless instructed to do so.
Inhalation	Carbon monoxide from heating fixtures, fumes from certain solvents, fumes produced by mixing cleaning products together	Get the victim to fresh air right away. Then call poison control. Be prepared to perform rescue breathing if necessary.
Through the eyes	Any strong chemical that enters the eye	Flush the eye with fresh water for 15 to 20 minutes. Call poison control.
Through the skin	Caustic chemicals such as drain cleaner or rust remover; certain pesticides.	Remove clothing the poison has touched. Rinse skin with running water for 15 to 20 minutes. Call poison control.

Snakebite. Certain types of snakes can inject **venom**, *a poisonous secretion,* into the victim's body. In the United States, poisonous snakes include rattlesnakes, copperheads, water moccasins (also known as cottonmouths), coral snakes, and cobras. You should treat any snakebite seriously unless you are absolutely sure of the species. Follow these steps:

READING CHECK

Identify List one thing you can do in the event of a snakebite.

- Call 911 for medical help and follow the dispatcher's instructions.
- Try to keep the victim from moving. Keep the affected body part below chest level to reduce the flow of venom to the heart.
- Remove rings and other constricting items, since the affected area may swell up.
- Try using a snakebite suction kit, if one is available in your first-aid kit.
- Do *not* apply a tourniquet, use cold compresses, cut into the bitten area with a blade, suck the venom out by mouth, or give the victim any medications without being advised to do so by a doctor or 911 dispatcher.

Insect and Spider Bites or Stings. The stings of insects such as bees, hornets, and wasps, as well as the bites of certain spiders, are painful but usually not dangerous. If someone allergic to the venom of these insects or spiders has been stung or bitten, call 911. For other cases, follow these steps:

- Remove the stinger by scraping it off with a firm, straight-edged object such as a credit card. Do not use tweezers, since they may squeeze the stinger and release more venom.

- Wash the site thoroughly with mild soap and water to help prevent infection.

- Apply ice (wrapped in a cloth) to the site for ten minutes to reduce pain and swelling. Alternate ten minutes on, ten minutes off.

- Antihistamines and anti-itch creams may help reduce itching.

- If the victim shows signs of severe reaction, such as weakness, difficulty breathing, or swelling of the face, call 911 immediately.

Poisonous Plants. Most people are allergic to poison ivy, poison oak, and poison sumac. Exposure to these plants will cause itching, swelling, redness, burning, and blisters at the site of contact. If you brush up against one of these plants, do not rub your skin, because that will spread the plant oils that cause an allergic reaction. Washing the area immediately with soap and water may prevent a reaction. Take care to wash any clothing or other objects that have touched the plant as well. If an allergic reaction develops, an over-the-counter cream or oral antihistamine may ease the itching.

■ **Figure 27.11** Exposure to poison ivy, poison oak, and poison sumac can cause itching, swelling, and blisters. *What should you do if you accidentally brush against one of these plants?*

LESSON 3 ASSESSMENT

After You Read

Reviewing Facts and Vocabulary

1. What are the symptoms of a fracture or dislocation?

2. Why is the recovery position the safest position for an unconscious person?

3. What is the first step in any case of suspected poisoning?

Thinking Critically

4. **Evaluate.** Why should you always seek professional medical care for fractures and dislocations?

5. **Analyze.** List the items you would need to treat a bee sting in a victim who is not allergic to the venom.

Applying Health Skills

6. **Practicing Healthful Behaviors.** Make a poster illustrating the steps of the P.R.I.C.E. procedure.

Writing Critically

7. **Narrative.** Write a newspaper-style article about a child or teen who is bitten by a wild animal. The article should describe the steps the victim and his or her parents take to treat the wound and prevent rabies and other diseases.

Real Life Issues ....

After completing the lesson, review and analyze your response to the Real Life Issues question on page 771.

LESSON 4

GUIDE TO READING

BIG Idea *Planning ahead and knowing what to expect can help you survive severe weather and natural disasters.*

Before You Read

Organize Information. Write "Weather Emergencies & Natural Disasters" in a circle. Surround this circle with the following terms: "Thunderstorms," "Hurricanes," "Tornadoes," "Blizzards," "Floods," "Earthquakes," and "Wildfires." As you read, add information about each type of emergency.

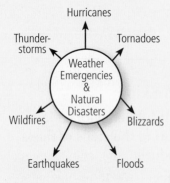

New Vocabulary

▶ hurricane
▶ tornado
▶ blizzard
▶ flash floods
▶ earthquake
▶ emergency survival kit

Emergency Preparedness

Real Life Issues

Safe in a Storm. Dean is at home looking after his younger brother and sister when it begins to snow heavily. Turning on the radio, he hears that it is a blizzard and everyone is advised to stay indoors. However, Dean's younger brother and sister want to go out and play in the snow. Dean doesn't want to spoil their fun, but he knows he's responsible for keeping them safe.

Writing *Write a dialogue between Dean and his siblings in which he explains the need to stay indoors during the snowstorm and proposes an alternative activity for them to enjoy.*

Storm Safety

Main Idea It is important to pay attention to weather warnings and follow safety guidelines during a severe storm.

When a severe storm or any other type of harsh or dangerous weather condition may occur, the National Weather Service will issue a severe weather alert. A *watch* indicates that severe weather is possible during the next few hours. A *warning* means that severe weather has already been observed or is expected soon. Watches and warnings go out over radio, television, and the Internet to let the public know about the dangers and take steps to protect themselves.

Severe Thunderstorms

Thunderstorms typically produce heavy rain and are accompanied by lightning, strong winds, and sometimes hail or tornadoes. If a thunderstorm is forecast, or if you see signs

©Hola Images/Alamy

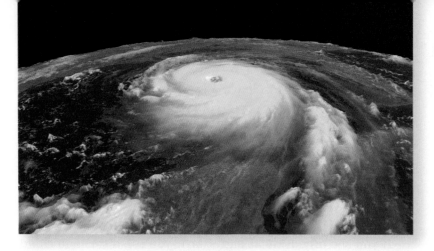

Figure 27.12 Radio and TV broadcasts will warn you if a hurricane is expected in your area. *What steps should you take to protect yourself during a hurricane?*

that one is approaching, get indoors as soon as you can. The lightning produced by thunderstorms can result in electrocution. If you are in a car and cannot reach secure shelter, stay in the car and avoid touching anything made of metal. If you are on open water, return to shore. If you are in a forest, seek shelter under shorter trees. If you are trapped in an open area, try to get into a low-lying spot such as a valley or ravine; however, be on the alert for flash floods.

If you see lightning or hear thunder, use the 30/30 rule for lightning safety. Get or stay indoors if you hear thunder within 30 seconds of seeing lightning, and stay there for 30 minutes after hearing the last peal of thunder. Avoid bathing or showering since plumbing and bathroom fixtures can conduct electricity. Unplug all electrical appliances. Avoid using a corded telephone, except in an emergency. Cordless and cellular phones are safe to use.

Hurricanes

A **hurricane** is *a powerful storm that generally forms in tropical areas, producing winds of at least 74 miles per hour, heavy rains, and sometimes tornadoes.* In the United States, hurricanes strike mainly along the eastern and southern coasts. Hurricanes cause **major** flooding, and flying debris can injure or kill people and cause property damage. High winds can topple trees, power lines, and even buildings. The deadliest part of a hurricane is the storm surge, a massive wave that sweeps in from the ocean, occasionally advancing up to hundreds of miles inland.

If a hurricane is predicted in your area, take steps to secure your property. Listen to radio or TV reports for information on the progress of the storm. Follow their instructions, and be prepared to evacuate if ordered to do so by local authorities.

Academic Vocabulary

major *(adjective):* notable in effect or scope

Tornadoes

Hurricanes and severe thunderstorms can produce a **tornado**—*a whirling, funnel-shaped windstorm that causes destruction as it advances along the ground in a narrow path.*

Tornadoes are most common east of the Rocky Mountains. The whirling winds of a tornado can reach speeds of 300 miles per hour and can leave a trail of damage a mile wide.

Although tornadoes can strike without warning, there are danger signs you can look for, such as

- darkened or greenish looking skies.
- a hailstorm that produces large hailstones.
- a large, dark, low-lying cloud that may be rotating.
- a loud roar like that of a freight train.

If you see any of these signs, or if you see a tornado cloud, take shelter immediately. Safe places to go include the lowest level in a house or other building, or the center of an interior room, such as a bathroom or closet. You should be as far away as possible from windows, doors, and outer walls. To protect yourself, crouch down as close to the floor as possible and use your arms and hands to shield your head. If possible, cover yourself with a mattress or blankets to protect yourself from falling debris.

If you are in a car or a mobile home when a tornado strikes, leave the vehicle and get into a secure shelter as quickly as possible. Never try to outrun a tornado in your car. If you are caught out in the open, lie flat in a ditch or other low-lying area and cover your head with your hands.

Winter Storms

Severe winter storms can block roads, knock down power lines, and cause floods. One type of hazardous winter storm is a **blizzard**, *a snowstorm with winds that reach 35 miles an hour or more.* To protect yourself during a winter storm, follow these guidelines:

- Stay indoors. It's the safest place to be.
- If you must go outdoors, wear layers of loose-fitting, lightweight clothing. Choose an outermost layer that will repel wind and water. Wear a hat, a scarf to protect your mouth and neck, and mittens or gloves. Wear insulated, water-resistant boots to keep your feet warm and dry.
- Whenever you are outside in a winter storm, watch out for signs of frostbite and hypothermia. (See Chapter 12.)
- Avoid driving during a severe winter storm unless it is absolutely necessary. If you must go out, use main roads.
- If you are caught in a blizzard while driving, pull off the road and turn on your emergency flashers. Stay in the car until help arrives or the storm ends. Turn on the engine and run the heater for about ten minutes each hour to help you stay warm. Roll down a window slightly to avoid carbon monoxide poisoning.

 READING CHECK

Compare and Contrast What is the difference between a storm watch and a storm warning?

Natural Disasters

Main Idea Know what to expect during natural disasters.

All natural disasters, from floods to earthquakes to wild-fires, have one thing in common: knowing what to expect is your best defense.

Floods

Some floods develop slowly as heavy rain raises the level of rivers and lakes. **Flash floods**, however, are *floods in which a dangerous volume of water builds up in a short time*. Listen to radio and TV broadcasts for instructions if a flood is expected. If ordered to evacuate, secure your home and move essential items to an upper floor. Shut off utilities and disconnect electrical appliances. Don't walk through moving water or drive into a flooded area. If floodwaters surround your car, leave the car and flee to higher ground.

After a flood, return home only when authorities tell you it is safe to do so. Clean and disinfect everything in your home that got wet. Floods can contaminate the water supply, so drink bottled water until authorities tell you the water is safe to drink.

Earthquakes

An **earthquake** is *a series of vibrations in the earth caused by a sudden movement of the earth's crust*. Earthquakes are most common in western states. In the event of an earthquake, take the following precautions.

- **If you are indoors:** Drop to the ground. Take cover under a sturdy table or desk and hold on until the shaking stops. If there is no nearby table, crouch in a corner and cover your head with your arms.

- **If you are outdoors:** Stay clear of buildings, trees, streetlights, and power lines.

- **If you are in a car:** Stop the car and stay inside. Avoid stopping near or under trees, buildings, freeway overpasses, and power lines.

■ **Figure 27.13** Floods are one of the most common natural disasters in the United States. *Why should you avoid drinking tap water after a flood?*

 READING CHECK

Explain What is the best defense against all natural disasters?

After an earthquake, be prepared for aftershocks—smaller tremors that occur after the main quake. Use caution when opening overhead cabinets, and be aware that utilities such as gas, power, and sewer lines may be damaged.

Wildfires

Wildfires are most likely to occur in especially dry regions. People who live in areas where wildfires are common can create a "safety zone" around their homes that is free of most vegetation and other flammable materials. If you spot a wildfire, call 911 to report it, then evacuate before the fire reaches your home. Before you leave, shut off gas and oil supplies at their source and clear away any flammable materials near the house. Close all doors and windows, but don't lock the house since firefighters may need to get inside.

Being Prepared for Emergencies

■ **Figure 27.14** An emergency survival kit can help you wait out a disaster at home or travel with you if you must evacuate your area. *Why might you need each of the items shown here?*

Main Idea **Emergency supplies can help you survive.**

In an emergency, you may need to evacuate your home in a hurry, or "shelter in place"—stay in a secure location in your home until the crisis has passed. In either case, you'll need supplies to get you through the disaster. An **emergency survival kit** is *a set of items you will need in an emergency situation.* These items may include

- a three-day supply of food and water for your family. Choose shelf-stable, ready-to-eat foods. Store at least 3 gallons of water per person (1 gallon per person per day).
- a battery-powered radio or television (with extra batteries).
- a change of clothing for each family member.
- sleeping bags or bedrolls for each family member.
- first-aid supplies, including any necessary medications.
- duct tape and plastic sheeting, in case you need to seal off the windows in your home.
- copies of important documents, such as passports and birth certificates (if you need to leave your home).
- money (if you need to leave your home).

Keep a list of phone numbers for each member of your family so that you can reach one another if you are separated. Identify an out-of-town contact person to call if you can't get through to one another. Choose a meeting place for family members to go if you have to evacuate your area.

READING CHECK

Classify Which two items would you need in your emergency survival kit if you had to evacuate your home?

Sheltering in Place

During certain disasters, including terrorist attacks, people in the area may need to "shelter in place" until it's safe to go outside. Select a small, interior room with as few windows as possible for your shelter in place. A room at or above ground level is best, and it should contain a landline phone, since cell phone systems can be overwhelmed in an emergency.

Follow these steps when taking shelter:

▶ Close and lock all windows and exterior doors. Also close window shades or blinds if there is a risk of explosions.

▶ Turn off all fans and heating and air-conditioning systems. Close fireplace dampers.

▶ Gather all family members and pets in your safe room. Bring your emergency survival kit with you. You should also have a plastic bucket with a tight lid to use for personal waste, along with soap, toilet paper, and disinfectant.

▶ Use duct tape and plastic sheeting to seal off the room you are in, including all vents and cracks around the door.

▶ Keep listening to your radio or television until you hear that it is safe to leave.

Activity

Role-play a scene involving a family sheltering in place. Show the steps the family takes to stay safe during the crisis.

LESSON 4 ASSESSMENT

 ### After You Read

Reviewing Facts and Vocabulary

1. Explain the 30/30 rule for lightning safety.

2. Identify two warning signs of an approaching tornado.

3. What should you do if you are in a car during an earthquake?

Thinking Critically

4. **Synthesize.** Suppose you hear on the radio that a tornado watch has been issued for your area. How would you respond?

5. **Evaluate.** What are some of the possible consequences of not having an emergency survival kit?

Applying Health Skills

6. **Goal Setting.** Develop an emergency plan for your family. Make a list of the items you will gather for your emergency kit and the steps you will take in case of an emergency. Then set a deadline for completing your emergency preparedness goal.

Writing Critically

7. **Expository.** Choose one of the emergencies discussed in this lesson. Write an informational handout for families about what steps to take in this emergency.

Real Life Issues

After completing the lesson, review and analyze your response to the Real Life Issues question on page 778.

Hands-On HEALTH

Activity First-Aid Station

In this activity you will set up a first-aid learning station. You will develop a creative presentation demonstrating how to respond to a common emergency.

What You'll Need

- pen or pencil
- notebook paper
- poster board and markers
- props (optional)

What You'll Do

Step 1

Working in small groups, select a common emergency discussed in this chapter. Identify how to recognize the emergency, appropriate steps to take, and what to do after first aid has been provided.

Step 2

Decide on a creative way to present your material at a learning station. Ideas might include a poster, a board game, a quiz show, a puzzle, a news story, or demonstrations with props such as bandages.

Step 3

Set up your learning station and give your presentation to the class.

Apply and Conclude

Write a brief essay discussing the importance of learning first-aid procedures for common emergencies.

Checklist: Practicing Healthful Behaviors

☑ Identify how to recognize a common emergency.

☑ List appropriate first-aid steps for responding to this emergency.

☑ Explain what action to take after first aid has been provided.

LESSON 1

Providing First Aid

Key Concepts

▶ When responding to an emergency, remember three steps: check, call, and care.
▶ Take universal precautions when providing first aid.
▶ Major burns require professional medical care.

Vocabulary

▶ first aid (p. 758)
▶ Good Samaritan laws (p. 759)
▶ universal precautions (p. 759)

LESSON 2

CPR and First Aid for Shock and Choking

Key Concepts

▶ CPR combines rescue breathing and chest compressions.
▶ In cases of choking, abdominal thrusts can be used to dislodge the object blocking the windpipe.
▶ Treatment for shock involves elevating the legs and trying to keep the victim warm and calm.

Vocabulary

▶ chain of survival (p. 764)
▶ defibrillator (p. 765)
▶ CPR (p. 765)
▶ rescue breathing (p. 767)
▶ shock (p. 770)

LESSON 3

Responding to Other Common Emergencies

Key Concepts

▶ Use the P.R.I.C.E. procedure to treat minor sprains and strains.
▶ Fractures and dislocations are medical emergencies that require professional care.
▶ Animal bites that break the skin require medical attention.
▶ In all cases of suspected poisoning, call a poison control center for emergency assistance.

Vocabulary

▶ fracture (p. 772)
▶ dislocation (p. 772)
▶ unconsciousness (p. 773)
▶ concussion (p. 774)
▶ poison (p. 775)
▶ poison control center (p. 775)
▶ venom (p. 776)

LESSON 4

Emergency Preparedness

Key Concepts

▶ It is important to pay attention to weather warnings.
▶ During severe storms, stay indoors and follow guidelines for protecting yourself and your home.
▶ To protect yourself during an earthquake, stay close to the ground, protect your head, and avoid objects that might fall.

Vocabulary

▶ hurricane (p. 779)
▶ tornado (p. 779)
▶ blizzard (p. 780)
▶ flash floods (p. 781)
▶ earthquake (p. 781)
▶ emergency survival kit (p. 782)

LESSON 1

Vocabulary Review

Use the vocabulary terms listed on page 785 to complete the following statements.

1. During an emergency, _____ can mean the difference between life and death.

2. Statutes that protect rescuers from being sued for giving emergency care are called _____.

3. You can protect yourself from disease by following _____.

Understanding Key Concepts

After reading the question or statement, select the correct answer.

4. Universal precautions require you to wear sterile gloves whenever you
   a. encounter an emergency.
   b. treat a burn.
   c. perform rescue breathing.
   d. come into contact with someone's blood.

5. When treating a minor burn, you should *not*
   a. cool the burned area with running water.
   b. apply ice to the burned area.
   c. cover the burn with a sterile gauze bandage.
   d. give the victim pain relievers.

Thinking Critically

After reading the question or statement, write a short answer using complete sentences.

6. **Predict.** What are the possible consequences of treating a wound without following universal precautions?

7. **Summarize.** Describe the procedure for treating an open wound.

8. **Evaluate.** How can you tell if a burn is minor enough to be treated at home?

Rick Brady/McGraw-Hill Education

LESSON 2

Vocabulary Review

Correct the sentences below by replacing the italicized term with the correct vocabulary term.

9. A *shock machine* is a device that delivers an electric shock to the heart to restore its normal rhythm.

10. *First aid* is a lifesaving procedure that can replace a patient's normal heartbeat and breathing when these body functions have stopped.

11. *Fainting* is a life-threatening condition in which the heart is not delivering an adequate supply of blood to the body.

Understanding Key Concepts

After reading the question or statement, select the correct answer.

12. Before beginning rescue breathing, you should check to see
    a. whether the victim is breathing.
    b. whether the victim has a pulse.
    c. whether there is something in the victim's mouth.
    d. whether the victim has any open wounds.

13. A person who clutches his or her throat is most likely experiencing
    a. a heart attack.
    b. a stroke.
    c. choking.
    d. shock.

14. You should wrap your arms around the rib cage, rather than the abdomen, when assisting a choking victim who is
    a. an infant.
    b. seated.
    c. unconscious.
    d. pregnant.

Thinking Critically

After reading the question or statement, write a short answer using complete sentences.

15. **Cause and Effect.** What is the likely consequence of keeping automated external defibrillators in public places?

16. **Describe.** What is the correct position in which to place your hands for performing chest compressions?

17. **Identify.** List three symptoms of shock.

LESSON 3

Vocabulary Review

Choose the correct term in the sentences below.

18. A *fracture / dislocation* is a separation of a bone from its normal position in a joint.

19. Fainting is a form of temporary *concussion / unconsciousness.*

20. *Poison / Venom* is a harmful substance secreted by some types of snakes, spiders, and insects.

Understanding Key Concepts

After reading the question or statement, select the correct answer.

21. You should always seek professional medical care for
    a. strains.
    b. sprains.
    c. fractures.
    d. animal bites.

22. When treating a nosebleed, you should *not*
    a. squeeze your nostrils shut.
    b. breathe through your mouth.
    c. try to swallow the blood.
    d. apply a cold compress to the nose.

23. The first step in any case of suspected poisoning is to
    a. find out what poison has been taken.
    b. call a poison control center.
    c. induce vomiting.
    d. see if the victim develops symptoms.

Thinking Critically

After reading the question or statement, write a short answer using complete sentences.

24. **Explain.** How can you tell if someone is unconscious?

25. **Describe.** When should you suspect that a victim has a concussion?

26. **Evaluate.** Under what circumstances are insect bites and stings medical emergencies?

LESSON 4

Vocabulary Review

Choose the correct term in the sentences below.

27. A *hurricane / tornado* is a powerful storm that generally forms in tropical areas, producing strong winds and heavy rains.

28. In a *blizzard / hurricane,* falling and blowing snow reduces visibility to less than a quarter mile, making it very easy to lose your way.

29. You should stay indoors and take cover under a sturdy table or desk during a(n) *earthquake / flash flood.*

Assessment

Understanding Key Concepts

After reading the question or statement, select the correct answer.

30. You should *not* stay in your car if you are out on the road during
   a. a severe thunderstorm that includes hail and sleet.
   b. a tornado.
   c. an earthquake.
   d. a wildfire.

31. If you are caught in a blizzard while driving, you should
   a. keep driving at a slow speed.
   b. pull off the road and turn on your emergency flashers.
   c. leave your car and attempt to find your way on foot.
   d. have the heater turned on the entire time and keep the windows tightly closed.

32. Earthquakes are most common in
   a. western states.
   b. eastern states.
   c. summer.
   d. winter.

Thinking Critically

After reading the question or statement, write a short answer using complete sentences.

33. **Compare and Contrast.** How are hurricanes and tornadoes alike? How are they different?

34. **Describe.** How should you respond to a wildfire?

35. **Explain.** Why is it necessary to clean and disinfect items that have been through a flood?

Technology PROJECT-BASED ASSESSMENT

Administering First Aid

Background

First aid is the immediate care given to someone who is injured or ill. First aid is provided until professional medical care can be reached. Proper first-aid procedures can help reduce further injury or even prevent death. Always remember if immediate care is needed, please call 911.

Task

Create a video that effectively demonstrates proper first-aid procedures.

Audience

Fellow students and adults in the community.

Purpose

Show the steps in first-aid procedures for specific injuries and medical conditions.

Procedure

1 Choose several of the first-aid procedures discussed in the chapter to demonstrate in the video. Review the steps that are required in the procedures.

2 Collaborate as a group to write a script to accompany each first-aid procedure that you will demonstrate.

3 Make a storyboard of your video, in which you show what will happen in each scene in the video.

4 Work on special features that will appear in the video, such as props and titles.

5 Assign roles in the video to the members of your group, finalize the script, and rehearse the scenes.

6 Record your video, and present it to the class.

Math Practice

Solve Word Problems. Use the passage below to answer Questions 1–3.

> While hiking, Antonio and his younger brother saw that a young woman had collapsed on the hiking trail. The woman was unconscious, not breathing, and had no heartbeat.
> "Here, take my cell phone," he told his brother. "Go back to the beginning of the trail entrance and call 911. Tell them that I'm starting CPR."
> As his brother ran for help, Antonio, who had been trained to perform CPR, began the procedure. Antonio began by doing 30 chest compressions followed by two rescue breaths. He repeated these two steps for four continuous cycles until paramedics arrived to take over.

1. Imagine that x represents the total number of chest compressions Antonio had to do until the paramedics arrived. Which expression below represents how many total minutes Antonio had to perform CPR on the woman?
   A. $x(4 \times 30)$
   B. $4x/30$
   C. $30x/4$
   D. $x/(4 \times 30)$

2. If Antonio performed CPR steadily as described, how much time has passed in three cycles of compressions and breaths?
   A. 45 seconds
   B. About 1 minute
   C. 3 minutes
   D. 3 minutes 45 seconds

3. How many total chest compressions did Antonio have to perform if the paramedics took 15 minutes to arrive?

Reading/Writing Practice

Understand and Apply. Read the passage below, and then answer the questions.

> BROWNWOOD, TEXAS—At South Elementary today, a tornado touched down, injuring a teacher. Tyrone Rasco, a third-grade teacher who was standing outside the building, suffered minor injuries from the storm.
> Thirty children under the supervision of Ann Katz, a physical education teacher, were outside on a playground adjacent to the building. Because the playground was next to the building, Mrs. Katz rushed the children into the building as soon as she heard the tornado alarm. The kids hurried to hallways in the center of the building before the tornado hit. The storm broke the school's front door and most of its windows.
> "We are thankful that no students were injured in the storm," said Principal Jennifer Rodriguez.

1. What is the purpose of this article?
   A. To describe the damage that a tornado caused at a school
   B. To report about Mrs. Katz's actions
   C. To explain why people should stay inside during tornadoes
   D. To promote tornado warning systems

2. Which word or phrase has the same meaning as the words *adjacent to* in the second paragraph?
   A. next to
   B. nearby
   C. far from
   D. underneath

3. Write an article about how to stay safe during a tornado.

National Education Standards
Math: Operations and Algebraic Thinking
Language Arts: LACC.910.RI.1, LACC.910.RL.2.4

CHAPTER 28 Community and Environmental Health

Lesson 1
Community and Public Health

BIG Idea *Many people and organizations work together to promote individual and public health.*

Lesson 2
Air Quality and Health

BIG Idea *Both outdoor and indoor air quality can affect your health.*

Lesson 3
Protecting Land and Water

BIG Idea *Human actions can either damage or protect land and water.*

Activating Prior Knowledge

Using Visuals Look at the picture on this page. How are these teens contributing to community and environmental health? Why do you think it is important to protect the environment? Write a paragraph explaining your thoughts.

Chapter Launchers

Health in Action

Discuss the **BIG** Ideas

Think about how you would answer these questions:

▶ Why is community and public health important?

▶ How do your surroundings affect your personal health?

▶ How can your actions affect your environment?

Assess Your Health

Read each statement. On a separate sheet of paper, write "yes," "sometimes," or "no" based on your typical behavior.

1. I conserve water by turning off the faucet while brushing my teeth.

2. I reduce waste by recycling paper and plastic bottles and containers.

3. I donate gently used items to a friend or charity instead of throwing it away.

4. I turn off the television when it is not in use.

5. I switch off lights whenever I leave a room.

A "yes" response shows that you practice healthy behaviors. "Sometimes" indicates that you should analyze and possibly modify your behavior. A "no" response means that you should modify the behavior.

LESSON 1

Tim Fuller Photography

GUIDE TO READING

BIG Idea *Many people and organizations work together to promote individual and public health.*

Before You Read

Create a Cluster Chart. Write "Health Care System" and circle it. Surround it with circles labeled "Health Care Professionals," "Health Care Facilities," "Health Insurance," and "Health Agencies." As you read, add details for each topic.

New Vocabulary

▸ health care system
▸ primary care physician
▸ specialists
▸ medical history
▸ health insurance
▸ public health

Review Vocabulary

▸ health fraud (Ch.2, L.4)

Community and Public Health

Real Life Issues ································

Choosing a New Doctor. Caleb's family is moving to a new town, so they need to choose a new family doctor. Their health insurer's Web site offers a "physician search" feature that looks for doctors within a given area. However, Caleb wants more from his new doctor than a convenient location. He wants someone who is easy to talk to, like his previous doctor. He isn't sure how he can find a new doctor he'll be comfortable with.

Writing *Brainstorm a list of questions Caleb could ask to help him choose a new doctor. Then write a dialogue between Caleb and a doctor he's considering.*

The Health Care System

Main Idea The health care system includes all the ways you receive and pay for medical care.

All the health care professionals you see on a regular basis—your doctor, your dentist, the pharmacist at your local drugstore—are part of the nation's health care system. A **health care system** includes *all the medical care available to a nation's people, the way they receive care, and the way they pay for it.* You use the health care system when you

• go for a checkup with a **primary care physician**, *a medical doctor who provides physical checkups and general care.*

• see the school nurse about an injury.

• have your teeth examined by a dentist.

• consult **specialists**, *medical doctors who focus on particular kinds of patients or on particular medical conditions.*

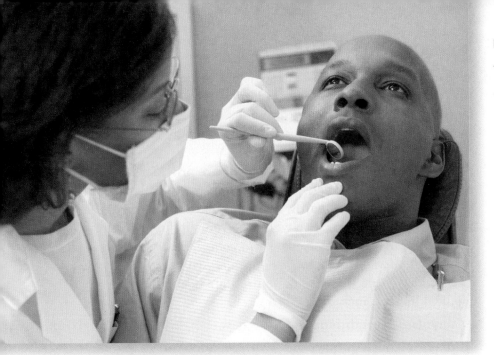

■ **Figure 28.1** Your dentist is just one of the people who may provide health care to you. *Who are other members of your health care "team"?*

©Kris Timken/Blend Images LLC

Examples of specialists include allergists (who treat allergies and asthma), dermatologists (who treat skin problems), gynecologists (who care for the female reproductive system), and pediatricians (who treat children).

Types of Health Care Facilities

You can receive health care in a variety of settings. For instance, if you become ill or injured at school, you might see the school nurse. If you need a checkup, you might visit a doctor's office or a clinic. A clinic is a community health facility where patients can receive *outpatient care,* which means being treated and returning home the same day.

Hospitals provide both outpatient care and *inpatient care,* which involves an overnight stay. Most hospitals have an emergency room where they handle urgent injuries or illnesses. Problems that are not life threatening may be treated at a facility called an urgent care center. Unlike ordinary doctor's offices, urgent care centers are typically open evenings and weekends and will see patients without an appointment.

Other types of health care facilities deal with specific problems or situations. Here are some examples:

- Birthing centers deliver babies in a homelike setting staffed by nurse-midwives.
- Drug treatment centers help people recover from drug and alcohol abuse.
- Assisted-living facilities provide care for people who need some help with everyday activities but do not need extended medical care.
- Hospices provide care for people who are terminally ill.

You and Your Health Care

A doctor shouldn't just be someone you call when you're sick or injured. Ideally, you should have an ongoing relationship with your health care provider to keep track of your health. Seeing a primary care physician regularly allows you to build up a relationship of trust.

Your relationship with your health care provider is a partnership. Your doctor can treat problems and make recommendations for your health, but you need to take an active role in promoting your own wellness. You should be aware of your **medical history**—*complete and comprehensive information about your immunizations and any health problems you have had to date*. Also, make sure your doctor knows your medical history, including

- any health conditions you have now.
- major physical or psychological problems you've had in the past.
- all medicines you are taking.
- any allergies you have to food or medication.
- any health problems that run in your family.
- your lifestyle and habits (for example, diet and exercise).

Your doctor should keep a record of your medical treatment on file. It will include information about health conditions, medications, and results of lab tests. A current trend in health care is to store these records in electronic form. This makes it easier for doctors to share information with each other and check for such problems as drug interactions. Whenever you have a doctor's appointment, note the reasons for your visit and list any questions you'd like to ask. During the visit, feel free to ask questions about the doctor's diagnosis or anything else you don't understand. Ask the doctor to write down any instructions so you won't forget them. If you have to fill a prescription at the pharmacy, you can also ask the pharmacist any questions you have about your medication.

■ Figure 28.2 Your pharmacist can be a good source of information about both prescription and over-the-counter medicines. *What kinds of questions might you ask a pharmacist?*

Paying Health Care Costs

Modern health care can be very expensive. Most people need some kind of health insurance to help pay their medical bills. **Health insurance** includes *private and government programs that pay for part or all of a person's medical costs.* People typically pay for health insurance with a monthly fee, known as a *premium.* There are two main forms of health insurance:

- **Fee-for-service.** Under these plans, patients must pay for all medical expenses up to a certain minimum amount, known as the *deductible.* After that, the insurance company covers a percentage of their costs. The portion that patients must still pay is called *coinsurance.* In many cases, patients must pay medical bills up front and then send a form to the insurance company to be reimbursed.

- **Managed care.** These plans hold down costs by limiting patients' choices and encouraging preventive care. Some plans, such as health maintenance organizations (HMOs), require patients to choose their doctors from a limited pool of physicians. Managed care plans may not cover certain types of medical care. However, they require less paperwork than fee-for-service plans. Rather than paying medical costs up front and applying for reimbursement, patients typically pay only a small fixed fee, known as a *copayment,* for each visit.

Most Americans receive insurance through their jobs (or through a family member's job). This is called group insurance. The employer may pay for part of the employee's premiums. People who cannot get insurance at work may buy individual policies. These vary widely in cost and in the benefits they offer, but generally cost more than group policies. People who cannot afford insurance at all may be covered under a federal government plan called Medicaid. All Americans over age 65 can receive coverage through a separate government program called Medicare.

Before selling a health policy to an individual, insurers generally require a medical exam to assess the person's level of health risk. People who are at a higher risk of developing health problems will be charged higher rates for insurance. A current **trend** is for group health plans to follow this practice as well. This reduces the risk that a plan will end up with more high-risk employees in its pool than it can afford.

The 2010 Patient Protection and Affordable Care Act (PP ACA) included a number of significant health care reforms. This law requires nearly all U.S. citizens to obtain health insurance and expands eligibility for Medicaid. The PPACA offers incentives for businesses to provide health care benefits. The act also prohibits insurers from using *pre-existing conditions* (health problems a person had before joining the insurance plan) as reasons to deny coverage or refuse claims.

 READING CHECK

Explain What costs must patients pay out of pocket with a fee-for-service plan? With a managed care plan?

Academic Vocabulary

trend *(noun):* a line of general direction or movement

Public Health Services

Blend Images/Ariel Skelley/Getty Images

Main Idea Agencies at all levels of government promote public health.

Your physician and other health care professionals help you take care of your personal health. However, you are also part of a community, and some health issues affect your community as a whole. To help deal with these issues, various agencies exist to promote public health. **Public health** includes *all efforts to monitor, protect, and promote the health of the population as a whole.* Public health agencies operate on all levels—local, state, national, and even worldwide. They work to make communities healthier by

- researching health problems.
- providing health services.
- educating the public.
- developing and enforcing policies that promote health.

Local Health Agencies

Local health departments are government agencies that operate at the city, county, or state level. They promote public health in various ways, including

- investigating threats to public health, such as outbreaks of disease.
- helping to plan responses to public health emergencies.
- enforcing local health regulations.
- providing information about health issues.

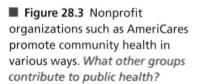

■ **Figure 28.3** Nonprofit organizations such as AmeriCares promote community health in various ways. *What other groups contribute to public health?*

Real World CONNECTION

Evaluating Health Care Services

Many communities offer a wide variety of health care options. People who live in these communities can choose from several different primary care physicians, dentists, and medical specialists. They may be able to receive care at a local hospital, an urgent care center, or a walk-in clinic.

Activity Technology

Conduct an Internet search for physicians using any online directories and hospital Web sites. Choose a physician near your home, and visit or call the office. Find out the answers to the following questions:

▸ When is the facility open?

▸ How much does an office visit cost? Which types of insurance are accepted?

▸ How soon can one get an appointment for a check-up? For a medical problem?

▸ Is the facility accepting new patients?

Write a brief report that evaluates this health care provider in terms of accessibility and cost.

National Health Agencies

Nonprofit agencies, such as the American Red Cross and the American Heart Association, work at the national level, but they may also have local chapters. These groups focus on specific health problems or goals. They may provide health services or help educate the public about specific health issues. They may also fund research into new treatments to fight disease.

Several departments of the U.S. government also promote public health at the national level.

- The **Environmental Protection Agency (EPA)** protects the country's land, air, and water. A major part of its job is enforcing environmental laws. The EPA also works to research issues related to the health of the environment and to educate the public about these issues.

- The **Occupational Safety and Health Administration (OSHA)** is part of the U.S. Department of Labor. It works to prevent injuries and other health problems in the workplace. OSHA sets safety standards for workplaces and helps train and educate workers.

- The **U.S. Department of Agriculture (USDA)** has several offices that promote public health. For example, the **Food Safety and Inspection Service** ensures the safety of meat, poultry, and eggs. The **Food and Nutrition Service** provides food to needy families.

Figure 28.4   **Health and Human Services Agencies**

The agencies in HHS oversee more than 300 health-related programs. *How does the FDA contribute to public health?*

National Institutes of Health (NIH)	Conducts and funds medical research
Food and Drug Administration (FDA)	Ensures the safety of foods and cosmetics and the safety and effectiveness of medicines
Centers for Disease Control and Prevention (CDC)	Works to track, prevent, and control outbreaks of disease
Indian Health Services (IHS)	Provides health care to Native Americans
Health Resources and Services Administration (HRSA)	Provides access to health care for low-income and uninsured people
Substance Abuse and Mental Health Services Administration (SAMHSA)	Funds programs to prevent and treat substance abuse and mental disorders
Agency for Healthcare Research and Quality (AHRQ)	Supports research on the health care system
Centers for Medicare and Medicaid Services (CMS)	Administers federal health insurance programs for elderly and low-income Americans
Administration for Children and Families (ACF)	Oversees programs to aid low-income families
Administration on Aging (AoA)	Provides services and support for older Americans

READING CHECK

Identify Which U.S. government agency works to fight health fraud?

- The **United Network for Organ Sharing (UNOS)** maintains data on potential organ recipients and donors. In many states, a person can indicate their choice to become an organ donor on a driver's license. When that person dies, his or her organs will be donated. Organ donation saves lives.

- The **Department of Health and Human Services (HHS)** includes ten agencies that promote public health in various ways. **Figure 28.4** shows the agencies that are part of HHS.

Global Health Organizations

Many countries don't have the same access to health care that the United States and other developed nations do. Disasters such as war, drought, flooding, or economic collapse can also harm the public health of a nation. Government agencies and private organizations from around the world work to help countries in such crises.

- The **World Health Organization (WHO)** is the health agency of the United Nations. Its goals include improving health care systems and fighting diseases such as AIDS and malaria.

- The **United Nations Children's Fund (UNICEF)** promotes children's health and well-being through immunization, disaster relief, and education.

- The **International Committee of the Red Cross** aids victims of war and other forms of violence. The organization also works to promote and strengthen humanitarian laws.

- The **U.S. Agency for International Development (USAID)** provides aid to foreign countries to promote health, economic growth, and democratic reforms.

- The **Peace Corps,** a U.S. government agency, sends volunteers to developing nations to promote such goals as health, education, and economic development.

- **Cooperative for Assistance and Relief Everywhere (CARE)** fights global poverty. Its work includes promoting education, improving sanitation, and fighting HIV/AIDS.

LESSON 1 ASSESSMENT

After You Read

Reviewing Facts and Vocabulary

1. What is the difference between a primary care physician and a specialist?

2. List three types of health care facilities.

3. Identify two organizations that work to promote global health.

Thinking Critically

4. **Synthesize.** Why is it important for your doctor to know your medical history?

5. **Compare and Contrast.** Compare the advantages of fee-for-service insurance and managed care.

Applying Health Skills

6. **Accessing Information.** Use reliable print and online resources to learn more about one of the public health agencies listed in this lesson. Write a report explaining how the agency promotes public health and prevents disease.

Writing Critically

7. **Expository.** Write a newspaper-style article advising other teens about how they can take an active role in their own health care. Include tips on how to get the most out of a visit to the doctor.

Real Life Issues ....

After completing the lesson, review and analyze your response to the Real Life Issues question on page 792.

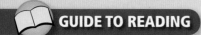
Air Quality and Health

Huntstock/Getty Images

GUIDE TO READING

BIG Idea *Both outdoor and indoor air quality can affect your health.*

Before You Read

Create a Comparison Chart. Make a three-column chart. Label the columns "Outdoor Air Pollution," "Indoor Air Pollution," and "Noise Pollution." As you read, use the chart to define each term, list causes and effects, and identify solutions.

Outdoor Air Pollution	Indoor Air Pollution	Noise Pollution

New Vocabulary

- air pollution
- smog
- Air Quality Index (AQI)
- greenhouse effect
- global warming
- noise pollution
- decibel

Real Life Issues

Saving Energy. Rachel has always been in the habit of turning on the television as soon as she comes home from school. Even if she's not watching it, she likes having it on in the background. Lately, though, she's started to wonder just how much electricity she's wasting by doing this—and how much air pollution she might be causing. This makes her think about what other habits she might have that waste energy and what she could do to cut back.

Writing *Evaluate your own energy usage. How might you reduce the amount of energy you use? Write your thoughts in a paragraph.*

Understanding Air Pollution

Main Idea Indoor and outdoor air pollutants can harm human health and damage the natural environment.

You normally can't see it, but air is all around you. The quality of the air you breathe has a significant impact on your health. **Air pollution**, *the contamination of the earth's atmosphere by harmful substances,* poses serious health concerns. In fact, numerous studies have linked it to a wide variety of health problems, including lung disease, cardiovascular disease, and cancer.

Air Quality

In the United States, the Environmental Protection Agency (EPA) sets air quality standards to prevent and correct problems related to environmental air pollution.

The EPA has placed limits on the levels of six pollutants that harm human health and the environment.

- **Ozone (O_3)** forms at ground level when certain other pollutants react chemically in the presence of sunlight. Ground-level ozone is a major component of smog, *a brownish haze that sometimes forms in urban areas.* Ozone irritates the lungs and makes breathing difficult. It can worsen respiratory problems such as asthma, bronchitis, and emphysema.

- **Particulate matter (PM)** is a general term for small particles found in the air, such as dust, soil, soot, smoke, mold, and droplets of liquid. PM can cause breathing difficulties, certain lung diseases, and even heart attacks.

- **Carbon monoxide (CO)** is a colorless, odorless gas that forms when carbon in fuel is not burned completely. Outdoor sources of CO include automobile exhaust and industrial processes. CO harms the body by preventing oxygen from reaching body tissues. At high enough levels, CO can be deadly.

- **Sulfur dioxide (SO_2)** comes chiefly from power plants, especially those that burn coal. In addition to harming respiratory health, SO_2 can combine with water to form acid rain, which is harmful to plants and animals.

- **Nitrogen oxides (NO_x)** are highly reactive gases that form when fuel is burned at high temperatures, as in motor vehicles and power plants. NO_x contributes to the formation of ground-level ozone, acid rain, PM, and a wide variety of toxic chemicals.

- **Lead** is a metal found naturally in the environment as well as in manufactured products. Exposure to lead can damage the kidneys, liver, brain, and nerves and can cause cardiovascular disease and anemia.

To track the levels of pollutants in the air, the EPA has created the **Air Quality Index (AQI)**, *an index for reporting daily air quality.* The AQI, shown in **Figure 28.5** on page 802, informs the public about local air quality and whether pollution levels pose health risks.

Greenhouse Gases

Air pollutants can also contribute to global climate change. The **greenhouse effect** is *the trapping of heat by gases in the earth's atmosphere.* These gases allow sunlight to enter our atmosphere but block radiation from escaping to outer space—much like the glass roof of a greenhouse. The chief greenhouse gas produced by human activity is carbon dioxide (CO_2). The burning of fossil fuels in power plants and motor vehicles is the chief source of CO_2 buildup.

Academic Vocabulary

component *(noun):* a constituent part or ingredient

Figure 28.5 **Air Quality Index (AQI)**

The EPA created this index to inform the public about daily air quality. *How can you use the AQI in your community to help protect your health?*

Range	Air Quality	Color Code
0 to 50	**Good:** There is little or no health risk.	Green
51 to 100	**Moderate:** Some pollutants may pose a moderate health concern for a very small number of people.	Yellow
101 to 150	**Unhealthy for Sensitive Groups:** Members of sensitive groups, such as people with lung disease, may experience health effects.	Orange
151 to 200	**Unhealthy:** Everyone may begin to experience health effects.	Red
201 to 300	**Very Unhealthy:** Everyone may experience more serious health effects.	Purple
301 to 500	**Hazardous:** Emergency conditions. The entire population is at risk.	Maroon

READING CHECK

Cause and Effect
How do human actions contribute to global warming?

The greenhouse effect is actually normal and necessary to support life on this planet. In the past 200 years, however, the concentration of greenhouse gases trapped in the earth's atmosphere has risen, resulting in **global warming**. This is *an overall increase in the earth's temperature.* Since 1900, the earth's average surface temperature has risen by 1.2 to 1.4 degrees F. If levels of greenhouse gases continue to rise, average temperatures could increase anywhere from 2.5 to 10.4 degrees F by the end of the twenty-first century.

The exact effects of global warming are hard to predict. Already, though, glaciers are beginning to melt, causing sea levels to rise. Global weather patterns could also shift. Areas might receive much less or much more rainfall than they do now. Plants and animals that cannot adapt to the new conditions could become extinct. The world's food supply could also be at risk if crop-growing areas are struck by drought.

Indoor Air Pollution

Research has found that in many cases, the air inside buildings contains more pollutants than the outdoor air, even in the biggest cities. Common sources of indoor air pollution include household chemicals, such as cleaning fluids and pesticides, and chemicals used in building and furnishing materials. Lack of ventilation makes the problem worse by trapping air pollutants inside.

Specific problems with indoor air quality include

- **carbon monoxide,** produced by fuel-burning equipment, such as stoves, furnaces, and fireplaces.

- **asbestos,** a mineral fiber. In the past, asbestos was often used as a fire retardant in insulation and building materials. Cutting or sanding these materials can release particles of asbestos into the air. Inhaling these particles can lead to lung cancer and other forms of lung damage.

- **radon,** an odorless, radioactive gas produced during the natural breakdown of the element uranium in soil and rocks. It can enter homes through dirt floors, cracks in concrete floors and walls, or floor drains. Exposure to high levels of radon can cause lung cancer.

Reducing Air Pollution

Main Idea Your choices can fight air pollution.

You can make choices to help reduce air pollution. Since power plants and home heating systems are sources of air pollution, reducing your use of energy is a good place to start. Here are some tips for saving energy:

- Switch off lights whenever you leave a room. Consider replacing regular incandescent lightbulbs with compact fluorescent bulbs, which use less energy and last longer.

- Turn off radios, computers, televisions, and other such appliances when they are not in use.

- In the winter, wear extra layers of clothing to stay warm so you can keep the thermostat at around 68 degrees F. Turn the thermostat down even lower at night. In the summer, set the thermostat at around 78 degrees F to keep the air conditioning from coming on too often. Instead, use a fan to cool rooms.

- Insulate your home to reduce your need for heating and cooling. Seal leaks around doors, windows, and electrical sockets to prevent heated or cooled air from escaping.

- Wash clothes in warm or cold water rather than hot water.

- When cooking, don't preheat the oven longer than necessary. Try cooking small amounts of food in a toaster oven or microwave rather than a full-size oven.

Cars are another major source of air pollution. Whenever you can, try walking, riding a bicycle, using public transportation, or carpooling to save gas. Another way to conserve gasoline is to reduce the use of motorized equipment, such as power mowers, chain saws, and leaf blowers. When possible, use hand tools to get the job done.

 **READING CHECK**

Identify List three actions you and your family can take to reduce air pollution.

Managing Indoor Air Pollution

To improve indoor air quality, you can identify sources of pollution and get rid of them. Home test kits and detectors can help you measure the levels of radon and carbon monoxide in your home. Depending on what you find, you may be able to eliminate the pollution sources yourself, or you may need the help of a professional.

If you can't get rid of all the sources completely, you may be able to reduce the pollutant levels in the air by increasing the ventilation in your home. Opening windows and turning on window or attic fans can help remove pollutants that build up in the short term. A long-term, more expensive solution is to modify your home's ventilation system. You can also try using air cleaners to filter out particle pollution. However, these devices cannot eliminate most gaseous pollutants.

■ **Figure 28.6** Riding mass transit is one way to reduce the air pollution associated with car use. *What are other transportation options that help reduce air pollution?*

 **READING CHECK**

Define *What is noise pollution?*

Noise Pollution

Main Idea **Exposure to loud noises can harm your health.**

Traffic, loud music, construction equipment, and power tools are all sources of **noise pollution**. This is *harmful, unwanted sound loud enough to damage hearing.* To better understand what types of noise levels can harm your hearing, take a look at **Figure 28.7**. This graph shows the decibel levels of some common sounds. A **decibel** is *a unit that measures the intensity of sound.* A level of 0 decibels represents the lowest level of sound the human ear can detect. Noise levels of 130 decibels or higher can cause pain.

If you are exposed to loud noise, you may experience a temporary hearing loss, which may be accompanied by ringing in the ears. In most cases, you will recover your normal hearing shortly after the noise stops. However, repeated exposure to noise at levels around 90 decibels or higher can lead to permanent hearing loss.

If you are going someplace where you are likely to be exposed to loud noise, wear earplugs or earmuffs. You can also avoid contributing to noise pollution by keeping the volume down on stereos and TV sets. Use manual tools instead of power tools, and avoid using your car horn unnecessarily.

Figure 28.7 Decibel Levels of Common Sounds

A decibel is not a fixed unit of sound. Instead, each 10-decibel increase roughly doubles the loudness of a sound.

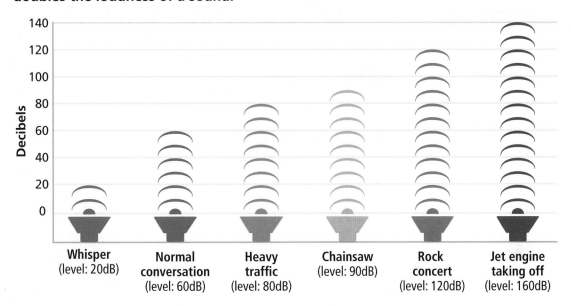

Whisper (level: 20dB)	
Normal conversation (level: 60dB)	
Heavy traffic (level: 80dB)	
Chainsaw (level: 90dB)	
Rock concert (level: 120dB)	
Jet engine taking off (level: 160dB)	

LESSON 2 ASSESSMENT

 ### After You Read

Reviewing Facts and Vocabulary

1. Name the six outdoor air pollutants for which the EPA sets limits.

2. How is radon harmful to human health?

3. What might result from repeated exposure to sounds of 90 decibels or louder?

Thinking Critically

4. **Evaluate.** How might your health be affected if the EPA stopped regulating common air pollutants?

5. **Analyze.** Explain the connection between the greenhouse effect and global warming.

Applying Health Skills

6. **Decision Making.** The morning news has reported an AQI of 145. Paul, who has asthma, was planning to go in-line skating with his friends. Use the decision-making process to determine what you would do in Paul's place.

Writing Critically

7. **Persuasive.** Write a script for a public service announcement urging teens and others to take steps to reduce air pollution. List specific steps in your announcement and show how each is tied to the goal of reducing air pollution.

Real Life Issues ....

After completing the lesson, review and analyze your response to the Real Life Issues question on page 800.

Protecting Land and Water

BIG Idea *Human actions can either damage or protect land and water.*

Before You Read

Create a Word Web. Write "Healthy Environment" in the center of a sheet of paper. Around it, jot down characteristics of a healthy environment. As you read, add more notes to your word web.

Healthy Environment

New Vocabulary

▶ biodegradable
▶ landfill
▶ hazardous wastes
▶ deforestation
▶ urban sprawl
▶ wastewater
▶ conservation
▶ precycling
▶ recycling

Real Life Issues

Wondering About Waste. Carlos has started to notice just how much trash his family throws out every day. A lot of it seems to be packaging, such as empty juice boxes and snack wrappers. He wonders whether changing some of their buying habits could make a big dent in the amount of waste they create. However, he's not sure how to convince his family that it's worth the effort.

Writing *Brainstorm a list of strategies to reduce waste. Then write a dialogue in which Carlos persuades his family to try some of these ideas.*

Waste Disposal

Main Idea Wastes need to be disposed of safely.

Getting rid of the waste we produce is a big problem for our society. If wastes aren't properly contained or destroyed, they can pollute the land and water we rely on to live. This can harm human health by making water supplies unsafe to drink or contributing to the spread of pathogens.

Many types of waste are **biodegradable**, or *able to be broken down by microorganisms in the environment.* Biodegradable wastes will not break down if they are disposed of in ways that do not expose them to the environment. Many other wastes are not biodegradable and need to be disposed of in ways that will do the least possible damage to the environment.

Solid Waste

Municipal solid waste (MSW) is another term for trash or garbage. There are two ways to dispose of MSW:

- **Landfills.** A **landfill** is *a specially engineered area where waste can be buried safely.* Modern landfills have a variety of safeguards in place to prevent wastes from damaging the nearby land and water. For example, they must be located away from sensitive natural areas and operated in ways that control odors and pests. They also must use special liners to prevent leakage that could contaminate groundwater.

- **Incineration.** Burning wastes in specially designed incinerators reduces the volume of trash that otherwise would go into landfills. Waste incinerators use "scrubbers" and filters on their smokestacks to reduce air pollution. Some incinerators use the energy from burning waste to produce electricity, reducing the need for fuel.

■ **Figure 28.8** Some MSW is burned in specially designed incinerators. *What is another way to dispose of solid waste?*

Hazardous Wastes

Hazardous wastes are *waste materials with properties that make them dangerous to human health or the environment.* Wastes may be considered hazardous because they are corrosive, chemically unstable, highly flammable, or toxic. Most hazardous wastes must be treated to make them less harmful before being discarded at special disposal sites. Types of hazardous waste include

- **industrial wastes.** These include solvents used for cleaning and degreasing as well as sludge and wastewater from certain industries, such as oil refining.

- **household wastes.** Products such as pesticides, paints, cleaning fluids, and batteries may be hazardous when discarded. Some household wastes are banned from landfills and must be disposed of at special collection sites.

- **radioactive wastes.** Sources such as nuclear power plants produce wastes that emit radiation. Exposure to radiation can increase the risk of cancer. It can also cause mutations, harmful changes in the body's DNA, which may be passed on to offspring. Extremely high levels of radiation can cause burns and radiation sickness.

READING CHECK

Explain What properties make certain wastes hazardous?

Because of these risks, such wastes must be isolated in secure storage sites until the radioactive materials decay, or cease to be radioactive.

- **mercury.** This naturally occurring substance is highly toxic to humans and other animals. It is found in some medical and dental wastes and in certain parts of cars.

Problems of Development

 **Main Idea** Urban development impacts the environment.

Throughout human history, the number of people on the planet has been growing at a faster and faster rate. As the world population grows, so does our use of resources. Wilderness areas give way to new urban developments, with drastic impacts on the environment.

Disappearing Forests

In many parts of the world, forests are being cleared away to make room for new developments. Timber from these forests is also used for fuel and manufacturing. **Deforestation**, or *destruction of forests*, causes a variety of problems.

- It destroys the habitats of plant and animal species.
- The loss of trees puts these areas at risk for soil erosion and flooding.
- It alters the local climate, making it hotter and drier.
- It contributes to global warming because trees absorb carbon dioxide, a greenhouse gas, from the atmosphere.

READING CHECK

Cause and Effect How does deforestation contribute to global warming?

■ **Figure 28.9** Forests play a vital role in the environment, providing oxygen, homes for a variety of living organisms, and natural beauty. *What are some of the consequences of destroying forests?*

Urban Sprawl

The spreading of city development (houses, shopping centers, businesses, and schools) onto undeveloped land is called **urban sprawl**. Sprawl contributes to several environmental problems. For example, paved areas do not filter rainwater the way soil does. As a result, runoff can carry pollutants into the water supply. Sprawl also destroys wildlife habitats and reduces air quality.

To combat these problems, some city planners have embraced a concept called *smart growth*. It involves planning communities in ways that use fewer resources and create less pollution. For example, planners may incorporate *mixed-use development*. In this type of development, businesses, homes, and schools are located close together, making it easier for people to walk from place to place instead of driving. Smart growth also involves building more compactly, which preserves open land and provides more **transit** choices.

Academic Vocabulary

transit *(noun):* local transportation

Water: A Limited Resource

Main Idea Pollution threatens our limited water supply.

Less than 1 percent of the earth's water is in a form humans can use. This limited supply makes water a precious resource. The EPA is responsible for protecting the water supply in the United States. It sets standards for the quality of drinking water and seeks to protect the health of oceans, rivers, and other water systems.

Sources of Water Pollution

Rivers, lakes, and aquifers (bodies of groundwater) provide much of our water supply. Pollutants can enter the supply in several different ways:

- **Runoff.** When rainwater or melting snow flows across the ground and into the water supply, it can pick up pesticides, fertilizers, salt, and other wastes.
- **Wastewater,** *used water from homes, communities, farms, and businesses.* Wastewater can contain pollutants, such as human wastes, metals, and pathogens. The EPA requires that wastewater be treated to remove pollutants before it is released back into the environment.
- **Sediment.** Runoff can carry soil and other sediments into the water supply. This can clog lakes and rivers.
- **Oil.** Spills from oil tankers and offshore drilling rigs can pollute the water. Oil poured down drains or onto the ground can also enter the water supply.

 **READING CHECK**

Identify Name four sources of water pollution.

Water Scarcity

Not only is the world's supply of water limited, but this supply is also unevenly distributed. An estimated 500 million people around the world have trouble getting the water they need to live. As the world population continues to grow, the demand for water increases. Disputes over water resources could lead to major conflicts between nations.

Protecting the Environment

Main Idea Conservation helps protect the environment.

People can help protect the environment by practicing **conservation**. This means *avoiding waste through careful management of natural resources,* such as energy, water, and materials. In Lesson 2, you learned about ways to conserve energy. You can also conserve water and land by reducing your water use and minimizing pollution:

- Repair leaky faucets, and never leave water running unnecessarily.
- Wait until you have a full load before doing laundry or running the dishwasher.
- Consider installing low-flow showerheads, faucet aerators, and low-flush toilets.
- Avoid overwatering your lawn and garden. Consider landscaping with plants that require less water.
- Try reducing your use of household chemicals or looking for versions that are less hazardous.
- Check with local authorities to find out how to discard hazardous wastes such as oil, paint, and batteries safely.

You also can reduce the amount of waste you produce by practicing the three Rs:

- **Reduce.** Source reduction, or **precycling**, means *reducing waste before it is generated.* For example, you can choose reusable products (such as cloth napkins) and products with less packaging.
- **Reuse.** Next to precycling, the most efficient way to reduce waste is to reuse items. If you have an item that you no longer need, you can sell it, give it to a friend, or donate it to a charity instead of throwing it away. You can also repair broken items instead of buying new ones.
- **Recycle.** Recycling is *the processing of waste materials so that they can be used again.* Recycling conserves materials and energy and reduces the need for new landfills and incinerators.

 **READING CHECK**

List What are three steps you can take to help conserve natural resources?

Tim Fuller Photography

Health Skills Activity

Advocacy

Promoting Recycling

Kathy's family has just moved to a new town. In her old neighborhood, her family could leave recyclables at the curb to be picked up, but her new town doesn't offer this service. There is a local drop-off center, but not many people seem to use it. Most people put all their recyclables—newspapers, cans, bottles, and so on—in the regular garbage.

Kathy wonders what she can do to encourage more of her neighbors to recycle and maybe even get a curbside recycling program started. She decides to write a letter to the editor of her local paper.

Writing Write Kathy's letter. In it, discuss the benefits of recycling and encourage people to use the drop-off center for recyclables. The letter should also urge city leaders to start a curbside recycling program.

LESSON 3 ASSESSMENT

After You Read

Reviewing Facts and Vocabulary

1. Name four types of hazardous wastes.

2. How can wastewater harm the environment?

3. What is the difference between *reuse* and *recycle*?

Thinking Critically

4. **Analyze.** What rights do individuals and groups have when it comes to the environment?

5. **Explain.** How does population growth contribute to environmental problems?

Applying Health Skills

6. **Advocacy.** Create a comic book for elementary school students about a superhero who fights pollution. The comic should contain a strong message encouraging young people to reduce land and water pollution.

Writing Critically

7. **Descriptive.** Write a description of a real or imaginary town that is built according to the principles of smart growth.

Real Life Issues ....

After completing the lesson, review and analyze your response to the Real Life Issues question on page 806.

Hands-On HEALTH

Activity Twenty-Four Hours and One Brown Paper Bag

Your challenge is to make a positive difference in the world by reducing, reusing, and recycling as much as you can during a 24-hour period.

What You'll Need
- small brown paper lunch bag
- paper and pen/pencil

What You'll Do

Step 1

Review Chapter 28 to gain a better understanding of the benefits of reducing, reusing, and recycling waste products.

Step 2

Put your name on one small brown paper lunch bag and for 24 hours challenge yourself to reduce, reuse, and recycle everything you can. Anything you cannot must fit into the paper bag.

Step 3

Journal about your experience. Bring your paper bag and journal to class after 24 hours.

Apply and Conclude

Based on your experience, describe an environmental goal that will have a positive impact on your community.

Checklist: Advocacy, Goal Setting

✓ Support for the position with relevant information

✓ Encourage others to make healthful choices

✓ Identify realistic goals

✓ Plan for reaching the goals

✓ Evaluate or reflect on the plan

LESSON 1

Community and Public Health

Key Concepts

▸ Types of health care providers include primary care physicians, school nurses, dentists, and specialists.

▸ You may receive health care in settings such as doctor's offices, clinics, hospitals, and urgent care centers.

▸ It is important to develop an ongoing relationship with a primary care physician.

▸ The two main forms of health insurance are fee-for-service and managed care.

▸ Agencies at all levels of government promote public health.

Vocabulary

▸ health care system (p. 792)
▸ primary care physician (p. 792)
▸ specialists (p. 792)
▸ medical history (p. 794)
▸ health insurance (p. 795)
▸ public health (p. 796)
▸ health fraud (p. 798)

LESSON 2

Air Quality and Health

Key Concepts

▸ The Environmental Protection Agency (EPA) places limits on the levels of six air pollutants: ozone, particulate matter, carbon monoxide, sulfur dioxide, nitrogen oxides, and lead.

▸ Greenhouse gases such as carbon dioxide contribute to global warming, which can dramatically alter the earth's climate.

▸ Examples of indoor air pollutants include carbon monoxide, asbestos, and radon.

▸ You can help reduce air pollution by using less energy in your home and reducing car use.

▸ Exposure to loud noises can damage your hearing.

Vocabulary

▸ air pollution (p. 800)
▸ smog (p. 801)
▸ Air Quality Index (AQI) (p. 801)
▸ greenhouse effect (p. 801)
▸ global warming (p. 802)
▸ noise pollution (p. 804)
▸ decibel (p. 804)

LESSON 3

Protecting Land and Water

Key Concepts

▸ Solid waste can be disposed of in landfills or incinerators.

▸ Hazardous wastes require special treatment for safe disposal.

▸ The impacts of population growth on the environment include deforestation and urban sprawl.

▸ Water pollution from runoff, wastewater, sediment, and oil threaten the earth's limited water supply.

▸ People can reduce environmental problems by conserving resources, reducing waste, and reusing or recycling materials.

Vocabulary

▸ biodegradable (p. 806)
▸ landfill (p. 807)
▸ hazardous wastes (p. 807)
▸ deforestation (p. 808)
▸ urban sprawl (p. 809)
▸ wastewater (p. 809)
▸ conservation (p. 810)
▸ precycling (p. 810)
▸ recycling (p. 810)

LESSON 1

Vocabulary Review

Use the vocabulary terms listed on page 813 to complete the following statements.

1. All the health care professionals you see are part of the nation's _____.

2. The two main forms of _____ are fee-for-service and managed care.

3. _____ includes all efforts to monitor, protect, and promote the health of the population as a whole.

Understanding Key Concepts

After reading the question or statement, select the correct answer.

4. You would likely see a specialist for
   a. an annual checkup.
   b. a flu shot.
   c. a minor injury.
   d. a condition that does not respond to normal treatment.

5. How are urgent care centers different from ordinary doctor's offices?
   a. They handle only medical emergencies.
   b. They can provide inpatient care.
   c. They see patients without appointments.
   d. They have several doctors working in the same place.

6. Which of the following is not a duty of local health departments?
   a. Investigating threats to public health
   b. Enforcing local health regulations
   c. Educating the public about health issues
   d. Funding research for new treatments

Thinking Critically

After reading the question or statement, write a short answer using complete sentences.

7. **Evaluate.** What are the benefits of seeing a primary care physician on a regular basis?

8. **Describe.** What does your medical history include?

9. **Evaluate.** What are the advantages of receiving group insurance through your employer over buying an individual policy?

10. **Explain.** What is the distinction between *health care* and *public health?*

11. **Explain.** What role does the Federal Trade Commission play in promoting public health?

LESSON 2

Vocabulary Review

Choose the correct term in the sentences below.

12. Particulate matter, carbon monoxide, and sulfur dioxide are all components of *air pollution / smog.*

13. *The greenhouse effect / Global warming* is natural and is necessary to support life on earth.

14. The EPA created the *Air Quality Index / decibels* to track the levels of pollutants in the air.

15. Traffic, loud music, construction equipment, and power tools such as lawn mowers can all be sources of *air pollution / noise pollution.*

Understanding Key Concepts

After reading the question or statement, select the correct answer.

16. The pollutant that makes up the largest component of urban smog is
  a. ozone.
  b. carbon monoxide.
  c. sulfur dioxide.
  d. nitrogen oxides.

17. An AQI of 76 falls into the "moderate" range. This means that the quality of the air poses
  a. little or no health risk for anyone.
  b. a moderate concern for a very small number of people.
  c. a threat to members of sensitive groups, such as those with lung disease.
  d. a serious threat to everyone.

18. Which of the following is *not* a common indoor air pollutant?
  a. Asbestos
  b. Carbon monoxide
  c. Ozone
  d. Radon

19. Which of the following noises is loud enough to cause hearing damage?
  a. Whispered conversation
  b. Normal conversation
  c. Heavy traffic
  d. Rock concert

Thinking Critically

After reading the question or statement, write a short answer using complete sentences.

20. **Analyze.** How does reducing your use of energy combat global warming?

21. **Evaluate.** Suppose your school has old asbestos insulation in its walls. Is it a good idea to cut into the walls and remove it? Why or why not?

22. **Identify.** Give two examples of specific steps you can take to reduce air pollution.

23. **Synthesize.** Suppose your family wants to find out whether there is an unsafe level of carbon monoxide in your home, and fix the problem if there is. What steps would you take?

Vocabulary Review

Correct the sentences below by replacing the italicized term with the correct vocabulary term.

24. Many types of waste are *recyclable,* or able to be broken down by microorganisms in the environment.

25. Waste materials with properties that make them dangerous to human health or the environment are known as *municipal solid waste.*

26. Soil erosion, flooding, and an increase in global warming are all problems associated with *urban sprawl.*

Understanding Key Concepts

After reading the question or statement, select the correct answer.

27. Which of the following is an example of smart growth?
  a. Zoning regulations that require homes and businesses to be in separate parts of a city or town
  b. Buying farmland to build new housing developments and shopping malls
  c. Buildings that have shops on the bottom level and apartments on the upper levels
  d. Widening streets to make more room for parking

Assessment

28. Which of the following is *not* a problem associated with deforestation?
   a. Loss of habitat for plants and animals
   b. Increased soil erosion
   c. Changes in the local climate
   d. Destruction of old farms and ranches

29. The best way to dispose of used motor oil is to
   a. dump it down a storm drain.
   b. pour it out onto the ground.
   c. put it out with your regular trash.
   d. take it to a service station for recycling.

30. Which of the following is an example of precycling?
   a. Using cloth shopping bags
   b. Giving your old computer to a friend
   c. Repairing your broken radio instead of throwing it out and buying another
   d. Returning empty bottles to the manufacturer to be sterilized and refilled

Thinking Critically

After reading the question or statement, write a short answer using complete sentences.

31. Explain. Why do radioactive wastes need to be isolated in secure storage sites for a long time?

32. Identify. List three problems associated with urban sprawl.

33. Analyze. If 70 percent of the earth's surface is covered with water, why do 500 million people around the world have trouble getting the water they need?

34. Evaluate. Rank the strategies of precycling, reuse, and recycling in terms of their effectiveness in reducing waste. Give reasons for your answer.

Technology — PROJECT-BASED ASSESSMENT

Promoting Precycling

Background
Precycling is a strategy for reducing waste before it is generated. It is an effective and important way to conserve natural resources, but it is not as widely understood or as widely promoted as recycling.

Task
Create a blog that promotes two methods of precycling.

Audience
Students in your school.

Purpose
Show one strategy for precycling, and show the benefits of precycling.

Procedure

1 Review the concept of precycling (covered in Lesson 3 of this chapter). Conduct an Internet search to explore the benefits of precycling. Investigate effective strategies for precycling that could be practiced by students.

2 In a group, brainstorm specific ways that the students at your school could practice precycling.

3 Select one of these precycling strategies to be the focus of your blog.

4 Present your blog to your teacher for consideration for the school's Web site.

Standardized Test Practice

Math Practice

Calculating Costs. Read the paragraph below, and then answer the questions.

Two families have medical insurance policies through different employers. The Lopez family pays $250 a month, and the insurance company will pay 85 percent of the cost of hospital stays. Family members pay $20 for each doctor visit. This insurance plan does not cover any vision costs. The Perez family's plan costs $410 a month, and hospital stays are completely covered. Family members pay $15 for each doctor visit. This plan pays the entire cost of an eye exam and $100 toward a pair of glasses or contact lenses.

1. Pedro Lopez had a hospital stay that cost $4,000. Before that, he had three doctor's appointments, each of which cost $93. Which function describes what Pedro has to pay? (Hint: C is Pedro's cost, H is the cost of the hospital stay, and D is the cost of each doctor visit.)
   A. $C = H + D$
   B. $C = H + 3D$
   C. $C = 0.15H + 20D$
   D. $C = 0.15H + 3D$

2. Melissa Perez has an eye exam and finds out she needs glasses. The glasses cost $395, and the exam is $95. How much does she have to pay?
   A. $295
   B. $300
   C. $395
   D. $490

3. In one year, both families had the following medical expenses: a hospital stay that cost $12,000, 14 trips to the doctor, four eye exams at $100 each, and two pairs of glasses at $300 each. Which policy would be the best to have under these circumstances? Why?

Reading/Writing Practice

Understand and Apply. Read the passage below, and then answer the questions.

Did you know that Americans throw away more than 245 million tons of trash each year? Much of that garbage ends up in landfills. Before these landfills fill up, we need to come up with new ways to get rid of our garbage. One way to do it is recycling.

Some people worry that recycling costs too much, but it actually saves money because it uses less energy than manufacturing new items. Others say that people are too lazy to separate trash and wash out cans and bottles. This is not true.

The best thing about recycling is that it is something everyone can do to help our planet. No matter what our age or our economic or education level, we can all take part. Earth is where we all are living and so we should care.

1. Which sentence should be added at the end of paragraph 2 to support the topic?
   A. It's hard to peel the labels off jars.
   B. It's better to precycle instead.
   C. Across the country, recycling rates have risen steadily over the years.
   D. My school started a recycling program.

2. What is the most effective way to rewrite the last sentence?
   A. We should all care about where we live.
   B. Why don't people care about recycling?
   C. Earth is our home, and we should keep it safe and healthy.
   D. Our planet is beautiful from space.

3. Write a persuasive paragraph urging sports arenas to recycle bottles and cans.

National Education Standards

Math: Operations and Algebraic Thinking
Language Arts: LACC.910.L.3.6, LACC.910.RL.2.4

Chapter 28 Standardized Test Practice **817**

TEENS *Speak Out*

LWA/Dann Tardif/Blend Images LLC

Is Graduated Driver's Licensing Good for Teens?

*T*een drivers are much more likely to get into accidents than older drivers, who have more experience. A recent study found that among teens ages 16 to 19, there were 148 crashes for every 1,000 teen drivers. This rate was much higher than the rate for any other age group.

To protect young drivers, most states now have graduated licensing programs. These programs put various restrictions on young drivers, which gradually decrease as these drivers gain experience. Read on to find out what two teens have to say about graduated licensing programs.

Drawbacks of Graduated Licensing

With graduated licensing, newly licensed teen drivers cannot engage in certain types of driving. For instance, they may not be allowed to drive at night or with more than one passenger. These restrictions apply to all young drivers, not just the problem drivers. As a result, teens may have trouble getting to school, work, and after-school activities.

❝My state's graduated license program says that I can't drive after dark unless I have an adult with me. I don't think it's fair that they're assuming I'll be a bad driver just because I'm a young person."

—Nancy B., age 17

Benefits of Graduated Licensing

Graduated licensing gives teens a chance to gain experience driving in safer situations. They can improve their driving skills with adult supervision. They can also practice driving in the daytime and without passengers to distract them. Graduated licensing has been found to reduce the rate of accidents by 10 to 30 percent.

❝Learning to drive is kind of like learning to swim. You don't want to jump into the deep end until you've had some practice. I like the idea that with my provisional license, I can gradually become more comfortable with the toughest kinds of driving, instead of being thrown into the deep end."

—Ali S., age 16

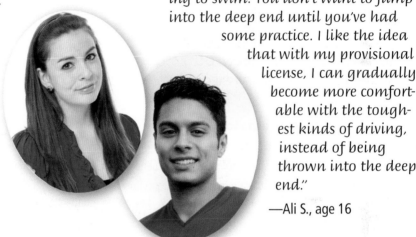

Activity Beyond the Classroom

1. **Research** graduated licensing requirements in your state. Do teens go directly from a learners' permit to full driving privileges, or is there an intermediate stage? If there is a special license for beginners, what restrictions does it impose?

2. **Discuss** these graduated licensing requirements with other teens and with adults. How do they feel about this issue? What arguments do they make for and against the program?

3. **Summarize** your findings in a newspaper article. Discuss the attitudes teens and adults have toward graduated licensing. In the last paragraph, draw your own conclusions about whether the benefits of graduated licensing outweigh the drawbacks.

Glossary / Glosario

Pronunciation Key

The following key will help you sound out words in the glossary.

a	back (BAK)	**yoo**	pure (PYOOR)
ay	day (DAY)	**yew**	few (FYEW)
ah	father (FAH thur)	**uh**	comma (CAH muh)
ow	flower (FLOW ur)	**u** (+ cons.)	rub (RUB)
ar	car (CAR)	**ur**	number (NUM bur)
e	less (LES)	**sh**	shelf (SHELF)
ee	leaf (LEEF)	**ch**	nature (NAY chur)
ih	trip (TRIHP)	**g**	gift (GIHFT)
i (i + cons. + e)	idea (i DEE uh)	**j**	gem (JEM)
oh	go (GOH)	**ing**	sing (SING)
aw	soft (SAWFT)	**zh**	vision (VIH zhun)
or	orbit (OR buht)	**k**	came (KAYM)
oy	coin (COYN)	**s**	cent (SENT)
oo	foot (FOOT)	**z**	zone (ZOHN)
ew	food (FEWD)		

Como usar el glosario en español:

1. Busca el término en inglés que desees encontrar.

2. El término en español, junto con la definición, se encuentran en la columna de la derecha.

Content vocabulary are words that relate to health content. They are boldface and highlighted in yellow in your text. Words below that have an asterisk (*) are academic vocabulary. They help you understand your school subjects and are blue boldfaced in your text.

English	A	Español

Abstinence A deliberate decision to avoid high-risk behaviors, including sexual activity and the use of tobacco, alcohol, and other drugs. (Ch. 1, 19)

Abstinencia Decisión deliberada de evitar conductas de alto riesgo, como la actividad sexual, el consumo de tabaco, alcohol y otras drogas.

Abuse The physical, mental, emotional, or sexual mistreatment of one person by another. (Ch. 9, 178)

Abuso Maltrato físico, mental, emocional o sexual que una persona le provoca a otra.

Accident chain A sequence of events that leads to an unintentional injury. (Ch. 26, 731)

Cadena de accidentes Serie de sucesos que generan una lesión no intencional.

Acquired Immunodeficiency Syndrome (AIDS) The final stage of the HIV infection. (Ch. 24, 669)

Síndrome de inmunodeficiencia adquirida (SIDA) Etapa final de la infección por VIH.

Action plan A multi-step strategy to identify and achieve your goals. (Ch. 2, 44)

Plan de acción Estrategia de varios pasos para identificar y lograr metas.

Active listening Paying close attention to what someone is saying and communicating. (Ch. 6, 154)

Escucha activa Escuchar atentamente lo que alguien dice o comunica.

Addiction A physiological or psychological dependence on a drug. (Ch. 22, 594)

Adicción Dependencia fisiológica o psicológica a una droga.

Addictive drug A substance that causes physiological or psychological dependence. (Ch. 20, 543)

Droga adictiva Sustancia que causa dependencia fisiológica o psicológica.

Additive interaction Occurs when medicines work together in a positive way. (Ch. 19, 528)

Interacción aditiva Situación en la cual los medicamentos interactúan de una manera positiva.

Adjust* To bring to a more satsfactory state. (Ch 22, 614)

Adolescence The period between childhood and adulthood. (Ch. 18, 496)

Adoption The legal process of taking a child of other parents as one's own. (Ch. 18, 505)

Adrenal glands Glands that help the body deal with stress and respond to emergencies. (Ch. 16, 444)

Advertising A written or spoken media message designed to interest consumers in purchasing a product or service. (Ch. 2, 46)

Advocacy Taking action to influence others to address a health-related concern or to support a health-related belief. (Ch. 2, 39)

Aerobic exercise All rhythmic activities that use large muscle groups for an extended period of time. (Ch. 12, 327)

Affect* To produce an effect upon. (Ch. 1, 19)

Affirmation Positive feedback that helps others feel appreciated and supported. (Ch. 7, 170)

Aggressive Overly forceful, pushy, or hostile. (Ch. 6, 152)

Air pollution The contamination of the earth's atmosphere by harmful substances. (Ch. 28, 800)

Air Quality Index (AQI) An index for reporting daily air quality. (Ch. 28, 801)

Alcohol abuse The excessive use of alcohol. (Ch. 21, 575)

Alcohol poisoning A severe and potentially fatal physical reaction to an alcohol overdose. (Ch. 21, 569)

Alcoholic An addict who is dependent on alcohol. (Ch. 21, 581)

Alcoholism A disease in which a person has a physical or psychological dependence on drinks that contain alcohol. (Ch. 21, 575)

Alienation Feeling isolated and separated from everyone else. (Ch. 5, 122)

Allergy A specific reaction of the immune system to a foreign and frequently harmless substance. (Ch. 25, 702)

Americans with Disabilities Act (ADA) A law prohibiting discrimination against people with physical or mental disabilities in the workplace, transportation, public accommodation, and telecommunications. (Ch. 25, 713)

Amniocentesis A procedure in which a syringe is inserted through a pregnant female's abdominal wall to remove a sample of the amniotic fluid surrounding the developing fetus. (Ch. 17, 481)

Anabolic-androgenic steroids Synthetic substances that are similar to male sex hormones. (Ch. 22, 601)

Ajustar Traer a un estado más satisfactorio.

Adolescencia Etapa entre la infancia y la edad adulta.

Adopción Proceso legal para tener como hijo a un niño de otros padres.

Glándulas suprarrenales Glándulas que ayudan al cuerpo a lidiar con el estrés y a reaccionar ante emergencias.

Publicidad Mensaje oral o escrito en los medios de comunicación, diseñado para incentivar a los consumidores a adquirir un producto o servicio.

Promoción Tomar medidas para influir en otras personas, con el propósito de abordar preocupaciones o apoyar creencias en relación con la salud.

Ejercicio aeróbico Toda actividad rítmica que use los grupos musculares grandes por un periodo prolongado.

Afectar Producir un efecto en algo o en alguien.

Afirmación Retroalimentación positiva que ayuda a que otras personas se sientan apreciadas y respaldadas.

Agresivo Excesivamente enérgico, insistente u hostil.

Contaminación atmosférica Contaminación de la atmósfera de la Tierra producto de sustancias peligrosas.

Índice de calidad del aire Indicador para informar sobre la calidad diaria del aire.

Abuso de alcohol Consumo excesivo de alcohol.

Intoxicación alcohólica Reacción física grave y potencialmente fatal a una sobredosis de alcohol.

Alcohólico Persona adicta al alcohol.

Alcoholismo Enfermedad en que la persona es adicta física o psicológicamente a las bebidas alcohólicas.

Alienación Sentirse solo y aislado de todo el mundo.

Alergia Reacción específica del sistema inmunológico a una sustancia extraña que usualmente es inofensiva.

Ley para estadounidenses discapacitados (ADA) Ley que prohíbe la discriminación de personas con discapacidades físicas o mentales en los lugares de trabajo, transporte, lugares públicos y telecomunicaciones.

Amniocentesis Procedimiento en el cual se inserta una jeringa a través de la pared abdominal de una embarazada hasta llegar al líquido amniótico que rodea al embrión en desarrollo.

Esteroides anabolizantes-androgénicos Sustancias sintéticas semejantes a las hormonas masculinas.

Glossary / Glosario

Glossary / Glosario

Anaerobic exercise Intense short bursts of activity in which the muscles work so hard that they produce energy without using oxygen. (Ch. 12, 328)

Angina pectoris Chest pain that results when the heart does not get enough oxygen. (Ch. 25, 690)

Anorexia nervosa An eating disorder in which an irrational fear of weight gain leads people to starve themselves. (Ch. 11, 300)

Antagonistic interaction Occurs when the effect of one medicine is canceled or reduced when taken with another medicine. (Ch. 19, 529)

Antibiotics A class of chemical agents that destroy disease-causing microorganisms while leaving the patient unharmed. (Ch. 24, 664)

Antibody A protein that acts against a specific antigen. (Ch. 23, 641)

Anticipate* To expect. (Ch. 4, 93)

Antibody screening test The first test to be run to detect HIV antibodies. (Ch. 24, 677)

Antigens Substances that are capable of triggering an immune response. (Ch. 23, 640)

Anxiety The condition of feeling uneasy or worried about what may happen. (Ch. 5, 114)

Anxiety disorder A condition in which real or imagined fears are difficult to control. (Ch. 5, 119)

Apathy A lack of strong feeling, interest, or concern. (Ch. 5, 116)

Appendicitis The inflammation of the appendix. (Ch. 15, 427)

Appetite The psychological desire for food. (Ch. 10, 255)

Approach* A particular manner of taking steps. (Ch. 2, 52)

Appropriate* Proper or fitting. (Ch. 1, 8)

Arrhythmias Irregular heartbeats. (Ch. 25, 690)

Arteriosclerosis Hardened arteries with reduced elasticity. (Ch. 25, 689)

Arthritis A group of more than 100 different diseases that causes pain and loss of movement in the joints. (Ch. 25, 707)

Aspect* A feature or phase of something. (Ch. 8, 202)

Assault An unlawful physical attack or threat of attack. (Ch. 9, 233)

Assertive Expressing your views clearly and respectfully. (Ch. 6, 153)

Asthma An inflammatory condition in which the trachea, the bronchi, and bronchioles become narrowed, causing difficulty breathing. (Ch. 15, 420)

Ejercicio anaeróbico Periodos cortos de actividad física intensiva, en los cuales los músculos trabajan tan arduamente que producen energía sin usar oxígeno.

Angina de pecho Dolor en el pecho causado porque el corazón no está recibiendo suficiente oxígeno.

Anorexia nerviosa Trastorno de la alimentación en la cual el miedo irracional a aumentar de peso provoca que las personas sigan una dieta de hambre.

Interacción antagónica Situación en la cual el efecto de un medicamento se elimina o reduce al interactuar con otro.

Antibióticos Tipo de agentes químicos que destruye los microorganismos que provocan enfermedades sin dañar al paciente.

Anticuerpo Proteína que ataca antígenos específicos.

Anticipar Esperar.

Prueba de detección de anticuerpos Primera prueba realizada para detectar los anticuerpos del VIH.

Antígenos Sustancias capaces de provocar una respuesta inmune.

Ansiedad Estado en el cual una persona se siente abrumada o preocupada acerca de lo que le pueda pasar.

Trastorno de ansiedad Estado en el cual el miedo, ya sea real o imaginario, es difícil de controlar.

Apatía Falta de sentimientos intensos, interés o preocupación.

Apendicitis Inflamación del apéndice.

Apetito Deseo psicológico de comer.

Acceso Una manera particular de tomar medidas.

Apropiado Apropiado o quedando bien.

Arritmia Palpitaciones irregulares del corazón.

Ateroesclerosis Arterias endurecidas con elasticidad reducida.

Artritis Grupo de más de 100 enfermedades que causan dolor y pérdida de movimiento en las articulaciones.

Aspecto Caracteristica o fase de algo.

Asalto Ataque o amenaza de ataque físico ilegal.

Asertivo Persona que expresa sus puntos de vista clara y respetuosamente.

Asma Condición inflamatoria en que la tráquea, los bronquios y los bronquiolos se estrechan provocando dificultad para respirar.

Asymptomatic People who are infected show no symptoms or the infections produce mild symptoms that disappear. (Ch. 24, 659)

Atherosclerosis A disease characterized by the accumulation of plaque on artery walls. (Ch. 25, 689)

Attribute* A quality or characteristic. (Ch. 8, 193)

Auditory ossicles Three small bones linked together that connect the eardrum to the inner ear. (Ch. 13, 371)

Authority* The right to make decisions and give commands. (Ch. 7, 168)

Autoimmune disease A condition in which the immune system mistakenly attacks itself, targeting the cells, tissues, and organs of a person's own body. (Ch. 25, 705)

Autonomy The confidence that a person can control his or her own body, impulses, and environment. (Ch. 17, 485)

Asintomático Persona infectada que no presenta síntomas o infecciones que producen síntomas leves que desaparecen.

Ateroesclerosis Enfermedad caracterizada por la acumulación de depósitos en las paredes de las arterias.

Atributo Cualidad o característica.

Osículos auditivas Tres huesos pequeños conectados juntos que unen el tímpano con el oido interno.

Autoridad Derecho para tomar decisiones y dar órdenes.

Enfermedad autoinmune Condición en la cual el sistema inmune se ataca a sí mismo por error, afectando las células, los tejidos y los órganos del cuerpo de una persona.

Autonomía Capacidad que tiene una persona para controlar su propio cuerpo, impulsos y medio ambiente.

English · B · Español

Bacteria Single-celled microorganisms. (Ch. 23, 629)

Behavior therapy A treatment process that focuses on changing unwanted behavior through rewards and reinforcement. (Ch. 5, 128)

Benign Noncancerous. (Ch. 25, 696)

Bile A yellow-green, bitter fluid important in the breakdown and absorption of fats. (Ch. 15, 424)

Binge drinking Drinking five or more alcoholic drinks at one sitting. (Ch. 21, 569)

Binge eating disorder An eating disorder in which people overeat compulsively. (Ch. 11, 301)

Biodegradable Able to be broken down by microorganisms in the environment. (Ch. 28, 806)

Biopsy The removal of a small piece of tissue for examination. (Ch. 25, 700)

Blended family A married couple and their children from previous marriages. (Ch. 7, 167)

Blizzard A snowstorm with winds that reach 35 miles an hour or more. (Ch. 27, 780)

Blood alcohol concentration (BAC) The amount of alcohol in a person's blood expressed as a percentage. (Ch. 21, 578)

Blood pressure A measure of the amount of force that the blood places on the walls of blood vessels, particularly large arteries, as it is pumped through the body. (Ch. 15, 413)

Bacteria Microorganismos compuestos de una sola célula.

Terapia del conducta Terapia que se enfoca en cambiar las conductas no deseadas a través de recompensas y refuerzos.

Benigno No canceroso.

Bilis Líquido amargo de color amarillo verdoso que es importante para la descomposición y absorción de las grasas.

Borrachera Consumo de cinco o más bebidas alcohólicas consecutivas.

Trastorno de atracones compulsivos Trastorno de la alimentación caracterizada por comer demasiado y de manera compulsiva.

Biodegradable Algo que los microorganismos del medio ambiente pueden descomponer.

Biopsia La extirpación diagnóstica de una pequeña muestra de tejido.

Familia mixta Pareja casada y sus hijos de matrimonios anteriores.

Ventisca Tormenta de nieve con vientos que superan las 35 millas por hora.

Concentración de alcohol en la sangre Cantidad de alcohol en la sangre de una persona expresada como porcentaje.

Presión arterial Medida de la presión que ejerce la sangre sobre las paredes de los vasos sanguíneos, especialmente en las arterias grandes, a medida que es bombeada por el cuerpo.

Body image The way you see your body. (Ch. 11, 297)

Body language Nonverbal communication through gestures, facial expressions, behaviors, and posture. (Ch. 6, 155)

Body mass index (BMI) A measure of body weight relative to height. (Ch. 11, 291)

Brain stem A three-inch-long stalk of nerve cells and fibers that connects the spinal cord to the rest of the brain. (Ch. 14, 396)

Bronchi The main airways that reach into each lung. (Ch. 15, 417)

Bulimia nervosa An eating disorder that involves cycles of overeating and purging, or attempts to rid the body of food. (Ch. 11, 301)

Bullying Deliberately harming or threatening another person who cannot easily defend himself or herself. (Ch. 6, 150)

Imagen corporal Forma en que uno ve su propio cuerpo.

Lenguaje corporal Comunicación no verbal a través de gestos, expresiones faciales, comportamientos y postura.

Índice de masa corporal (IMC) Medida de peso corporal en relación con la estatura.

Vástago cerebral Ramificación de neuronas y fibras de tres pulgadas de largo que conecta la médula espinal con el resto del cerebro.

Bronquios Vías aéreas principales que llegan a los pulmones.

Bulimia nerviosa Trastorno de la alimentación que implica ciclos en que la persona come en exceso y purga o intenta eliminar la comida del cuerpo.

Intimidación Daño o amenazas deliberadas hacia una persona que no se puede defender fácilmente.

English	C	Español

Calorie A unit of heat used to measure the energy your body uses and the energy it receives from food. (Ch. 10, 254)

Cancer Uncontrollable growth of abnormal cells. (Ch. 25, 695)

Capillaries Small vessels that carry blood from arterioles to small vessels called venules, which empty into veins. (Ch. 15, 412)

Carbohydrates Starches and sugars found in foods which provide your body's main source of energy. (Ch. 10, 259)

Carbon monoxide A colorless, odorless, and poisonous gas. (Ch. 20, 543)

Carcinogen A cancer-causing substance. (Ch. 20, 543)

Cardiac muscles A type of striated muscle that forms the wall of the heart. (Ch. 14, 388)

Cardiopulmonary resuscitation (CPR) A first-aid procedure that combines rescue breathing and chest compressions to supply oxygen to the body until normal body functions can resume. (Ch. 27, 765)

Cardiorespiratory endurance The ability of your heart, lungs, and blood vessels to send fuel and oxygen to your tissues during long periods of moderate to vigorous activity. (Ch. 12, 324)

Cardiovascular disease A disease that affects the heart or blood vessels. (Ch. 25, 688)

Cartilage A strong, flexible connective tissue. (Ch. 14, 383)

Caloría Unidad de calor que mide la energía que usa el cuerpo y la energía que la comida proporciona al cuerpo.

Cáncer Crecimiento incontrolable de células anormales.

Capilares Vasos sanguíneos delicados que transportan sangre desde las arteriolas hasta vasos pequeños conocidos como vénulas, las cuales terminan en las venas.

Carbohidratos Almidones y azúcares que se encuentran en los alimentos, los cuales proporcionan al cuerpo la fuente principal de energía.

Monóxido de carbono Gas incoloro, inodoro y venenoso.

Cancerígeno Sustancia que produce cáncer.

Músculo cardiaco Tipo de músculo estriado que forma las paredes del corazón.

Resucitación cardiopulmonar (RCP) Procedimiento de primeros auxilios que combina la respiración artificial con compresiones en el pecho a fin de proporcionar oxígeno hasta que las funciones vitales puedan reanudarse.

Resistencia cardiorrespiratoria Capacidad que tienen el corazón, los pulmones y los vasos sanguíneos de enviar energía y oxígeno a los tejidos durante largos periodos de tiempo con actividad moderada a enérgica.

Enfermedad cardiovascular Enfermedad que afecta el corazón o los vasos sanguíneos.

Cartílago Tejido conjuntivo fuerte y flexible.

Cerebellum The second largest part of the brain. (Ch. 14, 395)

Cerebral palsy A group of neurological disorders that are the result of damage to the brain before, during, or just after birth or in early childhood. (Ch. 14, 399)

Cerebrum The largest and most complex part of the brain. (Ch. 14, 394)

Cervix The opening to the uterus. (Ch. 16, 454)

Chain of survival A sequence of actions that maximize the victim's chances of survival. (Ch. 27, 764)

Character The distinctive qualities that describe how a person thinks, feels, and behaves. (Ch. 3, 73)

Child abuse Domestic abuse directed at a child. (Ch. 7, 179)

Cholesterol A waxy, fatlike substance. (Ch. 10, 262)

Chorionic villi sampling (CVS) A procedure in which a small piece of membrane is removed from the chorion, a layer of tissue that develops into the placenta. (Ch. 17, 482)

Chromosomes Threadlike structures found within the nucleus of a cell that carry the codes for inherited traits. (Ch. 17, 478)

Chronic disease An ongoing condition or illness. (Ch. 1, 10)

Chronic stress Stress associated with long-term problems that are beyond a person's control. (Ch. 4, 97)

Circumstances* An event that influences another event. (Ch. 7, 175)

Cirrhosis Scarring of the liver tissue. (Ch. 23, 635)

Citizenship The way you conduct yourself as a member of the community. (Ch. 6, 144)

Clique A small circle of friends, usually with similar backgrounds or tastes, who exclude people viewed as outsiders. (Ch. 8, 195)

Closure Acceptance of a loss. (Ch. 4, 103)

Cluster suicides A series of suicides occurring within a short period of time and involving several people in the same school or community. (Ch. 5, 123)

Cognition The ability to reason and think out abstract solutions. (Ch. 18, 498)

Cognitive therapy A treatment method designed to identify and correct distorted thinking patterns that can lead to feelings and behaviors that may be troublesome, self-defeating, or self-destructive. (Ch. 5, 129)

Commitment A promise or a pledge. (Ch. 18, 504)

Communicable disease A disease that is spread from one living organism to another or through the environment. (Ch. 23, 628)

Cerebelo La segunda parte más grande del cerebro.

Parálisis cerebral Grupo de trastornos neurológicos que son el resultado de daños al cerebro antes, durante o inmediatamente después del nacimiento o durante la niñez temprana.

Corteza cerebral La parte más grande y compleja del cerebro.

Cuello del útero La entrada del útero.

Cadena de supervivencia Secuencia de acciones que tiene como objetivo maximizar las posibilidades de supervivencia de una víctima.

Carácter Características distintivas que describen cómo una persona piensa, siente y actúa.

Maltrato infantil Abuso doméstico dirigido hacia los niños.

Colesterol Sustancia cerosa de apariencia grasa.

Cromosomas Estructuras parecidas a hilos que se encuentran dentro del núcleo de una célula y que tienen los códigos de los rasgos heredados.

Enfermedad crónica Afección o enfermedad permanente.

Estrés crónico Estrés relacionado con problemas de largo plazo y fuera del control de una persona.

Biopsia de vellosidades coriónicas Procedimiento en el cual se saca una pequeña muestra de membrana del corion, una capa de tejido que se desarrolla en la placenta.

Circunstancia Acontecimiento que influye en otro acontecimiento.

Cirrosis Lesiones en el tejido del hígado.

Ciudadanía Manera de comportarse como miembro de la comunidad.

Pandilla Grupo pequeño de amigos, generalmente con gustos y experiencias similares, que excluyen a otras personas consideradas ajenas a ellos.

Resignación Aceptación de una pérdida.

Serie de suicidios Varios suicidios que ocurren en un periodo de tiempo corto y que involucran a personas de un mismo colegio o comunidad.

Cognición Capacidad de razonar y generar soluciones abstractas.

Terapia cognoscitiva Terapia diseñada para identificar y corregir patrones de pensamiento distorsionados, los cuales pueden generar sentimientos y comportamientos problemáticos, contraproducentes o autodestructivos.

Compromiso Una promesa.

Enfermedad contagiosa Enfermedad que se quede transmitir de un ser vivo a otro o a través del medio ambiente.

Community* A population of individuals in a common location. (Ch. 23, 648)

Comparison shopping Judging the benefits of different products by comparing several factors, such as quality, features, and cost. (Ch. 2, 47)

Competence Having enough skills to do something. (Ch. 3, 68)

Component* A constituent part or ingredient. (Ch. 28, 801)

Compromise A problem-solving method in which each participant gives up something to reach a solution that satisfies everyone. (Ch. 6, 146)

Computer* A device that can store, retrieve, and process data. (Ch. 6, 147)

Concussion A jarring injury to the brain that can cause unconsciousness. (Ch. 27, 774)

Conduct disorder* Patterns of behavior in which the rights of others or basic social rules are violated. (Ch. 5, 121)

Confidentiality Respecting the privacy of both parties and keeping details secret. (Ch. 9, 227)

Confine* To keep within limits. (Ch. 24, 671)

Conflict Any disagreement, struggle, or fight. (Ch. 9, 220)

Conflict resolution The process of ending a conflict through cooperation and problem solving. (Ch. 2, 36)

Conservation Avoiding waste through careful management of natural resources. (Ch. 28, 810)

Consistent* Free from variation or contradiction. (Ch. 1, 25)

Constructive* Promoting improvement or development. (Ch. 5, 129)

Constructive criticism Nonhostile comments that point out problems and encourage improvement. (Ch. 3, 76)

Consumer advocates People or groups whose sole purpose is to take on regional, national, and even international consumer issues. (Ch. 2, 51)

Contact* Union or junction of surfaces. (Ch. 23, 630)

Contract To draw together. (Ch. 15, 409)

Contradict* To imply the opposite of. (Ch. 6, 155)

Cool-down Low-level activity that prepares your body to return to a resting state. (Ch. 12, 335)

Cooperation Working together for the good of all. (Ch. 6, 146)

Coping Dealing successfully with difficult changes in your life. (Ch. 4, 103)

Cornea A transparent tissue that bends and focuses light before it enters the lens. (Ch. 13, 368)

Comunidad Población de personas que viven en el mismo lugar.

Compras informadas Evaluar los beneficios de diferentes productos comparando diversos factores, como calidad, características y precio.

Competencia Capacidad suficiente para realizar algo.

Componente Parete o ingrediente constitutivo.

Acuerdo Método para resolver problemas en que cada participante debe sacrificar algo para llegar a una solución satisfactoria para todos.

Computadora Recuperar datos.

Conmoción cerebral Lesión violenta en el cerebro que puede conducir a la pérdida de conocimiento.

Trastorno de conducta Patrón de comportamiento en el cual se infringen los derechos de los demás o las reglas sociales básicas.

Confidencialidad Respetar la vida privada de ambas partes y mantener en secreto los detalles.

Confinar Encerrar en un lugar.

Conflicto Cualquier desacuerdo, pelea o enojo.

Resolución de conflictos Proceso de resolver un conflicto a través de métodos de cooperación y solución de problemas.

Conservación Evitar el desperdicio de recursos a través de un manejo correcto de los recursos naturales.

Coherente Sin variaciones ni contradicciones.

Constructivo Que promeuve mejoras o progresos.

Crítica constructiva Comentarios no hostiles que señalan los problemas y fomentan su mejoramiento.

Defensores del consumidor Gente o grupos cuyo único propósito es confrontar los problemas regionales, nacionales y hasta internacionales del consumidor.

Contacto Unión o conexión de superficies.

Contraer Reducirse, disminuir.

Contradecir Decir o hacer lo contrario.

Recuperación Actividad liviana que prepara al cuerpo para volver a un estado de descanso.

Cooperación Trabajar juntos para el beneficio de todos.

Sobrellevar Encargarse exitosamente de los cambios difíciles de la vida.

Córnea Tejido transparente que refracta y enfoca la luz antes de pasar al cristalino.

Crisis center A facility that offers advice and support to people dealing with personal emergencies. (Ch. 7, 182)

Cross-contamination The spreading of pathogens from one food to another. (Ch. 10, 279)

Crucial* Important or essential. (Ch. 03, 76)

Culture The collective beliefs, customs, and behaviors of a group. (Ch. 1, 13)

Cumulative risks Related risks that increase in effect with each added risk. (Ch. 1, 18)

Custody The legal right to make decisions affecting children and the responsibility for their care. (Ch. 7, 174)

Cyberbullying Cruel or hurtful online contact. (Ch. 26, 729)

Cycle of violence Pattern of repeating violent or abusive behaviors from one generation to the next. (Ch. 7, 180)

Cystitis An inflammation of the bladder. (Ch. 15, 432)

Centro para crisis Plantel que maneja emergencias y envía a un individuo que necesita ayuda a especialistas.

Contaminación cruzada Transmisión de agentes patógenos de una comida a otra.

Crucial Importante o esencial.

Cultura Las creencias, costumbres y comportamientos colectivos de un grupo de personas.

Riesgos acumulativos Riesgos relacionados que aumentan en efecto con cada nuevo peligro.

Custodia Derecho legal de tomar decisiones que afecten a los niños y la responsabilidad de cuidarlos.

Intimidación cibernético Contacto cruel o dañino que se produce en línea.

Ciclo de violencia Patrón de comportamiento violento o abusivo que se repite de una generación a la siguiente.

Cistitis Inflamación de la vejiga.

English   D Español

Date rape One person in a dating relationship forces the other person to take part in sexual intercourse. (Ch. 9, 238)

Decibel A unit that measures the intensity of sound. (Ch. 28, 804)

Decision-making skills Steps that enable you to make a healthful decision. (Ch. 2, 41)

Defense mechanisms Mental processes that protect individuals from strong or stressful emotions and situations. (Ch. 3, 81)

Defensive driving Being aware of potential hazards on the road and taking action to avoid them. (Ch. 26, 746)

Defibrillator A device that delivers an electric shock to the heart to restore its normal rhythm. (Ch. 27, 765)

Deforestation Destruction of forests. (Ch. 28, 808)

Deoxyribonucleic acid (DNA) The chemical unit that makes up chromosomes. (Ch. 17, 479)

Depressant A drug that slows the central nervous system. (Ch. 21, 567)

Depression Prolonged feeling of helplessness, hopelessness, and sadness. (Ch. 5, 115)

Dermis The thicker layer of skin beneath the epidermis that is made up of connective tissue and contains blood vessels and nerves. (Ch. 13, 356)

Designer drug A synthetic drug that is made to imitate the effects of hallucinogens and other drugs. (Ch. 22, 605)

Violación durante una cita (violación por un conocido, violación a escondidas) Una persona que se encuentra en una cita obliga a la otra a participar en una actividad sexual.

Decibelios (Decibeles) Medida que se usa para expresar la intensidad del sonido.

Habilidades para tomar decisiones Pasos necesarios para tomar una decisión correcta.

Mecanismos de defensa Procesos mentales que protegen a los individuos de emociones y situaciones intensas o estresantes.

Conducción a la defensiva Estar consciente de posibles peligros en la carretera y tomar medidas para evitarlos.

Máquina de desfibrilación Un aparato que proporciona choques eléctricos al corazón para recuperar su ritmo normal.

Deforestación Destrucción de los bosques.

Ácido desoxirribonucleico (ADN) Unidad química que compone los cromosomas.

Depresor Sustancia que tiende a disminuir el funcionamiento (actividad) del sistema nervioso central.

Depresión Sentimiento prolongado de soledad, desesperación y tristeza.

Dermis La capa más gruesa de la piel que se encuentra debajo de la epidermis que está compuesta de tejidos conectivos y contiene vasos sanguíneos y nervios.

Droga de diseño Sustancias sintéticas que tratan de imitar los efectos de los alucinógenos y otras drogas peligrosas.

Glossary / Glosario

Developmental tasks Events that need to happen in order for a person to continue growing toward becoming a healthy, mature adult. (Ch. 17, 484)

Devote* To give time or effort to an activity. (Ch. 12, 321)

Diabetes A chronic disease that affects the way cells convert sugar into energy. (Ch. 25, 705)

Diaphragm A muscle that separates the chest from the abdominal cavity. (Ch. 15, 417)

Dietary Guidelines for Americans A set of recommendation about smart eating and physical activity for all Americans. (Ch. 10, 266)

Dietary Supplements Products that supply one or more nutrients as a supplement to, not as a substitute for, healthful foods. (Ch. 11, 304)

Disability Any physical or mental impairment that limits normal activities, including seeing, hearing, walking, or sleeping. (Ch. 25, 709)

Dislocation A separation of a bone from its normal position in a joint. (Ch. 27, 772)

Display* To make evident. (Ch. 5, 123)

Divorce A legal end to a marriage contract. (Ch. 7, 174)

Domestic* Relating to the household or family. (Ch. 7, 180)

Domestic violence Act of violence involving family members. (Ch. 7, 178)

Drug therapy The use of certain medications to treat or reduce the symptoms of a mental disorder. (Ch. 5, 129)

Drug watches Organized community efforts by neighborhood residents to patrol, monitor, report, and otherwise stop drug deals and drug abuse. (Ch. 22, 613)

Drug-free school zone Areas within 1,000 feet of schools and designated by signs, within which people caught selling drugs receive especially severe penalties. (Ch. 22, 613)

Drugs Substances other than food that change the structure, function of the body, or mind. (Ch. 19, 524)

Tareas requeridas para el desarrollo Sucesos necesarios para que una persona continúe creciendo y se convierta en un adulto saludable y maduro.

Dedicar Destinar tiempo o esfuerzo a una actividad.

Diabetes Una enfermedad crónica que afecta el modo en que las células del cuerpo convierten los alimentos en energía.

Diafragma El músculo que separa la cavidad toráxico de la cavidad abdominal.

Guías alimentarias para los estadounidenses Conjunto de recomendaciones acerca de alimentarse inteligentemente y de la actividad física para todos los estadounidenses.

Suplementos alimentarios Productos que suministran uno o más nutrientes en forma de suplementos, no de sustitutos, a los alimentos saludables.

Discapacidad Cualquier impedimento físico o mental que limita el desarrollo de actividades normales tales como ver, oír, caminar o dormir.

Dislocación Separación del hueso de su posición normal en una articulación.

Exponer Hacer evidente.

Divorcio Fin legal de un contrato de matrimonio.

Doméstico Relativo al hogar o a la familia.

Violencia doméstica (intrafamiliar) Acto de violencia que incluya a los miembros de una familia.

Terapia farmacológica Uso de ciertos medicamentos para tratar o reducir los síntomas de una enfermedad mental.

Vigilantes de la droga Un grupo de personas de un vecindario organizadas para supervisar, controlar, denunciar o directamente frenar el abuso y la venta de drogas.

Zona de escuela libre de drogas Un área que comprende 1,000 pies alrededor de una escuela y se encuentra señalizada, en la cual las personas que son atrapadas vendiendo drogas son gravemente penalizadas o castigadas.

Drogas Sustancias distintas de los alimentos, que cambian la estructura o el funcionamiento del cuerpo o la mente de las personas.

English	**E**	Español

Earthquake A series of vibrations in the earth caused by sudden movements of the earth's crust. (Ch. 27, 781)

Eating disorders Extreme, harmful eating behaviors that can cause serious illness or even death. (Ch. 11, 300)

Terremoto Serie de vibraciones en la tierra provocada por movimientos repentinos de la superficie de la tierra.

Trastorno alimentario Un comportamiento que se caracteriza por comer en forma extrema y dañina lo que causa que la persona se pueda enfermar o morir.

Eggs Female gametes. (Ch. 16, 452)

Elder abuse The abuse or neglect of older family members. (Ch. 7, 179)

Embryo A cluster of cells that develops between the third and eighth week of pregnancy. (Ch. 17, 470)

Emergency survival kit A set of items you will need in an emergency. (Ch. 27, 782)

Emerging infection Communicable diseases whose occurrence in humans has increased within the past two decades or threatens to increase in the near future. (Ch. 23, 645)

Emotional abuse A pattern of attacking another person's emotional development and sense of worth. (Ch. 9, 237)

Emotional maturity The state at which the mental and emotional capabilities of an individual are fully developed. (Ch. 18, 502)

Emotions Signals that tell your mind and body how to react. (Ch. 3, 78)

Empathy The ability to imagine and understand how someone else feels. (Ch. 3, 80)

Emphysema A disease that progressively destroys the walls of the alveoli. (Ch. 15, 421)

Empty-nest syndrome The feelings of sadness or loneliness that accompany children's leaving home and entering adulthood. (Ch. 18, 510)

Encounter* To experience. (Ch. 26, 728)

Endocrine glands Ductless or tubeless organs or groups of cells that secrete hormones directly into the bloodstream. (Ch. 16, 442)

Environment The sum of your surroundings. (Ch. 1, 12)

Environmental tobacco smoke (ETS) Air that has been contaminated by tobacco smoke. (Ch. 20, 553)

Epidemic An occurrence of a disease in which many people in the same place at the same time are affected. (Ch. 24, 648)

Epidermis The outer, thinner layer of the skin that is composed of living and dead cells. (Ch. 13, 356)

Epilepsy A disorder of the nervous system that is characterized by recurrent seizures—sudden episodes of uncontrolled electric activity in the brain. (Ch. 14, 399)

Escalate Become more serious. (Ch. 9, 221)

Estimate* To determine roughly the size or extent of. (Ch. 24, 674)

Ethanol The type of alcohol in alcoholic beverages. (Ch. 21, 566)

Euphoria A feeling of intense well-being or elation. (Ch. 22, 605)

Óvulos Gametos femeninos.

Abuso de mayores El abuso o la negligencia de miembras ancianos de la familia.

Embrión Grupo de células que se desarrolla entre la tercera y la octava semana del embarazo.

Botiquín de emergencia Conjunto de elementos necesarios para una situación de emergencia.

Infección emergente Una enfermedad infecciosa cuya incidencia en humanos ha aumentado durante las últimas dos décadas o que amenaza con aumentar en el futuro cercano.

Abuso emocional Patrón de ataque al desarrollo emocional y al sentido de estima de otra persona.

Madurez emocional Un estado en el cual las capacidades mentales y emocionales de una persona se encuentran totalmente desarrolladas.

Emociones Señales que le comunican a la mente y al cuerpo cómo actuar.

Empatía La habilidad para imaginar y entender cómo siente otra persona.

Enfisema Una enfermedad que destruye progresivamente las paredes de los alvéolos.

Síndrome del nido vacío Sentimiento de tristeza y soledad que ocurre cuando los hijos, quienes ya se convirtieron en adultos, se van de la casa de sus padres.

Encuentro Expriencias.

Glándulas endocrinas Órganos o grupos de células sin conductos o tubos que secretan hormonas directamente al torrente sanguíneo.

Medio ambiente Todo lo que nos rodea.

Ambiente con humo de cigarro Aire que ha sido contaminado por el humo de cigarrillos.

Epidemia Una situación en la cual mucha gente contrae una enfermedad al mismo tiempo y en el mismo lugar.

Epidermis La capa más fina y externa de la piel la cual se encuentra compuesta de células vivas y muertas.

Epilepsia Trastorno del sistema nervioso caracterizado por convulsiones continuas—repentinos episodios de actividad eléctrica incontrolable en el cerebro.

Intensificar Situación que se hace más grave.

Estimar Determinar aproximadamente el tamaño o la extension de algo.

Etanol Tipo de alcohol que se encuentra en las bebidas alcohólicas.

Euforia Sentimiento de un intenso bienestar o alegría.

Glossary / Glosario

Eventually* At an unspecified later time. (Ch. 17, 471)

Exclude* To prevent or restrict the entrance of. (Ch. 8, 195)

Exercise Purposeful physical activity that is planned, structured, and repetitive, and that improves or maintains physical fitness. (Ch. 12, 319)

Expand* To open up. (Ch. 15, 417)

Exposure* The condition of being unprotected. (Ch. 12, 340)

Extended family A family that includes additional relatives beyond parents and children. (Ch. 7, 167)

Extensor The muscle that opens a joint. (Ch. 14, 388)

Eventual Sin especificar un rato más después.

Excluir Para prevenir o restringir la entrada de.

Ejercicio Actividad física dirigida que es planeada, estructurada y repetitiva y que tiene como objetivo el mantenimiento o el mejoramiento del estado físico de una persona.

Expandir Ampliar.

Exposición Falta de protección.

Familia extendida Familia que incluye a otros parientes, distintos a padres e hijos.

Extensor Músculo que abre una articulación.

English · F · Español

Factor* An element that contributes to a particular result. (Ch. 1, 13)

Fad diet Weight-loss plans that tend to be popular for only a short time. (Ch. 11, 298)

Fallopian tubes A pair of tubes with fingerlike projections that draw in the ovum. (Ch. 16, 453)

Family therapy Helping the family function in more positive and constructive ways by exploring the patterns in communication and providing support and education. (Ch. 5, 129)

Fermentation The chemical action of yeast on sugars. (Ch. 21, 566)

Fertilization The union of a male sperm cell and a female egg. (Ch. 17, 470)

Fetal alcohol syndrome A group of alcohol-related birth defects that includes both physical and mental problems. (Ch. 17, 474)

Fetus Group of developing cells after about the eighth week of pregnancy. (Ch. 17, 470)

Fiber A tough, complex, carbohydrate that the body cannot digest. (Ch. 10, 259)

Fire extinguisher A portable device for putting out small fires. (Ch. 26, 732)

First aid The immediate, temporary care given to an ill or injured person until professional care can be provided. (Ch. 27, 758)

Flash floods Floods in which a dangerous volume of water builds up in a short time. (Ch. 27, 781)

Flexibility The ability to move your body parts through their full range of motion. (Ch. 12, 325)

Flexor The muscle that closes a joint. (Ch. 14, 388)

Factor Elemento que contribuye a un resultado en particular.

Dietas de moda Planes para perder peso que son populares por poco tiempo.

Trompas de falopio Un par de conductos con terminaciones en forma de dedos que atrae el ovario.

Terapia familiar Ayudar a que la familia funcione de maneras más constructivas y positivas mediante la exploración de los patrones de comunicación y en proporcionar apoyo y educación.

Fermentación Reacción química de la levadura en los azúcares.

Fertilización La unión del espermatozoide y el óvulo.

Síndrome de alcoholismo fetal Un grupo de defectos de nacimiento causados por el alcohol y que incluyen problemas físicos y mentales.

Feto Grupo de células en desarrollo después de las ocho semanas de embarazo.

Fibra Un carbohidrato complejo y duro que el cuerpo no puede digerir.

Extintor de incendios Aparato portátil para apagar fuego.

Primeros auxilios La atención inmediata y temporal que se le proporciona a una persona hasta que se le puede otorgar atención profesional.

Inundaciones rápidas Inundaciones en las cuales se acumula un volumen peligroso de agua en poco tiempo.

Flexibilidad La capacidad de mover una parte del cuerpo fácilmente y en muchas direcciones.

Músculo flexor Músculo que abre una articulación.

Food additives Substances added to a food to produce a desired effect. (Ch. 10, 275)

Aditivos alimentarios Sustancias que son adicionadas a los alimentos en forma intencional para generar un efecto deseado.

Food allergy A condition in which the body's immune system reacts to substances in some foods. (Ch. 10, 281)

Alergia alimentaria Una condición en la cuál el sistema inmunológico del cuerpo reacciona a sustancias contenidas en algunos alimentos.

Foodborne illness Food poisoning. (Ch. 10, 278)

Enfermedad producida por alimentos Intoxicación alimentaria.

Food intolerance A negative reaction to food that does not involve the immune system. (Ch. 10, 281)

Intolerancia alimentaria Reacción negativa a los alimentos (o un elemento particular del alimento) en la cual no participa el sistema inmunológico.

Foster care The temporary placement of children in the homes of adults who are not related to them. (Ch. 7, 167)

Cuidados temporales Colocación provisoria de niños en hogares de adultos que no son sus parientes.

Fracture A break in a bone. (Ch. 27, 772)

Fractura Ruptura de un hueso.

Friendship A significant relationship between two people that is based on trust, caring, and consideration. (Ch. 6, 143)

Amistad Una relación importante entre dos personas que está basada en solidaridad, confianza y consideración.

Frostbite Damage to the skin and tissues caused by extreme cold. (Ch. 12, 339)

Congelación Daño a la piel y a los tejidos provocados por frío extremo.

English **G** **Español**

Gastric juices Secretions from the stomach lining that contain hydrochloric acid and pepsin, an enzyme that digests protein. (Ch. 15, 424)

Jugos gástricos La secreciones que provienen del revestimiento del estómago y que contienen ácido clorhídrico y pepsina, una enzima que digiere la proteína.

Gene therapy The process of inserting normal genes into human cells to correct genetic disorders. (Ch. 17, 482)

Terapia genética Un proceso que consiste en introducir genes normales en las células humanas para corregir trastornos genéticos.

Genes The basic units of heredity. (Ch. 17, 479)

Genes Unidades básicas de la herencia.

Genetic disorders Disorders caused partly or completely by a defect in genes. (Ch. 17, 481)

Trastorno genético Trastorno causado parcial o completamente por defectos en los genes.

Giardia A microorganism that infects the digestive system. (Ch. 23, 647)

Giardia Microorganismo que infecta el sistema digestivo.

Global warming An overall increase in the earth's temperature. (Ch. 28, 802)

Calentamiento global Aumento general de la temperatura de la Tierra.

Goals Those things you aim for that take planning and work. (Ch. 2, 42)

Metas Las cosas de que te esfuerzas que necesita planificación y trabajo.

Good Samaritan laws Statutes that protect rescuers from being sued for giving emergency care. (Ch. 27, 759)

Leyes del Buen Samaritano Estatutos que protegen a los rescatistas de ser demandados por otorgar atención de urgencia.

Graduated licensing A system that gradually increases driving privileges over time. (Ch. 26, 745)

Licencia graduada Sistema que gradualmente aumenta los privilegios de conducción en el transcurso del tiempo.

Greenhouse effect The trapping of heat by gases in the earth's atmosphere. (Ch. 28, 801)

Efecto invernadero Calor atrapado por gases en la atmósfera de la Tierra.

Group therapy Treating a group of people who have similar problems and who meet regularly with a trained counselor. (Ch. 5, 129)

Terapia de grupo Tratamiento de un grupo de personas que tienen problemas similares.

Hair follicles Sacs or cavities that surround the roots of hairs. (Ch. 13, 360)

Folículos pilosos Sacos o cavidades que rodean las raíces de los pelos.

Halitosis Bad breath. (Ch. 13, 365)

Halitosis Mal aliento.

Hallucinogens Drugs that alter moods, thoughts, and sense perceptions including vision, hearing, smell, and touch. (Ch. 22, 605)

Alucinógenos Drogas que alteran el estado de ánimo, el pensamiento y la percepción, lo que incluye vista, oído, olfato y tacto.

Harassment Persistently annoying others. (Ch. 8, 199)

Acoso Molestar continuamente a otra persona.

Hazardous wastes Waste materials with properties that make them dangerous to human health or the environment. (Ch. 28, 807)

Desechos peligrosos Materiales de desecho que se caracterizan por ser peligrosos para la salud humana o el medio ambiente.

Hazing Making others perform certain tasks in order to join the group. (Ch. 6, 150)

Acoso personal Forzando otras a complir ciertas tareas para formar parte del grupo.

Health The combination of physical, mental/emotional, and social well-being. (Ch. 1, 6)

Salud Combinación de bienestar físico, mental-emocional y social.

Health consumer Someone who purchases or uses health products or services. (Ch. 2, 46)

Consumidor de salud Cualquier persona que adquiere o consume productos o servicios de salud.

Health disparities Differences in health outcomes among groups. (Ch. 1, 23)

Desigualdades de salud Diferencias de los resultados de salud entre distintos grupos.

Health education Providing accurate health information and health skills teaching to help people make healthy decisions. (Ch. 1, 22)

Educación en salud Proveer a las personas información adecuada y enseñar destrezas de salud para que puedan tomar decisiones saludables.

Health fraud The sale of worthless products or services that claim to prevent diseases or cure other health problems. (Ch. 2, 52)

Fraude de salud Venta de productos o servicios inútiles que supuestamente sirven para prevenir enfermedades o mejorar otros problemas de la salud.

Health insurance Private and government programs that pay for part or all of a person's medical costs. (Ch. 28, 795)

Seguro de salud Programas privados y gubernamentales que financian total o parcialmente los costos médicos de una persona.

Health literacy A person's capacity to learn about and understand basic health information and services and to use these resources to promote one's health and wellness. (Ch. 1, 25)

Conocimientos de salud Capacidad que tiene una persona para aprender y comprender información básica de salud y los servicios relacionados, y usar esos conocimientos para mejorar su propia salud y bienestar.

Health skills Specific tools and strategies to maintain, protect, and improve all aspects of your health. (Ch. 2, 34)

Habilidades de salud Herramientas y estrategias específicas que ayudan a mantener, proteger y mejorar todos los aspectos de la salud.

Health care system All the medical care available to a nation's people, the way they receive care, and the way they pay for it. (Ch. 28, 792)

Sistema de atención de salud Toda atención médica disponible para los habitantes de un país, la forma en que la reciben y el sistema de pago.

Healthy People A nationwide health promotion and disease prevention plan designed to serve as a guide for improving the health of all people in the United States. (Ch. 1, 22)

Healthy People Plan de promoción de la salud y prevención de enfermedades diseñado para que sirva como guía en el mejoramiento de la salud de todos los habitantes de Estados Unidos.

Heat exhaustion A form of physical stress on the body caused by overheating. (Ch. 12, 340)

Agotamiento debido al calor Forma de estrés físico provocado por el sobrecalentamiento del cuerpo.

Heatstroke A dangerous condition in which the body loses its ability to cool itself through perspiration. (Ch. 12, 340)

Insolación Estado peligroso en el cual el cuerpo pierde su capacidad de enfriarse mediante la transpiración.

Hemodialysis A technique in which an artificial kidney machine removes waste products from the blood. (Ch. 15, 433)

Hemodiálisis Técnica en la cual una máquina de diálisis limpia los desechos de la sangre.

Hemoglobin The oxygen-carrying protein in blood. (Ch. 15, 410)

Hemoglobina Proteína que lleva el oxígeno en la sangre.

Herbal supplements Dietary supplements containing plant extracts. (Ch. 11, 308)

Suplementos hierbas Suplementos alimentarios que contienen extractos vegetales.

Heredity All the traits that were biologically passed on to you from your parents. (Ch. 1, 11)

Herencia Todo rasgo biológicamente transmitido de padres a hijos.

Hernia Occurs when an organ or tissue protrudes through an area of weak muscle. (Ch. 14, 390)

Hernia Cuando un órgano o tejido sobresale en un área de músculos débiles.

Hierarchy of Needs A ranked list of those needs essential to human growth and development, presented in ascending order, starting with basic needs and building toward the need for reaching your highest potential. (Ch. 3, 70)

Jerarquización de necesidades Lista priorizada de aquellas necesidades esenciales para el desarrollo óptimo del ser humano, presentada en orden ascendente, comenzando con las necesidades básicas y ascendiendo hacia la necesidad de alcanzar los potenciales máximos.

Histamines Chemicals that can stimulate mucus and fluid production. (Ch. 25, 703)

Estaminas Sustancias químicas que pueden estimular la producción de mucosidades y líquidos corporales.

HIV (Human Immunodeficiency Virus) The virus that causes Acquired Immune Deficiency Syndrome (AIDS). (Ch. 24, 669)

VIH (Virus de la Inmunodeficiencia Humana) Virus que provoca el Síndrome de Inmunodeficiencia Adquirida (SIDA).

Homicide The willful killing of one human being by another. (Ch. 9, 233)

Homicidio Cuando una persona mata intencionalmente a otra.

Hormones Chemicals produced by your glands that regulate the activities of different body cells. (Ch. 3, 78)

Hormonas Produciónes químicas secretadas por las glándulas que regulan las actividades de diferentes células corporales.

Hostility The intentional use of unfriendly or offensive behavior. (Ch. 3, 79)

Hostilidad Comportamiento intencional que es antipático, desagradable u ofensivo.

Hunger The natural physical drive to eat, prompted by the body's need for food. (Ch. 10, 255)

Hambre Impulso físico natural de comer, provocado por la necesidad del cuerpo de obtener alimento.

Hurricane A powerful storm that generally forms in tropical areas, producing winds of at least 74 miles per hour, heavy rains, and sometimes tornadoes. (Ch. 27, 779)

Huracán Una tormenta muy fuerte que se origina en áreas tropicales y que se caracteriza por vientos de al menos 74 millas por hora, fuertes lluvias, inundaciones y, algunas veces, tornados.

Hypertension High blood pressure. (Ch. 25, 689)

Hipertensión Presión arterial alta.

Hypothermia Dangerously low body temperature. (Ch. 12, 340)

Hipotermia Descenso peligroso de la temperatura corporal.

English　　Español

"I" message A statement that focuses on your feelings rather than on someone else's behavior. (Ch. 6, 154)

Mensaje en primera persona Una declaracion enfocada en sus propias sensaciones mas bien que en el comportamiento de alguien mas.

Illegal drugs Chemical substances that people of any age may not lawfully manufacture, possess, buy, or sell. (Ch. 22, 592)

Drogas ilegales Sustancias químicas que ninguna persona, cualquiera sea su edad, puede legalmente producir, poseer, comprar o vender.

Illicit drug use The use or sale of any substance that is illegal or otherwise not permitted. (Ch. 22, 592)

Uso ilegal de drogas El uso o venta de cualquier sustancia que es ilegal o no permitida.

Immune system A network of cells, tissues, organs, and chemicals that fight off pathogens. (Ch. 23, 639)

Sistema de defensas (inmunológico) Una combinación de células, tejidos, órganos y sustancias químicas que combaten a los agentes patógenos.

Immunity The state of being protected against a particular disease. (Ch. 23, 640)

Inmunidad Estado de protección contra una enfermedad en particular.

Implantation The process by which the zygote attaches to the uterine wall. (Ch. 17, 470)

Implantación El proceso en que el cigoto se adhiere a la pared uterina.

Infatuation Exaggerated feelings of passion. (Ch. 8, 206)

Enamoramiento Sentimientos exagerados de pasión.

Infection A condition that occurs when pathogens in the body multiply and damage body cells. (Ch. 23, 628)

Infección Una condición que ocurre cuando agentes patógenos entran al cuerpo, se multiplican y dañan las células.

Inflammatory response A reaction to tissue damage caused by injury or infection. (Ch. 23, 639)

Respuesta inflamatoria Reacción al daño de tejidos causada por una lesión o infección.

Inhalants Substances whose fumes are sniffed or inhaled to give a mind-altering effect. (Ch. 22, 601)

Inhalantes Sustancias cuyos gases se aspiran o inhalan para alcanzar un estado que altera la mente.

Insecure* Not confident or sure. (Ch. 9, 230)

Inseguro Que no tiene confianza o seguridad.

Instance* To mention as a case or example. (Ch. 12, 334)

Caso Para mencionar como un caso o ejemplo.

Integrity A firm observance of core ethical values. (Ch. 3, 74)

Integridad Adherencia firme a los valores éticos fundamentales.

Intense* Existing in an extreme degree. (Ch. 22, 600)

Intenso Que existe en grado extremo.

Intermediate* Being at the middle place or stage. (Ch. 16, 444)

Intermedio Que está en medio de un lugar o de una etapa.

Interpersonal communication The exchange of thoughts, feelings, and beliefs between two or more people. (Ch. 2, 35)

Comunicación interpersonal Intercambio de pensamientos, sentimientos y creencias entre dos o más personas.

Interpersonal conflict Conflicts between people or groups of people. (Ch. 9, 220)

Conflicto interpersonal Desacuerdo entre personas o grupos de personas.

Intimacy Closeness between two people that develops over time. (Ch. 8, 206)

Intimidad Cercanía entre dos personas que se desarrolla en el transcurso del tiempo.

Intoxication The state in which the body is poisoned by alcohol or another substance and the person's physical and mental control is significantly reduced. (Ch. 21, 567)

Intoxicación Estado en el cual el cuerpo se encuentra envenenado por el alcohol u otra sustancia, y el control físico y mental de la persona se encuentra reducido significativamente.

Involve* To require as a necessary accompaniment (Ch. 15, 423)

Implicar Que incluye algo.

Isolation* The act of being withdrawn or separated. (Ch. 21, 575)

Aislamiento Acción de retirar o separar.

English	J	Español

Jaundice A yellowing of the skin and eyes. (Ch. 23, 635)

Ictericia Estado en el cual la piel y los ojos se ponen de color amarillo.

English	L	Español

Labyrinth The inner ear. (Ch. 13, 371)

Laberinto Oído interno.

Landfill A specially engineered area where waste can be buried safely. (Ch. 28. 807)

Vertedero Área diseñada especialmente para enterrar los desechos en forma segura.

Legal* Permitted by law. (Ch. 26, 740)

Leukoplakia Thickened, white, leathery-looking spots on the inside of the mouth that can develop into oral cancer. (Ch. 20, 544)

Lifestyle factors The personal habits or behaviors related to the way a person lives. (Ch. 1, 20)

Ligament A band of fibrous, slightly elastic connective tissue that attaches one bone to another. (Ch. 14, 383)

Link* A connecting element or factor. (Ch. 25, 697)

Long-term goal A goal that you plan to reach over an extended period. (Ch. 2, 43)

Lymph The clear fluid that fills the spaces around body cells. (Ch. 15, 412)

Lymphocyte Specialized white blood cell that coordinates and performs many of the functions of specific immunity. (Ch. 23, 640)

Legal Permitido por la ley.

Leucoplaquia Granos con apariencia de piel blanca dura y espesa que se encuentran dentro de la boca y que pueden llegar a producir un cáncer oral.

Factores del estilo de vida Hábitos o conductas personales relativos a la forma de vivir de las personas.

Ligamento Tejido conjuntivo fibroso y levemente elástico que une dos huesos.

Enlace Elemento o factor de conexión.

Meta a largo plazo Meta que planeas alcanzar en un periodo prolongado.

Linfa Líquido transparente que llena los espacios entre las células del cuerpo.

Linfocito Glóbulo blanco especializado que coordina y realiza muchas de las funciones de inmunidad específica.

English — M — Español

Mainstream smoke The smoke exhaled from the lungs of a smoker. (Ch. 20, 553)

Malignant Cancerous. (Ch. 25, 696)

Malocclusion A misalignment of the upper and lower teeth. (Ch. 13, 365)

Malpractice Failure by a health professional to meet accepted standards. (Ch. 2, 52)

Manipulation An indirect, dishonest way to control or influence others. (Ch. 8, 200)

Marijuana A plant whose leaves, buds, and flowers are usually smoked for their intoxicating effects. (Ch. 22, 599)

Mastication The process of chewing. (Ch. 15, 423)

Media Various methods for communicating information. (Ch. 1, 14)

Mediation Bringing in a neutral third party to help others resolve their conflicts peacefully. (Ch. 9, 226)

Medical history Complete and comprehensive information about your immunizations and any health problems you have had to date (Ch. 28, 794)

Medicines Drugs that are used to treat or prevent diseases or other conditions. (Ch. 19, 524)

Megadoses Very large amount. (Ch. 11, 308)

Melanin A pigment that gives the skin, hair, and iris of the eyes their color. (Ch. 13, 356)

Melanoma The most serious form of skin cancer. (Ch. 13, 360)

Menstruation The shedding of the uterine lining. (Ch. 16, 454)

Humo directo Humo exhalado por los pulmones de un fumador.

Maligno Canceroso.

Oclusión defectuosa Alineación defectuosa de los dientes superiores e inferiores.

Mala práctica médica (mala praxis) Condición en la que un profesional de la salud no cumple con los estándares aceptados.

Manipulación Controlar o influenciar a otros de manera indirecta y deshonesta.

Marihuana Una planta cuyas hojas, brotes y flores son generalmente fumados por su efecto intoxicante.

Masticación Proceso de masticar.

Medios de comunicación Diversos métodos para comunicar información.

Mediación Proceso en el cual una tercera parte neutra ayuda a otros a resolver sus conflictos pacíficamente.

Historial médico Información completa acerca de las vacunas recibidas y los problemas de salud que una persona ha teido hasta la fecha.

Medicamentos Fármacos para tratar o prevenir una enfermedad u otro problema de salud.

Megadosises Gran cantidad.

Melanina Pigmento que da el color a la piel, el cabello y el iris del ojo.

Melanoma El cáncer de la piel más grave de todos.

Menstruación El eliminación del revestimiento del útero.

Mental* Of or relating to the mind. (Ch. 3, 67)

Mental disorder An illness of the mind that can affect the thoughts, feelings, and behaviors of a person, preventing him or her from leading a happy, healthful, and productive life. (Ch. 5, 118)

Mental retardation A below-average intellectual ability present from birth or early childhood and associated with difficulties in learning and social adaptation. (Ch. 25, 712)

Mental/emotional health The ability to accept yourself and others, express and manage emotions, and deal with the demands and challenges you meet in your life. (Ch. 3, 66)

Metabolism The processes by which the body breaks down substances and gets energy from food. (Ch. 11, 290)

Metastasis The spread of cancer from the point where it originated to other parts of the body. (Ch. 25, 696)

Minerals Elements found in food that are used by the body. (Ch. 10, 262)

Minor* Not serious or involving risk to life. (Ch. 27, 760)

Misinterpret* To understand wrongly. (Ch. 9, 221)

Monitor* To watch or keep track of. (Ch. 15, 430)

Mood disorders Illness that involves mood extremes that interfere with everyday living. (Ch. 5, 120)

Mourning The act of showing sorrow or grief. (Ch. 4, 104)

Mucous membrane The lining of various body cavities, including the nose, ears, and mouth. (Ch. 23, 634)

Muscle cramps Sudden and sometimes painful contractions of the muscles. (Ch. 12, 341)

Muscular endurance The ability of your muscles to perform physical tasks over a period without tiring. (Ch. 12, 325)

Muscular strength The amount of force your muscles can exert. (Ch. 12, 324)

MyPyramid An interactive guide to healthful eating and active living. (Ch. 10, 267)

Mental Relativo a la mente.

Trastorno mental Enfermedad mental que puede afectar la manera de pensar, los sentimientos y el comportamiento de una persona, y que le impide tener una vida feliz, saludable y productiva.

Retardo mental Capacidad intelectual inferior al promedio que se presenta desde el nacimiento o la niñez temprana y que se relaciona con dificultades de aprendizaje y adaptación social.

Salud mental-emocional Habilidad para aceptarse a sí mismo y a otras personas, expresar, manejar las emociones y hacer frente a las exigencias y desafíos de la vida.

Metabolismo Proceso mediante el cual el cuerpo procesa las sustancias y obtiene energía de los alimentos.

Metástasis Extensión del cáncer desde el punto de origen a otras partes del cuerpo.

Minerales Elementos que se encuentran en los alimentos y son utilizados por el cuerpo.

Menor Que no es grave ni representa riesgo vital.

Malinterpretar Entender equivocadamente.

Controlar Vigilar o comprobar.

Trastornos del ánimo Enfermedad que involucran estados de ánimo extremos, los cuales interfieren con la vida diaria.

Luto Acto de mostrar pena o dolor.

Membrana mucosa Revestimiento de diferentes cavidades corporales que incluyen la nariz, los oídos y la boca.

Calambres musculares Contracciones repentinas y, algunas veces, dolorosas de los músculos.

Resistencia muscular Capacidad de los músculos para hacer actividades físicas durante un periodo de tiempo sin fatigarse.

Fuerza muscular Fuerza que puedan ejercer los músculos.

Mi Pirámide Guía interactiva para alimentarse sanamente y llevar una vida activa.

English	N	Español

Neglect The failure to provide for a child's basic needs. (Ch. 7, 179)

Negotiation The use of communication and, in many cases, compromise to settle a disagreement. (Ch. 9, 225)

Nephrons The functional units of the kidneys. (Ch. 15, 430)

Abandono No satisfacer las necesidades básicas de un niño.

Negociación Uso de la comunicación y, frecuentemente, el compromiso para resolver un desacuerdo.

Nefronas Unidades funcionales de los riñones.

Neurons Nerve cells. (Ch. 14, 393)

Neutralize* To counteract the effect of. (Ch. 19, 525)

Nicotine The addictive drug found in tobacco. (Ch. 20, 543)

Nicotine substitute A product that delivers small amounts of nicotine into the user's system while he or she is trying to give up the tobacco habit. (Ch. 20, 551)

Nicotine withdrawal The process that occurs in the body when nicotine, an addictive drug, is no longer used. (Ch. 20, 551)

Noise pollution Harmful, unwanted sound loud enough to damage hearing. (Ch. 28, 804)

Noncommunicable disease A disease that is not transmitted by another person, a vector, or the environment. (Ch. 25, 688)

Nuclear family Two parents and one or more children living in the same space. (Ch. 7, 167)

Nutrient-dense A high ratio of nutrients to calories. (Ch. 10, 269)

Nutrients Substances in food that your body needs to grow, to repair itself, and to supply you with energy. (Ch. 10, 254)

Nutrition The process by which your body takes in and uses food. (Ch. 10, 254)

Neuronas Células nerviosas.

Neutralizar Contrarrestar el efecto de algo.

Nicotina Droga adictiva que se encuentra en el tabaco.

Sustituto de la nicotina Producto que libera pequeñas cantidades de nicotina en el cuerpo de una persona que está tratando de dejar de fumar.

Reacción al retiro de la nicotina Proceso que ocurre en el cuerpo cuando la nicotina, una droga adictiva, deja de ser consumida.

Contaminación acústica Nivel de ruido perjudicial y no deseado que es lo suficientemente alto como para dañar la audición de las personas.

Enfermedad no contagiosa Enfermedad que no se transmite entre las personas o por un vector, y que tampoco proviene del medio ambiente.

Familia nuclear Ambos padres y uno o más hijos que viven en el mismo espacio.

Rico en nutrientes Que tiene una relación alta de nutrientes a calorías.

Nutrientes Sustancias presentes en los alimentos que el cuerpo necesita para crecer, regenerarse y producir energía.

Nutrición Proceso mediante el cual el cuerpo absorbe y usa los alimentos.

English **O** **Español**

Obese Having an excess of body fat. (Ch. 11, 292)

Occupational Safety and Health Administration (OSHA) The agency within the federal government that is responsible for promoting safe and healthful conditions in the workplace. (Ch. 26, 738)

Opiates Drugs like those derived from the opium plant that are obtainable only by prescription and are used to relieve pain. (Ch. 22, 609)

Ossification The process by which bone is formed, renewed, and repaired. (Ch. 14, 383)

Osteoarthritis A disease of the joints in which cartilage breaks down. (Ch. 25, 707)

Osteoporosis A condition in which the bones become fragile and break easily. (Ch. 10, 264)

Ovaries The female sex glands that store the ova, eggs, and produce female sex hormones. (Ch. 16, 452)

Over the counter (OTC) Medicines you can buy without a prescription. (Ch. 19, 531)

Obeso Que tiene exceso de grasa corporal.

Administración de Seguridad y Salud Ocupacional (OSHA) Agencia del gobierno federal que es responsable de promover condiciones de trabajo seguras y saludables.

Opiáceos Drogas, como aquellas derivadas del opio, que sólo se obtienen con prescripción médica y se utilizan para aliviar el dolor.

Osificación Proceso mediante el cual el hueso se forma, renueva y repara.

Osteoartritis Enfermedad de las articulaciones en la cual el cartílago se deteriora.

Osteoporosis Enfermedad en la cual los huesos se vuelven frágiles y se rompen con facilidad.

Ovarios Glándulas sexuales femeninas que contienen los óvulos y producen hormonas sexuales.

Medicamentos de venta libre (OTC) Medicamentos que se pueden comprar sin prescripción médica.

Glossary / Glosario

Overdose A strong, sometimes fatal reaction to taking a large amount of a drug. (Ch. 22, 593)

Overexertion Overworking the body. (Ch. 12, 340)

Overload Exercising at a level that's beyond your regular daily activities. (Ch. 12, 333)

Overweight Heavier than the standard weight range for your height. (Ch. 11, 291)

Ovulation The process of releasing a mature ovum into the fallopian tube each month. (Ch. 16, 452)

Sobredosis Reacción fuerte, y algunas veces fatal, al consumir una droga en grandes cantidades.

Esfuerzo excesivo Cuando el cuerpo trabaja demasiado.

Sobrecarga Ejercitarse a un nivel que sobrepasa las actividades diarias normales.

Sobrepeso Peso superior al rango normal según la estatura.

Ovulacion El proceso de saltar un ovario maduro cada mes en las trompas de falopio.

English — P — Español

Pancreas A gland that serves both the digestive and endocrine systems. (Ch. 16, 443)

Pandemic A global outbreak of an infectious disease. (Ch. 23, 648)

Paranoia An irrational suspiciousness or distrust of others. (Ch. 22, 600)

Parathyroid glands Produce a hormone that regulates the body's balance of calcium and phosphorus. (Ch. 16, 443)

Partner* One associated with another. (Ch. 9, 238)

Passive Unwilling or unable to express thoughts and feelings in a direct or firm manner. (Ch. 6, 152)

Pasteurization Treating a substance with heat to kill or slow the growth of pathogens. (Ch. 10, 279)

Pathogen A microorganism that causes disease. (Ch. 15, 412)

Peer mediation Processes in which specially trained students help other students resolve conflicts peacefully. (Ch. 9, 228)

Peer pressure The influence that people your age may have on you. (Ch. 8, 198)

Peers People of the same age who share similar interests. (Ch. 1, 13)

Penis A tube-shaped organ that extends from the trunk of the body just above the testes. (Ch. 16, 447)

Peptic ulcer A sore in the lining of the digestive tract. (Ch. 15, 427)

Percent* One part in a hundred. (Ch. 5, 119)

Perception The act of becoming aware through the senses. (Ch. 4, 92)

Performance enhancers Substitutes that boost athletic ability. (Ch. 11, 307)

Periodontium The area immediately around the tooth. (Ch. 13, 363)

Period* The completion of a cycle. (Ch. 12, 327)

Páncreas Glándula utilizada tanto por el sistema digestivo como el endocrino.

Pandemia Brote global de una enfermedad infecciosa.

Paranoia Sospecha o desconfianza irracionales hacia otras personas.

Glándulas paratiroides Glándulas que producen una hormona que regula el equilibrio de calcio y fósforo en el cuerpo.

Socio Persona asociada con otra.

Pasivo Persona que no está dispuesto o no es capaz de expresar sus pensamientos y sentimientos.

Pasteurización Tratamiento de una sustancia con calor para matar organismos patógenos o para hacer más lento su desarrollo.

Patógeno Microorganismo que causa enfermedades.

Mediación entre compañeros Proceso en el cual estudiantes especialmente entrenados ayudan a otros a resolver sus conflictos pacíficamente.

Presión de los compañeros La influencia que gente de tu misma edad puede tener en ti.

Compañeros Personas cuya edad e intereses son similares a los tuyos.

Pene Órgano con forma de tubo que se extiende desde el tronco del cuerpo, justo arriba de los testículos.

Úlcera péptica Herida en el revestimiento del tracto digestivo.

Porcentaje Una parte de cada cien.

Percepción Acto de tomar conciencia de algo a través de los sentidos.

Potenciadores del rendimiento Sustitutos que mejoran las capacidades atléticas.

Periodontio Área que rodea los dientes.

Periodo Ciclo de tiempo.

Periodontal disease An inflammation of the periodontal structures. (Ch. 13, 365)

Peristalsis A series of involuntary muscle contractions that move food through the digestive tract. (Ch. 15, 424)

Personal flotation device Life jacket. (Ch. 26, 742)

Personal identity Your sense of yourself as a unique individual. (Ch. 3, 72)

Personal safety The steps you take to prevent yourself from becoming the victim of crime. (Ch. 26, 726)

Personality A complex set of characteristics that make you unique. (Ch. 3, 73)

Phagocyte White blood cells that attack invading pathogens. (Ch. 23, 640)

Physical abuse A pattern of intentionally causing bodily harm or injury to another person. (Ch. 9, 237)

Physical activity Any form of movement that causes your body to use energy. (Ch. 12, 318)

Physical fitness The ability to carry out daily tasks easily and have enough reserve energy to respond to unexpected demands. (Ch. 12, 319)

Physical maturity The state at which the physical body and all its organs are fully developed. (Ch. 18, 502)

Physiological dependence A condition in which the user has a chemical need for the drug. (Ch. 21, 572)

Pituitary gland Regulates and controls the activities of all other endocrine systems. (Ch. 16, 443)

Plaque A combination of bacteria and other particles, such as small bits of food, which adheres to the outside of a tooth. (Ch. 13, 364)

Plasma The fluid in which other parts of the blood are suspended. (Ch. 15, 410)

Platelets Types of cells in the blood that cause blood clots to form. (Ch. 15, 410)

Platonic friendship A friendship with a member of the opposite gender in which there is affection but the two people are not considered a couple. (Ch. 8, 194)

Pneumonia An infection of the lungs in which the air sacs fill with pus and other liquids. (Ch. 23, 634)

Poison Any substance—solid, liquid, or gas—that causes injury, illness, or death when it enters the body. (Ch. 27, 775)

Poison control center A round-the-clock service that provides emergency medical advice on how to treat poisoning victims. (Ch. 27, 775)

Portion* A part set off from the whole. (Ch. 13, 368)

Pose* To put or set forth. (Ch. 11, 298)

Enfermedad periodontal Inflamación de la estructura de soporte dental.

Movimientos peristálticos Serie de contracciones musculares involuntarias que mueven la comida a través del tracto digestivo.

Dispositivo de flotación personal Chaqueta salvavidas.

Identidad personal Sentirse como un individuo único.

Seguridad personal Medidas que se toman para evitar ser víctima de un delito.

Personalidad Conjunto complejo de características que hacen que una persona sea única.

Fagocito Glóbulo blanco que combate la invasión de patógenos.

Abuso físico Patrón intencionalmente de causar daño o lesión corporal a otra persona.

Actividad física Cualquier forma de movimiento que provoque que el cuerpo consuma energía.

Buen estado Capacidad de realizar fácilmente las tareas diarias y tener suficiente energía para responder a exigencias inesperadas.

Madurez física Estado en el cual el cuerpo y todos sus órganos se encuentran totalmente desarrollados.

Dependencia fisiológica Enfermedad en la cual el usuario tiene una necesidad física de consumir una droga.

Glándula pituitaria Glándula que regula o controla las actividades de todas las glándulas endocrinas.

Placa Combinación de bacterias y otras partículas, tales como pequeños trozos de alimentos, que se adhieren a la parte externa de los dientes.

Plasma Líquido en el cual están suspendidos los componentes de la sangre.

Plaquetas Tipos de células sanguíneas que provocan la formación de coágulos.

Amistad platónica Amistad con una persona del sexo opuesto en la cual hay sentimientos mutuos de afecto, pero que no se consideran una pareja.

Neumonía Infección de los pulmones en la cual los alvéolos se llenan de pus y otros líquidos.

Veneno Cualquier sustancia, sea sólida, líquida o gaseosa, que al entrar en el cuerpo causa una herida, una enfermedad o la muerte.

Centro para el control de envenenamientos Servicio que funciona las 24 horas del día para dar consejos médicos de emergencia sobre el tratamiento de víctimas de intoxicaciones.

Porción Parte separada de un todo.

Plantear Exponer o presentar.

Precycling Reducing waste before it is generated. (Ch. 28, 810)

Reciclaje previo Reducir la basura antes de generarla.

Prejudice An unfair opinion or judgment of a particular group of people. (Ch. 6, 150)

Prejuicio Opinión o juicio injusto acerca de un grupo específico de personas.

Prenatal care Steps that a pregnant female can take to provide for her own health and the health of her baby. (Ch. 17, 473)

Cuidado prenatal Todas las medidas que una mujer embarazada puede tomar para cuidar su propia salud y la de su bebé.

Prescription medicines Medicines that cannot be used without the written approval of a licensed physician or nurse practitioner. (Ch. 19, 531)

Medicamentos con prescripción Medicamentos que no se pueden utilizar sin la aprobación escrita de un médico o enfermera profesionales.

Prevention Taking steps to keep something from happening or getting worse. (Ch. 1, 18)

Prevención Tomar medidas para evitar que algo ocurra o empeore.

Primary care physician A medical doctor who provides physical check-ups and general care. (Ch. 28, 792)

Médico de atención primaria Médico que realiza chequeos físicos y se encarga del cuidado general de los pacientes.

Priorities The goals, tasks, values, and activities that you judge to be more important than others. (Ch. 8, 205)

Prioridades Metas, tareas, valores y actividades que se consideran más importantes que otras.

Process* A series of actions geared toward an end result. (Ch. 8, 206)

Proceso Serie de acciones para obtener un resultado final.

Profound deafness Hearing loss so severe that a person affected cannot benefit from mechanical amplification such as a hearing aid. (Ch. 25, 710)

Sordera profunda Pérdida de la audición tan grave que no se beneficia con la amplificación mecánica (por ejemplo, los audífonos).

Progression Gradually increasing the demands on your body. (Ch. 12, 333)

Progresión Aumento gradual de las exigencias corporales.

Promote* To contribute to the growth of. (Ch. 2, 38)

Promover Contribuir al crecimiento de algo.

Proteins Nutrients the body uses to build and maintain its cells and tissues. (Ch. 10, 260)

Proteínas Nutrientes que el cuerpo utiliza para generar y mantener las células y los tejidos.

Psychoactive drugs Chemicals that affect the central nervous system and alter activity in the brain. (Ch. 22, 603)

Drogas psicoactivas Sustancias químicas que afectan el sistema nervioso central y alteran la actividad del cerebro.

Psychological* Directed toward the mind. (Ch. 10, 255)

Psicológico Relativo a la mente.

Psychological dependence A condition in which a person believes that a drug is needed in order to feel good or to function normally. (Ch. 21, 572)

Dependencia psicológica Enfermedad en la cual una persona cree que una droga es necesaria para sentirse bien o para funcionar normalmente.

Psychosomatic response A physical reaction, which results from stress rather than from an injury or illness. (Ch. 4, 95)

Respuesta psicosomática Reacción física producida por el estrés en lugar de corresponder a una lesión o una enfermedad.

Psychotherapy An ongoing dialogue between a patient and a mental health professional. (Ch. 5, 128)

Psicoterapia Diálogo continúo entre un paciente y un profesional de la salud mental.

Puberty The time when a person begins to develop certain traits of adults of his or her gender. (Ch. 18, 496)

Pubertad Periodo en el cual una persona comienza a desarrollar ciertos rasgos de adultez que son característicos de su sexo.

Public health All efforts to monitor, protect, and promote the health of the population as a whole. (Ch. 28, 796)

Salud pública Todo esfuerzo para supervisar, proteger y promover la salud de la población como una unidad.

Pulp The tissue that contains the blood vessels and nerves of a tooth. (Ch. 13, 364)

Pulpa Tejido que contiene los vasos sanguíneos y los nervios de un diente.

Random violence Violence committed for no particular reason. (Ch. 9, 233)

Range* The distance between possible extremes. (Ch. 11, 292)

Rape Any form of sexual intercourse that takes place against a person's will. (Ch. 9, 234)

Rapid test Used in situations where the infected person might not come back to learn the results to the test. A blood sample is collected and analyzed immediately. (Ch. 24, 677)

Reaction* A response to a stimulus or influence. (Ch. 10, 264)

Recovery The process of learning to live an alcohol-free life. (Ch. 21, 583)

Recycling The processing of waste materials so that they can be used again. (Ch. 28, 810)

Refusal skills Communication strategies that can help you say no when you are urged to take part in behaviors that are unsafe or unhealthful, or that go against your values. (Ch. 2, 36)

Regulate* To fix the time, amount, degree, or rate of. (Ch. 25, 690)

Rehabilitation Process of medical and psychological treatment for physiological or psychological dependence on a drug or alcohol. (Ch. 22, 614)

Relationship A bond or connection you have with other people. (Ch. 6, 142)

Relaxation response A state of calm. (Ch. 4, 99)

Remission A period of time when symptoms disappear. (Ch. 25, 701)

Remove* To get rid of. (Ch. 13, 358)

Require* To demand as necessary. (Ch. 5, 115)

Rescue breathing Breathing for a person who is not breathing on his or her own. (Ch. 27, 767)

Resilient The ability to adapt effectively and recover from disappointment, difficulty, or crisis. (Ch. 3, 67)

Resolve* To deal with successfully. (Ch. 4, 104)

Resource* A source of supply or support. (Ch. 3, 82)

Respiratory tract The passageway that makes breathing possible. (Ch. 23, 633)

Respond* To react in response. (Ch. 25, 705)

Resting heart rate The number of times your heart beats per minute when you are not active. (Ch. 12, 336)

Retina The inner layer of the eye wall. (Ch. 13, 368)

Violencia de azar Violencia infringida sin una razón en especial.

Rango Distancia entre posibles extremos.

Violación Cualquier tipo de acto sexual que ocurre contra la voluntad de una persona.

Prueba rápida Prueba que se realiza cuando la persona infectada podría no regresar para conocer los resultados. Se toma y analiza una muestra de sangre en forma inmediata.

Reacción Respuesta a un estimulo o influencia.

Recuperación Proceso de aprender a vivir sin consumir alcohol.

Reciclaje Proceso en el cual los desechos se pueden utilizar nuevamente.

Habilidades de negación Estrategias de comunicación que ayudan a decir no cuando te presionan a participar en actividades peligrosas, no saludables o que van en contra de tus valores.

Regular Organizar el tiempo, la cantidad, el grado o el rtimo de algo.

Rehabilitación Proceso de tratamiento médico y psicológico para la dependencia fisiológica o psicológica de una droga o del alcohol.

Relación Lazo que una persona tiene con los demás.

Respuesta de relajación Estado de calma.

Remisión Periodo en el cual desaparecen los síntomas.

Eliminar Deshacerse de algo o de alguien.

Requerir Solicitar como necesario.

Respiración de rescate Respiración artificial para una persona que no está respirando por sí misma.

Resiliencia Capacidad de adaptarse eficazmente y recuperarse después de una decepción, dificultad o crisis.

Resolver Manejar algo con éxito.

Recurso Fuente de suministro o apoyo.

Tracto respiratorio Vías que hacen posible la respiración.

Reaccionar Actuar en respuesta a algo o alguien.

Frecuencia cardiaco en reposo Número de latidos por minuto que se producen cuando la persona se encuentra en estado pasivo.

Retina Capa interna de la pared ocular.

Rheumatoid arthritis A disease characterized by the debilitating destruction of the joints due to inflammation. (Ch. 25, 708)

Risk behaviors Actions that can potentially threaten your health or the health of others. (Ch. 1, 16)

Road rage Responding to a driving incident with violence. (Ch. 26, 746)

Role The parts you play in your relationships. (Ch. 6, 144)

Role model Someone whose success or behavior serves as an example for you. (Ch. 3, 73)

Artritis reumatoide Enfermedad caracterizada por la destrucción debilitadora de las articulaciones debido a la inflamación.

Comportamiento riesgoso Acciones que pueden poner en peligro tu salud o la de otras personas.

Agresividad al volante Reaccionar de manera violenta a un incidente de conducción.

Función Papel que desempeña una persona en una relación.

Modelo de conducta Alguien cuyo éxito o comportamiento sirve de ejemplo para otros.

English — S — Español

Sclera The white part of the eye. (Ch. 13, 367)

Scoliosis A lateral, or side-to-side, curvature of the spine. (Ch. 14, 384)

Scrotum An external skin sac, which holds the testes. (Ch. 16, 447)

Sebaceous glands Structures within the skin that produce an oily secretion called sebum. (Ch. 13, 358)

Secure* To make fast or seal. (Ch. 27, 761)

Sedentary Involving little physical activity. (Ch. 12, 321)

Seek* To go in search of. (Ch. 27, 774)

Self-actualization To strive to become the best you can be. (Ch. 3, 70)

Self-control A person's ability to use responsibility to override emotions. (Ch. 8, 207)

Self-defense Any strategy for protecting yourself from harm. (Ch. 26, 727)

Self-directed Able to make correct decisions about behavior when adults are not present to enforce rules. (Ch. 18, 506)

Self-esteem How much you value, respect, and feel confident about yourself. (Ch. 3, 68)

Semen A thick fluid containing sperm and other secretions from the male reproductive system. (Ch. 16, 447)

Separation A decision by two married people to live apart from each other. (Ch. 7, 174)

Sexual abuse A pattern of sexual contact that is forced upon a person against his or her will. (Ch. 9, 237)

Sexual assault Any intentional sexual attack against another person. (Ch. 9, 234)

Sexual violence Any form of unwelcome sexual contact directed at an individual. (Ch. 9, 234)

Sexually transmitted diseases (STDs) Infectious diseases spread from person to person through sexual contact. (Ch. 8, 208)

Esclerótica Parte blanca del ojo.

Escoliosis Desviación lateral, o de lado a lado, de la columna.

Escroto Saco de piel externa que sostiene los testículos.

Glándulas sebáceas Estructuras dentro de la piel que producen una secreción aceitosa llamada sebo.

Asegurar Dejar firme y seguro.

Sedentar Poca actividad física.

Buscar Hacer algo para encontrar a alguien o algo.

Realización personal Esfuerzo para lograr lo mejor de uno mismo.

Dominio de sí mismo Capacidad de una persona para controlar sus emociones por medio de la responsabilidad.

Defensa propia Cualquier estrategia para protegerse de un daño.

Auto-dirigido Capaz de tomar decisiones correctas acerca de su comportamiento en la ausencia de adultos que impongan las reglas.

Autoestima Valor, respeto y sentimiento de confianza que uno tiene de sí mismo.

Semen Líquido espeso que contiene los espermatozoides y otras secreciones del aparato reproductor masculino.

Separación Cuando una pareja casada decide vivir aparte uno del otro.

Abuso sexual Patrón de contacto sexual realizado a la fuerza o contra la voluntad de una persona.

Agresión sexual Cualquier ataque sexual intencional en contra de otra persona.

Violencia sexual Cualquier forma de contacto sexual no deseado dirigido a una persona.

Enfermedades de transmisión sexual (ETS) Enfermedades que se transmiten a través del contacto sexual entre dos personas.

Sexually transmitted infections (STIs) Infections spread from person to person through sexual contact. (Ch. 24, 658)

Shock A life-threatening condition in which the heart is not delivering an adequate supply of blood to the body. (Ch. 27, 770)

Short-term goal A goal that you can reach in a short period of time. (Ch. 2, 43)

Siblings Brothers or sisters. (Ch. 7, 167)

Side effects Reactions to medicine other than the one intended. (Ch. 19, 528)

Sidestream smoke The smoke from the burning end of a cigarette, pipe, or cigar. (Ch. 20, 553)

Significant* Having meaning. (Ch. 18, 497)

Skeletal muscles Muscles attached to bone that cause body movements. (Ch. 14, 388)

Smog A brownish haze that sometimes forms in urban areas. (Ch. 28, 801)

Smoke alarm A device that produces a loud warning noise in the presence of smoke. (Ch. 26, 732)

Smooth muscles Muscles that act on the lining of the body's passageways and hollow internal organs. (Ch. 14, 388)

Sobriety Living without alcohol. (Ch. 21, 582)

Specialists Medical doctors who focus on particular kinds of patients or particular medical conditions. (Ch. 28, 792)

Specificity Choosing the right types of activities to improve a given element of fitness. (Ch. 12, 333)

Sperm Male gametes. (Ch. 16, 446)

Spiritual health A deep-seated sense of meaning and purpose in life. (Ch. 1, 8)

Spirituality A deep-seated sense of meaning and purpose in life. (Ch. 1, 8)

Spousal abuse Domestic violence or any other form of abuse directed at a spouse. (Ch. 7, 178)

Sprains Injuries to the ligaments around a joint. (Ch. 12, 342)

Stages of grief A variety of reactions that may surface as an individual makes sense of how a loss affects him or her. (Ch. 4, 103)

Stalking Repeatedly following, harassing, or threatening an individual. (Ch. 9, 237)

Stereotype Exaggerated or oversimplified beliefs about people who belong to a certain group. (Ch. 6, 150)

Sterility The inability to reproduce. (Ch. 16, 450)

Stigma A mark of shame or disapproval that results in an individual being shunned or rejected by others. (Ch. 5, 118)

Infecciones de transmisión sexual Infecciones que se transmiten a través del contacto sexual entre dos personas.

Shock Enfermedad que puede ser fatal en la cual el corazón no envía suficiente sangre al cuerpo.

Meta a corto plazo Meta que se puede lograr en un periodo breve.

Hermanos Hermanos o hermanas.

Efectos secundarios Reacciones inesperadas a un medicamento.

Humo indirecto Humo que proviene de una colilla de cigarrillo, pipa o cigarro.

Significativo Que es importante.

Músculos del esqueleto Músculos unidos a los huesos y que producen el movimiento del cuerpo.

Smog Bruma de color café que se forma algunas veces en las áreas urbanas.

Alarma de humo Dispositivo que emite un sonido de advertencia que se activa en presencia del humo.

Músculos lisos Músculos que actúan sobre el revestimiento de los conductos y órganos internos.

Sobriedad Vivir sin consumir alcohol.

Especialistas Médicos que se dedican a un tipo particular de pacientes o condiciones medicas particulares.

Especificar Elegir el tipo correcto de actividades que mejoren un elemento específico del bienestar físico.

Espermatozoides Gametos masculinos.

Salud espiritual Un sentido profundo de significado y proposito en vida.

Espiritualidad Sentido profundamente asentado del significado y propósito de la vida.

Abuso conyugal Violencia doméstica o cualquier forma de abuso dirigida hacia el esposo o la esposa.

Esguince Daño a los ligamentos que rodean una articulación.

Etapas del duelo Variedad de reacciones que pueden aparecer a medida que una persona entiende cómo le afecta una pérdida.

Acecho Seguimiento, acoso o amenaza repetidos que una persona hace a otra.

Estereotipo Creencias exageradas o demasiado simplificadas acerca de las personas que pertenecen a un grupo específico.

Esterilidad Incapacidad de reproducirse.

Estigma Señal de vergüenza o desaprobación que da como resultado el rechazo de una persona por parte de los demás.

Glossary / Glosario

Stimulant A drug that increases the action of the central nervous system, the heart, and other organs. (Ch. 20, 543)

Strains Overstretching and tearing a muscle. (Ch. 12, 341)

Stress The reaction of the body and mind to everyday challenges and demands. (Ch. 2, 38)

Stress management The use of healthy ways to reduce and manage stress in your life. (Ch. 2, 38)

Stress management skills Skills that help you reduce and manage your stress. (Ch. 2, 38)

Stressor Anything that causes stress. (Ch. 4, 93)

Stroke An acute injury in which blood flow to the brain is interrupted. (Ch. 25, 692)

Substance abuse Any unnecessary or improper use of chemical substances for non-medical purposes. (Ch. 22, 592)

Suicide The act of intentionally taking one's own life. (Ch. 5, 122)

Survival* The continuation of life or existence. (Ch. 27, 768)

Suspend* To bar temporarily. (Ch. 26, 737)

Synergistic effect* The interaction of two or more medicines that results in a greater effect than when the medicines are taken alone. (Ch. 19, 528)

Estimulante Droga que acelera el funcionamiento del sistema nervioso central, el corazón y otros órganos.

Distensión muscular Estirar en exceso y rasgar un músculo.

Estrés Reacción del cuerpo y la mente a las exigencias y desafíos de la vida diaria.

Manejo del estrés Utilización de formas saludables para reducir y manejar el estrés en tu vida.

Habilidades para controlar el Habilidades que le ayudan a reducir y manejar el estrés.

Estresante Cualquier cosa que produce estrés.

Accidente cerebrovascular Lesión aguda que interrumpe el flujo de sangre al cerebro.

Abuso de sustancias Cualquier uso inapropiado o excesivo de sustancias químicas con propósitos no médicos.

Suicidio Acto de quitarse la vida intencionalmente.

Supervivencia Continuación de la vida o de la existencia.

Suspender Prohibir temporalmente.

Efecto sinérgico Interacción entre dos o más medicamentos que produce un efecto más fuerte que si se toman por separado.

English	T	Español

Tar A thick, sticky, dark fluid produced when tobacco burns. (Ch. 20, 543)

Technique* A method of accomplishing a desired aim. (Ch. 4, 99)

Technology Radio, television, and the Internet. (Ch. 1, 14)

Tendon A fibrous cord that attaches muscle to the bone. (Ch. 14, 383)

Tendonitis The inflammation of a tendon. (Ch. 14, 389)

Testes Two small glands that secrete testosterone and produce sperm. (Ch. 16, 447)

Testosterone The male sex hormone. (Ch. 16, 446)

Thyroid gland Produces hormones that regulate metabolism, body heat, and bone growth. (Ch. 16, 443)

Tinnitus A condition in which a ringing, buzzing, whistling, roaring, hissing, or other sound is heard in the ear in the absence of external sound. (Ch. 13, 373)

Tolerance The ability to accept others' differences. (Ch. 6, 150)

Alquitrán Líquido espeso, pegajoso y oscuro que se forma al quemarse el tabaco.

Técnica Método para lograr el propósito deseado.

Tecnología Radio, televisión e internet.

Tendón Tejido fibroso que une los músculos a los huesos.

Tendinitis Inflamación de un tendón.

Testículos Par de pequeñas glándulas que secretan testosterona y producen espermatozoides.

Testosterona Hormona sexual masculina.

Tiroides Glándula que produce las hormonas que regulan el metabolismo, el calor del cuerpo y el crecimiento de los huesos.

Tinnitus Enfermedad en la cual se oye un timbre, zumbido, silbido, siseo, rugido u otro sonido en ausencia de ruidos externos.

Tolerancia Capacidad para aceptar las diferencias de los demás.

Tornado A whirling, funnel-shaped windstorm that extends from a storm to the ground and advances along the ground. (Ch. 27, 779)

Tornado Tormenta de viento en forma de embudo giratorio que se extiende desde una tormenta hasta el suelo y avanza por la tierra.

Toxin A substance that kills cells or interferes with their functions. (Ch. 23, 629)

Toxina Sustancia que mata células o que interfiere con su funcionamiento.

Trachea The windpipe. (Ch. 15, 417)

Tráquea Vía respiratoria principal.

Transit* Conveyance of persons or things from one place to. (Ch. 28, 809)

Tránsito Desplazamiento de personas o cosas de un lugar a otra.

Transitions Critical changes that occur in all stages of life. (Ch. 18, 508)

Transiciones Cambios críticos que ocurren en todas las etapas de la vida.

Transmit* To send from one person or place to another. (Ch. 24, 660)

Transmitir Trasladar desde una persona o lugar a otro.

Traumatic event Any event that has a stressful impact sufficient to overwhelm your normal coping strategies. (Ch. 4, 105)

Suceso traumático Todo suceso que tiene un efecto estresante suficiente como para sobrepasar las estrategias normales para lidiar con un hecho.

Trend* A line of general direction or movement. (Ch. 28, 795)

Tendencia Linea general de dirección o movimiento.

Trigger* To initiate or set off. (Ch. 11, 305)

Desencadenar Comenzar o iniciar algo.

Tuberculosis A contagious bacterial infection that usually affects the lungs. (Ch. 15, 421)

Tuberculosis Infección bacteriana contagiosa que comúnmente afecta los pulmones.

Tumor An abnormal mass of tissue that has no natural role in the body. (Ch. 25, 695)

Tumor Masa de tejido anormal que no cumple ninguna función natural en el cuerpo.

English	U	Español

Unconditional love Love without limitation or quantification. (Ch. 18, 507)

Amor incondicional Amor sin límite ni cuantificación.

Unconsciousness The condition of not being alert or aware of your surroundings. (Ch. 27, 773)

Inconciencia Condición en la cual una persona no está alerta o consciente de lo que lo rodea.

Underweight Below the standard weight range for your height. (Ch. 11, 293)

Bajo peso Peso inferior al rango normal según la estatura.

Unintentional injury An injury resulting from an unexpected event. (Ch. 26, 731)

Lesión no intencional Lesión resultante de un suceso inesperado.

Universal precautions Steps taken to prevent the spread of disease through blood and other body fluids when providing first aid or health care. (Ch. 27, 759)

Precauciones universales Medidas que se toman para evitar el contagio de enfermedades a través de la sangre y otros fluidos corporales cuando se proporcionan primeros auxilios o atención médica.

Urban sprawl The spreading of city development (houses, shopping centers, businesses, and schools) onto undeveloped land. (Ch. 28, 809)

Expansión urbana Crecimiento del desarrollo de una ciudad (casas, centros comerciales, negocios y escuelas) hacia zonas no desarrolladas.

Ureters Tubes that connect the kidneys to the bladder. (Ch. 15, 430)

Uréter Cada uno de los dos canales que conectan los riñones y la vejiga.

Urethra The tube that leads from the bladder to the outside of the body. (Ch. 15, 431)

Uretra Canal que nace en la vejiga y se extiende hacia el exterior del cuerpo.

Urethritis The inflammation of the urethra. (Ch. 15, 432)

Uretritis Inflamación de la uretra.

Uterus The hollow, muscular, pear-shaped organ that nourishes and protects a feterilized ovum until birth. (Ch. 16, 452)

Útero El órgano baso, muscular, en forma de pera que alimienta y protege el ovario hasta nacimiento.

Glossary / Glosario

Vaccine A preparation of dead or weakened pathogens that are introduced into the body to stimulate and immune response. (Ch. 23, 641)

Vagina A muscular, elastic passageway that extends from the uterus to the outside of the body. (Ch. 16, 454)

Valid* Well-grounded or justifiable. (Ch. 2, 48)

Values The ideas, beliefs, and attitudes about what is important, that help guide the way you live. (Ch. 2, 41)

Vector An organism that carries and transmits pathogens to humans or other animals. (Ch. 23, 630)

Vegetarian A person who eats mostly, or only, plant-based foods. (Ch. 11, 304)

Vehicular safety Obeying the rules of the road and exercising common sense and good judgment while driving. (Ch. 26, 744)

Venom A poisonous secretion. (Ch. 27, 776)

Verbal abuse The use of words to mistreat or injure another person. (Ch. 9, 237)

Violence The threatened or actual use of physical force or power to harm another person and to damage property. (Ch. 9, 229)

Virus A piece of genetic material surrounded by a protein coat. (Ch. 23, 629)

Visualize* To form a mental image of. (Ch. 10, 270)

Vitamins Compounds found in food that help regulate many body processes. (Ch. 10, 262)

Volume* The degree of loudness. (Ch. 25, 711)

Vacuna Preparación de agentes patógenos muertos o debilitados que se introducen en el cuerpo para estimular el sistema inmune.

Vagina Conducto muscular y elástico que va desde el útero hasta la parte externa del cuerpo de una mujer.

Válido Correctamente fundamentado o justificable.

Valores Ideas, creencias y actitudes sobre lo que es importante que guíen la vida de una persona.

Vector Organismo que lleva y transmite agentes patógenos a personas y otros animales.

Vegetariano Persona que come principal o exclusiva-mente alimentos que provienen de las plantas.

Seguridad vehicular Usar el sentido común, un buen criterio y obedecer las reglas de tránsito mientras se conduce.

Veneno Secreción venenosa.

Abuso verbal El uso de palabras para maltratar o dañar a otras personas.

Violencia Amenaza o uso de la fuerza física o el poder para maltratar a una persona o dañar una propiedad.

Virus Partícula de material genético rodeada de una capa proteica.

Visualizar Formarse una imagen mental.

Vitaminas Compuestos que se encuentran en los alimentos y que ayudan a regular muchos procesos corporales.

Volumen Intensidad del sonido.

Warm-up Gentle cardiovascular activity that prepares the muscles for work. (Ch. 12, 334)

Warranty A company's or a store's written agreement to repair a product or refund your money if the product does not function properly. (Ch. 2, 48)

Wastewater Used water from homes, communities, farms, and businesses. (Ch. 28, 809)

Weight cycling A repeated pattern of losing and regaining body weight. (Ch. 11, 298)

Wellness An overall sense of well-being or total health. (Ch. 1, 9)

Western blot Used to both detect HIV antibodies and to confirm the results of earlier tests. (Ch. 24, 677)

Workout The part of an exercise session when you are exercising at your highest peak. (Ch. 12, 335)

Precalentamiento Actividad cardiovascular liviana que prepara los músculos para el ejercicio.

Garantía Acuerdo escrito en el cual una empresa o tienda se compromete a reparar un producto o a devolver el dinero si el producto no funciona correctamente.

Aguas residuales Agua ya utilizada que proviene de casas, comunidades, granjas e industrias.

Ciclo de peso Patrón repetido de subir y bajar de peso.

Bienestar Sensación general de gozar de buena salud.

Western blot Prueba utilizada tanto para detectar anticuerpos de VIH como para confirmar el resultado de pruebas anteriores.

Entrenamiento Parte de un programa de actividad física en la que los ejercicios se realizan en el nivel más alto de rendimiento.

Index

Page numbers in italics refer to pictures or features.

Index

Index

Index

Index

Index

Self-defense, 727–728
Self-destructive behaviors
    and substance abuse, 123
    suicide, 122–125
    therapy for, 129
Self-directed children, 506
Self-esteem, 67–69
    definition of, 68
    improving, 69
    and participation in violent crimes, 230
    physical activity enhancing, 320, 321
    tips for improving, 84
Self-examinations
    breast, 455
    to detect cancer, 700
    for male reproductive health, 449
    testicular, 450–451
Self-management skills
    anger management, 82–83
    building resiliency, 100–101
    coping with grief and loss, 102–105
    as health skill, 38
    personal safety, 726–730
Self-respect, 148–149
Self-sufficiency, 67
Self-talk, 68, *81*
Sell-by dates (on food labels), 277
Semen, 447
Seminal vesicles, 448, *449*
Sense organs
    ears, 371–373
    in embryo, 471
    eyes, 367–371
    nose, 417
    skin, 356–360
    tongue, 423
Sensitivities, food, 280, 281
Sensorineural hearing loss, 373
Sensory neurons, 393, 397
Separation, changes in family related to, 174
Septum (heart), 409
Serving sizes, 273, *276*
Setting limits. *See* Limits, setting
Severe Acute Respiratory Syndrome (SARS), 648
Sex (of a person). *See* Gender
Sex characteristics, 497
Sexual abuse, 234–239
Sexual activity
    abstinence from, 206–207
    and alcohol use, 574
    consequences of, 208–209
    and risk of STDs, 666
    risk situations for, 208
    and spread of HIV, 671
Sexual assault, 234, 574
Sexual attack
    avoiding, 234
    responding to, 234–235
Sexual harassment, 234
Sexual violence, 234–235, 238–239
Sexually transmitted diseases (STDs), 658–679
    abstinence for prevention of, 664–667
    and alcohol use, 574
    as asymptomatic, 659
    cancer caused by, 698
    chlamydia, 660, 661
    definition of, 208, 658
    diagnosing and treating, 667–668
    as epidemic, 663

    in female reproduction system, 456
    genital herpes, 661
    genital HPV infections, 660, 661
    gonorrhea, 661, 662
    HIV/AIDS, 669–679
    most common, 660
    as risk of sexual activity, 208, 209
    spread of, 658
    symptoms and long-term effects of, *661*
    syphilis, 662
    trichomoniasis, 662
Sexually transmitted infections (STIs), 208, 658. *See also* Sexually transmitted diseases
Shared interests, in positive friendships, 194
Sheltering in place, 783
Shock, first aid for, 770
Shopping
    comparison, 47, 48
    compulsive, *120*
Short-term goals, 43
Siblings, 167
Sickle-cell anemia, *481*
Side effects of medicines, 528
Sidestream smoke, 553
SIDS (sudden infant death syndrome), 554
Sight
    impairment of, 709, 710
    in middle adulthood, 509
Simple carbohydrates, 250
Single-parent families, 167
Sinusitis, 419, 420
Sit-and-reach test, *326,* 327
Skateboarding safety, 338
Skating safety, 338, 748
Skeletal muscles, 388, 390
Skeletal system, 382–386
    caring for, 383–384
    problems with, 384, 386
Skiing safety, 741
Skin cancers, 340, 374, *696,* 697
Skin care, 356–360
Skin-fold testing, 292
Sledding safety, 741
Sleep
    for building resiliency, 100–101
    cycles of, 404
    for endocrine health, 445
    physical activity enhancing, 320
Small intestine, 424, 425
Small motor vehicles, safety with, 748–749
Smog, 475, 801
Smoke alarms, 732
Smokeless tobacco, 544
Smoking
    antismoking organizations, 558
    carcinogens from, 697–698
    consequences of, 542
    effects of, 420
    emphysema from, *555*
    health risks of, 553–555
    heart damage from, 434
    making decisions about, 563
    marijuana, 600
    during pregnancy, 474, 544, 554
    in public places, 556
    reasons for starting, 548
    and respiratory health, 419
    restricting places for, 622–623

Smooth muscles, 388, 413
Snack foods, calories in, *291*
Snacks, 272
Snakebites, 776
Snell, 48
Snowboarding safety, 741
Snowmobiling safety, 741
Sobriety, 582–583
Social and human services assistants, 248
Social environment, 13
Social health, 8
    and alcohol use, 569
    benefits of physical activity to, 320, 321
    conflict and harm to, 222
    and conflict resolution, 222
    consequences of sexual activity on, 209
    and drug use, 594–596
    family in promotion of, 170–171
    influence of mental/emotional health on, 67
    in middle adulthood, 510
Social Security system, 511
Social services careers, 248–249
Society, consequences of drug use on, 596
Solid waste, 806–807
Somatic nervous system, 397
Somatotropic hormone, 443
Speaking skills, 153, 154
Special Olympics, 624
Specialists (doctors), 792
Specificity, in building fitness, 332
Sperm, 446, 447, 497
    chromosomes in, 479
    definition of, 446
Sphincter muscle, 424
Spider bites/stings, 776–777
Spinal cord, *395*
    in central nervous system, 393
    in nervous system, 392, 393
Spinal cord injury, 405
Spinal injuries, 398, 399
Spinal meninges, 393
Spiritual health, 8
Spleen, 412, 413
Sports
    dealing with conflict in, *225*
    safety equipment for, 338
    winter, 740–741
Sportsmanship, 320, 337
Spousal abuse, 178–179
Sprains, 342, 389
Stages of grief, 103
Stalking, 237
Starches. *See* Complex carbohydrates
STDs. *See* Sexually transmitted diseases
Step test, 325
***Step*tember campaign,** 2
Stereotypes, 150
    in cliques, 195
    definition of, 150
Sterility, male, 450
Steroids, 307, 601
Stigma, 118
Stillbirths, 475, 476
Stimulants, 606–607
    amphetamines, 607
    caffeine, 606
    cocaine, 607
    crack, 607
    definition of, 543

Index